Neurobiological and Clinical Consequences of Stress

From Normal Adaptation to Post-Traumatic Stress Disorder

Neurobiological and Clinical Consequences of Stress

From Normal Adaptation to Post-Traumatic Stress Disorder

Editors

Matthew J. Friedman, M.D., Ph.D.
Professor, Departments of Psychiatry and Pharmacology
Dartmouth Medical School
Hanover, New Hampshire; and
Executive Director
National Center for Post-Traumatic Stress Disorder
Veterans Administration Medical Center
White River Junction, Vermont

Dennis S. Charney, M.D.
Professor and Associate Chairman for Research
Department of Psychiatry
Yale University School of Medicine
New Haven, Connecticut; and
Chief, Psychiatry Service
National Center for Post-Traumatic Stress Disorder
Veterans Affairs Medical Center
West Haven, Connecticut

Ariel Y. Deutch, Ph.D.
Associate Professor, Departments of Psychiatry and Pharmacology
Yale University School of Medicine
New Haven, Connecticut; and
National Center for Post-Traumatic Stress Disorder
Veterans Affairs Medical Center
West Haven, Connecticut

Lippincott - Raven
PUBLISHERS
Philadelphia • New York

Lippincott-Raven Publishers, 227 East Washington Square
Philadelphia, Pennsylvania 19106–3780

Made in the United States of America

Library of Congress Cataloging-in-Publication Data

Neurobiological and clinical consequences of stress : from normal adaptation to PTSD / edited by Matthew J. Friedman, Dennis S. Charney, Ariel Y. Deutch.

p. cm.

Includes bibliographical references and index.

ISBN 0-7817-0177-5

1. Post-traumatic stress disorder—Pathophysiology. 2. Stress (Psychology)—Physiological aspects. 3. Stress (Physiology) I. Friedman, Matthew J. II. Charney, Dennis S. III. Deutch, Ariel Y.

[DNLM: 1. Stress, Psychological—physiopathology. 2. Stress Disorders, Post-Traumatic—physiopathology. 3. Neurobiology. WM 172 N494 1995]

RC552.P67N47 1995

616.9′8—dc20

DNLM/DLC

for Library of Congress 95-16250

CIP

The material contained in this volume was submitted as previously unpublished material, except in the instances in which credit has been given to the source from which some of the illustrative material was derived.

Great care has been taken to maintain the accuracy of the information contained in the volume. However, neither Lippincott-Raven Publishers nor the editors can be held responsible for errors or for any consequences arising from the use of the information contained herein.

9 8 7 6 5 4 3 2 1

To our wives, Gayle Smith, Andrea Charney, and Laura Young

We could not have completed this book without their love, wisdom, patience, and sense of humor

To our children, Abby, Alex, Allison, Becky, Bo, Danielle, Jessie, Lauren, Mattis, and Meredith. Each made a significant contribution in his or her unique fashion

Contents

Part IV. Clinical Issues

Part V. Summary

Contributing Authors

Alexandre Bennett, Ph.D
Department of Psychiatry
Yale University School of Medicine
34 Park Street
New Haven, Connecticut 06510; and
National Center for Post-Traumatic Stress Disorder
Veterans Affairs Medical Center
950 Campbell Avenue
West Haven, Connecticut 06516

J. Douglas Bremner, M.D.
Assistant Professor of Psychiatry
Department of Psychiatry
Yale University School of Medicine
34 Park Street
New Haven, Connecticut 06510; and
National Center for Post-Traumatic Stress Disorder
Veterans Affairs Medical Center
950 Campbell Avenue
West Haven, Connecticut 06516

Dennis S. Charney, M.D.
Professor and Associate Chairman for Research
Department of Psychiatry
Yale University School of Medicine
333 Cedar Street
New Haven, Connecticut 06510; and
Chief, Psychiatry Service
National Center for Post-Traumatic Stress Disorder
Veterans Affairs Medical Center
950 Campbell Avenue
West Haven, Connecticut 06516

William E. Cullinan, Ph.D.
Research Fellow
Mental Health Research Institute
University of Michigan School of Medicine
205 Zina Pitcher Place
Ann Arbor, Michigan 48109

Jonathan R. T. Davidson, M.B., B.S. F.R.C.Psych
Professor of Psychiatry
Department of Psychiatry and Behavioral Medicine
Duke University Medical Center
P. O. Box 3812
Durham, North Carolina 27710

Michael Davis, Ph.D.
Professor of Psychiatry and Psychology
Department of Psychiatry
Yale University School of Medicine
34 Park Street
New Haven, Connecticut 06508

Ariel Y. Deutch, Ph.D.
Associate Professor of Psychiatry and Pharmacology
Departments of Psychiatry and Pharmacology
Yale University School of Medicine
34 Park Street
New Haven, Connecticut 06508; and
National Center for Post-Traumatic Stress Disorder
Veterans Affairs Medical Center
950 Campbell Avenue
West Haven, Connecticut 06516

Ronald S. Duman, Ph.D.
Associate Professor of Psychiatry and Pharmacology
Department of Psychiatry
Yale University School of Medicine
34 Park Street
New Haven, Connecticut 06508

John A. Fairbank, Ph.D.
Research Triangle Institute
P. O. Box 12194
Research Triangle Park, North Carolina 27709

William A. Falls, Ph.D.
Assistant Professor of Psychology
Department of Psychology
Northern Illinois University
DeKalb, Illinois 60115

Janet M. Finlay, Ph.D.
Research Assistant Professor
Department of Neuroscience
University of Pittsburgh
446 Crawford Hall
Pittsburgh, Pennsylvania 15260

Edna B. Foa, Ph.D.
Professor and Director
Center for the Treatment and Study of Anxiety
Department of Psychiatry
Medical College of Pennsylvania and Hahnemann University
3200 Henry Avenue
Philadelphia, Pennsylvania 19129

Matthew J. Friedman, M.D., Ph.D.
Professor of Psychiatry and Pharmacology
Departments of Psychiatry and Pharmacology
Dartmouth Medical School
Hanover, New Hampshire 03755; and
Executive Director
National Center for Post-Traumatic Stress Disorder
Veterans Administration Medical Center
White River Junction, Vermont 05009

Earl L. Giller, Jr., M.D., Ph.D.
Associate Director
CNS Clinical Research
Pfizer, Inc.
Eastern Point Road
Groton, Connecticut 06340

Paul E. Gold, Ph.D.
Director, Neuroscience Graduate Program
Department of Psychology
University of Virginia
102 Gilmer Hall
Charlottesville, Virginia 22903

Philip W. Gold, M.D.
Branch Chief
Clinical Neuroendocrinology Branch
National Institute of Mental Health
Building 10, Room 3S231
10 Center Drive
Bethesda, Maryland 20892

Dana L. Helmreich, Ph.D.
Research Fellow
Mental Health Research Institute
University of Michigan School of Medicine
205 Zina Pitcher Place
Ann Arbor, Michigan 48109

George R. Heninger, M.D.
Director
Laboratory of Clinical and Molecular Neurobiology
Department of Psychiatry
Yale University School of Medicine
34 Park Street
New Haven, Connecticut 06508

James P. Herman, Ph.D.
Assistant Professor
Department of Anatomy and Neurobiology
University of Kentucky Medical Center
800 Rose Street
Lexington, Kentucky 40536

Brian A. Horger, Ph.D.
Postdoctoral Associate
Department of Pharmacology
Yale University School of Medicine
333 Cedar Street
New Haven, Connecticut 06520

David R. Johnson, Ph.D.
Associate Professor of Psychiatry
Department of Psychiatry
Yale University School of Medicine
34 Park Street
New Haven, Connecticut 06510; and
National Center for Post-Traumatic Stress Disorder
Veterans Affairs Medical Center
950 Campbell Avenue
West Haven, Connecticut 06516

Peter W. Kalivas, Ph.D.
Director of Washington State University Alcoholism and Drug Abuse Program; and
Professor of Pharmacology
Department of Veterinary and Comparative Anatomy, Pharmacology, and Physiology
Washington State University
Pullman, Washington 99164

Danny G. Kaloupek, Ph.D.
Associate Clinical Professor of Psychiatry
Department of Psychology
Tufts University School of Medicine
National Center for Post-Traumatic Stress Disorder
Boston Veterans Affairs Medical Center
150 South Huntington Avenue
Boston, Massachusetts 02130

Terrence M. Keane, Ph.D.
Professor of Psychiatry
Tufts University School of Medicine; and
Director, National Center for Post-Traumatic Stress Disorder (Behavioral Science Division)
Boston Veterans Affairs Medical Center
150 South Huntington Avenue
Boston, Massachusetts 02130

Clinton D. Kilts, Ph.D.
Associate Professor of Psychiatry
Assistant Professor of Pathology
Department of Psychiatry and Behavioral Sciences
Emory University School of Medicine
1639 Pierce Drive
Atlanta, Georgia 30322

Thomas R. Kosten, M.D.
Professor of Psychiatry
Department of Psychiatry
Yale University School of Medicine
34 Park Street
New Haven, Connecticut 06510; and
National Center for Post-Traumatic Stress Disorder
Veterans Affairs Medical Center
950 Campbell Avenue
West Haven, Connecticut 06516

John H. Krystal, M.D.
Associate Professor of Psychiatry
Department of Psychiatry
Yale University School of Medicine
34 Park Street
New Haven, Connecticut 06510; and
National Center for Post-Traumatic Stress Disorder
Veterans Affairs Medical Center
950 Campbell Avenue
West Haven, Connecticut 06516

Joseph E. LeDoux, Ph.D.
Professor
Center for Neural Science
New York University
6 Washington Place
New York, New York 10003

Robert A. Levengood, M.D.
Associate Clinical Professor of Psychiatry
Department of Psychiatry
Mount Sinai School of Medicine
Bronx Veterans Affairs Medical Center
130 West Kingsbridge Road
Bronx, New York 10468

Julio Licinio, M.D.
Clinical Neuroendocrinology Branch
National Institute of Mental Health
Building 10, Room 3S231
10 Center Drive
Bethesda, Maryland 20892

Hadar Lubin, M.D.
Assistant Professor of Psychiatry
Department of Psychiatry
Yale University School of Medicine
34 Park Street
New Haven, Connecticut 06510; and
National Center for Post-Traumatic Stress Disorder
Veterans Affairs Medical Center
950 Campbell Avenue
West Haven, Connecticut 06516

Charles R. Marmar, Ph.D.
Department of Psychiatry
University of California, San Francisco
Box F-0984
San Francisco, California 94143

John W. Mason, M.D.
Professor Emeritus of Psychiatry
Yale University School of Medicine
34 Park Street
New Haven, Connecticut 06510; and
National Center for Post-Traumatic Stress Disorder
Veterans Affairs Medical Center
950 Campbell Avenue
West Haven, Connecticut 06516

Richard C. McCarty, Ph.D.
Professor and Chair
Department of Psychology
University of Virginia
102 Gilmer Hall
Charlottesville, Virginia 22903

Bruce S. McEwen, Ph.D.
Professor and Head
Laboratory of Neuroendocrinology
Rockefeller University
1230 York Avenue
New York, New York 10021

David Michelson, M.D.
Clinical Neuroendocrinology Branch
National Institute of Mental Health
Building 10, Room 3S231
10 Center Drive
Bethesda, Maryland 20892

Chris Molnar, B.A.
Senior Research Assistant
Center for the Treatment and Study of Anxiety
Department of Psychiatry
Medical College of Pennsylvania and Hahnemann University
3200 Henry Avenue
Philadelphia, Pennsylvania 19129

C. A. Morgan III, M.D.
Assistant Professor of Psychiatry
Department of Psychiatry
Yale University School of Medicine
34 Park Street
New Haven, Connecticut 06510; and
National Center for Post-Traumatic Stress Disorder
Veterans Affairs Medical Center
950 Campbell Avenue
West Haven, Connecticut 06516

Charles B. Nemeroff, M.D., Ph.D.
Professor and Chairman
Department of Psychiatry and Behavioral Sciences
Emory University School of Medicine
1639 Pierce Drive
Atlanta, Georgia 30322

Annabel Prins, Ph.D.
Instructor of Psychology
Department of Psychology
Smith College
Northampton, Massachusetts 01060

Robert M. Post, M.D.
Chief
Biological Psychiatry Branch
National Institute of Mental Health
Building 10, Room 3N212
10 Center Drive, MSC 1272
Bethesda, Maryland 20892

Robert S. Pynoos, M.D., M.P.H.
Associate Professor
Department of Psychiatry and Biobehavioral Sciences
University of California, Los Angeles
Neuropsychiatric Institute
300 UCLA Medical Plaza
Los Angeles, California 90024

Sherry J. Riney, M.S.W.
Research Coordinator
National Center for Post-Traumatic Stress Disorder
Veterans Affairs Medical Center
3801 Miranda Avenue
Palo Alto, California 94304

Robert H. Roth, Ph.D.
Professor
Departments of Pharmacology and Psychiatry
Yale University School of Medicine
333 Cedar Street
New Haven, Connecticut 06520

Barbara Olasov Rothbaum, Ph.D., A.B.P.P.
Assistant Professor
Department of Psychiatry
Emory University School of Medicine
The Emory Clinic
1365 Clifton Road
Atlanta, Georgia 30322

Philip A. Saigh, Ph.D.
Professor
Ph.D. Program in Educational Psychology
City University of New York
33 West 42nd Street
New York, New York 10036

William E. Schlenger, Ph.D.
Research Triangle Institute
P. O. Box 12194
Research Triangle Park, North Carolina 27709

Paula P. Schnurr, Ph.D.
Research Associate Professor
Department of Psychiatry
Dartmouth Medical School
Lebanon, New Hampshire 03756; and
Deputy to the Executive Director
National Center for Post-Traumatic Stress Disorder
Veterans Administration Medical Center
White River Junction, Vermont 05009

Larry J. Siever, M.D.
Professor
Department of Psychiatry
Mount Sinai School of Medicine
Bronx Veterans Affairs Medical Center
130 West Kingsbridge Road
Bronx, New York 10468

Mark A. Smith, M.D., Ph.D.
Medical Officer
Biological Psychiatry Branch
National Institute of Mental Health
Building 10, Room 3N212
9000 Rockville Pike
Bethesda, Maryland 20892

Barbara A. Sorg, Ph.D.
Assistant Professor of Pharmacology
Department of Veterinary and Comparative Anatomy, Pharmacology, and Physiology
Washington State University
Pullman, Washington 99164

Steven M. Southwick, M.D.
Associate Professor
Department of Psychiatry
Yale University School of Medicine
333 Cedar Street
New Haven, Connecticut 06510; and
National Center for Post-Traumatic Stress Disorder
Veterans Affairs Medical Center
950 Campbell Avenue
West Haven, Connecticut 06516

Susan M. Stine, M.D., Ph.D.
Assistant Professor
Department of Psychiatry
Yale University School of Medicine
34 Park Street
New Haven, Connecticut 06510; and
National Center for Post-Traumatic Stress Disorder
Veterans Affairs Medical Center
950 Campbell Avenue
West Haven, Connecticut 06516

Steven C. Stout, B.A.
Department of Psychiatry and Behavioral Sciences
Emory University School of Medicine
1639 Pierce Drive
Atlanta, Georgia 30322

Alan F. Sved, Ph.D.
Professor of Neuroscience and Psychiatry
Department of Neuroscience
University of Pittsburgh
446 Crawford Hall
Pittsburgh, Pennsylvania 15260

Sheila Wang, Ph.D.
Research Associate
Department of Psychiatry
Yale University School of Medicine
34 Park Street
New Haven, Connecticut 06510; and
National Center for Post-Traumatic Stress Disorder
Clinical Neuroscience Division
Veterans Affairs Medical Center
950 Campbell Avenue
West Haven, Connecticut 06516

Stanley J. Watson, Jr., M.D., Ph.D.
Professor of Psychiatry
Department of Psychiatry
Mental Health Research Institute
University of Michigan School of Medicine
205 Zina Pitcher Place
Ann Arbor, Michigan 48109

Daniel S. Weiss, Ph.D.
Department of Psychiatry
University of California, San Francisco
Box F-0984
San Francisco, California 94143

Susan R. B. Weiss, Ph.D.
Chief, Unit on Behavioral Biology
Biological Psychiatry Branch
National Institute of Mental Health
Building 10, Room 3N212
9000 Rockville Pike
Bethesda, Maryland 20892

Redford B. Williams, M.D.
Professor of Psychiatry and Psychology
Department of Psychiatry
Duke University Medical Center
2212 Elmer Street
Durham, North Carolina 27710

Steven H. Woodward, Ph.D.
Psychology Service
Veterans Affairs Medical Center
3801 Miranda Avenue
Palo Alto, California 94304

Rachel Yehuda, Ph.D.
Assistant Professor
Department of Psychiatry
Mount Sinai School of Medicine
Bronx Veterans Affairs Medical Center
130 West Kingsbridge Road
Bronx, New York 10468

Cheryl D. Young, Ph.D.
Departments of Psychiatry and Pharmacology
Yale University School of Medicine
34 Park Street
New Haven, Connecticut 06508; and
National Center for Post-Traumatic Stress Disorder
Veterans Affairs Medical Center
950 Campbell Avenue
West Haven, Connecticut 06516

Michael J. Zigmond, Ph.D.
Professor of Neuroscience and Psychiatry
Department of Neuroscience
University of Pittsburgh
570 Crawford Hall
Pittsburgh, Pennsylvania 15260

Foreword

Fifteen years ago most of the information contained in this book did not exist. The onrush of research, particularly related to the consequences of massive emotional stress, has been notable. This research has influenced the directions of study in animal laboratories and has led to the present widespread interest in the long-term consequences of experiencing life-threatening assaults, including war, rape, or industrial and natural disasters.

Each successive war has laid its burden of human misery. Often from that suffering there has emerged some new knowledge, understanding of illness, or therapeutic approach. For the most part, such insights were gleaned during the immediate period of aftercare of the wounded and acutely suffering. During the last century, increasing attention has been drawn to the more long-lasting disabilities and disorders that appear in the aftermath of brutal struggles. The American Civil War with its many casualties and incredible number of amputations and other long-lasting physical disabilities focused attention in the United States on the plight of the disabled survivors and on the need to aid them to live effectively.

Such concerns expanded in this century. During the two world wars, huge populations were pressed into the armed forces. Innumerable people became acutely incapacitated on the front lines with poorly understood behavioral and emotional disturbances. The explanations for these conditions varied from those who espoused a purely psychological cause to others who considered the disorders a result of shock of the nervous system. Surprisingly, in spite to the large number of military affected worldwide, little research was carried out on these conditions during and following the world wars.

As the United States prepared to enter World War II, Kardiner's clinical study of the long-term behavioral disturbances of World War I combat veterans was published as a research monograph. Noting the persistence of the startle reaction through many years among these veterans, Kardiner hypothesized that it was the nuclear symptom of this chronic condition and postulated that all other symptoms and personality deviations were organized, over time, around the abnormal startle reaction. Interestingly, Kardiner hypothesized that this symptom had resulted from conditioning of the basic startle reflex. He combined the emerging psychophysiological laboratory research of his time with psychoanalytical theories of personality development. Between World War II and America's major military involvement in Southeast Asia, only one investigation was carried out to test his hypothesis. Nevertheless, new information related to the impact of exposure to various forms of stress emerged from the laboratories of behavioral scientists.

It was the war in Vietnam and its living victims, particularly the combat veterans with similar conditions, that exploded the indifference of researchers and clinicians. Veterans seeking help for their symptoms began to attract the attention of a few investigators who initiated (largely unfunded) psychophysiological investigations. Often demeaned in their attempts to obtain aid and even turned away by the Veterans Administration health system from which they sought such assistance, Vietnam veterans organized politically to make their case. They demanded from the federal government and the U.S. Veterans Administration that more clinical attention and more research be dedicated to alleviating the suffering and long-term consequences of combat-induced stress disorders and other service-connected disabilities. They have been eminently successful in achieving their political goals.

Neurobiological and Clinical Consequences of Stress: From Normal Adaptation to Post-

Traumatic Stress Disorder, is an example of that remarkable success. Many of the chapters in this book are written by members of the staff of the National Center for Post-Traumatic Stress Disorder, that came into being as a result of the political activity of the Vietnam Veterans of America. Matthew Friedman directs that center and Dennis Charney directs one of its several divisions. With Ariel Deutch, another member of the National Center staff, they have recognized the need to bring together and integrate the latest results from laboratory and clinical research on stress. Therefore, they have brought together and organized reviews and reports on the most recent experimental data from both sources. They have prepared summaries, presented their interpretations of these exciting new findings, and reflected on possible new avenues for future investigations. Throughout this book, one also finds superb bibliographies related to the contents of each chapter.

This book also reflects the moving focus of research studies over the past 15 years. First came investigations on the abnormal psychophysiological arousal found in patients with post-traumatic stress disorder. Shortly thereafter, neurohumoral abnormalities were detected implicating dysfunction of the pituitary-adrenal axis. Within the past 5 years, attention has shifted to improving our understanding of the central neural processes that underlie impairments in perception, memory, and habituation seen in post-traumatic stress disorder patients.

When reading the first section of this book, my fascination reached its height with Bruce S. McEwen's chapter "Adrenal Steroid Actions on Brain: Dissecting the Fine Line between Protection and Damage." I was at first interested in his review section on noradrenergic innervation of the hippocampus and became even more engaged by his descriptions of recent work demonstrating the unusual sensitivity of the hippocampus to adrenal steroids. His discussion of the concept of "allostatsis" in contrast to "homeostasis" as an important working dynamic of the central nervous system, as well as the consequent "allostatic load" that develops under circumstance, is most thought provoking. It brings to the forefront the usefulness of the now loosely used term "stress" and the need to relate its impact to systemic changes in internal bodily functions.

The reader is likely to discover new or reformulated previous information, even in areas where one has tried to keep up with the emerging literature. For those who have not previously been acquainted with explanatory models on brain responses to stress, the book's second section will be most informative.

One may infer that through the studies and findings reported herein that neurobiologists are now approaching detailed studies of a major and specific brain system—the agonistic or alarm system necessary to sustain life for all organisms. The system includes a complex cortical perceptual segment which evaluates each new threat and compares it with previously stored memories produced by distressing emotional experiences. There are also genetic factors that influence this process. Following this evaluation phase, the system determines appropriate (active or inhibitory) response and electrochemically initiates the relevant organ actions.

I certainly expect that those who are seriously interested in the field of stress disorders will place this book on their shelves.

Lawrence C. Kolb, M.D.
Emeritus Professor of Psychiatry
College of Physician and Surgeons
Present address:
232 Van Wies Point
Glenmont, New York 12077

Preface

This book is about the neurobiological consequences of stress. To our knowledge, it represents the first attempt to review the entire spectrum of stress responses from normal to pathological reactions. The capacity to cope with stress is a crucial theme in human evolution. Humans are well equipped to respond biologically and psychologically to the everyday stressors that punctuate life on this planet. It can even be argued that stress may have a salutary as well as an adverse impact on individuals, and that efforts to master such challenging situations promote human achievements in scholarship, athletics, creativity, and psychological growth.

Sometimes, the response to stress is not successful, and adaptation is not achieved. This may occur because the stressor is so extreme and overwhelming that the adaptive capacities of most persons are exceeded. In a clinical context, such stressors are labeled "traumatic," and they may precipitate post-traumatic stress disorder (PTSD) in susceptible individuals. We know much less about the neurobiology of the traumatic (as compared with the normal) stress response since there are far fewer studies with PTSD patients than with volunteer subjects exposed to nontraumatic laboratory stress paradigms. Since the introduction of PTSD as a psychiatric diagnosis in 1980, however, we have learned that PTSD is not uncommon and that exposure to traumatic events, including war, rape, child abuse, natural disasters, is even less uncommon (see Chapter 23 for the relevant epidemiologic data). For this reason it is imperative that we try to understand the pathophysiology of PTSD. In our opinion, we can begin to achieve this goal by extrapolating from research on the normal stress response in humans.

It has been recognized that people exposed to stress can suffer from profound and sometimes irreversible biological consequences. As early as 1864, reports began appearing in the medical literature suggesting that military stress could result in serious cardiovascular abnormalities, variously labeled as soldier's heart, nervous exhaustion, DeCosta's syndrome, irritable heart, effort syndrome, and neurocirculatory asthenia. In 1915, Walter Cannon proposed the fight-or-flight reaction to stress, the proposition that the sympatho-adrenal-medullary axis is mobilized in response to threat. This represents, perhaps, the first neurobiological theory of the stress response. Hans Selye, the second great pioneer in this field, postulated in 1950 that both immunologic and hypothalamic-pituitary-adrenocortical (HPA) systems are affected following persistent exposure to stress. What has evolved since then is a great body of modern research identifying a neurobiologic "stress system" that mediates the organism's response to threat. Major components of this stress system are the HPA and central and peripheral adrenergic systems that activate a complex cascade of physiological, pharmacological, endocrinological, and immunological responses under stressful conditions.

These neurobiological processes do not occur *in vacuo*. There is a major psychological component, the alerting response, (see Chapters 7, 9, 10, and 11) that sets all of this activity into motion after the organism recognizes a specific situation as stressful. The cognitive process by which a potential stressor is identified as either a threat or a challenge is called appraisal. A change in appraisal can produce a significant change in the subsequent neurobiological response to a given stressor. A number of brain structures, especially the amygdala (see Chapters 5, 7, 11, and 15) participate in this process by which specific situations are designated as either safe or potentially dangerous.

The literature on the neurobiology of the traumatic stress response begins with the seminal work of Abram Kardiner (1941). Along with the other physiological abnormalities he observed among World War I veterans with war-related "traumatic neurosis," Kardiner was especially impressed by their autonomic hyperarousal, exaggerated startle reflexes, and disturbed sleep. He called this disorder a "physioneurosis." Lawrence Kolb (1987) has revived Kardiner's important observations for contemporary researchers and suggested that PTSD is associated with alterations in brain function, especially in the limbic system. More recently a number of fundamental bio-behavioral animal models have been proposed for PTSD including inescapable stress, fear conditioning, behavioral sensitization, and kindling (Charney, et al. 1993; Davis 1990; Post, et al. 1988; van der Kolk, et al. 1985).

The guiding principle of this book is that humans exposed to catastrophic stressors utilize the same neurobiological mechanisms that are activated following exposure to a less severe "normal" stressor. We believe, however, that failure to successfully cope with traumatic stress has significant neurobiological consequences. Whereas successful adaptation is followed by a restoration of normal homeostatic balance, unsuccessful adaptation may result in an equilibrium state which, though stable, deviates significantly from normative neurobiological standards. McEwen (see Chapter 8) has called such an equilibrium "allostatis." He gives as a concrete example of the difference between the homeostatic balance of a see-saw with two 50-pound 6-year olds at either end compared with the allostatic horizontality of the same see-saw with two 5-ton elephants at either end. In both cases, the see-saw is perfectly balanced, but in the case of allostatis, such balance places severe demands on the organism and may be impossible to sustain indefinitely. Yehuda and associates (see Chapter 19) indicate that there is strong evidence to suggest that such an allostatic equilibrium is present in the HPA system of patients with PTSD.

The book is divided into five sections, each of which is preceded by a brief overview to maintain context and continuity. Part I includes eight chapters that emphasize basic science studies of stress from a variety of approaches. Part II moves from primary data to synthesis. These seven chapters concern animal models of neurobiological processes that have heuristic value for clinical theory on the pathophysiology of PTSD. Part III consists of seven chapters concerning laboratory and clinical abnormalities detected in human subjects exposed to normal stressors and in PTSD patients exposed to traumatic stressors. The fourth part contains seven chapters that synthesize findings presented previously and attempt to show their relevance for the diagnosis and treatment of PTSD. Part V, the final chapter, is an overview of the work presented throughout this book. It includes our speculations on its significance and on its implications for future research and treatment. We believe that our emphasis on the homology between the normal stress response and the pathophysiology of PTSD is fundamental to understanding this disorder.

This book will help PTSD investigators to familiarize themselves with basic research on stress. It will provide PTSD clinicians with a theoretical context and some practical tools to assist them in their efforts to alleviate the suffering, immobility, and despair caused by this disabling disorder.

Matthew J. Friedman, M.D., Ph.D.
Dennis S. Charney, M.D.
Ariel Y. Deutch, Ph.D.

Acknowledgments

In 1989, the Department of Veterans Affairs established the National Center for Post-Traumatic Stress Disorder as mandated by the U.S. Congress. It brought us together. The seven divisions of the National Center provided a context and impetus through which basic neurobiologists and clinical neuroscientists developed a research agenda that is reflected by the wide spectrum of topics covered in this book. Indeed, most of our thirty chapters were written by members of the National Center's nuclear or extended family. Therefore, we wish to thank the veterans, veterans service organizations, and Congressional legislators who were instrumental in creating the National Center. We thank the officials in the Department of Veterans Affairs who have supported us during our first 5 years of operation, especially Paul Errera, Thomas Horvath, Laurent Lehmann, and Gay Koerber.

We thank a number of key individuals whose efforts were essential in the actual production of this book. They include Peggy Pearson, Laura Ocker, and Betsy Carlucci, in White River Junction, Vermont, and Evelyn Testa in West Haven, Connecticut.

Finally, we thank Mark Placito and Jill Barnowski of Lippincott-Raven Publishers who provided the professional guidance, logistic support, and technical expertise needed to make this book a reality.

Neurobiological and Clinical Consequences of Stress

From Normal Adaptation to Post-Traumatic Stress Disorder

PART I

From Molecules to Behavior: Stress in Laboratory Animals

As one reads the contributions to this section, it becomes apparent that research on stress encompasses a broad range of studies that require a correspondingly broad range of techniques and concepts. Nonetheless, certain common themes can be seen to emerge.

Historically, a considerable amount of attention has been devoted to both peripheral and central sequelae to exposure to a stressor, and thus to the hypothalamo-hypophyseal-adrenal (HPA) axis. Attention has now focused on understanding the regulatory features that govern the activity of the HPA axis. This has led to a strong emphasis on studies that reveal the functional activation of neurons regulating HPA function. A method that has been used with considerable success is examination of the functional activation of neurons as reflected by immediate-early gene induction. Chapter 1 shows how this approach has been used to demonstrate those neurons that are activated by stress and contribute to functional circuits regulating the HPA axis. The utility of studying immediate-early gene induction as a marker for neuronal activation is amplified when one considers that c-fos is a transcriptional regulator; expression of the immediate-early gene is indicative not only of activation, but also of a long-term molecular change in neurons. Chapter 2 outlines identification of changes in gene expression that result from stress. The elucidation of the genes that are regulated and the transduction mechanisms that determine induction of a particular gene have only recently been addressed, but are certain to be a major focus of future studies, particularly in view of disorders such as post-traumatic stress disorder (PTSD).

The functional localization of stress effects as revealed by molecular and anatomical methods is consistent with the increasing attention being devoted to understanding the neurotransmitter systems involved in stress. Localization studies have revealed that a large number of CNS regions are involved in the stress response, and it is likely that there are a similarly large number of transmitter systems that respond to stress. There have been remarkable strides in uncovering the ways in which various stressors impact different transmitter systems in distinct regions of the CNS. For example, there has been an interest in norepinephrine (NE) dynamics in response to stress since the turn of the century. Chapter 3 covers new information concerning the regionally and functionally distinct responses of NE neurons to stress, including target areas such as the hippocampus (where NE changes may be related to certain forms of memory) and the neocortex. Chapter 5 describes parallel studies of the central dopamine systems, which are very sensitive to stress-induced perturbation and which are functionally heterogenous. These authors focus on how sensitization and cross-sensitization of the dopamine (DA) responses occur. These changes may be relevant to a number of clinical conditions that are marked by co-morbidity with substance abuse. Chapter 4 reviews recent studies on the involvement of amino acid systems in stress. These studies have revealed that neurons once considered to be homogeneously distributed sets of excitatory (e.g., glutamate) and inhibitory γ-aminobutyric acid (GABA) cells are more complex. When one considers the striking regional and functional heterogeneity of amino acid systems as reflected by the receptors for these transmitters, it is apparent that pharmacological interventions mediated by inhibitory or excitatory amino acids provide both current and future opportunities for pharmacotherapy for anxiety and panic disorders, and PTSD. Similarly, Chapter 6 shows that there have been remarkable strides in uncovering the involvement of peptidergic transmitters in stress. The identification and cloning of these peptides and their receptors have allowed us to examine in detail the functional regulation of the HPA axis, and to identify peptides that are likely to be important targets for pharmacotherapy of stress-related disorders (e.g., neuropeptide Y and neurotensin).

In addition to the issue of localization of function in studies of stress, another common theme that emerges is the relationship between stress and different forms of learning and memory. Current studies have shown that certain of the neuronal systems that respond to stress are key stations in the distributed networks that subserve different forms of memory. Indeed, as reviewed in Chapter 7, we have moved far past Lashley's engram to a view that the response to stress, including the emotional salience and cognitive elaboration of stress, occurs in key subcortical as well as cortical regions. The interrelationship of stress and memory functions may shed important new light on clinical conditions such as PTSD.

A common theme addressed in virtually every chapter in this section is that we must attend to both the acute and chronic sequelae of stress. The effects of chronic stress involve a number of distinct adaptations, ranging from molecular to behavioral changes. As shown in Chapter 8, some of these adaptations may be beneficial (e.g., the development of coping strategies) whereas others may be detrimental (e.g., dystrophic changes and loss of neurons in the hippocampus). As studies describing the differences in neuronal function in response to chronic stress have accumulated, we have also begun to appreciate that there are large differences in the stress response that depend upon the type of stressor (anticipated versus unexpected, or escapable versus nonescapable).

As one reads the various contributions in this section, there is one final theme that becomes apparent: investigation of stress has moved past the simple (and simplistic) dichotomy of basic as opposed to clinical research. Investigators who have traditionally been called basic scientists have focused their efforts, which are increasingly molecular and cellular in orientation, on integrating their findings into system themes and generalizing the relevance of this work to broader clinical issues. And while basic science drives clinical science, both in terms of treatment and understanding underlying pathophysiology, there is a reciprocating relationship in which clinical science has also provided important new findings that have directed attention in basic laboratories to novel mechanisms and paradigms.

Neurobiological and Clinical Consequences of Stress: From Normal Adaptation to PTSD, edited by M. J. Friedman, D. S. Charney, and A. Y. Deutch, Lippincott-Raven Publishers, Philadelphia © 1995.

1

A Neuroanatomy of Stress

William E. Cullinan, **James P. Herman, Dana L. Helmreich, and †Stanley J. Watson, Jr.*

*Mental Health Research Institute, †Department of Psychiatry, University of Michigan School of Medicine, Ann Arbor, Michigan 48109, and *Department of Anatomy and Neurobiology, University of Kentucky Medical Center, Lexington, Kentucky 40536*

Although the concept of "stress" remains ill-defined, it generally refers to physical or psychological alterations capable of disrupting homeostasis. In mammals, the response to stress involves a number of processes comprising what was termed by Selye the "General Adaptation Syndrome" (1); a major neuroendocrine component of this response is mediated by the hypothalamic-pituitary-adrenocortical (HPA) axis. Various stress-related inputs converge upon the neurons of the hypothalamic paraventricular nucleus (PVN) (see Fig. 1). The PVN neurons that synthesize corticotropin-releasing hormone (CRH) (2), arginine vasopressin (AVP), and other secretagogues project to the external layer of the median eminence, where they release their products into the portal circulation, resulting in the synthesis and release of adrenocorticotropic hormone (ACTH) from anterior pituitary corticotrophs. The ACTH, in turn, activates the biosynthesis and release of glucocorticoids from the adrenal cortex, which act broadly throughout the body to mediate changes in processes (e.g., metabolic, immune, inflammatory) required for adaptation. Due to their potency and wide range of action, glucocorticoid levels must be maintained at appropriate levels; either too much or too little is deleterious to an organism. This regulation is accomplished through complex multiple feedback loops operating at pituitary and brain levels.

In accordance with its role in coordinating responses to various neuronal and hormonal influences indicative of multiple forms of "stress," the PVN is known to receive a diverse array of inputs, many of which are beginning to come into sharper focus. In the following sections we will briefly summarize some of what is known regarding the organization of neural circuitry underlying integration of the stress response, focusing initially on putative activational pathways, followed by a review of possible negative feedback mechanisms. We will also review recent evidence from functional/anatomical studies aimed at disclosing neural pathways mediating responses to acute stressors.

Although the present chapter chiefly concerns neural influences directed upon the HPA axis, it is important to bear in mind that such effects represent only a portion of the complex physiological and behavioral responses to stress. The brain's responses to stressful stimuli are clearly widespread and dramatic, and recent evidence, some of which is reviewed here, has served to reinforce this fact. Thus, while activation of the HPA axis represents one primary manifestation of "stress," additional aspects of the brain's response to stress, and their potential interplay,

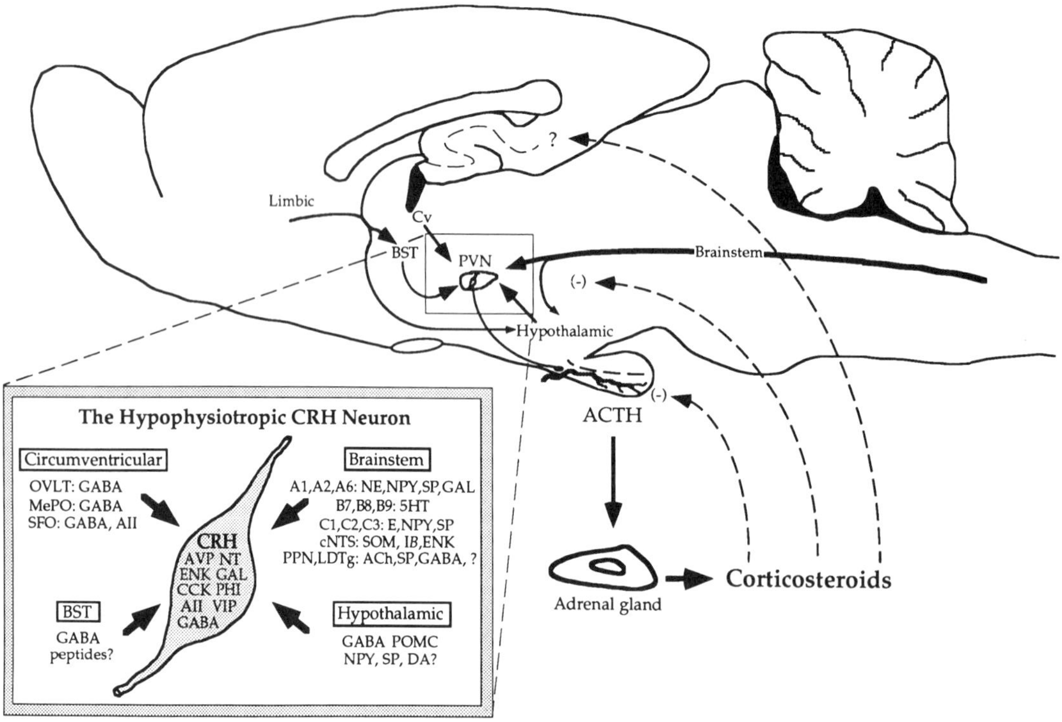

FIG. 1. Summary diagram in saggital plane illustrates components of the rat hypothalamic-pituitary-adrenocortical (HPA) axis, some sources of afferents, and potential sites of glucocorticoid feedback. Boxed region lists various neuropeptides colocalized in hypophysiotropic CRH neurons, as well as the neurochemical specificity of a number of inputs. Data for boxed region drawn from references: 13–16,20,35–38,42,67,68,79,101,106,128,129,167,168,173,189–193. Abbreviations: A1, A2, A6, brainstem noradrenergic cell groups A1, A2, A6; AII, angiotensin II; ACh, acetylcholine; ACTH, adrenocorticotropic hormone; AVP, arginine vasopressin; B7–B9, serotoninergic cell groups B7–B9; BST, bed nucleus of the stria terminalis; C1–C3, brainstem adrenergic cell groups C1–C3; CCK, cholecystokinin; cNTS, nucleus of the solitary tract, caudal portion; CRH, corticotropin-releasing hormone; Cv, circumventricular organs; DA, dopamine; E, epinephrine; ENK, enkephalin; GABA, gamma-aminobutyric acid; Gal, galanin; *IB*, inhibin; LDTg, laterodorsal tegmental nucleus; MePO, median preoptic nucleus; NE, norepinephrine; NPY, neuropeptide Y; NT, neurotensin; PHI, peptide histidine-isoleucine; POMC, pro-opiomelanocortin; PPN, pedunculopontine nucleus; SFO, subfornical organ; SOM, somatostatin; SP, substance P; VIP, vasoactive intestinal peptide.

will need to be considered for a complete understanding of these effects.

THE PARAVENTRICULAR HYPOTHALAMIC NUCLEUS

The final common pathway controlling activity of the HPA axis originates in the PVN. This wing-shaped nucleus positioned at the rostral end of the hypothalamus along the dorsal border of the third ventricle can be divided into distinct subterritories on the basis of cytoarchitectural, cytochemical, and connectional features (3). Initially, the PVN can be separated into magnocellular and parvicellular divisions. Magnocellular neurons are large neurosecretory cells that primarily synthesize AVP or oxytocin (OT) (as well as a number of neuropeptides), project to the posterior pituitary, and release their products into the peripheral circulation. The classic role of AVP released from magnocellular neurons is in maintenance of water and salt balance, while

OT released from the posterior pituitary is critical for lactation and parturition. Parvicellular elements of the PVN can be divided into five distinct subregions: the periventricular region, anterior parvicellular region, medial parvicellular region (which is further divided into dorsal and ventral components), dorsal parvicellular region, and lateral parvicellular region (4). Neurons located mainly within the dorsal portion of the medial parvicellular region, which synthesize CRH, project to the external lamina of the median eminence, and release CRH into the hypophyseal portal vasculature, comprise the origin of the final common pathway for ACTH release. The periventricular and dorsal medial parvicellular regions contain neurons that synthesize somatostatin, growth hormone releasing hormone, thyrotropin-releasing hormone (TRH), and dopamine; these neurons also project to the external lamina of the median eminence and control the release of growth hormone, thyrotropin, and prolactin from the anterior pituitary (5). The anterior, ventral medial parvicellular, and lateral parvicellular regions of the PVN contain cells that project to the intermediolateral column of the spinal cord and/or dorsal vagal complex and other parasympathetic cell groups, while dorsal parvicellular neurons project almost exclusively to the intermediolateral column of the spinal cord; the descending projections of these parvicellular neurons impart an integral function in the regulation of the autonomic nervous system. Coupled with the endocrine regulation afforded by the magnocellular and neuroendocrine parvicellular neurons, the PVN is thus considered a critical locus for integration and regulation of pathways concerned with maintaining homeostasis.

CRH is acknowledged as the primary ACTH secretagogue due to its potent intrinsic ACTH-releasing activity (6), coupled with the observation that pharmacological blockade of CRH can eliminate the ACTH response to various stress conditions (7,8). However, a number of additional peptides present within the dorsal medial parvicellular region of the PVN, or "hypophysiotropic zone," have been found to be co-stored with CRH. Most notable among these is AVP, which was detected in 50% of CRH neurons in colchicine-treated male rats (9). Interestingly, terminals of CRH neurons are apparently segregated into vasopressin-rich and vasopressin-deficient subpopulations that are differentially regulated by glucocorticoids and stress (10–13). In addition to AVP, cholecystokinin (CCK), galanin, vasoactive intestinal peptide (VIP), peptide histidine-isoleucine (PHI), GABA angiotentin II, enkephalin, and neurotensin have all been found to be variably colocalized in CRH-containing neurons (14–16). In terms of secretagogue activity, AVP has been shown to have weak effects on intrinsic ACTH release, but can greatly potentiate the ability of CRH to stimulate ACTH secretion (6–8,17). Weak activity of CCK, VIP, PHI, and Angiotensin II has also been found on ACTH release (18–22).

The multiplicity of neuropeptides located within hypophysiotropic CRH neurons allows for differential patterns of secretagogue release in response to a wide range of physiological conditions, dependent upon factors such as the corticosteroid environment and the category of afferent information. For example, the regulation of AVP within the hypophysiotropic zone appears to be much more sensitive to circulating corticosterone levels than CRH (23,24). Different categories of stressors may also target relatively distinct populations of hypophysiotropic neurons through different projections. Plotsky et al. (7,8) reported that hemorrhage causes an increase in portal levels of CRH and AVP, while insulin-induced hypoglycemia causes an increase in portal levels of AVP only. Similarly, ether stress but not cold or swim stress has been shown to stimulate an increase in enkephalin mRNA levels within the mpPVN (25,26). Results such as these suggest a marked capacity for neuronal plasticity and precise regulation of the ACTH secretagogue signal reaching the anterior pituitary corticotrophs. However, further investigation will be needed to ascertain the anatomical basis for and functional implications of multiple peptide "activation" and release, particularly after different stress conditions.

It should also be noted that in addition to the hypophysiotropic CRH neuron, magnocellular neurosecretory neurons represent another potential source of ACTH-releasing activity. A largely magnocellular origin of AVP and OT in the hypophyseal portal blood has been reported (27) and OT, similar to AVP, has been shown to have weak intrinsic ACTH-releasing activity and can potentiate the ability of CRH to stimulate ACTH secretion (28). Various mechanisms have been proposed for magnocellular influences on pituitary corticotrophs based on anatomical and functional data (29–31), and a role for magnocellular neuropeptides in the regulation of ACTH secretion in response to at least some forms of stress is supported by a recent study (32). However, the manner and extent to which magnocellular neuropeptides contribute to stress-induced ACTH secretion remain to be clearly defined.

AFFERENT REGULATION

Hypophysiotropic neurons of the PVN are known to receive a diverse set of inputs that have been previously defined according to class (5,33,34): 1) Brainstem pathways that are largely catecholaminergic are thought to be involved in transmitting visceral or interoceptive information to the PVN (35,36). A pivotal role in this circuitry is played by the nucleus of the solitary tract (NTS), which serves as the primary relay for information coursing via the glossopharyngeal and vagus nerves. In addition to its catecholaminergic projections, the NTS provides various nonaminergic inputs to the PVN (37,38), and projects to several brainstem regions which, in turn, innervate the PVN. These regions include the A1 and C1 cell groups, and the lateral parabrachial nucleus (39–41), and may represent routes by which stress-related information influences the HPA axis. Additionally, several midbrain and pontine cell groups have been suggested to be involved in the relay of somatosensory information to the PVN, based on connectional data (33,42). These areas include various mesencephalic nuclei, the laterodorsal tegmental (LDTg) and pedunculopontine nuclei, and the periaqueductal and central gray area. 2) A number of sites within the limbic telencephalon are thought to mediate cognitive/emotional influences upon the HPA axis. These regions include the septum, amygdala, and hippocampus, as well as a number of limbic cortical areas, and the bed nucleus of the stria terminalis (BST). However, with the exceptions of the BST and perhaps restricted amygdaloid territories, most of these regions appear to lack significant projections to the PVN. The majority are thus thought to influence the HPA axis indirectly, and the BST has been frequently viewed as a potential relay for many of such effects. 3) Numerous portions of the hypothalamus itself are known to project to the hypophysiotropic zone of the PVN. Since most of these hypothalamic regions also receive dense innervation from limbic and brainstem structures, they are in a position to provide integration of a broad array of behavioral and motivational responses. 4) A final class of inputs emanates from circumventricular regions such as the organum vasculosum of the lamina terminalis (OVLT) and subfornical organ (SFO), which relay blood-borne chemosensory signals to the PVN, including the hypophysiotropic zone.

The extent to which these various afferent systems are engaged in response to specific stressors, and the precise pathways by which they influence HPA activity, are currently the focus of intensive study. The following sections review some of what is known about a number of these pathways in the context of stress responsiveness.

Brainstem Pathways

Catecholaminergic Inputs

Perhaps the most extensively studied neuronal system in the regulation of HPA activity is the brainstem catecholaminergic system. An enhancement of catecholaminergic activity following exposure to different forms of stress has long been established (43–48). While initial studies indicated that catecholamines have inhibitory effects on ACTH secretion (49,50), subsequent data involving chemical or mechanical interruption of ascending catecholaminergic pathways strongly suggested a stimulatory action that is essential for stress-induced HPA activation (51–58). Early activation of brainstem catecholaminergic cell groups in response to stress has also been suggested from biochemical studies (59) as well as experiments detecting induction of cellular immediate-early genes as markers for neuronal activation (60–64).

The brainstem noradrenergic and adrenergic inputs to the PVN have been mapped in detail, and ultrastructural evidence has confirmed noradrenergic and adrenergic synapses on CRH neurons (35,36,65,66). The noradrenergic innervation of the PVN emanates from the A1 cell group of the ventrolateral medulla, the A2 cell group, corresponding to the nucleus of the NTS, and the A6 cell group of the locus coeruleus. However, while the A1 cell group appears to project primarily to the magnocellular portions of the PVN (36), it is the A2 catecholaminergic cell group that provides the CRH zone of the PVN with its major noradrenergic innervation, along with a minor contribution from the locus coeruleus (A6) (35). Consistent with this, recent lesion data have underscored the importance of A2 in stimulating adrenocortical responses to ether stress (58). The adrenergic innervation of the PVN originates from the C1 cell group of the ventrolateral medulla, the C2 of the NTS, and the C3 cell group of the dorsomedial medulla. These projections are directed to the dorsal and dorsomedial parvocellular portions of the nucleus (36). The PVN-projecting brainstem catecholaminergic neurons also display a number of peptides that are variably colocalized in these cells. For example, neuropeptide Y (NPY) is found in most adrenergic neurons that project to the PVN, as well as in a subpopulation of PVN-projecting neurons in the A1 region (67), and galanin is co-localized in PVN-projecting neurons within A1 and A6 (68).

Whereas noradrenergic afferents to the PVN appear to provide a primarily stimulatory signal to the CRH neurons, the functional impact of adrenergic afferents is less clear. Pharmacological data have suggested that adrenaline can have stimulatory effects on the HPA axis (69–71), however, evidence suggesting an inhibitory role has also been reported (72). The functional significance of brainstem peptidergic inputs also remains to be established.

While brainstem catecholaminergic neurons appear to be important in the early activation of the HPA axis, at least for several forms of stress, a number of important issues remain unresolved. Among them is the question of whether such influences are exerted at the level of the PVN, the median eminence, or both. Thus, an intact catecholaminergic innervation of the PVN has been shown to be necessary for the HPA response to a conditioned emotional stimulus (57), although catecholaminergic effects at the level of the median eminence also appear to be involved in ether stress-induced ACTH release (54). In addition, the receptor mechanisms by which the effects of noradrenaline and adrenaline are mediated remain to be firmly established, although the stimulatory actions of noradrenaline at least may operate via $\alpha-1$ receptors (see reviews in 73,74). An understanding of the functional roles of neuropeptides that are either co-localized or parallel these catecholaminergic projections are also likely to be of importance in understanding regulation of HPA activity.

Serotoninergic Inputs

Similar to the effects of catecholamines, the issue of serotonin effects on stress-induced HPA activity is the subject of controversy. Early pharmacological studies suggested an inhibitory role (75,76), but subsequent studies challenged this view (77,78). Anatomical data have revealed a modest serotoninergic projection to the CRH zone of the PVN, originating from the brainstem serotoninergic cell groups B7-B9 (79), and ultrastructural evidence for synaptic contact between serotoninergic terminals and CRH neurons has been advanced (80). A number of in vivo and in vitro studies suggest that serotoninergic agents may exert stimulatory effects at the level of the PVN (81–85), however, the picture is complicated by data suggesting that serotonin may also increase corticosterone secretion via effects at the adrenal level (86,87). Also, while depletion of hypothalamic serotonin appears to reduce the adrenocortical response to photic stimuli and stimulation of various limbic and cortical sites, the lesions apparently have no effect on the adrenocortical response to stressors such as ether, restraint, hypertonic saline administration, or insulin-induced hypoglycemia (88–92). It has thus been suggested that the role of serotoninergic pathways in stress-induced HPA activation may be stressor-specific (89–90). The diffuse nature of the serotoninergic projection system offers numerous possibilities for sites of neurotransmitter action that might explain these differential responses.

Mesopontine Afferents

The neural pathways by which somatosensory information accesses the HPA axis have remained elusive, as the PVN appears to lack direct input from somatosensory cortical or thalamic regions. Sawchenko and colleagues have identified a number of mesencephalic and pontine cell groups that may be in a position to relay somatosensory information to the PVN (42, see review 33). Included are a series of mesencephalic nuclei (e.g., posterior intralaminar, peripeduncular) that relay auditory information (93) and have been suggested to be involved based on defined PVN projections (42). Similarly, projections to the hypophysiotropic zone have been reported from the periaqueductal and pontine central gray (42), representing a pathway by which sensory and nociceptive information might be conveyed. Additionally, two cell groups located in the mesopontine tegmentum, the pedunculopontine nucleus and laterodorsal tegmental nucleus, appear to be in a position to relay visual and auditory information, as well as somesthetic inputs from the spinal cord (42,94–97). These cell groups are largely cholinergic and variably colocalize substance P (98–101), and the proposed pathway is consistent with previous evidence suggesting stimulatory effects of acetylcholine on CRH secretion (102,103). Conversely, the PVN appears to receive a relatively sparse cholinergic innervation in comparison to surrounding hypothalamic regions (104,105), and a recent study reported that the projections of the LDTg to the PVN are only partially cholinergic (106). In any case, possible functional roles for the aforementioned projections in HPA activation will clearly require investigation. It is notable, however, that experiments detecting induction of cellular immediate-early genes have suggested neuronal activation in many of the same regions (e.g., peripeduncular nucleus, LDTg) in response to various stress paradigms (64).

Limbic Inputs

Numerous studies have indicated that HPA activity can be influenced by portions of the limbic telencephalon, including the hippocampus, medial prefrontal cortex, septum, and amygdala (107–122). The precise pathways by which such effects are mediated have remained unclear because, with the exception of a modest input from the central (and possibly the medial) amygdaloid nucleus, recent anatomical data have indicated that these regions lack substantial direct projections to the PVN (106,123–128) (Fig. 2D–E). Findings such as these suggest that

FIG. 2. A: An injection site within the ventral portion of the hippocampal formation (ventral subiculum) of the anterograde neuronal tracer PHA-L. **B–D:** Darkfield photomicrographs illustrate the distribution of axons resulting from the PHA-L injection site shown in (A) at forebrain levels of the anterior portion of the BST (B), caudal portion of the BST (C), and PVN (D). Note that while a heavy innervation of the BST is seen, this projection avoids the PVN. **E:** A similar distribution pattern is seen at the level of the PVN in darkfield photomicrograph, following an injection of the tracer in the lateral septal nucleus. **F:** Darkfield photomicrograph illustrates the distribution of PHA-L–labeled fibers in the PVN following injection of the tracer in the caudal BST within the posterior intermediate division of Moga et al. (241). Abbreviations: ac, anterior commissure; AHA, anterior hypothalamic area; AM, anteromedial thalamic nucleus; BST, bed nucleus of the stria terminalis; CA1, field CA1 Ammon's horn; Ent, entorhinal cortex; f, fornix; ic, internal capsule; MPA, medial preoptic area; ot, optic tract; pi, posterior intermediate division of BST; PVN, hypothalamic paraventricular nucleus; Sub, subiculum; VP, ventral pallidum; ZI, zona incerta; 3V, third ventricle.

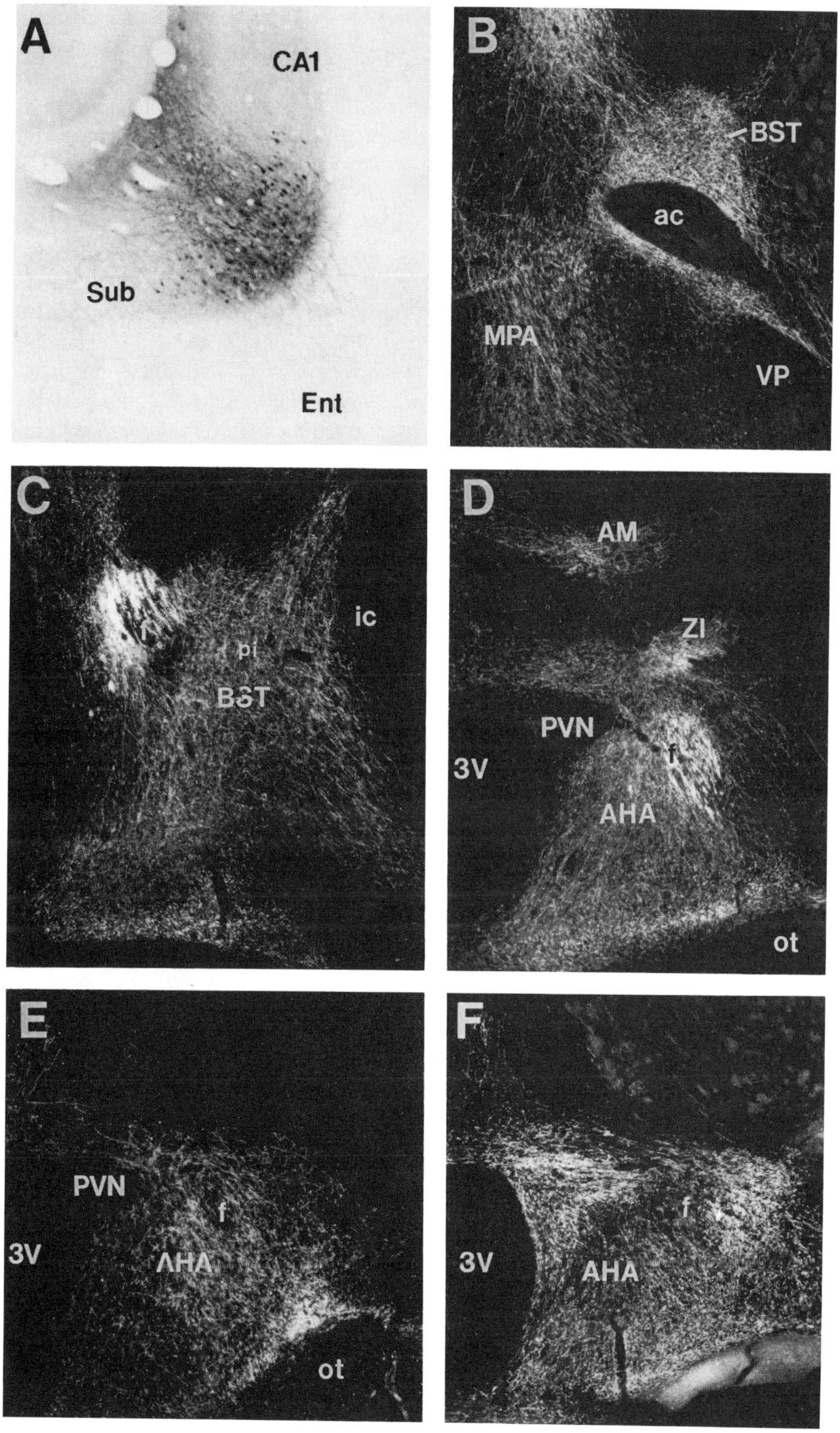
A
CA1
Sub
Ent
B
BST
ac
MPA
VP
C
f
pl
BST
ic
D
AM
ZI
PVN
3V
f
AHA
ot
E
PVN
f
3V
AHA
ot
F
f
3V
AHA

stress-related inputs from limbic regions are mediated by neuronal relays. The BST has been proposed to subserve such a function (129,130), based on its projections to the medial parvicellular PVN (mpPVN) (Fig. 2F) and its interconnectivity with limbic brain regions implicated in HPA regulation, including the hippocampus, lateral septum, and amygdala (123,124,131,132). Further support for this notion has come from stimulation and lesion experiments, which have shown that the BST can influence HPA output measures, and have implicated this structure as a site of convergence of stress-related information (133–135).

We have recently investigated the BST as a possible relay for hippocampal influences on the mpPVN in combined neuroanatomical studies (128). These experiments suggested the existence of a hippocampal–BST–PVN circuit (see Fig. 3), and indicated that a significant proportion of the BST projection to the PVN may be GABAergic. In view of the fact that the output of the hippocampus appears to be excitatory (136,137), such a pathway might explain the fact that hippocampal stimulation results in reductions in circulating levels of corticosteroids, or that ablation of the hippocampus or its major efferent pathways leads to corticosteroid hypersecretion and induction of CRH mRNA in the mpPVN (113,118,119,138–141). However, the role of the BST in mediating the effects of stress on the HPA axis is likely to be complex. Recent evidence has indicated that different BST subsectors may have opposing influences on HPA activity (142,143), and BST ablation experiments have suggested that effects on stress-induced corticosterone secretion may be dependent upon the nature of the stressor (144).

While the BST may relay limbic information to the PVN, it should be noted that portions of the hypothalamus are in a position to serve in a similar capacity. Although not well-studied to date, numerous hypothalamic areas are known to receive afferents from limbic-related regions (e.g., lateral septum, amygdala, hippocampus, medial prefrontal cortex) and have been reported

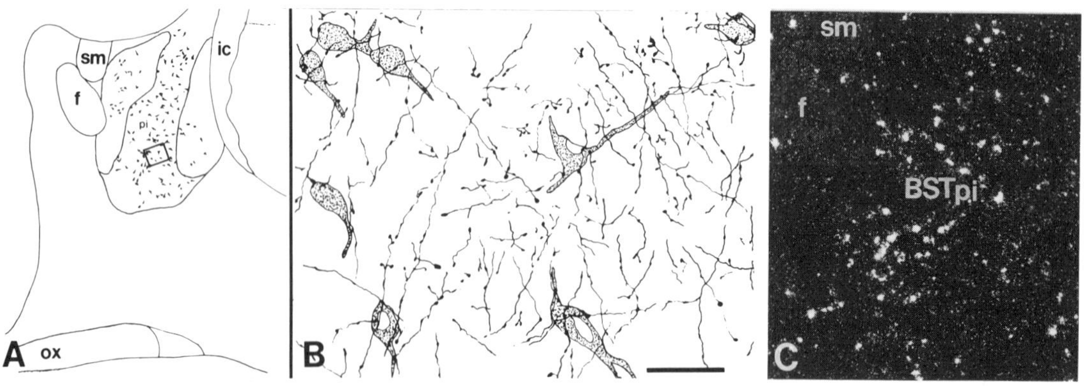

FIG. 3. A: Camera lucida drawing illustrates the distribution of neurons within the BST that were retrogradely labeled from the PVN with the tracer Fluoro-gold (242). Visualization of retrogradely labeled neurons was enhanced using a Fluoro-gold antibody (243). (Only retrogradely neurons within the BST are shown for simplicity). **B:** Camera lucida drawing from boxed region in (A). Section was double immunolabeled for Fluoro-gold and PHA-L following injections of the retrograde tracer in the PVN, and the anterograde tracer (PHA-L) in the ventral subiculum. Fluoro-gold–labeled neurons in the BST are abutted by PHA-L–labeled fibers and terminals, suggesting that hippocampal influences on the PVN may be relayed via the BST. Scale bar: 50 μm. **C:** Darkfield photomicrograph illustrating c-fos mRNA expression within the BST 30′ after swim stress. Note that labeling within the BST at this relatively caudal level is selective for the posterior intermediate subdivision. Abbreviations: BSTpi, bed nucleus of the stria terminalis, posterior intermediate division; f, formix; ic, internal capsule; ox, optic chiasm; sm, stria medullaris.

to project to the mpPVN (145–148). Moreover, the high degree of interconnectivity between the hypothalamus and BST raises the possibility of multisynaptic pathways involving these structures. Of course, whether limbic influences on HPA activity are mediated via monosynaptic pathways involving the BST or hypothalamus, or polysynaptic circuits involving these structures are not mutually exclusive possibilities. However, as noted in a recent discussion of the matter (33), it is questionable how the integrity of any such limbic signals could be maintained over the course of multisynaptic circuits that must certainly participate in additional functions. Clearly, further careful functional/anatomical data will be required to approach this issue.

Hypothalamic Inputs

The PVN has been reported to receive inputs from numerous hypothalamic sources. Indeed, based on retrograde tracing data (145–148), at least some PVN-projecting neurons can be localized within virtually all hypothalamic regions except the medial and lateral mammillary nuclei (145). While more recent studies involving the use of the sensitive anterograde tracer PHA-L have aided in refining the relative contributions of some hypothalamic regions to the innervation of the PVN (68,106,149–153), the potential remains for direct or indirect afferent regulation via numerous hypothalamic cell groups. Several possibilities are considered in the next paragraphs.

Considerable anatomical and physiological evidence has been advanced for GABAergic modulation of the hypophysiotropic CRH neuron. The PVN has been shown to receive a prominent GABAergic innervation, and GABA-containing terminals have been confirmed to contact CRH neurons (154–157). A recent anatomical study has demonstrated that this input comes in part from local hypothalamic sources such as the anterior portion of the PVN, anterior hypothalamic area, and perinuclear region of the supraoptic nucleus (106). The dorsomedial hypothalamic nucleus may also contribute to the GABAergic innervation of the PVN to some extent, based on its substantial projections to the hypophysiotropic zone and the high number of neurons expressing GABA markers (106,149, 155,158,159). Hypothalamic GABAergic inputs to the PVN may serve to relay limbic inputs to the CRH neurons, or convey information coursing via brainstem pathways, which ramify extensively in the hypothalamus. Support for GABA-mediated inhibition of HPA activity has come from in vitro studies (160–162), as well as in vivo experiments demonstrating GABAergic inhibition of CRH release and ACTH secretion (163,164). The precise role of GABA in basal and stress-induced HPA activity remains to be determined, however.

The hypothalamus is also involved in mediating opioidergic influences on HPA activity. A number of immunocytochemical studies have revealed ACTH/β-End-containing fibers in the PVN that originate from cells in the arcuate nucleus (165–168), and ultrastructural data has confirmed ACTH-IR synapses on CRH-containing neurons (169). The relevance of this projection to HPA regulation is suggested by a recent study that found ACTH-IR varicosities localized within the hypophysiotropic zone in close approximation to neurons that displayed Fos-IR in response to a stressful stimulus (170). While conflicting data exist, particularly from experiments examining the effects of morphine, recent evidence from in vivo and in vitro studies have suggested an inhibitory action of β-End on PVN neuronal activity and CRH secretion (171,172). The ACTH/β-End arcuate-PVN pathway appears to be paralleled by an NPY-containing projection (173); this peptide has been similarly confirmed at the electron microscopic level to contact CRH neurons (169), and appears to have a stimulatory effect on HPA activity (174). The arcuate nucleus may also be part of the origin of the reported dopaminergic innervation of the PVN (175,176), which appears to stimulate corticosterone secretion (177,178). While the possible functional roles for these pathways remain to be clarified, some suggestion that they might participate in mediating steroid feedback effects upon the HPA axis has been raised by the finding of relatively prominent glucocorticoid receptor expression in the arcuate nucleus (179–181).

A substance P innervation of the PVN also originates in part from the hypothalamus, particularly from the lateral hypothalamic area, and to a lesser extent from the anterior periventricular, dorsomedial, and ventromedial hypothalamic nuclei (101). Substance P has been reported to inhibit CRH secretion in vitro (182), although it apparently stimulates AVP secretion when centrally administered (183).

Finally, a number of hypothalamic cell groups appear to be in a position to mediate the effects of the suprachiasmatic nucleus (SCN) on basal and stress-induced HPA activity. The SCN, which is known to be the primary source of the diurnal rhythm of ACTH and corticosterone secretion (184,185), apparently lacks significant direct projections to the mpPVN (150,186), but appears to innervate several hypothalamic cell groups that have been confirmed to provide afferents to the hypophysiotropic zone. These areas include the periventricular PVN and the dorsomedial hypothalamic nucleus (186).

Circumventricular Inputs

A final class of afferents to the PVN arises from circumventricular regions such as the SFO, as well as a series of contiguous forebrain regions associated with the lamina terminalis: the median preoptic nucleus and the organum vasculosum of the lamina terminalis. These structures are known to be involved in the regulation of water balance, and while their projections appear to principally affect magnocellular components of the PVN, anatomical and physiological data suggest that parvicellular elements are also influenced (187,188). Projections to the PVN, including the hypophysiotropic zone, have been reported for all of these areas (147,189–191), and some evidence for GABA in the projections from these regions to magnocellular, and presumably parvicellular, divisions of the PVN has also been advanced (192). Angiotensin II (AII) has been localized in the projections from the SFO and median preoptic nucleus to the PVN (193), and the SFO sends an additional AII-containing projection to the median preoptic nucleus. As the SFO has been shown to be important in sensing circulating AII (194), these pathways are thought to mediate the effects of blood-borne AII on the HPA axis.

NEGATIVE FEEDBACK

The HPA axis is tightly controlled to produce rapid optimal glucocorticoid responses that can be promptly terminated upon the cessation of stress. This regulation is thought to be achieved through glucocorticoids acting via multiple inhibitory feedback loops involving pituitary, hypothalamic, and suprahypothalamic sites (195). The effects of glucocorticoid hormones are mediated via low-affinity (type II; glucocorticoid receptor, GR) and/or high-affinity (type I; mineralocorticoid receptor, MR) receptors. The precise roles of GR and MR on HPA function are unclear. Binding data indicate that stress-induced glucocorticoid secretion affects primarily type II receptors, suggesting that negative feedback is a GR-mediated event (196). Recent data have shown an ability of the GR to inhibit FOS-JUN induced gene transcription (197), indicating that high glucocorticoid levels may exert a negative influence on stimulus-transcription coupling. This type of interaction may underlie the genomic component of negative feedback. Conversely, MR appears to have a role in modulation of basal HPA tone and diurnal glucocorticoid rhythms (198), in that: 1) occupancy of type I receptors appears to mediate basal inhibition of ACTH secretion at both the peak and nadir of the circadian cycle (199); 2) MR shows a marked circadian rhythm in binding and biosynthesis (196,200); and 3) blockade of MR using a specific antagonist (RU28318) increases basal corticosterone secretion assayed during the circadian nadir (201). The majority of functional data suggest prominent involvement of both receptors in HPA regulation. However, the precise CNS loci at which these effects are relayed to neuronal regulators of the HPA remain to be definitively determined.

The ability of individual neurons to bind glucocorticoids was first demonstrated by McEwen as early as 1968, using in vivo uptake of tritiated corticosterone (202). In ensuing years, localization of receptors in brain has been visualized by in vitro and in vivo autoradiography, immunocytochemistry, and in situ hybridization (179–181,203–208). The results of these endeavors have revealed a widespread distribution of these molecular species across numerous brain regions. The GR is particularly widely distributed throughout the neuraxis (Fig. 4), and is present in both neurons and glia (see 181 for review). Areas of abundance include the hippocampus, PVN, arcuate nucleus, central amygdaloid nucleus, locus coeruleus, and cortex, although numerous additional regions express GR at low to moderate levels. Fewer regions show prominent expression of MR (e.g., hippocampus, cortical amygdaloid nucleus, several cranial nerve nuclei), yet it too is expressed at low to moderate levels throughout many brain loci (Fig. 4). Pre-

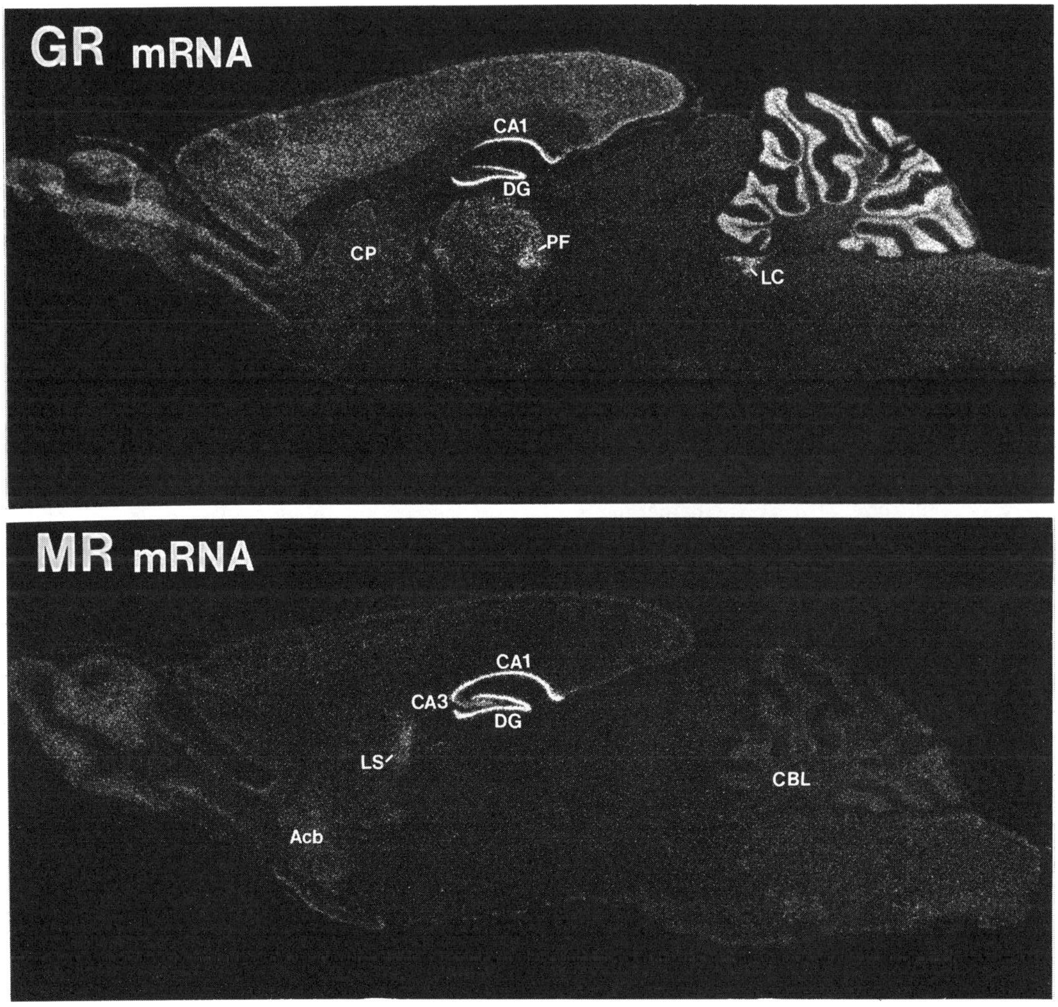

FIG. 4. Darkfield photomicrographs illustrate the distributions of mRNAs encoding the glucocorticoid receptor (*upper panel*) and mineralocorticoid receptor (*lower panel*) detected by in situ hybridization using cRNA probes. Abbreviations: Acb, accumbens nucleus; CA1, CA3, fields CA1, CA3 Ammon's horn; CBL, cerebellum; CP, caudate putamen; DG, dentate gyrus; LC, locus coeruleus; LS, lateral septum; PF, parafascicular thalamic nucleus.

sumably, adrenocorticosteroid receptor species act to modulate activity in a wide range of CNS systems, in accordance with glucocorticoid status by way of general modulation of gene transcription and/or actions at the cell membrane (either directly or as metabolites).

One brain region frequently implicated in mediating the feedback effects of glucocorticoids is the hippocampus. The hippocampus contains very high levels of GR and MR, and has been suggested to play a role in feedback based on data involving hippocampal lesions and/or corticosteroid administration (140,209–213). However, the precise role of the hippocampus in HPA regulation has been the subject of considerable debate and, indeed, strong cases can be made for prominent involvement or non-involvement of this structure in negative feedback control of the HPA axis (214,215). Data from our studies suggest that only a subset of neurons within the ventral hippocampal/subicular region may mediate feedback effects on the HPA axis (119,141); discrete lesions of this area result in delayed termination of the stress response. However, these results do not permit us to conclude that the hippocampus is the site of action of glucocorticoids, since the integrity of the hippocampus may be required for feedback to occur at other brain sites (216). Thus whether the hippocampus plays a *permissive* role or a *direct* role in steroid-mediated inhibition of HPA activity remains to be determined. Hippocampal GR and MR may also be important for modulation of circadian influences on the axis, as well as other functions (e.g., controlling expression of genes essential for cognitive integration).

Glucocorticoids are also thought to act directly at the level of the PVN to inhibit ACTH secretagogue release. Previous studies have shown that steroid implants (dexamethasone or corticosterone) in the PVN region are able to inhibit the ACTH and corticosterone responses to neural stimuli (217) and reverse adrenalectomy-induced up-regulation of CRH and AVP (218,219). These effects are presumably mediated via type II receptors, as the PVN contains high amounts of GR immunoreactivity and mRNA (200,203,207). However, other evidence has shown that extrahypothalamic afferents to the PVN are required for maintenance of basal CRH and AVP mRNA expression (220). Additionally, local application of glucocorticoids or antagonists into extrahypothalamic sites can modulate HPA activity (213,221,222). Thus, while glucocorticoids can apparently act at the level of the PVN, it is likely that other glucocorticoid-sensitive regions within the CNS are capable of modulating HPA axis activity via neuronal projections to the PVN. In the context of stress termination, such effects could be mediated via different mechanisms (genomic and/or nongenomic) that result in an inhibition of stimulatory afferents, activation of neuronal circuits that are ultimately inhibitory to the HPA axis, or both.

FUNCTIONAL NEUROANATOMY: STRESS-INDUCED IMMEDIATE-EARLY GENE (IEG) EXPRESSION IN THE CNS

While there remains much to be learned regarding the neural pathways by which stress stimulates the HPA axis, recent advances have been made through the use of IEGs as markers for neuronal activity. IEGs are rapidly and transiently induced in many brain regions, and their protein products, many of which are transcription factors, are thought to mediate the long-term responses of cells to external stimuli (223, 224). A number of recent experiments have used IEGs as markers for cellular activation in response to various stressors (60,62,225–233); these studies have reported stress-induced IEG expression in a variable number of brain areas, and support the utility of this approach, with certain caveats (234), to determine cellular markers for neuronal activity.

We have recently undertaken a systematic investigation of IEG mRNA expression (c-fos, c-jun, zif/268) in brain following acute swim stress (64), which has been generally viewed as a combined physical/psychological form of stress. The results indicate a broad pattern of neuronal activation in response to this stressor (Figs. 5 and 6, Table 1). Among areas demonstrating prominent IEG induction were: 1) brainstem regions including the cells of origin of catecholaminergic inputs to the PVN (A1/C1,

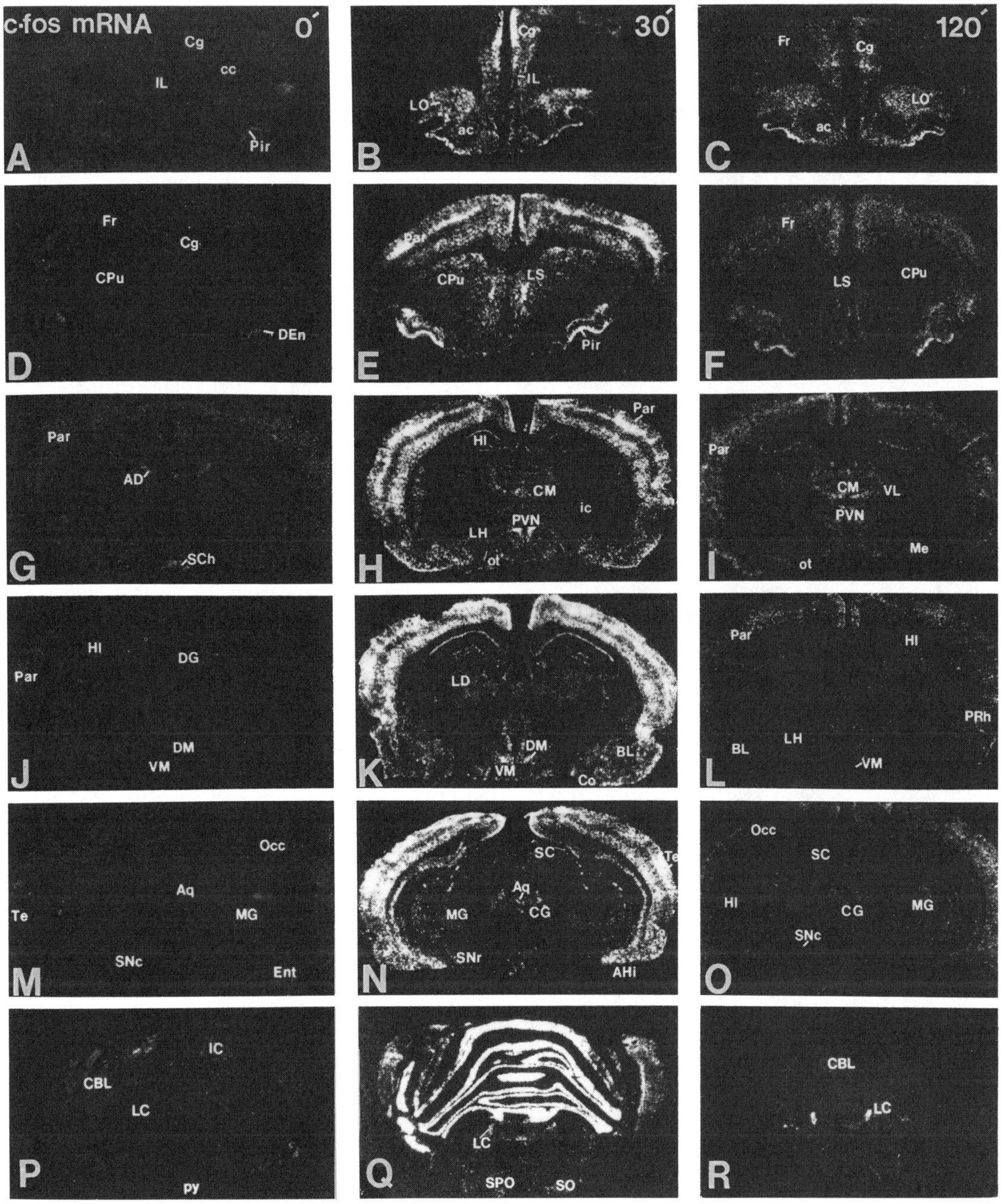

FIG. 5. Darkfield photomicrographs illustrating the expression of c-fos mRNA basally (**A,D,G,J,M,P**) and 30′ (**B,E,H,K,N,Q**) and 120′ (**C,F,I,L,O,R**) following onset of 10′ forced swim stress. Abbreviations: ac, anterior commissure; AD, anterodorsal thalamic nucleus; AHi, amygdalohippocampal area; Aq, cerebral aqueduct; BL, basolateral amygdaloid nucleus; CBL, cerebellum; cc, corpus callosum; Cg, cingulate cortex; CG, central gray; CM, central medial thalamic nucleus; Co, cortical amygdaloid nucleus; CPu, caudate putamen; DEn, dorsal endopiriform nucleus; DG, dentate gyrus; DM, dorsomedial hypothalamic nucleus; Fr, frontal cortex; HI, hippocampal formation; ic, internal capsule; IC, inferior colliculus; IL, infralimbic cortex; LC, locus coeruleus; LD, laterodorsal thalamic nucleus; LH, lateral hypothalamic area; LO, lateral orbital cortex; LS, lateral septal nucleus; Me, medial amygdaloid nucleus; MG, medial geniculate nucleus; Occ, occipital cortex; ot, optic tract; Par, parietal cortex; Pir, piriform cortex; PRh, perirhinal cortex; PVN, paraventricular hypothalamic nucleus; py, pyramidal tract; SCh, suprachiasmatic nucleus; SNc, substantia nigra, pars compacta; SNr, substantia nigra, pars reticulata; SO, superior olivary complex; SPO, superior preolivary nucleus; Te, temporal cortex; VL, ventrolateral thalamic nucleus; VM, ventromedial hypothalamic nucleus.

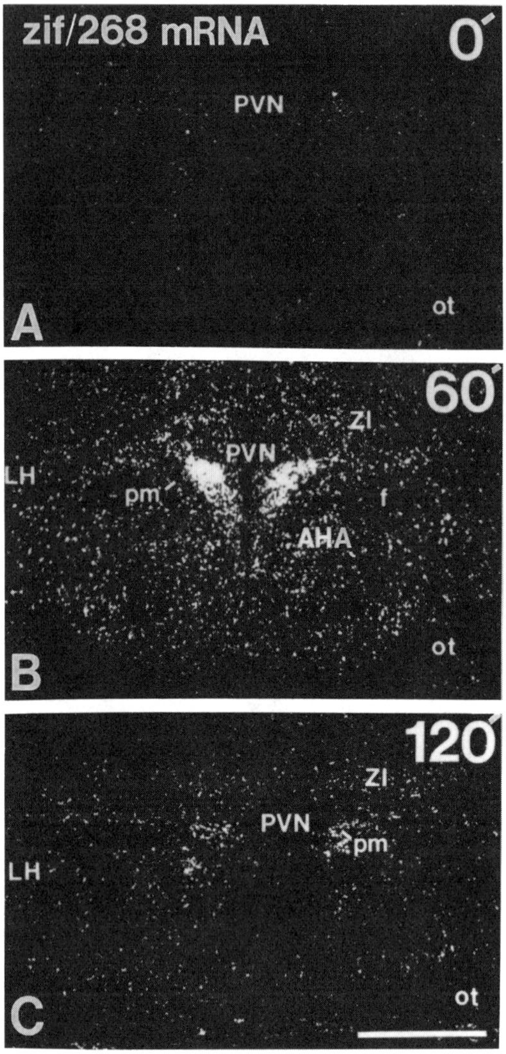

FIG. 6. Darkfiled photomicrographs illustrating the distribution of neurons expressing zif/268 mRNA basally (**A**), and 60′ (**B**) and 120′ (**C**) following swim stress. Note the heavy labeling within the PVN and adjacent anterior hypothalamic area at 60′ post-stress. Scale bar: 1.0 mm.

A2/C2, locus coeruleus); 2) some of the mesopontine regions proposed to relay somatosensory information (peripeduncular and laterodoral tegmental nuclei); 3) specific portions of the limbic telencephalon inferred to mediate the psychological aspects of stress (231) (medial prefrontal cortex, lateral septal nucleus, BST, medial and cortical amygdaloid nuclei); and 4) numerous hypothalamic nuclei and widespread portions of the neocortex. Interestingly, we have found that this pattern of IEG induction was generally similar to that seen in response to restraint stress, which has been viewed as a primarily psychological stressor. At present it is unclear whether such findings reflect commonalities in the neural pathways mediating physical versus psychological aspects of stress, similarities among the stress paradigms themselves, or perhaps represent non–stress-related neuronal activation. These distinctions are important for our understanding of the organization of stress activation pathways, since although different stressors can be expected to access the PVN from distinct starting points, integration may occur at common sites proximal to the level of the PVN. Our studies have also suggested a more widespread neuronal activation pattern in response to these acute stressors than has generally been reported. Such distinctions may well be due to technical differences in detection methods (i.e., mRNA versus protein detection). In any case, further comparison of the patterns of IEG induction following these and other classes of stressors may be used to provide an entry point to dissecting out key components of the stress-activation circuitry.

Examination of the time course of activation of various brain loci may also provide insight into their possible roles in the stress response. For example, we have found that while mRNA levels of the IEGs c-fos and zif/268 peak near 30 minutes in the parvicellular PVN following swim or restraint stress, little expression is evident in magnocellular neurons; peak levels of magnocellular c-fos and zif/268 mRNA expression occurred at 120 minutes post-stress, a time point at which corticosterone levels are known to have returned to baseline. These data may suggest delayed or secondary activation of magnocellular elements following these stressors, rather than a primary effect in stimulating pituitary corticotrophs.

While several recent studies have detected induction of IEGs in brain regions thought to provide stimulatory effects of HPA activity (60,62,63), it is interesting that stress-induced IEG expression has also been detected in brain regions thought to provide inhibitory inputs to

TABLE 1. *Relative expression of C-fos mRNA and zif/268 mRNA basally and 30′ following acute swim stress*

Region	c-fos 0′	c-fos 30′	zif 0′	zif 30′
Telencephalon				
Ant. Olfactory n.	+/o	+++	+++	++++
D. Endopiriform n.	++	++++	+++	++++
Neocortex				
Frontal	+	+++	++	+++
Parietal	+	++++	+++	++++
Temporal	+	++++	+++	++++
Occipital	+	+++	+++	++++
Allocortex				
Piriform	++	++++	+++	++++
Orbital	+/o	++++	+++	++++
Infralimbic	+/o	++++	+++	++++
Ant. Cingulate	+/o	++++	+++	++++
Entorhinal	+/o	+++	++	+++
Hippocampus				
CA1	+/o	++	+++	++++
CA2	+/o	+	+++	+++
CA3–4	+	+	+	++
Dentate gyrus	+	++	+	++
Subiculum	+/o	++	+++	++++
Accumbens n.	+/o	++	++	+++
Lateral septal n.	+/o	+++	++	+++
BST (medial)	+/o	++	+/o	+/o
BST (lateral)	+/o	+/o	+/o	++
Parastrial n.	+/o	+++	o	o
Septohypopthal. n.	+/o	++++	+	++
Amygdaloid body				
Ant. amyg. area	+/o	++	+	++
Medial n.	+/o	++++	+	++
Cortical n.	+/o	++++	+	+++
Central n.	o	+	+	++
Basomedial n.	o	++	++	+++
Basolateral n.	o	+	++	++
Diencephalon				
Thalamus				
Anterodorsal n.	++	++	++	+++
Paraventricular n.	+/o	++++	o	++
Central medial n.	+/o	++++	+	++
Ventrolateral n.	+/o	+++	+/o	+
Reticular n.	+/o	+/o	+/o	+/o
Med. habenula	o	o	o	+
Hypothalamus				
Medial preoptic a.	+/o	++	+/o	++
Suprachiasmatic a.	+	+	+	++
Paraventricular n. (parvi)	o	++++	+	++++
Paraventricular n. (magno)	o	+	+/o	+

Region	c-fos 0′	c-fos 30′	zif 0′	zif 30′
Supraoptic n.	o	+/o	+/o	+/o
Anterior hypothal. area	o	+++	+/o	+++
Ventromedial n.	o	+/o	+	+
Dorsomedial n.	o	++++	+	++
Lateral hypothal. area	+/o	+++	+	+++
Dorsal premammillary n.	+/o	+++	+/o	++
Ventral premammillary n.	+/o	++	+/o	++
Mesencephalon				
Substantia nigra, compacta	o	o	+/o	+/o
Substantia nigra, reticulata	o	+	+/o	+/o
Periaqueductal gray	+/o	++++	+	++
Peripeduncular n.	+/o	+	+/o	+++
Deep mesen-cephalic n.	+/o	+++	+	+
Pons				
Laterodorsal tegmental n.	+/o	++	+	+
Locus coeruleus	+/o	+++	+/o	+/o
Posterodorsal tegmental n.	+/o	+++	+/o	+/o
Pontine central gray	+/o	++	+/o	+/o
Dorsal raphe n.	+/o	++	+/o	+
Median raphe n.	+/o	++	+/o	+
Pontine n.	+/o	++	++	++
Medulla				
Vestibular complex	+/o	++	+/o	+/o
Lateral reticular n.	+/o	+++	+/o	+/o
N. solitary tract	+/o	+++	+	+
Dorsal cochlear n.	+	++	+/o	+/o
Raphe magnus	+/o	+++	+	+
Cuneate n.	+/o	+	+/o	+/o
Gracile n.	+/o	++	+/o	+/o
A1/C1	+/o	+++	+/o	+/o
Lat. paragiganto-cellular n.	+/o	+++	+/o	+/o
Inferior olivary complex	+/o	+++	+	++
Cerebellum				
Purkinje cell layer	+/o	++	++	+++
Granule cell layer	+/o	+++	++	+++
Molecular layer	+/o	+	o	o

BST, bed nucleus of the stria terminalis

the PVN (64,235). Included among such areas are portions of the BST (Fig. 3C) and several hypothalamic regions that are the origin of GABAergic projections to the hypophysiotropic zone (106) (Fig. 6). Although this requires confirmation, such data may suggest a GABAergic neuronal mechanism involved in termination of the stress response, particularly if it can be shown that relevant GABAergic inputs to the PVN are engaged temporally *following* activation of putative stimulatory pathways.

The results of experiments detecting IEG expression in response to stress must be interpreted with caution, since not all neuronal populations can be expected to utilize any single IEG, and part of the activation pattern revealed using these markers is likely to reflect neural events unrelated to the stress axis. However, the use of this methodology, particularly when multiple IEG markers are applied in tandem, provides an important starting point with which to initiate experimental evaluation of putative stress-related circuitries.

SUMMARY

As additional data regarding the neuroanatomy of the stress response have come to the fore, a number of key issues have emerged that may ultimately be important to our understanding of the system. While significant progress has been made in delineating pathways by which modality-specific inputs access the HPA axis, relatively little is understood about possible points of integration proximal to the hypophysiotropic PVN neuron. Such information may provide insight into the possibility that a unique circuitry exists, perhaps involving predominantly limbic brain regions, which identifies a particular stimulus as being "stressful" to an organism. Also requiring clarification is the nature of the glucocorticoid feedback signal in the context of the acute response to stress. Although ample evidence exists for genomic actions of glucocorticoids, the potential exists for rapid nongenomic effects that may be involved in mediating these processes (236,237). Furthermore, the neuronal pathways mediating negative feedback remain to be firmly established. To complicate matters, although activational and negative feedback components of these circuitries are often distinguished, it has become increasingly apparent that a continuous interaction between these functions is operative in maintaining optimal glucocorticoid levels both basally and after stress.

Future attempts at elucidating these issues may be aided by activity-based IEG mapping, which appears to represent a powerful new approach to delineating circuit components responsible for orchestrating HPA responses. Application of this method in combined studies may also help reveal the extent to which various neuropeptide systems are involved in integrating responses to stress, as previously suggested for extrahypothalamic CRH-containing pathways (238–240). Finally, though not considered in the present review, a clearer picture of the interplay between various blood-borne molecules (e.g., cytokines, adrenomedullary hormones) and the HPA axis will be required for a full understanding of adaptational responses to stress. Further insight into mechanisms of regulatory control over activity of the HPA axis will aid our understanding of the impact of stress on neuronal functioning in health and disease.

ACKNOWLEDGMENTS

This work was supported by NIDA grants DA05452 (W.E.C.) and DA00265-12 (S.J.W.) and NIMH grants MH42251-06 (S.J.W.) and MH49698 (J.P.H.). The authors wish to acknowledge the contributions of Professor Huda Akil in several of the concepts discussed in this paper.

REFERENCES

1. Selye H. *The stress of life.* New York: McGraw-Hill, 1956.
2. Vale W, Speiss J, Rivier C, Rivier J. Characterization of a 41-residue ovine hypothalamic peptide that stimulates the secretions of corticotropin and B-endorphin. *Science* 1981;213:1394–1397.
3. Swanson LW, Sawchenko PE. Hypothalamic integration: organization of the paraventricular and supraoptic nuclei. *Ann Rev Neurosci* 1983;6:275–335.
4. Swanson LW, Kuypers HGJM. The paraventricular nucleus of the hypothalamus: cytoarchitectonic subdivisions and organization of projections to the pituitary, dorsal vagal complex, and spinal cord as demonstrated by fluorescence double-labeling methods. *J Comp Neurol* 1980;194:555–570.
5. Swanson LW, Sawchenko PE, Lind RW. Regulation of multiple peptides in CRF parvicellular neurosecretory neurons. *Prog Brain Res* 1986;68:169–190.
6. Vale W, Vaughan J, Smith M, Yamamoto G, Rivier J, Rivier C. Effects of synthetic ovine corticotropin-releasing factor, glucocorticoids, catecholamines, neurohypophyseal peptides, and other substances on cultured corticotropic cells. *Endocrinology* 1983;113:1121–1131.
7. Plotsky PM, Bruhn TO, Vale W. Hypophysiotropic regulation of adrenocorticotropin secretion in response to insulin-induced hypoglycemia. *Endocrinology* 1985;117:323-329.
8. Plotsky PM, Bruhn TO, Vale W. Evidence for multifactor regulation of the adrenocorticotropin secretory re-

sponse to hemodynamic stimuli. *Endocrinology* 1985; 116:633–639.
9. Whitnall MH. Distributions of pro-vasopressin expressing and pro-vasopressin deficient CRH neurons in the paraventricular hypothalamic nucleus in colchicine-treated normal and adrenalectomized rats. *J Comp Neurol* 1988;275:13–28.
10. Whitnall MH, Smyth D, Gainer H. Vasopressin coexists in half of the corticotropin-releasing factor axons present in the external zone of the median eminence in normal rats. *Neuroendocrinology* 1987a;45:420–424.
11. Whitnall MH, Key S, Gainer H. Vasopressin-containing and vasopressin-deficient subpopulations of corticotropin-releasing factor axons are differentially affected by adrenalectomy. *Endocrinology* 1987b;120: 2180–2182.
12. Whitnall MH. Stress selectively activates the vasopressin-containing subset of corticotropin-releasing hormone neurons. *Neuroendocrinology* 1989;50:702–707.
13. de Goeij DCE, Kvetnansky R, Whitnall MH, Jezova D, Berkenbosch F, Tilders FJH. Repeated stress-induced activation of corticotropin-releasing factor neurons enhances vasopressin stores and colocalization with corticotropin-releasing factor in the median eminence of rats. *Neuroendocrinology* 1991;53:150–159.
14. Lind RW, Swanson LW, Chin DA, Bruhn TO, Ganten D. Angiotensin II: an immunohistochemical study of its distribution in the paraventriculo-hypophysial system and its co-localization with vasopressin and CRF in parvicellular neurons. *Soc Neurosci Abstr* 1984;10:88.
15. Meister B, Hökfelt T, Geffard M, Oertel W. Glutamic acid decarboxylase- and gamma-aminobutyric acid-like immunoreactivities in corticotropin-releasing factor-containing parvicellular neurons of the hypothalamic paraventricular nucleus. *Neuroendocrinology* 1988; 48:516–526.
16. Ceccatelli S, Erikson M, Hökfelt T. Distribution and coexistence of corticotropin-releasing factor-, neurotensin-, enkephalin-, cholecystokinin-, galanin-, and vasoactive intestinal polypeptide/peptide histidine isoleucine-like peptides in the parvicellular part of the paraventricular nucleus. *Neuroendocrinology* 1989;49: 309–323.
17. Rivier C, Vale W. Modulation of stress-induced ACTH release by corticotropin-releasing factor, catecholamines, and vasopressin. *Nature* 1983;305:325–327.
18. Tilders F, Tatemoto K, Berkenbosch F. The intestinal peptide PHI-27 potentiates the action of corticotropin-releasing factor on ACTH release from rat pituitary fragments in vitro. *Endocrinology* 1984;115:1663–1635.
19. Antoni FA. Hypothalamic control of adrenocorticotropin secretion: advances since the discovery of 41-residue corticotropin-releasing factor. *Endocrine Rev* 1986;7:351–378.
20. Mezey E, Reisine T, Skirboll L, Beinfeld M, Kiss JZ. Role of cholecystokinin in corticotropin release: coexistence with vasopressin and corticotropin-releasing factor in cells of the rat hypothalamic paraventricular nucleus. *Proc Natl Acad Sci* 1986;83:3510–3512.
21. Reisine T, Jensen R. Cholecystokinin-8 stimulates adrenocorticotropin release from anterior pituitary cells. *J Pharmacol Exp Ther* 1988;236:621–626.
22. Watanabe T, Orth DN. Effects of several in vitro systems on the potencies of putative adrenocorticotropin secreatgogues on rat pituitary cells. *Endocrinology* 1988;122:2299–2308.
23. Swanson LW, Simmons DM. Differential steroid hormone and neuronal influences on peptide mRNA levels in CRH cells of the paraventricular nucleus: a hybridization histochemical study in the rat. *J Comp Neurol* 1989;285:413–435.
24. Schafer MK, Herman JP, Watson SJ. In situ hybridization analysis of gene expression in the HPA stress axis: regulation by glucocorticoids. Wenner-Grenn International Symposium Series, *Visualization of Brain Function,* The Macmillan Press, 1990;11–22.
25. Watts AG. Ether anesthesia differentially affects the content of prepro-corticotropin-releasing hormone, prepro-neurotensin/neuromedin N, and prepro-enkephalin mRNAs in the hypothalamic paraventricular nucleus of the rat. *Brain Res* 1991;544:353–357.
26. Ceccatelli S, Orazzo C. Effect of different types of stressors on peptide messenger ribonucleic acids in the hypothalamic paraventricular nucleus. *Acta Endocrinologica* 1993;128:485–492.
27. Antoni FA, Fink G, Sheward WJ. Corticotrophin-releasing peptides in rat hypophysial portal blood after paraventricular lesions: a marked reduction in the concentration of corticotrophin-releasing factor-41, but no change in vasopressin. *J Endocrinology* 1990;125: 175–183.
28. Linke H, Dayanithi G, Fohr KJ, Gratzi M. Oxytocin at physiological concentrations evokes adrenocorticotropin (ACTH) release from corticotrophs by increasing intracellular free calcium mobilized mainly from intracellular stores. Oxytocin displays synergistic or additive effects on ACTH-releasing factor or arginine vasopressin-induced ACTH secretion, respectively. *Endocrinology* 1992;130:2183–2191.
29. Holmes MC, Antoni FA, Aguiera G, Catt KJ. Magnocellular axons in passage through the median eminence release vasopressin. *Nature* 1986;319:326–329.
30. Page RB. The pituitary portal system. In: Ganten D, Pfaff D eds. *Current topics in neuroendocrinology, vol. 7. Morphology of the hypothalamus and its connections.* Berlin: Springer Verlag, 1986;1–47.
31. Buma P, Nieuwenhuys R. Ultrastructural demonstration of oxytocin and vasopressin release sites in the neural lobe and median eminence of the rat by tannic acid and immunogold methods. *Neurosci Lett* 1987; 74:151–157.
32. Dohanics J, Hoffman GE, Verbalis JC. Hyponatremia-induced inhibition of magnocellular neurons causes stressor-selective impairment of stimulated adrenocorticotropin secretion in rats. *Endocrinology* 1991;128: 331–340.
33. Sawchenko PE. The final common path: issues concerning the organization of central mechanisms controlling corticotropin secretion. In: Brown MR, Koob GF, Rivier C eds. *Stress Neurobiology and Neuroendocrinology.* New York: Marcel Dekker, 1991, 55–71.
34. Swanson LW. Biochemical switching in hypothalamic circuits mediating responses to stress. *Prog Brain Res* 1991;87:181–200.
35. Cunningham ET Jr, Sawchenko PE. Anatomical specificity of noradrenergic inputs to the paraventricular and supraoptic nuclei of the rat hypothalamus. *J Comp Neurol* 1988;274:60–76.

36. Cunningham ET Jr, Bohn MC, Sawchenko PE. The organization of adrenergic projections to the paraventricular and supraoptic nuclei of the hypothalamus in the rat. *J Comp Neurol* 1990;292:651–667.
37. Sawchenko PE, Benoit R, Brown MR. Somatostatin 28-immunoreactive inputs to the paraventricular and supraoptic nuclei: principal origin from non-aminergic neurons in the nucleus of the solitary tract. *J Chem Neuroanatomy* 1988;1:81–94.
38. Sawchenko PE, Arias C, Bittencourt JC. Inhibin β, somatostatin, and enkephalin immunoreactivities coexist in caudal medullary neurons that project to the paraventricular nucleus of the hypothalamus. *J Comp Neurol* 1990;291:269–280.
39. Norgren R. Projections from the nucleus of the solitary tract in the rat. *Neuroscience* 1978;3:207–218.
40. Saper CB, Loewy AD. Efferent connections of the parabrachial nucleus in the rat. *Brain Res* 1980;197: 291–317.
41. Hancock MB. Evidence for direct projections from the nucleus of the solitary tract onto medullary adrenaline cells. *J Comp Neurol* 1988;276:460–467.
42. Levin MC, Cunningham ET Jr, Sawchenko PE. The organization of mesencephalic and pontine afferents to the paraventricular and supraoptic nuclei in the rat. *Soc Neurosci Abstr* 1987;13:1166.
43. Corrodi H, Fuxe K, Hökfelt T. The effect of immobilization stress on the activity of central monoamine neurons. *Life Sci* 1968;7:107–112.
44. Thierry AM, Javoy F, Glowinski J, Kety SS. Effects of stress on the metabolism of norepinephrine, dopamine, and serotonin in the central nervous system of the rat. I. Modification of norepinephrine turnover. *J Pharmacol Exp Ther* 1968;163:163–171.
45. Hedge GA, Van Ree JM, Versteeg DHG. Correlation between hypothalamic catecholamine synthesis and ether stress-induced ACTH secretion. *Neuroendocrinology* 1976;21:236–246.
46. Saavedra JM, Kvetnansky R, Kopin IJ. Adrenaline, noradrenaline and dopamine levels in specific brain stem areas of acutely immobilized rats. *Brain Res* 1979;160:271–280.
47. Smythe GA, Bradshaw JE, Vining RF. Hypothalamic monoamine control of stress-induced adrenocorticotropin release in the rat. *Endocrinology* 1983;113: 1062–1071.
48. Johnston CA, Spinedi EJ, Negro-Vilar A. Effect of acute ether stress on monoamine metabolism in the median eminence and discrete hypothalamic nuclei of the rat brain and on anterior pituitary hormone secretion. *Neuroendocrinology* 1985;41:83–88.
49. Di Renzo G, Gillham B, Holmes MC, Jones MT. The effect of pretreatment on with intraventricular 6-hydroxydopamine on the hypothalamo-pituitary-adrenocortical function of the rat. *J Physiol* 1975;293:50–56.
50. Weiner RI, Ganong WF. Role of brain monoamines and histamine in regulation of anterior pituitary secretion. *Physiol Rev* 1978;58:904–976.
51. Darlington D, Shinsako J, Dallman MF. Medullary lesions eliminate ACTH responses to hypotensive hemorrhage. *Am J Physiol* 1986;251:R106–R115.
52. Gibson A, Hart SL, Patel S. Effects of 6-hydroxydopamine-induced lesions of the paraventricular nucleus, and of prazosin, on the corticosterone response to restraint in rats. *Neuropharmacology* 1986;25:257–260.
53. Szafarczyk A, Alonso G, Ixart G, Malaval F, Assenmacher I. Diurnal-stimulated and stress-induced ACTH release is mediated by ventral noradrenergic bundle. *Am J Physiol* 1985;249:E219–E226.
54. Szafarczyk A, Guillame V, Conte-Devolx B, Alonso G, Malaval F, Pares-Herbute N, Oliver C, Assenmacher I. Central catecholaminergic system stimulates secretion of CRH at different sites. *Am J Physiol* 1988;255:E463–E468.
55. Feldman S, Conforti N, Melamed E. Norepinephrine depletion in the paraventricular nucleus inhibits adrenocortical responses to neural stimuli. *Neurosci Lett* 1986;64:191–195.
56. Feldman S, Conforti N, Melamed E. Involvement of ventral noradrenergic bundle in corticosterone secretion following neural stimuli. *Neuropharmacology* 1988;27:129–133.
57. Feldman S, Conforti N, Melamed E. Involvement of ventral noradrenergic bundle in corticosterone secretion following neural stimuli. *Neuropharmacology* 1988;27:129–133.
58. Gaillet S, Alonso G, Le Borgne, Barbanel G, Malaval F, Assenmacher I, Szafarczyk A. Effects of discrete lesions in the ventral noradrenergic ascending bundle on the corticotropic stress response depend on the site of the lesion and on the plasma levels of adrenal steroids. *Neuroendocrinology* 1993;58:408–419.
59. Laucher J, Gaillet S, Barbagli B, Buda M, Tappaz M. Differential early time course activation of the brainstem catecholaminergic groups in response to various stresses. *Neuroendocrinology* 1991;53:589–596.
60. Ceccatelli S, Vilar MJ, Goldstein M, Hökfelt T. Expression of c-Fos immunoreactivity in transmitter-characterized neurons after stress. *Proc Natl Acad Sci* 1989;86:9569–9573.
61. Chan RKW, Brown ER, Kovacs KJ, Sawchenko PE. A comparison of two immediate-early genes, c-fos and NGFI-B, as markers for functional activation in stress-related neuroendocrine circuitry. *J Neurosci* 1993;13: 5126–5138.
62. Pezzone MA, Lee W-S, Hoffman GE, Pezzone KM, Rabin BS. Activation of brainstem catecholaminergic neurons by conditioned and unconditioned aversive stimuli as revealed by c-Fos immunoreactivity. *Brain Res* 1993;608:310–318.
63. Senba E, Matsunaga K, Tohyama M, Noguchi K. Stress-induced c-fos expression in the rat brain: activation mechanism of sympathetic pathway. *Brain Res Bull* 1993;31:329–344.
64. Cullinan WE, Herman JP, Battaglia DF, Akil H, Watson SJ. Pattern and time course of immediate early gene activation following acute stress. *Neuroscience* 1995;64:477–505.
65. Liposits Zs, Phelix C, Paull WK. Electron microscopic analysis of tyrosine hydroxylase, dopamine-B-hydroxylase and phenylethanolamine-N-methyl-transferase immunoreactive innervation of the hypothalamic paraventricular nucleus of the rat. *Histochemistry* 1986; 84:105–120.
66. Liposits Zs, Phelix C, Paull WK. Adrenergic innervation of the corticotropin releasing factor (CRF)-synthesizing neurons in the hypothalamic paraventricular nucleus of the rat. A combined light and electron microscopic study. *Histochemistry* 1986;84:201–205.
67. Sawchenko PE, Swanson LW, Grzanna R, Howe PRC,

Polak J, Bloom SR. Co-localization of neuropeptide Y-immunoreactivity in brainstem catecholaminergic neurons that project to the paraventricular nucleus of the hypothalamus. *J Comp Neurol* 1985;241:138–153.
68. Levin MC, Sawchenko PE, Howe PRC, Bloom SR, Polak JM. The organization of galanin-immunoreactive inputs to the paraventricular nucleus with special reference to their relationship to catecholaminergic afferents. *J Comp Neurol* 1987;261:562–582.
69. Plotsky P. Facilitation of immunoreactive corticotropin-releasing factor secretion into the hypophysealportal circulation after activation of catecholaminergic pathways or central norepinephrine injection. *Endocrinology* 1987;121:924–930.
70. Szafarczyk A, Malaval F, Laurent A, Gibaud R, Assenmacher I. Further evidence for a central stimulatory action of catecholamines on adrenocortical release in the rat. *Endocrinology* 1987;121:883–892.
71. Widmaier EP, Lim AT, Vale W. Secretion of corticotropin-releasing factor from cultured rat hypothalamic cells: effects of catecholamines. *Endocrinology* 1989; 124:583–590.
72. Mezey E, Kiss JZ, Skirboll LP, Goldstein M, Axelrod J. Increase of corticotropin releasing factor staining in the rat paraventricular nucleus neurones by depletion of hypothalamic adrenaline. *Nature* 1984;310:140–141.
73. Plotsky PM, Cunningham ET Jr, Widmaier EP. Catecholaminergic modulation of corticotropin-releasing factor and adrenocorticotropin secretion. *Endocrine Rev* 1989;10:437–458.
74. Plotsky PM. Neural coding of stimulus-induced ACTH secretion. In: Brown MR, Koob GF, Rivier C, eds. *Stress Neurobiology and Neuroendocrinology*. New York: Marcel Dekker, 1991, 137–150.
75. Vernikos-Danellis J, Berger P, Barchas JD. Brain serotonin and pituitary-adrenal function. *Prog Brain Res* 1973;39:301–309.
76. Kovacs GL, Kishonti J, Lissak K, Telegdy G. Inhibitory action of midbrain raphe stimulation on stress-induced elevation of plasma corticosterone levels in rats. *Neurosci Lett* 1976;3:305–310.
77. Szafarczyk A, Alonso G, Ixart G, Malaval J, Nougyier-Soule J, Assenmacher I. Serotonergic system and circadian rhythms of ACTH and corticosterone in rats. *Am J Physiol* 1980;239:482–489.
78. Fuller RW, Snoddy HD. Elevation of plasma corticosterone by swim stress and insulin-induced hypoglycemia in control and fluoxetine pre-treated rats. *Endocrin Res Commun* 1977;4:11–23.
79. Sawchenko PE, Swanson LW, Steinbusch HWM, Verhofstad AAJ. The distribution and cells of origin of serotonergic inputs to the paraventricular and supraoptic nuclei of the rat. *Brain Res* 1983;277:355–360.
80. Liposits Zs, Phelix C, Paull WK. Synaptic interaction of serotonergic axons and corticotropin releasing factor (CRF) synthesizing neurons in the hypothalamic paraventricular nucleus of the rat. *Histochemistry* 1987; 86:541–549.
81. Jones MT, Hillhouse E, Burden J. Effects of various neurotransmitters on the secretion of corticotrophin-releasing hormone from the rat hypothalamus in vitro: a model of the neurotransmitters involved. *J Endocrinol* 1976;69:1–10.
82. Hashimoto K, Ohno N, Murakami K, Kageyama J, Aoki YTJ. The effect of serotonin agonist 1-(trifluoromethylphenyl) piperazine on corticotropin releasing factor and arginine vasopressin in rat hypothalamic nuclei. *Endocrinol Japan* 1982;29:383–388.
83. Gibbs DM, Vale W. Effect of serotonin reuptake inhibitor fluoxetine on corticotropin-releasing factor and vasopressin secretion into hypophysial portal blood. *Brain Res* 1983;280:176–179.
84. Hillhouse EW, Milton NGN. Effect of acetylcholine and 5-hydroxytryptamine on the secretion of corticotrophin-releasing factor-41 and arginine vasopressin from the rat hypothalamus in vitro. *J Endocrinol* 1989;122:713–718.
85. Calogero AE, Bernardini R, Margioris AN et al. Effects of serotonergic agonists and antagonists of corticotropin-releasing hormone secretion by explanted rat hypothalami. *Peptides* 1989;10:189–200.
86. Van de Kar LD, Wilkinson CW, Skrobik Y, Brownfield MS, Ganong WF. Evidence that serotoninergic neurons in the dorsal raphe exert a stimulatory effect on the secretion of renin but not corticosterone. *Brain Res* 1982;235:233–243.
87. Calogero AE, Bagdy G, Szemeredi K, Tartagli ME, Gold PW, Chrousos GP. Mechanisms of serotonin receptor agonist-induced activation of the rat hypothalamic-pituitary-adrenal axis in the rat. *Endocrinology* 1990;126:1888–1894.
88. Fuller RW, Snoddy HD. Elevation of plasma corticosterone by swim stress and insulin-induced hypoglycemia in control and fluoxetine pre-treated rats. *Endocrin Res Commun* 1977;4:11–23.
89. Feldman S, Melamed E, Conforti N, Weidenfeld J. Effect of central serotonin depletion on adrenocortical responses to neural stimuli. *Exp Neurol* 1984;85: 661–666.
90. Feldman S, Conforti N, Melamed E. Paraventricular nucleus serotonin mediates neurally stimulated adrenocortical secretion. *Brain Res Bull* 1987;18:165–168.
91. Feldman S, Weidenfeld J, Conforti N, Saphier D. Differential recovery of adrenocortical responses to neural stimuli following administration of 5,7-dihydroxytryptamine into the hypothalamus. *Exp Brain Res* 1991;85: 144–148.
92. Harbuz MS, Chalmers J, DeSouza L, Lightman SL. Stress-induced activation of CRF and c-fos mRNAs in the paraventricular nucleus are not affected by serotonin depletion. *Brain Res* 1993;609:167–173.
93. LeDoux JE, Ruggiero DA, Reis DJ. Projections to the subcortical forebrain from anatomically defined region of the medial geniculate body in the rat. *J Comp Neurol* 1985;242:182–213.
94. Schroeder DM, Jane JA. Projections of dorsal column nuclei and spinal cord to brainstem and thalamus in the tree shrew *tupaia glis*. *J Comp Neurol* 1971;142: 309–350.
95. Henkel CK. Afferent sources of a lateral midbrain tegmental zone associated with the pinna in the cat as mapped by retrograde transport of horseradish peroxidase. *J Comp Neurol* 1981;203:213–226.
96. Moon-Edley S, Graybiel AM. The afferent and efferent connections of the feline nucleus tegmenti pedunculopontinus, pars compacta. *J Comp Neurol* 1983;217: 187–215.
97. Massopust LC, Hauge DG, Ferneding JC, Doubek WG, Taylor JJ. Projection systems and terminal localization of dorsal column afferents: an autoradiographic and

horseradish peroxidase study in the rat. *J Comp Neurol* 1985;237:533–544.

98. Armstrong DM, Saper CB, Levey AI, Wainer BH, Terry RD. Distribution of cholinergic neurons in rat brain demonstrated by the immunocytochemical localization of choline acetyltransferase. *J Comp Neurol* 1983;216:53–68.
99. Vincent SR, Satoh K, Armstrong DM, Fibiger HC. Substance P in the ascending cholinergic reticular system. *Nature* 1983;306:688–691.
100. Rye DB, Saper CB, Lee HJ, Wainer BH. Pedunculopontine tegmental nucleus of the rat: cytoarchitecture, cytochemistry, and some extrapyramidal connections of the mesopontine tegmentum. *J Comp Neurol* 1987; 259:483–528.
101. Bittencourt JC, Benoit R, Sawchenko PE. Distribution and origins of substance P-immunoreactive projections to the paraventricular and supraoptic nuclei: partial overlap with catecholaminergic projections. *J Chem Neuroanat* 1991;4:63–78.
102. Calogero AE, Gallucci WT, Bernardini R, Saoutis C, Gold PW, Chrousos GP. Effect of cholinergic agonists and antagonists on rat hypothalamic corticotropin-releasing hormone secretion in vitro. *Neuroendocrinology* 1988;47:303–308.
103. Tsagarakis S, Holly JMP, Rees LH, Besser GM, Grossman A. Acetylcholine and norepinephrine stimulate the release of corticotropin releasing factor-41 from the rat hypothalamus in vitro. *Endocrinology* 1988;123: 1962–1969.
104. Meeker RB, Swanson DJ, Hayward JN. Local synaptic organization of cholinergic neurons in the basolateral hypothalamus. *J Comp Neurol* 1988;276:157–168.
105. Ruggiero DA, Giuliano R, Anwar M, Stornetta R, Reis DJ. Anatomical substrates of cholinergic-autonomic regulation in the rat. *J Comp Neurol* 1990;292:1–53.
106. Roland BL, Sawchenko PE. Local origins of some GABAergic projections to the paraventricular and supraoptic nuclei of the hypothalamus in the rat. *J Comp Neurol* 1993;332:123–143.
107. Matheson KG, Branch BJ, Taylor AN. Effects of amygdaloid stimulation on pituitary-adrenal activity in conscious cats. *Brain Res* 1971;32:151–167.
108. DuPont A, Bastarache E, Endroczi E, Fortier C. Effect of hippocampal stimulation on plasma thyrotropin (THS) and corticosterone responses to acute cold exposure in the rat. *Can J Physiol Pharmacol* 1972;50: 364–367.
109. Redgate ES, Fahringer EE. A comparison of the pituitary adrenal activity elicited by electrical stimulation of preoptic, amygdaloid and hypothalamic sites. *Neuroendocrinology* 1973;12:334–343.
110. Seggie J, Uhlir I, Brown GI. Adrenal responses following septal lesions in the rat. *Neuroendocrinology* 1973;16:225–236.
111. Wilson MM, Greer SE, Greer MA, Roberts L. Hippocampal inhibition of pituitary-adrenocortical function in female rats. *Brain Res* 1980;197:433–441.
112. Feldman S, Conforti N. Amygdalectomy inhibits adrenocortical responses to somatosensory and olfactory stimulation. *Neuroendocrinology* 1981;32:330–334.
113. Dunn J, Orr S. Differential plasma corticosterone responses to hippocampal stimulation. *Exp Brain Res* 1984;54:1–6.
114. Dunn J, Whitener J. Plasma corticosterone responses to electrical stimulation of the amygdaloid complex: cytoarchitectural specificity. *Neuroendocrinology* 1986; 42:211–217.
115. Beaulieu S, Di Paolo T, Barden N. Control of ACTH secretion by the central nucleus of the amygdala: implication of the serotoninergic system and its relevance to the glucocorticoid delayed negative feedback mechanism. *Neuroendocrinology* 1986;44:247–254.
116. Saphier D, Feldman S. Effects of septal and hippocampal stimulation on paraventricular nucleus neurons. *Neuroscience* 1987;20:749–755.
117. Seggie J. Differential responsivity of corticosterone and prolactin to stress following lesions of the septum or amygdala: implications for psychoneuroendocrinology. *Prog Neuropsychopharmacol Biol Psychiatry* 1987;11:315–324.
118. Herman JP, Schäfer MK-H, Young EA, Thompson R, Douglass J, Akil H, Watson SJ. Evidence for hippocampal regulation of neuroendocrine neurons of the hypothalamic-pituitary-adrenocortical axis. *J Neurosci* 1989;9:3072–3082.
119. Herman JP, Cullinan WE, Young EA, Akil H, Watson SJ. Selective forebrain fiber-tract lesions implicate ventral hippocampal structures in tonic regulation of paraventricular nucleus CRH and AVP mRNA expression. *Brain Res* 1992;592:228–238.
120. Sapolsky RM, Armanini DR, Sutton SW, Plotsky PM. Elevations of hypophyseal portal concentrations of adrenocorticotropin secretagogues after fornix transection. *Endocrinology* 1989;125:2881–2887.
121. Van de Kar LD, Piechowski RA, Rittenhouse PA, Gray TS. Amygdaloid lesions: differential effect on conditioned stress and immobilization-induced increases in corticosterone and renin secretion. *Neuroendocrinology* 1991;54:89–95.
122. Diorio D, Viau V, Meaney MJ. The role of the medial prefrontal cortex (cingulate gyrus) in the regulation of hypothalamic-pituitary-adrenal responses to stress. *J Neurosci* 1993;13:3839-3847.
123. Swanson LW, Cowan WM. An autoradiographic study of the organization of the efferent projections of the hippocampal formation in the rat. *J Comp Neurol* 1977;172:49–84.
124. Swanson LW, Cowan WM. The connections of the septal region in the rat. *J Comp Neurol* 1979;186: 621–656.
125. Gray TS, Carney ME, Magnuson DJ. Direct projections from the central amygdaloid nucleus to the hypothalamic paraventricular nucleus: possible role in stress-induced adrenocorticotropin release. *Neuroendocrinology* 1989;50:433–446.
126. Kohler C. Subicular projections to the hypothalamus and brainstem: some novel aspects revealed in the rat by the anterograde Phaseolus vulgaris leukoagglutinin (PHA-L) tracing method. *Prog Brain Res* 1990;83: 59–69.
127. Canteras NS, Swanson LW. Projections of the ventral subiculum to amygdala, septum, and hypothalamus: a PHA-L anterograde tract-tracing study in the rat. *J Comp Neurol* 1992;324:180–194.
128. Cullinan WE, Herman JP, Watson SJ. Ventral subicular interaction with the hypothalamic paraventricular nucleus: evidence for a relay in the bed nucleus of the stria terminalis. *J Comp Neurol* 1993;332:1–20.
129. Sawchenko PE, Swanson LW. The organization and

biochemical specificity of afferent projections to the paraventricular and supraoptic nuclei. *Prog Brain Res* 1983;60:19–29.
130. Swanson LW, Sawchenko PE, Lind RW. Regulation of multiple peptides in CRF parvocellular neurosecretory neurons: implications for the stress response. *Prog Brain Res* 1986;68:169–190.
131. Weller KL, Smith DA. Afferent connections to the bed nucleus of the stria terminalis. *Brain Res* 1982;232: 255–270.
132. Krettek JE, Price JL. Amygdaloid projections to subcortical structures within the basal forebrain and brainstem in the rat and cat. *J Comp Neurol* 1978;178: 225–254.
133. Henke PG. The bed nucleus of the stria terminalis and immobilization stress: unit activity, escape behaviour, and gastric pathology in rats. *Behav Brain Res* 1984;11:35–45.
134. Feldman S, Conforti N, Saphier D. The preoptic area and bed nucleus of the stria terminalis are involved in the effects of the amygdala on adrenocortical secretion. *Neuroscience* 1990;37:775–779.
135. Mor G, Saphier D, Feldman D. Neural pathways that mediate the effects of afferent stimuli on paraventricular multiunit activity in freely moving rats. *J Neurosci Res* 1987;17:452–458.
136. Walaas I, Fonnum F. Biochemical evidence for glutamate as a transmitter in hippocampal efferents to the basal forebrain and hypothalamus in the rat brain. *Neuroscience* 1980;5:1691–1698.
137. Finch DM, Wong EE, Derian EL, Babb TL. Neurophysiology of limbic system pathways in the rat: projections from the subicular complex and hippocampus to the entorhinal cortex. *Brain Res* 1986;397:205–213.
138. Knigge KM. Adrenocortical response to stress in rats with lesions of the hippocampus and amygdala. *Proc Soc Exp Biol Med* 1961;108:67–69.
139. Fischette CT, Komisaruk BR, Edinger HM, Siegel A. Differential fornix ablations and the circadian rhythmicity of adrenal corticosteroid secretion. *Brain Res* 1980;195:373–387.
140. Margariños AM, Somoza G, DeNicola AF. Glucocorticoid negative feedback and glucocorticoid receptors after hippocampectomy in rats. *Horm Metab Res* 1987;19:105–109.
141. Herman JP, Cullinan WE, Morano MI, Watson SJ. Involvement of the ventral hippocampus in regulation of the hypothalamo-pituitary-adrenocortical axis. *Soc Neurosci Abstr* 1992;22:1541.
142. Dunn J. Plasma corticosterone responses to electrical stimulation of the bed nucleus of the stria terminalis. *Brain Res* 1987;407:327–331.
143. Herman JP, Cullinan WE, Watson SJ. Involvement of the bed nucleus of the stria terminalis in tonic regulation of paraventricular hypothalamic CRH and AVP mRNA expression. *J Neuroendocrinology* 1994;6: 433–442.
144. Gray TS, Piechowski RA, Yracheta JM, Rittenhouse PA, Bethea CL, Van de Kar LD. Ibotenic acid lesions in the bed nucleus of the stria terminalis attenuate conditioned stress-induced increases in prolactin, ACTH and corticosterone. *Neuroendocrinology* 1993; 57:517–524.
145. Swanson LW. The hypothalamus. In: Björklund A, Hökfelt T, Swanson LW, eds. *Handbook of chemical neuroanatomy, vol. 5. Integrated Systems of the CNS, Part I.* Amsterdam: Elsevier, 1987;1–124.
146. Silverman AJ, Hoffman DL, Zimmerman. The descending afferent connections of the paraventricular nucleus of the hypothalamus. *Brain Res Bull* 1981;6: 47–61.
147. Tribollet E, Dreifuss JJ. Localization of neurones projecting to the hypothalamic paraventricular nucleus area of the rat: a horseradish peroxidase study. *Neuroscience* 1981;6:1315–1328.
148. Sawchenko PE, Swanson LW. The organization of forebrain afferents to the paraventricular and supraoptic nuclei of the rat. *J Comp Neurol* 1983;218:121–144.
149. Ter Horst GJ, Luiten PGM. The projections of the dorsomedial hypothalamic nucleus in the rat. *Brain Res Bull* 1986;16:231–248.
150. Watts AG, Swanson LW, Sanchez-Watts G. Efferent projections of the suprachiasmatic nucleus: I. Studies using anterograde transport of *Phaseolus vulgaris*-Leucoagglutinin in the rat. *J Comp Neurol* 1987;258: 204–229.
151. Simerly RB, Swanson LW. Projections of the medial preoptic nucleus: a *Phaseolus vulgaris* Leucoagglutinin anterograde tract-tracing study in the rat. *J Comp Neurol* 1988;270:209–242.
152. Cullinan WE, Zaborszky L. Organization of ascending hypothalamic projections to the rostral forebrain with special reference to the innervation of cholinergic projection neurons. *J Comp Neurol* 1991;306:631–667.
153. Canteras NS, Simerly RB, Swanson LW. Projections of the ventral premammillary nucleus. *J Comp Neurol* 1992;324:195–212.
154. Tappaz ML, Wassef M, Oertel WH, Paut L, Pujol JF. Light- and electron-microscopic immunocytochemistry of glutamic acid decarboxylase (GAD) in the basal hypothalamus: morphological evidence for neuroendocrine γ-aminobutyrate (GABA). *Neuroscience* 1983;9: 271–287.
155. Mugnani E, Oertel WH. An atlas of the distribution of GABAergic neurons and terminals in the rat CNS as revealed by GAD immunohistochemistry. In: Björklund A, Hökfelt T, eds. *Handbook of chemical neuroanatomy, vol. 4, GABA and neuropeptides in the CNS, Part I.* Amsterdam: Elsevier, 1985;436–622.
156. Olschowka JA. GABAergic innervation of corticotropin releasing factor neurons in the rat paraventricular nucleus: an electron microscopic immunocytochemical study. *Soc Neurosci Abstr* 1987;460:15.
157. Decavel C, Van Den Pol AN. GABA: a dominant neurotransmitter in the hypothalamus. *J Comp Neurol* 1990;302:1019–1037.
158. Ferraguti FM, Zoli M, Arinsson M, Agnati LF, Goldstein M, Filer D, Fuxe K. Distribution of glutamic acid decarboxylase messenger RNA-containing nerve cell populations in the male rat brain. *J Chem Neuroanat* 1990;3:377–396.
159. Okamura H, Abitbol M, Julien J-F et al. Neurons containing messenger RNA encoding glutamate decarboxylase in rat hypothalamus demonstrated by in situ hybridization, with special emphasis on cell groups in medial preoptic area, anterior hypothalamic area, and dorsomedial hypothalamic nucleus. *Neuroscience* 1990; 39:675–699.
160. Jones MT, Hillhouse EW, Burden J. Effect of various putative neurotransmitters on the secretion of cortico-

tropin-releasing hormone from the rat hypothalamus *in vitro:* a model of the neurotransmitters involved. *J Endocrinology* 1976;69:1–10.
161. Calogero AE, Galluci WT, Chrousos GP, Gold PW. Interaction between GABAergic neurotransmission and rat hypothalamic corticotropin-releasing hormone secretion in vitro. *Brain Res* 1988;463:28–36.
162. Hillhouse EW, Milton NGN. Effect of noradrenaline and γ-aminobutyric acid on the secretion of corticotropin-releasing factor-41 and arginine vasopressin from the rat hypothalamus in vitro. *J Endocrinology* 1989;122:719–723.
163. Makara GB, Stark E. Effect of gamma-aminobutyric acid (GABA) and GABA antagonist drugs on ACTH release. *Neuroendocrinology* 1974;16:178–190.
164. Plotsky PM, Otto S, Sutton S. Neurotransmitter modulation of corticotropin releasing factor secretion into the hypophyseal-portal circulation. *Life Sci* 1987;41: 1311–1317.
165. Watson SJ, Richard CW III, Barchas JD. Adrenocorticotropin in rat brain: immunocytochemical localization in cells and axons. *Science* 1978;200:1180–1181.
166. Joseph SA. Immunocreative adrenocorticotropin in rat brain: a neuroanatomical study using antiserum generated against synthetic ACTH[1-39]. *Am J Anat* 1980;158: 533–548.
167. Sawchenko PE, Swanson LW, Joseph SA. The distribution and cells of origin of ACTH (1-39)-stained varicosities in the paraventricular and supraoptic nuclei. *Brain Res* 1982;232:365–374.
168. Kiss JZ, Cassell MD, Palkovits M. Analysis of the ACTH/β-End/α MSH-immunoreactive afferent input to the hypothalamic parventricular nucleus of rat. *Brain Res* 1984;324:91–99.
169. Liposits Zs, Sievers L, Paull WK. Neuropeptide-Y and ACTH-immunoreactive innervation of corticotropin releasing factor (CRF)-synthesizing neurons in the hypothalamus of the rat. *Histochemistry* 1988;88:227–234.
170. Pretel S, Piekut DT. ACTH and enkephalin axonal input to paraventricular neurons containing c-fos-like immunoreactivity. *Synapse* 1991;8:100–106.
171. Plotsky PM. Opioid inhibition of immunoreactive corticotropin-releasing factor secretion into the hypophysial-portal circulation of rats. *Regul Peptides* 1986;16: 235–242.
172. Tsagarakis S, Rees LH, Besser GM, Grossman A. Opiate receptor subtype regulation of CRF-41 release from rat hypothalamus in vitro. *Neuroendocrinology* 1990; 51:599–605.
173. Bai FL, Yamano M, Shiotani Y, Emson PC, Smith AD, Powell JF, Tohyama M. An arcuato-paraventricular and -dorsomedial hypothalamic neuropeptide Y-containing system which lacks noradrenaline in the rat. *Brain Res* 1985;331:172–175.
174. Wahlestedt C, Skagerberg G, Ekman R, Heilig M, Sundler F, Hakanson R. Neuropeptide Y (NPY) in the area of the hypothalamic paraventricular nucleus activates the pituitary-adrenocortical axis in the rat. *Brain Res* 1987;417:33–38.
175. Buijs RM, Geffard M, Pool CW, Hoorneman EMD. The dopaminergic innervation of the supraoptic and paraventricular nucleus. A light and electron microscopical study. *Brain Res* 1984;323:65–72.
176. Decavel C, Geffard M, Calas A. Comparative study of dopamine- and noradrenaline-immunoreactive terminals in the paraventricular and supraoptic nuclei of the rat. *Neurosci Lett* 1987;77:149–154.
177. Fuller RW, Snoddy HD. Elevation of serum corticosterone concentrations in rats by pergolide and other dopamine agonists. *Endocrinology* 1981;109:1026–1032.
178. Kitchen I, Kelly M, Turner M. Dopamine receptor modulation of corticosterone secretion in neonatal and adult rats. *J Pharm. Pharmacol* 1988;40:580–581.
179. Arriza JL, Simerly RB, Swanson LW, Evans RM. Neuronal mineralocorticoid receptor as a mediator of glucocorticoid response. *Neuron* 1988;1:887–900.
180. Whitfield HJ, Brady LS, Smith MA, Mamalaki E, Fox RJ, Herkenham M. Optimization of cRNA probe in situ hybridization methodology for localization of glucocorticoid receptor mRNA in rat brain: a detailed protocol. *Cell Mol Neurobiol* 1990;10:145–157.
181. Herman JP. Regulation of adrenocorticosteroid receptor mRNA expression in the central nervous system. *Cell Mol Neurobiol* 1993;13:349–372.
182. Faria M, Navarra P, Tsagarakis S, Besser GM, Grossman A. Inhibition of CRF-41 release by substance P, but not substance K, from rat hypothalamus in vitro. *Brain Res* 1990;538:76–78.
183. Cantalamessa F, De Caro G, Pefumi M. Water intake inhibition and vasopressin release following intracerebroventricular administration of tachykinins to rats of the Wistar and Brattleboro strains. *Pharmacol Res Comm* 1981;13:641–655.
184. Moore RY, Eichler VB. Loss of a circadian adrenal corticosterone rhythm following suprachiasmatic lesions in the rat. *Brain Res* 1972;42:201–206.
185. Cascio CS, Shinsako J, Dallman MF. The suprachiasmatic nuclei stimulate evening ACTH secretion in the rat. *Brain Res* 1987;423:173–178.
186. Buijs RM, Markman M, Nunes-Cardoso B, Hou Y-X, Shinn S. Projections of the suprachiasmatic nucleus to stress-related areas in the rat hypohtalamus: a light and electron microscopic study. *J Comp Neurol* 1993;335: 42–54.
187. Ferguson AV, Day TA, Reanud LP. Subfornical organ efferents influence the excitability of neurohypophyseal and tuberoinfundibular paraventricular nucleus neurons in the rat. *Neuroendocrinology* 1984;39:423–428.
188. Kovacs KJ, Sawchenko PE. Mediation of osmoregulatory influences on neuroendocrine corticotropin-releasing factor expression in the ventral lamina terminalis. *Proc Natl Acad Sci* 1993;90:7681–7685.
189. Swanson LW. An autoradiographic study of the efferent connections of the preoptic area of the rat. *J Comp Neurol* 1976;167:227–256.
190. Miselis RR. The efferent projections of the subfornical organ of the rat: a circumventricular organ within a neural network subserving water balance. *Brain Res* 1981;230:1–23.
191. Lind RW, Van Hoesen GW, Johnson AK. An HRP study of the connections of the subfornical organ of the rat. *J Comp Neurol* 1982;210:265–277.
192. Roland BL, Sawchenko PE. The lamina terminalis is a source of the GABAergic innervation of the magnocellular neurosecretory system. *Soc Neurosci Abstr* 1992;18:823.
193. Lind RW, Swanson LW, Ganten D. Angiotensin II immunoreactivity in the neural afferents and efferents

of the subfornical organ of the rat. *Brain Res* 1984;321: 209–215.

194. Simpson JB. The circumventricular organs and the central actions of angiotensin. *Neuroendocrinology* 1981;32:248–256.
195. Keller-Wood ME, Dallman MF. Corticosteroid inhibition of ACTH secretion. *Endocrinol Rev* 1984;5:1–24.
196. Reul JMHM, de Kloet ER. Two receptor systems for corticosterone in rat brain: microdistribution and differential occupation. *Endocrinology* 1985;117:2502–2511.
197. Pearce D, Yamamoto KR. Mineralocorticoid and glucocorticoid receptor activities distinguished by nonreceptor factors at a composite response element. *Science* 1993;259:1161–1165.
198. de Kloet ER, Reul JMHM. Feedback action and tonic influence of glucocorticoids on brain function: a concept arising from heterogeneity of brain receptor systems. *Psychoneuroendocrinology* 1987;12:83–105.
199. Dallman MF, Levin N, Cascio CS, Akana SF, Jacobson L, Kuhn RW. Pharmacological evidence that the inhibition of diurnal adrenocorticotropin secretion by corticosteroids is mediated via type I corticosterone-prefering receptors. *Endocrinology* 1989;124:2844–2850.
200. Herman JP, Watson SJ, Chao HM, Coirini HM, McEwen BS. Diurnal regulation of glucocorticoid receptor and mineralocorticoid receptor mRNAs in the rat hippocampus. *Mol Cell Neurosci* 1993;4:181–190.
201. Ratka A, Sutanto W, Bloemers M, de Kloet ER. On the role of brain mineralocorticoid (type I) and glucocorticoid (type II) receptors in neuroendocrine regulation. *Neuroendocrinology* 1989;50:117–123.
202. McEwen BS, Weiss JM, Schwartz LS. Selective retention of corticosterone by limbic structures in rat brain. *Nature* 1968;228:911–912.
203. Füxe K, Wikstrom AC, Okret S et al. Mapping of glucocorticoid immunoreactive neurons in the rat tel- and diencephalon using a monoclonal antibody against rat liver glucocorticoid receptor. *Endocrinology* 1985; 117:1803–1812.
204. Reul JM, de Kloet ER. Anatomical resolution of two types of corticosterone receptor sites in rat brain with in vitro autoradiography and computerized image analysis. *J Steroid Biochem* 1986;24:269–272.
205. Aronsson M, Füxe K, Dong Y, Agnati LF, Okret S, Gustafsson J-A. Localization of glucocorticoid receptor mRNA in the male rat brain by in situ hybridization. *Proc Natl Acad Sci* 1988;55:9331–9335.
206. Sousa RJ, Tannery NH, Lafer EM. In situ hybridization mapping of glucocorticoid receptor messenger ribonucleic acid in rat brain. *Mol Endocrinol* 1989;3: 481–494.
207. Ahima RS, Harlan RE. Charting of type II glucocorticoid receptor-like immunoreactivity in the rat central nervous system. *Neuroscience* 1990;39:579–604.
208. Ahima RS, Krozowski Z, Harlan RE. Type I corticosteroid receptor-like immunoreactivity in the rat CNS: distribution and regulation by corticosteroids. *J Comp Neurol* 1991;313:522–538.
209. Feldman S, Conforti N. Participation of the dorsal hippocampus in the glucocorticoid feedback effect on adrenocortical activity. *Neuroendocrinology* 1980;30: 52–55.
210. Sapolsky RM, Krey LC, McEwen BS. Glucocorticoid-sensitive hippocampal neurons are involved in terminating the adrenocortical stress response. *Proc Natl Acad Sci* 1984;81:6174–6177.
211. Sapolsky RM, Armanini MP, Packan DR, Sutton SW, Plotsky PM. Glucocorticoid feedback inhibition of adrenocorticotropic hormone secretagogue release. *Neuroendocrinology* 1990;51:328–336.
212. Plotsky PM, Otto S, Sapolsky RM. Inhibition of immunoreactive corticotropin-releasing factor secretion into the hypophyseal-portal circulation by delayed glucocorticoid feedback. *Endocrinology* 1987;119:1126–1130.
213. Kovacs KJ, Makara G. Corticosterone and dexamethasone act at different brain sites to inhibit adrenalectomy-induced adrenocorticotropin hypersecretion. *Brain Res* 1988;474:205–210.
214. Jacobson L, Sapolsky RM. The role of the hippocampus in feedback regulation of the hypothalamo-pituitary-adrenocortical axis. *Endocrine Rev* 1991;12:118–134.
215. Bradbury MJ, Strack AM, Dallman MF. Lesions of the hippocampal efferent pathway (fimbria-fornix) do not alter sensitivity of adrenocorticotropin to feedback inhibition by corticosterone in rats. *Neuroendocrinology* 1993;58:396–407.
216. Feldman S, Weidenfeld J. The dorsal hippocampus modifies the negative feedback effect of glucocorticoids on the adrenocortical and median eminence CRF-41 responses to photic stimulation. *Brain Res* 1993; 614:227–232.
217. Feldman S, Saphier D, Weidenfeld J. Corticosterone implants in the paraventricular nucleus inhibit ACTH and corticosterone responses and the release of corticotropin-releasing factor following neural stimuli. *Brain Res* 1992;578:251–255.
218. Kovacs K, Kiss JZ, Makara GB. Glucocorticoid implants around the hypothalamic paraventricular nucleus prevent the increase in corticotropin-releasing factor and arginine vasopressin immunostaining induced by adrenalectomy. *Neuroendocrinology* 1986;44:229–234.
219. Sawchenko PE. Evidence for a local site of action for glucocorticoids in inhibiting CRF and vasopressin expression in the paraventricular nucleus. *Brain Res* 1987;403:213–223.
220. Herman JP, Wiegand SJ, Watson SJ. Regulation of basal corticotropin-releasing hormone and arginine vasopressin messenger ribonuclei acid expression in the paraventricular nucleus: effects of selective hypothalamic deafferentations. *Endocrinology* 1990;127:2408–2417.
221. Slusher MA. Effects of cortisol implants in the brainstem and ventral hippocampus on diurnal corticosteroid levels. *Exp Brain Res* 1966;1:184–194.
222. Bradbury MJ, Dallman MF. Effects of type I and type II glucocorticoid receptor antagonists on ACTH levels in the PM. *Soc Neurosci Abstr* 1989;15:176.
223. Morgan JI, Curran T. Stimulus-transcription coupling in neurons: role of cellular immediate early genes. *Trends Neurosci* 1989;12:459–462.
224. Sheng M, Greenberg ME. The regulation and function of c-fos and other immediate early genes in the nervous system. *Neuron* 1990;4:477–485.
225. Chastrette N, Pfaff DW, Gibbs RB. Effects of daytime and nighttime stress on Fos-like immunoreactivity in the paraventricular nucleus of the hypothalamus, the

habenula, and the posterior paraventricular nucleus of the thalamus. *Brain Res* 1991;563:339–344.

226. Deutch AY, Lee MC, Gillham MH, Cameron DA, Goldstein M, Iadorola MJ. Stress selectively increases Fos protein in dopamine neurons innervating the prefrontal cortex. *Cerebral Cortex* 1991;1:273–292.
227. Schreiber SS, Tocco G, Shors TJ, Thompson RF. Activation of immediate early genes after acute stress. *Neuroreport* 1991;2:17–20.
228. Sharp FR, Sagar SM, Hicks K, Lowenstein D, Hisanga K. Fos and Fos-related antigen induction by hypertonic saline and stress. *J Neurosci* 1991;11:2321–2331.
229. Honkaniemi J, Kainu T, Ceccatelli S, Rechardt L, Hökfelt T. Fos and jun in rat central amygdaloid nucleus and paraventricular nucleus after stress. *Neuroreport* 1992;3:849–852.
230. Kononen J, Honkaniemi J, Alho H, Koistinaho J, Iadarola M, Pelto-Huikko M. Fos-like immunoreactivity in the rat hypothalamic-pituitary axis after immobilization stress. *Endocrinology* 1992;130:3041–3047.
231. Pezzone MA, Lee W-S, Hoffman GE, Rabin BS. Induction of c-Fos immunoreactivity in the rat forebrain by conditioned and unconditioned aversive stimuli. *Brain Res* 1992;597:41–50.
232. Duncan GE, Johnson KB, Breese GR. Topographic patterns of brain activity in response to swim stress: assessment by 2-deoxyglucose uptake and expression of Fos-like immunoreactivity. *J Neurosci* 1993;13:3932–3943.
233. Imaki T, Shibasaki T, Hotta M, Demura H. Intracerebroventricular administration of corticotropin-releasing factor induces c-fos mRNA expression in brain regions related to stress responses: comparison with pattern of c-fos mRNA induction after stress. *Brain Res* 1993;616:114–125.
234. Hoffman GE, Smith MS, Verbalis JG. c-Fos and related immediate early gene products as markers of activity in neuroendocrine systems. *Front Neuroendocrinol* 1993;14:173–214.
235. Cullinan WE, Herman JP, Watson SJ. Pattern of immediate early gene activation in rat brain following acute stress. *Soc Neurosci Abstr* 1992;18:1541.
236. Hua SY, Chen YZ. Glucocorticoid's fast effects on neuronal membrane in hippocampal slices of rat and isolated coeliac ganglion of the guinea pig. *Neurosci Lett* 1988;31:s58.
237. Chen YZ, Hua SY, Wang CA, Wu LG, Gu Q, Xing BR. An electrophysiological study on the membrane receptor-mediated actions of glucocorticoids in mammalian neurons. *Neuroendocrinology* 1991;53(suppl 1):25–30.
238. Koob GF, Bloom FE. Corticotropin-releasing factor and behavior. *Fed Proc* 1985;44:259–263.
239. Vale W, Greer M, eds. Corticotropin-releasing factor. *Fed Proc* 1985;44.
240. Dunn AJ, Berridge CW. Physiological and behavioral responses to corticotropin-releasing factor administration: is CRF a mediator of anxiety and stress responses? *Brain Res Rev* 1990;15:71–100.
241. Moga MM, Saper CB, Gray TS. Bed nucleus of the stria terminalis: cytoarchitecture, immunocytochemisty, and projection to the parabrachial nucleus in the rat. *J Comp Neurol* 1989;283:315–332.
242. Schmued LC, Fallon JH. Fluoro-gold: a new fluorescent retrograde axonal tracer with numerous unique properties. *Brain Res* 1986;377:147–154.
243. Chang HT, Kuo H, Whittaker JA, Cooper NG. Light and electron microscopic analysis of projection neurons retrogradely labeled with Fluoro-gold: notes on the application of antibodies to Fluoro-gold. *J Neurosci Meth* 1990;1:31–37.

Neurobiological and Clinical Consequences of Stress: From Normal Adaptation to PTSD, edited by M. J. Friedman, D. S. Charney, and A. Y. Deutch, Lippincott-Raven Publishers, Philadelphia © 1995.

2

Regulation of Intracellular Signal Transduction and Gene Expression by Stress

Ronald S. Duman

Department of Psychiatry, Yale University School of Medicine, New Haven, Connecticut 06508

The influence of environmental or psychological stressors on central nervous system function is mediated by activation of specific neurotransmitter and neuropeptide pathways. Rapid neuronal responses to stress are mediated by activation of excitatory or inhibitory receptor/ion channel complexes that directly control neuronal excitability. In addition, the actions of stress on many neurotransmitter and neuropeptide systems result in regulation of receptor-coupled intracellular signal transduction pathways. These signal transduction pathways mediate many of the short- and long-term effects of stress on neuronal function. Continued exposure to stressful stimuli may lead to additional modulatory changes, some of which may be beneficial. However, other chronic effects of stress may be maladaptive and may thereby underlie, or contribute to, stress-related psychiatric disorders such as post-traumatic stress disorder (PTSD) and depression. The present chapter provides an overview of the intracellular pathways that mediate the actions of acute stress. In addition, examples of adaptations of signal transduction pathways and gene expression to chronic stress are discussed.

OVERVIEW OF INTRACELLULAR SIGNAL TRANSDUCTION PATHWAYS

Acute stress leads to activation of many brain signaling systems, including those for certain neurotransmitters and neuropeptides, as well as those for some nerve growth factors. The rapid actions of stress are mediated by amino acid neurotransmitters that regulate neuronal activity via gating of ion channels, and thereby directly inhibit or stimulate cell firing rate. However, most neurotransmitter and neuropeptide receptors belong to the G protein-coupled receptor family. These receptors are coupled, via guanosine 5′-triphosphate (GTP)-binding proteins or G proteins, to second messenger systems. Regulation of these second messengers represents the first step in a cascade of intracellular events mediated in most cases by protein phosphorylation. Second messenger-coupled intracellular pathways regulate most aspects of cellular function, including neuronal activity, metabolism, and gene expression. Thus, these intracellular signal transduction systems represent the primary pathways by which stress, as well as other stimuli, regulate neuronal function.

Receptor-Coupled Second Messenger Systems

With the exception of receptors that form an ion channel complex, the G protein-coupled receptor superfamily mediates synaptic transmission to most other neurotransmitter and neuropeptide systems. These receptors couple to second messenger systems via four major types of G proteins: Gs, Gi, Go, and Gq. There are also multiple subtypes for each of these G proteins (1,2). The G proteins are heterodtrimers composed of α, β, and γ subunits. The α subunits confer specific functional activity on the different types of G proteins. There appear to be some βγ subunits shared by G proteins, although there is also some diversity of the type and function of these subunits. The G protein-coupled receptors regulate multiple intracellular effector proteins. The best characterized second messengers regulated by these receptors in brain include adenosine 3′,5′-cyclic monophosphate (cAMP), calcium, and the phosphatidylinositol system, citric guanosine 3′,5′-cyclic monophosphate (cGMP) and nitric oxide (NO). These second messenger pathways and the neurotransmitter receptors that regulate these pathways are briefly discussed.

Stress and Receptor Regulation of the cAMP System

Stress influences several neurotransmitter receptors that regulate cAMP formation. Receptors expected to be activated by stress that stimulate the cAMP system include β-adrenergic, D_1- and D_5-dopaminergic, subtypes of serotonin receptors (5-HT_4, 5-HT_6, and 5-HT_7), and corticotrophin-releasing factor (CRF) receptors. Receptors that negatively regulate the cAMP system and are thought to be influenced by stress include α_2-adrenergic, D_2- and D_4-dopaminergic, 5-HT_{1A}, and neuropeptide Y (NPY). Gs and Gi couple stimulatory and inhibitory receptors, respectively, to adenylyl cyclase, the enzyme responsible for synthesis of cAMP (Fig. 1). Receptor inhibition of adenylyl cyclase may also be mediated by Go, and a variant form of Gi, termed Gz (1,3). There are at least six forms of adenylyl cyclase that have different regional distributions in brain, as well as distinct regulatory properties (2). In addition to regulation by Gs and Gi α-subunits, the enzyme isoforms of adenylyl cyclase differ in their ability to be activated or inhibited upon binding to free βγ complexes or Ca^{2+}/calmodulin complexes (2,4).

In addition to regulation of cAMP, many of the receptors which couple via Gi and/or Go lead to activation of an inward rectifying K^+ channel and to inhibition of a voltage-dependent Ca^{2+} channel (5) (Fig. 1). These effects on ion channels lead to hyperpolarization of neurons and may thereby work in concert with receptor-inhibition of cAMP formation in some neurons where cAMP is stimulatory. The exact mechanisms subserved by each receptor-G protein interaction probably depends on the neuronal cell type, the brain region being studied, and expression level of the specific G proteins and effectors in that particular neuron.

Stress and Regulation of Calcium and the Phosphatidylinositol System

Neurotransmitters can influence levels of Ca^{2+} by either activation of receptors that influence the flux of extracellular Ca^{2+} into neurons or by regulation of the release of Ca^{2+} from intracellular stores (6) (Fig. 1). Extracellular Ca^{2+} flux can be regulated by receptors that inhibit voltage-sensitive Ca^{2+} channels, via interactions with G proteins as just discussed. In addition, Ca^{2+} channel conductances can be influenced by other second messenger systems, such as cAMP, via phosphorylation of Ca^{2+} channels. Finally, extracellular Ca^{2+} can pass through some ligand-gated channels, such as the *N*-methyl-D-aspartate (NMDA)-glutamate receptor.

Free Ca^{2+} levels can also be regulated by receptor regulation of the phosphatidylinositol second messenger system and release of intracellular Ca^{2+} from storage sites (Fig. 1). Stress activates several receptors that stimulate phospholipase C, including α_1-adrenergic, 5-HT_2, neurotensin, and neurokinin receptors. Neurotransmitter receptors couple with Gq (or subtypes of Gi and Go in some cell types) to activate phospholipase C (also referred to as phosphoinositidase), which catalyzes the breakdown of phosphatidylinositol 4,5-bisphosphate to inositol triphosphate (IP_3) and diacylglycerol (DG). There are multiple forms of phospholipase C in brain that demonstrate different regional distributions and regulatory properties (4,6). Gq and subtypes of Gi/Go can directly couple receptors

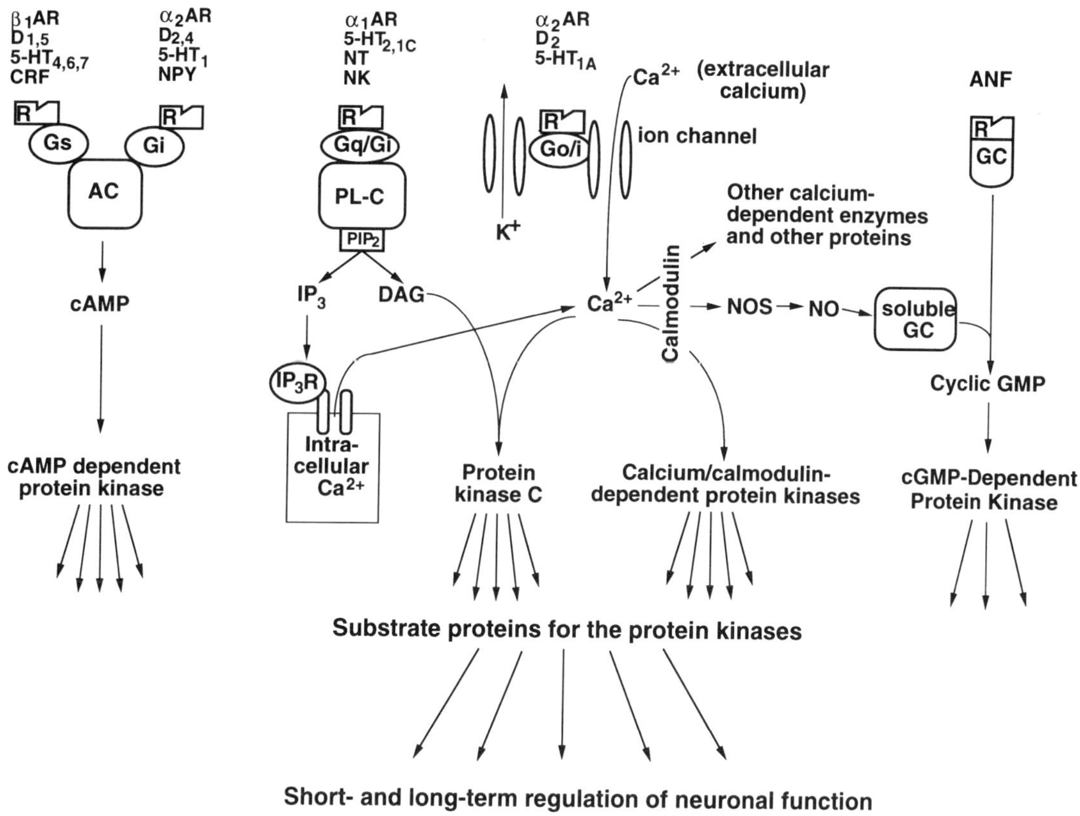

FIG. 1. Schematic illustration of major intracellular second messenger-dependent signal transduction pathways regulated by stress in the brain. Stress results in activation of several different neurotransmitter and neuropeptide receptors, including subtypes of α- and β-adrenergic receptors (αAR and βAR, dopamine (D), serotonin (5-HT), corticotropin releasing factor (CRF), neuropeptide Y (NPY), neurotensin (NT), and neurokinin (NK) receptors. These receptors are specifically coupled to intracellular signal transduction pathways via G protein subtypes. Neurotransmitter receptors (R) that either stimulate or inhibit adenylyl cyclase, the enzyme that catalyzes the synthesis of adenosine 3′,5′-cyclic monophosphate (cAMP), couple with Gs and Gi/o, respectively. Receptors interact with Gq/Gi to stimulate phospholipase C (PLC), which metabolizes phosphatidylinositol 4,5-bisphosphate (PIP_2) into the second messengers inositol triphosphate (IP_3) and diacylglycerol (DAG). IP_3 then acts on specific IP_3 receptors (IP_3R) to increase intracellular levels of free Ca^{2+} (also a second messenger in brain) by releasing Ca^{2+} from internal stores. Increased levels of intracellular Ca^{2+} also result from the flux of Ca^{2+} across the plasma membrane through Ca^{2+} channels. Ca^{2+} flux is stimulated by nerve impulses, and certain neurotransmitter receptors are capable of regulating these channels via interactions with Go/Gi. Elevation of Ca^{2+} may result in regulation of several different enzymes. One such enzyme is nitric oxide synthetase (NOS), which leads to increased levels of nitric oxide (NO) and the activation of cytoplasmic guanylyl cyclase (the enzyme that catalyzes the synthesis of guanosine 3′,5′ cyclic monophosphate (cGMP)) and of ADP-ribosyltransferases (not shown). Other forms of guanylyl cyclase reside in specific plasma membrane receptors, such as those for atrial natriuretic factor (ANF).

The actions of these second messengers on neuronal function are mediated largely by regulation of specific protein kinases. There is one major class of protein kinase for cAMP and one for cGMP, although subtypes of these enzymes are differentially expressed throughout the brain. Activation of these enzymes results in phosphorylation of specific substrate proteins that mediate most of the numerous second messenger actions of cAMP and cGMP in the nervous system. There are two major classes of Ca^{2+}-dependent protein kinase in brain. One is activated by Ca^{2+} and calmodulin and is referred to as Ca^{2+}/calmodulin-dependent protein kinase; there are at least five distinct subtypes of this enzyme in brain. The other major class is activated by Ca^{2+} in conjunction with DAG and various phospholipids, and is referred to as Ca^{2+}/DAG-dependent protein kinase or protein kinase C; there are at least 7 closely-related variants of this enzyme present in the brain. Activation of these second messenger-dependent protein kinases alters the phosphorylation and function of substrate proteins and thereby leads to the biological responses of the extracellular messengers. Modified from Duman and Nestler, (4).

to activation of certain subtypes of phospholipase C (β_1) via free G protein α-subunits. Alternatively, receptor-G protein coupling can activate other subtypes of phospholipase C (β_2) via the release of free G protein $\beta\gamma$-subunits.

Receptors for IP_3 form Ca^{2+} channels located on intracellular storage vesicles, and these receptors mediate the release of Ca^{2+} in response to elevated levels of IP_3 (6). The Ca^{2+} is required for activation of many regulatory enzymes, including Ca^{2+}/calmodulin-dependent protein kinases and protein kinase C (Fig. 1). Diacylglycerol is also involved in activation of protein kinase C (see next section). The IP_3 receptors, like ryanodine receptors, are also sensitive to Ca^{2+}, a sensitivity that may underlie the formation of "Ca^{2+} waves or oscillations" in neurons as well as other cell types (6). These Ca^{2+} waves represent a positive feedback that allows the continued neuronal spread of the Ca^{2+} signal.

Stress and Regulation of cGMP and Nitric Oxide (NO)

Neurotransmitters and neuropeptides can regulate cGMP levels via two major pathways (2,7). The first is where the enzyme guanylyl cyclase and the ligand recognition site reside within the same protein (e.g., the atial natriuretic peptide receptor). The second major pathway involves the formation of NO, which activates soluble guanylyl cyclase. The NO is formed upon activation of receptors that increase intracellular Ca^{2+}, which then activates NO synthetase. Stress activates many neutotransmitter receptors that increase intracellular Ca^{2+} levels via activation of the phosphatidylinositol system (see previous sentence), and would thereby be expected to result in elevated cGMP via activation of NO synthetase.

Nitric oxide has also been shown to regulate endogenous adenosine 5′-diphosphate (ADP)-ribosylation via activation of ADP-ribosyltransferases found in brain (8). Cholera and pertussis toxins can be considered ADP-ribosyltransferases in that they catalyze the ADP-ribosylation of Gs and Gi/o, respectively, although they are not regulated by NO. While a few of the physiological substrates for NO-stimulated ADP-ribosylation have been identified, most remain unknown (4,8). Further studies are required to identify the nature of other substrates of endogenous ADP-ribosylation and to ascertain the physiological consequences of this covalent modification.

Stress and Regulation of Arachidonic Acid Metabolites

Arachidonic acid is a precursor for the intracellular messengers known as the prostaglandins and leukotrienes (9). Phospholipase A_2 (PLA_2) cleaves membrane phospholipids to yield free arachidonic acid, which is the substrate for two major enzymatic pathways: 1) cleavage by cyclooxygenase, and several additional enzymatic steps, to yield prostaglandins and other cyclic endoperoxides, and 2) by lipoxygenase to yield leukotrienes. The prostaglandins and leukotrienes influence cellular function by actions on other cellular proteins, including adenylyl cyclase, guanylyl cyclase, ion channels, protein kinases, and other proteins (9).

Increasing evidence supports a role for neurotransmitter regulation of PLA_2 via coupling with G proteins (9). A PLA_2-activating protein (PLAP), which shares some homology with G protein β-subunits, has also been identified (10). The enzymatic activity of PLA_2 is also dependent on Ca^{2+}, and neurotransmitter regulation of Ca^{2+}, via the mechanisms just discussed, could also lead to activation of this enzyme. For example, α-adrenergic receptors may exert some actions on PLA_2 via mobilization of Ca^{2+} (11). Finally, PLA_2 activity can be inhibited by another cellular protein, macrocortin, that is induced by glucocorticoids (12). Elevation of glucocorticoids in response to stress may thereby regulate PLA_2 activity. These regulatory pathways provide several different mechanisms for short- and long-term regulation of PLA_2 activity in response to stress and glucocorticoids.

Stress and Regulation of Protein Phosphorylation

With the exception of direct regulation of ion channels, nearly all actions of neurotransmitter receptors are mediated largely by activation of their corresponding protein kinases (13,14). These kinases mediate the transfer of phosphate from adenosine 5′-triphosphate (ATP) to serine and threonine residues of specific substrate proteins or phosphoproteins. Several types of protein kinases and the pathways leading to their regulation will be discussed.

Second Messenger-Dependent Protein Kinases

There are two major classes of cyclic nucleotide regulated kinases: cAMP- and cGMP-dependent protein kinases (13,14) (Figs. 1, 2). In addition, two major classes of Ca^{2+}-regulated protein kinases have been characterized, which require additional factors for activation (Fig. 1). One of these is activated by Ca^{2+} in conjunction with calmodulin and is named accordingly Ca^{2+}/calmodulin-dependent protein kinase. The other is activated by Ca^{2+} in conjunction with diacylglycerol and is referred to as protein kinase C. There are several subtypes of each of these Ca^{2+}-dependent kinases in brain. These protein kinases would be expected to be involved in most of the short- and long-term actions of neurotransmitters and neuropeptides that are activated by stress.

Second Messenger-Independent Protein Kinases

The best example of second messenger-independent protein kinases are the microtubule-associated protein (MAP) kinases (also referred to as ERKs for *e*xtracellular signal-*r*egulated *k*inases). These kinases were first identified by their involvement in the phosphorylation of microtubule-associated proteins (15,16). The MAP kinases can be indirectly regulated by extracellular signals, including growth factors and neurotransmitters (Fig. 2). Activation of MAP kinases by growth factors or neurotrophins is mediated by activation of Ras and a cascade of kinases, including MAP kinase kinase (also referred to as MEK or *M*AP kinase/*E*RK *K*inase) and MAP kinase kinase kinases (also referred to as MEK kinases or ERK kinase kinases; one example is Raf kinase). Neurotransmitters can indirectly

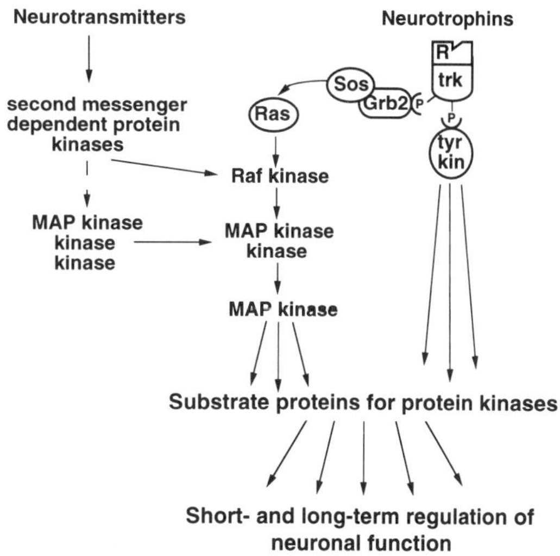

FIG. 2. Schematic illustration of the intracellular pathways that mediate the actions of receptor-associated protein tyrosine kinases in the brain. Stress alters the expression of certain neurotrophins in brain and activation of neurotrophin receptors (trk) stimulates protein tyrosine kinase activity intrinsic to the receptor. Activation of related receptor protein tyrosine kinases, and by analogy trk, can result in phosphorylation of the receptor itself as well as adaptor proteins with src homology (SH) domains. The SH-adaptor proteins, or phosphorylated receptor in some cases, may then associate other proteins (Grb2 and Sos) to activate the GTP-binding protein Ras. This in turn leads to activation of Raf, a protein serine/threonine kinase, by an unknown mechanism. Raf phosphorylates and activates microtubule-associated-proteins (MAP) kinase kinase, which then activates MAP-kinase by phosphorylating it on threonine and tyrosine residues. Also shown is a mechanism by which second messenger-dependent protein kinases may interact with and regulate Raf and possibly other MAP-kinase kinases. In addition to the MAP kinase pathway, receptor-protein tyrosine kinases can also influence cellular responses by phosphorylation and activation of soluble protein tyrosine kinases (tyr kin) that lack a receptor domain. These pathways have only recently been identified, and it is likely that future studies will identify additional pathways that mediate the biological actions of protein tyrosine kinases.

regulate these kinases via activation of second messenger-dependent protein kinases (Fig. 2). Thus, stress could regulate the MAP kinase pathway by activation of both second messenger-dependent and -independent protein kinases.

Protein Tyrosine Kinases

Protein tyrosine kinases can be subdivided into two major classes: 1) the receptor-protein tyrosine kinases, which contain the receptor and kinase domains within the same protein, and 2) the soluble protein tyrosine kinases, which lack a receptor domain (14,17). Examples of the receptor protein tyrosine kinases include the neurotrophin receptors, also referred to as the trk receptor family. Stress may influence the expression of neurotrophins, in part via activation of monoamine neurotransmitter systems, and thereby activates neurotrophin receptor signaling pathways. Activation of receptor-tyrosine kinases may result in phosphorylation of the receptor-kinase itself, as well as other proteins (Fig. 2). Phosphorylation of proteins with SH (for *s*rc *h*omology) adaptor domains can result in activation of intracellular signaling pathways. For example, such an interaction leads to activation of Ras: this occurs via SH interactions between Grb2 and Sos, a guanine nucleotide exchange factor. Activation of Ras, in turn, leads to activation of Raf kinase and the pathway that regulates MAP kinase (15,18) (Fig. 2).

Activation of receptor-tyrosine kinases may also result in regulation of another intracellular pathway. This pathway appears to involve phosphorylation and activation of the soluble protein tyrosine kinases (18). Examples of soluble protein tyrosine kinases include src, yes, fes, and many others (15,16). Targets of these soluble protein tyrosine kinases, which may mediate some actions of receptor tyrosine kinases, include cytoplasmic transcription factors. Phosphorylation of these cytoplasmic transcription factors results in translocation of the transcription factor to the nucleus.

Protein Phosphatases

In addition to regulation of protein kinases, neurotransmitters can also regulate protein phosphatases that remove phosphate groups from either serine/theonine or tyrosine residues of target proteins (14). One type of serine/threonine phosphatase, calcineurin or phosphatase type 2B, can be activated by Ca^{2+}/calmodulin, and could therefore be activated by neurotransmitter-receptor pathways that increase Ca^{2+}, as just discussed. One other known mechanism for regulation of phosphatases of the serine/threonine type is by regulation of phosphatase inhibitors. The best characterized examples are phosphatase inhibitors-1 and -2, and DARPP-32, which are inhibitors of phosphatase type 1 and are regulated by cAMP-dependent kinases and other protein kinases. The mechanisms underlying the regulation of tyrosine phosphatases have not been well characterized.

Stress and Regulation of Phosphoproteins

Nearly all actions of the neurotransmitter second messenger systems already discussed are mediated by regulation of specific phosphoproteins. Thus, protein kinases are known to phosphorylate and thereby regulate most neuronal proteins, including receptors, G proteins, effector enzymes, ion channels, other protein kinases, transporters, transmitter biosynthetic enzymes, metabolic enzymes, transcription factors, neurofilaments, and many others (4). Acute stress activation of neurotransmitters would be expected to result in diverse biological actions that would be dependent on the brain region and transmitter system activated, as well as the intracellular machinery present in each cell type.

Stress and Regulation of Gene Expression

Stress Regulation of Immediate-Early Gene (IEG) Transcription Factors

One of best characterized actions of acute stress and subsequent protein phosphorylation is induction of IEG transcription factors, such as c-fos (19–26) (Fig. 3). The protein products of IEGs, FOS and JUN, form a dimer that bind to AP-1 DNA sites and thereby regulate transcriptional activity at these sites (27,28). Expression of IEGs is known to be regulated by a variety of extracellular stimuli and a corresponding diverse array of intracellular signal transduction pathways (Fig. 3). The best characterized are medi-

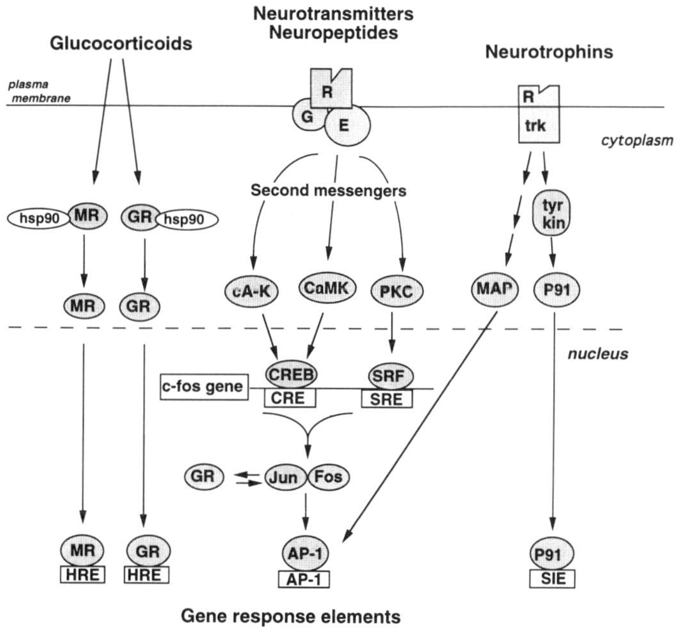

FIG. 3. Schematic illustration of the mechanisms by which stress regulates gene expression in brain. The longterm actions of stress on brain may be mediated by regulation of the transcriptional activity and expression of specific genes. There are several pathways by which stress could influence gene expression, including those regulated by second messengers, glucocorticoid receptors (GR and MR), and protein tyrosine kinases. The actions of second messengers are mediated by activation of protein kinases, including adenosine 3′,5′-cyclic monophosphate (cAMP)- (cA-K) and Ca/calmodulin- (CaMK) dependent protein kinases and protein kinase C (PKC). These kinases lead to regulation of gene expression via phosphorylation of specific transcription factors: for instance, the actions of cA-K and CaMK are mediated by phosphorylation of the CREB (cAMP response element binding protein), while PKC influences gene expression by phosphorylation of serum response factor (SRF). CREB and SRF recognize specific DNA regulatory elements, CRE and SRE, respectively, and upon phosphorylation they regulate the expression of genes containing these elements. One example is the immediate early gene c-fos, which is induced by these second-messenger–dependent pathways via CRE and SRE regulatory elements. Induction of c-fos also demonstrates a second pathway for regulation of gene transcription factors in addition to phosphorylation (i.e., altered expression). The protein product of c-fos (Fos) may also be regulated by second messenger-dependent phosphorylation (*not shown*). Fos interacts with Jun to form a dimer referred to as the AP-1 complex that binds to DNA regulatory elements of the same name.

The actions of glucocorticoids on gene expression are mediated by two receptors, GR and MR, which have different affinities for glucocorticoids. Heat shock proteins (hsp90) bind to and help maintain GR and MR in their inactive state. Upon binding of glucocorticoid, the receptors dissociate from hsp90 and are then translocated to the nucleus. Activated GR and MR recognize specific hormone response elements (HRE) and thereby regulate gene expression. In addition, GR may interact with other transcription factors, such as Jun, which may result in mutual inhibition of the functional activity of each factor. GR may also inhibit AP-1 function by binding to HRE sites located in close proximity (*not shown*). Interactions of GR and AP-1 proteins would be likely to occur during stress when levels of glucocorticoids are elevated and expression of Fos is increased in many brain regions.

Activation of receptor-protein tyrosine kinases, and by analogy the neurotrophin receptors, can result in regulation of gene expression by two separate pathways. The first pathway involves activation of Ras and the pathway leading to activation of microtubule-associated-protein (MAP) kinase (Fig. 2). MAP kinase can then influence gene expression by phosphorylation of transcription factors, such as Fos and Jun. The second pathway involves phosphorylation of cytoplasmic transcription factors such as P91. This appears to occur not by the receptor-protein tyrosine kinase itself, but by soluble tyrosine kinases (tyr kin) that are activated by the former. Upon phosphorylation, P91 is translocated to the nucleus, where it recognizes specific DNA sequences such as the sis-inducible element (SIE).

ated by cAMP- and Ca^{2+}-dependent protein kinases, and phosphorylation of another type of transcription factor known as CREB (*c*AMP *r*esponse *e*lement *b*inding protein) (29). Activation of CREB by phosphorylation leads to regulation of inducible CRE and CaRE (*c*alcium *r*esponse *e*lements) sites located on the promoter region of the genes for c-fos and other IEGs (Fig. 3). In addition, fos and other IEG proteins are themselves regulated by protein phosphorylation via several different protein kinases.

Expression of c-fos and related IEGs has provided a useful tool for identification of brain regions activated by pharmacological and environmental stimuli, including stress. Thus, acute stress increases the levels of c-fos in brain, and the pattern of c-fos expression is dependent on the type of stress being studied (19–26). Regulation of IEGs could also provide a mechanism whereby acute stimuli lead to changes in gene expression and long-term adaptations of neuronal function. Adaptations to chronic stress (see next section) could also result from regulation of IEG as well as CREB transcription factors. The genes for many neuronal proteins regulated by stress contain CRE- or AP-1–binding sites (Fig. 3). These include genes for neuropeptides, receptors, G proteins, and neurotransmitter synthetic enzymes, just to list a few. However, only a few target genes have been directly demonstrated to be regulated by IEG transcription factors. Further characterization of the genes regulated by IEG transcription factors in brain is currently a subject of great interest.

Regulation of Gene Expression by Glucocorticoids

In addition to regulation by second messenger-activated transcription factors, gene expression is regulated during stress by adrenal-glucocorticoids via binding to cytoplasmic glucocorticoid receptors. Two types of cytoplasmic receptors for glucocorticoids in brain display different affinities for glucocorticoids and have distinct regional distributions (30). The type I or mineralcorticoid receptor (MR) has high affinity for corticosterone and aldosterone, while the type II or glucocorticoid receptor (GR) has lower affinity for corticosterone. Upon binding of steroid, both the GR and MR are translocated to the nucleus where they control transcription via binding to hormone response elements (HRE) (Fig. 3).

In addition to regulating HRE sites, GR and MR may also influence the activity of other transcription factors. First, HRE sites located in proximity to certain gene response elements can result in a mutual increase in the activity of each element (31). Second, GR can modulate the transcriptional activity of other transcription factors, such as the IEGs (32,33). Activated GR and AP-1 proteins (i.e., Jun or Fos) combine to form a complex that then represses the binding of each transcription factor to its respective promoter element (Fig. 3). Repression of DNA binding is dependent on protein-protein interactions between the two transcription factors, and not on competition for the DNA binding domain. The GR, but not the MR, is also able to inhibit Fos/Jun activation of AP-1 response elements located in close proximity to an HRE (31). Such interactions between GR and AP-1 would be likely to occur during stress, when GR is activated by elevated levels of glucocorticoids and expression of c-fos is increased by activation of intracellular signal transduction pathways. This demonstrates the complexity of the mechanisms that control gene expression, and the ability of separate pathways to converge and modulate each other. Moreover, these examples most likely represent only an elementary description of the many and varied forms of modulatory interactions that occur between different types of transcription factors.

Another class of proteins, the heat shock proteins, may also be involved in regulation of GR and MR (34). Several classes of heat shock protein are induced in response to cell injury or stress. One type, HSP90, appears to play a role in the control of GR and MR (Fig. 3). By binding to GR or MR, HSP90 helps to maintain the receptors in an inactive state. Upon binding of steroid, the receptor is released from HSP90 and becomes activated and able to regulate gene expression.

Regulation of Gene Expression by Nerve Growth Factors

Nerve growth factors and cytokines regulate gene expression by at least two pathways (15,18) (Fig. 3). The first pathway to be identified is indirect and involves activation of Ras and eventually activation of the MAP kinases (Figs. 1, 3). These kinases are then able to regulate transcriptional activity by phosphorylation of Fos, Jun, and other transcription factors. A second, more direct pathway involves tyrosine phosphorylation of cytoplasmic transcription factors that are then translocated to the nucleus (Fig. 3). These transcription factors appear to be phosphorylated not by the receptor protein tyrosine kinase but by a second, soluble protein tyrosine kinase that is activated by a receptor protein tyrosine kinase.

REGULATION OF INTRACELLULAR PATHWAYS AND GENE EXPRESSION BY CHRONIC STRESS

While stress has certain acute actions mediated by the activation of intracellular pathways discussed in the previous section, extended exposure to stressful stimuli results in additional molecular changes that could mediate the behavioral and functional consequences of chronic stress. Such changes due to chronic stress might be most relevant to the pathophysiology of stress-related affective illnesses such as PTSD and some forms of depression. Long-term changes may represent adaptive responses to the actions of acute stress or may represent new responses observed only after chronic exposure to stress. Highlighted in the next section are some actions of chronic stress on receptor-coupled intracellular signal transduction pathways and gene expression.

Molecular Actions of Chronic Stress in Locus Coeruleus

The norepinephrine neurotransmitter system has been implicated in the actions of stress on brain and in the etiology of stress-related illnesses such as PTSD and depression. These studies demonstrate alterations in levels of norepinephrine and its metabolites in response to acute and chronic stress, and in PTSD and depressed patients (35,36). This work has led to further examination of the molecular and cellular actions of stress on neurons of the locus coeruleus (LC), the major noradrenergic nucleus in brain. Acute stress increases the firing rate of LC neurons, in part via activation of CRF pathways, and this activation of LC neurons is maintained even after repeated stress exposure (36,37). This sustained activation of LC firing results in increased expression of tyrosine hydroxylase (TH), the rate-limiting enzyme for catecholamine synthesis, including increased levels of TH enzyme activity, mRNA, and protein immunoreactivity (36,38–40) (Fig. 4). The induction of TH in response to chronic stress is blocked by CRF antagonists, which also block the activation of LC neurons in response to stress, and repeated intracerebralventricular (ICV) administration of CRF increases levels of TH (41). Increased firing rate and TH expression appear to result from sustained activation of stimulatory inputs to LC neurons and increased demand for norepinephrine.

Recent studies have identified some of the intracellular sites and molecular mechanisms that mediate the actions of chronic stress on LC neurons. Vasoactive intestinal peptide (VIP) and presumably CRF increase the firing rate of LC neurons by activation of a Na^+-dependent inward current (42) (Fig. 4). The Na^+ current is activated by cAMP-dependent protein kinase, via phosphorylation of a Na^+ channel. CRF activation of the cAMP intracellular pathway probably accounts for the increased firing rate of LC neurons in response to stress. Chronic exposure to stress leads to increased levels of adenylyl cyclase and cAMP-dependent protein kinase in LC (36). Up-regulation of these components of the cAMP-pathway may contribute to increased firing of LC neurons, via cAMP activation of the Na^+-dependent current, in response to chronic stress.

In addition to regulation of firing rate, up-regulation of the cAMP second messenger path-

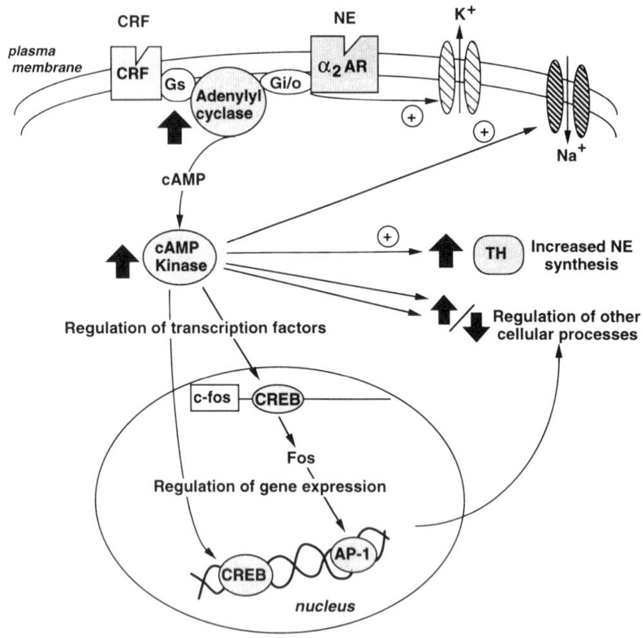

FIG. 4. Schematic illustration of the actions of stress on intracellular signal transduction and gene expression in locus coeruleus (LC). Acute stress increases the firing rate of LC neurons, in part by activation of corticotropin-releasing factor (CRF) receptors that stimulate the adenosine 3′,5′-cyclic monophosphate (cAMP) second messenger system. Stimulation of cAMP increases LC firing by activation of a Na^+-dependent inward current. This Na^+ current is activated by cAMP-dependent protein kinase, via phosphorylation of a Na^+ channel or pump. The cAMP pathway is inhibited by α_2-adrenergic receptors (α_2AR), as well as opiate receptors (*not shown*), resulting in inhibition of LC cell firing rate. α_2AR and opiate receptors inhibit cell firing by a second mechanism in LC: the receptors interact with Gi/o and increase the conductance of a K^+ channel.

Chronic stress results in sustained activation of LC neurons, and up-regulation of the cAMP system and expression of tyrosine hydroxylase (TH), the rate-limiting enzyme for norepinephrine (NE) synthesis. Thus, levels of adenylyl cyclase and cAMP-dependent protein kinase are increased by chronic stress. The up-regulated cAMP system could contribute to increased firing rate of LC neurons by continued activation of the Na^+-dependent current. Chronic stress also leads to elevated levels of TH enzyme activity and protein immunoreactivity in LC. In addition, levels of TH mRNA are increased, raising the possibility that TH gene expression is increased by chronic stress. Up-regulation of the cAMP system could mediate such an increase in TH gene expression. This could occur via the transcription factor CREB (cAMP response element binding protein), which is activated by cAMP-dependent protein kinase. In addition, the expression of c-fos is increased by stress, in part via the cAMP system. Induction of c-fos and other immediate early genes could also contribute to the regulation of TH during stress. It is notable that chronic antidepressant treatments result in actions opposite to those of stress: decreased firing rate of LC neurons and down-regulation of the cAMP system and expression of TH (*not shown*).

way by stress could also contribute to increased expression of TH in LC. This could occur via cAMP activation of CREB and/or induction of IEG transcription factors (Fig. 4). While the influence of stress on the phosphorylation and expression of CREB has not been examined, acute stress is reported to increase the expression of c-fos in LC neurons (26). Up-regulation of the cAMP pathway and activation of CREB and IEG transcription factors could also play a role in the expression of other regulatory proteins, such as adenylyl cyclase and cAMP-dependent protein kinase, influenced by chronic stress in LC.

Chronic stress would also be expected to influence LC neurons via elevated glucocorticoid levels and GR regulation of gene expression. In fact, expression of TH in adrenal cells is reported to be increased by glucocorticoids (43), and a similar mechanism might be in effect in LC. However, treatment with glucocorticoids does not influence the expression of TH in LC (36, 40), even though GRs are reported to be expressed in this brain region (44). This suggests that the expression of TH in response to glucocorticoids may be dependent on additional factors, the expression of which may vary between cell types.

It is notable that chronic administration of antidepressant drugs, which are used for the treatment of PTSD, results in effects opposite to those described for chronic stress in LC. Thus, chronic administration of most classes of antidepressant drugs decreases the expression of TH and cAMP-dependent protein kinase in LC (36). Moreover, chronic antidepressant pretreatment can block the induction of TH in response to chronic stress (45). Down-regulation of TH and the cAMP pathway could result from sustained inhibition of LC cell firing, via activation of α_2-adrenergic inhibitory autoreceptors, resulting in decreased demand for norepinephrine. Interestingly, antidepressants are also reported to increase levels of glucocorticoid receptors in LC, although the consequences of this effect have not been determined (44).

Molecular Actions of Chronic Stress in Dorsal Raphe

Serotonin is another monoamine neurotransmitter that has been implicated in some of the central nervous system actions of stress, as well as in PTSD and depression. The influence of stress on tryptophan hydroxylase, a key serotonin biosynthetic enzyme, has been examined in serotonergic cell bodies located in the midbrain dorsal raphe (46,47). These studies demonstrate a role for phosphorylation of tryptophan hydroxylase, as well as adaptations of total amount of enzyme protein, in the regulation of this enzyme by stress. Acute stress results in a transient elevation of tryptophan hydroxylase enzyme activity. This effect of acute stress on enzyme activity is reversible by pretreatment of midbrain extracts with alkaline phosphatase, which removes phosphate groups from tryptophan hydroxylase. These results indicate that acute stress increases enzyme activity by a phosphorylation-dependent mechanism. Chronic stress results in a more long-lasting increase in levels of tryptophan hydroxylase activity; this effect is insensitive to pretreatment with alkaline phosphatase, suggesting that levels of enzyme protein are increased by chronic stress.

While the precise mechanisms underlying these in vivo effects of stress on tryptophan hydroxylase are not known, in vitro studies indicate that Ca^{2+}/calmodulin-dependent protein kinase may be responsible for the acute stress-activation of enzyme activity (47). Stress is reported to increase the expression of c-fos in midbrain serotonin neurons, and may contribute to altered expression of tryptophan hydroxylase, as well as other regulatory proteins (26). Interestingly, the regulation of tryptophan hydroxylase in midbrain raphe neurons by both acute and chronic stress is dependent on the presence of an intact adrenal-glucocorticoid system (47). These findings suggest that glucocorticoid levels play a permissive role in the regulation of tryptophan hydroxylase by stress. Additional studies are required to further characterize the mechanisms underlying the regulation of tryptophan hydroxylase in response to stress, and to identify other molecular actions of stress in serotonin neurons.

Molecular Actions of Chronic Stress in Ventral Tegmentum

The dopamine neurotransmitter system also plays a role in mediating the actions of stress in brain (22). Moreover, the mesolimbic dopamine system has also been implicated in the etiology and treatment of affective disorders (48). Acute stress is reported to increase the activity of ventral tegmental neurons, where the cell bodies for the mesolimbic dopamine system are found. This has been demonstrated by elevated levels of dopamine and dopamine metabolites in prefrontal cortex and nucleus accumbens, terminal field regions for ventral tegmental neurons. While there have been no direct measures of tyrosine hydroxylase, stress is reported to increase in vivo tyrosine hydroxylation in ventral tegmentum (22), suggesting that levels of tyrosine hydroxylase enzyme activity are increased. Such an effect probably results from a phosphorylation-dependent increase of tyrosine hydroxylase activity, since there are no reports that stress, either acute or chronic, influences the expression of tyrosine hydroxylase mRNA or immunoreactivity. Acute stress is also reported to increase the expression of c-fos in a subpopulation of ventral tegmental dopaminergic neurons (22). Induction of c-fos in ventral tegmentum further demonstrates that stress activates these neurons, but the functional consequences of c-fos expression are not known.

While the actions of stress on mesolimbic dopamine intracellular signal transduction pathways are not known, studies of psychotropic drugs may provide some insights. Treatment with psychostimulants and antidepressants, as well as stress, are reported to enhance or "sensitize" locomotor responses to dopamine agonists. These actions of psychostimulants and antidepressants are associated with increased expression of the cAMP system in the nucleus accumbens (48,49). It is possible that stress is associated with similar alterations of the cAMP system in this brain region. The mesolimbic dopamine system is involved in drug reward and craving and is referred to as a "pleasure center" of the brain. Because of the anhedonia associated with depression and stress-related disorders, it is interesting to speculate on whether the mesolimbic dopamine system is involved in the pathophysiology of depression and therapeutic actions of antidepressant treatments.

Molecular Actions of Chronic Stress on Cerebral Cortex

Stress influences several neurotransmitter and neuropeptide systems in the cerebral cortex, but one of the best characterized is the norepinephrine-stimulated cAMP second messenger system. One action of chronic stress is to decrease the ability of norepinephrine to stimulate cAMP production in cerebral cortex (50). The mechanism by which stress influences norepinephrine-stimulated cAMP production involves α-, but not β-adrenergic receptor function. The cAMP response to norepinephrine is mediated by both α-adrenergic receptors, which have little stimulatory effect alone but enhance the cAMP response to β-adrenergic receptors, which directly activate adenylyl cyclase by coupling with Gs (11,51). It is the α-adrenergic receptor augmentation of cAMP that is influenced by chronic stress (50). This effect is mimicked by chronic administration of glucocorticoids or adrenocorticotrophin hormone (ACTH), while adrenalectomy has the opposite effect (51–53). These results indicate that down-regulation of norepinephrine-stimulated cAMP formation in response to stress is mediated by elevated glucocorticoid levels. Adrenalectomy may also influence receptor-stimulated cAMP formation in a region- and receptor specific manner (54). In addition to these actions of chronic stress and glucocorticoid treatments, more long-term administration of ACTH or the potent glucocorticoid analogue, dexamethasone, increases receptor-stimulated cAMP production in brain (55).

The mechanisms by which stress and glucocorticoid treatments regulate norepinephrine-stimulated cAMP formation are unclear. Chronic antidepressant treatments also decrease norepinephrine-stimulated cAMP production, but this effect can be explained largely by antidepressant-induced down-regulation of β-adrenergic receptor sites (52). In contrast, stress does not decrease levels of α- or β-adrenergic receptor sites that mediate the regulation of cAMP. This suggests that some postreceptor component of the system is modified by stress. One possibility is that stress or glucocorticoid treatments influence the expression of G protein subtypes that mediate receptor-regulation of cAMP. Glucocorticoid treatment has been reported to increase levels of Gsα, but decrease levels of Giα, mRNA, and protein in cerebral cortex, and similar effects have been observed in cultured cells (56). Such regulation of Gsα and Giα could explain some of the actions of glucocorticoids just discussed on receptor-regulation of cAMP responses.

It is also possible that regulation of the cAMP system by stress and glucocorticoids involves sites beyond the level of receptors and G proteins. McEwen and colleagues have reported that glucocorticoid treatment decreases Ca^{2+}/calmodulin-stimulated adenylyl cyclase in cerebral cortex (57). This effect could be involved in down-regulation of α-adrenergic receptor augmentation of cAMP responses, which is dependent on Ca^{2+} and therefore could involve Ca^{2+}/calmodulin (11). Another possibility is that stress and glucocorticoids regulate the expression of adenylyl cyclase itself; there are several subtypes of the enzyme that could be differentially regulated by stress or glucocorticoid treatments. In addition, glucocorticoids could influence the expression of other proteins that are capable of modulating the cAMP response. Glucocorticoid treatment has been reported to regulate endogenous ADP-ribosylation and expression of ADP-ribosylation factor which appear to play a role in endogenous and toxin catalyzed ADP-ribosylation (8,58). In addition, stress and glucocorticoids may influence the function or expression of other effectors such as phospholipase A_2, which could indirectly influence the cAMP system (11,53). These are just a few of the many potential actions of chronic stress and glucocorticoids on the expression of neural proteins in cerebral cortex.

The consequences of stress regulation of the cAMP system on the function of cerebral cortical neurons are not well known. Acute stress increases the expression of c-fos and other IEGs in frontal cortex, primarily via activation of α- and β-adrenergic receptors (19,21). Chronic stress may result in desensitization of the c-fos response to the same stress in some brain regions, and this desensitization may be specific to the stress stimulus presented during chronic treatment (22). Chronic stress may also result in expression of novel c-fos related antigens (FRAs), which are expressed only after chronic exposure to the same stimulus (see 59). Such alterations of the cAMP system and IEG transcription factors may result in altered phosphorylation and expression of other regulatory proteins in response to chronic stress. Chronic stress does result in a small decrease in levels of cAMP-dependent protein kinase in frontal cortex (36). Moreover, the potential interaction between IEG transcription factors (i.e., c-fos and c-jun) and glucocorticoid receptors provides a mechanism for additional points of regulation of gene expression.

Molecular Actions of Chronic Stress on Hypothalamic-Pituitary-Adrenal Axis

The most notable response to stress is activation of the hypothalamic-pituitary-adrenal (HPA) axis, which results in elevated levels of adrenal glucocorticoids. Feedback inhibition of the HPA axis involves actions of glucocorticoids on the hippocampus, as well as the hypothalamus and pituitary (see Chapter 2 and 60,61). In the hippocampus, glucocorticoids regulate a pathway that results in inhibition of hypothalamic neurons that synthesize and secrete CRF. Long-term exposure to glucocorticoids results in down-regulation of type I and II receptor mRNA and ligand binding in hippocampus, which presumably represents a compensatory down-regulation in response to receptor activation (62). Sustained elevation of glucocorticoids results in neurotoxic effects and cell death, which could contribute to loss of negative feedback control of the HPA axis (see Chapter 10 for further discussion).

Stress is also reported to influence the expression of neurotrophic factors in hippocampus (63). Acute stress increases the expression of neurotrophic factor-3 (NT-3), but decreases levels of brain-derived neurotrophic factor (BDNF) in hippocampus. Expression of neurotrophins is also reported to be regulated by acute and chronic administration of antidepressants in cerebral cortex (64). While the actions of neurotrophic factors were originally thought to be limited to neuronal development and injury, the induction of these factors by stress, neurotransmitters, and psychotropic drugs suggests that neurotrophic factors are involved in many aspects of neuronal function. The precise role of neurotrophins in response to stress remains to be determined, but they are conceivably involved in the induction of axonal sprouting in response to stress (65).

The actions of stress and glucocorticoids on hypothalamic neuropeptides have been extensively studied. Acute stress stimulates the synthesis and release of CRF and arginine vasopressin (AVP) from parvicellular neurons of the hypothalamic paraventricular nucleus (66,67). In addition, acute stress increases the expression of the immediate early gene c-fos in hypothalamus (20,24–26). However, there is no direct evidence for a causative relationship between the expression of c-fos and these neuropeptides. Moreover, the intracellular pathways that lead to elevation of c-fos and CRF have not been identified. Both MR and GR are found in the hypothalamus, and stress or glucocorticoids may also regulate the expression of CRF via activation of these receptors.

Expression of CRF mRNA in hypothalmic neurons remains elevated during chronic stress, even though glucocorticoid levels are elevated. This indicates that the mechanisms that stimulate the expression and release of CRF override the inhibitory actions of glucocorticoids on the HPA axis. One possibility is that stress activation of CRF expression involves induction of c-fos and other IEGs. As just discussed, Fos protein is capable of interacting with GR and could thereby influence its transcriptional activity.

In addition to hypothalamus, CRF is expressed in several areas throughout the brain and may play a role in coordinating central responses to stress (66,67). Stress is reported to increase CRF expression in amygdala and Barringtons nucleus and to decrease expression of CRF in olfactory bulb. The neurotransmitter systems and intracellular pathways involved in the regulation of CRF expression in these brain regions remain to be determined.

Regulation of CRF mRNA suggests that stress alters the gene transcription rate for this neuropeptide. However, studies of gene transcription rate in vivo are extremely difficult, and stress regulation of CRF gene transcription has not been demonstrated. Therefore, it is possible that alterations in levels of CRF mRNA, as well as regulation of mRNA levels for other neuropeptides and regulatory proteins, result from regulation of mRNA stability. Regulation of mRNA stability and mRNA processing represent additional steps that could be potentially regulated by stress.

SUMMARY

Stress influences many different neurotransmitter and neuropeptide systems that influence neuronal function largely via regulation of intracellular signal transduction pathways. In addition, stress also exerts actions on neuronal function via activation of the HPA axis and elevated levels of adrenal glucocorticoids. Regulation of intracellular signal transduction pathways and glucocorticoid receptors may also result in more long-term effects of stress via regulation of gene expression. In addition, second messenger-regulated gene transcription factors, such as c-fos, and glucocorticoid receptors may interact with each other, providing an additional level for regulation of gene expression.

Many of the acute and chronic actions of stress most likely evolved as "positive" adaptive responses that should increase the survival and behavioral performance of the animal in response to subsequent environmental stressors. In this regard, it is notable that many of the acute actions of stress are similar to those of antidepressants, which are used for the treatment of stress-related disorders such as PTSD and depression. Acute stress and antidepressant treatments increase synaptic levels of monoamines that regulate intracellular signal transduction pathways and increase the expression of IEG transcription factors.

Other actions of chronic stress could represent "negative" adaptations that have adverse side effects on neuronal function and behavior. These adverse effects of stress could result from sustained activation of certain neurotransmitter and intracellular pathways, or could result from the chronic exposure to hormones such as glucocorticoids or other bioactive substances such as cytokines released by stress. These negative adaptations could contribute to the underlying etiology of psychiatric disorders such as PTSD and depression. In addition, psychiatric disorders may result from molecular-genetic alterations that interfere with the formation of positive adaptations. For example, many individuals have significant stressful experiences, but such experiences only result in PTSD or depression in a small percentage of the population. Thus, certain individuals may have a genetic predisposition, which results in stress-associated psychiatric abnormalities. Further characterization of the molecular actions of acute and chronic stress will ultimately contribute to the identification of the etiological determinants of PTSD and depression.

REFERENCES

1. Simon MI, Strathman MP, Gautam N. Diversity of G proteins in signal transduction. *Science* 1991;252: 802–808.
2. Nestler EJ, Duman RS. G-proteins and cyclic nucleotides in the nervous system. In: Siegel G, Agranoff B, Albers RW, Molinoff P, eds. *Basic neurochemistry*. 5th Ed. New York: Raven Press, 1994;427–448.
3. Taussig R, Iniguez-Lluhi JA, Gilman AG. Inhibition of adenylyl cyclase by Gia. *Science* 1993;261:218–221.
4. Duman RS, Nestler EJ. Signal transduction pathways for catecholamine receptors. In: Bloom FE, Kupfer DJ, eds. *Psychopharmacology, 4th Generation of Progress,* 1944.
5. Nicoll RA. The coupling of neurotransmitter receptors to ion channels in the brain. *Science* 1988;241:545–551.
6. Berridge MJ. Inositol triphosphate and calcium signaling. *Nature* 1993;361:315–325.
7. Yuen PST, Garbers DL. Guanylyl cyclase-linked receptors. *Annu Rev Neurosci* 1992;15:193–225.
8. Duman RS, Terwilliger RZ, Nestler EJ. Endogenous ADP-ribosylation in brain: initial characterization of substrate proteins. *J Neurochem* 1991;57:2124–2132.
9. Piomelli D, Greengard P. Lipoxygenase metabolites of arachidonic acid in neuronal transmembrane signalling. *TiPS* 1990;11:367–373.
10. Peitsch MC, Borner C, Tschopp J. Sequence similarity of phospholipase A2 activating protein and the G protein b-subunits: a new concept of effector protein activation in signal transduction. *TiBS* 1993;18:292–293.
11. Duman RS, Karbon EW, Harrington C, Enna SJ. An examination of the involvement of phospholipases A_2 and C in the a-adrenergic and g-aminobutyric acid receptor modulation of cyclic AMP accumulation in rat brain slices. *J Neurochem* 1986;47:800–810.
12. Blackwell GJ, Carnuccio R, Di Rosa M, Flower RJ, Parente L, Persioco P. Macrocortin: A polypeptide causing the antiphospholipase effect of glucocorticoids. *Nature* 1980;287:147–149.
13. Nestler EJ, Greengard P. *Protein phosphorylation in the nervous system.* New York: J. Wiley and Sons, 1984.
14. Hunter T, Sefton BM, eds. Protein phosphorylation. Parts A and B. *Methods in Enzymology*. Vols. 200 and 201. New York: Academic Press, 1991.
15. Blenis J. Signal transduction via the MAP kinases: proceed at your own RSK. *Proc Natl Acad Sci USA* 1993;90:5889–5892.
16. Davis RJ. The mitogen-activated protein kinase signal transduction pathway. *J Biol Chem* 1993;268: 14553–14556.
17. Korsching S. The neurotrophic factor concept: a reexamination. *J Neurosci* 1993;13:2739–2748.
18. Montminy M. Trying on a new pair of SH2s. *Science* 1993;261:1694–1695.
19. Gubits RM, Smith TM, Fairhurst JL, Yu H. Adrenergic receptors mediate changes in c-fos mRNA levels in brain. *Mol Brain Res* 1989;6:39–45.
20. Ceccatelli S, Villar MJ, Goldstein M, Hökfelt T. Expression of c-fos immunoreactivity in transmitter-characterized neurons after stress. *Proc Natl Acad Sci USA* 1989;86:9569–9573.
21. Bing G, Filer D, Miller JC, Stone EA. Noradrenergic activation of immediate early genes in rat cerebral cortex. *Mol Brain Res* 1991;11:43–46.
22. Campeau S, Hayward MD, Hope BT, Rosen JB, Nestler EJ, Davis M. Induction of the c-fos proto-oncogene in rat amygdala during acute and chronic stress. *Brain Res* 1991;565:349–352.
23. Deutch AY, Lee MC, Gillham MH, Cameron DA, Goldstein M, Iadarola MJ. Stress selectively increases fos protein in dopamine neurons innervating the prefrontal cortex. *Cerebral Cortex* 1991;1:273–292.
24. Sharp FR, Sagar SM, Hicks K, Lowenstein D, Hisanaga K. c-fos mRNA, Fos, and Fos-related antigen induction by hypertonic saline and stress. *J Neurosci* 1991;11: 2321–2331.
25. Duncan GE, Johnson KB, Breese GR. Topographic patterns of brain activity in response to swim stress: assessment by 2-deoxyglucose uptake and expression of fos-like immunoreactivity. *J Neurosci* 1993;13:3932–3934.
26. Senba E, Matsunaga K, Tohyama M, Noguchi K. Stress-induced c-fos expression in the rat brain: activation mechanism of sympathetic pathway. *Brain Res Bull* 1993;31:329–344.
27. Sheng M, Greenberg ME. The regulation and function of c-fos and other immediate early genes in the nervous system. *Neuron* 1990;4:477–485.
28. Morgan JI, Curran T. Stimulus-transcription coupling in the nervous system: involvement of the inducible proto-oncogenes fos and jun. *Annu Rev Neurosci* 1991;14:421–451.
29. Armstrong RC, Montminy MR. Transsynaptic control of gene expression. *Annu Rev Neurosci* 1993;16:17–29.
30. Reul JHM, de Kloet ER. Two receptor systems for corticosterone receptors in rat brain: microdistribution and differential occupation. *Endocrin* 1985;117:2505–2511.
31. Pearce D, Yamamoto KR. Mineralocorticoid and glucocorticoid receptor activities distinguished by nonreceptor factors at a composite response element. *Science* 1993;259:1161–1165.
32. Yang-Yen HF, Chambard JC, Sun YL, Smeal T, Schmidt TJ, Drouin J, Karin M. Transcriptional interference between c-jun and the glucocorticoid receptor: mutual inhibition of DNA binding due to direct protein-protein interaction. *Cell* 1990;62:1205–1215.
33. Schule R, Rangarajan P, Kliewer S, Ransone LJ, Bolado

J, Yang N, Verma IM, Evans RM. Functional antagonism between oncoprotein c-jun and the glucocorticoid receptor. *Cell* 1990;62:1217–1226.

34. Welch WJ. How cells respond to stress. *Scientific American* 1993;56–64.
35. Gold PW, Goodwin FK, Chrousos GP. Clinical and biochemical manifestations of depression: relation to the neurobiology of stress. *N Engl J Med* 1988;319: 348–353.
36. Melia KR, Rasmussen K, Terwilliger RZ, Haycock JW, Nestler EJ, Duman RS. Coordinate regulation of the cyclic AMP system with firing rate and expression of tyrosine hydroxylase in the rat locus coeruleus: effects of chronic stress and drug treatments. *J Neurochem* 1992;58:494–502.
37. Valentino RJ, Wehby RG. Corticotropin-releasing factor: evidence for a neurotransmitter role in the locus coeruleus during hemodynamic stress. *Neuroendocrin* 1988;48:674–677.
38. Zigmond RE, Schon F, Iversen LL. Increased tyrosine hydroxylase activity in the locus coeruleus of rat brain stem after reserpine treatment and cold stress. *Brain Res* 1974;70:547–555.
39. Richard F, Faucon-Biguet N, Labatut R, Rollet D, Mallet J, Buda M. Modulation of tyrosine hydroxylase gene expression in rat brain and adrenals by exposure to cold. *J Neurosci Res* 1988;20:32–37.
40. Mamalaki E, Kvetnansky R, Brady LS, Gold PW, Herkenham M. Repeated immobilization stress alters tyrosine hydroxylase, corticotropin-releasing hormone and corticosteroid receptor messenger ribonucleic acid levels in rat brain. *J Neuroendocrin* 1993;4(6): 689–699.
41. Melia KR, Duman RS. Involvement of corticotropin-releasing factor in chronic stress regulation of the brain noradrenergic system. *Proc Natl Acad Sci USA* 1991; 88:8382–8386.
42. Alreja M, Aghajanian GK. Pacemaker activity of locus coeruleus neurons: whole-cell recordings in brain slices show dependence on cAMP and protein kinase A. *Brain Res* 1991;556:339–343.
43. Fossom LH, Sterling CR, Tank AW. Regulation of tyrosine hydroxylase gene transcription rate and tyrosine hydroxylase mRNA stability by cyclic AMP and glucocorticoid. *Mol Pharm* 1992;42:898–908.
44. Kitayama I, Janson AM, Cintra A, Fuxe K, Agnati LF, Ogren S-O, Harfstrand A, Eneroth P, Gustafsson J-A. Effects of chronic imipramine treatment on glucocorticoid receptor immunoreactivity in various regions of the rat brain. *J Neural Trans* 1988;73:191–203.
45. Melia KR, Nestler EJ, Duman RS. Chronic imipramine treatment normalizes levels of tyrosine hydroxylase in the locus coeruleus of chronically stressed rats. *Psychopharmacology* 1992;108:23–26.
46. Boadle-Biber MC, Corley KC, Graves L, Phan TH, Rosecrans J. Increase in the activity of tryptophan hydroxylase from cortex and midbrain of male Fischer 344 rats in response to acute or repeated sound stress. *Brain Res* 1989;482:306–316.
47. Singh VB, Corley KC, Phan TH, Boadle-Biber MC. Increases in the activity of tryptophan hydroxylase from rat cortex and midbrain in response to acute or repeated sound stress are blocked by adrenalectomy and restored by dexamethasone treatment. *Brain Res* 1990;516: 66–76.
48. Rosin DL, Melia K, Knorr AM, Nestler EJ, Roth RH, Duman RS. Chronic imipramine administration alters the activity and phosphorylation state of tyrosine hydroxylase in dopaminergic regions of rat brain. *Neuropsychopharmacology* 1995;12:113–121.
49. Nestler EJ. Molecular mechanisms of drug addiction. *J Neurosci* 1992;12:2439–2450.
50. Stone EA, Platt JE, Herrera AS, Kirk KL. Effect of repeated restraint stress, desmethylimipramine or adrenocorticotropin on the alpha and beta adrenergic components of the cyclic AMP response to norepinephrine in rat brain slices. *J Pharmacol Exp Ther* 1986;237: 702–707.
51. Duman RS, Enna SJ. Modulation of receptor-mediated cyclic AMP production in brain. *Neuropharm* 1987; 26:981–986.
52. Mobley PL, Sulser F. Adrenal corticoids regulate the sensitivity of noradrenaline receptor coupled adenylate cyclase in brain. *Nature* 1980;286:608–609.
53. Duman RS, Strada SJ, and Enna SJ. Effect of imipramine and adrenocorticotropin administration on the rat brain norepinephrine-coupled cyclic nucleotide generating system: alterations in alpha and beta adrenergic components. *J Pharm Exp Ther* 1985;234:409–414.
54. Harrelson AL, McEwen BS. Hypophysectomy increases vasoactive intestinal peptide-stimulated cyclic AMP generation in the hippocampus of the rat. *J Neurosci* 1987;7:2807–2810.
55. Duman RS, Strada SJ, Enna SJ. Glucocorticoid administration increases receptor-mediated and forskolin-stimulated cyclic AMP accumulation in rat brain cerebral cortical slices. *Brain Res* 1989;477:166–171.
56. Saito N, Guitart X, Hayward M, Tallman JF, Duman RS, Nestler EJ. Corticosterone differentially regulates the expression of Gsa and Gia messenger RNA and protein in rat cerebral cortex. *Proc Natl Acad Sci USA* 1989;86:3906–3910.
57. Gannon MN, McEwen BS. Calmodulin involvement in stress- and corticosterone-induced down-regulation of cyclic AMP-generating systems in brain. *J Neurochem* 1990;55:276–284.
58. Duman RS, Winston SM, Clark JA, Nestler EJ. Corticosterone regulates the expression of ADP-ribosylation factor messenger RNA and protein in rat cerebral cortex. *J Neurochem* 1990;55:1813–1816.
59. Hope BT, Kelz M, Duman RS, Nestler EJ. Chronic administration of ECS produces a long-lasting AP-1 complex in rat cerebral cortex of altered composition and characteristics. *J Neurosci* 1994;14:4318–4328.
60. Sapolsky RM, Krey LC, McEwen BS. Glucocorticoid-sensitive hippocampal neurons are involved in terminating the adrenocortical stress response. *Proc Natl Acad Sci USA* 1984;81:6174–6177.
61. Herman JP, Schafer MKH, Young EA, Thompson R, Douglass J, Akil H, Watson SJ. Evidence for hippocampal regulation of neuroendocrine neurons of the hypothalamo-pituitary-adrenocortical axis. *J Neurosci* 1989;9: 3072–3082.
62. Sapolsky RM, Krey LC, McEwen BS. Stress down-regulates corticosterone receptors in a site-specific manner in the brain. *Endocrin* 1984b;114:287–292.
63. Smith MA, Makino S, Kvetnansky R, Post RM. Stress alters brain derived neurotrophic factor and neurotrophin-3 mRNA levels in the hippocampus. *Soc Neurosci Abst* 1993;19:866.

64. Morinobu S, Duman RS. Induction of BDNF mRNA by electroconvulsive seizure (ECS) and antidepressants in rat frontal cortex. *Soc Neurosci Abst* 1993;19:499.
65. Sakaguchi T, Nakamura S. Duration-dependent effects of repeated restraint stress on cortical projections of locus coeruleus neurons. *Neurosci Lett* 1990;118: 193–196.
66. Chappell PB, Smith MA, Kilts CD, Bissette G, Ritchie J, Anderson C, Nemeroff CB. Alterations in corticotropin-releasing factor-like immunoreactivity in discrete rat brain regions after acute and chronic stress. *J Neurosci* 1986;6:2908–2914.
67. Imaki T, Nahan JL, Rivier C, Sawchenko PE, Vale W. Differential regulation of corticotropin-releasing factor mRNA in rat brain regions by glucocorticoids and stress. *J Neurosci* 1991;11:585–599.

Neurobiological and Clinical Consequences of Stress: From Normal Adaptation to PTSD, edited by M. J. Friedman, D. S. Charney, and A. Y. Deutch, Lippincott-Raven Publishers, Philadelphia © 1995.

3

Neurochemical Studies of Central Noradrenergic Responses to Acute and Chronic Stress

Implications for Normal and Abnormal Behavior[1]

Michael J. Zigmond, Janet M. Finlay, and Alan F. Sved

Department of Neuroscience, University of Pittsburgh, Pittsburgh, Pennsylvania 15260

It was the work of Walter Cannon that provided the first unambiguous link between stress and norepinephrine (NE). Cannon noted that extensive removal of the sympathetic ganglia from cats produced little or no physiological change under basal laboratory conditions, but severely impaired the normal physiological response to an acute homeostatic challenge. He also observed that struggling cats showed an increase in heart rate even in the absence of a direct sympathetic innervation to that structure. From such findings Cannon concluded that the nerves and the adrenal gland of the sympathetic nervous system are involved in an animal's response to "fight or flight" situations (1).

When NE was first identified in the central nervous system it was assumed that the catecholamine was contained exclusively within the sympathetic innervation of that region. Yet in a seminal paper, Marthe Vogt (2) showed that the central distribution of NE was heterogenous, being particularly high in the brain stem and hypothalamus, and did not disappear after sympathectomy. She also observed that pharmacological treatments provoking adrenal medullary secretion caused the depletion of NE in the brain, as well. This led Vogt to suggest that the capacity of brain to replace NE through synthesis was less efficient than was the case in the periphery, where it was difficult to decrease tissue NE content despite prolonged stimulation. Finally, she noted that in some cases, such as following the systemic administration of apomorphine and morphine, there was a correlation between the extent of the behavioral excitation produced by a drug and the loss of central NE. Based on these observations, Vogt postulated that NE served either as a neurotransmitter, as had already been suggested for the peripheral postganglionic neurons, or as a modulator of other neurotransmitters. In either case, she suggested, central NE might play a role in the activities associated with the peripheral sympathetic nervous system.

Many of the studies that were performed in the years immediately following the work of Vogt were interpreted as indicating that NE or another of the biogenic amines had specific roles in regulating a physiological or behavioral function. However, evidence for the conceptualization first enunciated by Vogt continued to mount, leading numerous authors to describe a

[1]Dedicated to Marthe Vogt on her 90th birthday.

central noradrenergic system whose relation to behavior bore a strong resemblance to the relation between that of the peripheral noradrenergic system and physiology (3–7). This is our view, as well, and will be the focus of our review. We will begin by summarizing the basic characteristics of noradrenergic neurons in brain. Then we will discuss their response to acute and chronic stress. Finally, we will comment on the functional implications of these findings.

CHARACTERISTICS OF LOCUS COERULEUS NORADRENERGIC NEURONS

Several features of noradrenergic neurons are consistent with their putative role as elements within a diffuse, modulatory system participating in the response to stress. First, there are the *anatomical characteristics* of these neurons. The neurons are present within several cell groups in areas classically considered part of the reticular formation (8). Chief among these are the cells of the locus coeruleus (LC), located within the brain stem along the ventrolateral edge of the fourth ventricle. The terminal fields of LC neurons are widespread, showing extensive axonal collateralization and multiple synaptic varicosities en passage (9–11). These noradrenergic terminals engage in conventional synaptic transmission (12–14). However, like some other aminergic neurons, these terminals show a relative paucity of surface area devoted to synaptic contacts (15,16), suggesting that they may exert a portion of their influences through nonfunctional means as well. Thus, central noradrenergic neurons, like their peripheral counterpart, have the basic structure required of a system whose task is to influence simultaneously the activity of diverse regions of the CNS.

The *electrophysiological characteristics* of these neurons provide a second body of evidence in support of a diffuse, modulatory function. Noradrenergic axons typically have an extremely slow conduction velocity (0.4–0.6 m/second) (17), a characteristic that is better suited to a neuromodulatory role than to a rapid, point-to-point transfer of information. Under basal conditions, noradrenergic neurons in the LC exhibit a slow, steady, spontaneous firing rate (1–5 hz) and a relatively stable rate of transmitter release (18–20). In addition, synchronous firing between cells is frequently observed (21), suggesting that these neurons may function as an ensemble that enhances their ability to simultaneously influence diverse areas of the CNS. Moreover, when environmentally-induced fluctuations in LC firing are examined, the evidence suggests that a given cell can receive input from several different sensory systems (18,22). Finally, NE appears to modify the activity of target neurons not by a rapid, direct alteration in their ionic permeability, but instead by a more gradual influence at the level of second messenger systems such as cyclic adenosine monophosphate (AMP), often increasing the signal-to-noise ratio at such targets (23–25).

A third set of observations consistent with a modulatory function for central NE is that suggesting that noradrenergic neurons, like other aminergic systems, display a variety of *homeostatic regulatory properties* (26,27). For example, there are numerous feedback loops that appear designed to maintain a relatively constant level of communication between noradrenergic neurons and their targets. Terminal NE autoreceptors can inhibit the release of NE, which should serve to keep extracellular NE within narrow limits, and these receptors may be supplemented by additional local feedback loops that involve other neuronal terminals (28). Additionally, NE released from collateral branches and dendrites of LC neurons may act on the LC to regulate neuronal firing rates (11,29,30). These homeostatic control mechanisms have the capacity to provide marked regulation of NE release. They are supplemented by two other levels of regulation. One of these controls the rate at which NE is synthesized (see information that follows). The second regulatory process determines the impact that NE has on its targets. Thus, for example, the postsynaptic action of NE is significantly enhanced after a prolonged reduction of NE availability, as might occur following damage to the LC projection (31,32).

In summary, the characteristics of the LC system should allow a relatively small, localized number of neurons to simultaneously influence neuronal transmission in many parts of the CNS. Moreover, whereas this influence can be expected to change from time to time—for such is the nature of a communication system—the system seems to be designed for constancy.

ACUTE STRESS AND NORADRENERGIC NEURONS OF THE LC

With the advent of techniques for measuring extracellular NE levels such as in vivo microdialysis and voltammetry, many laboratories have now clearly demonstrated that acute stress increases extracellular NE in brain (33–38) (Fig. 1). This finding was predicted on the basis of results from numerous earlier reports in which NE turnover within noradrenergic neurons of the LC was found to increase in response to a wide variety of stressors and using a number of different biochemical measures. For example, earlier studies indicated that stressors such as foot shock, cold exposure, and prolonged exercise decreased the concentration of NE in brain tissue (39–42) and increased the concentration of its metabolite, methoxyhydroxyphenylglycol sulfate (MHPG-SO_4) (43–46). Stress-induced increases in NE synthesis also have been observed (43,45,47,48). Such findings suggest that stress increases NE release in the CNS as well as in the periphery (49,50).

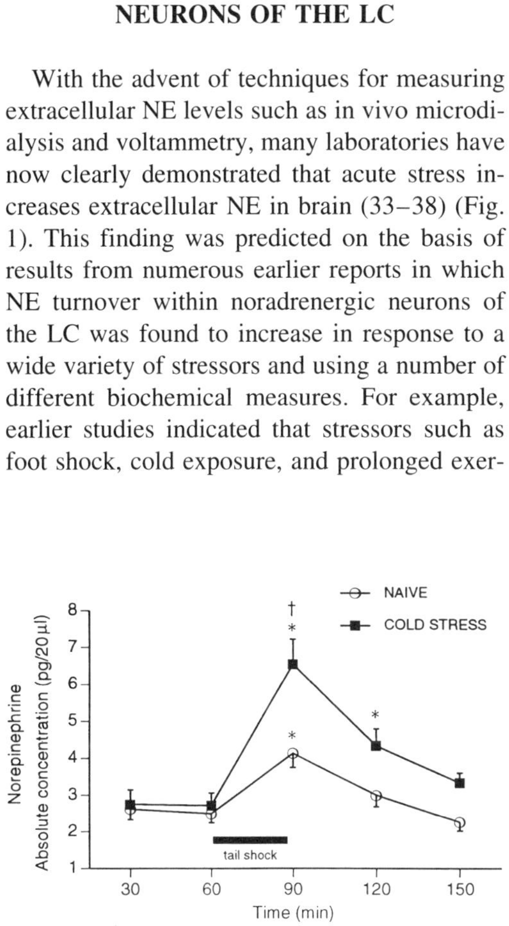

FIG. 1. Extracellular norepinephrine (NE) within the medial prefrontal cortex of rats exposed to a novel stressor. Naive rats and rats previously exposed to chronic cold (17–28 days @ 5°C) were subjected to 30 minutes of intermittent tail shock and extracellular NE was measured by microdialysis. Data are presented as pg/20 μl (mean±S.E.M.; n=10/group). Pairwise comparisons were performed using Tukey's HSD.* Significantly different from respective within-group baseline at 60 minutes ($p<0.05$). †Significantly different from naive control. Figure adapted from Gresch et al.[73]

Initial studies of the effects of stress on NE turnover did not specifically address whether these changes occurred within the neurons of the LC system as opposed to other NE projection systems. However, subsequent studies demonstrated that stress increased MHPG-SO_4 in brain regions receiving NE input derived largely from the LC as well as within the LC itself (51,52). Such observations have been confirmed with in vivo microdialysis that has demonstrated stress-induced increases in the extracellular levels of NE in both hippocampus and frontal cortex, two different terminals fields of LC neurons (33,53).

These neurochemical studies are consistent with the hypothesis that acute stress increases the synthesis, release, and catabolism of NE in neurons projecting from the LC and that these noradrenergic neurons may respond in a qualitatively homogeneous manner. Although these conclusions are limited by the poor spatial resolution of such methods, electrophysiological recordings in conscious animals support the same conclusion. Those studies have demonstrated that noradrenergic neurons of the LC increase their firing rate in response to all stimuli that elicit attention, with stressful or noxious stimuli eliciting larger and more sustained increases in firing than more mild stimuli (7,11,54). As anticipated by Vogt, these changes also correlated with sympathetic indices of arousal (7). Furthermore, based on these electrophysiological studies, it appears that all of the noradrenergic neurons of the LC respond to a variety of stressful stimuli, despite heterogeneity of these neurons with regard to their projections (55) and neuropeptide content (56–58). This conclusion is further reinforced by recent studies showing that virtually all LC neurons express the immediate-early gene c-fos in response to foot shock stress (59). Although the functional significance of the induction of c-fos in these cells is currently unclear, expression of c-fos in neurons is typically coupled with neuronal activation (60) and therefore expression of c-fos in LC neurons may serve as a useful marker for stress-evoked action.

The ability of stress to elicit activation of all cells within LC is reminiscent of the widespread effects of stress on noradrenergic neurons of the sympathetic nervous system. However, also by analogy to the sympathetic nervous system, it might be expected that more subtle stimuli would differentially activate specific populations of NE neurons and this is likely to be the case. Indeed, evidence is accumulating to suggest that noradrenergic neurons of the LC may exhibit a degree of heterogeneity (61). For example, in contrast to foot shock stress, which evokes the expression of c-fos in all LC neurons, a conditioned stressor has been shown to evoke the expression of c-fos in only a subset of LC neurons (59). Thus, the notion that acute stress induces a global activation of noradrenergic neurons in the LC may be further refined to reflect subtle heterogeneities within the cell group. Heterogeneity in the impact of LC neurons also could result from regional differences in the input to noradrenergic terminal fields, as has been suggested for dopaminergic systems (62,63). Consistent with the hypothesis of significant functional heterogeneity within the LC system is evidence of distinct behavioral syndromes following selective lesioning of different noradrenergic projections (see next section) (64).

CHRONIC STRESS AND NORADRENERGIC NEURONS OF THE LC

Whereas a considerable amount is known about the response of noradrenergic neurons of the LC to acute stress, less attention has been paid to the response of these neurons to chronic stress. Based on the available information, it is clear that the response to a persistent or repetitive stressor often is quite different from that seen during acute exposure. Since organisms are most often confronted not with a single isolated exposure to a stressor but rather with multiple or prolonged exposure to stressful events, understanding the impact of such events on the function of central noradrenergic neurons is critical in attempts to address the biological substrate of stress-induced psychiatric disorders. In this same regard, it is important to consider the impact of chronic stress not only on basal functioning of noradrenergic neurons, but also on the activity of the LC system when it is subsequently challenged by presentation of the same or a different stressor. In addition, it must be determined whether the specific alterations in the function of central noradrenergic neurons induced by chronic stress are adaptive, or are maladaptive and therefore lead to stress-induced pathologies (50,65).

Impact of Chronic Stress on Neurochemical Activity Under Basal Conditions and in Response to a Stressor

As noted in the previous section, acute stress can produce a reduction in tissue NE content when it is sufficiently severe. In contrast, repeated or protracted exposure to a stressor does not deplete tissue stores, and in many instances actually leads to increased tissue NE levels under basal conditions (42,66,67). For example, 1 hour of restraint per day for 24 days produces a 50% increase in NE levels in the hippocampus and a 35% increase in frontal cortex measured 24 hours following the final restraint period (66). Furthermore, rats exposed to 30 minutes of foot shock per day for several days did not exhibit the depletion of brain NE that occurs with acute exposure to the stressor (40). Similarly, chronic cold exposure increases brain NE levels (68,69). These observations indicate that under some conditions the effect of a stressor on NE release habituates with long-term exposure. Such an interpretation would be consistent with the observation that stress-evoked release of NE from sympathetic nerves becomes reduced with subsequent presentations of the stressor, and that a comparable decline in the stress-induced increase in LC firing rate also occurs (70–72).

Although animals may develop a tolerance to a given stressor after continued exposure, something very different has been observed when an animal that has been chronically exposed to a stressor is exposed to an acute novel stressor. Under these conditions, the response to the novel stressor is enhanced. This enhanced response, referred to as sensitization, has been demonstrated in projections of LC neurons in a variety of paradigms (47,65–67,73). For example, in rats placed in a cold environment (4–5°C) for approximately 3 weeks, exposure to 30 minutes of tail shock elicits an increase in extracellular fluid NE as measured by microdialysis that is greater than the evoked response in naive rats (hippocampus, +82% versus +54%, respectively; prefrontal cortex, +141% versus +56%, respectively) (47,73) (Fig. 1). A similar stress-evoked sensitization of NE release from sympathetic neurons has also been observed (72,74).

The development and maintenance of stress-induced sensitization may be a function of the duration of exposure to the stressor (treatment-dependent) and/or the passage of time following the initial exposure to the stress (time-dependent). Few data are available that address the issue of the time course of development of stress-induced sensitization of NE release from LC neurons. However, we have recently observed that the enhancement of NE release in medial prefrontal cortex is present for 24 hours after 2 weeks but not after 1 week of cold exposure (J. Finlay, P. Gresch, M. Zigmond, and A. Sved, unpublished observations). It has been observed that other forms of sensitization are time-dependent, in that a single exposure to the stimulus will result in the development of sensitization as time passes and this sensitization is stable over time (75). However, in our experiment sensitized NE efflux was not present in rats that had been exposed to 1 week of cold followed by 2 weeks out of the cold, suggesting that the lack of sensitization observed following 1 week of cold exposure is not due to a process initiated by less chronic exposure to cold that requires additional time to become manifest.

While little information on the development of stress-evoked sensitization of NE release is available, nothing is known about the decay of this sensitized response, the relation between intensity of the stressor and the extent of sensitization, or the generality of the phenomenon across different stressors. We also do not know how chronic stress causes sensitization of NE release in response to a subsequent stressor. There are a great many possibilities, some of which we are beginning to explore (Fig. 2). The following sections will briefly review what has been learned over the past few years. Then we will turn to the possible functional significance of the phenomenon.

Changes in NE Synthesis

Few studies have addressed the mechanism underlying stress-induced sensitization in NE neurons. The factor that has been considered most thoroughly is the role of NE synthesis, since many studies have shown that stress increases the rate of NE synthesis. This increase in NE synthesis that occurs is due to, at least in part, an increase in the activity of tyrosine hydroxylase (TH), the first and rate-limiting enzyme involved in NE synthesis. As originally documented in sympathoadrenal catecholaminergic cells, there are two components in this response (76). One involves a rapid allosteric activation of the enzyme (77–81), and the other an increase in the number of TH molecules (78,82,83). Although virtually all of the attention on the regulation of NE synthesis has focused on the TH molecule itself, several other aspects of NE synthesis deserve consideration, including: 1) other components of the tyrosine hydroxylation step (e.g., availability of cofactor or of allosteric activators or inhibitors); 2) the decarboxylation of L-DOPA to dopamine by aromatic amino acid decarboxylase (AADC); and 3) the β-hydroxylation of dopamine to NE. The enzyme responsible for the latter of these steps, dopamine-β-hydroxylase (DβH), has been shown to increase LC neurons in chronically stressed rats (84).

Despite clear evidence for an increase in levels of TH activity and in mRNA and protein for TH in the LC in response to chronic cold exposure (78,82,83,85,86), no increase in basal TH activity was observed under these same conditions in the terminal fields of these neurons in either hippocampus (47) or cerebellum (L. Ni-

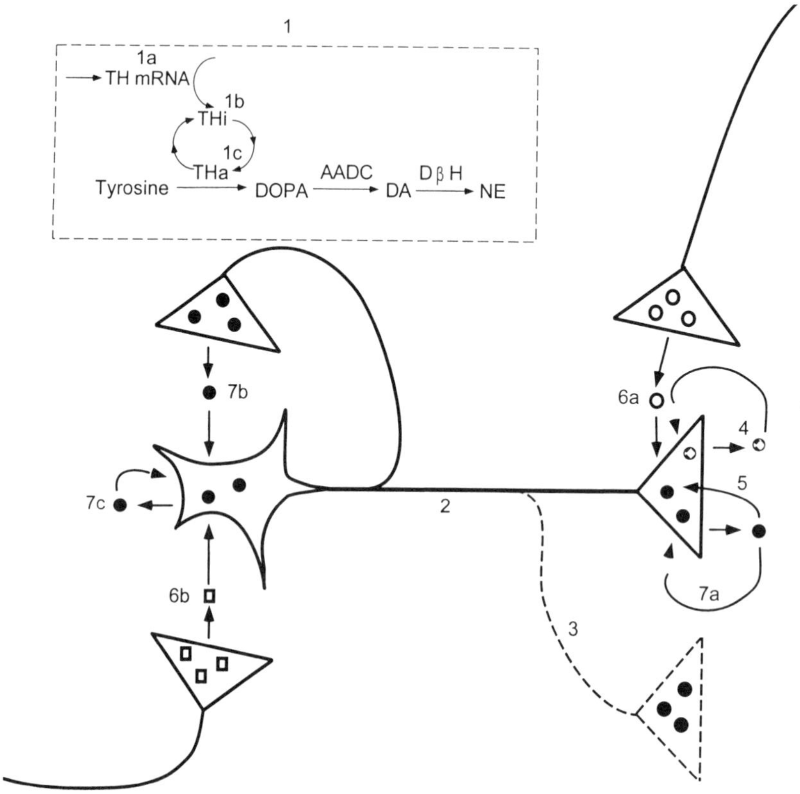

FIG. 2. Regulatory pathways influencing neurotransmitter dynamics that may contribute to stress-induced sensitization of norepinephrine (NE) efflux. Shown are some of the interactions that may support enhanced neurotransmitter release. Such mechanisms include: (*1*) increases in NE synthesis mediated by an increase in transcription (*1A*) or translation (*1B*) or the activation of existing TH molecules (*1C*); (*2*) increased firing of LC neurons; (*3*) sprouting of axon collaterals; (*4*) increased release of a cotransmitter that might facilitate NE release; (*5*) decreased reuptake of NE; (*6*) changes in afferent input to either the terminal (*6A*) or the cell body (*6B*); and (*7*) decreased autoreceptor inhibition occurring at either the terminal (*7A*) or the input of local collaterals (*7B*) or dendrites (*7C*) on the cell body.

senbaum, E. Abercrombie, and M. Zigmond, unpublished data). Chronic cold exposure also did not increase basal tyrosine hydroxylation in hippocampus as determined by the rate of DOPA accumulation following blockade of AADC (47). Consistent with this observation, local inhibition of AADC resulted in steady state levels of DOPA in the extracellular fluid of hippocampus that are not significantly increased in rats previously exposed to chronic cold (87). Similarly, extracellular levels of 3,4-dihydroxyphenylacetic acid (DOPAC), another index of NE synthesis, also were not increased in this structure in animals exposed chronically to cold (47). Thus, despite an increase in TH production in the cell bodies of LC neurons in animals exposed chronically to cold, this does not appear to result in a detectable increase in tyrosine hydroxylation in terminals of these neurons under basal conditions.

Although chronically stressed rats show no change in basal turnover or extracellular level of NE, chronic exposure to cold does give rise to an enhanced increase in NE synthesis expressed when an organism is subsequently challenged with a novel stressor. For example, prior exposure to chronic stress resulted in a greater increase in tissue and extracellular DOPA accumulation when these subjects were subsequently exposed to an acute novel stressor (47,87). Using extracellular fluid levels of DOPAC in the hippocampus as a measure of NE synthesis provides a similar conclusion (47). However, the relation between this enhanced stress-evoked increase in NE synthesis in the terminal region and increased production of TH in the cell body is unknown. Moreover, it is not clear whether this sensitized increase in NE synthesis during stress underlies the sensitized stress-evoked NE release, or is simply a parallel phenomenon.

Changes in LC Firing

The electrophysiological activity of LC neurons is also altered in rats that have been exposed to chronic stress. For example, repeated foot shock (88) or restraint (89), as well as long-term constant cold exposure (90), all increased the subsequent basal firing rate of LC neurons. More interestingly, repeated foot shock or chronic cold also increased the response of LC cells to noxious stimuli such as foot pinch or foot shock, suggesting that enhanced responsivity of these cells to noxious stimuli might contribute to stress-induced sensitization of NE release.

Repeated restraint (89) or chronic cold exposure (90) have also been reported to increase the responsiveness of LC neurons to the inhibitory effects of the α_2-adrenoreceptor agonist clonidine. Such an alteration is inconsistent with the observation that chronic stress enhances central noradrenergic activity, as it would seem to offset the stress-induced enhancement of the electrophysiological activity of these cells. However, this compensation complements the observation that chronic cold increases α_2-adrenoreceptor sensitivity of the terminals of LC cells (see information that follows), and raises the interesting possibility that chronic or repeated stress may elicit alterations in LC neurons that would *selectively* enhance their subsequent activity. This hypothesis becomes especially important when the effects of chronic or repeated stress on the electrophysiological activity of LC neurons are compared to the changes that have been reported in the release of NE from these cells. Although both techniques suggest that central noradrenergic activity is enhanced when a chronically stressed animal is subsequently faced with an acute stressor, electrophysiological measures indicate an increase in LC activity under basal conditions whereas neurochemical measures do not. This apparent paradox may reflect the fact that the electrophysiological studies were conducted in anesthetized rats, whereas the dialysis studies were performed using conscious, freely moving animals. Alternatively, it may reflect the inhibition of NE release due to supersensitive α-2 adrenoreceptors at the terminals of the LC neurons (see next section). The α-2 adrenoreceptors may compensate for the inability of similar changes in α-2 adrenoreceptors at the cell body to completely suppress alterations that result in an increase in the basal electrophysiological activity of these cells. This form of compensation would increase the contrast in NE release seen when the organism is subsequently stressed beyond that which would be produced if the enhanced evoked release of NE were superimposed on elevated basal levels of the transmitter.

Impact of Chronic Stress on Postsynaptic Receptors

In addition to there being changes in presynaptic events related to NE involved in chemical neurotransmission, the postsynaptic response to NE may also be altered as a result of exposure to chronic stress. Chronic stress has been found to decrease the binding of ligands to β-adrenergic receptors, due perhaps to a decrease in the number of receptors (91) and/or the sensitivity of these receptors (92). Some controversial findings regarding the impact of chronic stress on the β-adrenergic receptors have been noted and may be related to the stressor or β-adrenergic receptor ligands utilized in particular studies (50). Nonetheless, a hypothesis regarding the impact of chronic stress on central noradrenergic neurons must take into account changes in both the presynaptic and postsynaptic elements of neurotransmission. Specifically: 1) why does chronic stress increase the capacity for NE release and at the same time decrease the sensitivity of the postsynaptic elements upon which the neurotransmitter is acting?; 2) does the increase in neurotransmitter efflux result in a compensatory down-regulation of the postsynaptic receptor?; 3) is the decrease in sensitivity of the postsynaptic receptor responsible for a decrease in feedback inhibition of the LC neurons that then allows the expression of enhanced responsiveness of these neurons?; and 4) in any case, does the down-regulation of β-adrenergic receptors offset the impact of stress-induced sensitization of NE release?

Changes in Afferent Input

Although it is well known that LC neurons receive both excitatory and inhibitory input, and can be influenced by glucocorticoids as well, little information is available about the alteration of these inputs or their receptors during chronic stress. We have observed that the ability of diazepam to decrease extracellular NE in prefrontal cortex is no longer present after 3 weeks of cold (53). It also has been shown that benzodiazepine (93,94) and γ-aminobutyric acid ($GABA_A$) receptors (95,96) are reduced in animals exposed to chronic stress. Thus, it seems likely that one effect of stress is to reduce the impact of an inhibitory input to LC neurons (53). Other changes in afferent input also may occur under these conditions.

Changes in NE Terminals

In addition to changes in afferent regulation of the LC, chronic stress-induced changes at the level of the nerve terminal may also occur. A sensitized release of NE in the hippocampus was observed with local infusion of depolarizing concentrations of potassium directly into the hippocampus (87) although, we have been unable to see a comparable increase in the response to electrical field stimulation using hippocampal minces incubated in vitro (Fig. 3). Moreover, local administration of clonidine to the hippocampus via a microdialysis probe produced a greater decrease in extracellular fluid levels of NE in the hippocampus after exposure to chronic stress than in control animals. Similarly, the increase in extracellular fluid NE levels elicited by the α_2-adrenergic receptor antagonist idazoxan also was enhanced (87). Finally, although acute

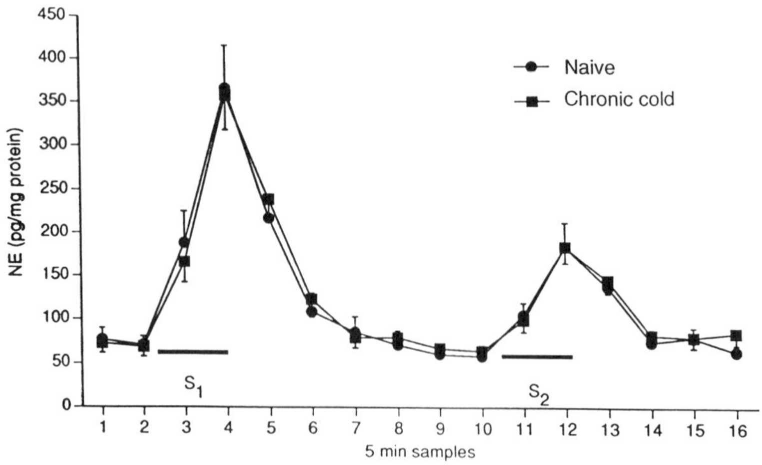

FIG. 3. Effect of electrical field depolarization on the overflow of endogenous norepinephrine (NE) from hippocampal minces. Tissue was prepared from naive rats and rats previously exposed to cold for 3 weeks. Tissue was stimulated with bipolar pulses at 10 Hz for 10 minutes (*horizontal bar*). Data are expressed as mean ±S.E.M. (n=9/group). (P.J. Gresch, J.M. Finlay, A.F. Sved, and M.J. Zigmond, unpublished data.)

stress does not appear to affect NE uptake (97), in mice exposed to overcrowding an increase in the K_m of NE uptake accompanied by an increase in the V_{max} has been reported (98). Thus, whereas some of the changes at the terminal level may promote stress-induced sensitization, others may actually counteract its effects.

Proliferation of NE Terminals

If changes at the level of the NE terminal do contribute to the sensitization, one possibility is that animals exposed to chronic stress develop more NE terminals. For example, it has been reported that exposure of rats to 2 weeks of intermittent restraint for 1 hour daily resulted in an increase in the percentage of LC neurons that could be antidromically activated from cerebral cortex, whereas if the restraint was provided for 6 hours per day the percentage of activated neurons was decreased (99,100). It is noteworthy that cortical NE levels increase when rats are exposed to a stress paradigm similar to that which increased the percentage of antidromically activated LC neurons (66). Although indirect, this evidence suggests that the density of LC innervation of terminal fields may change in response to chronic stress, and this change may depend on the severity of the stress.

Changes in Cotransmitter Release

Another plastic response that may occur in LC neurons in response to stress is a change in the amount or nature of the neuropeptides that they produce. Although the physiological significance of neuropeptides in LC neurons is not clear at present, a majority of LC neurons in the rat synthesize the neuropeptide galanin, and many produce neuropeptide Y (58). Other peptides have been shown to be produced in some portion of LC neurons. Whereas it is known that certain pharmacological treatments can alter the expression of these peptides in LC neurons (101), the effects of stress are unknown. Changes in the expression of neuropeptides in LC neurons could contribute to stress-evoked changes in NE release if these peptides modulate NE release from LC nerve terminals, as has been shown for neuropeptide Y in peripheral noradrenergic neurons (102).

Stress and Other NE Neurons

Noradrenergic neurons present in brainstem loci other than the LC also respond to a variety of stressful stimuli (39,103,104). Despite the fact that stress-evoked increases in activity are not unique to noradrenergic neurons in LC, there are several reasons to focus attention on this cell group in examining the role of noradrenergic neurons in stress: 1) noradrenergic neurons of LC, in contrast to other noradrenergic cell groups, innervate the entire neuroaxis (10) and are therefore in a position to influence widespread areas of the CNS; 2) LC neurons receive multisensory input and respond to virtually all stressful stimuli (54); and 3) most important, studies involving selective stimulation or inhibition of the LC support a role of the LC in responses to stress (see next section).

ROLE OF NORADRENERGIC NEURONS OF THE LC IN THE RESPONSE TO STRESS

Given that stressful stimuli increase NE release from LC neurons, it is appropriate to inquire about the role of these central noradrenergic neurons in the response to a stressful stimulus. Two complimentary approaches have been used to address this issue: 1) examination of the behavioral and physiological responses elicited by increased release of NE in the brain and 2) examination of the effects of destruction of noradrenergic neurons on the behavioral and physiological responses to stressful stimuli. Studies taking the first of these approaches have yielded reasonably consistent results: stimulation of LC neurons in conscious animals elicits a constellation of responses similar to the responses evoked by stressful stimuli. However, this literature is rather limited in scope, and is based primarily on two experimental paradigms. Most of the studies have relied on administration of corticotropin-releasing hormone (CRH) directly into the LC as a means of stimulating these neurons (105,106). When administered into the LC, CRH elicits behavioral and physiological responses resembling those produced by stressful stimuli (107,108). Stimulation of the LC also has been examined in nonhuman primates, although these studies have focused on the effects of electrical stimulation. Such stimulation elicits behavioral and physiological responses that resemble anxiety or fear (109-111). Although studies relying on electrical stimulation of the LC are problematic because of the potential for stimulating non-LC neurons (112), these studies also used stimulation of the ascending projections of the LC cells, as well as various pharmacological probes, to support the role of noradrenergic LC neurons in eliciting the evoked responses.

The second approach to understanding the role of NE in stress is to disrupt the function of the LC system and examine the consequences. Here, the idea has been to either destroy the LC with electrical current or selective neurotoxins, or temporarily impair LC function with pharmacological antagonists. There are at least two problems with these paradigms: 1) neuronal plasticity and functional redundancy of neural circuits make it difficult to interpret negative effects in lesion studies (5,113) and 2) most such lesions affect noradrenergic function throughout the brain and not simply that associated with the LC, and pharmacological intervention is still less specific, influencing peripheral as well as central function. This may help to explain the relatively mild impairments that have been observed after pharmacological treatments, as well as the contradictory nature of some of the results. Examples of these contradictory results are provided by studies measuring stress-evoked increases in adrenocorticotropin hormone (ACTH) or corticosterone levels, a hallmark of the stress response. Some of these studies show that destruction of noradrenergic neurons of the LC eliminate or attenuate the release of these hormones in response to stressful stimuli (114,115), whereas seemingly similar studies show these lesions to be ineffective (116–119).

Despite these caveats, some limited consensus may be emerging from investigations of LC functioning. Most such studies have investigated the effects in the rat of lesions of the LC or of the dorsal noradrenergic bundle, which projects from the LC to a variety of forebrain structures. In general, such lesions do not impair the acquisition, performance, or extinction of a learned response (120–122), although there is some evidence of impairments in acquisition (123). In addition, there appear to be no significant effects of such lesions on basal motor activity (120). However, lesions of the LC system often alter the response to being placed in a novel situation. Most investigators have reported a significant lesion-induced decrease in the hyperactivity normally associated with exposure to novel stimuli (117,120,124). Interestingly, lesions of the ventral noradrenergic bundle, which projects to hypothalamus but not to the more dorsal or rostral terminal fields of LC, may have very different effects, possibly because of the autonomic or endocrine changes that accompany such lesions (64,125). Studies of LC lesions in nonhuman primates provide a complementary perspective on the role of these neurons. Such animals appear to be less affected by situations that would usually be expected to cause mild stress or anxiety. As a result, they are less responsive to threats and appear less fearful of humans or other, more dominant, monkeys (111).

Thus, stimulation of LC neurons produces the appearance of behaviors normally associated with stress, whereas lesions seem to reduce such behaviors. Such studies further support the hypothesis that activation of the noradrenergic neurons of the LC during stress plays a role in at least some aspects of the generalized stress response.

FUNCTIONAL IMPLICATIONS FOR NORADRENERGIC RESPONSE TO CHRONIC STRESS: ADAPTATION OR MALFUNCTION?

If noradrenergic neurons of the LC are important in the response to acute stress, what are the implications of the impact of chronic stress on noradrenergic function? On the one hand, prior exposure to chronic stress has been shown to increase the availability of TH, at least within cell bodies, and to prevent subsequent stress-induced depletion of tissue NE. This would seem to be adaptive, permitting a wider range of subsequent stress-induced increases in NE release before failure of transmission. Yet, if the relation between arousal and performance is represented by an inverted U-shaped curve, can the sensitization of NE release that occurs with chronic stress represent too much of a good thing? Several investigators have recently called attention to this possibility, further suggesting that posttraumatic stress disorder may be one manifestation of a hyperactive central noradrenergic systems (65,126–129).

The evidence for functional impairments in animals previously exposed to a period of stress comes primarily from studies of motor activity and escape or avoidance performance (130–133). In general, these studies suggest that a period of exposure to an aversive stimulus can lead to a syndrome of "learned helplessness" in which animals show a reduction in motor activity, a decrease in avoidance, and an impairment in previously learned escape performance. These functional deficits have been likened to some of the negative symptoms seen in posttraumatic stress disorder (134).

SOME QUESTIONS FOR THE FUTURE

It has been 40 years since Marthe Vogt provided the first clear indication that noradrenergic neurons are present in the brain and that they are influenced by stress. We have learned much since that time. Yet in the years to come it will be important to ask a number of additional questions if we are to fully understand the relationship between central noradrenergic neurons and stress, particularly as it pertains to health and disease. These questions include:

1. Are stress-induced increases in TH gene expression necessarily accompanied by changes in TH protein? In TH activity? Does that TH get transported to nerve terminals and, if so, does it result in an increase in NE synthesis?
2. Does an increase in the capacity for NE synthesis result in increased NE release? If so, when does the increased release occur? For example, does it occur under basal conditions or only after a long period of intense stress that might otherwise lead to the exhaustion of releasable NE pools?
3. If stress-induced changes in the synthesis and release of NE do occur, are they accompanied by persistent changes in the electrophysiological activity of LC neurons? If so, do these changes play a role in the generation of the neurochemical changes or is there a dissociation between electrophysiological responses measured in LC cells and neurochemical responses measured in LC terminals?
4. If increases in NE synthesis and release occur as a result of chronic stress, either under basal conditions or in response to subsequent stress, are these increases associated with changes in the functional consequences of LC activation? For example, do they permit LC neurons to function more effectively in response to prolonged, intense stimulation, or do they result in hyperactivity of LC neurons leading to dysfunction?
5. What are the parameters of stress-induced sensitization? Which stressors produce it, at which intensities, and over how long a period? Is there a difference between continuous stress and intermittent stress? Is there an intensity of stress that is so severe that it can lead to the gradual (or immediate) development of sensitization after a single exposure?
6. If stress leads to sensitization of NE release and then to dysfunction at the level of central neural systems, might this underlie the behavioral aspects of clinical conditions such as anxiety, depression, and post-traumatic stress disorder? How does it relate to the autonomic dysfunctions that accompany these conditions; are these symptoms a manifestation of parallel events happening in the periphery or are they related more directly to central events?

These unresolved issues range from the most molecular to the most clinical, from those that are overlooked because it is assumed that they have been long-answered to those that have only recently been asked. And, of course, there will be other questions still. An understanding of these and related issues will lead us to appreciate which stress-induced changes in LC noradrenergic neurons are adaptive and which are maladaptive, and then to new, more effective approaches to treating and ultimately preventing the disorders that stress can precipitate.

ACKNOWLEDGMENTS

Our thanks to Elizabeth D. Abercrombie and Edward M. Stricker for discussions through the years and to Michael J. Mana for helpful comments on this manuscript. Findings resulting from the doctoral dissertation of Laura K. Nisenbaum (135) played an important role in shaping our thinking as well. We thank Paul J. Gresch for technical assistance and Susan D. Giegel for her help in preparing the manuscript. This work was supported in part by U.S. Public Health Service Grants NS19608, MH29670, MH45156, and MH00058.

REFERENCES

1. Cannon WB. *The wisdom of the body.* New York: Norton, 1932.
2. Vogt M. The concentration of sympathin in different parts of the central nervous system under normal conditions and after the administration of drugs. *J Physiol (London)* 1954;123:451–481.
3. Brodie BB, Shore PA. A concept for a role of serotonin and norepinephrine as chemical mediators in the brain. *Ann NY Acad Sci* 1957;66:631–642.
4. Amaral DG, Sinnamon HM. The locus coeruleus: neurobiology of a central noradrenergic nucleus. *Prog Neurobiol* 1977;9:147–196.
5. Stricker EM, Zigmond MJ. Brain monoamines, homeostasis, and adaptive behavior. In: Bloom FE, ed. *Handbook of physiology: the nervous system IV.* Bethesda: American Physiological Society, 1986;677–700.
6. Svensson TH. Peripheral, autonomic regulation of locus coeruleus noradrenergic neurons in brain: putative implications for psychiatry and psychopharmacology. *Psychopharmacol* 1987;92:1.
7. Abercrombie ED, Jacobs BL. Single-unit response of noradrenergic neurons in the locus coeruleus of freely moving cats. I. Acutely presented stressful and non-stressful stimuli. *J Neurosci* 1987;7:2837–2843.
8. Dahlström A, Fuxe K. Evidence for the existence of monoamine-containing neurons in the central nervous system. I:demonstration of monoamines in the cell bodies of brain stem neurons. *Acta Physiol Scand Suppl* 1964;62:5–55.
9. Ungerstedt U. Stereotaxic mapping of the monoamine pathways in the rat brain. *Acta Physiol Scand Suppl* 1971;367:1–48.
10. Moore RY, Bloom FE. Central catecholamine neuron systems: anatomy and physiology of the norepinephrine and epinephrine systems. *Ann Rev Neurosci* 1979;2:113–166.
11. Foote SL, Bloom FE, Aston-Jones G. Nucleus locus coeruleus: new evidence of anatomical and physiological specificity. *Physiol Rev* 1983;63:844–914.
12. Koda LY, Bloom FE. A light and electron microscopic study of noradrenergic terminals in the rat dentate gyrus. *Brain Res* 1977;120:327–335.
13. Parnavelas JG, Papadopoulos GC. The monoaminergic innervation of the cerebral cortex is not diffuse and nonspecific. *Trends Neurosci* 1989;12:315–320.
14. Bloom FE. An integrative view of information handling in the CNS. In: Fuxe K, Agnati LF, eds. *Volume transmission in the brain: novel mechanisms for neural transmission.* New York: Raven Press, 1991;11–23.
15. Beaudet A, Descarries L. The monoamine innervation of rat cerebral cortex: synaptic and nonsynaptic axon terminals. *Neuroscience* 1978;3:851–860.
16. Descarries L, Seguela P, Watkins KC. Nonjunctional relationships of monoamine-axon terminals in the cerebral cortex of adult rat. In: Fuxe K, Agnati LF, eds. *Volume transmission in the brain; novel mechanisms for neural transmission.* New York: Raven Press, 1991;53–62.
17. Aston-Jones G, Segal M, Bloom FE. Brain aminergic axons exhibit marked variability in conduction velocity. *Brain Res* 1980;195:215–222.
18. Aston-Jones G, Bloom FE. Norepinephrine-containing locus coeruleus neurons in behaving rats exhibit pronounced responses to non-noxious environmental stimuli. *J Neurosci* 1981;1:887–900.
19. Jacobs BL. Locus coeruleus neuronal activity in behaving animals. In: Heal DJ, Marsden CA, eds. *The pharmacology of noradrenaline in the central nervous system.* New York: Oxford University Press, 1990; 248–265.
20. Marsden CA. Methods for measuring noradrenaline release. In: Heal DJ, Marsden CA, eds. *The pharmacology of noradrenaline in the CNS.* Oxford: Oxford University Press, 1990;155–186.
21. Aston-Jones G, Bloom FE. Activity of norepinephrine containing locus coeruleus neurons in behaving rats anticipates fluctuations in the sleep-waking cycle. *J Neurosci* 1982;1:876–886.
22. Foote SL, Aston JG, Bloom FE. Impulse activity of locus coeruleus neurons in awake rats and monkeys is a function of sensory stimulation and arousal. *Proc Natl Acad Sci USA* 1980;77:3033–3037.
23. Waterhouse BD, Sessler FM, Cheng JT, Woodward DJ, Azizi A, Moises HC. New evidence for a gating action of norepinephrine in central neuronal circuits of mammalian brain. *Brain Res Bull* 1988;21:425–432.
24. Waterhouse BD, Sessler FM, Liu W, Lin C-S. Second messenger-mediated actions of norepinephrine on target neurons in central circuits: a new perspective on intracellular mechanisms and functional consequences. *Prog Brain Res* 1991;88:351–362.
25. Woodward DJ, Moises HC, Waterhouse BD, Hoffer BJ, Freedman R. Modulatory actions of norepinephrine in the central nervous system. *Fed Proc* 1979;38: 2109–2116.

26. Zigmond MJ, Stricker EM. Adaptive properties of monoaminergic neurons. In: Lajtha A, ed. *Handbook of neurochemistry, vol. 9.* New York: Plenum Press, 1985;87–102.
27. Zigmond MJ. Chemical transmission in the brain: homeostatic regulation and its functional implications. In: Bloom F. *Neuroscience: From the Molecular to the Cognitive. Prog Brain Res* 1994;100:115–122.
28. Starke K, Taube HD, Borowski E. Presynaptic receptor systems in catecholaminergic transmission. *Biochem Pharmacol* 1977;26:259–268.
29. Aghajanian GK, Cedarbaum JM, Wang RY. Evidence for norepinephrine-mediated collateral inhibition of locus coeruleus neurons. *Brain Res* 1977;136:570–577.
30. Groves PM, Wilson CJ. Monoaminergic presynaptic axons and dendrites in rat locus coeruleus seen in reconstructions of serial sections. *J Comp Neurol* 1980; 193:853–862.
31. Kalisker A, Rutledge CO, Perkins JP. Effect of nerve degeneration by 6-hydroxydopamine on catecholamine-stimulated adenosine 3:5′-monophosphate formation in rat cerebral cortex. *Mol Pharmacol* 1973;9: 619–629.
32. U'Prichard DC, Reisine TD, Mason ST, Fibiger HC, Yamarmura HI. Modulation of rat brain α and β-adrenergic receptor populations by lesions of the dorsal noradrenergic bundle. *Brain Research* 1980;187: 143–154.
33. Abercrombie ED, Keller RW, Zigmond MJ. Characterization of hippocampal norepinephrine release as measured by microdialysis perfusion: pharmacological and behavioral studies. *Neurosci* 1988;27:897–904.
34. Britton KT, Segal DS, Kvezenski R, Hauger R. Dissociation between in vivo hippocampal norepinephrine response and behavioral neuroendocrine responses to noise stress in rats. *Brain Res* 1992;574:125–130.
35. Kalén P, Rosengren E, Lindvall O, Björklund A. Hippocampal noradrenaline and serotonin release over 24 hours as measured by the dialysis technique in freely moving rats: correlation to behavioural activity state, effect of handling and tail-pinch. *Eur J Neurosci* 1989; 1:181–187.
36. Rossetti ZL, Portas C, Pani L, Carboni S, Gessa G. Stress increases noradrenaline release in the rat frontal cortex: Prevention by diazepam. *Eur J Pharmacol* 1990;176:229–231.
37. Tanaka T, Yokoo H, Mizoguchi K, Yoshida M, Tsuda A, Tanaka M. Noradrenaline release in the rat amygdala is increased by stress: studies with intracerebral microdialysis. *Brain Res* 1991;544:174–176.
38. Yokoo H, Tanaka M, Tanaka T, Tsuda A. Stress-induced increase in noradrenaline release in the rat hypothalamus assessed by intracranial microdialysis. *Experientia* 1990;46:290–292.
39. Maynert EW, Levi R. Stress-induced release of brain norepinephrine and its inhibition by drugs. *J Pharmacol Exp Ther* 1964;143:90–97.
40. Zigmond MJ, Harvey JA. Resistance to central norepinephrine depletion and decreased mortality in rats chronically exposed to electric foot shock. *J Neurovisc Rel* 1970;31:373–381.
41. Kvetnansky R, Palkovits M, Mitro A, Torda T, Mikulaj L. Catecholamines in individual hypothalamic nuclei of acutely and repeatedly stressed rats. *Neuroendocrin* 1977;23:257–267.
42. Roth KA, Mefford IM, Barchas JD. Epinephrine, norepinephrine, dopamine, and serotonin: differential effects of acute and chronic stress on regional brain amines. *Brain Res* 1982;239:417–424.
43. Thierry AM, Javoy J, Glowinski J, Kety SS. Effects of stress on the metabolism of norepinephrine, dopamine and serotonin in the central nervous system of the rat. I. modifications of norepinephrine turnover. *J Pharmacol Exp Ther* 1968;163:163–171.
44. Lidbrink P, Corrodi H, Fuxe K, Olson L. Barbiturates and meprobamate: decreases in catecholamine turnover of central dopamine and noradrenaline neuronal systems and the influence of immobilization stress. *Brain Res* 1972;45:507–513.
45. Korf J, Aghajanian GK, Roth RH. Increased turnover of norepinephrine in the rat cerebral cortex during stress: role of the locus coeruleus. *Neuropharmacol* 1973;12:933–938.
46. Stone EA. Stress and catecholamines. In: *Catecholamines and behavior, vol. 2.* Friedhoff AJ, ed. New York: Plenum Press. 1975;31–72.
47. Nisenbaum LK, Zigmond MJ, Sved AF, Abercrombie ED. Prior exposure to chronic stress results in enhanced synthesis and release of hippocampal norepinephrine in response to a novel stressor. *J Neurosci* 1991; 11:1478–1484.
48. Graham-Jones S, Fillenz M, Gray JA. The effects of footshock and handling on tyrosine-hydroxylase activity in synaptosomes and solubitized preparations from rat brain. *Neurosci* 1983;9:679–686.
49. Glavin GB. Stress on brain noradrenaline; a review. *Neurosci Biochem Rev* 1985;9:233–243.
50. Stanford SC. Monoamines in response and adaptation to stress. In: Stanford SC, Salmon P, eds. *Stress: from syndrome to synapse.* London: Academic Press. 1993; 281–331.
51. Tanaka M, Kohno Y, Nakawaga R, Ida Y, Takeda S, Nagasaki N, Noda Y. Regional characteristics of stress-induced increases in brain noradrenaline release in rats. *Pharmacology Biochem Behav* 1983;19:543–547.
52. Glavin GB, Tanaka M, Tsuda A, Kohno Y, Hoaki Y, Nagasaki N. Resional rat brain noradrenaline turnover in response to restraint stress. *Pharmacol Biochem Behav* 1983;19:287–290.
53. Finlay JM, Zigmond MJ, Abercrombie ED. Increased dopamine and norepinephrine release in medial prefrontal cortex induced by acute and chronic stress: effects of diazepam. *Neurosci* 1995;64:619–628.
54. Aston-Jones G, Chiang C, Alexinsky T. Discharge of noradrenergic locus coeruleus neurons in behaving rats and monkeys suggests a role in vigilance. *Prog Brain Res* 1991;85:47–75.
55. Loughlin SE, Foote SL, Bloom FE. Efferent projections of nucleus locus coeruleus: topographic organization of cells of origin demonstrated by three-dimensional reconstruction. *Neurosci* 1986;18:291–306.
56. Everitt BJ, Hökfelt T, Terenius L, Takemoto K, Mutt V, Goldstein M. Differential coexistence of neuropeptide Y (NPY)-like immunoreactivity with catecholamines in the central nervous system of the rat. *Neurosci* 1984;11:443–462.
57. Holets VR, Hökfelt T, Rokaeus A. Locus coeruleus neurons in the rat containing neuropeptide Y, tyrosine hydroxylase, galanin, and their efferent projections to

the spinal cord, cerebral cortex, and hypothalamus. *Neurosci* 1988;24:893–906.

58. Moore RY, Gustafson EL. The distribution of dopamine-β-hydroxylase, neuropeptide Y and galanin in locus coeruleus neurons. *J Chem Neuroanat* 1989; 2:95–106.
59. Pezzone MA, Lee W-S, Hoffman GE, Pezzone KM, Rabin BS. Activation of brainstem catecholaminergic neurons by conditional and unconditional aversive stimuli as revealed by c-Fos immunoreactivity. *Brain Res* 1993;607:310–318.
60. Morgan JK, Curran T. Stimulus-transcription coupling in the nervous system: involvement of the inducible proto-oncogenes *fos* and *jun*. *Ann Rev Neurosci* 1991;14:421–451.
61. Valentino RJ, Aston-Jones G. Physiological and anatomical determinants of locus coeruleus discharge: behavioral and clinical implications. In: *Psychopharmacology: fourth generation of progress,* Bloom FE, Kupfer DJ, (eds). Raven Press, New York, USA, 1995:373–385.
62. Abercrombie ED, Zigmond MJ. Striatal dopamine release: In vivo evidence for local initiation. *Ann NY Acad Sci* 1990;604:575–578.
63. Konarska M, Stewart RE, McCarty R. Sensitization of sympathetic adrenal medullar responses to a novel stressor in chronically-stressed laboratory rats. *Physiol Behav* 1989;46:129–135.
64. Everitt B, Robbins TW, Selden NRW. Functions of the locus coeruleus noradrenergic system: a neurobiological and behavioural synthesis. In: Heal DJ, Marsden DA, eds. *The pharmacology of noradrenaline in the central nervous system.* North Carolina: Oxford University Press, 1990;349–378.
65. Anisman H, Zacharko RM. Multiple neurochemical and behavioral consequences of stressors: implications for depression. *Pharm Ther* 1990;46:119–136.
66. Adell A, Garcia-Marquez C, Armario A, Gelpi E. Chronic stress increases serotonin and noradrenaline in rat brain and sensitizes their responses to a further acute stress. *J Neurochem* 1988;50:1678–1681.
67. Adell A, Garcia-Marquez C, Armario A, Gelpi E. Effects of chronic stress on serotonin and noradrenaline in the rat brain. *Biogenic Amines* 1990;7:19–26.
68. Bhagat B. Effect of chronic cold stress on catecholamine levels in rat brain. *Psychopharmaco* 1969;16: 1–4.
69. Ingenito AJ, Bunnycastle DB. The effect of exposure to heat and cold upon rat brain: catecholamine and serotonin levels. *Can J Physiol Pharmacol* 1962;45: 733–743.
70. Abercrombie ED, Jacobs BL. Single-unit response of noradrenergic neurons in the locus coerulus of freely moving cats. II. Adaptation to chronically presented stressful stimuli. *J Neurosci* 1987;7:2844–2848.
71. Konarska M, Stewart RE, McCarty R. Habituation of sympathetic-adrenal medullary responses following exposure to chronic intermittent stress. *Phys Behav* 1989;45:255–261.
72. Konarska M, Stewart RE, McCarty R. Habituation and sensitization of plasma catecholamine responses to chronic intermittent stress: effects of stressor intensity. *Phys Behav* 1990;47:647–652.
73. Gresch PJ, Sved AF, Zigmond MJ, Finlay JM. Stress-induced sensitization of dopamine and norepinephrine efflux in medial prefrontal cortex of the rat. *J Neurochem* 1994;63:575–583.
74. Konarska M, Stewart RE, McCarty R. Sensitization of sympathetic-adrenal responses to a novel stressor in chronically stressed laboratory rats. *Phys Behav* 1989; 46:129–135.
75. Antelman SM, Yehuda R. Time-dependent change following acute stress: relevance to the chronic and delayed aspects of PTSD. In: Spiegel D, Murburg MM, eds. *Catecholamine function in posttraumatic stress disorder: emerging concepts.* Washington, DC: American Psychiatric Press, 1994;87–98.
76. Zigmond RE, Schwarzschold MA, Rittenhouse AR. Acute regulation of tyrosine hydroxylase by nerve activity and by neurotransmitters via phosphorylation. *Ann Rev Neurosci* 1989;415–461.
77. Guidotti A, Costa E. Transsynaptic regulation of tyrosine-3-monooxygenase biosynthesis in rat adrenal medulla. *Biochem Pharmacol* 1977;26:817–823.
78. Thoenen H. Induction of tyrosine hydroxylase in peripheral and central adrenergic neurons by cold-exposure of rats. *Nature* 1970;228:861–862.
79. Zigmond RE. Long-term regulation of ganglionic tyrosine hydroxylase by preganglionic nerve activity. *Fed Proc* 1980;39:3003–3008.
80. Fluharty SJ, Snyder GL, Stricker EM, Zigmond MJ. Short- and long-term changes in adrenal tyrosine hydroxylase activity during insulin-induced hypoglycemia and cold stress. *Brain Res* 1983;267:384–387.
81. Fluharty SJ, Snyder GL, Zigmond MJ, Stricker EM. Tyrosine hydroxylase activity and catecholamine biosynthesis in the adrenal medulla of rats during stress. *J Pharm Exp Ther* 1985;233:32–38.
82. Richard F, Faucon-Biguet N, Labatut R, Rollet D, Mallet J, Buda M. Modulation of tyrosine hydroxylase gene expression in rat brain and adrenals by cold exposure. *J Neurosci Res* 1988;20:32–37.
83. Zigmond RE, Schon F, Iversen LL. Increased tyrosine hydroxylase activity in the locus coeruleus of rat brain stem after reserpine treatment and cold stress. *Brain Res* 1974;70:547–552.
84. Daiguji M, Okada F, Yamashita I. Dopamine-β-hydroxylase activity in the locus coeruleus and hypothalamus in cold-stressed rats. *Psychoneuroendocrinol* 1982;7:223–227.
85. Smith MA, Brady LS, Glowa J, Gold PW, Herkenham M. Effects of stress and adrenalectomy on tyrosine hydroxylase mRNA levels in the locus coeruleus by in situ hybridization. *Brain Res* 1991;544:26–32.
86. Melia KR, Rasmussen K, Terwilliger RZ, Haycock JW, Nestler EJ, Duman DS. Coordinate regulation of the cyclic AMP system with firing rate and expression of tyrosine hydroxylase in the rat locus coeruleus: effects of chronic stress and drug treatments. *J Neurochem* 1992;58:494–502.
87. Nisenbaum LK, Abercrombie ED. Enhanced tyrosine hydroxylation in hippocampus of chronically stressed rats upon exposure to a novel stressor. *J Neurochem* 1992;58:276–281.
88. Simson PE, Weiss JM. Altered activity of the locus coeruleus in an animal model of depression. *Neuropsychopharmacol* 1988;1:287–294.
89. Pavcovich LA, Ramirez OA. Time course of uncontrollable stress in the locus coeruleus neuronal activity. *Brain Res Bull* 1991;26:17–21.

90. Mana MJ, Grace AA. Chronic cold stress alters the basal and evoked electrophysiological activity of noradrenergic neurons in the locus coeruleus. *Soc Neurosci Abstr* 1993;19:1829.
91. Stanford SC. Central adrenoceptors in response and adaptation to stress. In: Meal DJ, Marsden CA, eds. *Pharmacology of noradrenaline in the central nervous system.* Oxford: Oxford University Press. 1991; 379–422.
92. Stone EA. Central cyclic-AMP-linked noradrenergic receptors: new findings on properties as related to the actions of stress. *Neurosci Biobehav Rev* 1987;11: 391–398.
93. Lippa AS, Critchett D, Sano MC, et al. Benzodiazepine receptors: cellular and behavioral characteristics. *Pharmacol Biochem Behav* 1979;10:831–843.
94. Weizman R, Weizman A, Kook DA, Vocci F, Deutsch SI, and Paul SM. Repeated swim stress alters brain benzodiazepine receptors measure in vivo. *J Pharmacol Exp Ther* 1989;249:701–707.
95. Biggio G, Concas A, Serra M, et al. Stress and β-carbolines decrease the density of low affinity GABA binding sites: an effect reversed by diazepam. *Brain Res* 1984;305:13–18.
96. Concas A, Corda MG, Biggio G. Involvement of benzodiazepine recognition sites in the foot shock-induced decrease of low affinity GABA receptors in the rat cerebral cortex. *Brain Res* 1985;341:50–55.
97. Thierry AM, Blanc G, Glowinski J. Effect of stress on the disposition of catecholamines localized in various intraneuronal storage forms in the brain stem of the rat. *J Neurochem* 1971;18:449–461.
98. Hendley ED, Moisset B, Welch BL. Catecholamine uptake in cerebral cortex: adaptive change induced by fighting. *Science* 1973;180:1050–1052.
99. Nakamura S, Sakaguchi T. Development and plasticity of the locus coeruleus: a review of recent physiological and pharmacological experimentation. *Prog Neurobiol* 1990;34:505–526.
100. Sakaguchi T, Nakamura S. Duration-dependent effects of repeated restraint stress on cortical projections of locus coeruleus neurons. *Neurosci Lett* 1990;118: 193–196.
101. Austin MC, Cottingham SL, Paul SM, Crawley JN. Tyrosine hydroxylase and galanin mRNA levels in locus coeruleus neurons are increased following reserpine administration. *Synapse* 1990;6:351–357.
102. Lundberg JM, Hökfelt T. Multiple co-existence of peptides and classical transmitters in peripheral autonomic and sensory neurons—functional and pharmacological implications. *Prog Brain Res* 1986;68:241–262.
103. Saavedra JM, Kvetnansky R, Kopin I. Adrenalin, noradrenaline and dopamine levels in specific brain stem areas of acutely immobilized rats. *Brain Res* 1979; 160:271–280.
104. Lachuer J, Gaillet S, Barbagli B, Buda M, Tappaz M. Differential early time course activation of the brainstem catecholaminergic groups in response to various stresses. *Neuroendocrin* 1991;53:589–596.
105. Valentino RJ, Foote SL. Corticotropin-releasing hormone increases tonic but not sensory-evoked activity of noradrenergic locus coeruleus neurons in unanesthetized rats. *J Neurosci* 1988;8:1016–1025.
106. Valentino RJ, Foote SL, Aston-Jones G. Corticotropin-releasing factor activates noradrenergic neurons of the locus coeruleus. *Brain Res* 1983;270:363–367.
107. Rassnick S, Sved AF, Rabin BS. Locus coeruleus stimulation by corticotropin-releasing hormone suppresses in vitro cellular immune responses. *J Neurosci* 1994; 14:6033–6040.
108. Butler PD, Weiss JM, Stout JC, Nemeroff CB. Corticotropin-releasing factor produces fear-enhancing and behavioral activating effects following infusion into the locus coeruleus. *J Neurosci* 1990;10:76–83.
109. Grant SJ, Huang YH, Redmond DE, Jr. Behavior of monkeys during opiate withdrawal and locus coeruleus stimulation. *Pharmacol Biochem Behav* 1988;30: 13–19.
110. Redmond DE Jr., Krystal JH. Multiple mechanisms of withdrawal from opioid drugs. *Ann Rev Neurosci* 1984;7:443–478.
111. Redmond DE. Studies of the nucleus locus coeruleus in monkeys and hypotheses for neuropsychopharmacology. In Meltzer H, ed. *The third generation of progress.* New York: Raven Press, 1987;967–975.
112. Sved AF, Felsten G. Stimulation of the locus coeruleus decreases arterial pressure. *Brain Res* 1987;414: 119–132.
113. Robbins TW, Everitt BJ, Cole BJ, Archer T, Mohammed A. Functional hypotheses of the coeruleo-cortical noradrenergic projection: a review of recent experimentation and theory. *Physiological Psychology* 1985;13:127–150.
114. Feldman S, Sigel RA, Weidenfeild J, Conforti N, Melamed E. Adrenocortical responses to either stress and neural stimuli in rats following the injection of 6-hydroxydopamine into the medial forebrain bundle. *Exp Neurol* 1984;83:215–220.
115. Gaillet S, Lachuer J, Maluval F, Assenmacher I, Szafarczyk A. The involvement of noradrenergic ascending pathways on the stress-induced activation of ACTH and corticosterone secretions is dependent on the nature of stressors. *Exp Brain Res* 1991;87:173–180.
116. Dooley DJ, Hunziker G, Hänsler A. Corticosterone secretion in the rat after DSP-4 treatment. *Neurosci Lett* 1984;46:271–274.
117. Velley L, Mormede P, Kempf E. Neurochemical lesion of the nucleus locus coeruleus increases neophobia in a specific exploration task but does not modify endocrine response to moderate stress. *Pharmacol Biochem Behav* 1988;29:1–7.
118. Bakke HK, Murison R, Walther B. Effect of central noradrenaline depletion on corticosterone levels and gastric ulcerations in rats. *Brain Res* 1986;368: 256–261.
119. Castagne V, Rivet JM, Mormede P. The integrity of the ventral noradrenergic bundle is not necessary for a normal neuroendocrine stress response. *Brain Res* 1990;511:349–352.
120. Britton DR, Ksir C, Thatcher Britton K, Young D, Koob GF. Brain norepinephrine depleting lesions selectively enhance behavioral responsiveness to novelty. *Physiology and Behavior* 1984;33:473–478.
121. Sessions GR, Kant GJ, Koob GF. Locus coeruleus lesions and learning in the rat. *Physiol Behav* 1976; 17:852–859.
122. Pisa M, Fibiger HC. Intact selective attention in rats with lesions of the dorsal noradrenergic bundle. *Behav Neurosci* 1983;97:519–529.

123. Cole BJ, Robbins TW. Dissociable effects of lesions to the dorsal or ventral noradrenergic bundle on the acquisition, performance and extinction of averse conditioning. *Behavioural Neuroscience* 1987;101: 476–488.
124. Cole BJ, Robbins TW, Everitt BJ. Lesions of the dorsal noradrenergic bundle simultaneously enhance and reduce responsivity to novelty in a food preference test. *Brain Res Rev* 1988;13:325–349.
125. Dunn LT, Everitt BJ. The effects of lesions to the noradrenergic projections from the locus ceruleus and lateral tegmental cell groups on conditioned taste aversion in the rat. *Behavioural Neuroscience* 1987;101: 409–422.
126. Aston-Jones G, Valentino RJ, Van Bockstaele EJ, Meyerson AT. Locus Coeruleus, stress, and PTSD: neurobiological and clinical parallels. In Murburg MM, ed. *Catecholamine function in posttraumatic stress disorder: emerging concepts.* Washington, DC: American Psychiatric Press, 1994;17–62.
127. Weiss JM, Goodman PA, Losito BG, Corrigan S, Charry JM, Bailey WH. Behavioral depression produced by an uncontrollable stressor: relationship to norepinephrine, dopamine, and serotonin levels in various regions of rat brain. *Brain Res Rev* 1981;3: 167–205.
128. Anisman H, Zacharko RM. Depression: the predisposing influence of stress. *Behav Brain Sci* 1982;5: 89–137.
129. Murburg MM, Asleigh EA, Hommer DW, Beith RC. Biology of catecholaminergic systems and their relevance to PTSD. In Murburg MM, ed. *Catecholamine function in posttraumatic stress disorder.* Washington, DC: American Psychiatric Press, 1994;3–16.
130. Weiss JM, Stone EA, Harrell N. Coping behavior and brain norepinephrine levels in rats. *J Comp Physiol Psych* 1970;72:153–160.
131. Anisman H, Sklar LS. Catecholamine depletion in mice upon re-exposure to stress: mediation of the escape deficits produced by inescapable shock. *J Compar Physiol Psych* 1979;93:610–625.
132. Weiss JM, Simson PG. Depression in an animal model: focus on the locus coeruleus. In Porter R, Bock G, Clark S, eds. *Antidepressants and receptor function.* John Wiley, New York, 1986;191–209.
133. Krystal JH, Kosten TR, Perry BD, et al. Neurobiological aspects of PTSD: review of clinical and preclinical studies. *Behavior Therapy* 1989;20:177–198.
134. Van der Kolk B, Greenberg M, Boyd H, et al. Inescapable shock, neurotransmitters, and addiction to trauma: toward a psychobiology of post-traumatic stress. *Biol Psychiatry* 1985;20:314–325.
135. Nisenbaum LK. *The effect of acute and chronic stress on the release and synthesis of norepinephrine in the locus coeruleus-noradrenergic system.* Dissertation. Pittsburgh; University of Pittsburgh, 1991.

Neurobiological and Clinical Consequences of Stress: From Normal Adaptation to PTSD, edited by M. J. Friedman, D. S. Charney, and A. Y. Deutch, Lippincott-Raven Publishers, Philadelphia © 1995.

4

Stress and Central Amino Acid Systems

*Brian A. Horger and *†Robert H. Roth

*Departments of *Pharmacology and †Psychiatry, Yale University School of Medicine, New Haven, Connecticut 06520*

There are a number of amino acids in the mammalian central nervous system (CNS) believed to have a role in synaptic transmission. Over the past decade several amino acids have gained recognition as major transmitter candidates. On the basis of neurophysiological studies, amino acids have been separated into two general classes, excitatory amino acids (EAA) (glutamate, aspartate, homocysteic acid, and cysteic acid) that depolarize neurons in the mammalian CNS and inhibitory amino acids (γ-aminobutyric acid (GABA), glycine, taurine, and β-alanine) that hyperpolarize mammalian CNS neurons. The major EAA transmitters in the CNS are glutamate and aspartate. Gamma-aminobutyric acid is the prototypical inhibitory amino acid transmitter. Both GABA and EAA transmitters are believed to influence emotive states in the organism. In particular, amino acid transmitters have been shown to have a role in the organism's response to physical (i.e., pain or discomfort) or psychological (i.e., fear) stress. The following chapter is a review of the responses of central amino acid systems to stress and the neurobiological consequences of these responses.

Inhibitory Amino Acid Transmitters (GABA)

The existence of a link between anxiety (the state of apprehension resulting from stress and/or fear) and GABAergic transmission was established independently 20 years ago by Costa and Haefely and their collaborators with the seminal findings that anxiolytic and hypnotic benzodiazepines exert their pharmacological effects by a selective facilitation of GABAergic transmission (1,2). The GABA field has progressed a long way in the past 2 decades with the onslaught of molecular biology and the development of new techniques to study and characterize receptors. The GABA receptors and transporters have been cloned, expressed, and sequenced. Two major types of GABA receptors have been identified: $GABA_A$ and $GABA_B$ receptors.

The $GABA_A$ receptors (Fig. 1) are the most prevalent form in mammalian brain and have been extensively studied and characterized. These receptors are believed to be comprised of five subunits that form a Cl^- ionophore. Binding of an agonist at the GABA recognition site causes a conformational change that opens the ionophore to Cl^-, resulting in membrane hyperpolarization and reduced neuronal excitability. Positive allosteric modulators enhance GABA-induced channel gating by facilitating the conformational transition from the closed to open form (see 3,4 for review). Specifically, benzodiazepines increase the frequency of GABA-induced Cl^- channel openings, while barbiturates increase the duration of channel openings (5). Neurosteroids augment GABA-stimulated Cl^- conductance by increasing benzodiazepine binding and augmenting the duration of Cl^- channel openings at a site distinct from barbiturates (6,7). In addition, these allosteric modulators (benzodi-

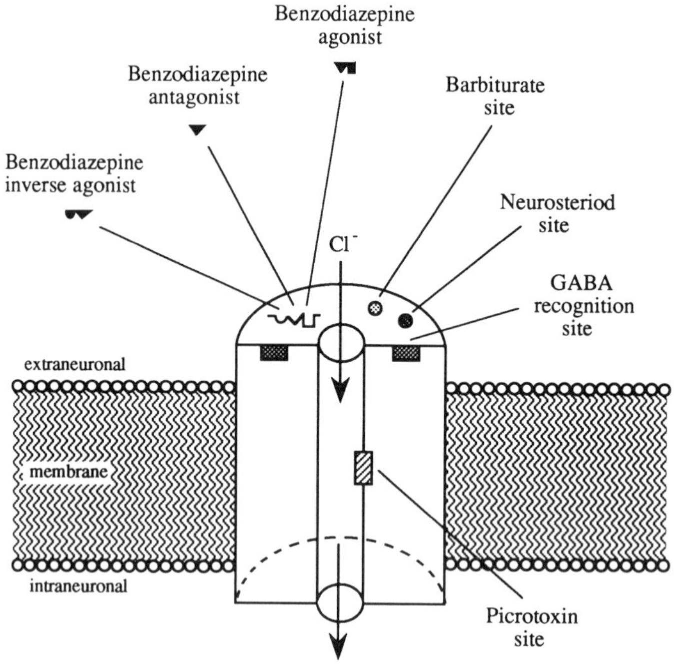

FIG. 1. Diagram depicting the $GABA_A$ receptor. Agonists at the benzodiazepine recognition site increase the frequency of GABA-induced channel openings, while barbiturates increase the duration of channel openings. GABA-elicited Cl^- conductance is also modulated by picrotoxin and neurosteroids.

azepines, barbiturates, and neurosteroids) have been shown to increase the affinity of the GABA recognition site (6,8,9). Picrotoxin is believed to block the Cl^- channel directly or via binding to a closely located site (10,11).

Although the $GABA_A$ receptor is the target of a number of anxiolytic drugs, the anatomical location, receptor subtypes, and/or subunits that are critically involved in anxiety have not been clearly defined. A family of at least fifteen subunits have been identified, making possible more than 2,000 subunit combinations of pentameric $GABA_A$ receptor subtypes, each with a unique structure and function (12–14). The affinity of benzodiazepine receptor ligands has been shown to vary as a function of subunit composition. Receptors expressed in oocytes without a γ subunit or those containing $\gamma1$ are less responsive to benzodiazepines relative to receptors containing a $\gamma2$ subunit (15–17). Subunit combinations are also a critical factor in determining the effectiveness of benzodiazepine receptor ligands, as demonstrated by the observation that flunitrazepam potentiates GABA-induced Cl^- currents in oocytes coexpressing $\gamma2$ and $\alpha3$ subunits, but not when either subunit was independently coexpressed with $\beta1$ (18). The $GABA_A$ receptors in the brain are largely heterogeneous, although distinct patterns have been observed. For example, serotonergic and GABAergic neurons in the raphe nuclei express distinct patterns of α subunits, indicating that these neurons exhibit unique $GABA_A$ receptor subtype profiles (19). Thus, subtle differences in subunit organization result in subpopulations of $GABA_A$ receptors that may have different regional and cellular locations, each with differential sensitivity to GABA and allosteric modulators.

Alterations in GABA receptors have been observed in response to various stressors. For example, acute exposure to cold-immobilization or footshock stress produced a decrease in the number of GABA receptor binding sites in the cerebral cortex and striatum (20–23). Similar effects were observed following a pharmacological stressor. Repeated administrations of the anxiogenic benzodiazepine inverse agonist FG 7142 decreased the number of cortical low-affinity GABA receptors (23). Even the mild stress of handling reduced the number of GABA receptor binding sites in the cerebral cortex, striatum, and cerebellum (20,21,23–25). However, $GABA_A$ receptor subunit ($\alpha 1$ and $\gamma 2$) mRNAs in the cortex have been reported to increase following acute exposure to social stress, which consisted of being attacked by other mice (26). In addition, acute exposure to forced swim stress significantly increased the number of benzodiazepine binding sites in the cortex (27,28). While these data are difficult to integrate they clearly indicate that stress produces compensatory changes in GABA receptors.

Stress has also been shown to alter GABA turnover. Repeated immobilization stress produced an increase in GABA accumulation following inhibition of its degradation with a GABA-transaminase inhibitor (29). In addition, repeated stress increased cortical glutamic acid decarboxylase (GAD) activity, a rate limiting enzyme in the metabolism of GABA from glutamate (29–31). Further work in this area are needed to follow up these initial findings indicating increased GABA turnover in the frontal cortex in response to stress.

Stress has also been found to alter the modulating effects of neurosteroids. Two of these neurosteroids, 3α-hydroxy-5α-pregnane-20-one (allopregnanolone) and 3α,21-dihydroxy-5α-pregnane-20-one (alloTHDOC), which are metabolites of progesterone and deoxycorticosterone, respectively, formed in the brain, are among the most potent known endogenous ligands of $GABA_A$ receptors in the CNS (32). AlloTHDOC has been shown to have anxiolytic effects at systemic doses lower than those required to induce sedation (33). The modulatory influence of neurosteroids on $GABA_A$ receptors has been shown to be augmented by stress. Acute exposure to swim stress produced robust increases in levels of the neurosteroids allopregnanolone and alloTHDOC in the cerebral cortex and hypothalamus (34). In addition, the pharmacologic action of neurosteroids are enhanced by stress in that the ability of alloTHDOC to enhance benzodiazepine agonist-binding in the cerebral cortex is augmented following acute swim stress (35). Swim stress also appears to alter the $GABA_A$ receptor-coupled Cl^- ionophore itself by enhancing muscimol-stimulated Cl^- uptake and increasing the binding sites and affinity of ^{35}S-*t*-butylbicyclophosphorothionate, a ligand that binds at or near the $GABA_A$-coupled Cl^- ionophore (36–38). These changes all lead to an augmentation of GABAergic inhibition and may function in CNS homeostasis, offsetting stress-induced activation of other transmitter systems (i.e., EAA transmitters).

Excitatory Amino Acid Transmitters

It has been recognized for many years that certain EAAs such as glutamate and aspartate occur in uniquely high concentrations in the brain and that they exert very powerful stimulatory effects on neuronal activity. However, their status as neurotransmitters in mammalian CNS was uncertain for many years due to the fact that glutamate and aspartate are compounds intimately involved in intermediary metabolism in neuronal tissue. Thus, it has been extremely difficult to dissociate the role glutamate plays in neuronal metabolism and as a GABA precursor from its possible role as a neurotransmitter. Our understanding of the role of EAAs as CNS neurotransmitters and their regulatory control and function has been greatly enhanced by studies directed at the identification, characterization, localization, and isolation of receptors for these EAAs. In the last decade, progress on the definition of receptor subtypes and the availability of more selective agonists and antagonists for EAA receptors have resulted in a quantum leap in knowledge concerning the possible role of EAA transmitters at synaptic sites throughout the vertebrate CNS.

Recently, studies have focused on the response of the EAA transmitter systems to stress. These studies became possible with the advent of the in vivo microdialysis procedure and the ability to recover local brain dialysate samples from extracellular compartments coupled with the availability of specific pharmacological probes for EAA receptors. Immobilization or restraint stress has been shown to increase extracellular glutamate levels in the medial prefrontal cortex (mPFC), hippocampus, striatum, and nucleus accumbens (NAS) (39–41). Restraint stress also produced an increase in extracellular levels of aspartate in the mPFC, hippocampus, and NAS (41). Stress-induced increases in extracellular levels of both glutamate and aspartate were significantly greater in the mPFC relative to the hippocampus and basal ganglia (41). These effects were attenuated by local perfusion with tetrodotoxin, suggesting that stress-induced glutamate release was impulse-dependent from neuronal pools (41). Consistent with this, increased glutamate release has been observed in synaptosomes prepared from the frontal cortex and hippocampus of immobilized subjects (42).

Other stress procedures have produced similar results. Increased extracellular levels of glutamate and aspartate have been observed in the mPFC and hippocampus following 20 minutes of swimming in room-temperature water (41). In addition, acute formalin (pain) stress produced an increase in tissue levels of glutamate and aspartate in cortical areas (43). These data suggest that the increases in EAA transmitter release following restraint stress was not due to factors unassociated with stress, such as the inability of the subject to engage in locomotor activity. Rather, the EAA response to stress appears to be common to stress-inducing procedures regardless of the motoric demand.

A possible mechanism by which EAA transmitters are released following stress is via glucocorticoid secretion. Stress induces secretion of glucocorticoids from the adrenal glands that produce varied effects throughout the CNS. Adrenalectomy markedly attenuates elevations in extracellular levels of glutamate in the mPFC and hippocampus following restraint stress; these effects are reversed by glucocorticoid replacement (40,44). These data suggest a critical role of the hypothalamic-pituitary-adrenal cortical axis in stress-induced increases in EAA release.

Interestingly, chronic exposure to stress appears to attenuate the stress-induced activation of glutamatergic neurons. As measured by lactography, acute restraint stress produced an increase in glutamatergic activity in the hippocampus (45). This effect was attenuated following several exposures to the stressor.

The synaptic actions of EAA transmitters are mediated via interactions with EAA receptors. Of the EAA receptor subtypes identified, the *N*-methyl-D-aspartate (NMDA) receptor (Fig. 2) has been the most fully characterized in terms of its pharmacology, physiology, anatomical distribution, and biophysical properties. The NMDA receptor can be subdivided into three functional domains, each with discrete ligand binding domains: 1) a transmitter recognition site; 2) an ion channel; 3) and an allosteric site activated by glycine.

The NMDA receptor regulates the conductance of cations through gated ion channels. Dissociative anesthetics such as dizocilpine (MK 801) and phencyclidine (PCP) block the ability of NMDA to regulate cation conductance in a use- and voltage-dependent manner (46–48) and are thus postulated to be acting at a site within the channel itself. In addition, Mg^{2+} and Zn^{2+} block the cation channel of the NMDA receptor in a voltage-dependent manner at a site independent of the dissociative anesthetics (49,50). The NMDA receptor is allosterically modulated by a strychnine-insensitive glycine receptor site that increases the frequency of NMDA receptor-induced channel openings (51). The dependence of glycine site activation in NMDA receptor function seems apparent, given numerous studies with (+)-1-hydroxy-3-aminopyrrolidone-2 (HA-966) and other glycine antagonists or partial agonists that demonstrate an attenuation of NMDA receptor activity. While it is still unclear as to whether glycine site activation constitutes an additional mechanism regulating NMDA receptor function under physiological conditions, there is some evidence to indicate that the glycine site is not always maximally activated in vivo. In view of the potential role played by endogenous glycine in the modulation of NMDA function, clearly there is a need to understand the endogenous mechanisms that regulate the synaptic levels of glycine in the vicinity of NMDA receptors and the potential role that endogenous glycine antagonists such as kynurenic acid may play in regulating the degree of glycine site acti-

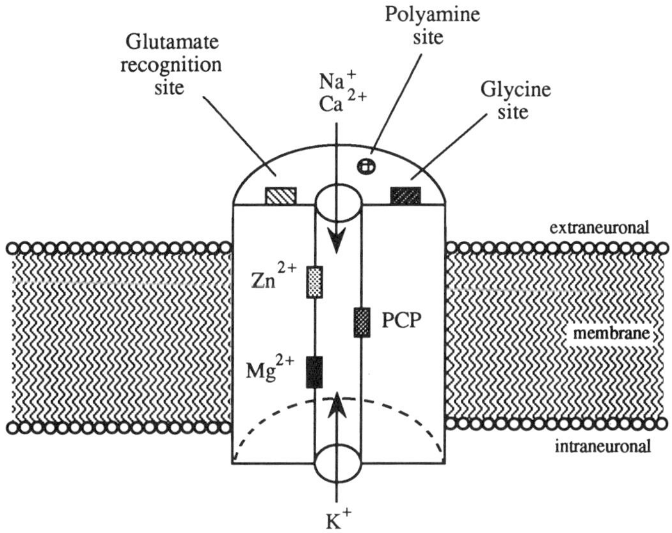

FIG. 2. Diagram depicting the NMDA receptor subtype. The glutamate recognition site is also activated by the selective agonist NMDA. PCP, Mg^{2+}, and Zn^{2+} block the gated cation channel in a voltage and use-dependent manner. Activation of strychnine-insensitive glycine sites not only increases the frequency of glutamate-induced channel openings, but appears to be an absolute requirement for glutamate receptor functioning. Polyamines also potentiate channel openings. However, contrary to glycine, the presence of polyamines is not a necessary requirement for glutamate receptor activation.

vation and, thus, NMDA receptor function in vivo.

Alpha-amino-3-hydroxy-5-methyl-4-isoxazole propionic acid (AMPA–formerly quisqualate) and kainate receptors are also believed to gate cation conductance (52). While there is sufficient evidence to warrant the existence of a separate kainate receptor subtype, currently there is a lack of an agonist selective for this site, as kainate is also active at the AMPA receptor. This dual agonist quality combined with the absence of a selective kainate antagonist makes it difficult to differentiate pharmacological activity of the kainate receptor from that of the AMPA receptor subtype.

Acute stress has been found to alter the binding properties of EAA receptor subtypes. Tail-shock stress has been shown to increase AMPA- but not NMDA-receptor binding in several areas of the hippocampus (53). Acute stress also appears to induce lasting effects on EAA receptor binding. Krugers et al. (54) observed an increase in NMDA binding and a decrease in AMPA binding in selected regions of the hippocampus 24 hours after exposure to dominant-opponent social stress. These studies indicate selective effects on EAA receptor subtype binding in response to stress.

Acute stressors have also been shown to increase endogenous levels of allosteric modulators of the NMDA receptor. Formalin stress produced an increase in cortical levels of glycine in the cortex (55). Restraint stress has been found to elevate extracellular levels of Zn^{2+} in the hippocampus as measured by in vivo microdialysis (39). In addition, swim stress increased MK 801 binding to NMDA receptors in the forebrain (56). Thus, acute stress appears to alter the functioning of EAA transmitter systems at several distinct levels.

AMINO ACID INFLUENCES ON MIDBRAIN DA NEURONS: A MODEL REPORTER SYSTEM

In order to provide some relevance to the data implicating a role for amino acid transmitter systems in stress, we chose the mesoprefrontal dopamine (DA) system as a model reporter system in terms of a neurochemical response to stress. Mesoprefrontal DA neurons projecting from the ventral tegmental area (VTA) to the mPFC appear to be particularly vulnerable to stress-induced activation. While acute stress increases DA metabolism and release from a number of specific brain areas, mild or brief exposure to stress produces a selective metabolic activation of mesoprefrontal DA neurons. Low-intensity electric footshock increases in vivo tyrosine hydroxylase (TH) activity and levels of the DA metabolite 3,4-dihydroxyphenylacetic acid (DOPAC) in the mPFC, but not in the NAS or striatum (57,58). Increased DA utilization in the mPFC occurs rapidly after stress but appears to develop tolerance and dissipates over time, even during the continued presence of the stressor (58). It is only after the presentation of more severe stress (i.e., increased shock intensity or duration) that a response from other mesotelencephalic DA systems is recruited.

Conditioned fear can also selectively activate the mesoprefrontal DA system (59,60). The ability of a previously neutral stimulus paired with a noxious stimulus to selectively increase DA utilization in the mPFC suggests that this effect is not a response to physical pain. Rather, activation of mesoprefrontal DA neurons appears to be associated with the psychological aspect of stress (i.e., anxiety) and may possibly be associated with the animal coping with the stressor (61–63). It should be noted that the responsiveness of the mesoprefrontal DA system to stress is heterogeneous, probably reflecting the presence of multiple cortical innervations imbedded in this region (64,65).

The unique response of the mesoprefrontal DA system to stressful stimuli could be due to singular characteristics of these neurons. The most notable characteristic of mesocortical DA neurons is the absence of specific DA autoreceptors. Presynaptic synthesis-regulating autoreceptors are either absent or markedly decreased in the prefrontal, cingulate, and entorhinal cortical terminal regions of ascending midbrain DA neurons (66–69). In addition, a majority of mesencephalic DA neurons projecting to the prefrontal and cingulate cortices lack, or have decreased densities of, somatodendritic impulse-modulating autoreceptors (69–70). Mesoprefrontal DA cells that do possess impulse-regulating autoreceptors appear limited to a small group of neurons located in the more lateral aspects of the VTA (71,72).

Some mesocortical DA neurons also exhibit a higher turnover rate of DA and a higher degree of burst firing. Mesoprefrontal and mesocingulate DA cells, for example, exhibit increased mean discharge rates, burst firing activity, and turnover rates of DA relative to mesopiriform and nigrostriatal DA neurons (70,73). It is possible that the absence of impulse- and synthesis-regulating autoreceptors contribute to this characteristic of mesoprefrontal and mesocingulate DA neurons. However, the lack of autoreceptors cannot in itself explain the selective activation of mesoprefrontal neurons to mild stress. Mesocingulate DA neurons also lack impulse- and synthesis-regulating autoreceptors, yet DA utilization is not elevated in the cingulate cortex following exposure to mild stress (58). Thus, other factors must also contribute to the unique responsiveness of mesoprefrontal DA neurons to mild stress.

Mesoprefrontal DA neurons appear to be regulated by a number of chemically distinct afferent systems (62,74). For example, immunoneutralization of substance P in the VTA, a source of the cortical DA innervation, prevents the DAergic response to stress in the mPFC (75). Similarly, opiate- or NMDA-receptor blockade in the VTA also prevents stress-induced activation of the mesoprefrontal DA system (76,77). In addition, systemic administration of $GABA_A$ receptor agonists or VTA administration of the $GABA_B$ receptor agonist baclofen can attenuate stress-induced activation of mesoprefrontal DA neurons (62,74,78). This suggests that both excitatory and inhibitory amino acid transmitter systems contribute to the selective modulation of mesoprefrontal DA neurons. The following sections review recent advances in defining the role of amino acid transmitters in the response of midbrain DA systems to stress.

GABAergic Modulation of the Mesoprefrontal DAergic Response to Stress

The GABA-containing projections to the ventral midbrain originate from the NAS, striatum, globus pallidus, and ventral pallidum (79–82). In addition, GABA interneurons are located in these regions. The GABAergic influences on VTA DA cells are mediated by both $GABA_A$ (Fig. 1) and $GABA_B$ receptors (83–85). As discussed in the previous section, benzodiazepines exert their anxiolytic effects through $GABA_A$ receptors. Benzodiazepine agonists prevent selective increases in DA utilization in the mPFC. Stress-induced activation of mesoprefrontal DA neurons is blocked by systemically or VTA-administered anxiolytic benzodiazepine agonists (58,86–91). Benzodiazepine agonists are also effective in blocking the mesoprefrontal DAergic response to conditioned fear (92). These effects are reversed by pretreatmet with the benzodiazepine antagonist RO 15-1788 suggesting that they are mediated via the benzodiazepine recognition site (58,92).

Benzodiazepine inverse agonists exert an opposite effect. Systemic administration of methyl-β-carboline-3-carboxamide (FG 7142), β-carboline-3-carboxylic acid ethylester (β-CCE), β-carboline-3-carboxylic acid methylester (β-CCM), and other benzodiazepine inverse agonists preferentially activate mesoprefrontal DA neurons (58,86,93–96). Increased DA utilization is not observed in other terminal fields of midbrain DA neurons such as the NAS or the cingulate and piriform corticies (58,95,96). In fact, DA utilization in the striatum is decreased following FG 7142 administration. Similar effects are observed in vitro, with FG 7142 selectively increasing TH activity in mPFC slice preparations and decreasing TH activity in striatal slices (62,97). The in vivo activation can be prevented by coadministration of benzodiazepine agonists or antagonists. These data suggest that mesoprefrontal DA neurons are preferentially influenced by allosteric modulation of the $GABA_A$ receptor and provide a possible mechanism by which these neurons could be selectively activated by stress.

Selectivity of Benzodiazepine Analogues for Midbrain DA Neurons

As noted in the previous section, FG 7142 increased DA utilization in the mPFC, but not other terminal areas of midbrain DA neurons (58,95,96). Similarly, in vivo microdialysis measures by our laboratory have revealed that FG 7142 administration increased extracellular levels of DA in the mPFC but not the striatum (94). Recently, however, FG 7142 administration was reported to increase extracellular levels of DA in the NAS as determined by in vivo microdialysis (98). These data are inconsistent with the earlier data derived from postmortem assays of DA metabolism. The discrepancy in the mesolimbic DAergic response to FG 7142 could be attributable to procedural differences. While postmortem tissue analyses of DA and metabolites are determined at a single time point subsequent to FG 7142 administration, in vivo microdialysis offers the assessment of extracellular levels of DA over several time points. Further, microdialysis is a more sensitive assay for detecting subtle alterations in DAergic activity. Thus, it is possible that postmortem tissue analyses failed to observe small yet significant alterations in DA utilization in the NAS following FG 7142 administration.

In addition to these procedural differences, different subdivisions of the NAS may have been assessed. Postmortem collection of the NAS was obtained using a punch that included the entire core subdivision of the NAS (58,96). While histological data of the probe placement was not provided in the microdialysis study (98), it is likely that recovery was predominantly from the shell region of the NAS, considering the listed coordinates. Given biochemical, pharmacological, behavioral, and anatomical differences between the core and shell of the NAS (99–103), consistent sampling of the NAS is an important experimental variable. Neuroanatomical data suggest that the shell subdivision of the NAS is more closely allied with mesolimbic sites such as the amygdala, while the core region is associated with the striatum. Biochemical characterization of the subdivisions of the NAS has indicated increased basal levels of DA in the shell region relative to the core (100). However, haloperidol-induced increases in DA utilization are greater in the core subdivision relative to that of the shell. By contrast, DA utilization is selectively increased in the shell region following short-duration immobilization stress (100).

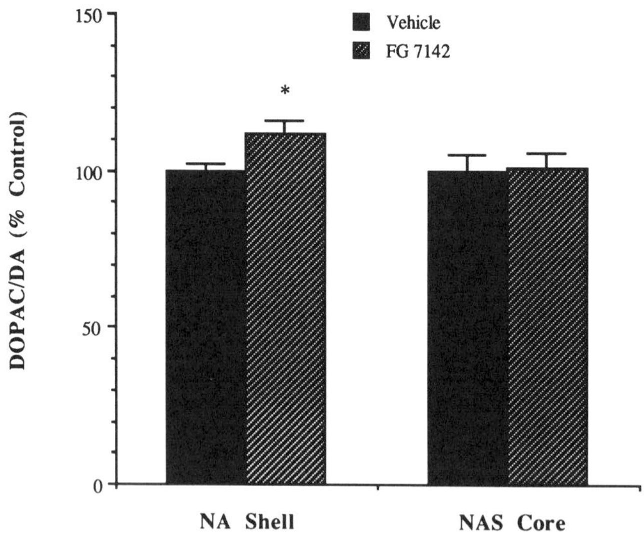

FIG. 3. DOPAC/DA ratios in the shell and core subdivisions of the NAS expressed as percent of mean of the vehicle controls (±SEM). Rats received 15 mg/kg (i.p.) FG 7142 or vehicle 30 minutes prior to sacrifice. Samples of the shell and core subdivisions of the NAS were collected according to Deutch and Cameron (79). ANOVA revealed a significant treatment effect in the shell ($F(1,9)=6.74$, $p<0.05$), but not the core ($F(1,9)=0.01$), NS), subdivision of the NAS.

Recently, we investigated whether sampling from differential subdivisions of the NAS could account for discrepant findings regarding the effects of FG 7142. Specifically, we assessed whether selective sampling from the shell subdivision of the NAS postmortem could yield data similar to that obtained by McCullough and Salamone (77) using in vivo microdialysis. As depicted in Fig. 3, FG 7142 produced a significant increase in DA utilization in the shell, but not core, subdivision of the NAS. These data reconcile earlier discrepancies and demonstrate the need to address the core and shell subdivisions of the NAS as unique and independent sites.

Site of Action of Benzodiazepine Inverse Agonists on Midbrain DA Neurons

Interestingly, FG 7142 does not increase extracellular levels of DA in the NAS when administered locally through the microdialysis probe (104). It is unclear which subdivision of the NAS the probe was sampling from specifically. However, the stereotaxic coordinates were similar to that of McCullough and Salamone (98) which reported increased DA release following systemic administration of FG 7142. It is possible that the lack of an observed effect when FG 7142 was administered through the probe was due to an insufficient concentration of drug. Alternatively, the facilitatory effects of systemically administered FG 7142 on DA release could be due primarily to somatodendritic effects. Consistent with this hypothesis, we have data indicating a somatodendritic action of benzodiazepine inverse agonists. Intra-VTA administration of the inverse agonist 6,7-dimethoxy-4-ethyl-β-carboline methylester (DMCM) produced a dose-related increase in DA turnover in the mPFC, but not the NAS (unpublished observations). These data suggest that the benzodiazepine inverse agonists may act to increase impulse flow in mesoprefrontal neurons by causing a reduction in GABAergic inhibition.

However, benzodiazepines also appear to exert local effects on DAergic terminals in the mPFC. It has been shown that FG 7142 increases TH activity in slices prepared from the mPFC, but not other terminal regions (62,97). This effect was blocked by coincubation with the benzodiazepine antagonist RO 15-1788. By contrast, incubation with GABA decreased TH activity in mPFC slice preparations (62). These data are consistent with a role for presynaptic $GABA_A$ receptors in the mPFC in the selective regulation of mesoprefrontal DA neurons.

Allosteric Modulation by Neurosteroids

As just discussed, neurosteroids formed in the brain act as allosteric modulators of the $GABA_A$ receptor and potentiate GABA-elicited Cl^- conductance (32). It has been suggested that these endogenous steroids may have anxiolytic actions. Recently, we have shown that intraventricular administration of alloTHDOC will antagonize stress-induced increases in DA metabolism in the mPFC at doses that do not induce sedation (105). The observation that both brain and plasma levels of allopregnanolone and alloTHDOC increase rapidly in the brain after stress (four- to twentyfold within less than 5 minutes) suggests that these neurosteroids may have a physiological role in stress and anxiety (34). Thus, neurosteroid modulation of $GABA_A$ receptors may provide another important adaptive response of the organism to stress. In addition, conditions that could lead to large increases in neuroactive steroid levels (e.g., puberty, pregnancy, or menstrual cycle) could alter neurochemical and behavioral adaptations to stress.

EAA Modulation of the Mesoprefrontal DAergic Response to Stress

The EAA neurotransmitters have been implicated in the selective modulation of mesoprefrontal DA neurons and control of their firing patterns. Midbrain DA neurons fire in two different patterns: 1) an irregular mode characterized by spike frequencies of 2–9 Hz, and 2) a bursting mode in which 3–7 spikes occur in groups with short interspike intervals and instantaneous frequencies of 10–20 Hz (71,106–108). In anesthetized animals, bursting of DAergic neurons correlates with dramatic increases in DA release in terminal regions (109,110). We have shown that electrical stimulation of the median forebrain bundle in a physiological bursting pattern, as contrasted to a tonic mode of the same number of pulses, increases release of both DA and neurotensin in the mPFC (111). Other studies have demonstrated that pressure injection of NMDA into the VTA results in burst firing in presumptive DAergic neurons (110) suggesting a role for NMDA receptors in burst firing-induced DA release.

Stimulation of the prefrontal and anterior cingulate cortices also elicits an induction of burst firing in VTA DA cells (112). Evidence suggests that corticofugal projections to the VTA utilize EAA neurotransmitters (113,114) and a direct monosynaptic input from mPFC efferents to DAergic neurons in the VTA has been demonstrated (115). Recent studies have suggested that this mPFC projection to the VTA may be physiologically relevant; glutamate injected into the mPFC selectively increased burst firing of DA neurons in the VTA and enhanced release of DA in the NAS (116). These data are complemented by observations that restraint stress preferentially increases extracellular levels of glutamate and aspartate in the mPFC (41).

The administration of glutamate into the VTA has been shown to increase DA release in both the mPFC and NAS terminal regions (77). However, intra-VTA infusions of NMDA selectively increase DA utilization in the mPFC (77). Similarly, VTA administration of the NMDA antagonist (±)-3-(2-carboxypiperazine-4-yl)propyl-1-phosphonic acid (CPP) blocks stress-induced increases in DA metabolism in the mPFC. By contrast, the mesoaccumbal DA system appears to be preferentially influenced by kainate receptors as intra-VTA kainate infusions increases DA utilization in the NAS, but not the mPFC (77). These data allude to selective regulation of midbrain DA neurons by EAA afferents to the VTA via EAA receptor subtypes and provide yet another possible mechanism by which mesoprefrontal DA neurons can be selectively activated by mild stress.

EAA Modulation of VTA DA Cell Firing

As discussed in the previous section, it is likely that glutamate administered into the mPFC increases burst firing of DA cells in the VTA (116) via EAA corticofugal projections (113, 114). While glutamate is active at all EAA receptor subtypes, aspartate is more selective for the NMDA receptor. Activation of midbrain DA neurons can be induced by iontophoretic application of either NMDA or non-NMDA receptor selective agonists into the VTA (110,117,118). The relative potencies for selective EAA receptor agonists on the activity of VTA DA neurons is AMPA > kainate > NMDA (118). It should be noted, however, that the endogenous EAA transmitter glutamate has a higher affinity for NMDA relative to non-NMDA receptor subtypes. This may account for data indicating that glutamate-induced excitation of VTA DA neurons is preferentially mediated by NMDA receptors. Wang and French (119) report that the excitatory effects of low concentrations of glutamate on VTA DA cell firing in vitro is completely blocked by coperfusion with competitive (CGS 19755) and noncompetitive (PCP) NMDA antagonists, but not the non-NMDA selective antagonist 2,3-dihydroxy-6-nitro-7-sulfamoyl-benzo(F) quinoxaline (NBQX). Non-NMDA receptors appear to be activated at higher concentrations of glutamate, as a combination of CGS 19755 and NBQX coperfusion was required to completely block the effects of glutamate when applied at concentrations of 300 μM or greater (119).

Burst firing was not induced by application of glutamate or selective EAA receptor agonists in these in vitro studies. Other studies have also failed to observe EAA-induced burst firing in vitro (120,121). Only Johnson et al. (122) have reported NMDA-induced burst firing in ventral midbrain slice preparations. However, these researchers utilized longer incubation periods of NMDA and added apamine, a bee venom neurotoxin with potent inhibitory effects on a Ca^{2+}-activated K^{+} channels, to the incubation medium.

Spontaneous burst activity of VTA DA neurons in vivo appears to be selectively modulated by NMDA receptors. Iontophoretic or i.c.v. administration of the nonspecific EAA receptor antagonist kynurenate has been shown to suppress spontaneous burst firing activity in vivo (123,124). Recently, Chergui et al. (125) investigated the subtypes of EAA receptors involved in spontaneous burst firing of midbrain DA neurons in vivo. Iontophoretic application of the NMDA selective antagonist (±) 2-amino, 5-phosphonopentanoic acid (AP5), but not the non-NMDA antagonist 6-cyano, 7-nitroquinoxaline-2,3-dione (CNQX), reduced the occurrence of burst firing and markedly regularized the firing pattern of these cells. While application of NMDA, kainate, and quisqualate stimulated the discharge rate of VTA DA cells, only NMDA administration elicited burst firing (110,125). These data suggest that EAA-induced burst firing of midbrain DA cells is selectively induced by activation of the NMDA receptor subtype.

Midbrain DA cells are often referred to and studied as if they were a homogeneous group of cells, despite the fact that they are quite heterogeneous. It is important, however, to be aware of and identify differences between subpopulations of DA cells in the VTA. Murase et al. (126) recently reported that dizocilpine had differential actions on the firing pattern of DA cells in the nucleus paranigralis (NP) and nucleus parabrachialis pigmentosus (PBP) subdivisions of the VTA. The DA cells located in the PBP were unique in exhibiting a significant reduction in burst firing activity. While there is limited overlap, the NP and PBP DA cells of the VTA appear to innervate different forebrain structures. The DA cells in the NP subdivision predominantly project to the areas involved in motor function, while PBP afferents innervate the mPFC (127–131). These data are consistent with initial evidence for a unique modulation of mesoprefrontal DA neurons by NMDA receptors accumulated by Kalivas and coworkers (77) as just discussed. Thus, it is possible that EAA afferents may have a role in the selective activation of mesoprefrontal DA neurons by mild stress via NMDA receptors.

Consistent with this hypothesis, we have recently observed a selective blockade of the mesoprefrontal DAergic response to immobilization stress in rats pretreated with NMDA receptor-selective antagonists. Systemic or VTA infusions of the glycine site NMDA receptor antagonist (partial agonist) (+)−HA 966 prevented stress-induced increased DA utilization in the mPFC, but not the NAS (Fig. 4). A similar selective blockade of stress-induced activation of the mesoprefrontal DA system was observed in subjects pretreated with low doses of the NMDA receptor antagonist dizocilpine, which by itself had no effect on basal DA utilization (132). By contrast, the negative enantiomer of HA 966 failed to block increased DA utilization in the mPFC following immobilization stress.

We have also found (+)−HA 966 pretreatment to selectively block activation of mesoprefrontal DA neurons produced by conditioned fear (133). As depicted in Fig. 5, presentation of a tone previously paired with moderate levels of footshock resulted in an increase in DA utilization in both the mPFC and NAS. Pretreatment with (+)−HA 966 blocked conditioned fear-induced increases in DA utilization in the mPFC, but did not attenuate the response of mesoaccumbal DA neurons.

We have additional data suggesting that pretreatment with (+)−HA 966 blocks the selective increase in DA utilization in the mPFC produced by the benzodiazepine inverse agonist FG 7142. As depicted in Fig. 6, administration of FG 7142 increased DA utilization in the mPFC, but not the NAS (samples contained both core and shell subdivisions). Pretreatment with (+)−HA 966 15 minutes prior to FG 7142 administration completely blocked the increase in DA utilization in the mPFC, with no significant effects in the NAS. These data suggest that NMDA receptors are involved in the selective activation of mesoprefrontal DA neurons induced by restraint stress, conditioned fear, and treatment with anxiogenic β-carbolines.

EAA Modulation of DA Release

The previous section showed that restraint stress has increased extracellular glutamate and aspartate levels in the mPFC and NAS (41). The

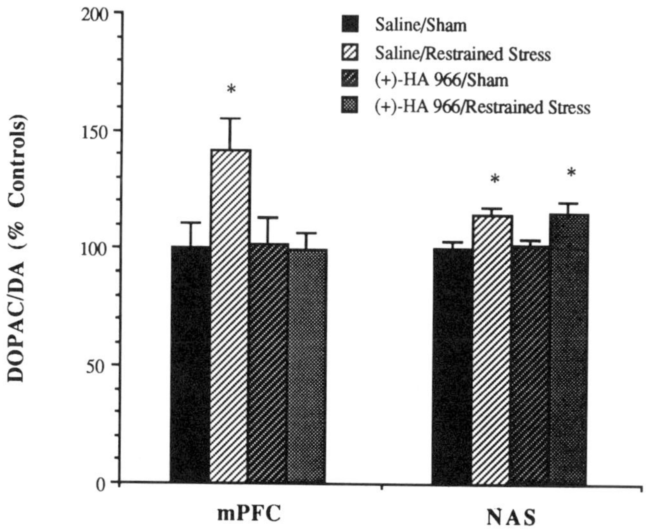

FIG. 4. DOPAC/DA ratios in the mPFC and NAS following 30 minutes of restraint stress or sham expressed as percent of mean of the Saline/Sham controls (±SEM). Rats were pretreated with (+)−HA 966 (15 mg/kg, i.p.) or saline vehicle 20 minutes prior to restraint stress or sham. ANOVA revealed significant interactions for both the mPFC ($F(3,31)=3.23$, $p<0.05$) and NAS ($F(3,32)=5.22$, $p<0.01$). Significant mean differences ($p<0.5$) relative to the Saline/Sham controls are indicated with an asterisk as determined by Duncan's multiple range post-hoc analyses. Modified from Morrow et al. (132).

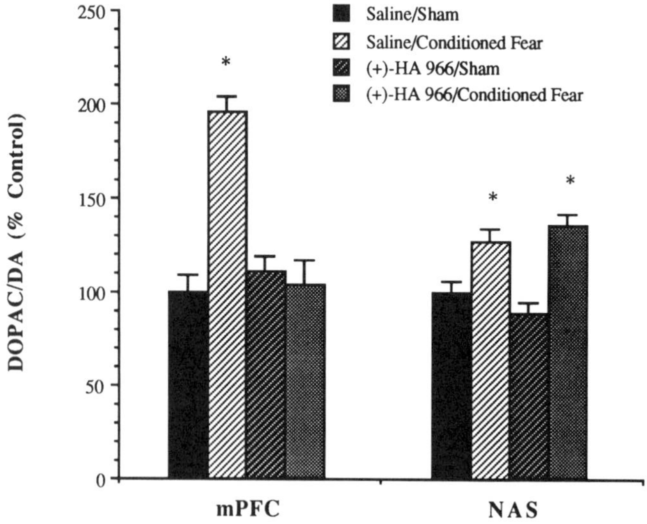

FIG. 5. DOPAC/DA ratios in the mPFC and NAS immediately following conditioned fear stress or sham expressed as percent of mean of the Saline/Sham controls (±SEM). Previously, subjects in the Saline/Conditioned Fear and (+)−HA 966/Conditioned Fear group received foot shocks (0.4 mA) paired with the presentation of a tone. Saline/Sham controls were subjected to the same training procedures without receiving foot shock. The next day rats were pretreated with (+)−HA 966 (15 mg/kg, i.p.) or saline vehicle 15 minutes prior to the presentation of the tone without foot shock. ANOVA revealed significant interactions for both the mPFC ($F(2,39)=40.53$, $p<0.0001$) and NAS ($F(2,20)=5.01$, $p<0.05$). Significant mean differences ($p<0.5$) relative to the Saline/Sham controls are indicated with an asterisk as determined by Student-Newman-Keuls post-hoc analyses. Modified from Goldstein et al. (133).

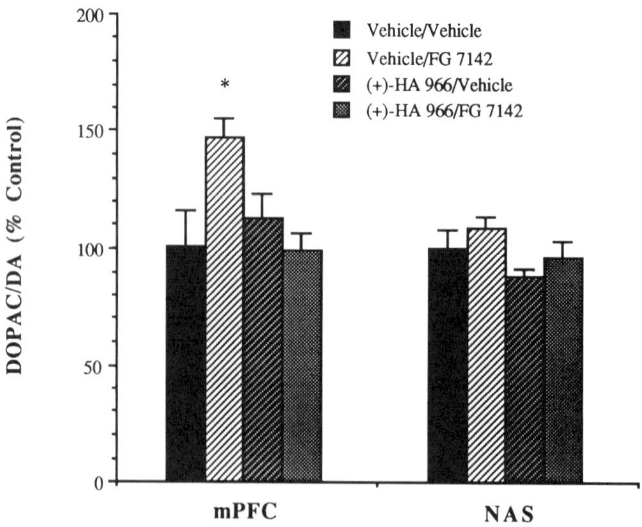

FIG. 6. DOPAC/DA ratios in the mPFC and NAS expressed as percent of mean of the Vehicle/Vehicle controls (±SEM). Rats received 15 mg/kg (i.p.) (+)−HA 966 or vehicle 15 minutes prior to FG 7142 (15 mg/kg, i.p.) or vehicle administration. Thirty minutes subsequent to the FG 7142 or vehicle injection, the rats were sacrificed. ANOVA revealed a significant interaction for the mPFC ($F(3,11)=4.27$, $p<0.05$), but not the NAS ($F(3,11)=0.22$), NS). Significant mean differences ($p<0.5$) relative to the Vehicle/Vehicle controls are indicated with an asterisk as determined by Fishers protected least significant difference (PLSD) post-hoc analyses.

EAA transmitters have been shown to modulate DA release. Local perfusions with the AMPA/kainate receptor antagonist CNQX through the microdialysis probe effectively blocked the increase in extracellular levels of DA in the mPFC produced by tail-pinch stress (134). This effect was not observed upon local perfusion of AP5, suggesting that stress-induced DA release in the mPFC is mediated by non-NMDA receptors. It should be noted, however, that in vitro glutamate-stimulated ^{3}H-DA release in mPFC slice preparations was effectively blocked by the addition of the NMDA antagonist AP5 to the superfusion (135). In addition, NMDA was more effective than kainate in stimulating ^{3}H-DA release. While these data clearly indicate a neuromodulatory role for EAA transmitters in the release of DA in the mPFC, the relative contributions of NMDA and non-NMDA receptors in mediating this effect are not clear.

Stress-induced increases in extracellular EAA transmitter levels have also been observed in the NAS and striatum (41). Excitatory corticofugal pathways to the basal ganglia are proposed to contain the EAA transmitters glutamate and aspartate (136–143). While histological studies in the basal ganglia note a lack of axo-axonal synapses formed by corticofugal afferents (144, 145), corticofugal neurons have been shown to come into direct apposition to converging mesolimbic and nigrostriatal neurons (115,145). This proximal location of terminals may be sufficient to allow for transsynaptic modulation of DAergic activity. Glutamate has been shown to facilitate the release of DA in the basal ganglia (146,147). It has been proposed that corticofugal projections tonically elevate DA release in the basal ganglia; this effect is directly modulated by the mesoprefrontal DA system. In support of this model, 6-hydroxydopamine (6-OHDA) lesions of mesoprefrontal DA neurons have been reported to increase DA turnover in the NAS and striatum (148–151). However, these results have been difficult to replicate in nonchallenged systems (152–154). Initial studies using in vivo microdialysis failed to observe decreases in recovered DA from the striatum after administration of NMDA and non-NMDA antagonists through the probe (155–157). In fact, high concentrations of selective EAA antagonists increase, rather than decrease, extracellular levels of DA. However, it is possible that a tonic excitatory influence on DA release may affect only a small fraction of DA terminals. Thus, a simple blockade of EAA receptors would not make a great enough contribution to the extracellular fluid to be detected by microdialysis in an unchallenged system. Consistent with this hypothesis, both AP5 and CNQX administered through the microdialysis probe have been found to decrease extracellular DA levels in the striatum in the presence of DA uptake blockers (158). Together, these data suggest limited tonic excitation of DA release by EAA afferents.

It is likely that under certain conditions mesoprefrontal DA neurons could exert an inhibitory influence on the activation of EAA corticofugal projections to the basal ganglia. Supporting this, we have found that stress-induced and pharmacologically induced increases in DA utilization in the NAS are augmented following 6-OHDA lesions of mesoprefrontal DA neurons (152, 154). However, it remains unclear as to whether this effect is the direct result of increased DA release via reduced inhibition of excitatory cortico-accumbal projections or increased impulse flow resulting from reduced inhibition of excitatory corticofugal projections to the VTA.

SUMMARY AND CONCLUSIONS

Acute stress has several effects on amino acid transmitter systems. Stress-inducing procedures reduce the number of GABA binding sites and increase benzodiazepine binding sites in the cortex. This is accompanied by increased levels of neurosteroids in the cortex, which act as potent allosteric modulators of the $GABA_A$ receptor. The enhancement of GABAergic transmission via allosteric modulation of the $GABA_A$ receptor that occurs during stress could serve as a possible coping response to the stressor.

Exposure to stress also produces an increase in extracellular levels of EAA transmitters in the cortex and other regions. These increases appear to be the result of impulse-dependent release from neuronal pools. Acute stress also induces selective increases in AMPA and MK 801 binding and increases the endogenous levels of glycine and Zn^{2+}, which can function as allosteric modulators of the NMDA receptor.

Stress-induced alterations in amino acid transmitter functioning could play a role in the enhanced responsivity of mesoprefrontal DA neurons to stress. Mesoprefrontal DA neurons have several singular properties that could contribute to their unique response to stress. Most notably, these DA neurons lack or have a decreased number of functional impulse- and synthesis-regulating autoreceptors. The absence of this important autoregulatory control makes the DA neurons more vulnerable to the modulatory effects of afferent regulatory inputs to the VTA, causing these neurons to exhibit an increase in burst firing and enhanced transmitter turnover during basal conditions.

Numerous data just summarized demonstrate that EAA input to the VTA is at least partly responsible for converting pacemaker-like firing in DA cells into burst-firing patterns. Burst firing of midbrain DA cells appears to be selectively mediated by NMDA receptors; AP5, but not CNQX, decreased burst firing in midbrain DA cells (125). In addition, (+)−HA 966 has been shown to suppress burst firing in DA cells in the zona compacta of the substantia nigra (159). Thus, it is possible that (+)−HA 966 is exerting its antagonistic effects on stress-induced DA turnover in the mPFC by a similar "regularization" of mesoprefrontal DA cell activity patterns.

Firing in a bursting mode results in an increase in the amount of DA released relative to metered single-spike firing rates (109–111). Neurochemical studies have also demonstrated that specific EAA receptors appear to be involved in the selective metabolic activation of mesoprefrontal DA neurons (77). As depicted by the model in Fig. 7, stimulation of NMDA receptors activates the DAergic projection to the mPFC, while stimulation of non-NMDA receptors activates mesoaccumbens and nigrostriatal neurons. Thus, NMDA receptors located on DA neurons in the VTA may have a role in the preferential activation of mesoprefrontal DA neurons by stress.

The unique role of NMDA receptors in modulating the mesoprefrontal DAergic response to stress suggests a potential target site for the development of agents that might have some utility in clinical treatment of stress-related disorders such as post-traumatic stress disorder (PTSD).

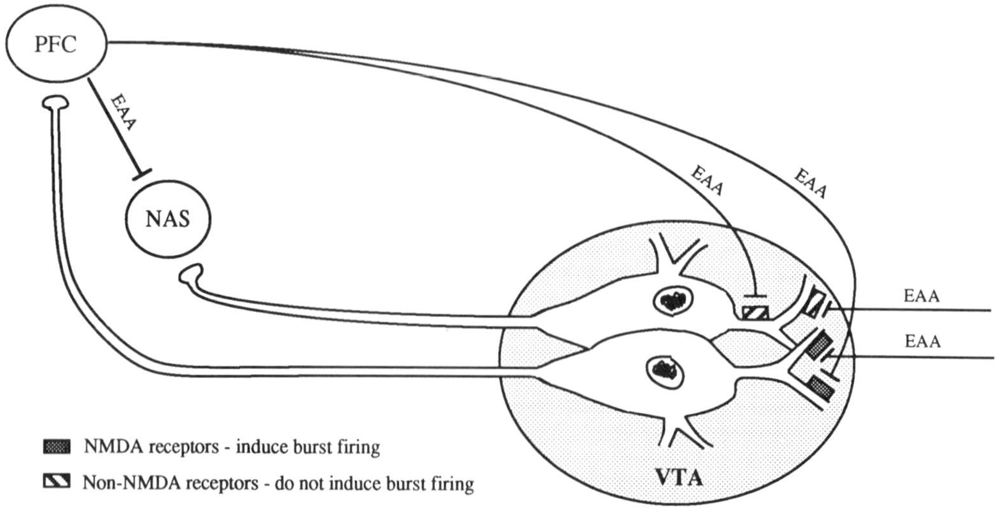

FIG. 7. Model depicting differential modulation of mesoprefrontal and mesoaccumbal DA neurons by EAA receptor subtypes.

Supporting this possibility, systemic administration of (+)−HA 966 blocks the mesoprefrontal DAergic response to conditioned fear stress (Fig. 5) and attenuates behavioral correlates of this stress paradigm (i.e., freezing) (133). The animal model of conditioned fear has been suggested to directly mimic the same psychophysiological processes that are etiologically relevant in the development of PTSD (160). A number of investigators have extensively commented on the correlation between aversive conditioning phenomena and the development of anxiety states (161). Recent studies utilizing the conditioned fear procedure in the rat suggest that this animal model shares many features with anxiety states in PTSD. Both paradigms elicit a diverse set of behavioral, neurochemical, and neuroendocrine reactions. In PTSD and the animal model of conditioned fear, anxiety states are generated quickly under stressful circumstances and can be subsequently elicited by associated discrete stimuli in the absence of pain. In addition, anxiety states associated with PTSD and animal models of conditioned fear are relatively long lasting. We have observed that conditioned psychological stress robustly activates a coordinated stress response in the rat that was observed one day, as well as one month, after training indicating the temporal stability of the conditioned response (unpublished observations). Thus, the behavioral, neuroendocrine, and neurochemical correlates of the conditioned fear stress response were shown to be stable over a time period roughly equal to 3 years of human life. In this respect, the conditioned fear paradigm in the rat may be useful for understanding the underlying neurobiology of PTSD and developing effective pharmacologic and behavioral treatment strategies for this stress disorder.

Mesoprefrontal DA neurons are also uniquely modulated by GABAergic systems. As depicted by the model in Fig. 8, systemic administration of the anxiogenic β-carboline FG 7142 increases DA utilization in the mPFC, but not the cingulate and entorhinal cortices. The administration of FG 7142 also appears to increase DA utilization in the shell, but not the core, subdivision of the NAS. Given the fact that $GABA_A$ receptors have been identified on DA neurons in the VTA (36,40), receptors located on midbrain DA neurons projecting to the mPFC and the shell subdivision of the NAS could be comprised of an enriched population of $GABA_A$ receptor subunits that are uniquely receptive to benzodiazepine inverse agonists.

The appreciation that endogenous neurosteroids can be formed rapidly and quickly accumulate in the brain following exposure to stress suggests that these endogenous agents may provide an important adaptive response of the organism to stress via an enhancement of GABAergic transmission. As just discussed, increased DA utilization in the mPFC occurs rapidly after stress but appears to develop tolerance and dissipates over time even during the continued presence of the stressor. Endogenous neurosteroids could have a role in the rapid dissipation of the DAergic response to stress. Further, anomalies in the functioning of endogenous neurosteroids could lead to deficits in the ability of the organism to adapt and cope with stressful stimuli, as typified in patients with PTSD.

In sum, both inhibitory and excitatory amino acid transmitters appear to have a role in the unique response of the mesoprefrontal DA system to stress. These opposing influences appear to functionally interact with regard to regulation of mesoprefrontal DA neurons. This is illustrated by data indicating a blockade of FG 7142-induced increases in DA utilization in the mPFC following (+)−HA 966 pretreatment (Fig. 6). Thus, it is likely that these amino acid transmitter systems function in concert to selectively modulate mesoprefrontal DA neurons.

It should be noted that, in addition to amino acid transmitters, several other systems have been suggested to have a role in modulating stress-induced activation of mesoprefrontal DA neurons. As just mentioned, 5-HT, opioids, and substance P also appear to differentially modulate mesoaccumbal and mesoprefrontal DA neurons. In addition, some mesoprefrontal DA neurons lack synthesis- and impulse-regulating autoreceptors. Thus, the selective effects of amino acid receptors are a subset of several dy-

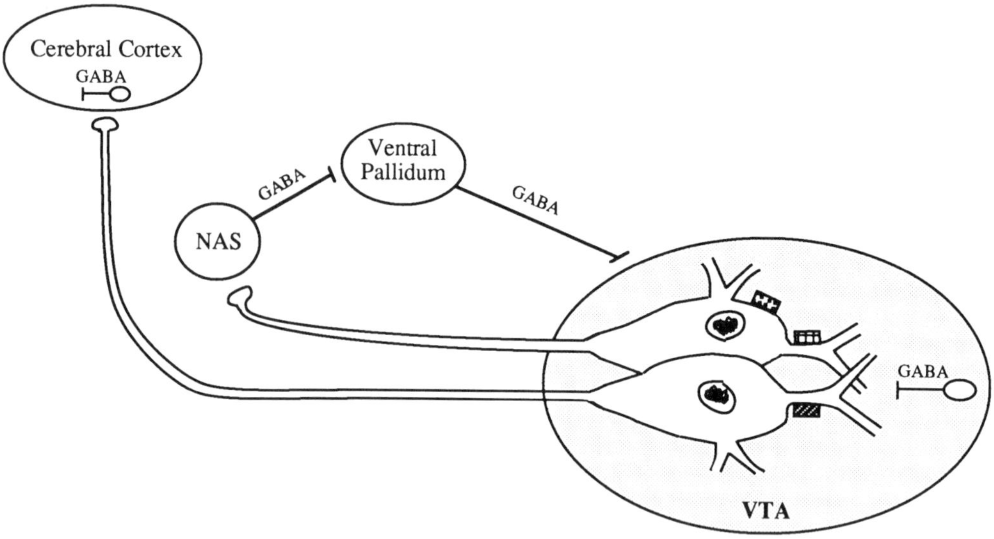

Benzodiazepine/GABA$_A$ receptor complex activated by benzodiazepine inverse agonists selectively modulating midbrain DA neurons projecting to the mPFC, but not the entorhinal or cingulate corticies.

Benzodiazepine/GABA$_A$ receptor complex activated by benzodiazepine inverse agonists located on mesoaccumbal DA neurons projecting to the shell subdivision of the nucleus accumbens.

Benzodiazepine/GABA$_A$ receptor complex not activated by benzodiazepine inverse agonists located on mesoaccumbal DA neurons projecting to the core subdivision of the nucleus accumbens.

FIG. 8. Model depicting differential modulation of ascending DA neurons in the VTA by GABA$_A$ receptors. Neurons depicted within the cerebral cortex and VTA regions represent GABAergic interneurons.

namic factors that govern the unique response of mesoprefrontal DA neurons to stress.

ACKNOWLEDGMENTS

The authors wish to thank Bret Morrow and Lee Goldstein for their work contributing to this review, and John Elsworth and Kati Piros for their technical assistance. This work was supported by U.S. Public Health Service (MH 14092 to RHR).

REFERENCES

1. Costa E, Guidotti A, Mao CC. Involvement of GABA in the action of benzodiazepine—studies on rat cerebellum. *Psychopharmacol Bull* 1975;11(4):59–60.
2. Haefely W, Kulcsár A, Möhler H. Possible involvement of GABA in the central actions of benzodiazepines. *Psychopharmacol Bull* 1975;11(4):58–59.
3. Macdonald RL, Olsen RW. GABA$_A$ receptor channels. *Annu Rev of Neurosci* 1994;17:569–602.
4. Sieghart W. GABA$_A$ receptors: ligand-gated Cl^- ion channels modulated by multiple drug-binding sites. *Trends Pharmacol Sci* 1992;13:446–450.
5. Study RE, Barker JL. Diazepam and (−)-pentobarbital: fluctuation analysis reveals different mechanisms for potentiation of γ-aminobutyric acid responses in cultured central neurons. *Proc Natl Acad Sci USA* 1981; 11:7180–7184.
6. Majewska MD, Harrison NL, Schwartz RD, Barker JL, Paul SM. Steroid hormone metabolites are barbiturate-like modulators of the GABA receptor. *Science* 1986; 232:1004–1007.
7. Turner DM, Ransom RW, Yang JS-J, Olsen RW. Steroid anesthetics and naturally occurring analogs modulate the γ-aminobutyric acid receptor complex at a site distinct from barbiturates. *J Pharmacol Exp Ther* 1989; 248:960–966.
8. Doble A, Martin IL. Multiple benzodiazepine receptors: no reason for anxiety. *Trends Pharmacol Sci* 1992; 13:76–81.
9. Hales TG, Lambert JJ. Modulation of GABAA and glycine receptors by chlormethiazole. *Eur J Pharmacol* 1992;210:239–246.
10. Akaike N, Hattori K, Oomura Y, Carpenter D. Bicuculline and picrotoxin block gamma-aminobutyric acid-

gated C1-conductance by different mechanisms. *Experientia* 1985;41:70–71.

11. Barker JL, McBurney RN, Mathers DA. Convulsant-induced depression of amino acid responses in cultured mouse spinal neurones studied under voltage clamp. *Br J Pharmacol* 1983;80:619–629.
12. Costa E. The Di Maseio lectures. The allosteric modulation of $GABA_A$ receptors. Seventeen years of research. *Neuropsychopharmacology* 1991;4:225–235.
13. Olsen RW, Tobin AJ. Molecular biology of $GABA_A$ receptors. *FASEB J* 1990;4:1469–1480.
14. Mohler H. GABAergic synaptic transmission. Regulation by drugs. *Arzneimittelforschung* 1992,42.211–214.
15. Pritchett DB, Sontheimer H, Shivers BD, Ymer S, Kettenmann H, Schofield PR, Seeburg PH. Importance of a novel $GABA_A$ receptor subunit for benzodiazepine pharmacology. *Nature* 1989;338:582–585.
16. Ymer S, Draghun A, Wisden W, Werner P, Keinanen K, Schofield PR, Sprengel R, Pritchett DB, Seeburg PH. Structural and functional characterization of the $\gamma 1$ subunit of $GABA_A$/benzodiazepine receptors. *EMBO J* 1990;9:3261–3267.
17. Wafford KA, Bain CJ, Whiting PJ, Kemp JA. Functional comparison of the role of gamma subunits in recombinant human gamma-aminobutyric $acid_A$/benzodiazepine receptors. *Mol Pharmacol* 1993;44: 437–442.
18. Knoflach F, Backus KH, Giller T, Malherbe P, Pflimlin JP, Mohler H, Trube G. Pharmacology and electrophysiological properties of recombinant $GABA_A$ receptors comprising the $\alpha 3$, $\beta 1$ and $\gamma 2$ subunits. *Eur J Neurosc* 1992;4:1–9.
19. Gao B, Fritshy JM, Benke D, Mohler H. Neuron-specific expression of $GABA_A$-receptor subtypes: differential association of the $\alpha 1$- and $\alpha 3$-subunits with serotonergic and GABAergic neurons. *Neuroscience* 1993; 54:881–892.
20. Biggio G, Corda MG, Concas A, Demontis G, Rossetti Z, Gessa GL. Rapid changes in GABA binding induced by stress in different areas of the rat brain. *Brain Res* 1981;229:363–369.
21. Concas A, Corda MG, Biggio G. Involvement of benzodiazepine recognition sites in the foot shock-induced decrease of low affinity GABA receptors in the rat cerebral cortex. *Brain Res* 1985;341:50–56.
22. Zhang M-Y, Hashimoto T, Kuriyama K. Stress induced alterations in striatal $GABA_A$ receptorcomplex. *Japan J Pharmacol* 1990;53:121–124.
23. Biggio G, Concas A, Mele S, Corda MG. Changes in GABAergic transmission induced by stress, anxiogenic and anxiolytic β-carbolines. *Brain Res Bull* 1987;19: 301–308.
24. Biggio G, Concas A, Serra M, Salis M, Corda MG, Nurchi V, Crisponi C, Gessa GL. Stress and β-carbolines decrease the density of low affinity GABA binding sites: an effect reversed by diazepam. *Brain Res* 1984;305:13–18.
25. Biggio G, Corda MG, Demontis G, Concas A, Gessa GL. Sudden decrease in cerebellar GABA binding induced by stress. *Pharmacol Res Comm* 1980;12(5): 489–493.
26. Kang I, Thompson ML, Heller J, Miller LG. Persistent elevation in $GABA_A$ receptor subunit mRNAs following social stress. *Brain Res Bull* 1991;26:809–812.
27. Motohashi N, Okamoto Y, Osada M, Yamawaki S. Acute swim stress increases benzodiazepine receptors, but not $GABA_A$ or $GABA_B$ receptors, in the rat cerebral cortex. *Neurochem Int* 1993;23(4):327–330.
28. Rägo L, Kiivet R-A, Harro J, Pold M. Central- and peripheral-type benzodiazepine receptors: similar regulation by stress and GABA receptor agonists. *Pharmacol Biochem Behav* 1989;32:879–883.
29. Otero Losada ME. Changes in central GABAergic function following acute and repeated stress. *Br J Pharmacol* 1988;93:483–490.
30. Acosta GB, Otero Losada ME, Rubio MC. Area-dependent changes in GABAergic function after acute and chronic cold stress. *Neurosci Lett* 1993;154:175–178.
31. Maroulakou IG, Stylianopoulou F. The effects of adrenalectomy and thermal stress on glutamic acid decarboxylase activity in different regions of the rat brain. *Neurochem Res* 1991;16:1265–1268.
32. Paul SM, Purdy RH. Neuroactive steroids. *FASEB J* 1992;6:2311–2322.
33. Crawley JN, Glowa JR, Majewska MD, Paul SM. Anxiolytic activity of an endogenous adrenal steroid. *Brain Res* 1986;398:382–385.
34. Purdy RH, Morrow AL, Moore PH Jr., Paul SM. Stress-induced elevations of gamma-aminobutyric acid type A receptor-active steroids in the rat brain. *Proc Natl Acad Sci USA* 1991;88:4553–4557.
35. Deutsch SI, Park CH, Hitri A. Allosteric effects of a GABA receptor-active steroid are altered by stress. *Pharmacology Biochemistry and Behavior* 1994;47: 913–917.
36. Havoundjian H, Paul SM, Skolnick P. Rapid, stress-induced modification of the benzodiazepine receptor-coupled chloride ionophore. *Brain Res* 1986;375: 401–406.
37. Schwartz RD, Wess, MJ, Labarca R, Skolnick P, Paul SM. Acute stress enhances the activity of the GABA receptor-gated chloride ion channel in brain. *Brain Res* 1987;411:151–155.
38. Trullas R, Havoundjian H, Zamir N, Paul S, Skolnick P. Environmentally induced modification of the benzodiazepine/GABA receptor coupled chloride ionophore. *Psychopharmacology* 1987;91:384–390.
39. Itoh T, Saito T, Fujimura M, Watanabe S, Saito K. Restraint stress-induced changes in endogenous zinc release from the rat hippocampus. *Brain Res* 1993;618: 318–322.
40. Lowy MT, Gault L, Yamamoto BK. Adrenalectomy attenuates stress-induced elevations in extracellular glutamate concentrations in the hippocampus. *J Neurochem* 1993;61:1957–1960.
41. Moghaddam B. Stress preferentially increases extraneuronal levels of excitatory amino acids in the prefrontal cortex: comparison to hippocampus and basal ganglia. *J Neurochem* 1993;60:1650–1657.
42. Gilad GM, Gilad VH, Wyatt RJ, Tizabi Y. Region-selective stress-induced increase of glutamate uptake and release in rat forebrain. *Brain Res* 1990;525: 335–338.
43. Palkovits M, Láng T, Patthy A, Elekes I. Distribution and stress-induced increase of glutamate and aspartate levels in discrete brain nuclei of rats. *Brain Res* 1986;373:252–257.
44. Moghaddam B, Bolinao ML, Stein-Behrens B, Sapolsky R. Glucocorticoids mediate the stress-induced ex-

tracellular accumulation of glutamate. *Brain Res* 1994; in press.
45. Schasfoort EMC, DeBruin LA, Korf J. Mild stress stimulates rat hippocampal glucose utilization transiently via NMDA receptors, as assessed by lactography. *Brain Res* 1988;475:58–63.
46. MacDonald JF, Miljkovic Z, Pennefather P. Use-dependent block of excitatory amino acid currents in cultured neurons by ketamine. *J Neurophysiol* 1987; 58:251–266.
47. Reynolds IJ, Miller RJ. Multiple sites for the regulation of the N-methyl-D-aspartate receptor. *Mol Pharmacol* 1988;33:581–584.
48. Wong EH, Kemp JA, Priestley T, Knight AR, Woodruff GN, Iversen LL. The anticonvulsant MK-801 is a potent N-methyl-D-aspartate antagonist. *Proc Natl Acad Sci USA* 1986;83:7104–7108.
49. Nowak L, Bregestovski P, Ascher P, Herbet A, Prochiantz A. Magnesium gates glutamate-activated channels in mouse central neurones. *Nature* 1984;307:462–465.
50. Legendre P, Westbrook GL. Noncompetitive inhibition of gamma-aminobutyric acidA channels by Zn. *Mol Pharmacol* 1991;39:267–274.
51. Johnson JW, Ascher P. Glycine potentiates the NMDA response in cultured mouse brain neurons. *Nature* 1987;325:529–531.
52. Young AB, Fagg GE. Excitatory amino acid receptors in the brain: membrane binding and receptor autoradiographic approaches. *Trends Pharmacol Sci* 1991;11: 126–133.
53. Tocco G, Shors TJ, Baudry M, Thompson RF. Selective increase of AMPA binding to the AMPA/quisqualate receptor in the hippocampus in response to acute stress. *Brain Res* 1991;559:168–171.
54. Krugers HJ, Koolhaas JM, Bohus B, Korf J. A single social stress-experience alters glutamate receptor-binding in rat hippocampal CA3 area. *Neurosci Lett* 1993;154:73–77.
55. Elekes I, Patthy A, Láng T, Palkovits M. Concentrations of GABA and glycine in discrete brain nuclei. *Neuropharmacology* 1986;25:703–709.
56. Akinci MK, Johnston GAR. Sex differences in acute swim stress-induced changes in the binding of MK-801 to the NMDA subclass of glutamate receptors in mouse forebrain. *J Neurochem* 1993;61:2290–2293.
57. Dunn AJ. Stress-related activation of cerebral dopaminergic systems. *Ann N Y Acad Sci* 1988;537: 188–205.
58. Roth RH, Tam S-Y, Ida Y, Yang J-X, Deutch AY. Stress and the mesocorticolimbic dopamine systems. *Ann N Y Acad Sci* 1988;537:138–147.
59. Deutch AY, Tam S-Y, Roth RH. Footshock and conditioned stress increase 3,4-dehydroxyphenylacetic acid (DOPAC) in the ventral tegmental area but not substantia nigra. *Brain Res* 1985;333:143–146.
60. Herman JP, Guillonneau D, Dantzer R, Scatton B, Semerdjian-Rouquier L, Le Moal M. Differential effects of inescapable footshocks and of stimuli previously paired with inescapable footshocks on dopamine turnover in cortical and limbic areas of the rat. *Life Sci* 1982;30:2207–2214.
61. D'Angio M, Serrano A, Driscoll P, Scatton B. Stressful environmental stimuli increase extracellular DOPAC levels in the prefrontal cortex of hypoemotional (Roman high-avoidance) but not hyperemotional (Roman low-avoidance) rats. An in vivo voltammetric study. *Brain Res* 1988;451:237–247.
62. Deutch AY, Roth RH. The determinants of stress-induced activation of the prefrontal cortical dopamine system. In Uylings HBM, Van Eden CG, DeBruin JPC, Corner MA, Feenstra MGP, eds. *Progress in Brain Research,* vol. 85. Cambridge: Elsevier Science Publishers BV, 1990;367–403.
63. Scatton B, D'Angio M, Driscoll P, Serrano A. An in vivo voltammetric study of the response of mesocortical and mesoaccumbens dopaminergic neurons to environmental stimuli in strains of rats with differing levels of emotionality. *Ann N Y Acad Sci* 1988;537:124–137.
64. Deutch AY. Heterogeneity of the prefrontal cortical dopamine system in responsiveness to stress. *Soc Neurosci Abstr* 1991;17:529.
65. Deutch AY, Lee MC, Gillham MH, Cameron DA, Goldstein M, Iadorola MJ. Stress selectively increases fos protein in dopamine neurons innervating the prefrontal cortex. *Cereb Cortex* 1991;1:273–292.
66. Bannon MJ, Roth RH. Mesocortical dopamine neurons. Lack of autoreceptors modulating dopamine synthesis. *Mol Pharmacol* 1981;19:270–275.
67. Bannon MJ, Roth RH. Pharmacology of mesocortical dopamine neurons. *Pharmacol Rev* 1983;35:53–68.
68. Bannon MJ, Wolf ME, Roth RH. Pharmacology of dopamine neurons innervating the prefrontal, cingulate and piriform cortices. *Eur J Pharmacol* 1983;91: 119–125.
69. Roth RH. CNS dopamine autoreceptors: distribution, pharmacology and function. *Ann N Y Acad Sci* 1984; 430:27–53.
70. Chiodo LA, Bannon MJ, Grace AA, Roth RH, Bunney BS. Evidence for the absence of impulse-regulating somatodendritic and synthesis-modulating nerve terminal autoreceptors on subpopulations of mesocortical dopamine neurons. *Neuroscience* 1984;12:1–16.
71. Gariano RF, Tepper JM, Sawyer SF, Young SJ, Groves PM. Mesocortical dopaminergic neurons. 1. Electrophysiological properties and evidence for somadendritic autoreceptors. *Brain Res Bull* 1989;22:511–516.
72. Shepard PD, German DC. A subpopulation of mesocortical dopamine neurons possesses autoreceptors. *Eur J Pharmacol* 1984;98:455–456.
73. Bannon MJ, Chiodo LA, Bunney EB, Wolf ME, Grace AA, Bunney BS, Roth RH. In vivo biochemical and electrophysiological studies on the distribution and pharmacology of dopamine autoreceptors. In Usdin E, Carlsson A, Dahlström, Engel J, eds. *Catecholamines: Neuropharmacology and Central Nervous System—Theoretical Aspects.* New York: Alan R. Liss, 1984; 25–36.
74. Roth RH, Tam S-Y. Regulatory control of midbrain dopamine neurons. In Kaufman S, ed. *Amino Acids in Health and Disease: New Perspectives.* New York: Alan R. Liss, 1987;159–178.
75. Bannon MJ, Elliot PJ, Alpert JE, Goedert M, Iverson SD, Iverson LL. Selective activation of mesocortical dopamine neurons by stress: role of endogenous substance P in stress-induced activation of mesocortical dopaminergic neurons. *Nature* 1983;306:791–792.
76. Kalivas PW, Abhold R. Enkephalin release into the ventral tegmental area in response to stress: modulation of mesocorticolimbic dopamine. *Brain Res* 1987;414: 339–348.

77. Kalivas PW, Duffy P, Barrow J. Regulation of the mesocorticolimbic dopamine system by glutamic acid receptor subtypes. *J Pharmacol Exp Ther* 1989;251: 378–387.
78. Kalivas PW, Duffy P, Eberhardt H. Modulation of A10 dopamine neurons by g-aminobutyric acid agonists. *J Pharmacol Exp Ther* 1990;253:858–866.
79. Araki M, McGeer PL, McGeer EG. Striatonigral and pallidonigral pathways studied by a combination of retrograde horseradish peroxidase tracing and a pharmacochistochemical method for g-aminobutyric acid transaminase. *Brain Res* 1985;331:17–24.
80. Fonnum F, Gottesfeld Z, Grofova I. Distribution of glutamate decarboxylase, choline acetyltransferase, and aromatic amino acid decarboxylase in the basal ganglia of normal and operated rats. Evidence for striatopallidal, striatoentopeduncular, and striatonigral GABAergic fibers. *Brain Res* 1978;143:125–138.
81. Ribak CE, Vaughn JE, Roberts E. GABAergic nerve terminals decrease in the substantia nigra following hemitransections of the striatonigral and pallidonigral pathways. *Brain Res* 1980;192:413–420.
82. Walaas I, Fonnum F. Biochemical evidence for gamma-aminobutyrate containing fibres from the nucleus accumbens to the substantia nigra and ventral tegmental area in the rat. *Neuroscience* 1980;5:63–72.
83. Grace AA, Bunney BS. The control of firing pattern in nigral dopamine neurons: burst firing. *J Neurosci* 1984;4:2877–2890.
84. Lacey MG, Mercuri NB, North RA. On the potassium conductance increase activated by $GABA_B$ and dopamine D2 receptors in rat substantia nigra neurones. *J Physiol* 1988;401:437–453.
85. Santiago M, Machado A, Cano J. In vivo release of dopamine from rat striatum, substantia nigra and prefrontal cortex: differential modulation by baclofen. *Br J Pharmacol* 1993;109:814–818.
86. Claustre Y, Rivy JP, Dennis T, Scatton B. Pharmacological studies on stress-induced increase in frontal cortical dopamine metabolism in the rat. *J Pharmacol Exp Ther* 1986;238:693–700.
87. D'Angio M, Serrano A, Rivy JP, Scatton B. Tail-pinch stress increases extracellular DOPAC levels (as measured by in vivo voltammetry) in the rat nucleus accumbens but not frontal cortex: antagonism by diazepam and zolpidem. *Brain Res* 1987;409:169–174.
88. Lavielle S, Tassin J-P, Theirry A-M, Blanc G, Hervé D, Barthelemy C, Glowinski J. Blockade by benzodiazepines of the selective high increase in dopamine turnover induced by stress in mesocortical dopaminergic neurons of the rat. *Brain Res* 1978;168:585–594.
89. Onel KK, Tam SY, Roth RH. Benzodiazepine receptor modulation of stress-induced activation of prefrontal dopamine neurons. *Soc Neurosci Abstr* 1984;10:881.
90. Reinhard JF Jr, Bannon MJ, Roth RH. Acceleration by stress of dopamine synthesis and metabolism in prefrontal cortex: antagonism by diazepam. *Naunyn Schmiedebergs Arch Pharmacol* 1982;318:374–377.
91. Fadda F, Argiolas A, Melis MR, Tissari A, Onali P, Gessa GL. Stress-induced increase in 3,4-dihydroxyphenylacetic acid (DOPAC) levels in the cerebral cortex and in n. accumbens: Reversal by diazepam. *Life Sci* 1978;23:2219–2224.
92. Kaneyuki H, Yokoo H, Tsuda A, Yoshida M, Mizuki Y, Yamada M, Tanaka M. Psychological stress increases dopamine turnover selectively in mesoprefrontal dopamine neurons of rats: reversal by diazepam. *Brain Res* 1991;557:154–161.
93. Biggio G, Concas A, Corda MG, Giorgi O, Sanna E, Serra M. GABAergic and dopaminergic transmission in the rat cerebral cortex: effect of stress, anxiolytic and anxiogenic drugs. *Pharmac Ther* 1990;48:121–142.
94. Bradberry CW, Lory JS, Roth RH. The anxiogenic beta-carboline FG 7142 selectively increases dopamine release in rat prefrontal cortex as measured by microdialysis. *J Neurochem* 1991;56:748–752.
95. Tam S-Y, Roth RH. Selective increase in dopamine metabolism in the prefrontal cortex by the anxiogenic beta-carboline FG 7142. *Biochem Pharmacol* 1985;34: 1595–1598.
96. Tam S-Y, Roth RH. Modulation of mesoprefrontal dopamine neurons by central benzodiazepine receptors: I. Pharmacological characterization. *J Pharmacol Exp Ther* 1990;252:989–996.
97. Knorr AM, Deutch AY, Roth RH. The anxiogenic β-carboline FG-7142 increases in vivo and in vitro tyrosine hydroxylation in the prefrontal cortex. *Brain Res* 1989;495:355–361.
98. McCullough LD, Salamone JD. Anxiogenic drugs beta-CCE and FG 7142 increase extracellular dopamine levels in nucleus accumbens. *Psychopharmacology* 1992;109:379–382.
99. Cools AR, de Beltrán KK, Prinssen EPM, Koshikawa N. Differential role of core and shell of the nucleus accumbens in jaw movements of rats. *Neurosci Res Comm* 1993;13:55–61.
100. Deutch AY, Cameron DS. Pharmacological characterization of dopamine systems in the nucleus accumbens core and shell. *Neuroscience* 1992;46:49–56.
101. Heimer L, Zahm DS, Churchill L, Kalivas PW, Wohltmann C. Specificity in the projection patterns of accumbal core and shell in the rat. *Neuroscience* 1991; 41:89–125.
102. Zahm DS. Compartments in rat dorsal and ventral striatum revealed following injection of 6-hydroxydopamine into the ventral mesencephalon. *Brain Res* 1991;552:164–169.
103. Zahm DS, Heimer L. Two transpallidal pathways originating in rat nucleus accumbens. *J Comp Neurol* 1990; 302:437–446.
104. Zetterström T, Fillenz M. Local administration of flurazepam has different effects on dopamine release in striatum and nucleus accumbens a microdialysis study. *Neuropharmacology* 1990;29:129–134.
105. Grobin AC, Roth RH, Deutch AY. Regulation of the prefrontal cortical dopamine system by the neuroactive steroid 3a, 21-dihydroxy-5a-pregnane-20-one. *Brain Res* 1992;578:351–356.
106. Bunney BS, Walters JR, Roth RH, Aghajanian GK. Dopaminergic neurons: effect of antipsychotic drugs and amphetamine on single cell activity. *J Pharmacol Exp Ther* 1973;185:560–571.
107. Grace AA, Bunney BS. Intracellular and extracellular electrophysiology of nigral dopaminergic neurons. I. Identification and characterization. *Neuroscience* 1983;10:301–315.
108. Wang RY. Dopaminergic neurons in the rat ventral tegmental area. I. Identification and characterization. *Brain Res Rev* 1981;3:123–140.
109. Gonon FG. Nonlinear relationship between impulse

flow and dopamine released by rat midbrain dopaminergic neurons as studied by in vivo electrochemistry. *Neuroscience* 1988;24:19–28.
110. Suaud-Chagny MF, Chergui K, Chouvet G, Gonon F. Relationship between dopamine release in the rat nucleus accumbens and the discharge activity of dopaminergic neurons during local in vivo application of amino acids in the ventral tegmental area. *Neuroscience* 1992;49:63–72.
111. Bean AJ, Roth RH. Extracellular dopamine and neurotensin in rat prefrontal cortex in vivo: effects of median forebrain bundle stimulation frequency, pattern and dopamine autoreceptors. *J Neuroscience* 1991;11: 2694–2702.
112. Gariano RF, Groves PM. Burst firing induced in midbrain dopamine neurons by stimulation of the medial prefrontal and anterior cingulate cortices. *Brain Res* 1988;462:194–198.
113. Christie MJ, Bridge S, James LB, Beart PM. Excitotoxic lesions suggest an aspartategic projection from rat medial prefrontal cortex to ventral tegmental area. *Brain Res* 1985;333:169–172.
114. Nieoullon A, Dusticier N. Glutamate uptake, glutamate decarboxylase and cholineacetyltransferase in subcortical areas after sensorimotor cortical ablations in the cat. *Brain Res Bull* 1983;10:287–293.
115. Sesack SR, Pickel VM. Prefrontal cortical efferents in the rat synapse on unlabeled neuronal targets of catecholamine terminals in the nucleus accumbens septi and on dopamine neurons in the ventral tegmental area. *J Comp Neurol* 1992;320:145–160.
116. Murase S, Grenhoff J, Ghouvet G, Gonon FG, Svensson TH. Prefrontal cortex regulates burst firing and transmitter release in rat mesolimbic dopamine neurons studied in vivo. *Neurosci Lett* 1993;157:53–56.
117. Seutin V, Verbanck P, Massotte L, Dresse A. Evidence for the presence of N-methyl-D-aspartate receptors in the ventral tegmental area of the rat: an electrophysiological in vitro study. *Brain Res* 1990;514:147–150.
118. Wang T, French ED. Electrophysiological evidence for the existence of NMDA and non-NMDA receptors on rat ventral tegmental dopamine neurons. *Synapse* 1993;13:270–277.
119. Wang T, French ED. L-Glutamate excitation of A10 dopamine neurons is preferentially mediated by activation of NMDA receptors: extra- and intracellular electrophysiological studies in brain slices. *Brain Res* 1993;627:299–306.
120. Grace AA, Onn SP. Morphology and electrophysiological properties of immunocytochemically identified rat dopamine neurons recorded in vitro. *J Neurosci* 1989;9:3463–3481.
121. Wu H-Q, Schwarcz R, Shepard PD. Excitatory amino acid-induced excitation of dopamine-containing neurons in the rat substantia nigra: modulation by kynurenic acid. *Synapse* 1994;16:219–230.
122. Johnson SW, Seutin V, North RA. Burst firing in dopamine neurons induced by N-methyl-D-aspartate: role of electrogenic sodium pump. *Science* 1992;258: 665–667.
123. Charlety PJ, Grenhoff J, Chergui K, DeLaChapelle B, Buda M, Svensson TH, Chouvet G. Burstfiring of mesencephalic dopamine neurons is inhibited by somatodendritic application of kynurenate. *Acta Physiol Scand* 1991;142:105–112.
124. Grenhoff J, Tung C-S, Svensson TH. The excitatory amino acid antagonist kynurenate induces pacemaker-like firing of dopamine neurons in rat ventral tegmental area in vivo. *Acta Physiol Scand* 1988;134:567–568.
125. Chergui K, Charléty PJ, Akaoka H, Saunier CF, Brunet J-L, Buda M, Svensson TH, Chouvet G. Tonic activation of NMDA receptors causes spontaneous burst discharge of rat midbrain dopamine neurons in vivo. *Eur J Neurosci* 1993;5:137–144.
126. Murase S, Mathé JM, Grenhoff J, Svensson TH. Effects of dizocilpine (MK-801) on rat midbrain dopamine cell activity: differential actions on firing pattern related to anatomical localization. *J Neural Transm* 1993;91: 13–25.
127. Deniau JM, Thierry AM, Feger J. Electrophysiologica identification of mesencephalic ventromedial tegmen tal (VMT) neurons projecting to the frontal cortex septum and nucleus accumbens. *Brain Res* 1980;198 315–326.
128. Fallon JH. Collateralization of monoamine neurons: mesotelencephalic dopamine projections to caudate, septum and frontal cortex. *J Neurosci* 1981;1: 1361–1368.
129. Lindvall O, Björklund A, Divac I. Organization of catecholamine neurons projecting to the frontal cortex in the rat. *Brain Res* 1978;142:1–24.
130. Simon H, Le Moal M, Calas A. Efferents and afferents of the ventral tegmental-A10 region studied after local injection of [^{3}H]leucine and horseradish peroxidase. *Brain Res* 1979;178:17–40.
131. Swanson LW. The projections of the ventral tegmental area and adjacent regions: a combined fluorescent retrograde tracer and immunofluorescence study in the rat. *Brain Res Bull* 1982;9:321–353.
132. Morrow BA, Clark WA, Roth RH. Stress activation of mesocorticolimbic dopamine neurons: effects of a glycine/NMDA receptor antagonist. *Eur J Pharmacol* 1993;238:255–262.
133. Goldstein LE, Rasmusson AM, Bunney BS, Roth RH. The NMDA Glycine site antagonist (+)−HA-966 selectively regulates conditioned stress-induced metabolic activation of the mesoprefrontal cortical dopamine but not serotonin systems: a behavioral, neuroendocrine, and neurochemical study in the rat. *J Neurosci* 1994;14:4937–4950.
134. Jedema HP, Moghaddam B. Glutamatergic control of dopamine release during stress in the rat prefrontal cortex. *J Neurochem* 1994;63:785–788.
135. Jones CA, Zempléni É, Davis B, Reynolds GP. Glutamate stimulates dopamine release from cortical and limbic rat brain in vitro. *E J Pharmacol* 1992;242: 183–187.
136. Carter CJ. Glutamatergic pathways from the medial prefrontal cortex to the anterior striatum, nucleus accumbens and substantia nigra. *Br J Pharmac* 1980;70: 50P.
137. Carter CJ. Topographical distribution of possible glutamatergic pathways from the frontal cortex to the striatum and substantia nigra in rats. *Neuropharmacology* 1982;21:379–383.
138. Christie MJ, Summers RJ, Stephenson JA, Cook CJ, Beart PM. Excitatory amino acid projections to the nucleus accumbens septi in the rat: a retrograde transport study utilizing D[^{3}H]aspartate and [^{3}H]GABA. *Neuroscience* 1987;22:425–439.

139. Divac I, Fonnum F, Storm-Mathisen J. High affinity uptake of glutamate in terminals of corticostriatal axons. *Nature* 1977;266:377–378.
140. Fagg GE, Foster AC. Amino acid neurotransmitters and their pathways in the mammalian central nervous system. *Neuroscience* 1983;9:701–719.
141. Girault JA, Barbeito L, Spampinato U, Gozlan H, Glowinski J, Besson MJ. In vivo release of endogenous amino acids from the rat striatum: further evidence for a role of glutamate and aspartate in corticostriatal neurotransmission. *J Neurochem* 1986;47:98–106.
142. Kim J-S, Hassler R, Haug P, Paik K-S. Effect of frontal cortex ablation on striatal glutamic acid level in rat. *Brain Res* 1977;132:370–374.
143. Rowlands GJ, Roberts PJ. Specific calcium-dependent release of endogenous glutamate from rat striatum is reduced by destruction of the cortico-striatal tract. *Exp Brain Res* 1980;39:239–240.
144. Bouyer JJ, Park DH, Joh TH, Pickel VM. Chemical and structural analysis of the relation between cortical inputs and tyrosine hydroxylase-containing terminals in rat neostriatum. *Brain Res* 1984;302:267–275.
145. Sesack SR, Pickel VM. In the rat medial nucleus accumbens, hippocampal and catecholaminergic terminals converge on spiny neurons and are in apposition to each other. *Brain Res* 1990;527:266–279.
146. Glowinski J, Chéramy A, Romo R, Barbeito L. Presynaptic regulation of dopaminergic transmission in the striatum. *Cell Mol Neurobiol* 1988;8:7–17.
147. Jhamandas K, Marien M. Glutamate-evoked release of endogenous brain dopamine: inhibition by an excitatory amino acid antagonist and an enkephalin analogue. *Br J Pharmacol* 1987;90:641–650.
148. Carter CJ, Pycock CJ. Studies on the role of catecholamines in the frontal cortex. *Br J Pharmacol* 1978;62: 402P.
149. Carter CJ, Pycock CJ. Lesions of the frontal cortex of the rat: changes in neurotransmitter systems in subcortical regions. *Br J Pharmacol* 1978;64:430P.
150. Carter CJ, Pycock CJ. Behavioral and biochemical effects of dopamine and noradrenaline depletion within the medial prefrontal cortex of the rat. *Brain Res* 1980;192:163–176.
151. Pycock CJ, Carter CJ, Kerwin RW. Effect of 6-hydroxydopamine lesions of the medial prefrontal cortex on neurotransmitter systems in subcortical sites in the rat. *J Neurochem* 1980;34:91–99.
152. Deutch AY, Clark WA, Roth RH. Prefrontal cortical dopamine depletion enhances the responsiveness of mesolimbic dopamine neurons to stress. *Brain Res* 1990;521:311–315.
153. Oades RD, Tagzhouti K, Rivet J-M, Simon H, Le Moal M. Locomotor activity in relation to dopamine and noradrenaline in the nucleus accumbens, septal, and frontal areas: a 6-hydroxydopamine study. *Neuropsychobiology* 1986;16:37–42.
154. Rosin DL, Clark WA, Goldstein M, Roth RH, Deutch AY. Effects of 6-hydroxydopamine lesions of the prefrontal cortex on tyrosine hydroxylase activity in mesolimbic and nigrostriatal dopamine systems. *Neuroscience* 1992;48:831–839.
155. Keefe KA, Zigmond MJ, Abercrombie ED. Extracellular dopamine in striatum: influence of nerve impulse activity in medial forebrain bundle and local glutamatergic input. *Neuroscience* 1992;47:325–332.
156. Keefe KA, Zigmond MJ, Abercrombie ED. In vivo regulation of extracellular dopamine in the neostriatum: influence of impulse activity and local excitatory amino acids. *J Neural Transm* 1993;91:223–240.
157. Moghaddam B, Gruen RJ. Do endogenous excitatory amino acids influence striatal dopamine release? *Brain Res* 1991;544:329–330.
158. Moghaddam B, Bolinao ML. Glutamatergic antago nists attenuate the ability of dopamine uptake blocker to increase extracellular levels of dopamine: Implica tions for a tonic influence of glutamate on dopamin release. *Synapse* 1994;18:337–342.
159. Shepard PD, Lehmann H. ()-1-hydroxy-3-aminopyrrolidone-2 (HA-966) inhibits the activity of substantia nigra dopamine neurons through a non-N-methyl-D-aspartate receptor-mediated mechanism. *J Pharmacol Exp Ther* 1992;261(2):387–394.
160. Charney DS, Deutch AY, Krystal JH, Southwick SM, Davis M. Psychobiologic mechanisms of posttraumatic stress disorder. *Arch Gen Psych* 1993;50:294–305.
161. Wolpe J, ed. *The practice of behavior therapy,* 4th ed. New York: Pergamon Press, 1990.

Neurobiological and Clinical Consequences of Stress: From Normal Adaptation to PTSD, edited by M. J. Friedman, D. S. Charney, and A. Y. Deutch, Lippincott-Raven Publishers, Philadelphia © 1995.

5

Stress and Neuronal Sensitization

Barbara A. Sorg and Peter W. Kalivas

Departments of Veterinary and Comparative Anatomy, Pharmacology, and Physiology, Washington State University, Pullman, Washington 99164

Stressful life experiences can initiate long-term behavioral and interoceptive changes that have clear adaptive value to an organism. The capacity to respond more readily and/or effectively to subsequent stressors increases the probability of survival. However, exposure to extraordinary life stressors can precipitate long-term maladaptive behavioral alterations referred to as post-traumatic stress disorder (PTSD). A classic example of behavioral sensitization, PTSD can be defined as the capacity of "a strong, potentially threatening stimulus to enhance the response to the same or a weaker agent presented at a later time" (1). One characteristic associated with behavioral sensitization to stressors is that sensitization is relatively long lasting. For example, PTSD can endure for the lifetime of an individual. Another characteristic is the presence of cross-sensitization. Thus, a different stressor than that involved in the induction of PTSD may precipitate the sensitized behavioral response.

Other neuropsychiatric disorders present sensitization-like characteristics. For example, panic disorder and substance abuse show characteristics of producing long-term alterations in behavioral responsiveness and cross-sensitization between environmental or pharmacological stimuli (see Chapter 12). In fact, sensitization brought about by excessive use of psychostimulants can elicit maladaptive behavioral changes that are indistinguishable from psychiatric disorders such as panic attack or paranoid psychosis (see review 1–3). In addition, individuals with PTSD are more prone to develop a pattern of substance abuse (see Chapter 25). Consistent with this, it has been proposed that a psychostimulant challenge acts as a stressor to induce behavioral sensitization (see review 1).

Elucidating the neural substrates responsible for the initiation and expression of sensitization-related psychiatric disorders requires the use of appropriate animal models where invasive measurements can be employed to reveal subtle alterations in function at a synaptic or cellular level. Animal models of sensitization-related psychiatric disorders have traditionally utilized the clinical link between chronic psychostimulant administration and the induction of psychiatric disorders. Thus, repeated administration of psychostimulants produces a marked augmentation in motor activity in experimental animals that is postulated to be analogous to sensitization-related psychiatric disease (4–6). The capacity of environmental stressors to substitute for psychostimulants in the induction of behavioral sensitization in rats further validates the presentation of repeated, salient challenges to experimental animals as a model of sensitization-related disorders. Furthermore, behavioral sensitization in rats produced by either repeated psychostimulants, stressors, or a combination of the two endures for weeks or months after discontinuing treatments, supporting the analogy of this animal model to sensitization-related psychiatric disorders.

Several years of research using this animal model indicate that the physiological substrates involved in stress- and psychostimulant-induced changes in behavior lie in the brain dopaminergic pathways. Specifically, the mesocorticolimbic dopamine system is activated by stress and psychostimulants, and is involved in behavioral sensitization and the reinforcing effects of psychostimulants (7–9). This chapter focuses on dopamine perikarya located in the ventral tegmental area (VTA) and its projections to the medial prefrontal cortex (mPFC) and nucleus accumbens (NAS) (10,11). We also address the potential role of noradrenergic and serotonergic systems that project to these brain areas, since psychostimulants and stress alter norepinephrine (NE) and serotonin (5-HT) transmission.

The overlapping effects of stress and psychostimulants may be facilitated by activation of the hypothalamic-pituitary-adrenal (HPA) axis. Blockade or removal of stress hormones such as corticotropin-releasing hormone (CRH) and corticosterone (CORT) blunts the behavioral responses to stress and behavioral sensitization induced by repeated stress and psychostimulants in animals. In humans, hypersecretion of CRH and cortisol are found in patients with panic disorder and PTSD. Knowledge of how the HPA axis governs the relevant neurotransmitter systems will be important to incorporate into an animal model of behavioral sensitization.

PSYCHOSTIMULANT- AND STRESS-INDUCED SENSITIZATION

Studies examining behavioral sensitization have most often used repeated exposure to psychostimulants such as cocaine and amphetamine. Both drugs promote an increase in extracellular levels of dopamine, NE, and 5-HT by increasing release and/or blocking the reuptake of these neurotransmitters (12,13). Environmental stress can "cross-sensitize" with psychostimulants to induce behavioral sensitization (i.e., repeated daily stress augments the behavioral response to an acute cocaine or amphetamine challenge (14–18). These findings have led to the hypothesis by Antelman (1) that psychostimulants act as pharmacological stressors to the organism. Similar to the acute administration of psychostimulants, acute exposure to an environmental stressor elicits an increase in dopamine, NE, and 5-HT content (19–23). The similar neurochemical effects of stressors and psychostimulants have led to the hypothesis that behavioral cross-sensitization arises via influences on overlapping substrates. Indeed, several parallel alterations occur in the mesocorticolimbic dopamine system associated with acute and repeated treatment of psychostimulants and stress.

Sensitization and Dopamine

VTA

Sensitization has been divided into two temporally distinct phases: 1) *initiation* of neuronal responses and 2) *expression* of neuronal responses (see review 6). Initiation refers to the early priming events associated with daily sensitizing treatments that lead to a permanently altered responsiveness of dopamine neurons. Expression refers to the augmentation of synaptic dopamine levels in the terminal fields (NAS) after withdrawal from daily treatment and subsequent drug challenge (see information that follows). The VTA, where dopamine perikarya are located, is the only site thus far discovered to be the origin of sensitization. Several studies have shown that microinjection of amphetamine or morphine directly into the VTA, but not into the NAS, leads to a behaviorally sensitized response to subsequent systemic amphetamine, cocaine, or intra-NAS amphetamine administration (6,24). The role of the VTA in initiating sensitization is further supported by observations that behavioral sensitization to an acute psychostimulant or morphine challenge is produced following intra-VTA treatments with neurotensin, substance P, and pertussis toxin (see 6). Kalivas and Stewart (6) have proposed that the common factor promoting the initiation of sensitization is an increase in somatodendritic release of dopamine in the VTA. Behavioral sensitization to psychostimulants or morphine is blocked by intra-VTA administration of a D1 antagonist, opioid antagonist, gamma-amino butyric acid (GABA) agonist, or N-methyl-D-aspartate (NMDA) antagonists, which prevent increased somatodentritic dopamine release or its effects (25), (see discussion 6).

While a primary site for the initiation of stress-induced sensitization has not been examined, acute stress also increases postmortem measurements of dopamine metabolites in the VTA (26), and Kalivas et al. (27) provide further evidence that the VTA may be an important component. The acute stress-induced increase in mPFC postmortem dopamine metabolites is blocked by intra-VTA infusion of the NMDA receptor antagonist, 3-[(±)-2-carboxypiperazine-4-yl]propyl-1-phosphonic acid (CPP), which also blocks behavioral sensitization to repeated cocaine (25).

Intrinsic to a two-stage analysis of neurochemical sensitization is the notion that there are time-dependent effects from repeated daily drug treatments (i.e., early changes associated with initiation may disappear while permanent changes giving rise to expression of sensitization do not appear until later). In the VTA, extracellular dopamine levels and the rate-limiting enzyme for dopamine synthesis, tyrosine hydroxylase (TH), are enhanced early after withdrawal (<1 week) from repeated high-dose cocaine, but the effects disappear by 3 weeks of withdrawal (28,29). In addition, repeated stress followed by an acute cocaine or stress challenge at a late withdrawal time (10 days) results in a decrease in postmortem dopamine levels in the VTA (30). Conversely, extracellular dopamine levels are diminished in the NAS early after withdrawal from repeated cocaine or amphetamine exposure (31–33), but are enhanced at later withdrawal times (31,34). The delay in expression of sensitization may be due to overriding effects of tolerance that become diminished with time, allowing sensitization to appear. Alternatively, since protein synthesis in the VTA is required for the appearance of behavioral sensitization to cocaine (B.A. Sorg and C. Ulibarri, unpublished results), the time needed for axonal transport of any altered gene products may delay the onset of neurochemical sensitization in the NAS.

NAS

Most of the studies examining psychostimulant-induced sensitization have focused on the mesoaccumbens dopamine projection because dopamine release in these terminals is associated with the rewarding as well as the motor-stimulant effects of these drugs (8,9,35,36). A primary effect of cocaine and amphetamine on the expression of behavioral sensitization is the augmentation of extracellular dopamine levels in the striatum and NAS (34,37–40). Kalivas and Duffy (37) observed a correlation between extracellular dopamine levels and locomotor activity in response to an acute cocaine challenge in cocaine-sensitized rats. Figure 1A shows an example of sensitization of both the locomotor and NAS dopamine response in rats pretreated with daily cocaine.

Far fewer studies have been done regarding the effects of repeated stress on NAS dopamine levels. This is largely due to the relative ineffectiveness of milder stressors in elevating NAS dopamine transmission (18,20,41–43). However, enhanced extracellular dopamine levels and measures of postmortem accumbal dopamine metabolism have been found in some studies, depending on the type of repeated stressor (26,30,44), while no changes were reported in other studies (45). Figure 1B shows that repeated exposure to footshock stress results in sensitization of the locomotor and NAS dopamine response to an acute cocaine injection. The behavioral and neurochemical responses are nearly identical to those shown in Fig. 1A, in which rats received repeated daily cocaine. It should be noted that subregions of the NAS referred to as the core and shell areas are differentially responsive to stress, the shell being more sensitive than the core with regard to increased dopamine utilization (46). However, the effects of repeated stress or psychostimulants on NAS core/shell areas are unknown.

Cross-sensitization between psychostimulants and stress leads to augmented NAS extracellular dopamine levels and dopamine metabolism (18,30). These results indicate that repeated exposure to stressful experiences or psychostimulants is able to modify either: 1) the effective-

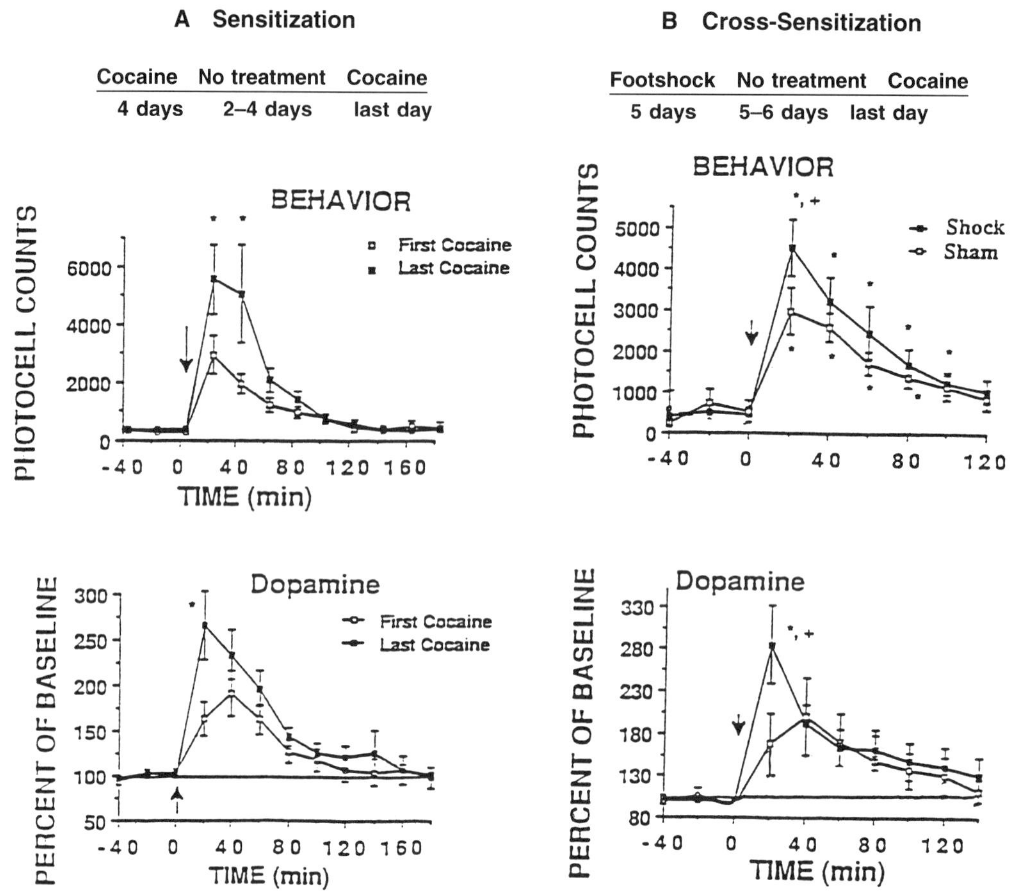

FIG. 1. Sensitization of locomotor activity and extracellular dopamine levels in the nucleus accumbens (NAS). **A.** Repeated daily cocaine was administered for 4 days followed by a withdrawal period and a cocaine challenge. Horizontal locomotor activity and extracellular dopamine levels in the NAS were measured after the first day of cocaine (first cocaine) and after an acute cocaine challenge (last cocaine), indicated by the arrow. Data are mean ±SEM, *$p<.05$, compared to first cocaine group. Data are taken from Kalivas and Duffy (37). **B.** Repeated 20-minute intermittent foot shock or sham shock sessions were administered daily, and following a withdrawal period, an acute cocaine was administered while measuring locomotor activity and extracellular dopamine levels in the NAS. Data are mean ±SEM, *$p<.05$, compared to sham shock group. Data are taken from Sorg and Kalivas (18).

ness of dopamine cell firing in the VTA and/or 2) the ability of the terminals to increase the ratio of release to reuptake of dopamine in the NAS. The observation that potassium- and amphetamine-stimulated dopamine release is augmented in in vitro slice preparations of NAS and striatum from psychostimulant-sensitized animals suggests that the permanent changes occur in the terminal fields (47–49). These effects have not always been observed in vitro after repeated cocaine (50,51); however, behavioral sensitization occurs in amphetamine- (52) and cocaine- (P.W. Kalivas, unpublished results) pretreated rats following in vivo amphetamine challenge into the NAS. These findings do not exclude the possibility that alterations in cell firing also contribute to the sensitized dopamine response, or that other neurotransmitter interactions in the NAS/striatum contribute to the robustness of the response (see next section).

mPFC

In contrast to the NAS, few studies have examined the effects of repeated stress or psychostimulants on mPFC dopamine. Evidence that cocaine is self-administered into the mPFC (53) suggests that this region may serve an important role in cocaine- and stress-induced sensitization. A number of studies have examined the effects of acute stress on mPFC dopamine and metabolites because of the high responsiveness of this pathway to stress compared to the NAS (7,20,43). Some studies have reported an enhancement of stress-induced dopamine metabolism and extracellular dopamine in the mPFC after repeated stress (21,30,54,55), while others have observed no change in response to stress after previous exposure (26,56). Other investigators have used cross-sensitization paradigms with psychostimulants and stress to demonstrate neurochemical sensitization. In general, postmortem measures of mPFC dopamine are increased by acute stimuli given following repeated exposure to stress or psychostimulants (30,57). More recently, however, tolerance was demonstrated in the mPFC when extracellular dopamine levels were measured following a cocaine or stress challenge in cocaine- or footshock stress-sensitized rats (58,59). In contrast, amphetamine challenge in cocaine-sensitized rats leads to a sensitized dopamine response (59). Hamamura and Fibiger (60) demonstrated a sensitized extracellular dopamine response to footshock stress challenge in a study employing pretreatment with repeated amphetamine, while amphetamine challenge produced no differences between the rise in extracellular dopamine levels. The reason for the discrepancies between studies is unknown, but stress, amphetamine, and cocaine produce a rise in extracellular dopamine via different mechanisms and exert their preferential actions in different brain regions (20,61,62). The effects of stress and cocaine on extracellular dopamine levels rely on impulse flow and vesicular dopamine release, while amphetamine's actions are impulse-independent and occur primarily via promotion of nonvesicular dopamine stores (63–65). Therefore, repeated exposure to different stimuli may induce dissimilar compensatory changes in the terminal fields. One interpretation of the data just presented is that repeated cocaine or stress may shift dopamine stores from vesicular to nonvesicular pools, resulting in decreased extracellular dopamine after cocaine or stress challenge, but elevated dopamine levels after amphetamine challenge. Repeated amphetamine administration, however, may produce elevated vesicular dopamine stores. Alternatively, local interactions with other neurotransmitter systems in the mPFC and/or VTA may account for differences in the dopamine response.

Interactions between mPFC and NAS

Much evidence supports a regulatory influence of mPFC dopamine in inhibiting dopamine release in the NAS via its inhibitory action on glutamatergic projections into this structure (66–68). Dopamine depletions in the PFC are correlated with an increase in spontaneous locomotor activity (69) and produce elevated amphetamine-induced stereotyped behavior (67, 70). Following 6-hydroxydopamine (6-OHDA) lesions of the mPFC, basal dopamine turnover (67,71) and footshock–induced increases in dopamine metabolism are found in the NAS (72). Tyrosine hydroxylase is also increased in the NAS after mPFC 6-OHDA lesions. Although basal levels of TH are not altered, haloperidol-stimulated increases in TH activity are potentiated in lesioned rats (73). Recently, Mitchell and Gratton (74) have reported that 6-OHDA lesions in the mPFC produce an augmentation in extracellular dopamine levels in the NAS compared to controls in response to the naturally reinforcing stimuli of food preparation and sex-related olfactory cues. Also, Schenk et al. (75) have demonstrated that 6-OHDA lesions of the PFC facilitate the acquisition of cocaine self-administration. These findings are consistent with the observation that dopaminergic activity is decreased in the mPFC but increased in the NAS of rats predisposed to acquire amphetamine self-administration (76).

It should also be pointed out that excitatory amino acids (EAA) project from the mPFC into the VTA (77). These projections may be important in modifying cell firing of mesoaccumbens dopamine neurons. Burst firing of VTA dopamine neurons, which significantly increases dopamine release in terminal fields (78), is produced by electrical stimulation of the PFC, and local cooling of the PFC blocks burst firing of VTA dopamine neurons (79,80).

Although the evidence just discussed strongly implicates a role for the PFC in regulating specific behaviors, the literature regarding the effects of cortical dopamine lesions on the responsiveness of subcortical dopamine regions and locomotor activity has been somewhat confusing. This arises partly from differences in the type of lesion, and the extent to which NE terminals are destroyed. Noradrenergic fibers overlap with dopamine terminals in the mPFC (81), and influence dopamine-mediated behaviors. Electrolytic lesions of the VTA destroy ascending dopamine neurons but leave noradrenergic neurons intact, and result in locomotor hyperactivity and D1 receptor supersensitivity in the PFC (69,82). On the other hand, 6-OHDA lesions in the VTA, which destroy ascending dopamine as well as NE fibers, result in neither behavioral deficits nor D1 receptor sensitivity changes (82,83). Carter and Pycock (67) observed similar responses when mPFC terminals were treated with 6-OHDA. Destruction of only dopamine terminals, but not NE terminals, in the mPFC produced an increase in spontaneous locomotion and amphetamine-induced stereotyped behavior, while destruction of both systems masked the effect of dopamine terminal loss, producing normal behavior. More recent evidence indicates that NE terminals in the mPFC influence extracellular dopamine levels. Treatment with blockers of NE uptake given peripherally (84) or locally into the mPFC (85) elevate extracellular dopamine concentrations in the mPFC. Together, the different evidence suggests that a balance between dopaminergic and noradrenergic systems in the mPFC provides normal responsiveness to the environment, although the nature of the relationship between these two systems at the ultrastructural level is unknown.

Sensitization and Norepinephrine

The dorsal NE system consists of NE perikarya in the locus coeruleus (LC) and their projections to the cerebral cortex, hippocampus, and other limbic structures, including the NAS and amygdala (86). The NE system has been postulated to function in the organism's attention to environmentally relevant stimuli (87,88). For example, LC neurons increase their discharge rate in the awake state and exhibit bursts of activity in anticipation of attention toward stimuli (87). As just mentioned, cocaine and amphetamine increase extracellular NE levels (89). Although NE has been speculated to play a role in amphetamine-induced locomotor activity (90, 91), repeated psychostimulant exposure does not produce alterations in NE turnover in the NAS or mPFC upon subsequent drug challenge (89).

Acute stress also elicits an increase in extracellular NE levels and NE metabolism in several brain regions (19,22,23,92), and increases the firing of LC neurons (93,94). Chronic or repeated stress has been shown to produce elevated levels of brain NE turnover after a stress challenge (95,96), and elevations in extracellular NE levels in the mPFC and hippocampus (21,97). It is important to note that these effects have generally been reported for early withdrawal from chronic or repeated stress, and therefore may represent a tolerance phenomenon rather than events associated with sensitization. Recently, Finlay et al. (98) have reported that enhanced extracellular NE in the mPFC levels are evoked by presentation of a novel stressor for 1 day, but not 14 days, after discontinuing chronic stress, suggesting that only short-term effects of chronic stress on NE release may occur.

Sensitization and Serotonin

Environmental stress increases the level of 5-HT turnover in the mPFC (99,100), and cocaine increases 5-HT levels in several brain areas due to its potent blockade of the 5-HT transporter (101). Serotonin neurons project from the median and dorsal raphe nuclei to the VTA (102,

103), and much evidence supports a role for 5-HT regulation of VTA dopamine neurons (see review 104). Repeated noncontingent cocaine enhances subsequent cocaine-induced increases in extracellular 5-HT in the VTA and NAS (39). It also increases the number of 5-HT uptake sites in the cell bodies in the dorsal raphe, as well as terminals in the mPFC (105). Repeated stress enhances the behavioral response to the 5-HT agonist, 5-methoxy-N,N-dimethyltryptamine (106), and increases 5-HT2 binding sites in the rat frontal cortex (107). Serotonin actions on the mesocorticolimbic dopamine system may also play a role in the therapeutic actions of the 5-HT2 antagonist, ritanserin, against the negative-type symptoms of schizophrenia (108). A reduction in prefrontal cortical dopamine activity has been related to the negative symptoms of schizophrenia, and it is proposed that this hypofrontality contributes to abnormal cognitive responses to stress in humans with schizophrenia (109) and to behavioral sensitization in cocaine-sensitized rats (58). Ritanserin given intravenously blocks the reduction of burst firing in VTA neurons produced by local cooling of the PFC (110), suggesting that the effects of the mPFC excitatory input to the VTA may be modulated by 5-HT levels.

Psychostimulant- and Stress-Induced Sensitization: Conclusions

Most of the emphasis in the study of behavioral and neurochemical sensitization has been on the mesocorticolimbic dopamine system because of its known role in the initiation and expression of behavioral sensitization. The precise neural substrates common to psychostimulant- and stress-induced sensitization are not known. However, similar neurochemical and behavioral responses to repeated stress and drugs indicate that the rapidly growing literature on psychostimulant-induced sensitization can form a point from which to investigate stress-induced alterations in specific brain pathways and further understand the cross-sensitization phenomenon. We discussed three important neurotransmitters, including dopamine, NE, and 5-HT, but several others are also known to be involved in stress- and psychostimulant-induced changes in dopamine transmission. For example, glutamate, GABA, enkephalin, and tachykinins modulate dopamine transmission in the VTA and/or the mPFC and NAS (review 6,7).

To begin to look at some of the common mechanisms mediating neuronal sensitization by stress and psychostimulants, several investigators have examined the role of stress hormones, including CRH and CORT, in producing sensitization.

ROLE OF THE HPA AXIS IN SENSITIZATION

Antelman and coworkers have long proposed that drugs in general act as stressors to the organism because of their foreign nature (1). Several similar neurochemical changes occur after exposure to psychostimulants and stress. These include increases in brain dopamine, 5-HT, and NE levels as well as increases in CRH, adrenocorticotropic hormone (ACTH), β-endorphin and CORT (111). More recently, studies examining the behavioral and neurochemical changes produced by psychostimulant drugs and environmental stressors have focused on the HPA axis because this pathway is activated by amphetamine (112), cocaine (113), and environmental stressors (114). In the typical stress response, CRH is produced in the paraventricular nucleus (PVN) of the hypothalamus, released in the median eminence, and, via the portal blood system, stimulates the release of ACTH in the anterior pituitary, in turn stimulating the release of CORT from the adrenal cortex.

Role of CRH in Sensitization

CRH and Behavior

Corticotropin-releasing factor is a 41-amino acid polypeptide that is released in the hypothalamic median eminence in response to stress and stimulates ACTH release from the anterior pituitary (115). Many reports demonstrate an increase in hypothalamic CRH and CRH mRNA levels following acute and chronic stress (116–122). Much evidence supports an extrahypothalamic role for CRH in the production of stresslike behavioral and autonomic responses. Administration of CRH directly into the brain (i.c.v.) produces behavioral responses similar to those induced by stress, including increased locomotor activity in a habituated environment (123,124), decreased activity in a novel environment (125), increased grooming (126) and decreased food consumption (127). Central administration of the CRH antagonist, alpha-helical CRH, decreases stress-induced reductions in exploratory behavior, footshock–induced fighting and freezing behaviors (128–130). Dissociation of this effect from CRH-induced ACTH release is evidenced by the lack of antagonism by hypophysectomy or dexamethasone-suppression (125,131). Corticotropin-releasing factor is also localized in the amygdala, hippocampus, and cortex (132–135). Changes in CRH-like immunoreactivity, CRH receptors, and/or CRH mRNA levels have been observed in discrete brain regions after acute and repeated stress (116,118) and following repeated cocaine exposure (136).

Since psychostimulants induce a stresslike response, several studies have addressed the question of whether CRH is a mediator of the motor-stimulant effect of cocaine and amphetamine. Pretreatment with either the CRH antibody or the CRH antagonist, alpha-helical CRH, prevents the acute cocaine-induced increase in locomotion (137). Corticotropin-releasing factor is also involved in psychostimulant-induced sensitization as well as cross-sensitization with stress. Peripheral pretreatment of animals with CRH antibody prevents amphetamine-induced behavioral sensitization to an amphetamine challenge (138), and central administration of a CRH antagonist prevents the development of stress-induced sensitization to amphetamine challenge (139). In addition, repeated central administration of CRH alone sensitizes the locomotor response to amphetamine challenge (140). Thus, stress- and psychostimulant-induced behaviors, including sensitization, may be mediated by the action of CRH at extrahypothalamic sites.

CRH and Neurochemistry

Further evidence for an effect of CRH on other brain regions is supported by the work of Dunn and Berridge (141) who found that i.c.v. administration of CRH into mice increased postmortem levels of dopamine and/or NE metabolism in the prefrontal cortex (as well as other brain areas), but not in the NAS or striatum, leading the authors to conclude that the overall pattern of neurochemical changes resembled that of footshock– or restraint-stressed animals. These changes in dopamine and NE metabolism did not occur following peripheral administration of ACTH or CORT, suggesting that i.c.v. CRH produced its effects by direct stimulation of extrahypothalamic sites. In another study, rats were given the same dose of CRH administered i.c.v. and demonstrated increased dopamine metabolism in several brain regions, including the frontal cortex, striatum, hippocampus, and amygdala, with increased NE metabolism restricted to the frontal cortex and hippocampus (126). However, the relationship between altered dopamine/NE metabolism and the behavioral effects of i.c.v. CRH is questionable. Koob et al. (142) could not prevent CRH-induced locomotion in rats with the dopamine antagonist, flupenthixol, and destruction of dopamine terminals by 6-OHDA in the NAS did not prevent the behavioral stimulation produced by i.c.v. CRH (143). Moreover, a dissociation between behavior and dopamine was reported by Dunn and Berridge (141), who reversed the behavioral, but not the neurochemical, effects of i.c.v. CRH with naloxone administration.

Kalivas et al. (144) applied CRH directly into the VTA to determine if CRH receptors located in this region (145) mediate CRH's effects on the locomotor response. Contrary to the stresslike increase in PFC dopamine utilization following i.c.v. CRH (141), intra-VTA CRH reduced PFC dopamine metabolism with accompanying increases in NAS dopamine utilization. As just described in detail, the decrease in PFC dopamine may in part mediate CRH-induced behavioral activity via EAA neurons projecting to the NAS, although behavioral activation from intra-VTA CRH was not blocked by haloperidol; this suggests that increased locomotion was independent of mesoaccumbal dopamine activation (144).

Activation of the noradrenergic system also increases behaviors related to the stress response, and CRH receptors located in the LC modulate stress-induced responses (146). Evidence suggests that an interaction between CRH and NE may occur in the LC in response to acute stress, such that CRH induces an increase in NE via increased TH biosynthesis in this brain area (147). The increase in TH levels in the LC after repeated stress is blocked by administration of alpha-helical CRH infused near the LC. In addition, CRH administered repeatedly into the lateral ventricals mimics the effects of repeated stress with regard to increased TH immunoreactivity in the LC, suggesting that repeated stress effects on TH are mediated by CRH.

Although we have focused on only a few of the brain areas that contain CRH receptors, it is apparent that CRH induces neurochemical changes that are similar to stress when infused into some of these discrete brain sites, and thus may be a contributing factor in stress-induced sensitization.

Role of Glucocorticoids in Sensitization

For many years, glucocorticoids have been proposed to play an important role in adaptation to stress (see review 148). In the rat, the naturally occurring glucocorticoid is corticosterone (CORT). Following its release from the adrenal cortex, CORT can enter neurons and bind to two types of receptors: 1) the mineralocorticoid receptor (MR, or type I) is confined to the hippocampus, septum, amygdala, olfactory nucleus, and regions of the cortex and brain stem sensory and motor neurons, while 2) the glucocorticoid receptor (GR, or type II) has been localized to several brain areas including the hippocampus, septum, PVN, thalamic nuclei, certain regions of the striatum, central amygdaloid nucleus, ascending monoaminergic neurons of the brainstem, and cortical hemispheres (149). Nearly all noradrenergic and 5-HT containing neurons in the lower brain stem contain GRs, while approximately one half of the dopamine-containing cells in the VTA and substantia nigra contain GRs (150). The affinity of MR for CORT is approximately 10 times higher than is GR for CORT. Thus, MRs are occupied by CORT under basal conditions, while GRs become increasingly occupied after stress and at the circadian peak (148). An increase in CORT levels provides negative feedback directly to the HPA axis to terminate stress-induced secretion of CRH and ACTH, and CORT additionally binds to extrahypothalamic sites, particularly the hippocampus, which regulates delayed negative feedback (148,151).

Effects of Stress and Psychostimulants on Glucocorticoids

It has long been known that acute stress increases plasma CORT levels (152–154). The predominant response of CORT to repeated intermittent stress is tolerance (153,155–157); however, sensitization has also been reported (55,158–161) as well as no change with repeated stress (162,163). These disparate observations demonstrate that several parameters of the treatment protocol are important in determining the direction of the CORT response, including the type and severity (intensity x duration) of the stressor, reexposure to the same versus a different stressor, interstressor interval, and the total number of daily treatments (155,160,162, 164). The overall pattern seen in these studies suggests that a transient sensitization of the CORT response to stress occurs after only a few repeated daily stress sessions (usually of higher intensity), while tolerance appears after several days. It has been suggested that this tolerance is due to familiarity with the same stressful environment (162), and that it is absent if animals are reexposed to a novel stimulus (162,165). The ability to demonstrate a switch in the biochemical response to different stressors shown in these studies is similar to studies using chronic stress in which hippocampal NE levels are elevated in response to a novel stressor (166).

Acute cocaine also increases plasma CORT levels (167,168), and this response is mediated by CRH (169). In contrast to most of the studies employing repeated stress discussed above, repeated cocaine produces neither sensitization nor tolerance of the CORT response to subsequent cocaine injections after various withdrawal times (170–172).

Glucocorticoids and Cross-Sensitization

Despite the differences in CORT responses shown between studies using repeated cocaine and stress, many investigators have found that behaviors such as behavioral sensitization produced by repeated stress or amphetamine and the acquisition of amphetamine self-administration are dependent on CORT secretion (173–178). The locomotor response to acute amphetamine is attenuated by adrenalectomy and enhanced in a dose-dependent fashion by implantation of CORT pellets (179). The behaviorally sensitized response to repeated amphetamine is also prevented by adrenalectomy with CORT replacement, but restored by replacement with dexamethasone, a GR agonist, suggesting that the effects of CORT on sensitization are mediated by GRs (178). In a cross-sensitization study, amphetamine challenge in stress-sensitized animals did not elicit a sensitized locomotor response in adrenalectomized rats given low levels of CORT replacement (173). Corticosterone itself has been repeatedly administered in doses that increase plasma CORT levels in the same range as that observed in response to stress, producing a behaviorally sensitized response to amphetamine (180). Thus, adrenalectomy prevents the locomotor response to amphetamine, low-level replacement CORT that mimics basal physiological CORT levels but not stress-induced levels restores the response, and higher CORT levels enhance the behavioral response to amphetamine. These data suggest that occupancy of MRs by low-level CORT therapy is sufficient for the normal response to acute amphetamine, but that GR occupancy may be required for the enhanced behavioral response. Additional evidence that CORT may mediate psychostimulant actions is that rats orally self-administer CORT, suggesting that drug reinforcing properties may be attributed to increases in CORT levels (181).

Piazza et al. (182) have recently shown that the stress-induced increase in extracellular dopamine levels in the NAS is attenuated in animals adrenalectomized and implanted with CORT pellets (to maintain basal physiological levels), while an additional CORT injection immediately prior to stress restores the stress-induced dopamine response. Only one study to date has examined the effects of acute cocaine on animals given adrenalectomy and CORT replacement that more closely mimics the circadian variation in CORT levels (183). In this study, implantation of CORT pellets was sufficient to allow for a normal locomotor response to acute cocaine, while the circadian fluctuation in CORT added to the drinking water during the dark cycle was necessary for acute morphine effects on behavior. However, the role of circadian variation in CORT levels has not been examined in stress- or psychostimulant-induced sensitization. Two major conclusions can be drawn from these data: 1) the *initiation* of sensitization is critically dependent on fluctuating levels of CORT, both circadian and those associated with the daily stress or drug pretreatment, and/or 2) the *expression* of sensitization to cocaine or amphetamine is dependent on the spike in CORT levels associated with the drug challenge.

Glucocorticoid Control of Neuronal Activity and Neurotransmitter Interactions

Considering the evidence that CORT serves a role in the development and/or expression of sensitization, what are the possible mechanisms by which CORT could influence neuronal activity in monoaminergic systems?

Corticosterone exerts its classic action via binding to intracellular GRs and MRs, which modulate transcription through regulatory elements on particular genes (184). Joels and de Kloet (185) observed that excitation of CA1 hippocampal neurons was altered by CORT, but only when the cells were depolarized electrically or by the addition of neurotransmitter. Neuronal excitability was increased by occupation of MRs, and decreased by occupation of GRs. The maximal effects of GR and MR occupancy on accommodation and after hyperpolarization of CA1 cells were observed following a delay of at least 20 minutes after removal of the steroid and remained effective for up to 90 minutes. Joels and de Kloet (185) proposed that these delayed effects are mediated by an action of CORT on the genome since they are dependent on protein synthesis (186). More immediate and shorter-acting effects of MR and GR occupancy on the synaptic response of CA1 cells have been recently demonstrated (187). Binding of MR increases, and GR decreases synaptic responses to amino acid-mediated transmission in these neurons.

Direct and rapid effects of CORT also occur via binding to specific receptors located in the plasma membrane. High-affinity binding has been shown in membrane preparations from amphibian brain that appear to modulate mating behavior (188). Although this high-affinity binding has not yet been reported in rat brain, binding of CORT in the plasma membrane provides a potential mechanism for rapid interactive effects with other membrane proteins. Indeed, CORT has been proposed to mediate some of its effects via G-protein receptor coupling in the rat hippocampus (185), and repeated CORT administration alters the levels of specific G-proteins and their mRNAs (189). This mechanism of action would permit indirect alteration of gene transcription in addition to its direct genomic effects. Thus, stress- and drug-induced increases in CORT levels may initiate a cascade of events that result in long-term changes in the responsiveness of neurons.

Several lines of evidence support the idea that CORT can modulate dopamine transmission in the mesocorticolimbic dopamine system, and that dopamine can alter CORT levels. One strategy for accumulating this evidence has been examination of the effects of circulating CORT levels on brain dopamine concentrations. Administration of CORT doses that mimic plasma CORT levels produced by stress increases the release of dopamine in the PFC (190) and NAS (182). The stress-induced increase in NAS dopamine levels is prevented by adrenalectomy plus basal-level CORT replacement, and blockade of CORT synthesis with metyrapone diminishes this dopamine response to stress (182). Other investigators have not observed a relationship between CORT and stress-induced extracellular dopamine levels in the mPFC or NAS (191) or postmortem dopamine metabolism (99). However, in these latter studies, measures of dopamine were collected 1–3 days after adrenalectomy, and a longer period may be necessary to observe the full effects resulting from adrenalectomy. This is supported by the observation that appearance of the reduced locomotor response to intra-NAS amphetamine occurs by 8 days, but not 4 days, after adrenalectomy (192). Also, electrophysiological evidence indicates that the effects of adrenalectomy are not observed earlier than 4 days in CA1 hippocampal neurons. Repeated stimulation of these neurons leads to a gradually diminished response to stimulation in adrenalectomized rats, which is apparent by 4 days, but not 2 days, after adrenalectomy (187).

A second approach to examining the interaction between CORT and mesocorticolimbic dopamine is to produce 6-OHDA lesions of dopamine neurons located in the ventral mesencephalon. Casolini et al. (193) have demonstrated a decrease in basal as well as restraint stress-induced CORT levels after lesioning these cells, indicating an excitatory effect of dopamine on plasma CORT levels. On the other hand, Rivet et al., (194) have shown no effect of 6-OHDA in the VTA on CORT levels. However, both NE and dopamine levels were decreased in the latter study, and the results may be due to counteractive effects similar to what is observed for dopamine and NE in the mPFC discussed earlier.

The neural mechanisms mediating mesocorticolimbic dopamine and CORT interactions may lie in the projections from limbic and cortical structures to the bed nucleus of stria terminalis, septum or amygdala, which in turn project to the PVN (195–197, also see discussion of this topic 151). In addition, dopamine may regulate CORT levels, and vice versa, indirectly by altering their respective receptors. Some studies have examined the relationship between dopamine and GR/MRs, as well as the interaction between CORT and dopamine receptors. Both MR and GRs appear to be involved in negative feedback on stress-induced CRH release (198), and the number of receptors, but not their affinity, alters basal and/or stress-induced CORT levels (199, 200). Since GRs rapidly turn over (201), and since repeated stress or corticosterone treatment generally reduces the number of GRs and MRs (202–204), while adrenalectomy increases the number of these receptors (204,205), the number or affinity of receptors present in discrete brain sites may change relatively rapidly and predict the sensitivity of neuronal activity. Animals that demonstrate an enhanced response to novelty (high responders, HR) and predisposition to self-administer cocaine have a longer duration of increased plasma CORT levels following exposure to novelty stress (177,206), which led these investigators to examine delayed negative feedback mediated by hippocampal GRs and MRs. Maccari et al. (175) reported a decreased affinity of both receptor types in the hippocampus of rats that readily acquire amphetamine self-administration. Dopamine depletion by 6-OHDA lesions in the VTA produces an increase in GR affinity in the ventral striatum, including the NAS, suggesting that dopamine is inhibitory on GRs (207). The effects of CORT on dopamine receptors has also been demonstrated; a long-term decrease in D1 and/or D2 receptors in the striatum and substantia nigra occur after adrenalectomy, while increases in D1 receptors in the NAS and striatum occur after dexamethasone treatment (208).

Both noradrenergic and serotonergic systems also interact with the HPA axis, in addition to acting directly to regulate CORT receptors in the brain independently of their regulation on adrenocortical function. Maccari et al. (209) have produced 6-OHDA lesions in ascending NE pathways in rats given adrenalectomy and adrenalectomy plus CORT replacement therapy. Both groups of rats demonstrated an increase in the number of MRs in the hippocampus, indicating regulation of HPA-axis activity via direct effects on MRs.

Plasma CORT modulates 5-HT transmission and 5-HT receptors, and 5-HT has been shown to modulate GR and MRs. Corticosterone alters the synthesis, turnover, and binding sites of 5-HT (210–214). Van de Kar et al. (215) reported that repeated cocaine exposure alters the responsiveness of 5-HT terminals, decreasing the ability of 5-HT to stimulate plasma ACTH levels. Therefore, it is apparent that stress- and psychostimulant-induced changes in NE and 5-HT function might also contribute to the observed changes in circulating CORT levels and their receptors.

Role of the HPA Axis in Sensitization: Conclusions

In summary, there appear to be numerous ways in which CORT interacts with neurotransmitters to produce the observed effects following repeated stress and psychostimulant drugs. These include altered CORT levels and changes in the number and affinity of GRs and MRs, and these receptors are probably regulated differentially in discrete brain regions by neurotransmitter systems both independently of and via the HPA axis. There is much evidence to support a role for CORT in stress- and drug-induced sensitization, but whether it is able to act as a primary mediator or in a permissive role is unknown. Most reports point to CORT as a permissive agent in altering mesolimbic or hippocampal neurons (99,148,214). The question of how CORT may translate stressful events into altered monoamine function may prove to be an important one in determining how repetition of stressful stimuli can produce behavioral and neurochemical sensitization.

The clinical data suggest that altered responsiveness to stress, as reflected by activity of the HPA axis, can lead to maladaptive behavioral responses to stressful stimuli. Hypersecretion of CRH and reduced ACTH responsiveness are associated with panic disorder and PTSD (216, 217). Elevated cortisol secretion or cortisol therapy can result in behavioral changes, including psychosis resembling that observed after repeated use of psychostimulants (1,218–220).

SUMMARY AND GENERAL CONCLUSIONS

Evidence to date shows that the neural substrates critical for development and expression of behavioral sensitization to psychostimulants include the VTA and NAS, respectively. An enhancement in NAS dopamine levels to subsequent drug challenge after at least 1 week of withdrawal from the drug is commonly observed, while a blunting of dopamine release appears in the mPFC after repeated cocaine exposure. Repeated stress, which produces behavioral cross-sensitization with psychostimulant drugs, produces dopamine responses in the NAS and mPFC similar to those of drug-induced sensitization. Expression of elevated dopamine levels in the NAS after drug challenge appears to be independent of dopamine perikarya, although it is possible that the magnitude of the sensitized response also depends on the rate and type of VTA dopamine cell firing, and local interactions with other neurotransmitters, including excitatory amino acid neurons projecting from the mPFC, hippocampus, and amygdala. A wealth of studies demonstrate an inhibitory action of mPFC dopamine on locomotor activity and NAS dopamine transmission. However, future work will need to focus on the role of mPFC dopamine in modulating excitatory amino acid output, specifically to the VTA versus the NAS.

Norepinephrine and 5-HT are also altered by stress and psychostimulants, and interactions between CRH and CORT with these neurotransmitter systems has been demonstrated. It will be necessary to assess NE and 5-HT transmission simultaneously with changes in dopamine transmission, and the role that CRH and CORT, acting at extrahypothalamic sites, play in modulating all three neurotransmitter systems.

The role of CRH and CORT in the initiation and/or expression of behavioral sensitization remains unclear. The fact that CORT can modulate cell firing in hippocampal neurons for extended periods suggests that activation of the HPA axis by drugs and stress may function in providing long-term modulation of neuronal activity in other GR- or MR-containing neurons. The abundance of evidence showing the interdependence of CORT and monoaminergic neurotransmitter systems suggests that an organism's assessment of fluctuations in its environment leads to the recruitment of these hormonal and neuronal responses, which are inextricably linked in translating environmental stimuli into behavior. Future studies will need to address whether CRH and CORT are involved in the initiation and/or expression of behavioral and neurochemical sensitization. Further, it will be important to demonstrate if CRH and CORT placed into the VTA can induce sensitization, whether they promote sensitization by acting as permissive agents, or whether activation of CRH and/or GRs in the VTA are necessary for inducing sensitization.

In humans, stress-precipitated psychoses such as panic disorder and PTSD resemble behavioral sensitization in rodents in the following manner: 1) there is a progression of the increased responsiveness to stressful stimuli over time; 2) the effects are long lasting and possibly permanent; and 3) hyperresponsiveness to a stimulus is not specific to the stressor but can crossreact with other stressful stimuli as well. There is overwhelming evidence that stress- and psychostimulant-induced alterations in the mesocorticolimbic dopamine system are critical for the emergence of behavioral sensitization in animals. Studies in animal models indicate that recruitment of hormonal systems involved in the stress response may importantly facilitate several behaviors, including sensitization to stress and psychostimulants. These models will provide information on neurotransmitter/hormone interactions at the behavioral, anatomical, and molecular levels to serve as the basis for future studies of human stress-induced psychopathologies.

ACKNOWLEDGMENTS

This work was supported in part by USPHS grants #DA-07827, DA-08212 (BAS), MH-

40817, DA 03906, and a Research Career Development Award DA-00158 (PWK).

REFERENCES

1. Antelman SM. Stressor-induced sensitization to subsequent stress: implications for the development and treatment of clinical disorders. In Kalivas PW, Barnes CD, eds. *Sensitization in the central nervous system,* New York: Academic Press, 1988;227–259.
2. Angrist B. Psychoses induced by central nervous system stimulants and related drugs. In Creese I, ed. *Stimulants: neurochemical, behavioral, and clinical perspectives,* New York: Raven Press, 1983;1–30.
3. Post RM, Weiss SRB. Sensitization and kindling: implications for the evolution of psychiatric symptomatology. In Kalivas PW, Barnes CD, eds. *Sensitization in the nervous system,* New Jersey: Telford Press, 1988;257–292.
4. Robinson TE, Becker JB. Enduring changes in brain and behavior produced by chronic amphetamine administration: a review and evaluation of animal models of amphetamine psychosis. *Brain Res* 1986;11:157–198.
5. Segal DS, Schuckit MA. Animal models of stimulant-induced psychosis. In Creese I., ed. *Stimulants: neurochemical, behavioral and clinical perspectives,* New York: Raven Press, 1983;131–167.
6. Kalivas PW, Stewart J. Dopamine transmission in the initiation and expression of drug- and stress-induced sensitization of motor activity. *Brain Res Rev* 1991; 16:223–244.
7. Deutch AY, Roth RH. The determinants of stress-induced activation of the prefrontal cortical dopamine system. *Prog Brain Res* 1990;85:357–393.
8. Fibiger HC, Phillips AG. Reward, motivation, cognition: psychobiology of mesotelencephalic dopamine systems, In Mountcastle VB, Bloom FE, Geiger SR, eds. *The handbook of physiology, vol IV, Intrinsic regulatory systems of the brain* Bethesda: American Physiology Society, 1986;647–675.
9. Wise RA, Bozarth MA. A psychomotor stimulant theory of addiction. *Psychol Rev* 1987;94:469–492.
10. Fallon JH, Moore RY. Catecholamine innervation of basal forebrain. IV. Topography of the dopamine projection to the basal forebrain and striatum. *J Comp Neurol* 1978;180:545–580.
11. Swanson LW. The projections of the ventral tegmental area and adjacent regions: A combined fluorescent retrograde tracer and immunofluorescence study in the rat. *Brain Res Bull* 1982;9:321–353.
12. Koe BK. Molecular geometry of inhibitors of the uptake of catecholamines and serotonin in synaptosomal preparations of rat brain. *J Pharmacol Exp Ther* 1976; 199:649–661.
13. Moore ME. Amphetamines: biochemical and behavioral actions in animals. In Iversen LL, Iversen SD, Snyder SH, eds. *Handbook of psychopharmacology,* vol 11, New York: Plenum Press, 1978;41–98.
14. Antelman SM, Eichler AJ, Black CA, Kocan D. Interchangeability of stress and amphetamine in sensitization. *Science* 1980;207:329–331.
15. Herman J-P, Stinus L, Le Moal M. Repeated stress increases locomotor response to amphetamine. *Psychopharmacol* 1984;84:431–435.
16. MacLennan AJ, Maier SF. Coping and the stress-induced potentiation of stimulant stereotypy in the rat. *Science* 1983;219:1091–1093.
17. Robinson TE, Angus AL, Becker JB. Sensitization to stress: the enduring effects of prior stress on amphetamine-induced rotational behavior. *Live Sci* 1985;37: 1039–1042.
18. Sorg BA, Kalivas PW. Effects of cocaine and footshock stress on extracellular dopamine levels in the ventral striatum. *Brain Res* 1991;559:29–36.
19. Abercrombie ED, Keller RW, Zigmond MJ. Characterization of hippocampal NE release as measured by microdialysis perfusion: pharmacological and behavioral studies. *Neurosci* 1988;27:897–904.
20. Abercrombie ED, Keefe KA, DiFrischia DS, Zigmond MJ. Differential effect of stress on in vivo dopamine release in striatum, nucleus accumbens, and medial frontal cortex. *J Neurochem* 1989;52:1655–1658.
21. Gresch PJ, Sved AF, Zigmond MJ, Finlay JM. Stress-induced sensitization of dopamine and norepinephrine efflux in medial prefrontal cortex of the rat. *J Neurochem* 1994;63:575–583.
22. Tanaka M, Kohno Y, Nakagawa R, Ida Y, Takeda S, Nagasaki N, Noda N. Regional characteristics of stress-induced increases in brain noradrenaline release in rats. *Pharmacol Biochem Behav* 1983;19:543–547.
23. Thierry AM, Javoy J, Glowinski J, Kety SS. Effects of stress on the metabolism of norepinephrine, dopamine and serotonin in the central nervous system of the rat. I. Modifications of norepinephrine turnover. *J Pharmacol Exp Ther* 1968;163:163–171.
24. Perugini M, Vezina P. Amphetamine administered to the ventral tegmental area sensitizes rats to the locomotor effects of nucleus accumbens amphetamine. *J Pharmacol Exp Ther* 1994;270:690–696.
25. Kalivas PW, Alesdatter JE. Involvement of N-methyl-D-aspartate in the ventral tegmental area and amygdala in behavioral sensitization to cocaine. *J Pharmacol Exp Ther* 1993;267:486–495.
26. Deutch AY, Tam SY, Roth RH. Footshock and conditioned stress increase 3,4-dihydroxyphenylacetic acid (DOPAC) in the ventral tegmental area but not substantia nigra. *Brain Res* 1985;333:143–146.
27. Kalivas PW, Duffy P, Barrow J. Regulation of the mesocorticolimbic dopamine system by glutamic acid receptor subtypes. *J Pharmacol Exp Ther* 1989;251: 378–387.
28. Kalivas PW, Duffy P. Time course of extracellular dopamine and behavioral sensitization to cocaine. II. Dopamine perikarya. *J Neurosci* 1993b;13:276–284.
29. Sorg BA, Chen S-Y, Kalivas P. Time course of tyrosine hydroxylase expression after behavioral sensitization to cocaine. *J Pharmacol Exp Ther* 1993;266:424–430.
30. Kalivas PW, Duffy P. Similar effects of daily cocaine and stress on mesocorticolimbic dopamine neurotransmission in the rat. *Biol Psychiat* 1989;25:913–928.
31. Kalivas PW, Duffy P. Time course of extracellular dopamine and behavioral sensitization to cocaine. I. Dopamine axon terminals. *J Neurosci* 1993a;13:266–275.
32. Segal DS, Kuczenski R. In vivo microdialysis reveals a diminished amphetamine-induced dopamine response

corresponding to behavioral sensitization produced by repeated amphetamine pretreatment. *Brain Res* 1992a; 521:330–332.
33. Segal DS, Kuczenski R. Repeated cocaine administration induces behavioral sensitization and corresponding decreased extracellular dopamine responses in caudate and accumbens. *Brain Res* 1992b;577:351–355.
34. Pettit HO, Pan H-T, Parsons LH, Justice JB Jr. Extracellular concentrations of cocaine and dopamine are enhanced during chronic cocaine administration. *J Neurochem* 1990;55:798–804.
35. DiChiara G, Imperato A. Drugs abused by humans preferentially increase synaptic dopamine concentrations in the mesolimbic system of freely moving rats. *Proc Natl Acad Sci USA* 1988;85:5274–5278.
36. Pettit HO, Justice JB Jr. Dopamine in the nucleus accumbens during cocaine self administration as studied by in vivo microdialysis. *Pharmacol Biochem Behav* 1989;34:899–904.
37. Kalivas PW, Duffy P. Effect of acute and daily cocaine treatment on extracellular dopamine in the nucleus accumbens. *Synapse* 1990;5:48–58.
38. Kazahaya Y, Akimoto K, Otsuki S. Subchronic methamphetamine treatment enhances methamphetamine- or cocaine-induced dopamine efflux in vivo. *Biol Psychiat* 1989;25:903–912.
39. Parsons LH, Justice JB Jr. Serotonin and dopamine sensitization in the nucleus accumbens, ventral tegmental area and dorsal raphe nucleus following repeated cocaine administration. *J Neurochem* 1993;61: 1611–1619.
40. Robinson TE, Jurson PA, Bennett JA, Bentgen KM. Persistent sensitization of dopamine neurotransmission in ventral striatum (nucleus accumbens) produced by prior experience with (+)-amphetamine: a microdialysis study in freely moving rats. *Brain Res* 1988;462: 211–222.
41. Cenci MA, Kalen P, Mandel RJ, Bjorklund A. Regional differences in the regulation of dopamine and noradrenaline release in medial frontal cortex, nucleus accumbens and caudate-putamen: a microdialysis study in the rat. *Brain Res* 1992;581:217–228.
42. Mantz J, Thierry AM, Glowinski J. Effect of noxious tail pinch on the discharge rate of mesocortical and mesolimbic dopamine neurons. Selective activation of the mesocortical system. *Brain Res* 1989;476:377–381.
43. Thierry AM, Tassin J-P, Blanc G, Glowinski J. Selective activation of the mesocortical dopamine system by stress. *Nature* 1976;263:242–243.
44. Doherty MD, Gratton A. High-speed chronoamperometric measurements of mesolimbic and nigrostriatal dopamine release associated with repeated daily stress. *Brain Res* 1992;586:295–302.
45. Imperato A, Angelucci L, Casolini P, Zocchi A, Puglisi-Allegra S. Repeated stressful experiences differently affect limbic dopamine release during and following stress. *Brain Res* 1992;577:194–199.
46. Deutch AY, Bourdelais AJ, Zahm DS. The nucleus accumbens core and shell: accumbal compartments and their functional attributes. In Kalivas PW, Barnes CD, eds. *Limbic Motor Circuits and Neuropsychiatry,* Boca Raton: CRC Press, 1993;45–88.
47. Castaneda E, Becker JB, Robinson TE. The long-term effects of repeated amphetamine treatment in vivo on amphetamine, KCl and electrical stimulation evoked striatal dopamine release in vitro. *Life Sci* 1988;42: 2447–2456.
48. Kolta MG, Shreve P, De Souza V, Uretsky NJ. Time course of the development of the enhanced behavioral and biochemical responses to amphetamine after pretreatment with amphetamine. *Neuropharmacol* 1985; 24:823–829.
49. Peris J, Zahniser N. One injection of cocaine produces a long-lasting increase in [3H] dopamine release. *Pharmacol Biochem Behav* 1987;27:533–535.
50. Kalivas PW, Duffy P. Effects of daily cocaine and morphine treatment on somatodendritic and terminal field dopamine release. *J Neurochem* 1988;50:1498–1504.
51. Yi S-J, Johnson KM. Effects of acute and chronic administration of cocaine on striatal uptake, compartmentalization and release of [3H]dopamine. *Neuropharmacol* 1990;29:475–486.
52. Paulson PE, Robinson TE. Sensitization to systemic amphetamine produces an enhanced locomotor response to a subsequent intra-accumbens amphetamine challenge in rats. *Psychopharmacol* 1991;104:140–141.
53. Goeders NE, Smith JE. Cortical dopaminergic involvement in cocaine reinforcement. *Science* 1983;221: 773–775.
54. Blanc G, Herve D, Simon H, Lisoprawski A, Glowinski J, Tassin J-P. Response to stress of mesocortical-frontal dopaminergic neurons after long-term isolation. *Nature* 1980;284:265–276.
55. Caggiula AR, Antelman SM, Aul E, Knopf S, Edwards DJ. Prior stress attenuates the analgesic response but sensitizes the corticosterone and cortical dopamine responses to stress 10 days later. *Psychopharmacol* 1989; 99:233–237.
56. Herman J-P, Guillonneau D, Dantzer R, Scatton B, Semerdjian-Rouquier L, Le Moal M. Differential effects of inescapable footshocks and of stimuli previously paired with inescapable footshock on dopamine turnover in cortical and limbic areas of the rat. *Life Sci* 1982;30:2207–2214.
57. Robinson TE, Becker JB, Moore CJ, Castaneda E, Mittleman G. The effects of footshock stress on regional brain dopamine metabolism and pituitary β-endorphin release in rats previously sensitized to amphetamine. *Neuropharmacol* 1987;26:679–691.
58. Sorg BA, Kalivas PW. Effects of cocaine and footshock stress on extracellular dopamine levels in the medial prefrontal cortex. *Neuroscience* 1993a;53:695–703.
59. Sorg BA, Kalivas PW. Effects of repeated cocaine on subsequent cocaine- and amphetamine-induced changes in extracellular dopamine levels in the medial prefrontal cortex. *Soc Neurosci Abstr* 1993b;19:1848.
60. Hamamura T, Fibiger HC. Enhanced stress-induced dopamine release in the prefrontal cortex of amphetamine sensitized rats. *Eur J Pharmacol* 1993;237: 65–71.
61. Kuczenski R, Segal DS. Differential effects of amphetamine and dopamine uptake blockers (cocaine, nomifensine) on caudate and accumbens dialysate dopamine and 3-methoxytyramine. *J Pharmacol Exp Ther* 1992; 262:1085–1094.
62. Robinson TE, Camp DM. Does amphetamine *preferentially* increase the extracellular concentration of dopa-

mine in the mesolimbic system of freely moving rats? *Neuropsychopharmacol* 1990;3:163–173.
63. Fischer JF, Cho AK. Chemical release of dopamine from striatal homogenates: evidence for an exchange diffusion model. *J Pharmacol Exp Ther* 1979;208: 203–209.
64. Hurd YL, Ungerstedt U. Ca^{2+} dependence of the amphetamine, nomifensine, and Lu 19-005 effect on in vivo dopamine transmission. *Eur J Pharmacol* 1989; 166:261–269.
65. Nomikos GG, Damsma G, Wenkstern D, Fibiger HC. In vivo characterization of locally applied dopamine uptake inhibitors by striatal microdialysis. *Synapse* 1990;6:106–112.
66. Carter CJ. Glutamatergic pathways from the medial prefrontal cortex to the anterior striatum, nucleus accumbens and substantia nigra. *Brit J Pharmacol* 1980; 70:50P–51P.
67. Carter CJ, Pycock CJ. Behavioral and biochemical effects of dopamine and noradrenaline depletion within the medial prefrontal cortex of the rat. *Brain Res* 1980;192:163–176.
68. Louilot A, LeMoal M, Simon H. Opposite influences of dopaminergic pathways to the prefrontal cortex or the septum on the dopaminergic transmission in the nucleus accumbens. An in vivo voltammetric study. *Neuroscience* 1989;29:45–56.
69. Tassin J-P, Stinus L, Simon H, Blanc G, Thierry A-M, LeMoal M, Card G, Glowinski J. Relationship between the locomotor hyperactivity induced by A10 lesions and the destruction of the frontocortical dopaminergic innervation in the rat. *Brain Res* 1978;141: 267–281.
70. Iversen SD, Wilkinson S, Simpson B. Enhanced amphetamine responses after frontal cortex lesions in the rat. *Eur J Pharmacol* 1971;13:387–390.
71. Pycock CJ, Carter CJ, Kerwin RW. Effect of 6-hydroxydopamine lesions of the medial prefrontal cortex on neurotransmitter systems in subcortical sites in the rat. *J Neurochem* 1980;34:91–99.
72. Deutch AY, Clark WA, Roth RH. Prefrontal cortical dopamine depletion enhances the responsiveness of mesolimbic dopamine neurons to stress. *Brain Res* 1990;521:311–315.
73. Rosin DL, Clark WA, Goldstein M, Roth RH, Deutch AY. Effects of 6-hydroxydopamine lesions of the prefrontal cortex on tyrosine hydroxylase activity in mesolimbic and nigrostriatal dopamine systems. *Neuroscience* 1992;48:831–839.
74. Mitchell JB, Gratton A. Partial dopamine depletion of the prefrontal cortex leads to enhanced mesolimbic dopamine release elicited by repeated exposure to naturally reinforcing stimuli. *J Neurosci* 1992;12:3609–3618.
75. Schenk S, Horger BA, Peltier R, Shelton K. Supersensitivity to the reinforcing effects of cocaine following 6-hydroxydopamine lesions to the medial prefrontal cortex in rats. *Brain Res* 1991;543:227–235.
76. Piazza PV, Rouge-Pont F, Deminiere JM, Kharoubi M, LeMoal M, Simon H. Dopaminergic activity is reduced in the prefrontal cortex and increased in the nucleus accumbens of rats predisposed to develop amphetamine self-administration. *Brain Res* 1991a;567: 169–174.
77. Christie MJ, Bridge S, James LB, Beart PM. Excitotoxin lesions suggest an aspartatergic projection from rat medial prefrontal cortex to ventral tegmental area. *Brain Res* 1985;333:169–172.
78. Gonon FG. Nonlinear relationship between impulse flow and dopamine released by rat midbrain dopaminergic neurons as studied by in vivo electrochemistry. *Neurosci* 1988;24:19–28.
79. Gariano RF, Groves PM. Burst firing induced in midbrain dopamine neurons by stimulation of the medial prefrontal and anterior cingulate cortices. *Brain Res* 1988;462:194–198.
80. Svensson TH, Tung, C-S. Local cooling of pre-frontal cortex induces pacemaker-like firing of dopamine neurons in rat ventral tegmental area in vivo. *Acta Physiol Scand* 1989;136:135–136.
81. Swanson LW, Hartman BK. The central adrenergic system. An immunofluorescence study of the location of cell bodies and their efferent connections in the rat utilizing dopamine-β-hydroxylase as a marker. *J Comp Neurol* 1975;163:467–506.
82. Tassin JP, Simon H, Herve D, Blanc G, LeMoal M, Glowinski J, Bockaert J. Nondopaminergic fibers may regulate dopamine-sensitive adenylate cyclase in the prefrontal cortex and the nucleus accumbens. *Nature* 1982;295:696–698.
83. LeMoal M, Galey D, Cardo B. Behavioural effects of local injection of 6-hydroxydopamine in the medial ventral tegmentum in the rat. Possible role of the mesolimbic dopaminergic system. *Brain Res* 1975;88:190–194.
84. Carboni E, Tanda GL, Frau R, DiChiara G. Blockade of the noradrenaline carrier increases extracellular dopamine concentration in the prefrontal cortex: evidence that dopamine is taken up in vivo by noradrenergic terminals. *J Neurochem* 1990;55:1067–1070.
85. Gresch PJ, Sved AF, Zigmond MJ, Finlay JM. Local application of desipramine increases basal and stress-induced dopamine efflux in medial prefrontal cortex. *Soc Neurosci Abstr* 1993;19:1828.
86. Lindvall O, Bjorklund A. Organization of catecholamine neurons in the rat central nervous system. In Iversen LL, Iversen SD, Snyder SH, eds. *Handbook of psychopharmacology,* New York: Plenum Press, vol 9, 1978;139–231.
87. Foote SL, Aston-Jones G, Bloom FE. Impulse activity of locus coeruleus neurons in awake rats and monkeys is a function of sensory stimulation and arousal. *Proc Natl Acad Sci USA* 1980;77:3033–3037.
88. Robbins TW, Everitt BJ. Psychopharmacological studies of arousal and attention. In Stahl SM, Iverson SD, Goodman E, eds. *Cognitive neurochemistry,* London: Oxford University Press, 1987;135–170.
89. Paulson PE, Camp DM, Robinson TE. Time course of transient behavioral depression and persistent behavioral sensitization in relation to regional brain monoamine concentrations during amphetamine withdrawal in rats. *Psychopharmacology* 1991;103:480–492.
90. Geyer MA. Behavioral effects of xylamine-induced depletions of brain norepinephrine: interaction with amphetamine. *Behav Brain Res* 1986;21:55–64.
91. Ogren SO, Archer T, Johansson C. Evidence for a selective brain noradrenergic involvement in the locomotor stimulant effects of amphetamine in the rat. *Neurosci Lett* 1983;43:327–331.

92. Rossetti ZL, Portas C, Pani L, Carboni S, Gessa GL. Stress increases noradrenaline release in the rat frontal cortex: prevention by diazapam. *Eur J Pharmacol* 1990;176:229–231.
93. Abercrombie ED, Jacobs BL. Single-unit response of noradrenergic neurons in the locus coeruleus of freely moving cats. I. Acutely presented stressful and non-stressful stimuli. *J Neurosci* 1981;7:2837–2843.
94. Aston-Jones G, Bloom FE. Norepinephrine-containing locus coeruleus neurons in behaving rats exhibit pronounced response to non-noxious environmental stimuli. *J Neurosci* 1981;1:887–900.
95. Adell A, Garcia-Marquez C, Armario A, Gelpi E. Chronic stress increases serotonin and noradrenaline in rat brain and sensitizes their responses to a further acute stress. *J Neurochem* 1988;50:1678–1681.
96. Kvetnansky R, Palkovits M, Mitro A, Torda T, Mikulaj L. Catecholamines in individual hypothalamic nuclei of acutely and repeatedly stressed rats. *Neuroendocrinology* 1977;23:257–267.
97. Nisenbaum LK, Zigmond MJ, Sved AF, Abercrombie ED. Prior exposure to chronic stress results in enhanced synthesis and release of hippocampal norepinephrine in response to a novel stressor. *J Neurosci* 1991;11: 1478–1484.
98. Finlay JM, Rabinovic AD, Gresch PJ, Sved AF, Zigmond MJ. Stress-induced sensitization of norepinephrine efflux in prefrontal cortex: further characterization. *Soc Neurosci Abstr* 1993;19:1828.
99. Dunn A. Stress-related changes in cerebral catecholamine and indoleamine metabolism: lack of effect of adrenalectomy and corticosterone. *J Neurochem* 1988; 51:406–412.
100. Torda T, Murgas K, Cechova E, Kiss A, Saavedra JM. Adrenergic regulation of 3[H]ketanserin binding sites during immobilization stress in the rat frontal cortex. *Brain Res* 1990;527:198–203.
101. Izenwasser S, Rosenberger JG, Cox BM. Inhibition of 3[H]dopamine and 3[H]serotonin uptake by cocaine; comparison between chopped tissue slices and synaptosomes. *Life Sci* 1992;50:541–547.
102. Azmitia EC, Segal M. An autoradiographic analysis of differential ascending projections of the dorsal and median raphe nuclei in the rat. *J Comp Neurol* 1978; 179:641–668.
103. Simon H, Le Moal M, Calas A. Efferents and afferents of the ventral tegmental A10 region studies after local injection of 3[H]leucine and horseradish peroxidase. *Brain Res* 1979;178:17–40.
104. Kalivas PW. Neurotransmitter regulation of dopamine neurons in the ventral tegmental area. *Brain Res Rev* 1993;18:75–113.
105. Cunningham KA, Paris JM, Goeders NE. Chronic cocaine enhances serotonin autoregulation and serotonin uptake binding. *Synapse* 1992;11:112–123.
106. Ohi K, Mikuni M, Takahashi K. Stress adaptation and hypersensitivity in 5-HT neuronal systems after repeated foot shock. *Pharmacol Biochem Beh* 1989; 34:603–608.
107. Torda T, Culaman J, Cechova E, Murgas K. 3H-Ketanserin (serotonin type 2) binding in rat frontal cortex: effect of immobilization stress. *Endocrinologia Experimentalis* 1988;22:99–105.
108. Gelders Y, Vanden Bussche G, Reyntjens A, Janssen P. Serotonin -S_2 receptor blockers in the treatment of chronic schizophrenia. *Clin Neuropharmacol* 1986;9, Suppl.4:325–327.
109. Weinberger DR. Implications of normal brain development for the pathogenesis of schizophrenia. *Arch Gen Psychiat* 1987;44:660–669.
110. Svensson TH, Tung C-S, Grenhoff J. The 5-HT2 antagonist ritanserin blocks the effect of pre-frontal cortex inactivation on rat A10 dopamine neurons in vivo. *Physiol Scand* 1989;136:497–498.
111. Antelman SM, Chiodo LA. Stress: its effect on interactions among biogenic amines and role in the induction and treatment of disease. In Iversen LL, Iversen SD, Snyder SH, eds. *Handbook of psychopharmacology,* vol. 18, New York: Plenum Press, 1990;279–341.
112. Knich ET, Eisenberg RM. Effect of amphetamine on plasma corticosterone in the conscious rat. *Neuroendocrinology* 1979;29:110–118.
113. Rivier C, Vale W. Cocaine stimulates adrenocorticotropin (ACTH) secretion through a corticotropin-releasing factor (CRF)-mediated mechanism. *Brain Res* 1987;422:403–406.
114. Munck A, Guyre PM. Glucocorticoid physiology, pharmacology and stress. *Adv Exp Med Biol* 1986;196: 81–96.
115. Vale W, Speiss J, Rivier C, Rivier J. Characterization of a 41-residue ovine hypothalamic peptide that stimulates secretion of corticotropin and β-endorphin. *Science* 1981;213:1394–1397.
116. Chappell PB, Smith MA, Kilts CD, Bissette G, Ritchie J, Anderson C, Nermeroff CB. Alterations in corticotropin-releasing factor-like immunoreactivity in discrete rat brain regions after acute and chronic stress. *J Neurosci* 1986;6:2908–2914.
117. Hiroshige T, Fujieda K, Kaneko M, Honma K. Assays and dynamics of corticotropin-releasing factor activity in rat hypothalamus. *Ann NY Acad Sci* 1977;297: 436–454.
118. Imaki T, Nahan J-L, Rivier C, Sawchenko PE, Vale W. Differential regulation of corticotropin-releasing factor mRNA in rat brain regions by glucocorticoids and stress. *J Neurosci* 1991;11:585–599.
119. Lightman SL, Young WS III. Influence of steroids on the hypothalamic corticotropin-releasing factor and preproenkephalin mRNA responses to stress. *Proc Natl Acad Sci USA* 1989;86:4306–4310.
120. Moldow RL, Fischman AJ. Physiological changes in rat hypothalamic CRF: circadian, stress and steroid suppression. *Peptides* 1982;3:837–840.
121. Moldow RL, Kastin AJ, Graf M, Fischman AJ. Stress mediated changes in hypothalamic corticotropin-releasing factor-like immunoreactivity. *Life Sci* 1986;40: 413–418.
122. Sato T, Sato M, Shinsako J, Dallman MF. Corticosterone-induced changes in hypothalamic corticotropin-releasing factor (CRF) content after stress. *Endocrinology* 1975;97:265–274.
123. Sutton RE, Koob GF, Le Moal M, Rivier J, Vale W. Corticotropin-releasing factor produces behavioral activation in rats. *Nature* 1982;297:331–333.
124. Veldhuis HD, de Wied D. Differential behavioral actions of corticotropin-releasing factor (CRF). *Pharmacol Biochem Behav* 1984;21:707–713.
125. Eaves M, Thatcher-Britton J, Rivier J, Vale W. Effects of corticotropin-releasing factor on locomotor activity in hypophysectomized rats. *Peptides* 1985;6:923–926.

126. Matsuzaki I, Takamatsu Y, Moroji T. The effects of intracerebro-ventricularly injected corticotropin-releasing factor (CRF) on the central nervous system: behavioral and biochemical studies. *Neuropeptides* 1989;13:147–155.
127. Levine AS, Rogers B, Kniep J, Grace M, Morley JE. Effects of centrally administered corticotropin releasing factor (CRF) on multiple feeding paradigms. *Neuropharmacology* 1983;22:337–339.
128. Berridge CW, Dunn AJ. A corticotropin-releasing factor antagonist reverses the stress-induced changes in exploratory behavior in mice. *Horm Behav* 1987;21: 393–401.
129. Kalin NH, Sherman JE, Takahashi LK. Antagonism of endogenous CRH systems attenuates stress-induced freezing behavior in rats. *Brain Res* 1988;457:130–135.
130. Tazi A, Dantzer R, Le Moal M, Rivier J, Vale W, Koob GF. Corticotropin-releasing factor antagonist blocks stress-induced fighting in rats. *Regul Peptides* 1987; 18:37–42.
131. Britton DR, Varela M, Garcia A, Rosenthal M. Dexamethasone suppresses pituitary-adrenal but not behavioral effects of centrally administered CRF. *Life Sci* 1986;38:211–216.
132. Bloom FE, Battenberg ELF, Rivier J, Vale W. Corticotropin-releasing factor (CRF): immunoreactive neurons and fibres in rat hypothalamus. *Regul Peptides* 1982;4:43–48.
133. De Souza EB, Perrin MH, Insel TR, Rivier J, Vale WW, and Kuhar MJ. Corticotropin-releasing factor receptors in rat forebrain: autoradiographic identification. *Science* 1984;224:1449–1451.
134. Olschowka JA, O'Donohue TL, Mueller GP, Jacobowitz DM. The distribution of corticotropin releasing factor-like immunoreactive neurons in rat brain. *Peptides* 1982;3:995–1015.
135. Swanson LW, Sawchenko PE, Rivier J, Vale WW. Organization of ovine corticotropin-releasing factor immunoreactive cells and fibers in the rat brain: an immunohistochemical study. *Neuroendocrinology* 1983; 36:164–186.
136. Goeders NE, Bienvenu OJ, De Souza EB. Chronic cocaine administration alters corticotropin-releasing factor receptors in the rat brain. *Brain Res* 1990;531: 322–328.
137. Sarnyai Z, Hohn J, Szabo G, Penke B. Critical role of endogenous corticotropin-releasing ractor (CRF) in the mediation of the behavioral action of cocaine in rats. *Life Sci* 1992a;51:2019–2024.
138. Cole BJ, Cador M, Stinus L, Rivier C, Rivier J, Vale W, Le Moal M, Koob G. Critical role of the hypothalamic pituitary adrenal axis in amphetamine-induced sensitization of behavior. *Life Sci* 1990a;47:1715–1720.
139. Cole BJ, Cador M, Stinus L, Rivier J, Vale W, Koob G, Le Moal M. Central administration of a CRF antagonist blocks the development of stress-induced behavioral sensitization. *Brain Res* 1990b;512:343–346.
140. Cador M, Cole BJ, Koob GF, Stinus L, Le Moal M. Central administration of corticotropin releasing factor induces long-term sensitization to D-amphetamine. *Brain Res* 1993a;606:181–186.
141. Dunn AJ, Berridge CW. Corticotropin-releasing factor administration elicits a stress-like activation of cerebral catecholaminergic systems. *Pharmacol Biochem Beh* 1987;27:685–691.
142. Koob GF, Swerdlow N, Seeligson M, Eaves M, Sutton R, Rivier J, Vale W. Effects of alpha-flupenthixol and naloxone on CRF-induced locomotor activation. *Neuroendocrinology* 1984;39:459–464.
143. Swerdlow N, Koob GF. Separate neural substrates of the locomotor-activating properties of amphetamine, heroin, caffeine and corticotropin releasing factor (CRF) in the rat. *Pharmacol Biochem Behav* 1985;23: 303–307.
144. Kalivas PW, Duffy P, Latimer LG. Neurochemical and behavioral effects of corticotropin-releasing factor in the ventral tegmental area of the rat. *J Pharmacol Exp Ther* 1987;242:757–763.
145. De Souza EB, Insel TR, Perrin MH, Rivier J, Vale WW, Kuhar MJ. Corticotropin-releasing factor receptors are widely distributed within the rat central nervous system: an autoradiographic study. *J Neurosci* 1985;5: 3189–3203.
146. Butler PD, Weiss JM, Stout JC, Nemeroff CB. Corticotropin-releasing factor produces fear-enhancing behavioral activating effects following infusion into the locus coeruleus. *J Neurosci* 1990;10:176–183.
147. Melia KR, Duman RS. Involvement of corticotropin-releasing factor in chronic stress regulation of the brain noradrenergic system. *Proc Natl Acad Sci USA* 1991; 88:8382–8386.
148. de Kloet ER. Brain corticosteroid receptor balance and homeostatic control. *Front Neuroendocrinol* 1991;12: 95–164.
149. de Kloet ER, Reul JMHM, Sutanto W. Corticosteroids and the brain. *J Steroid Biochem Molec Biol* 1990; 37:387–394.
150. Harfstrand A, Fuxe K, Cintra A, Agnati LF, Zini I, Wikstrom A-C, Okret S, Yu Z-Y, Goldstein M, Steinbusch H, Verhofstad A, Gustafsson J-A. Glucocorticoid receptor immunoreactivity in monoaminergic neurons of rat brain. *Proc Natl Acad Sci USA* 1986; 83:9779–9783.
151. Diorio D, Viau V, Meaney MJ. The role of the medial prefrontal cortex (cingulate gyrus) in the regulation of hypothalamic-pituitary-adrenal responses to stress. *J Neurosci* 1993;13:3839–3847.
152. Hennessy MB, Levine S. Sensitive pituitary-adrenal responsiveness to varying intensities of psychological stimulation. *Physiol Behav* 1978;21:295–297.
153. Kant GJ, Mougey EH, Pennington LL, Meyerhoff JL. Graded footshock stress elevates pituitary cyclic AMP and plasma β-endorphin, β-LPH, corticosterone and prolactin. *Life Sci* 1983;33:2657–2663.
154. Natelson BH, Tapp WN, Adamus JE, Mittler JC. Levine BE. Hormonal indices of stress in rats. *Physiol Behav* 1981;26:1049–1054.
155. De Boer SF, Koopmans SJ, Slangen JL, Van der Gugten J. Plasma catecholamine, corticosterone and glucose responses to repeated stress in rats: effect of interstressor interval length. *Physiol Behav* 1990;47:1117–1124.
156. Natelson BH, Ottenweller JE, Cook JA, Pitman DL, McCarty R, Tapp WN. Effect of stressor intensity on habituation of the adrenocortical stress response. *Physiol Behav* 1988;43:41–46.
157. Pitman DL, Ottenweller JE. Natelson BH. Plasma corticosterone levels during repeated presentation of two

intensities of restraint stress: Chronic stress and habituation. *Physiol Behav* 1988;43:47–56.
158. Hennessy MB. Sensitization of the plasma corticosterone response to novel environments. *Physiol Behav* 1991;50:1175–1179.
159. Levine S, Madden J IV, Conner RL, Moskal J. Anderson DC. Physiological and behavioral effects of prior aversive stimulation (preshock) in the rat. *Physiol Behav* 1973;10:467–471.
160. Orr TE, Meyerhoff JL, Mougey EH, Bunnell BN. Hyperresponsiveness of the rat neuroendocrine system due to repeated exposure to stress. *Psychoneuroendocrinology* 1990;15:317–328.
161. Vogel WH, Jensh R. Chronic stress and plasma catecholamine and corticosterone levels in male rats. *Neurosci Lett* 1988;87:183–188.
162. Kant GJ, Egglestone T, Landman-Roberts L, Kenion CC, Driver GC, Meyerhoff JL. Habituation to repeated stress is stressor specific. *Pharmacol Biochem Behav* 1985;22:631–634.
163. Ratner A, Yelvington DB, Rosenthal M. Prolactin and corticosterone response to repeated footshock stress in male rats. *Psychonueroendocrinology* 1989;14:393–396.
164. Ottenweller JE, Natelson BH, Pitman D, Drastal SD. Adrenocortical and behavioral responses to repeated stressors: Toward an animal model of chronic stress and stress-related mental illness. *Biol Psychiat* 1989;26:829–841.
165. Dobrakovova M, Juricovicova J. Corticosterone and prolactin responses to repeated handling and transfer of male rats. *Exp Clin Endocrinol* 1984;83:21–27.
166. Abercrombie ED, Nisenbaum LK, Zigmond MJ. Impact of acute and chronic stress on the release and synthesis of norepinephrine in brain: microdialysis studies in behaving animals. In Kvetnansky R, McCarty R, Axelrod J, eds. *Stress: neuroendocrine and molecular approaches,* New York: Gordon and Breach Science, 1992;29–42.
167. Borowsky B, Kuhn CM. Monoamine mediation of cocaine-induced hypothalamopituitary-adrenal activation. *J Pharmacol Exp Ther* 1991a;256:204–210.
168. Moldow RL, Fischman AJ, Cocaine induced secretion of ACTH, β-endorphin, and corticosterone. *Peptides* 1987;456:819–822.
169. Sarnyai Z, Biro E, Penke B, Telegdy G. The cocaine-induced elevation of plasma corticosterone is mediated by endogenous corticotropin-releasing factor (CRF) in rats. *Brain Res* 1992b;589:154–156.
170. Borowsky B, Kuhn CM. Chronic cocaine administration sensitizes behavioral but not neuroendocrine responses. *Brain Res* 1991b;543:301–306.
171. Levy AD, Li Q, Alvarez Sanz MC, Rittenhouse PA, Kerr JE, Van de Kar LD. Neuroendocrine responses to cocaine do not exhibit sensitization following repeated cocaine exposure. *Life Sci* 1992;51:887–897.
172. Torres G, Rivier C. Differential effects of intermittent or continuous exposure to cocaine on the hypothalamic-pituitary-adrenal axis and c-*fos* expression. *Brain Res* 1992;571:204–211.
173. Deroche V, Piazza PV, Casolini P, Maccari S, Le Moal M, Simon G. Stress-induced sensitization to amphetamine and morphine psychomotor effects depend on stress-induced corticosterone secretion. *Brain Res* 1992;598:343–348.
174. Deroche V, Piazza PV, Casolini P, Le Moal M, Simon H. Sensitization to the psychomotor effects of amphetamine and morphine induced by food restriction depends on corticosterone secretion. *Brain Res* 1993a; 611:352–356.
175. Maccari S, Piazza PV, Deminiere JM, Angelucci L, Simon H, Le Moal M. Hippocampal type I and type II corticosteroid receptor affinities are reduced in rats predisposed to develop amphetamine self-administration. *Brain Res* 1991;548:305–309.
176. Pauly JR, Robinson SF, Collins AC. Chronic corticosterone administration enhances behavioral sensitization to amphetamine in mice. *Brain Res* 1993;620: 195–202.
177. Piazza PV, Maccari S, Deminiere JM, Le Moal M, Mormede P, Simon H. Corticosterone levels determine individual vulnerability to amphetamine self-administration. *Proc Natl Acad Sci USA* 1991b;88:2088–2092.
178. Rivet J-M, Stinus L, Le Moal M, Mormede P. Behavioral sensitization to amphetamine is dependent on corticosteroid receptor activation. *Brain Res* 1989;498: 149–153.
179. Cador M, Dulluc J, Mormede P. Modulation of the locomotor response to amphetamine by corticosterone. *Neuroscience* 1993b;56:981–988.
180. Deroche V, Piazza PV, Maccari S, Le Moal M, Simon H. Repeated corticosterone administration sensitizes the locomotor response to amphetamine. *Brain Res* 1992;584:309–313.
181. Deroche V, Piazza PV, Deminiere J-M, Le Moal M, Simon H. Rats orally self-administer corticosterone. *Brain Res* 1993b;622:315–320.
182. Piazza PV, Rouge-Pont F, Deroche V, Kharouby M, Le Moal M, Simon H. Corticosterone and sensitivity to drugs of abuse: role of dopamine release. *Soc Neurosci Abstr* 1993;19:1860.
183. Marinelli M, Piazza PV, Deroche V, Maccari S, Le Moal M, Simon H. Corticosterone circadian secretion differentially facilitates dopamine-mediated psychomotor effect of cocaine and morphine. *J Neurosci* 1994; 14:2724–2731.
184. Yamamoto K. Steroid receptor regulated transcription of specific genes and gene networks. *Ann Rev Genet* 1985;19:209–252.
185. Joels M, de Kloet ER. Control of neuronal excitability by corticosteroid hormones. *Trends Neurosci* 1992;15: 25–30.
186. Karst H, Joels M. The induction of corticosteroid actions on membrane properties of hippocampal CA1 neurons requires protein synthesis. *Neurosci Lett* 1991; 130:27–31.
187. Joels M, de Kloet ER. Corticosteroid actions on amino acid-mediated transmission in rat CA1 hippocampal cells. *J Neurosci* 1993;13:4082–4090.
188. Orchinik M, Murray TF, Moore FL. A corticosteroid receptor in neuronal membranes. *Science* 1991;252: 1847–1851.
189. Saito N, Guitart X, Hayward M, Tallman JF, Duman RS, Nestler EJ. Corticosterone differentially regulates the expression of G_{salpha} and G_{ialpha} messenger RNA and protein in rat cerebral cortex. *Proc Natl Acad Sci USA* 1989;86:3906–3910.
190. Imperato A, Puglisi-Allegra P, Casolini P, Zocchi A, Angelucci L. Stress-induced enhancement of dopamine

and acetylcholine release in limbic structures: role of corticosterone. *Eur J Pharmacol* 1989;165:337–338.
191. Imperato A, Puglisi-Allegra S, Casolini P, Angelucci L. Changes in brain dopamine and acetylcholine release during and following stress are independent of the pituitary-adrenocortical axis. *Brain Res* 1991;538:111–117.
192. Cools AR. Differential role of mineralocorticoid and glucocorticoid receptors in the genesis of dexamphetamine-induced sensitization of mesolimbic, alpha1 adrenergic receptors in the ventral striatum. *Neuroscience* 1991;42:419–428.
193. Casolini P, Kabbaj M, Leprat F, Piazza PV, Rouge-Pont F, Angelucci L, Simon H, Le Moal M, Maccari S. Basal and stress-induced corticosterone secretion is decreased by lesion of mesencephalic dopaminergic neurons. *Brain Res* 1993a;622:311–314.
194. Rivet JM, Stinus L, Le Moal M, Mormede P. 6-Hydroxydopamine lesion of ventral tegmental area dopaminergic cell bodies does not impair neuroendocrine responses to environmental stimuli. *J Neuroendocrinol* 1990;2:733–736.
195. Bjorklund A, Lindvall O. Dopamine-containing systems in the CNS. In Bjorklund A, Hokfelt T, eds. *Handbook of chemical neuroanatomy, classical transmitters in the CNS,* vol. 2, Amsterdam: Elsevier, 1984; 55–122.
196. Feldman S, Conforti N, Saphier D. The preoptic area and bed nucleus of the stria terminalis are involved in the effects of the amygdala on adrenocortical secretion. *Neuroscience* 1990;37:775–779.
197. Swanson KW. The hypothalamus. In Bjorklund A, Hokfelt T, Swanson LW, eds. *Handbook of chemical neuroanatomy: integrated systems of the CNS,* vol. 5, Amsterdam: Elsevier, 1987;1–124.
198. Sapolsky RM, Armanini MP, Packan DR, Sutton WS, Plotsky PM. Glucocorticoid feedback inhibition of adrenocorticotropic hormone secretagogue release. *Neuroendocrinology* 1990;51:328–336.
199. Spencer RL, Young EA, Choo PH, McEwen BS. Adrenal steroid type I and type II receptor binding: estimates of in vivo receptor number, occupancy, and activation with varying level of steroid. *Brain Res* 1990;514:37–48.
200. Tornello S, Orti E, De Nicola AF, Rainbow TC, McEwen BS. Regulation of glucocorticoid receptors in brain by corticosterone treatment of adrenalectomized rats. *Neuroendocrinology* 1982;35:411–417.
201. Dong Y, Poellinger L, Gustafssono J-A, Okret S. Regulation of glucocorticoid receptor expression: evidence for transcriptional and posttranslational mechanisms. *Mol Endocrinol* 1988;2:1256–1264.
202. Sapolsky RM, Krey LC, McEwen BS. Glucocorticoid-sensitive hippocampal neurons are involved in terminating the adrenocortical stress response. *Proc Natl Acad Sci USA* 1984a;81:6174–6177.
203. Sapolsky RM, Krey LC, McEwen BS. Stress down-regulates corticosterone receptors in a site-specific manner in the brain. *Endocrinology* 1984b;114:287–292.
204. Spencer RL, Miller AH, Stein M, McEwen BS. Corticosterone regulation of type I and type II adrenal steroid receptors in brain, pituitary, and immune tissue. *Brain Re.* 1991;549:236–246.
205. Turner BB. Tissue differences in the up-regulation of glucocorticoid-binding proteins in the rat. *Endocrinology* 1986;118:1211–1216.
206. Piazza PV, Deminiere JM, Maccari S, Le Moal M, Mormede P, Simon H. Individual reactivity to novelty predicts vulnerability to amphetamine self-administration. *Behav Pharmacol* 1990;1:339–345.
207. Casolini P, Piazza PV, Kabbaj M, Leprat F, Angelucci L, Simon H, Le Moal M, Maccari S. The mesolimbic dopaminergic system exerts an inhibitory influence on brain corticosteroid receptor affinities. *Neuroscience* 1993b;55:429–434.
208. Biron D, Dauphin C, Di Paolo T. Effects of adrenalectomy and glucocorticoids on rat brain dopamine receptors. *Neuroendocrinology* 1992;55:468–476.
209. Maccari S, Mormede P, Piazza PV, Simon H, Angelucci L, Le Moal M. Hippocampal type I and type II corticosteroid receptors are modulated by central noradrenergic systems. *Psychoneuroendocrinology* 1992;17:103–112.
210. Azmitia EC, McEwen BS. Adrenalcortical influence on rat brain tryptophan hydroxylase activity. *Brain Res* 1974;78:291–302.
211. Biegon A, Rainbow TC, McEwen BS. Corticosterone modulation of neurotransmitter receptors in rat hippocampus: A quantitative autoradiographic study. *Brain Res* 1985;332:309–314.
212. de Kloet ER, Kovacs GL, Szabo G, Telegdy G, Bohus B, Versteeg DHG. Decreased serotonin turnover shortly after adrenalectomy: selective normalization after corticosterone substitution. *Brain Res* 1982;239:659–663.
213. Joels M, Hesen W, de Kloet ER. Mineralocorticoid hormones suppress serotonin-induced hyperpolarization of rat hippocampal CA1 neurons. *J Neurosci* 1991;11:2288–2294.
214. Singh VB, Corley KC, Phan T-H, Boadle-Biber MC. Increases in the activity of tryptophan hydroxylase from rat cortex and midbrain in response to acute or repeated sound stress are blocked by adrenalectomy and restored by dexamethasone treatment. *Brain Res* 1990;516:66–76.
215. Van de Kar LD, Bonadonna AM, Rittenhouse PA, Kerr JE, Levy AD, Iyer L, Herbert GB, Alvarez Sanz MC, Lent S, Carnes M. Prior chronic exposure to cocaine inhibits the serotonergic stimulation of ACTH and secretion of corticosterone. *Neuropharmacology* 1992;31:169–175.
216. Roy-Byrne PP, Uhde TW, Post RM, Gallucci W, Chrousos GP, Gold PW. The corticotropin-releasing hormone stimulation test in patients with panic disorder. *Am J Psychiat* 1986;143:896–899.
217. Smith MA, Davidson J, Ritchie JC, Kudler H, Lipper S, Chappell P, Nemeroff CB. The corticotropin-releasing hormone test in patients with post-traumatic stress disorder. *Biol Psychiat* 1989;26:349–355.
218. Krieger DT. Physiopathology of Cushing's disease. *Endocrinol Rev* 1983;4:22–53.
219. Ling MHM, Perry PJ, Tsuang MT. Side effects of corticosteroid therapy. *Arch Gen Psychiat* 1981;38:471–477.
220. Von Zerssen D. Mood and behavioural changes under corticosteroid therapy. In Itel TM, Laudahn G, Hermann WM, eds. *Psychotropic action of hormones,* New York: Spectrum, 1976;195–214.

Neurobiological and Clinical Consequences of Stress: From Normal Adaptation to PTSD, edited by M. J. Friedman, D. S. Charney, and A. Y. Deutch, Lippincott-Raven Publishers, Philadelphia © 1995.

6

Neuropeptides and Stress

Preclinical Findings and Implications for Pathophysiology

Steven C. Stout, Clinton D. Kilts, and Charles B. Nemeroff

Department of Psychiatry and Behavioral Sciences, Emory University School of Medicine, Atlanta, Georgia 30322

Depending on one's cutoff for the number of amino acids that comprise a protein versus a peptide, there are 50 to 75 neuropeptides that have been discovered in the mammalian central nervous system (CNS) in the last 3 decades. Functional roles for many neuropeptides have been demonstrated in the endocrine, autonomic, and central nervous systems, and we finally seem to be beyond the era of semantic arguments as to whether these substances are neurotransmitters, neuromodulators, or neuroregulators. Support for the role of neuropeptide systems in stress comes largely from preclinical laboratory animal studies examining changes in peptide concentrations, receptor binding, or preprohormone mRNA expression following exposure to an experimental stressor, or alternatively the behavioral response to an administered peptide. Based on such experiments, neuropeptides hypothesized to play some role in the stress response include corticotropin-releasing factor (CRF), vasopressin, oxytocin, opioid peptides, neurotensin, somatostatin, and neuropeptide Y. Although previous reviews have concentrated on one or another of these peptides, to our knowledge few if any authors have attempted a broad overview of neuropeptides in stress.

That a plethora of different neurohumoral mechanisms is involved in the response of animals to stress is not surprising when it is considered that there is clearly no such thing as a single, generic "stress response." On the contrary, a basic requirement for the survival of an organism is the ability to integrate a variety of environmental perturbations, or stressors, and to mount appropriate autonomic, immune, endocrine, and behavioral responses. Considering the neuroanatomical distribution of these neuropeptides, and their behavioral effects within the CNS, they represent preeminent candidates as chemical mediators of one or another aspect of the stress response. This statement is not made haphazardly, as considerable evidence supports this view. First, neuropeptides as a class account for neurotransmission at a very large percentage of synapses in the brain, perhaps only surpassed by the excitatory amino acids. However, their unique neuroanatomical distribution patterns may allow activation of a particular class of peptidergic neurons to have relatively precise and subtle effects. Second, it is not uncommon for a neuron to manufacture more than one neuropeptide, often colocalized with a nonpeptide neurotransmitter. Individual transcriptional control of the different precursor genes may allow stressor-specific activation of peptide synthesis; multiple second-messenger mechanisms may differentially alter peptide release and activation of peptide receptor mechanisms (signal transduction) as well. Finally, many neuropeptides are hypothalamic hypophysiotropic hormones, directly controlling the secretion of anterior pituitary hormones; they can therefore function both as hormones in the hypothalamic-hypophysial portal system and as neurotransmitters in the CNS. This dual functioning may allow neuropeptides to orchestrate both neural and endocrine responses to stress.

In this chapter we discuss salient evidence concerning those neuropeptides that appear to play a role in the stress response. Where possible, we will delineate the particular stressor that has been reported to alter peptidergic neurotranmission. However, the reader should be forewarned that attempts to make generalizations about correlating the severity, duration, and type (e.g., physical or psychological) of stressors and the resultant peptide alteration is premature at this time considering the extant data base. It is a testament to the variety of challenges to which an animal must respond and the multitude of neuropeptide systems that a simple categorization scheme is currently impossible. More optimistically, the continued outpouring of stress and neuropeptide research portends a clearer understanding of these issues in years to come.

Our discussion begins with CRF, for which the most overwhelming evidence exists for a widespread role in the stress response. Next we consider the neurohypophysial peptides that interact with CRF in certain functions. The remaining peptides are then discussed, from those with well-established roles in mediating stress responses to a few neuropeptides for which the evidence is slight. These latter peptidergic systems may be affected only by a small subset of stressors, whereas CRF is involved in responses to most stressors, although even this conclusion is speculative. Finally, we consider briefly the application of preclinical data to the study of neuropeptide involvement in clinical states such as posttraumatic stress disorder. Note that unless otherwise specified, all data come from studies in rats. Interspecies variation almost certainly precludes the direct application of these rodent findings to humans (e.g., oxytocin has different effects on adrenocorticotropin hormone (ACTH) release in rodents and primates). Clearly space limitations preclude a comprehensive treatise, but we and others have reviewed parts of this literature in detail elsewhere (1,2,3).

CORTICOTROPIN RELEASING FACTOR

In 1981, Vale and colleagues isolated CRF from roughly one half million ovine hypothalami and determined its 41-amino-acid sequence (4). Immunohistochemical studies (5,6) soon revealed that CRF-containing neurons are heterogeneously distributed throughout the brain in hypothalamic as well as extrahypothalamic areas. A concatenation of studies on the neuroanatomical distribution of CRF, the behavioral and physiological responses to centrally administered CRF, and the effects of experimental stress on indices of CRF neuronal activity reveal that CRF appears to orchestrate the endocrine, autonomic, immunological, and behavioral responses of an organism to stress (see review 7).

The most widely studied CRF neurons in the brain are undoubtedly the parvicellular neurons of the paraventricular nucleus (PVN) of the hypothalamus (8). CRF-containing PVN neurons project to the median eminence, where CRF is released into the hypothalamo-hypophysial portal system. CRF-containing perikarya are also present in other hypothalamic nuclei and in the magnocellular division of the PVN (9–11). A high density of CRF-containing cells are located in extrahypothalamic areas, including the cerebral cortex, amygdaloid complex, bed nucleus of the stria terminalis, central gray area, dorsal tegmentum, locus coeruleus, parabrachial nucleus, dorsal vagal complex, and inferior olive. Thus, CRF is neuroanatomically positioned to mediate limbic, cortical, and autonomic responses to stress, and to influence monoamine neurotransmitter systems also known to be stress responsive.

In laboratory animals, i.c.v. administration of CRF produces many of the behavioral and autonomic sequelae of the "stress response." Stress and i.c.v. CRF produce similar endocrine responses, adrenomedullary activation, cardiovascular and gastrointestinal changes, and behavioral responses such as increased grooming (see review 12,13,1). Administration of i.c.v. CRF also elicits dose-dependent, anxiety-like behaviors, including an increased startle response (14), decreased exploratory behavior (15,16), and increased locomotor activity in familiar environments (17,18). These behavioral effects can be attenuated by pretreatment with the CRF receptor antagonist, α-helical CRF_{9-41} (19,20).

Of particular importance in establishing CRF as a central neurotransmitter in stress pathways is the fact that administration of dexamethasone, a synthetic glucocorticoid, does not block the effect of i.c.v. CRF on locomotor activity; hence, the behavioral response is independent of the hypothalamic-pituitary-adrenal (HPA) axis (21). Moreover, CRF-induced locomotor activation is not reduced by destruction of dopaminergic nerve terminals (22) or by treatment with opioid receptor antagonists (23). The behavioral effect of i.c.v. CRF may depend somewhat on an increase in sympathetic tone, as ganglionic blocking drugs partially attenuate CRF-induced locomotor activation (24). However, the partial nature of this blockade taken together with the evidence from other experiments discussed in this section suggests that CRF exerts many of its behavioral actions directly within the CNS. Attenuation of the behavioral effects of CRF by benzodiazepines (14,15) implicates the involvement of γ-aminobutyric acid (GABA)-A receptor mechanisms; the effects of benzodiazepines on regional brain CRF concentrations also support such an interaction (vide infra).

The CRF has been microinjected directly into different brain regions in order to identify neuroanatomical loci where CRF produces its behavioral effects. Butler et al. (26) reported that CRF microinjected bilaterally into the locus coeruleus of the rat produces a decrease in exploratory behavior in a novel environment, interpreted as a fearful or anxiogenic response. Recently using a conflict test, Weiss et al. (27) demonstrated anxiogenic effects of CRF after bilateral injection into the locus coeruleus (LC) or parabrachial nucleus but not after injection into the nearby dorsal tegmentum. Moreover, direct application of CRF in the LC increases the firing rate of noradrenergic neurons (28). Like chronic stress, locally infused CRF increases tyrosine hydroxylase activity in the LC, and the CRF receptor antagonist α-helical CRF_{9-41} blocks induction of tyrosine hydroxylase by stress (29). Together, these results indicate that stressful stimuli may exert their effects, in part, by increasing CRF release in the LC, which in turn increases norepinephrine synthesis and release. Such an interaction may in part underlie both the classical findings of HPA axis and noradrenergic alterations in anxiety and affective disorders (30,31).

The amygdaloid complex is another site strongly implicated as a site of action of CRF. Microinjection of CRF into the amygdala produces a decrease in exploratory behavior (32) and attenuates stress-induced gastric ulcer formation (33). Furthermore, local injection of α-helical CRF_{9-41} into the central amygdala reduces defensive freezing behavior following conditioned stress (34). Lesioning studies have also revealed that the central nucleus of the amygdala plays a central role in the HPA axis response to various stressors (35,36). The amygdala is likely also to play a role in emotional memory processes, as amygdaloid lesions block the conditioned fear response in the potentiated startle paradigm (37). Thus, like the LC, the amygdaloid complex is a CRF-responsive site that appears to coordinate a number of different responses to stress.

A final line of evidence demonstrating the involvement of CRF in the stress response and most directly relevant to this chapter is that exposure to stress produces alterations in CRF synthesis, release, and receptor density in a number of brain regions. Chappell et al. (38) demonstrated that acute (3-hour cold immobilization) and chronic (14-day unpredictable stress regimen) stress exposure altered the CRF concentration in various microdissected brain regions. Both acute and chronic stress resulted in a twofold increase in the concentration of CRF in the LC, a finding of particular interest in light of the behavioral effect of CRF in the LC. Acute and chronic stress also led to a 50% decrease in median eminence CRF concentrations, reflecting a demand for CRF release into portal blood that cannot be met by increased CRF synthesis. Surprisingly, the concentration of CRF in the

PVN was not significantly altered in this study. It is of particular interest that administration of benzodiazepines such as alprazolam or adinazolam exerts effects on CRF concentrations that are opposite to those of stress (e.g., an increase in median eminence CRF and a decrease in LC CRF concentrations (39) (Fig. 1). Finally, a significant decrease in CRF was observed in the dorsal vagal complex following chronic stress, a result consistent with a proposed role for CRF in autonomic responsiveness to stress. We have recently demonstrated persistent CRF hypersecretion in nonhuman primates years after exposure to neonatal stress, as assessed by measurement of CSF CRF concentrations (40).

CRF transcriptional activity may be a more sensitive measure of the responsiveness of CRF neurons to stress than CRF concentrations, which depend simultaneously on rates of synthesis, transport, and release. Whereas no changes in CRF immunoreactivity were detected in the PVN following stress (vide supra), in situ hybridization histochemistry has revealed alterations in gene expression following a variety of stressors. Lightman and coworkers have observed increases in CRF mRNA as early as 4 hours after the onset of restraint, and hypertonic saline or swim stress, but not cold stress (41). However, this finding was not replicated in another study (42). This evidence as well as the finding that acute footshock stress does not alter hypothalamic CRF mRNA as measured by Northern blot analysis (43) indicate that acute stressors do not consistently alter CRF gene expression. Unlike acute footshock, chronic footshock did result in a significant increase in CRF mRNA expression (43) which can be reversed by the anxiolytic, chlordiazepoxide (44). It is plausible that acute stress results in a release of stored CRF in hypothalamic nerve terminals, whereas prolonged stress causes a compensatory increase in CRF synthesis. However, certain types of acute or chronic stressors may cause a more rapid or dramatic induction of CRF mRNA expression than others. These factors, though speculative, might explain the varied results in studies of stress-induced CRF gene expression in hypothalamic nuclei.

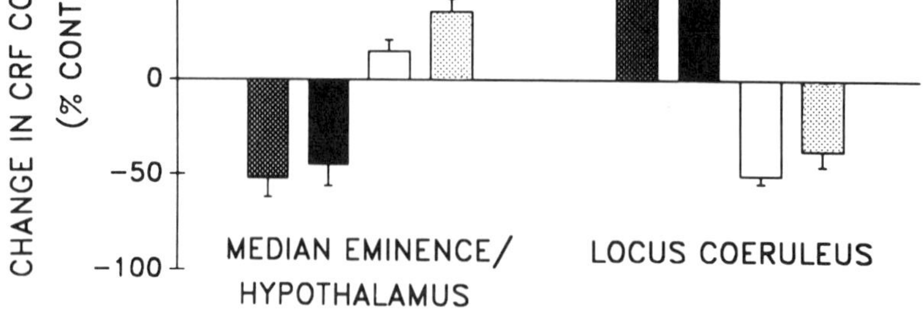

FIG. 1. Alprazolam and adinazolam administration exert opposite effects on corticotropin-releasing factor (CRF) concentrations in the hypothalamus/median eminence compared to the locus coeruleus. These effects are opposite of the CRF concentration changes seen in these brain regions after acute or chronic stress. Reprinted with permission from Owens et al (39). Copyright John Wiley & Sons, Inc. 1989.

Stress-induced CRF gene expression in extrahypothalamic brain regions has been investigated to a more limited extent. An increase in CRF mRNA expression in Barrington's nucleus (a pontine micturition center) and a decrease in CRF mRNA expression in the olfactory bulb were reported following acute and chronic stress (43). Moreover, the relationship between CRF and stress neurocircuity is strengthened by the demonstration that i.c.v. CRF produces a pattern of induced c-fos mRNA expression in the rat brain similar to the pattern following immobilization or footshock stress (45). It would be of interest to determine the time course of the effects of various stressors on heteronuclear RNA as detected with intronic probes to the pro-CRF gene.

Changes in synaptic availability of a neurotransmitter often produce converse changes in postsynaptic receptor densities. Anderson and colleagues (46) observed a decrease in the density of CRF receptors (i.e., downregulation) in response to chronic stress. Rats receiving intermittent foot shock for 3–4 days were found to have decreased [^{125}I]oCRF binding to membranes prepared from frontal cortex, anterior pituitary, and hypothalamus. Our group demonstrated that 24-hour maternal deprivation of rat pups on neonatal day 18 resulted in a decrease in pituitary CRF-binding density (47). Moreover, it appears that the downregulation of anterior pituitary CRF receptors produced by maternal deprivation persists into adulthood (48).

A simple receptor downregulation model may not apply to all CRF systems, however. Maternal deprivation produces upregulation of CRF receptors in 9- and 12-day-old rats in a variety of regions, including the LC and hippocampus (49). Recently, Fuchs and Flügge (50) have confirmed that CRF-binding density is modulated in response to stress. In their study, the chronic psychosocial stress of a subordinate tree shrew confronted by a dominant animal produced a complex set of effects on the density and affinity of CRF-binding sites. The number of binding sites was, not surprisingly, decreased in the anterior pituitary and hippocampal formation, but *increased* in the amygdala, frontal cortex, and other cortical areas. The authors suggested that an increased number of presynaptic CRF autoreceptors in regions such as the amygdaloid complex could act as a compensatory response to increased CRF release.

In summary, CRF appears to mediate a variety of anxiogenic and depressogenic behavioral effects, and is the preeminent regulator of the adrenocortical response to stress. Furthermore, the effects of a variety of stressors on CRF concentrations, CRF gene expression, and CRF receptor binding indicate that CRF neuronal activation is a final common pathway in the response to virtually every type of stressor thus far examined. However, the exact neuroanatomical, neurochemical, and temporal pattern of CRF effects vary according to the nature of the stressor.

Vasopressin

Vasopressin (VP) is a nonapeptide released from nerve terminals in the neurohypophysis, whose major endocrine role is to regulate fluid and electrolyte balance, particularly in response to dehydration. Its water-conserving effect on renal collecting tubules is reflected in its original name, antidiuretic hormone. Other peripheral effects of the peptide include vasoconstriction (51), and stimulation of hepatic glycogenolysis and gluconeogenesis (52). Vasopressin also contributes to HPA axis regulation, potentiating the effects of CRF on ACTH release but not induction of proopiomelanocortin transcription (see reviews 2,53). Finally, the existence of extrahypothalamic VP projections, and the effects of VP on the firing rate of hippocampal neurons, suggest that this peptide may function as a neurotransmitter (54).

With regards to its action as a neurohypophyseal hormone, the primary sites of VP synthesis are the magnocellular neurons of the supraoptic and paraventricular nuclei, which project by separate pathways to the posterior lobe of the pituitary (55). Vasopressin is also synthesized in other regions of the hypothalamus, including the suprachiasmatic nucleus and the CRF-containing cells of the medial parvicellular PVN

(56,57). Vasopressin that acts on the adenohypophysis may derive from these parvicellular sources, from *en passage* release into the median eminence by magnocellular projections (58), or from short portal vessels connecting the anterior and posterior pituitary (59). The relative contribution of parvicellular and magnocellular VP stores to stimulating ACTH secretion in basal and stress states is a controversial and as yet unresolved issue. Further complicating matters, VP may be synthesized in the anterior pituitary (60,61); a paracrine action of VP may therefore represent yet another pathway for stimulating ACTH secretion. Other putative extrahypothalamic sites of VP synthesis include the bed nucleus of the stria terminalis (62), LC and amygdala (63), cerebellum and dentate gyrus (64), pineal gland (65), and dorsal root ganglia (66).

Most reports of stress-induced changes in VP concentration or expression have focused on the hypothalamus, and in particular the PVN. Vasopressin synthesis and release are not only increased in response to hypertonic or hypotensive stress (67), during which its systemic hormonal effects predominate, but also in response to other stressors such as insulin-induced hypoglycemia, in which the role of VP in augmenting ACTH secretion is critical (68,69). In a chronic hyponatremic rodent model, hypertonic stress activates both parvicellular and magnocellular VP systems, whereas insulin-induced hypoglycemia only activates the parvicellular PVN system (70). Footshock is another stressor unrelated to water and mineral balance that has been demonstrated to produce alterations in hypothalamic VP concentrations. In a study by van Dijken et al. (71), a single session of repeated footshock produced a long-term increase in VP, but not CRF, concentration in the median eminence.

Increases in plasma VP concentrations have been well documented in response to physical but not psychological stressors (72,73). Nevertheless, recent studies have revealed increased VP mRNA expression, particularly in parvicellular PVN neurons, following exposure to psychological stressors. Pre-VP heteronuclear RNA expression in the rat parvicellular PVN is elevated for 2 hours following the onset of isolation stress (74). Vasopressin content and transcription are altered in the rat PVN following repeated immobilization stress (75) and VP transcription increased following a single 30-minute immobilization stress (76,77). Interestingly, coexpression of VP in CRF-containing parvicellular cells in the PVN increases dramatically following stress. It is possible that VP and CRF concentrations in the PVN are regulated independently or in concert by different types of stressors, allowing for a range of HPA axis responsiveness. Vasopressin may function to reduce the feedback inhibition of ACTH secretion by glucocorticoids during prolonged stress (78).

One source of the discrepancy between the studies of mRNA transcription rates and plasma VP concentrations following acute stress is the duration of immobilization in the experiments. Additional unexplored factors include efficiency of mRNA translation, independent regulation of synthesis versus release, and potentially different contributions of portal versus systemic VP release after emotional or physical stressors. A method for sampling hypophysial portal blood in awake, freely moving animals would be of clear benefit in resolving such issues.

The role of extrahypothalamic VP systems in stress responsiveness has been investigated to a much more limited extent. Like CRF, VP injected into the central nucleus of the amygdala (CEA) produces some stresslike effects. Bilateral microinfusion of low-dose (20 pg) VP into the CEA potentiated the bradycardia and immobility following footshock in a rat line selectively bred for a passive coping response to stress (79).

OXYTOCIN

Like vasopressin, oxytocin (OT) has a relatively limited extrahypothalamic distribution and is an important neurohypophysial hormone. The OT-immunoreactive cells are confined largely but not exclusively to the supraoptic and paraventricular nuclei (80). Systemic effects of OT include milk letdown and uterine contractility at time of parturition. Oxytocin also stimulates prolactin secretion, and potentiates CRF-stimulated ACTH release in a fashion similar to VP in rats (81); in contrast, OT inhibits ACTH secretion in primates (82, see review 83).

Oxytocin release has been implicated in responses to a variety of stressors. One minute of immobilization stress has produced increased plasma OT and decreased hypothalamic OT concentrations; plasma and hypothalamic VP concentrations were not altered by this stressor (72,84). Increased OT release has also been demonstrated in response to tail-hang and ether stressors, but not to the stress of a novel environment (73).

In summary, there is considerable evidence that CRF and to a lesser extent VP and OT are not only intimately involved in the regulation of ACTH secretion in response to stress, but that extrahypothalamic CRF-, and perhaps OT- and VP-containing neurons are major regulators of stress responsiveness. These findings have not surprisingly led to the consideration of novel pharmacotherapies directed towards one or another aspect of peptidergic neurotransmission. There is considerable evidence that a CRF receptor antagonist would likely be anxiolytic and perhaps antidepressant. Clinical findings also support the hypothesis that CRF hypersecretion at distinct extrahypothalamic sites occurs in a variety of affective and/or anxiety disorders (vide infra).

OPIOID PEPTIDES

The opioid peptides comprise three sets of homologous peptide classes, differentially expressed throughout the CNS and with a panoply of different functions (85). β-endorphin is a 31-amino-acid peptide derived from the ACTH precursor, proopiomelanocortin (POMC); the enkephalins and dynorphins are shorter peptides each with their own precursor genes. Met-enkephalin is an opioid pentapeptide that also constitutes the N-terminus of β-endorphin; the Leu-enkephalin sequence contains a single amino acid substitution and is identical to a part of the dynorphin sequence. The dynorphins can be degraded to Leu-enkephalin or other fragments that have a spectrum of bioactivity (86). Complicating matters even further are the anatomically separate opioid systems (e.g., the adenohypophyseal synthesis of POMC and the secretion of β-endorphin from the pituitary as contrasted with the extensive β-endorphin projections in the CNS from the arcuate nucleus of the hypothalamus (87). There are three major opioid receptor types: mu, delta, and kappa (88) (the epsilon receptor is less studied and the "sigma receptor" is not a true opiate receptor, with high-affinity binding to drugs such as haloperidol, (+)3-PPP, and BMY 14802 (89)). Naloxone or naltrexone have commonly been used as moderately selective mu receptor antagonists, though they also have some affinity for other receptor types. In recent years, the design of more selective antagonists for each of the opiate receptor subtypes has facilitated a more accurate dissection of the role of different opioid systems.

Despite the overwhelming complexity of the multiple β-endorphin, enkephalin, and dynorphin systems, one common theme of these endogenous opioid systems is mediation of stress responsiveness. Not surprisingly, the most intensively studied of the opioid mechanisms influenced by stress is pain modulation. A variety of stressful stimuli including electrical shock, forced swimming, and restraint have been documented to produce diminished responsiveness to pain, or "stress-induced analgesia" (SIA) (90,91). The reversal of some forms of SIA by naloxone suggests that opioid peptides play an important role in this response (91,93).

While it is beyond the scope of this chapter to describe in detail the anatomy and neurochemistry of nociceptive systems, as a first approximation the midbrain periaqueductal grey (PAG) projects to the rostroventral medulla, including the nucleus raphe magnus (NRM), which in turn sends noradrenergic and serotonergic projections to the dorsal horn of the spinal cord (92). It is believed that multiple opioid systems modulate pain through actions in the PAG, the NRM, and the spinal cord (93,94). It appears that different types and durations of stressors may produce analgesia by influencing separate pathways. For example, the analgesia produced after two tail shocks is not altered by decerebration or anesthesia, whereas the analgesia following 80–100 shocks is a distinct, conditioned reaction that can be prevented by anesthesia. Watkins et al (95) have provided evidence that both of the above SIA responses depend on descending projections and on kappa receptors in the spinal cord.

Some forms of SIA are not reversed by naloxone or naltrexone. While this type of analgesia has often been termed ''nonopioid,'' a more accurate description may be non-mu opioid receptor-mediated SIA, though true nonopioid systems or a combination of systems may be involved (96). The conditioned SIA produced after many shocks can be blocked by naloxone, whereas a shorter footshock paradigm produces naloxone-insensitive SIA (97). Moreover, hind paw versus front paw shock may produce apparent naloxone-insensitive or -sensitive SIA, respectively (98), although Cannon et al (99) were able to produce either type of analgesia in each region by varying stimulus intensity. The naloxone/naltrexone-sensitive and -independent forms of SIA may in fact be mutually inhibitory (100,101). To add further complexity, gender differences in the naloxone sensitivity of restraint stress-induced hypoalgesia in mice have been reported (102). In summary, opioid modulation of pain is a common response to many stressful stimuli, but the pairing of a stimulus type to a particular opioid system has not yet been fully elucidated.

A second physiological response to stress in which the opioid peptides appear to play a role is inhibition of the hypothalamic-pituitary-gonadal (HPG) axis. Intensive exercise or psychological stress in women may result in amenorrhea or other disruptions in fertility (103,104). β-endorphin antagonists in primates (105), and CRF antagonists in rats (106) can block stress-induced inhibition of HPG axis function; thus, both CRF and β-endorphin may be mediators in HPG axis suppression. The CRF is postulated to stimulate opioid peptide-containing interneurons, which in turn inhibit GnRH release from the hypothalamus (107). CRF-opioid interactions appear to be common to many stress responses (11).

Stressors produce various changes in opioid concentration in the hypothalamus and other brain areas. Nabeshima et al. (108) reported decreased enkephalin and increased dynorphin concentrations in the hypothalamus following conditioned suppression of motility- or forced-swim–induced immobility. Changes in the concentration of these peptides were also reported in the striatum and pituitary, showing further dissociation between the enkephalins and dynorphin. In some cases, Met- and Leu-enkephalin concentration changes differed in response to the two types of stressors.

The midbrain ventral tegmental area (VTA) contains dopamine perikarya that project to the prefrontal cortex; these cells are activated by stressors such as footshock and conditioned stress (see review 109). One or more opioids participate in activation of these neurons. Naloxone prevents stress-induced activation of the mesoprefrontal dopamine system (110), and a 20 minute footshock causes depletion of dynorphin concentrations in the VTA (111).

Opioid peptide gene expression in the hypothalamus is differentially induced in response to different stressors. Lightman and coworkers have reported elevated proenkephalin A gene expression in parvicellular neurons of the PVN after acute hypertonic saline injection, but not after restraint or swimming stress (39). Elevated enkephalin expression has been reported in a more circumscribed, dorsal region of the medioparvicellular PVN following ether but not cold or swim stress; in contrast to the first study, enkephalin mRNA expression was also increased following 1-hour restraint stress (40). In addition, chronic hypertonic saline stress was reported to increase enkephalin expression in magnocellular neurons of the PVN and SON (112). It has recently been reported that the increase in enkephalin expression following hypertonic saline is mediated by the transcription factor, adenosine 3′,5′-cyclic phosphate (cAMP)/Ca^{++} response element-binding protein (CREB) (113). Dehydration was reported to increase prodynorphin mRNA expression in magnocellular neurons of the PVN and SON in sheep (114). In the same study, prodynorphin, proenkephalin, and CRF mRNA expression increases were also observed in the PVN following isolation stress.

The litany of enkephalin and dynorphin changes following stressors may represent, in part, a fine-tuning mechanism for the regulation of OT, VP, and CRF secretion. Dynorphin is believed to be coreleased with VP, and acts at adjacent kappa receptors on magnocellular neurons to inhibit inappropriate OT secretion (115). In support of this model, dynorphin and VP are colocalized in terminals of magnocellular neurons (116); furthermore, dynorphin inhibits in vitro stimulated release of OT, but not VP (117,118). Enkephalin may also inhibit OT release, similar to its presynaptic inhibition of substance P in terminals of sensory neurons in the substantia gelatinosa (119). Unlike dynorphin, Leu-enkephalin and β-endorphin have been reported to inhibit VP release from the neurohypophysis (120). Although the opioid peptide increases following hypertonic and dehydration stressors are for the most part concordant with the proposed roles of opioids in VP and OT release, the remaining stress effects on hypothalamic opioid expression may well act through undiscovered effector pathways in the hypothalamus to modulate neurohypophysial and other peptide release or action.

Finally, different stressors also alter CNS opioid receptor binding in a complex fashion. A single footshock that produced SIA also produced reductions in mu receptor binding in the septum. In contrast, chronic water deprivation resulted in an increase in septal mu receptor binding, and increases in delta receptor binding in the caudate and nucleus accumbens (121). Surprisingly, no changes in receptor binding were observed in the hypothalamus, given the evidence for enkephalinergic responses to hyperosmolar stress (vide supra). The brain regions that exhibit altered opioid receptor binding in this study are of interest because they all comprise part of the limbic system. In addition, the contrasting results following the two behavioral paradigms highlight the recurrent theme of this chapter that different stressors are seldom, if ever, identical in their effects on neuropeptide circuits.

NEUROTENSIN

The tridecapeptide neurotensin (NT) fulfills all of the requisite criteria supporting its role as a neurotransmitter in CNS. The human and rat brain NT receptor have been cloned and sequenced, and their function studied in transfected cell systems and clonal cell lines (122–124). Neuroendocrine, biochemical, and molecular evidence exists in support of the involvement of NT in the response of the CNS to stress. The anatomical distribution of NT and the NT receptor is associated with neurocircuitry of stress as defined by c-fos mapping (43). Both NT and the NT receptor are distributed in hypothalamic, septal, amygdaloid, and hippocampal nuclei or subdivisions, and are proximal to many of the cell bodies of origin of the stress responsive classical neurotransmitters (125–129). In further support of the involvement of NT in the response to stress is the finding that the i.c.v. administration of NT results in a robust increase in serum corticosterone concentration in rats (130). The NT-induced increase in this measure of HPA axis activity is apparently independent of the well-documented effects of NT on brain dopamine neurons. A role for NT in the coping response to stress is suggested by the cytoprotective effects of i.c.v. injection of NT on restraint stress-induced gastric ulcers in rats (131); this effect has also been observed after microinjection of NT into the central amygdaloid nucleus (132).

A comparison of the effects of exposure to different stressors on distinct neuropeptide mRNA expression in the hypothalamic paraventricular nucleus of adult male rats revealed that stress associated with ether anesthesia and immobilization resulted in a significant increase in the NT mRNA level of the PVN (40). These results confirm previous findings of ether stress-induced increases in NT mRNA in the rat PVN (133) and suggest that the PVN may be one site where NT may activate the HPA axis. The CRF mRNA levels of the PVN, which are under negative feedback regulation by the stress associated increase in glucocorticoids (134), were not affected by any of the stressors tested. Thus NT mRNA expression in the PVN is not under similar feedback regulation by glucocorticoids, a distinction supported by the lack of effect of adrenalectomy on PVN NT concentration (134).

The involvement of NT in the response of the CNS to stress is suggested by the intimate functional and anatomical association between NT and dopamine-containing neurons in the brain; the mesocortical dopamine system has long been known to be particularly stress-responsive (135,136). The dendrites and perikarya of dopaminergic neurons in the rat tegmental area VTA make direct contacts with NT immunopositive axon terminals (137) and exhibit a highly localized distribution of NT receptors at cell-surface– and intracellular sites (125). The activity of dopamine neurons emanating from the VTA is enhanced by the central administration of NT (138,139). Deutch et al. (140) demonstrated a selective increase in the NT concentration of the VTA following footshock stress. These results suggested a possible involvement of NT in the stress-induced activation of dopamine neurons emanating from the VTA. In a subsequent examination of the effects of footshock and conditioned stress on NT immunoreactivity in discrete cell body groups of the VTA, Kilts et al. (139) demonstrated a significant increase in the concentration of NT in the nucleus paranigralis and nucleus parabrachialis pigmentosis that comprise the lateral VTA. Foot shock did not alter the concentration of NT in the medial VTA or the discrete subdivisions of the substantia nigra. Conditioned stress resulted in a significant increase in NT concentration in the zona compacta of the substantia nigra that was attenuated in a dose-dependent manner by pretreatment with the anxiolytic benzodiazepine, diazepam. Conditioned stress did not alter NT content in the lateral or medial VTA, or the remaining subdivisions of the substantia nigra. Collectively, these findings support the hypothesized role for specific subsets of NT-containing neurons in the VTA in the patterned activation of mesotelencephalic dopamine neurons by stress.

SOMATOSTATIN

Somatostatin (somatotropin release-inhibiting factor, SRIF) was discovered serendipitously during attempts to purify growth hormone-releasing factor (GRF) (141), and the tetradecapeptide was subsequently isolated and sequenced in 1973 (142). As the name implies, SRIF is the major physiological inhibitor of growth-hormone secretion. Somatostatin is widely distributed in brain, with high concentrations found in the hypothalamus and especially in the median eminence, amygdala, hippocampus, cerebral cortex, medial preoptic area, and nucleus accumbens (143). Several SRIF receptor subtypes have now been identified (144). Both inhibitory and excitatory neuronal responses to iontophoretic application of SRIF in various brain regions have been observed (145). Behavioral effects of centrally administered SRIF include changes in sleep patterns, food consumption, locomotor activity, and memory processes (146–148). It is notable that each of the behavioral responses just mentioned resemble symptoms of clinical depression.

It is now well established that in rats, acute stress results in suppression of growth hormone (GH) secretion and an increase in prolactin secretion (149–151). In contrast, in humans stress produces elevations in plasma GH concentrations (152). Administration of an antiserum to SRIF in rats blocks the suppression of GH secretion (153). Central administration of α-helical CRF_{9-41} also reverses stress-induced GH suppression; SRIF appears to be released in response to CRF (154), analagous to the putative CRF-activated opioid interneurons in stress-induced suppression of GnRH secretion. Using push-pull perfusion of the median eminence in unanesthesized rats, Benyassi et al. (155) recently demonstrated an increase in SRIF release within 15 minutes of immobilization stress, providing direct evidence of the involvement of SRIF in the stress response. Chronic, daily immobilization results in increased basal, as well as stress-induced, SRIF release and decreased GH release; in vitro responsiveness of the pituitary somatotroph response to SRIF following the chronic stress condition is not diminished (156,157). The concentration of GRF in anterior pituitary perfusates has also been reported to increase after 10 minutes of ether stress; however, a prolonged increase in SRIF release under these conditions appears to counter the increase in GRF and suppress GH secretion

(158). There is little data available on the response of extrahypothalamic SRIF neuronal systems to stress.

CHOLECYSTOKININ

Cholecystokinin (CCK) is a 33-amino-acid–containing hormone that is synthesized by the small intestine and exerts complex effects on gastrointestinal physiology. The C-terminal octapeptide form of CCK (CCK_8) is present in high concentrations in the brain, especially in the cortex, amygdala, hippocampus, and septum, and fulfills virtually all of the requisite criteria for designation as a neurotransmitter (see review 159). Cholecystokinin is extensively colocalized with opioid peptides and certain monoamine neurotransmitters, particularly dopamine, and may modulate the activity of these other neurotransmitter systems. Two CCK receptor subtypes have been cloned (160,161), of which the CCK_B receptor subtype predominates in the brain, whereas the pancreatic ("alimentary") CCK_A receptor subtype, pharmacologically distinguished by its high affinity for sulfated CCK, is found in more restricted regions such as medial aspects of the nucleus tractus solitarius (NTS), and the substantia nigra of primates (162–164).

Preclinical and clinical evidence (review 165) support an anxiogenic role for CCK acting at CCK_B receptors. Fekete et al. (166) demonstrated that injection of femtomolar quantities of either sulfated or nonsulfated CCK into the central nucleus of the amygdala produces fear and behavioral arousal. A number of investigators have since shown that peripheral or i.c.v. administration of CCK agonists in rodents causes decreased exploratory behavior (167,168), and it has been recently reported that i.v. CCK_4 administration produces fearful behavior in African green monkeys (169). Selective CCK_B receptor antagonists (e.g., L365260, CI-988) show anxiolytic effects in models such as the rat elevated plus-maze test (170), and are antidepressant-like in the mouse conditioned suppression of motility model (171). Bradwejn and colleagues have demonstrated that CCK is also anxiogenic in humans. Injection of CCK_4 in panic disorder patients reliably precipitates panic attacks, similar in nature to the patients' naturally occurring panic attacks as well as those induced by enriched CO_2 inhalation (172). Furthermore, CCK_4 injection sometimes induces panic attacks in healthy volunteers (173). Finally, CCK_B antagonists are efficacious in preventing CCK_4-induced panic attacks (174), consistent with the preclinical literature supporting CCK_B receptor-mediated anxiogenesis. These findings have served as an impetus for testing of numerous CCK antagonists as potential novel agents for the treatment of one or another anxiety disorder.

Whereas the concentrations or receptor densities of other neuropeptides discussed in this chapter have been measured in discrete brain regions following stress, such data is not yet available for CCK. Accordingly, the behavioral site(s) of action of CCK remain(s) unclear. Both limbic and brainstem sites such as the NTS have been proposed, the latter candidate site likely due to its incomplete blood-brain barrier and consequent accessibility to peripherally administered CCK (159); furthermore, increases in Fos-like immunoreactivity in the NTS have been demonstrated in response to peripheral CCK administration (175). The NTS is an important autonomic relay center, and is plausible as a site of CCK action in producing increases in blood pressure and heart rate during a panic attack. In summary, it is unclear whether CCK plays a preeminent role in the stress response. However, the reproducible effects of CCK in clinical studies support the view that a CCK receptor antagonist might truly represent a novel anxiolytic mechanism of action.

NEUROPEPTIDE Y

Neuropeptide Y (NPY) is a 36-amino-acid–containing peptide present in high concentrations in brainstem nuclei, the amygdaloid complex, hypothalamus, and cerebral cortex (176). Recent research suggests that NPY may act as an endogenous anxiolytic, perhaps by diminishing the actions of other stress-released peptides such as CRF (see review 177). In support of this hypothesis, i.c.v. administration of NPY produced an increase in open-arm exploration in the elevated plus-maze test (178). Direct microinjection of NPY into the central nucleus of the amygdala also produced an anxiolytic effect, without having an effect on food intake (179). Central administration of NPY protects against restraint stress-induced gastric ulceration in rats (180,181), further suggesting a "buffering" role of NPY in the stress response. The lack of effect of C-terminal fragments of NPY in behavioral tests indicates that the Y1 subclass of NPY receptors is important in its anxiolytic effects (177).

Little data are available on the effects of stress on regional NPY concentrations. Chronic immobilization stress produces a decrease in NPY concentrations in the arcuate nucleus of the hypothalamus and an increase in the PVN, effects reversed by synthetic glucocorticoid administration (182). The significance of this finding, and the exact role of NPY in neuroendocrine regulation, remain to be resolved. It is attractive to consider the hypothesis that hypothalamic NPY may oppose the actions of CRF following stress, similar to its proposed action in the amygdala (177), though no such firm evidence exists at this time. Finally, decreased CSF NPY concentrations in both the CSF of depressed patients (183) and in postmortem brain tissue of suicide victims (184) warrant further study of NPY in mood and anxiety disorders.

OTHER PEPTIDES

Other neuropeptides and their roles in stress responsiveness have been investigated to a much more limited extent than those already discussed; brief mention of a few will be made here. Increases in substance P concentrations in the brainstem periaqueductal grey matter and nucleus accumbens have been reported following the mild stress of subcutaneous saline injection plus 1-minute restraint (185). Vasoactive intestinal polypeptide (VIP) concentrations in the median eminence increase following chronic, but not acute, immobilization (186). Thyrotropin-releasing hormone (TRH), like NPY, has a cytoprotective effect on gastric ulcer formation following cold-restraint stress (187) thyroid-stimulating hormone (TSH), of which TRH is the principal secretagogue, is elevated in rat serum following stress (151). Chapter 20 presents new data that further illuminate the role of TRH in the stress response.

IMPLICATIONS OF PRECLINICAL FINDINGS TO CLINICAL DISORDERS

It is clear from the preceding discussion that exposure to stress has marked effects on the activity of a number of CNS neuropeptide systems. These stress-responsive neuropeptide systems are not restricted to those peptides that function at the apex of pituitary-target endocrine end organ axes, but include peptides distributed in extrahypothalamic sites in the CNS. Although this field is in its infancy, it is evident that there is a different response of specific neuropeptide circuits to distinct stressors. These results suggest that the distinct impact on a response to different stressors by the CNS involves a differential contribution of discrete neuropeptides in the underlying neurocircuitry. At the present time, there exist insufficient data to categorize peptide systems based upon their responses to a variety of physical and psychological stressors. Clearly, neuropeptides play a preeminent role in the neurobiology of stress. The stress-related roles exist in addition to, and probably in part as a result of, the stress responsive classical monoamine neurotransmitters with which neuropeptides exhibit close functional associations.

An obvious extension of the evolving role of neuropeptides in stress neurobiology is scrutiny of their potential role in the pathophysiology of clinical stress disorders. Post-traumatic stress disorder (PTSD) is a syndrome composed of characteristic symptoms following exposure to an extreme traumatic stressor. The initial stress response involves intense fear, helplessness, or horror, with subsequent symptoms involving persistent intrusive reexperiencing of the traumatic event, avoidance of stimuli associated with the trauma, and increased arousal. The diagnosis of PTSD is dependent upon the symptom picture being present for more than 1 month and being associated with clinically significant distress or impairment in social, occupational, or other important areas of functioning (188). This is reviewed in detail elsewhere in this volume.

Of importance to establishing the involvement of specific neuropeptides in PTSD is the identification of specific properties of the traumatic stressor and its sequelae to be used in developing and validating animal models of PTSD. Yehuda and Antelman (189) have identified five specific criteria for evaluating the relevance of individual stress paradigms to the phenomenology of PTSD: 1) the ability of even very brief stressors to induce biological and behavioral sequelae of PTSD; 2) the stressors should be capable of eliciting such sequelae in a dose-dependent manner; 3) the stressors should produce biological alterations that persist over time or become more pronounced with the passage of time; 4) such stress-induced alterations should have the potential for a bidirectional expression; and 5) the response to a stressor should be characterized by marked interindividual variability.

Although these criteria are not of equal value for establishing biological versus behavioral models of PTSD, it is apparent that several neuropeptides satisfy many of these criteria. Brief ether anesthesia (2 minutes) results in a significant increase in the levels of NT mRNA and enkephalin mRNA in the hypothalamic RVN (42). However, as just noted above, responses of neuropeptide systems to stressors are not limited to physical stressors. For example, the emotional stress associated with exposure of rats to a novel environment results in a significant increase in the level of VP mRNA in the hypothalamic PVN (74). Brief exposure to footshock stress results in persistent increases in the content of VP in the external zone of the median eminence, as well as longlasting changes in the HPA axis of rats (71).

It can be argued that the specific criteria of a potential for bidirectional expression of the response to a stressor would be more appropriate for behavioral rather than biological sequela. As such, these criteria may be of greater value for validating behavioral models of PTSD than in testing a biological basis of PTSD. The criteria of interindividual variability in the stress response incorporate the probability that factors other than the traumatic event itself contributes to the induction of PTSD. Such quantitative variability is characteristic of studies of the effects of stressors on brain neuropeptide systems. While at least several neuropeptide systems satisfy many of these criteria, it is obvious that the relationship between the response of neuropeptide systems to stressors in animal models is still quite distant from the pathophysiology of PTSD. For this reason, data demonstrating alterations in neuropeptide systems in patients with PTSD would represent more direct evidence for their involvement in PTSD. The relatively small database is reviewed in the next section.

Stress, Neuropeptides, and PTSD

A subset of the candidate stress neuropeptides has been investigated in patients with one or another psychiatric disorder. In particular, many studies have documented increased CSF concentrations of CRF in drug-free depressed patients and suicide victims (190,191). There is also a burgeoning literature indicating that depressed patients have a higher incidence of untoward life events, particularly in the early years. Whether it is these depressed patients who hypersecrete CRF remains an important unexplored area. Stressful life events may predispose individuals not only to depression but to other psychiatric disorders as well, with CRF and possibly other neuropeptides serving as biological substrates; we have previously reviewed this area in detail (192).

PTSD is the disorder in which there is the most clear evidence of a precipitating stressor. However, due in part to its relatively recent clinical description, research on the neurobiology of PTSD in general and its relationship to neuropeptides in particular has been limited. As reviewed by Southwick et al. (193), research to date has focused largely on noradrenergic and adrenergic dysfunction, including most notably the precipitation of panic attacks by yohimbine (194), with fewer studies involving amino acid neurotransmitters, opioid peptides and stress-induced analgesia, and CRF/HPA axis function. To our knowledge no brain-imaging studies investigating the neuroanatomy of PTSD or studies of the CSF concentrations of neuropeptides in PTSD patients have been published. We have recently completed a pilot study with Charney and colleagues at Yale on PTSD patients; CSF concentrations of CRF were elevated in the PTSD patients, similar to our previous findings in depression (Bremner et al, submitted for publication). However, CSF SRIF concentrations were elevated in the PTSD patients, a finding opposite to the decreased CSF SRIF concentrations reported by many investigators in depressed patients (195).

Indirect evidence for the involvement of stress and neuropeptide alterations in a clinical syndrome was obtained in the course of using the "neuroendocrine window" strategy. These studies revealed HPA axis hyperactivity in major depression as evidenced by hypercortisolemia, early escape from suppression of plasma cortisol concentrations by dexamethasone, and blunting of the ACTH response to i.v. administered CRF (see review 196). A hypothesis that could explain many of these findings is that CRF is hypersecreted from the median eminence in depressed patients, causing increases in secreted ACTH and cortisol as well as a diminished responsiveness of pituitary corticotrophs to further stimulation by exogenous CRF challenge. Further studies led to the hypothesis that CRF is hypersecreted not only from hypothalamic neurons but in other regions that may contribute to the signs and symptoms of depression. As noted earlier, CSF CRF concentrations have been repeatedly demonstrated to be elevated in depressed patients (190,191), and several lines of evidence indicate that CSF CRF is derived from extrahypothalamic brain sources (197,198). Moreover, there is a 23% decrease in the density of CRF receptors in the frontal cortex of suicide victims relative to control brains, presumably representing receptor downregulation in response to increased CRF synaptic availability (199).

Distinct HPA axis abnormalities have also been documented in PTSD patients, supporting the generally accepted belief that PTSD is a separate biological entity. Baseline urinary cortisol excretion has been reported to be decreased, rather than increased, in PTSD patients when compared to healthy control groups and other psychiatric patients (200). However, higher fluctuations in cortisol over a 24-hour period have been reported in PTSD patients than in normal and depressed patients, suggesting that the HPA axis may be "hyperdynamic" in PTSD (201), consistent with the clinical findings of hypervigilance and hyperarousal. Yehuda et al. (202) reported an increased number of lymphocyte glucocorticoid receptors in PTSD patients, believed to be secondary to receptor upregulation in the presence of low-circulating cortisol concentrations. PTSD patients were reported to respond to the dexamethasone suppression test similarly to normals (203), or, upon a metaanalysis of such studies, to be supersensitive to dexamethasone (204). Virtually all of these findings indicate opposite HPA axis activity measures in PTSD and depressed patients, although both diagnostic groups are reported to exhibit a blunted ACTH response to CRF challenge (205), and as just noted we have observed elevated CSF CRF concentrations in PTSD patients. One possible explanation for the differences in the HPA axis activity measures in PTSD patients and depressed patients may be differences in SRIF secretion. SRIF, which inhibits HPA axis activity, is elevated in the CSF of PTSD patients (vide supra) but reduced in depressed patients (205). A confound of many of the PTSD studies, including our original report of a blunted ACTH response to CRF, is that half of the patients in our study had comorbid major depression; the sample size of nondepressed PTSD patients was inadequate for significance, but the data tended towards the finding of a blunted ACTH response within this subgroup. A major finding relevant to the present discussion has recently been published. Debellis et al. (206) reported that girls (mean age, 11 years) who had been sexually abused as young children had a blunted ACTH response to CRF when compared to age- and sex-matched controls.

The CSF CRF as well as TRH concentrations have been measured in panic disorder patients, with no difference found between these patients and control (207,208). Panic disorder and PTSD are, of course, distinct clinical entities and it is not possible to draw conclusions about the biological basis of one disorder from the other, although the two reportedly share similar noradrenergic and possibly other mechanisms (194). It is likely that different patterns of stress in susceptible individuals may produce distinct CRF and/or other neurotransmitter system changes, producing predominately affective or anxiety symptoms, or both types of symptoms, in PTSD patients.

A second area of progress in the study of stress peptides and PTSD concerns endogenous opioids. In one study, Vietnam veterans with PTSD developed increased heat tolerance upon viewing a combat videotape (209). This response, similar to stress-induced analgesia in laboratory animals, was reversible with the mu opioid receptor antagonist naloxone. However, these investigators did not find a significant difference in either plasma β-endorphin or met-enkephalin concentrations pre- versus postcombat videotape, or between PTSD and control groups. Another group reported an increased rate of plasma met-enkephalin degradation in PTSD patients relative to controls, and a decreased release of met-enkephalin into the circulation; the net effect of these two alterations appears to be similar plasma met-enkephalin concentrations in both the PTSD and control groups (210). The blood samples were, however, obtained from these patients under basal, drug-treated conditions; perhaps met-enkephalin concentrations exhibit an increased ratio of stress-induced to tonic values (''signal-to-noise ratio''), as suggested for cortisol concentrations in PTSD patients (179).

Elucidation of the role of neuropeptides in the stress response has been only partially accomplished; considerably more research must be conducted considering the obvious complexity of the stress response. The role of neuroendocrine mechanisms and extrahypothalamic peptidergic neurons in the pathophysiology of PTSD requires more intensive scrutiny. Newer techniques of brain imaging, both structural and functional, as well as methods derived from the revolution in molecular biology will surely be applied to the study of PTSD and will undoubtedly shed more light on this perplexing area of basic and clinical research.

REFERENCES

1. Hayden-Hixson DM, Nemeroff CB. Role(s) of neuropeptides in responding and adapting to stress: a focus on corticotropin-releasing factor and opioid peptides. In Stanford SC, Salmon P, eds. *Stress: from synapse to syndrome.* London: Academic Press, 1993;355–391.
2. Whitnall MH. Regulation of the hypothalamic corticotropin-releasing hormone neurosecretory system. *Progress in Neurobiology* 1993;40:573–629.
3. Terenius L. Opioid peptides, pain and stress. *Progress in Brain Research* 1992;92:375–383.
4. Vale W, Speiss J, Rivier C, Rivier J. Characterization of a 41-residue ovine hypothalamic peptide that stimulates secretion of corticotropin and beta endorphin. *Science* 1981;213:1394–1397.
5. Swanson LW, Sawchenko PE, Rivier J, Vale W. Organization of ovine corticotropin-releasing factor immunoreactive cells and fibers in the rat brain: an immunohistochemical study. *Neuroendocrinology* 1983;36:165–186.
6. Merchenthaler I. Corticotropin-releasing factor (CRF)-like immunoreactivity in the rat central nervous system. Extrahypothalamic distribution. *Peptides* 1984;5:53–69.
7. Owens MJ, Nemeroff CB. The physiology and pharmacology of corticotropin-releasing factor. *Pharmaco Rev* 1991;43:425–473.
8. Bloom FE, Battenberg ELF, Rivier J, Vale W. Corticotropin releasing factor (CRF) immunoreactive neurons and fibers in rat hypothalamus. *Reg Peptides* 1982;4: 43–48.
9. Daikoku S, Okamura Y, Kawano H, Tsuruo Y, Maegawa M, Shibasaki T. CRF-containing neurons of the rat hypothalamus. *Cell Tissue Res* 1985;240:575–584.
10. Piekut DT, Joseph SA. Relationship of CRF-immunostained cells and magnocellular neurons in the paraventricular nucleus of rat hypothalamus. *Peptides* 1985;6: 873–882.
11. Kawata M, Hashimoto K, Takahara J, Sano Y. Immunohistochemical demonstration of the localization of corticotropin releasing factor-containing neurons in the hypothalamus of mammals including primates. *Anat Embryol* 1982;165:303–313.
12. Cole BJ, Koob GF. Corticotropin-releasing factor, stress and animal behavior. In McCubbin JA, Kaufmann PG, Nemeroff CB, eds. *Stress, neuropeptides, and systemic disease.* New York: Academic Press, 1991;119–148.
13. Koob GF, Heinrichs SC, Pich EM, Merzaghi F, Baldwin H, Miczek K, Britton KT. The role of corticotropin-releasing factor in behavioral responses to stress. In *Corticotropin-releasing factor Ciba Foundation*

Symposium. New York: John Wiley, 1993;172:227–295.
14. Swerdlow NR, Geyer MA, Vale WW, Koob GF. Corticotropin-releasing factor potentiates acoustic startle in rats: blockade by chlordiazepexide. *Psychopharmacology* 1986;88:147–152.
15. Berridge CW, Dunn AJ. Corticotropin-releasing factor elicits naloxone sensitive stress-like alterations in exploratory behavior in mice. *Reg Peptides* 1986;16:83–93.
16. Lee EHY, Tang YP, Chai CT. Stress and corticotropin-releasing factor potentiate center region activity of mice in an open field. *Psychopharmacology* 1987;93:320.
17. Sutton RE, Koob GF, Le Moal M. Corticotropin releasing factor produces behavioural activation in rats. *Nature* 1982;297:333–334.
18. Eaves M, Thatcher-Britton K, Rivier J, Vale W, Koob GF. Effects of corticotropin releasing factor on locomotor activity in hypophysectomized rats. *Peptides* 1985;6:923–926.
19. Britton KT, Lee G, Vale W, Rivier J, Koob GF. Corticotropin-releasing factor (CRF) receptor antagonist blocks activating and anxiogenic actions of CRF in the rat. *Brain Res* 1986;369:303.
20. Swerdlow NR, Britton KT, Koob GF. Potentiation of acoustic startle by corticotropin-releasing factor (CRF) and by fear are both reversed by α-helical CRF (9-41). *Neuropsychopharmacology* 1989;2:285–292.
21. Britton KT, Lee G, Dana R, Risch SC, Koob GF. Activating and "anxiogenic" effects of corticotropin releasing factor are not inhibited by blockade of pituitary-adrenal system with dexamethasone. *Life Science* 1986;39:1281–1286.
22. Swerdlow NR, Koob GF. Separate neural substates of the locomotor activating properties amphetamine, heroin, caffeine and corticotropin-releasing factor (CRF) in the rat. *Pharmacol Biochem Behav* 1985;23:303–307.
23. Koob GF, Swerdlow N, Seeligson M, et al. Effects of alpha-flupenthixol and naloxone on CRF-induced locomotor activation. *Neuroendocrinology* 1984;39:459–464.
24. Britton DR, Indyk E. Effects of ganglionic blocking agents on behavioral responses to centrally administered CRF. *Brain Res* 1989;478:205–210.
25. Britton KT, Morgan J, Rivier J, Vale W, Koob GF. Chlordiazepoxide attenuates response suppression induced by corticotropin-releasing factor in the conflict test. *Psychopharmacology* 1985;83:170–174.
26. Butler PD, Weiss JM, Stout JC, Nemeroff CB. Corticotropin-releasing factor produces fear-enhancing and behavioral activating effects following infusion into the locus coeruleus. *J Neurosci* 1990;10:176–183.
27. Weiss JM, Stout J, Aaron M, Owens MJ, Nemeroff CB. Experimental studies of depression and anxiety: role of locus coeruleus and corticotropin-releasing factor. *Brain Res Bull* 1994;35:561–572.
28. Valentino RJ, Foote SL, Aston-Jones G. Corticotropin-releasing factor activates noradrenergic neurons of the locus coeruleus. *Brain Res* 1983;270:363–367.
29. Melia KR, Duman RS. Involvement of corticotropin-releasing factor in chronic stress regulation of the brain noradrenergic system. *Proc Natl Acad Sci USA* 1991;88:8382–8386.
30. Carroll BJ. Pituitary-adrenal function in depression. *Lancet* 1968;556:1373–1374.
31. Bunney WE, Davis JM. Norepinephrine in depressive reactions. *Arch Gen Psychiatry* 1965;13:483–494.
32. Liang KC, Lee EHY. Intra-amygdala injections of corticotropin releasing factor facilitate inhibitory avoidance learning and reduce exploratory behavior rats. *Psychopharmacology* 1988;96:232–236.
33. Ray A, Henke PG, Gulati K, Sen P. The amygdaloid complex, corticotropin releasing factor and stress-induced gastric ulcerogenesis in rats. *Brain Res* 1993;624:286–290.
34. Swiergiel AH, Takahashi LK, Kalin NH. Attenuation of stress-induced behavior by antagonism of corticotropin-releasing factor receptors in the central amygdala in the rat. *Brain Res* 1993;623:229–234.
35. Beaulieu S, DiPaolo T, Barden N. Control of ACTH secretion by the central nucleus of the amygdala: implication of the serotoninergic systems and its relevance to the glucocorticoid delayed negative feedback mechanism. *Neuroendocrinology* 1986;44:247–254.
36. Van de Kar LD, Peichowshi RA, Ritterhouse PA, Gray TS. Amygdaloid lesions: differential effect of conditioned stress and immobilization-induced increases in corticosterone and renin secretion. *Neuroendocrinology* 191;54:89–95.
37. Hitchcock J, Davis M. Lesions of the amygdala, but not of the cerebellum or red nucleus, block conditioned fear as measured with the potentiated startle paradigm. *Behav Neurosci* 1986;100:11–22.
38. Chappell PB, Smith MA, Kilts CD, et al. Alterations in corticotropin-releasing factor-like immunoreactivity in discrete rat brain regions after acute and chronic stress. *J Neuroscience* 1986;6:2908–2914.
39. Owens MJ, Bissette G, Nemeroff CB. Acute effects of alprazolam and adinazolam on the concentrations of corticotropin-releasing factor in the rat brain. *Synapse* 1989;4:196–202.
40. Coplan JD, Rosenblum LA, Owens MJ, Andrews MW, Gorman JM, Nemeroff CB. Increased cerebrospinal fluid CRF concentrations in adult non-human primates previously exposed to adverse experiences as infants. *Soc Neurosci Abst* 1994;:17.
41. Harbuz MS, Lightman SL. Responses of hypothalamic and pituitary mRNA to physical and psychological stress in the rat. *J Endocrinology* 1989;122:705–711.
42. Ceccatelli S, Orazzo C. Effect of different types of stressors on peptide messenger ribonucleic acids in the hypothalamic paraventricular nucleus. *Acta Endocrinologica* 1993;128:485–492.
43. Imaki T, Nahan J-L, Rivier C, Sawchenko PE, Vale W. Differential regulation of corticotropin-releasing factor mRNA in rat brain regions by glucocorticoids and stress. *J Neuroscience* 1991;11:585–599.
44. Imaki T, Vale W. Chlordiazepoxide attentuates stress-induced accumulation of corticotropin-releasing factor mRNA in the paraventricular nucleus. *Brain Research* 1993;623:223–228.
45. Imaki T, Shibasaki T, Hotta M, Demura H. Intracerebroventricular administration of corticotropin-releasing factor induces c-fos mRNA expression in brain regions related to stress responses: comparison with pattern of c-fos mRNA induction after stress. *Brain Research* 1993;616:114–125.
46. Anderson SM, Kant GJ, DeSouza EB. Effects of

chronic stress on anterior pituitary and brain corticotropin-releasing factor receptors. *Pharm Biochem Behav* 1993;44:755–761.

47. Pihoker C, Owens MJ, Kuhn CM, Schanberg SM, Nemeroff CB. Maternal separation in neonatal rats elicits activation of the hypothalamic-pituitary-adrenocortical axis: a putative role for corticotropin-releasing factor. *Psychoneuroendocrinology* 1993;18:485–493.
48. Owens CA, Owens MJ, Thrivikraman KV, Nemeroff CB. The effects of early maternal deprivation on neurochemical and endocrine responses to stress in adult rats. *Soc Neuroscience Abst* 1994;20:372.
49. Nemeroff CB, Owens MJ, Plott SJ, Levine S. Increased density of regional brain CRF binding sites after maternal deprivation. *Soc Neurosci Abst* 1993;19:1.
50. Fuchs E, Flügge G. Modulation of binding sites for corticotropin-releasing hormone by chronic psychosocial stress. *Psychoneuroendocrinology* 1995;20:33–51.
51. Altura BM, Altura BT. Vascular smooth muscle and neurohypophyseal hormones. *Fed Proc* 1977;36:1853–1860.
52. Hems DA. Short-term hormonal control of hepatic carbohydrate and lipid catabolism. *FEBS Letters* 1977; 80:237–245.
53. Antoni FA. Vaspressingergic control of pituitary adrenocorticotropin secretion comes of age. *Frontiers in Neuroendocrinology* 1993;14:76–122.
54. Muhlethaler M, Dreifuss JJ, Gahwiler BH. Vasopressin excites hippocampal neurons. *Nature* 1982;296:749–751.
55. Morris JF, Chapman DB, Sokol HW. Anatomy and function of the classic vasopressin-secreting hypothalamo-neurohypophysial system. In Gash DM, Boer GJ, eds. *Vasopressin: principles and properties.* New York: Plenum Press, 1987;1–90.
56. Tramu G, Croix C, Pillez A. Ability of the CRF immunoreactive neurons of the paraventricular nucleus to produce a vasopressin-like material. *Neuroendocrinology* 1983;37:467–469.
57. Whitnall MH. Subpopulations of corticotropin-releasing hormone neurosecretory cells distinguished by presence or absence of vasopressin: confirmation with multiple corticotropin-releasing hormone antisera. *Neuroscience* 1990;36:201–205.
58. Holmes MC, Antoni FA, Aguilera G, Catt KJ. Magnocellular axons in passage through the median eminence release vasopressin. *Nature* 1986;319:326–329.
59. Baertschi AJ. Portal vascular route from hypophysial stalk/neural lobe to adenohypohysis. *Am J Physiol* 1980;239:R463–R469.
60. Watkins WB, Moore LG, Spiess J. Characterization of neurophysin in the anterior pituitary gland of sheep. *Neuroendocrinology* 1988;47:142–148.
61. Terrier C, Chabot JG, Autrat G, et al. Arginine-vasopressin in anterior pituitary cells; in situ hybridization of mRNA and ultrastructural localization of immunoreactivity. *Neuroendocrinology* 1991;54:303–311.
62. Sofroniew MV. Morphology of vasopressin and oxytocin neurons and their central and vascular projections. *Prog Brain Res* 1983;60:101–114.
63. Caffé AR, van Leeuwen FW. Vasopressin-immunoreactive cells in the dorsomedial hypothalamic region, medial amygdaloid nucleus and locus coeruleus of the rat. *Cell Tissue Res* 1983;233:23–34.
64. Lepetit P, Fevre-Montange M, Gay N, Belin M-F, Bobillier P. Vasopressin mRNA in the cerebellum and circumventricular organs: a quantitative in situ hybridization study. *Neuroscience Letters* 1993;159:171–174.
65. Buijs RM, Pevet P. Vasopressin- and oxytocin-containing fibres in the pineal gland and subcommissural organ of the rat. *Cell Tissue Res* 1980;205:11–17.
66. Kai-Kai M, Anderton BH, Keen P. A quantitative analysis of the interrelationships between subpopulations of rat sensory neurons containing arginine vasopressin or oxytocin and those containing substance P, fluoride-resistant acid phosphatase or neurofilament protein. *Neuroscience* 1986;18:475–486.
67. Burbach JP, DeHoop MJ, Schmale H, et al. Differential responses to osmotic stress of vasopressin-neurophysin mRNA in hypothalamic nuclei. *Neuroendocrinology* 1984;39:582–584.
68. Plotsky PM, Bruhn TO, Vale W. Hypophysiotropic regulation of adrenocorticotropin secretion in response to insulin-induced hypoglycemia. *Endocrinology* 1985;117:323–329.
69. Caraty A, Grino M, Locatelli A, et al. Insulin-induced hypoglycemia stimulates corticotropin-releasing factor and arginine vasopressin secretion into hypophysial portal blood of conscious, unrestrained rams. *J Clin Invest* 1990;85:1716–1721.
70. Dohanics J, Verbalis J. Hyponatremia-induced inhibition of magnocellular vasopressin secretion selectively impairs ACTH secretion in response to stress. *Proc 71st Ann Meeting Endocrin Soc Abstract,* Seattle, Washington; 1989;1727.
71. van Dijken HH, de Goeij DCE, Sutanto W, Mos J, de Kloet ER, Tilders FJH. Short inescapable stress produces long-lasting changes in the brain-pituitary-adrenal axis of adult male rats. *Neuroendocrinology* 1993;58:57–64.
72. Lang RE, Heil JWE, Ganten D, Hermann K, Unger T, Rascher W. Oxytocin unlike vasopressin is a stress hormone in the rat. *Neuroendocrinology* 1983;37: 314–316.
73. Gibbs DM. Stress-specific modulation of ACTH secretion by oxytocin. *Neuroendocrinology* 1986;42:456–458.
74. Priou A, Oliver C, Grino M. In situ hybridization of arginine vasopressin (AVP) heteronuclear ribonucleic acid reveals increased AVP gene transcription in the rat hypothalamic paraventricular nucleus in response to emotional stress. *Acta Endocrinologica* 1993;128: 466–472.
75. Bartanusz V, Jezova D, Bertini LT, Tilders FJH, Aubry J-M, Kiss JZ. Stress-induced increase in vasopressin and corticotropin-releasing factor expression in hypophysiotrophic paraventricular neurons. *Endocrinology* 1993;132:895–902.
76. Bartanusz V, Aubry J-M, Jezova D, Baffi J, Kiss JZ. Up-regulation of vasopressin mRNA in paraventricular hypophysiotrophic neurons after acute immobilization stress. *Neuroendocrinology* 1993;58:625–629.
77. Herman JP, Sherman TG. Acute stress upregulates vasopressin gene expression in parvocellular neurons of the hypothalamic paraventricular nucleus. *Ann NY Acad Sci* 1993;689:546–549.
78. Scaccianoce S, Muscolo LAA, Cigliana G, Navarra D, Nicolai R, Angelucci L. Evidence for a specific role of vasopressin in sustaining pituitary-adrenocortical

stress response in the rat. *Endocrinology* 1991;128: 3138–3143.

79. Roozendaal B, Wiersma A, Driscoll P, Koolhaas JM, Bohus B. Vasopressinergic modulation of stress responses in the central amygdala of the Roman high-avoidance and low-avoidance rat. *Brain Res* 1992; 596:35–40.
80. Buijs RM, Swaab DF, Dogterom J, van Leeuwen FW. Intra and extrahypothalamic vasopressin and oxytocin pathways in the rat. *Cell Tissue Res* 1978;186: 423–433.
81. Antoni FA, Holmes MC, Jones MT. Oxytocin as well as vasopressin potentiate ovine CRF in vitro. *Peptides* 1983;4:411–415.
82. Legros JJ, Chiodera P, Demey-Ponsart E. Inhibitory influence of exogenous oxytocin on adrenocorticotropin secretion in normal human subjects. *J Clin Endocrinol Metab* 1982;55:1035–1039.
83. Gibbs DM. Vasopressin and oxytocin: hypothalamic modulators of the stress response: a review. *Psychoneuroendocrinology* 1986;11:131–140.
84. Miaskowski C, Ong GL, Lukic D, Haldar J. Immobilization stress affects oxytocin and vasopressin levels in hypothalamic and extrahypothalamic sites. *Brain Res* 1988;458:137–141.
85. Olson GA, Olson RD, Kastin AJ. Endogenous opioids: 1992. *Peptides* 1993;14:1339–1378.
86. Nyberg F, Terenius L. Enzymatic inactivation of neuropeptides. In Henriksen JH, ed. *Degradation of bioactive substances: physiology and pathophysiology*. Boca Raton, Ann Arbor, Boston, London: CPC Press, 1991;189–200.
87. Wilkes MM, Watkins SB, Stewart RD, Yen SS. Localization and quantification of beta-endorphin in human brain and pituitary. *Neuroendocrinology* 1980;30:113–121.
88. Loh HH, Smith AP. Molecular characterization of opioid receptors. *Ann Rev Pharmacol Toxicol* 1990;30: 123–147
89. Walker JM, Bowen WD, Walter FD, et al. Sigma receptors: biology and function. *Pharmacol Rev* 1990;42: 355–402.
90. Terman GW, Shavit Y, Lewis JW, Cannon JT, Liebeskind JC. Intrinsic mechanisms of pain inhibition: activation by stress. *Science* 1984;226:1270–1277.
91. Kelly DD, ed. *Stress-induced analgesia.* New York: New York Academy of Sciences, 1986.
92. Willis WD Jr. Anatomy and physiology of descending control of nociceptive responses of dorsal horn neurons: comprehensive review. *Prog Brain Res* 1988; 77:1–29.
93. Millan MJ. Multiple opioid systems and pain. *Pain* 1986;27:303–348.
94. Yaksh TL, Al-Rodhan NF, Jensen TS. Sites of action of opiates in production of analgesia. In Fields, HL, Besson, JM, eds. *Progress in brain research, vol. 77.* Amsterdam: Elsevier, 1988;371–393.
95. Watkins LR, Wiertelak EP, Maier SF. Kappa opiate receptors mediate tail-shock induced antinociception at spinal levels. *Brain Res* 1992;582:1–9.
96. Watkins LR, Wiertelak EP, Grisel JE, Silbert LH, Maier SF. Parallel activation of multiple spinal opiate systems appears to mediate non-opiate stress-induced analgesias. *Brain Res* 1992;594:99–108.
97. Lewis JW, Cannon JT, Liebeskind JC. Opioid and nonopioid mechanisms of stress analgesia. *Science* 1980;208:623–635.
98. Watkins LR, Cobelli DA, Faris P, Aceto MD, Mayer DJ. Opiate vs non-opiate footshock-induced analgesia (FSIA): the body region shocked is a critical factor. *Brain Res* 1982;242:299–308.
99. Cannon JT, Terman GW, Lewis JW, Liebeskind JC. Body region shocked need not critically define the neurochemical basis of stress analgesia. *Brain Res* 1984;323:316–319.
100. Kirchgessner AL, Bodnar RJ, Pasternak GW. Naloxone and pain-inhibitory systems: Evidence for a collateral inhibition model. *Pharmacol Biochem Behav* 1982;17:1175–1179.
101. Grisel JE, Fleshner M, Watkins LR, Maier SF. Opioid and nonopioid interactins in two forms of stress-induced analgesia. *Pharmacol Biochem Behav* 1993;45: 161–172.
102. Wong C-L. The effect of naloxone on restraint-induced antinociception in mice. *Meth Find Exp Clin Pharmacol* 1992;14:695–700.
103. Shangoid MM. Exercise and amenorrhea. *Sem Reprod Endocrinol* 1985;3:35–41.
104. Cumming DC, Vickovic MM, Wall SR, Fluker MR, Belcastro AN. The effect of acute exercise on pulsatile release of luteinizing hormone in women runners. *Am J Obstet Gynecol* 1985;153:482–485.
105. Norman RL, Smith CJ. Restraint inhibits luteinizing hormone and testosterone secretion in intact male rhesus macaques: effects of concurrent naloxone administration. *Neuroendocrinology* 1992;55:405–415.
106. Rivier C, Rivier J, Vale W. Stress-induced inhibition of reproductive functions: role of endogenous corticotropin-releasing factor. *Science* 1986;231:607–609.
107. Nikolarakis KE, Almeida OFX, Herz A. Concomitant changes in the in vitro and in vivo release of opioid peptides and luteinizing hormone-releasing hormone in the hypothalamus following blockade of receptors for corticotropin-releasing factor. *Neuroendocrinology* 1988;47:545–550.
108. Nabeshima T, Katoh A, Wada M, Kameyama T. Stress-induced changes in brain Met-Enkephalin, Leu-Enkephalin and dynorphin concentrations. *Life Sciences* 1992;51:211–217.
109. Bannon MJ, Roth RH. Pharmacology of mesocortical dopamine neurons. *Pharmacol Rev* 1983;35:53–68.
110. Miller JD, Speciale SG, McMillen BA, German DC. Naloxone antagonism of stress-induced augmentation of frontal cortex dopamine metabolism. *Eur J Pharmacol* 1984;98:437–439.
111. Bannon MJ, Deutch AY, Tam S-Y, et al. Mild footshock stress dissociates substance P from substance K and dynorphin from Met- and Leu-enkephalin. *Brain Res* 1986;381:393–396.
112. Young III WS, Lightman SL. Chronic stress elevates enkephalin expression in the rat paraventricular and supraoptic nuclei. *Mol Brain Res* 1992;13:111–117.
113. Borsook D, Konradi C, Falkowski O, Comb M, Hyman SE. Molecular mechanisms of stress-induced proenkephalin gene regulation: CREB interacts with the proenkephalin gene in the mouse hypothalamus and is phosphorylated in response to hyperosmolar stress. *Mol Endocrin* 1994;8:240–248.

114. Matthews SG, Parrott RF, Sirinathsinghji DJS. Isolation- and dehydration-induced changes in neuropeptide gene expression in the sheep hypothalamus. *J Mol Endocrin* 1993;11:181–189.
115. Summy-Long JY, Rosella-Dampman LM, McLemore GL, Koehler E. Kappa opiate receptors inhibit release of oxytocin from the magnocellular system during dehydration. *Neuroendocrinology* 1990;51:376–384.
116. Watson S, Akil H, Fischili W, et al. Dynorphin and vasopressin: common localization in magnocellular neurons. *Science* 1982;216:85–87.
117. Falke N. Dynorphin (1-8) inhibits stimulated release of oxytocin but not vasopressin from isolated neurosecretory endings of the rat neurohypophysis. *Neuropeptides* 1988;11:163–167.
118. Bondy CA, Gainer H, Russell JT. Dynorphin A inhibits and naloxone increases the electrically stimulated release of oxytocin but not vasopressin from the terminals of the neural lobe. *Endocrinology* 1988;122:1321–1327.
119. Clarke G, Wood P, Merrick L, Lincoln DW. Opiate inhibition of peptide release from the neurohumoral terminals of hypothalamic neurones. *Nature* 1979; 282:746–748.
120. Iversen LL, Iversen SD, Bloom FE. Opiate receptors influence vasopressin release from nerve terminals in rat neurohypophysis. *Nature* 1980;284:350–351.
121. Stein EA, Hiller JM, Simon EJ. Effects of stress on opioid receptor binding in the rat central nervous system. *Neuroscience* 1992;51:683–690.
122. Tanaka K, Masu M, Nakanishi S. Structure and functional expression of the cloned rat neurotensin receptor. *Neuron* 1990;4:847–854.
123. Di Paola ED, Cusack B, Yamada M, Richelson E. Desensitization and down-regulation of neurotensin receptors in murine neuroblastoma clone N1E-115 by [d-Lys^8] Neurotensin(8-13). *J Pharmacol Expt Ther* 1993;264:1–5.
124. Vita N, Laurent P, Lefort S, et al. Cloning and expression of a complementary DNA encoding a high affinity human neurotensin receptor. *FEBS Letters* 1993; 317:139–142.
125. Szigethy E, Beaudet A. Correspondence between high-affinity ^{125}I-neurotensin binding sites and dopaminergic neurons in the rat substantia nigra and ventral tegmental area: a combined radioautographic and immunohistochemical light microscopic study. *J Comp Neurol* 1989;279:128–137.
126. Honkaniemi J. Colocalization of peptide- and tyrosine hydroxylase-like immunoreactivities with Fos-immunoreactive neurons in rat central amygdaloid nucleus after immobilization stress. *Brain Res* 1992;598: 107–113.
127. Dana C, Vial M, Leonard K, et al. Electron microscopic localization of neurotensin binding sites in the midbrain tegmentum of the rat. I. Ventral tegmental area and interfascicular nucleus. *J Neurosci* 1989;9:2247–2257.
128. Jennes L, Stumpf WE, Kalivas PW. Neurotensin: topographical distribution in rat brain by immunohistochemistry. *J Comp Neurol* 1982;210:211–224.
129. Nicot A, Rostene W, Berod A. Neurotensin receptor expression in the rat forebrain and midbrain: a combined analysis by in situ hybridization and receptor autoradiography. *J Comp Neurol* 1994;341:407–419.
130. Gudelsky GA, Berry SA, Meltzer HY. Neurotensin activates tuberoinfundibular dopamine neurons and increases serum corticosterone concentrations in the rat. *Neuroendocrinology* 1989;49:604–609.
131. Nemeroff CB, Hernandez DE, Orlando RC, Prange AJ Jr. Cytoprotective effect of centrally administered neurotensin in rats. *Dev Brain Res* 1982;3:497–501.
132. Henke PG, Sullivan RM, Ray A. Interactions of thyrotropin-releasing hormone (TRH) with neurotensin and dopamine in the central nucleus of the amygdala during stress ulcer formation in rats. *Neurosci Letters* 1988; 91:95–100.
133. Watts AG. Ether anesthesia differentially affects the content of prepro-corticotropin-releasing hormone, prepro-neurotensin-neuromedin N and prepro-enkephalin mRNAs in the hypothalamic paraventricular nucleus of the rat. *Brain Res* 1991;544:353–357.
134. Swanson LW, Sawchenko PE, Lind RW. Regulation of multiple peptides in CRF parvocellular neurosecretory neurons: implications for the stress response. *Prog Brain Res* 1986;68:1692.
135. Deutch AY, Roth RH. The determinants of stress-induced activation of the prefrontal cortical dopamine system. In Uylings HBM, Van Eden CG, de Bruin JPC, Corner MA, Feenstra MGP, eds. *Progress in brain research, vol. 85.* New York: Elsevier Press, 1990;367–403.
136. Coco ML, Kuhn CM, Ely TD, Kilts CD. Selective activation of mesoamygdaloid dopamine neurons by conditioned stress: alteration by diazepam. *Brain Res* 1992;590:39–47.
137. Wolfe J, Beaudet A. Immunocytochemical evidence for direct connections between neurotensin-containing axons and dopaminergic neurons in the rat ventral midbrain tegmentum. *Brain Res* 1989;479:402–406.
138. Seutin V, Massotte L, Dresse A. Electrophysiological effects of neurotensin on dopaminergic neurons of the ventral tegmental areas of the rat in vitro. *Neuropharmacology* 1989;28:949–954.
139. Kilts CD, Coco ML, Ely TD, Bissette G, Nemeroff CB. Differential effects of conditioned and unconditioned stress on the neurotensin content of dopamine cell body groups of ventral mesencephalon. *Ann New York Acad Sci* 1992;668:266–276.
140. Deutch AY, Bean AJ, Bissette G, Nemeroff CB, Robbins RJ, Roth RH. Stress-induced alterations in neurotensin, somatostatin and corticotropin-releasing factor in mesotelencephalic dopamine systems regions. *Brain Res* 1987;417:350–354.
141. Krulich L, Dhariwal APS, McCann SM. Stimulatory and inhibitory effects of purified hypothalamic extracts on growth hormone release from rat pituitary in vitro. *Endocrinology* 1968;83:783–790.
142. Brazeau P, Vale W, Burgus R, et al. Hypothalamic polypeptide that inhibits the secretion of immunoreactive pituitary growth hormone. *Science* 1973;197: 77–79.
143. Krisch B. Somatostatin-immunoreactive fiber projections into the brainstem and spinal cord of the rat. *Cell Tissue Res* 1981;217:531–552.
144. Bell GI, Reisine T. Molecular biology of somatostatin receptors. *Trends Neurosci* 1993;16:34–38.
145. Olpe HR, Balcar VJ, Bittinger H, Rink H, Sieber P.

Central actions of somatostatin. *Eur J Pharmacol* 1980;63:127–133.
146. Rezek M, Havlicek V, Hughes KR, Havlicek V, Hughes KR, Friesen H. Cortical administration of somatostatin (SRIF): effects on sleep and motor behavior. *Pharmacol Biochem Behav* 1976;5:73–77.
147. Aponte G, Leung P, Gross D, Yamada T. Effects of somatostatin on food intake in rats. *Life Sci* 1984; 35:741–746.
148. Vecsei L, Kiraly C, Bollok I, et al. Comparative studies with somatostatin and cysteamine in different behavior tests on rats. *Pharmacol Biochem Behav* 1984;21: 833–837.
149. Terry LC, Saunders A, Audet J, Willoughby JO, Brazeau P, Martin JB. Physiologic secretion of growth hormone and prolactin in male and female rats. *Clin Endocrin* 1977;6:19S–28S.
150. Takahashi K. Effect of various traumatic stresses on growth hormone release in pentobarbital anesthetized rats. *Neuroendocrinology* 1978;26:1–7.
151. Armario A, Jolín T. Influence of intensity and duration of exposure to various stressors on serum TSH and GH levels in adult male rats. *Life Sci* 1989;44:215–221.
152. Reichlin S. Prolactin and growth hormone secretion in stress. In Chrousos GP, Loriaux DL, Gold PW, eds. *Mechanisms of physical and emotional stress.* New York: Plenum Press, 1988;353–376.
153. Terry LC, Willoughby JO, Braseau P, Martin JB, Patel Y. Antiserum to somatostatin prevents stress-induced inhibition of growth hormone secretion in the rat. *Science* 1976;192:565–567.
154. Rivier C, Vale W. Involvement of corticotropin-releasing factor and somatostatin in stress-induced inhibition of growth hormone secretion in the rat. *Endocrinology* 1985;117:2478–2482.
155. Benyassi A, Gavalda A, Armario A, Arancibia S. Role of somatostatin in the acute immobilization stress-induced GH decrease in rat. *Life Sci* 1993;52:361–370.
156. Armario A, Marti O, Gavaldà A, Giralt M, Jolin T. Effects of chronic immobilization stress on GH and TSH secretion in the rat: response to hypothalamic regulatory factors. *Psychoneuroendocrinology* 1993;18: 405–413.
157. Benyassi A, Roussel J, Rougeot C, Gavald A, Astier H, Arancibia S. Chronic stress affects in vivo hypothalamic somatostatin release but not in vitro GH responsiveness to somatostatin in rats. *Neurosci Let* 1993; 159:166–170.
158. Aguila MC, Pickle RL, Yu WH, McCann SM. Roles of somatostatin and growth hormone-releasing factor in ether stress inhibition of growth hormone release. *Neuroendocrinology* 1991;54:515–520.
159. Moran TH, Schwartz GJ. The neurobiology of cholecystokinin. *Crit Rev Neurobiology* (in press).
160. Wank SA, Harkins R, Jensen RT, et al. Purification, molecular cloning, and functional expression of the cholecystokinin receptor from rat pancreas. *Proc Natl Acad Sci* 1992;89:3125–3129.
161. Wank SA, Pisegna JR, de Weerth A. Brain and gastrointestinal cholecystokinin receptor family: structure and functional expression. *Proc Natl Acad Sci* 1992; 89:8691–8695.
162. Carlberg M, Gundlach AL, Mercer LD, Beart PM. Autoradiographic localization of cholecystokinin A and B receptors in rat brain using [^{125}I] D-Tyr^{25} (Nle^{28,31})-CCK 25-33S. *Eur J Neurosci* 1992;4: 563–573.
163. Hill RG, Shaw TM, Graham W, Woodruff GN. Autoradiographic detection of cholecystokinin (CCK-A) receptors in primate brain using [125I]-Bolton Hunter CCK8 and [3H]-MK-329. *J Neurosci* 1990;10:1070–1081.
164. Hill DR, Campbell NJ, Shaw TM. Woodruff GN. Autoradiographic localization and biochemical characterization of peripheral type CCK receptors in rat CNS using highly selective non-peptide CCK antagonists. *J Neurosci* 1987;7:2967–2976.
165. Harro J, Vasar E, Bradwejn J. CCK in animal and human research on anxiety. *Trends Pharmacol Sci* 1993;14:244–249.
166. Fekete M, Lengyel A, Hegedus B, et al. Further analysis of the effects of cholecystokinin octapeptide on avoidance behavior in rats. *Eur J Pharmacol* 1984; 98:79–91.
167. Harro J, Vasar E. Evidence that CCK_B receptors mediate the regulation of exploratory behavior in the rat. *Eur J Pharmacol* 1991;193:379–381.
168. Dauge V, Böhme GA, Crawley JN, et al. Investigation of behavioral and electrophysiological responses induced by selective stimulation of CCKB receptors by using a new highly potent CCK analog, BC 264. *Synapse* 1990;6:73–80.
169. Palmour R, Bradwejn B, Ervin F. The anxiogenic effects of CCK_4 in monkeys are reduced by CCK-B antagonists, benzodiazepines or adenosine A2 agonists. *Eur Neuropsychopharmacol* 1992;2:193–195.
170. Singh L, Field MJ, Hughes J. The behavioural properties of CI 988, a selective $cholecystokinin_B$ receptor antagonist. *Br J Pharmacol* 1991;104:239–245.
171. Derrien M, Durieux C, Roques BP. Antidepressant-like effects of CCK_B antagonists in mice: antagonism by naltrindole. *Br J Pharmacol* 1994;111:956–960.
172. Bradwejn J, Koszycki D, Meterissian G. Cholecystokinin tetrapeptide-induced panic attacks in patients with panic disorder. *Can J Psychiatry* 1990;35:83–85.
173. Bradwejn J, Koszycki D, Bourin M. Dose ranging study of the effect of CCK_4 in healthy volunteers. *J Psychiatry Neurosci* 1991;16:260–264.
174. Bradwejn J, Koszycki D, Couetoux du Tertre A, et al. The panicogenic effects of cholecystokinin-tetrapeptide are antagonized by L-365,260, a central cholecystokinin receptor antagonist, in patients with panic disorder. *Arch Gen Psychiatry* 1994;51:486–493.
175. Luckman SM, Hamamura M, Antonijevic I, Dyes. Involvement of cholecystokinin receptor types in pathways controlling oxytocin secretion. *Br J Pharm* 1993;110:378–384.
176. Adrian TE, Allen JM, Bloom SR, et al. Neuropeptide Y distribution in human brain. *Nature* 1983;306: 584–586.
177. Heilig M, Koob GF, Ekman R, Britton KT. Corticotropin-releasing factor and neuropeptide Y: role in emotional integration. *Trends Neurosci* 1994;17:80–85.
178. Heilig M, Söderpalm B, Engel JA, Widerlöv E. Centrally administered neuropeptide Y (NPY) produces anxiolytic-like effects in animal anxiety models. *Psychopharmacology* 1989;98:524–529.
179. Heilig M, McLeod S, Brot M, et al. Anxiolytic-like action of neuropeptide Y: mediation by Y1 receptors

in amygdala, and dissociation from food intake effects. *Neuropsychopharmacology* 1993;8:357–363.
180. Heilig M, Murison R. Intracerebroventricular neuropeptide Y protects against stress-induced gastric erosion in the rat. *Eur J Pharmacol* 1987;137:127–129.
181. Penner SB, Smyth DD, Glavin GB. Effects of neuropeptide Y and [Leu31,Pro34] neuropeptide Y on experimental gastric lesion formation and gastric secretion in the rat.
182. Pralong FP, Corder R, Gaillard RC. The effects of chronic glucocorticoid excess, adrenalectomy and stress on neuropeptide Y in individual rat hypothalamic nuclei. *Neuropeptides* 1993;25:223–231.
183. Widerlöv E, Lindström LH, Wahlestedt C, Ekman R. Neuropeptide Y and peptide YY as possible cerebrospinal fluid markers for major depression and schizophrenia, respectively. *J Psychiatr Res* 1988;22:69–79.
184. Widdowson PS, Ordway GA, Halaris AE. Reduced neuropeptide Y concentrations in suicide brain. *J Neurochem* 1992;59:73–80.
185. Rosén A, Brodin K, Eneroth P, Brodin E. Short-term restraint stress and s.c. saline injection alter the tissue levels of substance P and cholecystokinin in the periaqueductal grey and limbic regions of rat brain. *Acta Physiol Scand* 1992;146:341–348.
186. Gavaldà A, Benyassi A, Arancibia S, Armario A. Chronic but not acute exposure to stress is associated with hypothalamic vasoactive intestinal polypeptide (VIP) release into median eminence. *J Neuroendocrin* 1993;5:421–425.
187. Hernandez DE, Arredondo ME, Xue BG, Jennes L. Evidence for a role of brain thyrotropin-releasing hormone (TRH) on stress gastric lesion formation in rats. *Brain Res Bull* 1990;24:693–695.
188. Diagnostic and Statistical Manual of Mental Disorders, Fourth Edition. Washington, DC: American Psychiatric Association, 1994.
189. Yehuda R, Antelman SM. Criteria for rationally evaluating animal models of posttraumatic stress disorder. *Biol Psychiatry* 1993;33:479–486.
190. Nemeroff CB, Widerlov E, Bissette G. Elevated concentration of CSF corticotropin-releasing factor-like immunoreactivity in depressed patients. *Science* 1984; 226:1342–1344.
191. Arato M, Banki SC, Bissette G, Nemeroff CB. Elevated CSF CRF in suicide victims. *Biol Psychiatry* 1989; 25:355–359.
192. Stout SC, Nemeroff CB. Stress and psychiatric disorders. *Seminars in Neurosci* 1994;6:271–280.
193. Southwick SM, Krystal JH, Johnson DR, Charney DS. Neurobiology of posttraumatic stress disorder. *Ann Rev Psychiatry* 1992;11:347–364.
194. Southwick SM, Krystal JH, Morgan CA, et al. Abnormal noradrenergic function in posttraumatic stress disorder. *Arch Gen Psychiatry* 1993;50:266–274.
195. Bissette G, Widerlöv E, Walléus H, et al. Alterations in cerebrospinal fluid concentrations of somatostatinlike immunoreactivity in neuropsychiatric disorders. *Arch Gen Psychiatry* 1986;43:1148–1151.
196. Musselman DL, Nemeroff CB. The role of corticotropin-releasing factor in the pathophysiology of psychiatric disorders. *Psychiatric Ann* 1993;23:676–681.
197. Post RM, Gold P, Rubinow DR. Peptides in cerebrospinal fluid of neuropsychiatric patients: an approach to central nervous system peptide function. *Life Science* 1982;31:1–15.
198. Kalin NH, Shelton SE, Barksdale CM, Brownfield MS. A diurnal rhythm in cerebrospinal fluid corticotropin-releasing hormone different from the rhythm of pituitary-adrenal activity. *Brain Res* 1987;426:385–391.
199. Nemeroff CB, Owens MJ, Bissette G, Andorn AC, Stanley M. Reduced corticotropin-releasing factor binding sites in the frontal cortex of suicide victims. *Arch Gen Psychiatry* 1988;45:577–579.
200. Mason J, Giller E, Kosten T, Harkness T. Elevation of urinary norepinephrine/cortisol ratio in post-traumatic stress disorder. *J Nerv Ment Dis* 1988;176:498–502.
201. Yehuda R, Levengood RA, Teicher M. Circadian release of cortisol in combat PTSD. *Proc 146th Ann Meeting Am Psychiat Assn* 1993;116.
202. Yehuda R, Lowy MT, Southwick SM, Shaffer S, Giller EL. Increased number of glucocorticoid receptors in post-traumatic stress disorder. *Am J Psychiatry* 1991; 149:499–504.
203. Kosten TR, Mason JW, Giller EL, Harness L. Sustained urinary norepinephrine and epinephrine evaluation in post-traumatic stress disorder. *Psychoneuroendocrinology* 1987;12:13–20.
204. Yehuda R, Giller EL, Mason JW. Psychoneuroendocrine assessment of posttraumatic stress disorder: current progress and new directions. *Prog Neuro-Psychopharmacol & Biol Psychiat* 1993;17:541–550.
205. Smith MA, Davidson J, Ritchie JC. The corticotropin releasing hormone test in patients with posttraumatic stress disorder. *Biol Psychiatry* 1989;26:349–355.
206. De Bellis MD, Chrousos GP, Dorn LD, et al. Hypothalamic-pituitary-adrenal axis dysregulation in sexually abused girls. *J Clin Endocrin Metab* 1994;78:249–255.
207. Jolkkonen J, Lepola U, Bissette G, Nemeroff C, Riekkinen P. CSF corticotropin-releasing factor is not affected in panic disorder. *Biol Psychiatry* 1993;33: 136–138.
208. Fossey MD, Lydiard RB, Ballenger JC, Laraia MT, Bissette G, Nemeroff CB. Cerebrospinal fluid thyrotropin-releasing hormone concentrations in patients with anxiety disorder. *J Neuropsychiatry and Clinical Neurosciences* 1993;5:335–337.
209. Van der Kolk BA, Greenberg MS, Orr SP, Pitman RK. Endogenous opioids, stress induced analgesia, and posttraumatic stress disorder. *Psychopharm Bull* 1989; 25:417–421.
210. Wolf ME, Mosnaim AD, Puente J, Ignacio R. Plasma methionine-enkephalin in PTSD. *Biol Psychiatry* 1991;29:295–308.

Neurobiological and Clinical Consequences of Stress: From Normal Adaptation to PTSD, edited by M. J. Friedman, D. S. Charney, and A. Y. Deutch, Lippincott-Raven Publishers, Philadelphia © 1995.

7

Setting "Stress" into Motion

Brain Mechanisms of Stimulus Evaluation

Joseph E. LeDoux

Center for Neural Science, New York University, New York, New York 10003

"Stress," however we choose to define this ambiguous term, is a common occurrence for many people and a major health issue in the modern world. A good deal of progress has been made in the effort to understand the biology of the stress reaction, the cascade of physiological events that occur as a consequence of exposure to stressful events in the environment. In this large and growing body of research, stress is usually depicted as a response to environmental challenges. While this focus on the "effects" of stress is valid, it is incomplete, as it emphasizes the events that follow the onset of the stress reaction and largely ignores how the reaction itself is initiated.

In experimental studies, intense physical stimulation (i.e., physical restraint or prolonged exposure to extreme temperatures or to painful stimulation) is often used as a means of eliciting stress reactions. However, in contemporary life, stress is more likely to depend on symbolic stimulus meanings than on physical stimulus features. It is our personal interpretation of real or imagined objects as actually or potentially harmful to our physical or psychological well-being that is most likely to cause us to be anxious and worried and ultimately depressed. In these cases it is not the stimulus that causes stress so much as the interpretation we apply to it. We will not understand the psychology or biology of stress until we understand how environmental events come to acquire symbolic significance and thereby gain access to circuits that control so-called stress responses. At the same time, from a clinical point of view, we will not fully understand how to control stress until we understand the mental events that initiate stress responses.

Although the process through which stress reactions are set into motion by stimulus evaluations has not been a major focus of stress researchers, the mechanisms of stimulus evaluation have been intensely investigated within the related area of research concerned with the neural basis of emotion, and emotional learning and memory. In this chapter I will describe research that has identified neural systems that underlie the evaluation of the emotional significance of environmental stimuli, especially research that has used the techniques of classical fear conditioning to study emotional mechanisms. Complementing the vast literature on the biological response to stress, this work illustrates how the brain learns and stores information about the emotional significance of external stimuli, thus providing insights into the manner in which aversive emotional reactions, including stress reactions, are set into motion.

FEAR CONDITIONING

Through fear conditioning, a meaningless stimulus such as a sound (tone) of moderate intensity

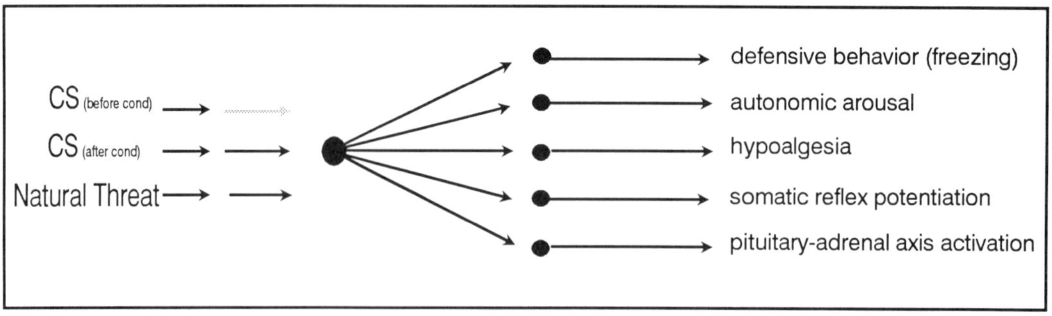

FIG. 1. In the presence of threatening events, a variety of "hard-wired" emotional defense responses are activated. The threat can be unlearned (natural) or learned (conditioned). As a result of emotional learning, a conditioned stimulus (CS) is able to enter into the circuits that control the hard-wired defense repertoire. (See Davis (8) for a similar diagram.)

can be transformed into an emotionally significant event. This occurs as a result of the association in time of the sound with some aversive stimulus. In studies with rats, the aversive stimulus is often the delivery of a brief electric shock to the feet. The shock is called an unconditioned stimulus (US) and the tone a conditioned stimulus (CS). After a few such pairings (as few as one in some instances), rats will exhibit emotional reactions in the presence of the tone CS (and in the absence of the shock US) that are typically exhibited when members of its species are faced with a natural threat (Fig. 1). For example, laboratory rats exhibit "freezing" behavior and elevations in arterial pressure and heart rate when they hear the tone CS after it has been paired with the US (1–3). In addition, in the presence of the CS, a number of reflex reactions are potentiated (4) and pain responses are diminished (5). Although these reactions are elicited by the CS as a result of learning, the responses themselves are not learned. The rat learns the association between the tone and shock and, on the basis of this association, responses are expressed. The responses are evolutionarily programmed reactions to threat. The learning allows novel stimuli to gain control over these responses.

NEURAL BASIS OF FEAR CONDITIONING

Over the past decade, several laboratories have made significant contributions to the elucidation of the neural circuitry through which conditioned stimuli gain control over conditioned fear responses as a result of CS–US pairings. In particular, studies by Kapp, Davis, Fanselow, and their colleagues are especially noteworthy (see reviews 4, 6–10; and Chapter 11). I will focus on research from my laboratory using the elicitation of species-typical defensive behavior (freezing) and increases in arterial pressure as conditioned responses by a tone CS paired with a shock US (Fig. 2), but the findings obtained in several laboratories using different procedures and responses are mostly in agreement.

Neural System Underlying Simple Conditioning

In simple conditioning, a single tone is paired with a shock. The rat has to learn only this association. That an association has been learned is determined through control studies, conducted in a separate group, showing that the so-called conditioned fear responses are not expressed in the presence of the CS when it is randomly related to, rather than paired with, the US.

The neural system underlying the learning of such associations involves transmission of the tone CS through the auditory system to the medial geniculate body (MGB). Conditioning can be mediated by projections from MGB directly to the amygdala or from MGB to auditory cortex and from there to the amygdala. Within the amygdala, the lateral nucleus (AL) receives the

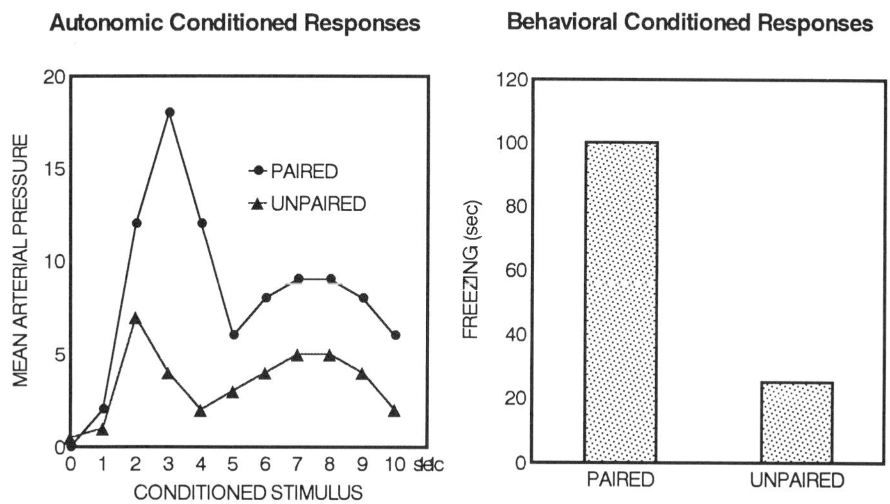

FIG. 2. Examples of autonomic and behavioral conditioned fear responses. The left panel shows arterial pressure responses and the right freezing behavior elicited by a tone to animals previously given paired or unpaired presentations of the tone with a shock. Only with systematic pairing does conditioning take place.

auditory signal from both the auditory thalamus and from the auditory cortex, and is thus the sensory interface of the amygdala in fear conditioning (Fig. 3). The AL is also a site where thalamic and cortical information can be integrated. Conditioning, as expressed at the level of behavior, most likely depends on the synergistic action of these two CS transmission pathways in AL.

The thalamo-amygdala pathway provides a rapid channel for activating the amygdala, but it does not provide the amygdala with a very accurate representation of the external event. The cortical pathway provides the amygdala with a very accurate representation of the stimulus, but it requires more synaptic links and thus more time to do so. The thalamic pathway may be particularly useful in situations requiring rapid responses, where it is more important to respond quickly than it is to be accurate in whether a response is required. The cortical pathway can then take over, either allowing the protective response to continue or aborting the reaction if it has, on further analysis of the stimulus, proven to be inappropriate. If thalamic and cortical systems do not remain in synchrony, such that thalamic reactions go unchecked by cortical processing, then aversive reactions might occur excessively and for reasons that are not fully understood by the conscious organism.

From AL, the stimulus appears to be transmitted within the amygdala to the basolateral and/or the basomedial nuclei (11). These nuclei in turn project to the central nucleus (ACe), an important waystation that connects the amygdala with areas in the brainstem involved in the control of various emotional response modalities (see review 12). For example, projections from ACe to the nucleus of the solitary tract are involved in the control of parasympathetic nervous system responses (6,13), whereas projections from ACe to the lateral hypothalamus and then to the rostral ventral medulla are involved in the control of sympathetic nervous system responses (14,15). In contrast, projections from ACe to the midbrain central gray region control the expression of species-typical behavioral responses such as freezing (14), as well as the inhibition of the behavioral response to pain (16). Projections to the reticulo pontis caudalis are involved in the potentiation of startle reflexes by conditioned fear stimuli (4,8,9). Finally, and most importantly for the current discussion, projections from ACe to the bed nucleus of the stria

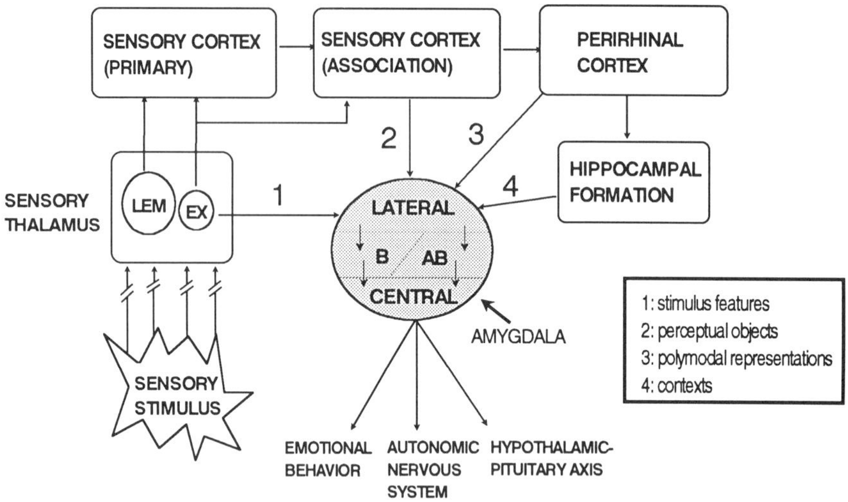

FIG. 3. Fear conditioning circuits. A sensory stimulus with emotional significance is transmitted through the brain to the sensory thalamus. Both lemniscal (LEM) and extralemniscal (Ex) processing regions in the thalamus send their outputs to the sensory processing areas of the cortex. In addition, Ex projects directly to the lateral nucleus of the amygdala (LATERAL). LATERAL also receives inputs from sensory association cortex and from higher-order association regions (PERIRHINAL CORTEX and HIPPOCAMPAL FORMATION). LATERAL projects to the Basal (B), Accessory Basal (AB), and CENTRAL nuclei. Outputs of the CENTRAL control behavioral, autonomic, and endocrine response systems.

terminalis mediate the release of adrenocorticotropic hormone (ACTH) from the pituitary gland and the resulting release of steroid hormones from the adrenal cortex (17–19). Thus, outputs of ACe can control behavioral, autonomic, and endocrine responses elicited by aversive stimuli in the environment.

These various findings characterize, from sensory to motor neurons, a brain system involved in learning that auditory stimuli are predictive of danger. A similar set of circuits exists for visual conditioned stimuli (8,9,20), but the anatomical connections linking the visual system to the amygdala in fear conditioning have not been elucidated as clearly.

Davis and colleagues, studying visual and auditory conditioning, recently suggested on the basis of posttraining lesions of a cortical area (anterior perirhinal cortex) that the thalamo-amygdala pathway is not normally used during learning, but used only when the cortical pathway to the amygdala is damaged, and that the perirhinal cortex is the crucial CS transmission route to the amygdala (71). However, their results cannot distinguish between an interpretation of the effects in terms of retrieval/retention deficits (due to cortical damage) and a failure to initially learn through the thalamo-amygdala pathway (72). Although anterior perirhinal cortex projects to the amygdala (21–25), there is no evidence to indicate that this is a CS transmission route to amygdala, much less *the* CS transmission route. Perirhinal cortex does receive information from sensory cortex (21,26,27), but is not an obligatory relay between sensory cortex and the amygdala (27). In the absence of anatomical and physiological evidence demonstrating that CS information is actually relayed to the amygdala by way of this cortical region, the argument that the region is a CS transmission route is one (and not necessarily the most obvious or plausible) of several interpretations of the lesion effects. In contrast, considerable anatomical and physiological evidence is consistent with the hypothesis that the thalamo-amygdala pathway transmits sensory information to the amygdala in a manner consistent with a role in CS transmission (12,28–32).

It seems likely, as already noted, that both thalamo-amygdala and cortico-amygdala pathways contribute to the overall learning process. However, the important point is to understand how sensory information is transformed by the amygdala during conditioning, regardless of where that information comes from. This can best be achieved by studies that pay careful attention to the anatomical and physiological organization of the system.

Neural System Underlying Differential Conditioning

In differential conditioning, one stimulus is paired with US and another remains unpaired. The subjects thus come to exhibit conditioned responses to the paired but not to the unpaired CS. Jarrell et al. (33) showed that the auditory cortex (and presumably cortico-amygdala connections) is required for differential fear conditioning. The animals still exhibit fear reactions, but to both stimuli. Thus, the lesions do not interfere with fear conditioning so much as with the ability to distinguish between the two stimuli. Through the thalamic pathways, the animal expresses responses to both stimuli.

Neural System Underlying Contextual Conditioning

When rats receive a tone and shock US in a conditioning chamber, they develop fear reactions to the chamber as well as the tone. This is called background or contextual fear conditioning. Damage to the hippocampal formation interferes with contextual conditioning (34,35), without affecting conditioning to the tone CS (36,37). In contrast, damage to the amygdala interferes with conditioning to both the tone CS and to the context (36). The amygdala is a crucial link in the fear conditioning pathway, regardless of the type of stimulus. The hippocampus is most likely required to process the kinds of complex stimulus representations that make up the context. Interestingly, the subiculum and ventral CA1 regions of the hippocampal formation project to the lateral amygdala and several other amygdala regions (Fig. 3). These pathways may be crucially involved in the processing of the emotional significance of contextual stimuli.

The lateral amygdala is thus the sensory and cognitive gateway to the amygdala and may well be involved in the coordination of stimulus evaluation, which involves a determination not only of whether a given stimulus is potentially harmful, but also under what conditions it might be so. A bear in the zoo is not threatening, but the same stimulus on a deserted trail in the woods is. Thus far, research on emotional learning has focused on questions about specific stimuli and their significance. However, contextual factors are an essential part of what gives a stimulus its meaning, and future work will have to explore these relations in more detail at both the psychological and biological levels.

Cellular Mechanisms of Conditioning

It has long been believed that classical conditioning involves the integration in the brain of information between neural pathways processing the CS and pathways processing the US (38,39). An important step towards the elucidation of the mechanisms underlying fear conditioning thus involves the identification of possible sites of CS–US convergence.

In recent studies, we demonstrated that neurons in AL are highly responsive to both auditory (CS) and somatosensory (US) stimuli (31,40,41) (Fig. 4). Further, essentially every cell that responded to an auditory stimulus also responded to footshock (41). However, many cells that responded to footshock did not respond to the auditory stimulus. Presumably, the shock stimulus can affect cells responsive to a number of CS modalities, and the auditory stimuli are ineffective in driving these cells.

What is also particularly important is the demonstration that the response latencies recorded in the amygdala to auditory (CS) stimulation are consistent with transmission from both the thalamus and the cortex. Thus, the earliest latencies found in AL were about the same as the

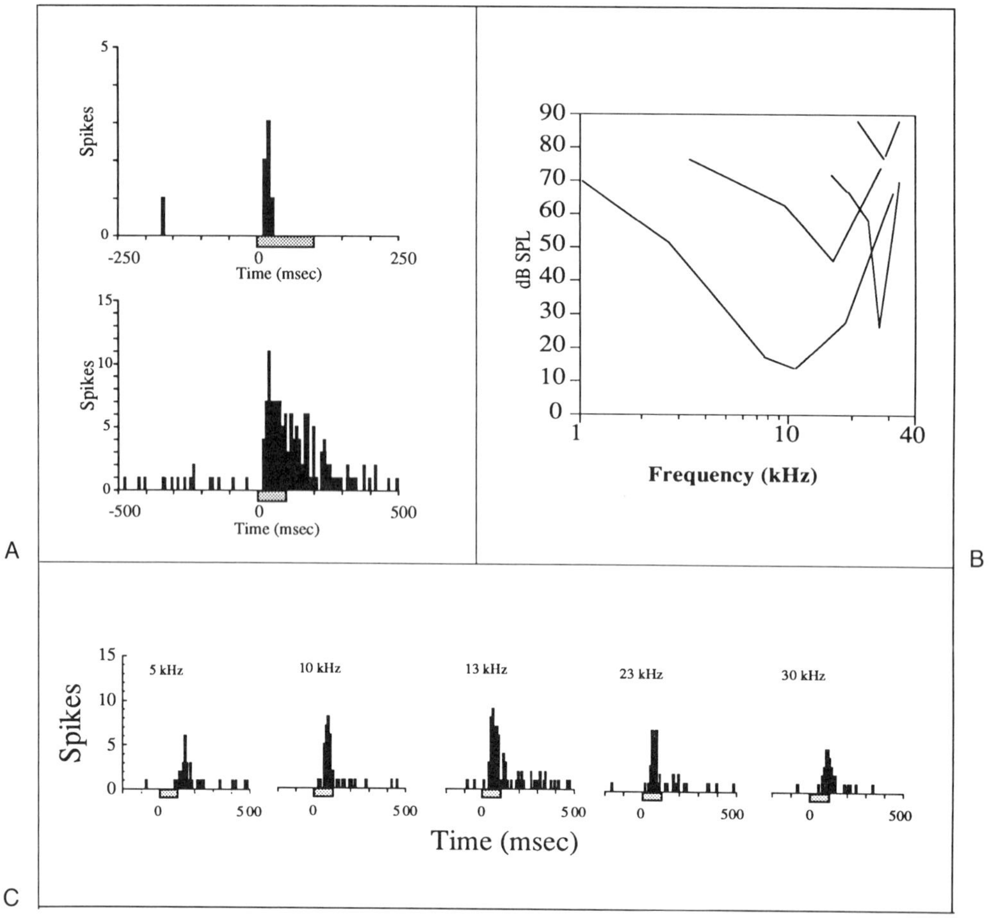

FIG. 4. Auditory Processing in the Lateral Amygdala: **A.** Typical, short-latency responses of lateral amygdala cells to auditory stimuli (bursts of white noise). **B.** Frequency-threshold (timing) curves of lateral amygdala cells. Note the relatively high intensity thresholds of the cells, the preference for higher frequencies, and the fairly broad bandwidth of responsivity, especially at higher intensities. **C.** Histograms of responses summed over 10 sweeps of pure tones of different frequencies at about 30 dB above threshold illustrating the preference of this cell for frequencies around 13 kHz. (Based on findings reported in Bordi and LeDoux (31) and Bordi et al (40).

earliest latencies in auditory cortex (40) and these latencies would seem to necessarily involve transmission from the thalamus.

Additional studies have shown that the CS pathway to the amygdala from the auditory thalamus contains the neurotransmitter glutamate (42,43). Further, it has been shown that blockade of excitatory amino acid postsynaptic receptors interferes with the transmission of stimulus information from the auditory thalamus to the amygdala (44). This is of interest given three additional facts: 1) the most well-studied form of longterm potentiation (LTP), a candidate mechanism for memory formation, involves excitatory amino acid receptors, in particular N-methyl-D-aspartate (NMDA) receptors, in the hippocampus (45–47); 2) LTP has been demonstrated in both thalamic (48) and cortical (49) projections to the amygdala; and 3) blockade of NMDA receptors in the amygdala inteferes with fear conditioning (50). Although the neural mechanisms of LTP in the amygdala remain unclear (51,52), as does the relationship between LTP and memory, the thalamo-amygdala system offers a new model network for examining these issues. One note of caution is suggested by our recent findings indicating that synaptic transmission in AL is partly mediated by NMDA receptors (44). If so, then the effects of NMDA blockade on conditioning (50) may have to be reconsidered.

FROM FEAR CONDITIONING TO STIMULUS EVALUATION AND FROM STIMULUS EVALUATION TO THE STRESS RESPONSE

These studies have mapped pathways involved in fear conditioning. The conclusion from this work is that the amygdala is situated in such a way as to be an interface between sensory systems, on the one hand, and motor systems, on the other. The amygdala is thus involved in fear conditioning independent of the sensory modality of the CS and the motor modality of the unconditioned response. Therefore, the amygdala is likely to be a site where sensory information is transformed into emotional control signals, allowing the determination of whether environmental threats exist and, if so, determining what they are and how they should be responded to. This conclusion is consistent with a variety of unit recording studies that have shown that neurons in the amygdala respond to the emotional significance of sensory stimuli, including complex social stimuli (53–55). The amygdala thus appears to play a crucial role in stimulus evaluation.

Once the aversive emotional significance of a stimulus is assessed by the amygdala, the classical stress reaction can be set into emotion. Stress reactions involve, initially, activation of the peripheral sympathetic nervous system, including the adrenal medulla, and activation of the hypothalamic-pituitary-adrenal cortex (HPA) axis. The pathways by which aversively conditioned stimuli such as a tone paired with shock elicit these responses are fairly clear. Although just reviewed here in some detail, the basic conclusions will be stated again. The amygdala receives information about the tone in the lateral nucleus and by way of intraamygdala connections activates the central nucleus. Projections from the central nucleus to the lateral nucleus of the hypothalamus, and from there to autonomic control regions in the medulla, are involved in the control of sympathetic outflow (14,15,56). Projections from the central nucleus to the midbrain central gray region and from there to the spinal cord are involved in the expression of species-typical defensive behavior, especially freezing behavior (14). Projections from the central amygdala to the bed nucleus of the stria terminalis, and from there to the paraventricular nucleus of the hypothalamus (which controls ACTH release from the pituitary), appear to be involved in HPA activation (17–19). Fear conditioning is thus a very useful model for examining the mechanisms through which environmental events come to trigger stress responses.

It is important to note that stimulus evaluation is important not only in the initiation of the stress response, but also in the continued interpretation of the situation and the continued outflow of the sympathetic and HPA systems. A good deal of work has gone into understanding how steroids released by HPA activation provide feedback to and affect the hippocampus (57). In particular, this work has shown that the hippocampus is involved in negative feedback control of the HPA axis, such that steroid receptor activation in the hippocampus gates the subsequent release of ACTH. Additionally, it has been shown that chronic stress produces physiological and pathological changes in the hippocampus.

The amygdala, like the hippocampus, contains a high concentration of steroid receptors (58) and is thus potentially an important target of action of these hormones. Interestingly, steroid hormone receptors are colocalized with corticotropin-releasing hormone (CRH) in the central nucleus of the amgdyala (59,60) and CRH mRNA is induced in this region by either stress (61) or by systemic injections of corticosterone (62,63). Thus, circulating steroids, released by actions in the amygdala, can return to the brain to influence ongoing amygdala function. Steroid influences on the cognitive functions of the hippocampus have been intensively studied in the past few years (64–66), but little is known about the effects of steroids on amygdala-mediated functions. This is an important area for future work.

The present framework is consistent in many ways with the model of stress reactions promoted by Henke on the basis of research on stomach pathology produced by physical stress (prolonged restraint) (see summary 67). His studies have shown that damage to the amygdala decreases stress-induced ulceration, whereas damage to the hippocampus facilitates ulceration. Henke thus proposes that the amygdala is involved in the perception of threats in the environment and the initiation of defensive reactions in response to threats, whereas the hippocampus is involved in producing coping responses. This effect on coping may be related to the interruption of negative feedback control over the pituitary-adrenal axis that the hippocampus normally exerts (57).

Although the emphasis in this chapter has been on the role of the amygdala in processing aversive signals, a good deal of evidence suggests that the amygdala is also involved in processing positive valence stimuli, signals for reward (53,54,68–70). This raises the interesting question of how signals with different implications are directed through the amygdala so that the appropriate response systems can be activated. The same stimulus can be a signal for approach or avoidance, depending on the individual's past history with that stimulus. The elucidation of how the amygdala can make such determinations, and how on the basis of such determinations different response systems are activated, are other important areas for future research.

CONCLUSIONS

Fear conditioning has been used extensively as a model for understanding how the brain evaluates the emotional significance of external stimuli. The logic and major findings of this research are applicable to the problem of understanding how the stress reactions are set into motion by external events, especially events that trigger stress reactions because of the way the individual interprets the stimulus. This research reminds us that while stress reactions are responses to environmental challenges, often it is only through mental interpretations that environmental events become challenges. As a result, efforts to control stress reactions through pharmacological or psychological manipulations should be as concerned, if not more concerned, with the mechanisms through which stimuli are interpreted as with the mechanisms through which stress responses are expressed.

ACKNOWLEDGMENTS

Supported by USPHS Grants MH38774 and MH46516, NSFGrant IBN 9209646 and a Research Scientist Development Award from NIMH (USPHS Grant MH00956). The author is grateful to Jay Schulkin for pointing out the effects of corticosterone on CRF in cells in the amgdyala.

REFERENCES

1. LeDoux JE, Sakaguchi A, Reis A. Subcortical efferent projections of the medial geniculate nucleus mediate emotional responses conditioned by acoustic stimuli. *J Neuroscience* 1984;4:683–698.
2. Bouton ME, Bolles RC. Conditioned fear assessed by freezing and by the suppression of three different baselines. *Anim Learn Behav* 1980;8:429–434.
3. Fanselow MS. Conditional and unconditional components of postshock freezing. *Pavlovian Journal of Biological Science* 1980;15:177–182.
4. Davis M, Hitchcock JM, Rosen JB. Anxiety and the amygdala: pharmacological and anatomical analysis of the fear-potentiated startle paradigm. In Bower GH, ed. *The psychology of learning and motivation.* vol 21, San Diego: Academic Press, 1987;263–305.
5. Fanselow MS, Helmstetter FJ. Conditional analgesia, defense freezing, and benzodiozepines. *Behav Neurosci* 1988;102:233–243.
6. Kapp BS, Wilson A, Pascoe J, Supple W, Whalen PJ. A neuroanatomical systems analysis of conditioned bradycardia in the rabbit. In Gabriel M, Moore J, eds. *Learning and computational neuroscience: foundations of adaptive networks.* Cambridge: MIT Press, 1990; 53–90.
7. Kapp BS, Whalen PJ, Supple WF, Pascoe JP. Amygdaloid contributions to conditioned arousal and sensory information processing. In Aggleton JP, ed. *The amygdala: neurobiological aspects of emotion, memory, and mental dysfunction.* New York: Wiley-Liss; 1992; 229–254.
8. Davis M. The role of the amygdala in conditioned fear. In Aggleton JP, ed. *The amygdala: neurobiological aspects of emotion, memory, and mental dysfunction.* New York: Wiley-Liss; 1992;255–306.

9. Davis M. The role of the amygdala in fear-potentiated startle: implications for animal models of anxiety. *Trends Pharmacol Sci* 1992;13:35–41.
10. Fanselow MS. Analgesia as a response to aversive Pavlovian conditional stimuli: cognitive and emotional mediators. In Denny MR, ed. *Fear, avoidance, and phobias: a fundamental analysis*. Hillsdale, NJ: Erlbaum; 1991;61–86.
11. Stefanacci L, Farb CR, Pitkanen A, Go G, LeDoux JE, Amaral DG. Projections from the lateral nucleus to the basal nucleus of the amygdala: a light and electron microscope PHA-L study in the rat. *J Comp Neurol* 1992;323:2–17.
12. LeDoux JE. Emotion. In Plum F, ed. *Handbook of physiology. 1: The nervous system. vol V, Higher functions of the brain*. Bethesda: American Physiological Society, 1987;419–460.
13. Kapp BS, Pascoe JP, Bixler MA. The amygdala: a neuroanatomical systems approach to its contributions to aversive conditioning. In Buttlers N, Squire LR, eds. *Neuropsychology of memory*. New York: Guilford; 1984; 473–488.
14. LeDoux JE, Iwata J, Cicchetti P, Reis DJ. Different projections of the central amygdaloid nucleus mediate autonomic and behavioral correlates of conditioned fear. *J Neurosci* 1988;8:2517–2529.
15. Smith, OA, Astley CA, Devito JL, Stein JM, Walsh RE. Functional analysis of hypothalamic control of the cardiovascular responses accompanying emotional. *Behavior Fed Proc* 1980;39:2487–2494.
16. Helmstetter F. The amygdala is essential for the expression of conditioned hypoalgesia. *Behav Neurosci* 1992; 106:518–528.
17. Van de Kar LD, Piechowski RA, Rittenhouse PA, Gray TS. Amygdaloid lesions: differential effect on conditioned stress and immobilization-induced increases in corticosterone and renin secretion. *Neuroendocrinology* 1991;54:89–95.
18. van der Kar LD, Piechowski RA, Rittenhouse PA, Gray TS. Amygdaloid lesions: differential effect on conditioned stress and immobilization-induced increases in corticosterone and renin secretion. *Neuroendocrinology* 1991;54:89–95.
19. Roozendaal B, Koolhaas JM, Bohus B. Central amygdaloid involvement in neuroendocrine correlates of conditioned stress responses. *Journal of Neuroendocrinology* 1992;4:46–52.
20. LeDoux JE, Romanski, LM, Xagoaris, AE. Indelibility of subcortical emotional memories. *J Cog Neurosci* 1989;1:238–243.
21. Amaral DG, Price JL, Pitkanen A, Carmichael ST. Anatomical organization of the primate amygdaloid complex. In Aggleton JP, ed. *The amygdala: neurobiological aspects of emotion, memory, and mental dysfunction*. New York: Wiley-Liss; 1992;1–66.
22. Romanski LM, LeDoux JE. Information cascade from primary auditory cortex to the amygdala: cortico-cortical and and cortico-amygdaloid projections of temporal cortex in the rat. *Cerebral Cortex* 1993;3:515–532.
23. McDonald AJ, Jackson TR. Amygdaloid connections with posterior insular and temporal cortical areas in the rat. *J Comp Neurol* 1987;262:59–77.
24. Turner BH, Zimmer J. The architecture and some interconnections of the rat amygdala and lateral periallocortex. *J Comp Neurol* 1984;227:540–557.
25. Turner BH, Mishkin M, Knapp M. Organization of the amygdalopetal projections from modality-specific cortical association areas in the monkey. *J Comp Neurol* 1980;191:515–543.
26. Deacon TW, Eichenbaum H, Rosenberg P, Eckmann K. Afferent connections of the perirhinal cortex in the rat. *J Comp Neurol* 1983;220:168–190.
27. Romanski LM, LeDoux JE. Organization of rodent auditory cortex: anterograde transport of PHA-L from MGv to temporal neocortex. *Cerebral Cortex* 1993,3:499–514.
28. LeDoux JE. Sensory systems and emotion. *Integ Psychiat* 1986;4:237–248.
29. LeDoux JE. Information flow from sensation to emotion: plasticity in the neural computation of stimulus value. In Gabriel M, Moore J, eds. *Learning and computational neuroscience: Foundations of adaptive networks*. Cambridge: MIT Press; 1990;3–52.
30. LeDoux JE. Brain mechanisms of emotion and emotional learning. *Current Opinion in Neurobiology* 1992; 2:191–198.
31. Bordi F, LeDoux JE. Sensory tuning beyond the sensory system: an initial analysis of auditory properties of neurons in the lateral amygdaloid nucleus and overlying areas of the striatum. *J Neurosci* 1992;12:2493–2503.
32. Bordi F, LeDoux JE. Response properties of units in areas of rat auditory thalamus that project to the amygdala. II. Cells receiving convergent auditory and somatosensory inputs and cells ntidromically activated by amygdala stimulation. *Exp Brain Res* 1994;98:275–286.
33. Jarrell TW, Gentile CG, Romanski LM, McCabe PM, Schneiderman N. Involvement of cortical and thalamic auditory regions in retention of differential bradycardia conditioning to acoustic conditioned stimulii in rabbits. *Brain Res* 1987;412:285–294.
34. Blanchard RJ, Blanchard DC, Fial RA. Hippocampal lesions in rats and their effect on activity, avoidance, and aggression. *J Comp Physiol Psychol* 1970;71:92–102.
35. Selden NRW, Everitt BJ, Jarrard LE, Robbins TW. Complementary roles for the amygdala and hippocampus in aversive conditioning to explicit and contextual cues. *Neuroscience* 1991;42:335–350.
36. Phillips RG, LeDoux JE. Differential contribution of amygdala and hippocampus to cued and contextual fear conditioning. *Behav Neurosci* 1992;106:274–285.
37. Kim JJ, Fanselow MS. Modality-specific retrograde amnesia of fear. *Science* 1992;256:675–677.
38. Pavlov IP. *Conditioned reflexes*. New York: Dover, 1927.
39. Konorski J. Integrative *Activity of the brain*. Chicago: University of Chicago Press, 1967.
40. Bordi F, LeDoux JE, Clugnet MC, Pavlides C. Single-unit activity in the lateral nucleus of the amygdala and overlying areas of the striatum in freely behaving rats: rates, discharge patterns, and responses to acoustic stimuli. *Behav Neurosci* 1993;107:757–769.
41. Romanski LM, Clugnet M, Bordi F, LeDoux JE. Somatosensory and auditory convergence in the lateral nucleus of the amygdala. *Behavioral Neuroscience* 1993;107:444–450.
42. Farb CR, Aoki C, Milner T, Kaneko T, LeDoux JE. Glutamate immunoreactive terminals in the lateral amygdaloid nucleus: a possible substrate for emotional memory. *Brain Res* 1992;593:145–158.
43. LeDoux JE, Farb CR. Neurons of the acoustic thalamus

that project to the amygdala contain glutamate. *Neurosci Lett* 1991;134:145–149.
44. Xingfang L, Phillips RG, LeDoux JE. NMDA and non-NMDA receptors contribute to synaptic transmission between the medial geniculate body and the lateral nucleus of the amygdala. *Exp Brain Res* (in press).
45. Cotman CW, Monaghan DT, Ganong AH. Excitatory amino acid neurotransmission: NMDA receptors and hebb-type synaptic plasticity. *Annu Rev Neurosci* 1988;11:61–80.
46. Brown TH, Chapman PF, Kairiss EW, Keenan CL. Long-term synaptic potentiation. *Science* 1988;242:724–728.
47. Madison DV, Malenka RC, Nicoll RA. Mechanisms underlying long-term potentiation of synaptic transmission. *Annu Rev Neurosci* 1991;14:379–397.
48. Clugnet MC, LeDoux JE. Synaptic plasticity in fear conditioning circuits: induction of LTP in the lateral nucleus of the amygdala by stimulation of the medial geniculate body. *J Neurosci* 1990;10:2818–2824.
49. Chapman PF, Kairiss EW, Keenan CL, Brown TH. Long-term synaptic potentiation in the amygdala. *Synapse* 1990;6:271–278.
50. Miserendino MJD, Sananes CB, Melia KR, Davis M. Blocking of acquisition but not expression of conditioned fear-potentiated startle by NMDA antagonists in the amygdala. *Nature* 1990;345:716–718.
51. Chapman PF, Bellavance LL. NMDA receptor-independent LTP in the amygdala. *Synapse* 1992;11:310–318.
52. Gean P-W, Chang F-C, Huang C-C, Lin J-H, Way L-J. Long-term enhancement of EPSP and NMDA receptor-mediated synaptic transmission in the amygdala. *Brain Res Bull* 1993;31:7–11.
53. Rolls ET. Neurophysiology and functions of the primate amygdala. In Aggleton JP, ed. *The amygdala: neurobiological aspects of emotion, memory, and mental dysfunction.* New York: Wiley-Liss; 1992;143–165.
54. Ono T, Nishijo H. Neurophysiological basis of the Kluver-Bucy Syndrome: responses of monkey amygdaloid neurons to biologically signficant objects. In Aggleton JP, ed. *The amygdala: neurobiological aspects of emotion, memory, and mental dysfunction.* New York: Wiley-Liss; 1992;167–190.
55. Brothers L, Ring B, Kling A. Response of neurons in the macaque amygdala to complex social stimuli. *Behav Brain Res* 1990;41:199–213.
56. Iwata J, LeDoux JE, Meeley MP, Arneric S, Reis DJ. Intrinsic neurons in the amygdaloid field projected to by the medial geniculate body mediate emotional responses conditioned to acoustic stimuli. *Brain Res* 1986;383:195–214.
57. Jacobson L, Sapolsky R. The role of the hippocampus in feedback regulation of the hypothalamic-pituitary-adrenocortical axis. *Endocrine Reviews* 1991;12:118–134.
58. McEwen BS, DeKloet ER, Rostene W. Adrenal steroid receptors and actions in the nervous system. *Physiol Rev* 1986;66:1121–1188.
59. Cintra A, Fuxe K, Solfrini V, Agnati LF, Tinner B, Wilkstrom AC, Staines W, Okret S, Gustafsson JA. Central peptidergic neurons as targets for glucocorticoid action: evidence for the presence of glucocorticoid receptor immunoreactivity in various types of classes of peptidergic neurons. *Steroid Biochem Mol Biol* 1991;40:93–103.
60. Honkaniemi J, Pelto-Huikko M, Rechardi L, Isola J, Lammi A, Fuxe K, Gustafsson J-A, Wikstrom A, Hokfelt T. Colocalization of peptide and glucocorticoid receptor immunoreactivities in rat central amygdaloid nucleus. *Neuroendocrinology* 1992;55:451–459.
61. Pich EM, Koob GF, Sattler SC, Menzghi G, Heilig M, Heinrichs SC, Vale W, Weiss F. Stress-induced release of corticotropin releasing factor in the amygdala measured by in vivo microdialysis. *Soc Neurosci Abs* 1992;18:535.
62. Swanson LW, Simmons DM. Differential steroid hormone and neural influences on peptide mRNA levels in CRH cells of the paraventricular nucleus: a hybridization histochemical study in the rat. *J Comp Neurol* 1989;285:413–435.
63. Makino S, Gold PW, Schulkin J. Corticosterone effects on corticotropin-releasing hormone mRNA in the central nucleus of the amygdala and the parvocellular region of the paraventricular nucleus of the hypothalamus. *Brain Res* 1994;640:105–112.
64. Foy MR, Stanton ME, Levine S, Thompson RF. Behavioral stress impairs long-term potentiation in rodent hippocampus. *Behav Neural Biol* 1987;48:138–149.
65. Shors TJ, Foy MR, Levine S, Thompson RF. Unpredictable and uncontrollable stress impairs neuronal plasticity in the rat hippocampus. *Brain Res Bull* 1990;24:663–667.
66. Pavlides C, Watanabe Y, McEwen BS. Effects of glucocorticoids on hippocampal long-term potentiation. *Hippocampus* 1993;3:183–192.
67. Henke PG. Stomach pathology and the amygdala. In Aggleton JP, ed. *The amygdala: neurobiological aspects of emotion, memory, and mental dysfunction.* New York: Wiley-Liss; 1992;323–338.
68. Gaffan D. Amygdala and the memory of reward. In Aggleton JP, ed. *The amygdala: neurobiological aspects of emotion, memory, and mental dysfunction.* New York: Wiley-Liss; 1992;471–483.
69. Hiroi N, White NM. The lateral nucleus of the amygdala mediates expression of the amphetamine-produced conditioned place preference. *J Neurosci* 1991;11:2107–2116.
70. Everitt BJ, Robbins TW. Amygdala-Ventral striatal interactions and reward-related processes. In Aggleton JP, ed. *The amygdala: neurobiological aspects of emotion, memory, and mental dysfunction.* New York: Wiley-Liss, 1992;401–429.
71. Campeau S, Davis M. Involvement of subcortical and cortical afferents to the lateral nucleus of the amygdala in fear conditioning measured with fear-potentiated startle in rats trained concurrently with auditory and visual conditioned stimuli. *J Neurosci* 1995;15:2312–2327.
72. Corodimas KP, LeDoux JE. Disruptive affects of post-training perirhinal cortex lesions and conditioned fear: contributions of contextual cues. *Behav Neurosci* 1995;169:in press.

Neurobiological and Clinical Consequences of Stress: From Normal Adaptation to PTSD, edited by M. J. Friedman, D. S. Charney, and A. Y. Deutch, Lippincott-Raven Publishers, Philadelphia © 1995.

8

Adrenal Steroid Actions on Brain

Dissecting the Fine Line between Protection and Damage

Bruce S. McEwen

Laboratory of Neuroendocrinology, Rockefeller University, New York, New York 10021

From the work of Hans Selye (1) to the present time, we have been aware of the delicate balance between the protective effects of adrenal steroids secreted in response to stressful experiences, and the negative consequences that these same hormones may have for many processes. While, on the one hand, excess adrenal steroids suppress immune defense mechanisms and cause such negative consequences as neural damage, muscular atrophy, and calcium loss from bone (2,3), the insufficiency of adrenal steroids makes the organism much more vulnerable to inflammatory disturbances and autoimmune responses, fever, and damage from catecholamine metabolites and alcohol (4–8), as well as increasing fear responding and anxiety (9). While the initial work focused on immune responses and metabolism, and ignored behavior and brain mechanisms, the fact that fear and anxiety are neural events, that the brain governs the endocrine and autonomic nervous systems and their numerous effects on immune and metabolic processes, and finally that hormones feed back on the brain, have made it imperative to consider the impact of stress on the nervous system.

That the brain is involved directly in the response to stressors and to diurnal changes in the secretion of adrenal steroids was shown after it became possible to detect intracellular steroid hormone receptors. In 1968 we reported that the adrenal glucocorticoid in the rat, corticosterone, was taken up and retained in high levels by the hippocampal formation after rats had been adrenalectomized (ADX) in order to remove endogenous hormone from receptor sites (10). The hippocampal localization of corticosterone was a surprise, particularly since the hippocampus was not known to be directly associated with neuroendocrine function, whereas the hypothalamus and preoptic area were known as the "hypophysiotrophic area" and were the sites of uptake and retention of sex steroid hormones. The hippocampus plays a major role in learning and memory, and is also involved in the matching of expectations to actual events in situations where there may be punishment or reward (11,12). Both noradrenergic and serotonergic input play a major role in modulating hippocampal responsiveness. Noradrenergic innervation, acting via alpha and beta receptors, increases excitability of neurons and produces disinhibition, a process that also results in facilitation of neural activity, in the hippocampus (13,14). Serotonergic neurotransmission in the hippocampus, acting through postsynaptic 5HT1A receptors, has suppressive effects on longterm potentiation (15) and facilitates extinction or inhibits learning of aversive associations (16,17). Thus, disruption of hippocampal neuronal circuitry or alterations in noradrenergic or serotonergic activity of the hippocampus are likely to have profound effects on the role that the hippocampus plays in learning and memory, and in the response to novel, aversive events (12).

Recent research has shown that the circuitry, electrophysiology, and neurochemistry of the hippocampus are uniquely sensitive to circulating adrenal steroids. These hormones play an important role in maintaining hippocampal function during the diurnal rhythm. In this capacity, they appear to have a role in the phenomenon of jet lag. Adrenal steroids also stabilize the neuronal population of the dentate gyrus, while reduction in adrenal steroid levels enhances neuronal turnover; in this capacity, we speculate that the adrenal secretions may participate in seasonal adjustments to hippocampal function.

Finally, adrenal steroids play a vital role in containing the body and brain response to stressful events; and, as long as adrenal secretion is itself contained, adaptation occurs. However, when adrenocortical secretion is no longer self-regulated and becomes persistently elevated, there are several negative consequences (e.g., normal hippocampal neuronal circuitry is disrupted and serotonin receptor levels are altered). Both of these consequences may reduce effective cognitive processing of and coping with threatening events. This review will summarize these effects and comment on their relevance to the pathophysiology of two stress-related disorders: post-traumatic stress disorder (PTSD) and endogenous depressive illness.

ADRENAL STEROID RECEPTORS AND EFFECTS ON THE BRAIN

Intracellular Receptors

Uptake and retention of ^{3}H steroid hormones by a cell is the result of binding of the hormone by intracellular receptors that end up in the cell nuclear compartment, where these receptors bind to specific nucleotide sequences on DNA known as ''hormone responsive elements'' or ''HREs'' (18). The uptake and retention of adrenal steroids by the hippocampus, first found in the rat, has been demonstrated in species as divergent as the duck and the rhesus monkey, and this seems to be a widespread and evolutionarily stable trait of the hippocampal region (19). At the same time, we now know that adrenal steroid receptors are found throughout the nervous system in virtually every cell type of the brain, and that they mediate many neurochemical and behavioral actions (19).

Two Distinct Intracellular Receptors

Although it was first thought to reflect a single intracellular receptor type, we now know that two distinct adrenal steroid receptor types are present not only in hippocampus but also in other brain regions and other tissues of the body. However, realization of this was very gradual (19). Studies of the kidney in the early 1970s revealed two adrenal steroid receptor types, referred to as Type I and Type II (19). Clues that there might also be two such receptor types in the brain were obtained but not appreciated. Radioimmunoassay of cell nuclear corticosterone in rats exposed in vivo to low, intermediate, and high levels of corticosterone revealed high-affinity and lower-affinity nuclear retention of the steroid (20). What was lacking were selective steroid agonists and antagonists for the two receptors. When these became available, Type I and Type II receptors were definitively demonstrated in hippocampus using specific agonist and antagonist steroids in specific binding assays (21). Final proof of the existence of two distinct receptors came with the cloning of mineralocorticoid (Type I) and glucocorticoid (Type II) receptors (22–24). The cloning demonstrated each form to be the product of a distinct gene, and the mRNA for both receptors can be demonstrated in the hippocampus (25). It is the Type I (mineralocorticoid) receptor that is concentrated in the hippocampal formation; this receptor, because of its high affinity for ^{3}H corticosterone, picks up and retains tracer levels of steroid injected into ADX rats, as we found in 1968 (10).

Nongenomic Effects

Besides the genomic actions of steroid hormones, there are also nongenomic effects (26). First described by Selye in 1941 as anesthetic effects of progesterone, these actions involve direct effects on membrane receptor sites. The best example is the γ-aminobutyric acid (GABA)a-benzodiazepine receptor, which binds metabolites of progesterone and desoxycorticosterone, and facilitates opening of the chloride channel (26). It is through this mechanism that anxiolytic, as well as sedative-hypnotic and anesthetic effects, are produced (27–29). Other actions of adrenal steroids are known on membrane events, but these are less well understood, particularly in mammals (26).

ADRENAL STEROID EFFECTS ON HIPPOCAMPAL NEURONAL PLASTICITY RELATED TO NATURAL RHYTHMS

During the past 10 years, we have begun to achieve a real understanding of the role of adrenal steroid receptors in hippocampus (25). The first effects of adrenal steroids concerned destruction of neural tissue (see next section), but later findings pointed to additional effects that protected neurons and enhanced plasticity (Fig. 1).

Death and Replacement of Dentate Gyrus Granule Neurons

Granule neurons of the adult dentate gyrus depend on adrenal steroids for their survival, in contrast to pyramidal neurons of Ammons horn, as will be soon described. Moreover, continuous replacement of granule neurons occurs in the adult dentate gyrus, and both cell death and neurogenesis increase following adrenalectomy (30). Adrenal steroids regulate this process through Type I receptors, and yet these receptors are absent from newborn neurons, suggesting that the primary effect of adrenal steroids is to inhibit cell death, and that the death of neurons signals the formation of replacements (30).

The magnitude of the ADX effect on neuronal death in dentate gyrus was indicated by the finding that some, but not all, rats showed almost total loss of the dentate gyrus 3 months after adrenalectomy as adults (31). This finding attracted attention in part because it was the opposite to the story of stress- and glucocorticoid-induced neuronal death in Ammons horn (see next paragraph). It was also very puzzling because only some ADX rats showed the extensive loss of dentate gyrus granule neurons.

Several recent findings have enabled us to understand both of these unusual aspects. The selectivity of the ADX effect on killing of dentate gyrus granule neurons, and not pyramidal neurons, has been verified, and it appears to be related to the unusual nature of the dentate gyrus, which arises postnatally and continues to generate new neurons in adult life (30). Furthermore, the occurrence of cell death in only some rats is related to the presence of accessory adrenal tissue. Indeed, only tiny amounts of adrenal steroids are required to protect dentate gyrus neurons from death, and these steroids act via Type I adrenal steroid receptors (32). When not removed at the time of ADX, the accessory adrenal tissue can supply enough adrenal steroids to prevent neuronal loss (30).

What is the physiological role of this type of plasticity, since adrenalectomy is not a natural event? One explanation for why the dentate gyrus can make new cells in adult life, as well as eliminate them, is that the hippocampus may change during adult life in relation to its function in spatial orientation and memory (33). It has been demonstrated that species of birds that use space around them to hide and locate food, and voles that use their environment to find mates, have larger hippocampal volumes than species that do not engage in these behaviors; moreover, there are indications that hippocampal volume may change under some circumstances (e.g., with season of the year (33), although it remains to be seen whether it is the dentate gyrus that changes under these changing conditions.

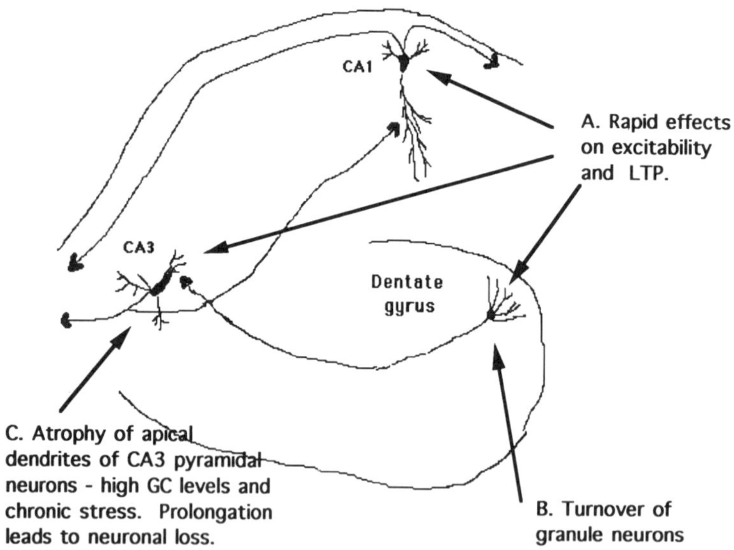

FIG. 1. Three types of plasticity regulated by adrenal steroids are: 1) Hippocampal neuronal excitability and long-term potentiation, over a time course of minutes to hours; both Type I and Type II adrenal steroid receptors are involved in producing a biphasic dose-response; 2) Dentate gyrus neuron death and neurogenesis, over a time course of days to weeks; this process is contained by low levels of adrenal steroids acting via Type I receptors; 3) Pyramidal neuron atrophy and death, over a time course of weeks and longer; Type II adrenal steroid receptors are implicated, along with excitatory amino acids and serotonin released by high levels of adrenal steroids and by stress.

Long-term Potentiation

Another type of plasticity that occurs in the hippocampus is longterm potentiation (LTP), in which a single burst of stimulation can alter the responsiveness of neurons to further stimulation over many hours (34). Recent work by Pavlides et al. (35) on the dentate gyrus has shown that LTP is biphasically modulated by adrenal steroids acting via Type I and Type II receptors; that is, Type I receptor stimulation in rats deprived of their own steroid by ADX enhances LTP within 1 hour, whereas Type II receptor stimulation in ADX rats suppresses LTP within the same time frame (35). These biphasic effects help to explain a biphasic dose response curve with respect to corticosterone levels and the induction of primed-burst potentiation (PBP), a type of response related to LTP, in the CA1 region of Ammon's horn, in which low levels of corticosterone within the daily variation of basal secretion facilitate PBP and stress levels of corticosterone inhibit PBP (36).

That the CA1 pyramidal neurons respond in a manner similar to the dentate gyrus granule neurons to corticosteroids is not surprising, for two reasons: 1) Type I and Type II receptors coexist in pyramidal neurons and dentate gyrus granule neurons (25) and 2) in CA1 pyramidal neurons, corticosteroids biphasically modulate excitability. In effect, Type I receptor stimulation facilitates excitability by reducing and inhibiting 5HT1A receptor input, while Type II receptor stimulation inhibits excitability by suppressing a beta receptor-mediated noradrenergic input (37).

The biphasic effects of adrenal steroids on LTP predict that rising levels of adrenal steroids during the diurnal rhythm, and as a result of stress, might have biphasic actions on functions of the hippocampus such as learning and memory. Thus, in situations such as jet airplane travel, the diurnal rise of adrenal steroids, which appears to facilitate hippocampal function, does not occur prior to waking, but at another time of day; this may be a contributing factor to impaired cognitive function in jet lag (38). On the other hand, elevation of adrenal steroids as a result of stress or as a consequence of the exogenous administration of steroids for inflammatory or autoimmune disorders may achieve levels that impair hippocampal function (39).

TOXIC EFFECTS OF GLUCOCORTICOIDS ON THE HIPPOCAMPUS

The first clue as to an effect of adrenal steroids in the hippocampus was the finding that adrenocorticotropic hormone (ACTH) or cortisone treatment of guinea pigs causes necrosis of pyramidal neurons of the hippocampus (40). Landfield and coworkers (41) then found that aging in the rat results in some loss of pyramidal neurons in hippocampus that can be retarded by adrenalectomy in midlife. Robert Sapolsky subsequently demonstrated that daily corticosterone injections into young adult rats over 12 weeks mimicked the pyramidal neuron loss seen in aging (42). He went on to demonstrate that excitatory amino acids play an important role in the cell loss by first showing that corticosterone exacerbates kainic acid-induced damage to hippocampus, as well as ischemic damage, and then showing that glucocorticoids potentiate excitatory amino acid-killing of hippocampal neurons in culture (3).

In order to examine the response of hippocampal neurons to high levels of glucocorticoids, we have recently used the Golgi technique to examine neuronal morphology after both corticosterone exposure and repeated stress. We have found that after 21 days of daily corticosterone exposure, apical dendrites of CA3 pyramidal neurons have atrophied (43); moreover, this atrophy is prevented by phenytoin, a blocker of excitatory amino acid release and action, given prior to corticosterone each day (44). The effect of corticosterone on dendritic length and branching is not found on basal dendrites of CA3 pyramidal neurons, nor on CA1 pyramidal neurons or dentate gyrus granule neurons (43,44). The sensitivity of the CA3 pyramidal neurons is reminiscent of the specific damage to CA3 neurons as a result of kainic acid infusion or as a consequence of seizure-inducing stimulation of the perforant pathway into the hippocampus; both of these effects are attributable to the mossy fiber input to CA3 from the dentate gyrus (44).

We next investigated whether the corticosteroid effect occurs physiologically by subjecting rats to repeated daily restraint stress, which is known to induce glucocorticoid secretion. Restraint stress for 21 days produced atrophy of apical dendrites of CA3 pyramidal neurons, without affecting basal dendritic length or branching (45); the pattern and specificity of change were identical to that produced by corticosterone treatment (43). Like corticosterone-induced atrophy, stress-induced atrophy was blocked by phenytoin, which prevents excitatory amino acid release and actions (44).

In addition, stress-induced atrophy of apical dendrites of CA3 pyramidal neurons was prevented by cyanoketone, a drug that virtually eliminates the stress-induced surge of corticosterone (46). Thus, stress-induced release of corticosterone synergizes with excitatory amino acid release to produce the atrophy, and there is evidence (46) that corticosterone actually facilitates the release of glutamate from nerve terminals in the hippocampus.

Serotonin also plays a role in the stress- and corticosterone-induced atrophy of dendrites of CA3 pyramidal neurons. This was shown by the finding that tianeptine, a tricyclic antidepressant that facilitates serotonin reuptake, attenuates the atrophy of apical CA3 dendrites caused by repeated restraint stress (47). The fact that tianeptine also blocks the atrophy caused by corticosterone treatment rules out any type of inhibition by tianeptine of corticosterone secretion as having a direct effect on dendritic atrophy. Rather, it appears that serotonin released during stress or as a result of injecting corticosterone may synergize with excitatory amino acid release in causing the atrophy. Indeed, some evidence for a serotonin-glutamate interaction has been found, and 5-HT2 receptor antagonists are known to attenuate ischemic damage to hippocampus, a process that also involves glutamate release and that is exacerbated by glucocorticoids (47).

More severe and prolonged stress (e.g., cold swim and social i.e., dominance/subordinance hierarchies) have been reported to produce actual loss of hippocampal neurons (48–50). Moreover, social stress in rats and tree shrews causes the same type of atrophy of apical dendrites of hippocampal CA3 pyramidal neurons as that caused by corticosterone and by repeated restraint stress (Magarinos, unpublished). Although dendritic atrophy precedes cell loss after corticosterone treatment and repeated restraint stress, we do not yet know the causal relationship of the dendritic atrophy to neuronal loss: whereas it is attractive to suppose that atrophy may represent the first stage of cell damage, the fact that this atrophy occurs on apical, but not on basal, dendrites indicates that it may be an adaptive process by viable neurons. We have determined that the atrophy caused by repeated restraint stress for 21 days is reversible (Magarinos, unpublished); thus, it may represent a mechanism whereby synaptic connectivity is reduced temporarily, perhaps in order to spare the CA3 pyramidal neurons from excessive bombardment by excitatory input.

Reversible and adaptive as it may be, the stress effect on hippocampal CA3 pyramidal neurons is accompanied by transient impairment of learning of a radial mask task (51); this impairment can be blocked by the same agents, namely, phenytoin and tianeptine, that prevent the stress-induced atrophy of CA3 apical dendrites (Villegas, Martinez, Luine, and McEwen, unpublished). A recent, preliminary report on human Cushing's disease patients indicates that some who show impairment of verbal memory also have decreased hippocampal volume, as determined by magnetic resonance imaging (52). Thus, the changes in hippocampal connectivity in relation to elevated glucocorticoids and stress may represent another mechanism, in addition to acute impairment of LTP or PBP (see previous sentence), by which learning and cognitive function may be impaired by excessive adrenal steroids.

ADAPTATION AND MALADAPTATION IN THE FACE OF STRESSFUL EVENTS

Homeostasis and Allostasis

What is stressful? In order to begin to understand the differences between adaptive and maladaptive responses to stressors, we must first define what is stressful. Stress is often defined as a perceived threat to homeostasis, and as an event or stimulus that causes an often abrupt but always large change in autonomic activity and hormone secretion, particularly hormones such as cortisol and prolactin. The term "perceived" emphasizes the extremely important point that each individual may react differently to an event or situation depending on physical status and prior experiences. Thus, trained athletes and sedentary individuals will react differently to physical exertion, and prior experiences will cause some individuals to be more or less anxious in the face of psychological challenges (e.g., examinations or events in the workplace).

Stress represents a physiological response to a large deviation from what is expected (S. Folkman, personal communication), and such a response perturbs the homeostatic balance. The term "homeostasis" captures the idea of a body setpoint that is reestablished after stress, but homeostasis fails to consider the dynamic aspect of anticipation of expected events (53,54). Normally, anticipation is a shortlived process; but, in psychopathology, expectancies can become self-sustaining and detached from reality in individuals suffering from anxiety disorders and depression.

An alternative to homeostasis is "allostasis." Allostasis is a term that incorporates change in physiological systems and the anticipation of change, and it promotes the idea of a variable setpoint for many physiological endpoints as a function of time of day and anticipated events (53). Allostasis, like homeostasis, stipulates that the body attempts to maintain whatever setpoint is operating at that time of day. However, the variability of the setpoint and adjustment of the setpoint in anticipation of future events are the important features that set allostasis apart from homeostasis.

Containment by Glucocorticoids

Glucocorticoid secretion is one of the most frequent responses to stressful events. Adrenal steroid secretion as a result of stress has multiple actions on the brain and body that have been characterized as "containing" or "counterregulating" other responses to stress and trauma such as inflammations, fever, edema, and immune responses (55). Containment implies that glucocorticoids prevent responses from being excessive. In the brain, glucocorticoids contain the synthesis of corticotropin-releasing hormone (CRH) and vasopressin, two neuropeptide-releasing hormones for ACTH; in doing this they prevent hypothalamic CRH, which has anxiogenic, arousing, immunosuppressive, and anorexigenic effects from becoming hyperactive (38). Other examples of containment will be discussed in the next section in conjunction with the noradrenergic and serotonergic systems.

Stress and Noradrenaline

Stressors of many kinds activate the release and turnover of noradrenaline, along with the release of catecholamines from the autonomic nervous system. Uncontrollable stressors tend to have more prolonged effects on noradrenaline (NA) turnover (56,57). One of the consequences of repeated stress is the induction of tyrosine hydroxylase (TH), the rate-limiting enzyme for noradrenaline and epinephrine formation, in both the locus coeruleus and the adrenal medulla (58). Stress-induced activation of catecholamine biosynthesis also increases the catalytic efficiency of TH and leads to increased catecholamine formation (see 58). Thus, in the face of increased TH amount and activity, there is a need for various forms of containment. Glucocorticoids play this role in several ways, reducing adenosine 3′,5′-cyclic phosphate (cAMP) formation in cerebral cortex in response to NA, (38), and decreasing catecholamine biosynthesis and release (59,60) (Fig. 2). In view of the importance of NA in promoting excitability and disinhibition of hippocampal neurons (13,14), it is interesting that, after repeated stress, noradrenergic terminals in the hippocampus retain the potential to release larger amounts of NA in response to a novel stressor (58). Yet, these terminals are contained to a large extent by presynaptic alpha adrenergic mechanisms (61) that may also be aided by circulating glucocorticoids (59,60). A similar situation exists with respect to ACTH and glucocorticoid secretion in response to novel stressors in animals that have been repeatedly stressed, as will be discussed shortly.

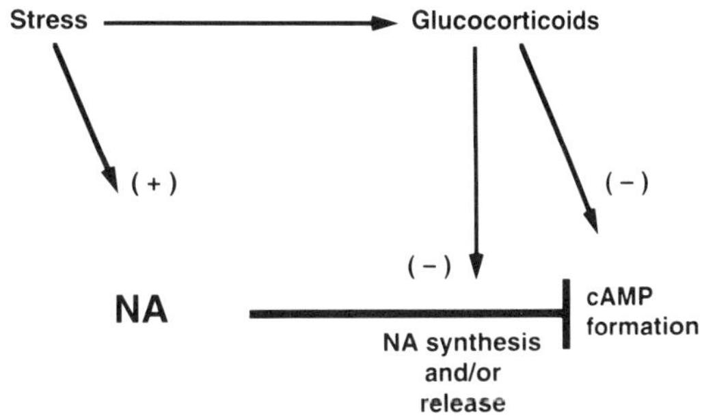

FIG. 2. Containment of noradrenergic activity by glucocorticoids.

Stress and Serotonin

Stressors also activate serotonin turnover and, in so doing, they activate a system that has both anxiogenic and anxiolytic pathways within the forebrain (17;47). Serotonin has a powerful role in the learning and retention of fear (62). A primary distinction in the qualitative nature of serotonin actions is between the dorsal raphe and median raphe nuclei. Dorsal raphe innervation of amygdala and hippocampus is believed to have anxiogenic effects, very likely via 5-HT$_2$ receptors. Median raphe innervation of the hippocampus reaches 5-HT$_{1A}$ receptors, stimulation of which facilitates the disconnection of previously learned associations with aversive events, or suppresses formation of new associations, thus providing a resilience in response to aversive events (16,17).

Glucocorticoids play a complex role in regulating the 5-HT system (Fig. 3). Circulating glucocorticoids acutely facilitate 5-HT turnover

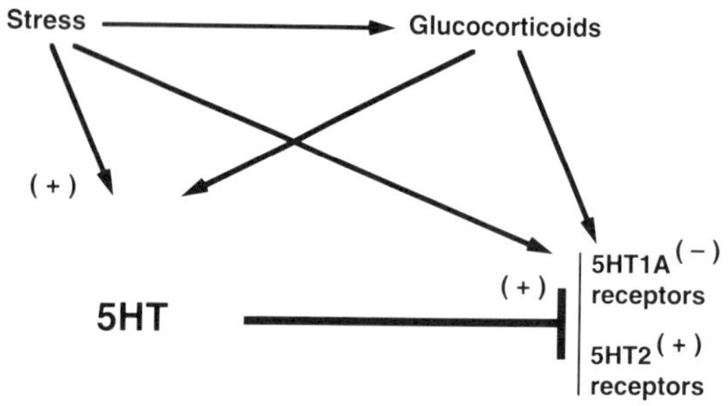

FIG. 3. Positive and negative effects of adrenal steroids on serotoninergic activity and receptor sensitivity.

provoked by a wide variety of stressors (63–65), and repeated ACTH or glucocorticoid treatment, or chronic stress, increase 5-HT_2 receptors in cerebral cortex (66–68) while reducing 5-HT_{1A} receptors in hippocampus (68–73). Thus, while glucocorticoids acutely facilitate the activity of the entire 5-HT system during stress, glucocorticoids chronically tip the balance in favor of 5-HT_2-mediated actions of 5-HT, at least in cerebral cortex, and suppress 5-HT_{1A}-mediated responses in hippocampus. These changes have implications for the pathophysiology of depression, which will be discussed in the next section.

Containment of HPA Axis

A key to successful adaptation as opposed to maladaptive effects of stress is the self-containment of the hypothalamo-pituitary-adrenal (HPA) axis (i.e., the ability of the HPA axis to shut itself down as well as contain other neurochemical responses to stress such as CRH and noradrenaline release (see previous section).

The paraventricular nuclei (PVN), which produce the CRH and vasopressin that stimulate ACTH release, are innervated by catecholaminergic and serotonergic inputs. Catecholaminergic input acts via alpha adrenergic receptors to facilitate HPA activity and via beta adrenergic receptors to inhibit it (74,75). In contrast, serotonergic input to PVN via 5-HT_{1A} receptors is reported to inhibit HPA activity (76).

In response to repeated stressors, HPA activity tends to habituate if the stressor is the same but become hypersensitive to novel stressors (77,78). Partial inhibition of glucocorticoid secretion in response to stress unmasks a strong facilitation, indicating that containment by glucocorticoids plays an important role (77,79). As noted, glucocorticoids contain the release of catecholamines by stress in the PVN (60). Moreover, the role of glucocorticoids in enhancing 5-HT release in stress may play a role in maintaining inhibition via 5-HT_{1A} receptors (76). It thus appears likely that the relative strengths of catecholaminergic and serotonergic input as a result of the same or different stressors may determine whether or not the HPA axis will habituate or be enhanced by repeated application of the same stressor. Moreover, the relative densities of the various adrenergic and serotonergic receptors types will play an important role in the net HPA response (i.e., to what degree it remains contained). Finally, the efficacy of glucocorticoid feedback on the containment mechanisms will be an important factor in the containment response. In this connection, it is interesting that depressed patients show a larger elevation of ACTH in response to Ru486 compared to control subjects (80). Ru486 is an antagonist of the Type II adrenal steroid receptors, and its effect in depressed subjects indicates that there is an increased drive to release ACTH that is held partially in check by glucocorticoid feedback acting via Type II receptors (80).

Allostatic Load

When we speak of repeated or chronic stress, we think of the body maintaining function in spite of an external load. As just noted, allostasis is a term that means stability through change, and ''allostatic load'' refers to the hidden price paid when an individual is under continuing stress. While we have seen that glucocorticoids are beneficial in the body's and brain's response to stressful situations, we have also noted that it is imperative that the HPA axis shut down again after the end of stress; if it does not, or if the stress persists without respite, then a vicious cycle may ensue in which elevated glucocorticoids cause insulin hypersecretion and insulin resistance. This is one form of allostatic load. The combination of elevated insulin and glucocorticoids is well-known to promote obesity and facilitate atherosclerotic plaque formation (81), such as has been reported in dominant monkeys on a normal diet in an unstable social situation (82).

Depending on the nature of the stressor, and individual differences in response to stress, allostatic load as a result of prolonged stress may take on different forms. One of the forms involves elevated insulin and cortisol, which increase the risk for obesity and atherosclerosis (Fig. 4). Interestingly, insulin resistance and elevated cortisol are hallmarks of endogenous depressive illness (83), and depressed individuals do have an increased risk for cardiovascular disease as well as shorter lifespans (84).

Another form of allostatic load is the hypervigilance and elevated sympathetic activity associated with posttraumatic stress disorder (PTSD). [See this volume.] Cortisol levels are not elevated, nor is insulin, and so the risk for obesity and atherosclerosis are not as high. Yet there may be adverse health consequences of the chronically elevated sympathetic tone associated with this form of allostatic load.

A third form of allostatic load is chronic exercise (85–87). With repeated exercise, cortisol is elevated, along with adrenal cortex and medulla size, but insulin is not. Because exercise increases glucose utilization and reduces insulin output, elevated cortisol is not able to cause insulin resistance as it does in sedentary individuals who experience endogenous depression. While regular exercise is regarded as beneficial, the effects of extreme physical exertion on a regular basis also entail some adverse health consequences related to elevated sympathetic and glucocorticoid activity.

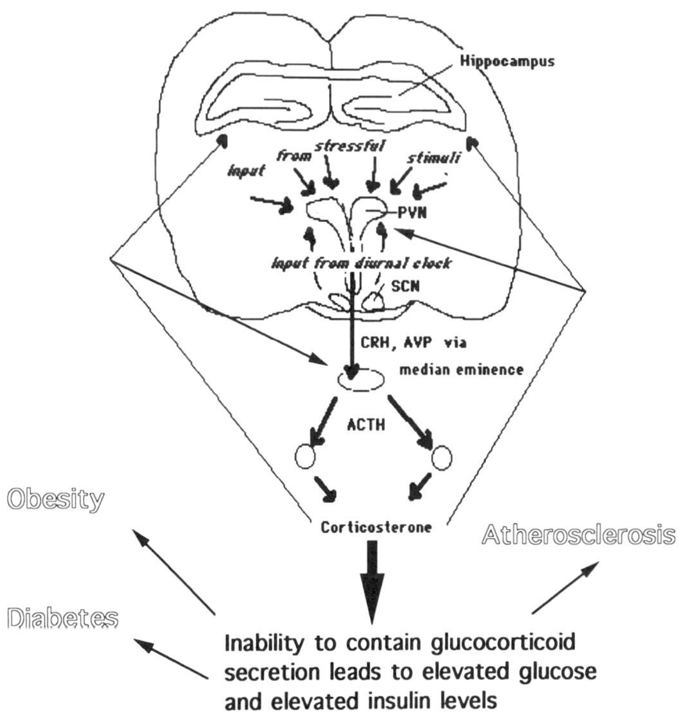

FIG. 4. When containment breaks down, one consequence is elevated basal glucocorticoid secretion. This form of allostatic load leads to elevated glucose levels and elevated insulin levels, with partial insulin insensitivity. This syndrome is characteristic of Cushing's disease and endogenous depressive illness, and is associated with obesity, diabetes, and atherosclerosis.

CONCLUSIONS: PROPOSED SEQUENCE OF EVENTS FOR BREAKDOWN OF CONTAINMENT

We have been arguing that containment of a variety of physiological responses to stress is a key feature of successful adaptation. This includes the HPA axis, because, when glucocorticoid secretion is elevated for prolonged periods, other actions of glucocorticoids besides the containment effects begin to accumulate that are not beneficial. It has been suggested that repeated stress may impair successful behavioral adaptation and may do so via glucocorticoid secretion. Evidence in support of this hypothesis has come from use of the adrenal steroid synthesis inhibitor, metyrapone, to increase behavioral adaptation to restraint stress (88).

One of the adverse events promoted by glucocorticoids, as we have seen, is altering the balance between 5-HT_2 and 5-HT_{1A} transmission in the forebrain. Another deleterious result of repeated, unremitting stress is the atrophy of aptical dendrites of hippocampal CA3 pyramidal neurons, a change associated with some cognitive impairment on a radial maze (see previous information). Whereas 5-HT_2 transmission elevated by glucocorticoid induction of increased 5-HT_2 receptors in cerebral cortex would enhance the anxiogenic side of stress inducement, the reduced 5-HT_{1A} transmission due to glucocorticoid actions in hippocampus would impair adaptation to stress by reducing the ability of the hippocampus to suppress negative associations (16,17). At the same time, synaptic connections through CA3 pyramidal neurons that are presumably disrupted by glucocorticoid-induced atrophy of dendrites would reduce the efficacy of the hippocampus in its ability to match expected outcomes with reality (12) and thus further increase the likelihood that fears and anxieties would become self-sustaining and less subject to matching with real events. This is a situation in which the concept of allostatic load, with its emphasis on anticipation, is of particular relevance to psychiatric disorders.

How might this sequence of events start in the first place? We have noted that the control of ACTH secretion through the PVN involves both facilitative and inhibitory influences via catecholamines and serotonin. It may be that the balance between these opposing influences breaks down in some individuals under stress, giving rise to persistently elevated cortisol levels. We have noted that depressed individuals show a high degree of stress, and considerable containment of the ACTH response as shown by challenge with Ru486 (80). Nevertheless, if containment breaks down enough so that the elevation in cortisol above baseline begins to alter neurochemical features of the brain by elevating 5-HT_2 receptors in cortex and suppressing 5-HT_{1A} receptors in hippocampus, as well as causing atrophy of dendrites in the hippocampus, then some of the negative features of glucocorticoid action on behavioral adaptation will occur.

Stressful life events are acknowledged as an important risk factor for major depressive illness (89;90) as well as PTSD (90). The fact that depression is a likely outcome in some individuals, resulting from the experience of a natural disaster, while PTSD may also occur in the same population and even in the same individuals (90), raises important questions. First, where they represent alternative responses to psychological trauma, what individual neurochemical and behavioral traits make a person susceptible to these disorders which, as we have seen, represent different trajectories along the pathway of allostatic load? Second, how can PTSD and major depression coexist in the same individuals, given their different pathophysiology? Do they occur at the same time, or do they alternate? Future studies should attempt to address these important questions.

SUMMARY

Under normal circumstances, adrenal steroids help to contain responses to stress and maintain stability within the brain, as well as perform important signaling roles (e.g., exerting biphasic effects on longterm potentiation, and regulating the balance between cell birth and cell death in the dentate gyrus, which may change according to season of the year). As long as steroid levels are themselves contained, the beneficial effects predominate; however, when adrenal corticosteroid levels are persistently elevated, due to a breakdown in negative feedback by adrenal steroids or a failure of neural containment of CRH and ACTH secretion, then a cascade of events begins in which neural and peripheral systems begin to show abnormalities resulting from excess exposure to corticosteroids. Then various types of pathology may ensue, including: 1) neuronal atrophy in hippocampus and associated cognitive impairment; 2) imbalances within the 5-HT system due to elevated 5-HT_2 and suppressed 5-HT_{1A} receptors, which may enhance anxiety and reduce the ability to suppress memories of aversive associations; and 3) elevated insulin and corticosteroids, which increase fat deposition and can increase atherosclerosis. It is interesting that these are all negative symptoms associated with major depression and/or with animal models of this disorder. The pathophysiology of PTSD is reputedly different, yet PTSD and major depression are both expressions of imbalances wrought by severe psychological trauma and may coexist in the same individuals. The mechanisms and consequences of this remain to be explored.

ACKNOWLEDGMENTS

Research in the author's laboratory related to this article is supported by NIH Grant MH 41256. The author also wishes to acknowledge intellectual and collegial support from members of the MacArthur Foundation Health and Behavior Network and from scientific colleagues who have participated in MacArthur Foundation supported activities. Other valuable information was gained in helpful discussions with Dr. Terrance M. Keane, U.S.V.A. Hospital, Boston, Massachusetts.

REFERENCES

1. Selye H. *The stress of life.* New York: McGraw-Hill; 1976.
2. Sapolsky R, Krey L, McEwen BS. Neuroendocrinology of stress and aging: the glucocorticoid cascade hypothesis. *Endocr Rev* 1986;7:284–301.
3. Sapolsky R. *Stress, the aging brain and the mechanisms of neuron death.* MIT Press: Cambridge, MA; 1992.
4. Leonard JP, MacKenzie FJ, Patel HA, Cuzner ML. Hypothalamic noradrenergic pathways exert an influence on neuroendocrine and clinical status in experimental autoimmune encephalomyelitis. *Brain Behav Immun* 1991;5:328–338.
5. Sternberg E, Young WS, Bernardini R, Calogero A, Chrousos G, Gold P, Wilder R. A central nervous system defect in biosynthesis of corticotropin-releasing hormone is associated with susceptibility to streptococcal cell wall-induced arthritis in Lewis rats. *Neurobiology* 1989;4771–4775.
6. Morrow L, McClellan J, Conn C, Kluger M. Glucocorticoids alter fever and IL-6 responses to psychological stress and to lipopolysaccharide. *Am J Physiol* 1993; R1010–1016.
7. Spencer R, McEwen BS. Adaptation of the hypothalamic-pituitary-adrenal axis to chronic ethanol stress. *Neuroendo.* 1990;52:481–489.
8. Ramey E, Goldstein M. The adrenal cortex and sympathetic nervous system. *Physiol Rev* 1957;37:155–195.
9. Weiss JM, McEwen BS, Silva M, Kalkut M. Pituitary-adrenal alterations and fear responding. *Am J Physiol* 1970;218:864–868.
10. McEwen BS, Weiss J, Schwartz L. Selective retention of corticosterone by limbic structures in rat brain. *Nature* 1968;220:911–912.
11. Eichenbaum H, Otto T. The hippocampus—what does it do? *Behav Neural Biol* 1992;57:2–36.
12. Gray J. Precis of the neuropsychology of anxiety: an enquiry into the functions of the septohippocampal system. *Behav Brain Sci* 1982;5:469–534.
13. Dunwiddie T, Taylor M, Heginbotham L, Proctor W. Long-term increases in excitability in the CA1 region of rat hippocampus induced by beta-adrenergic stimulation: possible mediation by cAMP. *J Neurosci* 1992; 506–517.
14. Doze VA, Coehn G, Madison D. Synaptic localization of adrenergic disinhibition in the rat hippocampus. *Neuron* 1991;6:889–900.
15. Sakai N, Tanaka C. Inhibitory modulation of long-term potentiation via the 5-HT1A receptor in slices of the rat hippocampal dentate gyrus. *Brain Res* 1993;613: 326–330.
16. Deakin W, Graeff F. 5-HT and mechanisms of defense. *J Psychopharm* 1991;5:305–315.
17. Graeff F. Role of 5-HT in defensive behavior and anxiety. *Rev Neurosci* 1993;4:181–211.
18. Miner J, Yamamoto K. Regulatory crosstalk at composite response elements. *TIBS* 1991;16:423–426.
19. McEwen BS, DeKloet R, Rostene W. Adrenal steroid receptors and actions in the nervous system. *Physiol Rev* 1986;66:1121–1188.
20. McEwen BS, Stephenson B, Krey L. Radioimmunoassay of brain tissue and cell nuclear corticosterone. *J Neurosci Meth* 1980;3:57–65.
21. Reul JM, DeKloet ER. Two receptor systems for corticosterone in rat brain: microdistribution and differential occupation. *Endocrinology* 1985;117:2505–2511.
22. Hollenberg S, Weinberger C, Ong E, Cerelli G, Oro A, Lebo R, Thompson B, Rosenfeld M, Evans R. Primary structure and expression of a functional human glucocorticoid receptor cDNA. *Nature* 1985;318:635–641.
23. Miesfeld R, Okret S, Wikstrom A-C, Wrange O, Gus-

tafsson J-A, Yamamoto K. Characterization of steroid receptor gene and mRNA in wild-type and mutant cells. *Nature* 1984;312:779–782.
24. Arriza J, Weinberger C, Cerelli G, Glaser T, Handelin B, Housman D, Evans R. Cloning of human mineralocorticoid receptor complementary DNA: structural and functional kinship with the glucocorticoid receptor. *Science* 1987;237:268–275.
25. McEwen BS, Cameron H, Chao H, Gould E, Luine V, Magarinos AM, Pavlides C, Spencer RL, Watanabe Y, and Woolley CS. Resolving a mystery: progress in understanding the function of adrenal steroid receptors in hippocampus. *Prog Brain Res* 1994;100:149–157.
26. McEwen BS. Non-genomic and genomic effects of steroids on neural activity. *TIPS* 1991;12:141–147.
27. Belelli D, Lan N, Gee K. Anticonvulsant steroids and the GABA/benzodiazepine receptor-chloride ionophore complex. *Neurosci Biobehav Rev* 1990;14:315–322.
28. Wieland S, Lan N, Mirasedeghi S, Gee K. Anxiolytic activity of the progesterone metabolite 5alpha-pregnan-3alpha-ol-20-one. *Brain Res* 1991;565:263–268.
29. Paul S, Purdy R. Neuroactive steroids. *Faseb* 1992; 6:2311–2322.
30. Gould E, McEwen BS. Neuronal birth and death. *Current Opinions in Neurobiology,* London: Current Biology; 1993; vol. 3/5.
31. Sloviter RS, Valiquette G, Abrams GM, Ronk C, Sollas AI, Paul LA, Neubort SL. Selective loss of hippocampal granule cells in the mature rat brain after adrenalectomy. *Science* 1989;243:535–538.
32. Woolley C, Gould E, Sakai R, Spencer R, McEwen BS. Effects of aldosterone or RU28362 treatment on adrenalectomy-induced cell death in the dentate gyrus of the adult rat. *Brain Res* 1991;554:312–315.
33. Sherry DF, Jacobs LF, Gaulin SJ. Spatial memory and adaptive specialization of the hippocampus. *Trends in Neurosciences* 1992;15:298–303.
34. Collingridge G. Synaptic plasticity: the role of NMDA receptors in learning and memory. *Nature* 1987;330: 604–605.
35. Pavlides C, Watanabe Y, Magarinos AM, McEwen BS, 1993. Opposing role of Type I and Type II adrenal steroid receptors in hippocampal long-term potentiation. *Neuroscience* (in press).
36. Diamond DM, Bennet MC, Fleshner M, Rose GM. Inverted-U relationship between the level of peripheral corticosterone and the magnitude of hippocampal primed burst potentiation. *Hippocampus* 1992;2:421–430.
37. Joels M, DeKloet ER. Control of neuronal excitability by corticosteroid hormones. *TINS* 1991;15:25–30.
38. McEwen BS, Angulo J, Cameron H, Chao H, Daniels D, Gannon M, Gould E, Mendelson S, Sakai R, Spencer R, Woolley C. Paradoxical effects of adrenal steroids on the brain: protection versus degeneration. *Biol Psychiatry* 1992;31:177–199.
39. Martignoni E, Costa A, Sinforiani E, Liuzzi A, Chiodini P, Mauri M, Bono G, Nappi G. The brain as a target for adrenocortical steroids: cognitive implications. *Psychoneuroendocrinology* 1992;17:343–354.
40. aus der Muhlen K, Ockenfels H. Morphologische veranderungen im diencephalon und telencephalon: storungen des regelkreises adenohypophysenebennierenrinde. Z. Zellforsch. *Mikrosck. Anat.* 1969;93:126–141.
41. Landfield P. Modulation of brain aging correlates by long-term alterations of adrenal steroids and neurally-active peptides. *Prog Brain Res* 1987;72:279–300.
42. Sapolsky R, Krey L, McEwen BS. Prolonged glucocorticoid exposure reduces hippocampal neuron number: implications for aging. *J Neurosci* 1985;5:1222–1227.
43. Woolley C., Gould E, McEwen BS. Exposure to excess glucocorticoids alters dendritic morphology of adult hippocampal pyramidal neurons. *Brain Res* 1990;531:225–231.
44. Watanabe Y, Gould E, Cameron H, Daniels D, McEwen BS. Phenytoin prevents stress- and corticosterone-induced atrophy of CA3 pyramidal neurons. *Hippocampus* 1992b;2:431–436.
45. Watanabe Y, Gould E, McEwen BS. Stress induces atrophy of apical dendrites of hippocampal CA3 pyramidal neurons. *Brain Res* 1992a;588:341–345.
46. Magarinos AM, McEwen BS. Stress-induced atrophy of apical dendrites of hippocampal CA3c neurons: Involvement of glucocorticoid secretion and excitatory amino acid receptors. *Neuroscience* (submitted).
47. Watanabe Y, Gould E, Daniels D, Cameron H, McEwen BS. Tianeptine attenuates stress-induced morphological changes in the hippocampus. *Eur J Pharm* 1992c;222: 157–162.
48. Mizoguchi K, Kunishita T, Chui D-H, Tabira T. Stress induces neuronal death in the hippocampus of castrated rats. *Neurosci Lett* 1992;138:157–160.
49. Uno H, Ross T, Else J, Suleman M, Sapolsky R. Hippocampal damage associated with prolonged and fatal stress in primates. *J Neurosci* 1989;9:1705–1711.
50. Uno H, Flugge G, Thieme C, Johren C, Fuchs E. Degeneration of the hippocampal pyramidal neurons in the socially stressed tree shrew. *Abstr Soc Neurosci* 1991; 17:#52.20.
51. Luine, V, Villegas M, Martinez C, McEwen BS. Repeated stress causes reversible impairments of spatial memory performance. *Brain Res* 1994;639:167–170.
52. Starkman M, Gebarski S, Berent S, Schteingart D. Hippocampal formation volume, memory dysfunction and cortisol levels in patients with Cushing's syndrome. *Biol Psych* 1992;32:756–765.
53. Sterling P, Eyer J. Allostasis: A new paradigm to explain arousal pathology. In Fisher S, Reason J, eds. *Handbook of life stress, cognition and health.* New York: John Wiley 1988;629–649.
54. McEwen BS, Stellar E. Stress and the Individual: Mechanisms leading to disease. *Arch. Int. Med.* 1993;153: 2093–2101.
55. Munck A, Guyre P, Holbrook N. Physiological functions of glucocorticoids in stress and their relation to pharmacological actions. *Endocr Rev* 1984;5:25–44.
56. Weiss J, Goodman P, Losito B, Corrigan S, Charry J, Bailey W. Behavioral depression produced by an uncontrollable stressor: Relationship to norepinephrine, dopamine and serotonin levels in various regions of rat brain. *Brain Res Rev* 1981;3:167–205.
57. Tsuda A, Tanaka M. Differential changes in noradrenaline turnover in specific regions of rat brain produced by controllable and uncontrollable shocks. *Behav Neurosci* 1985;802–817.
58. Nisenbaum L, Zigmond M, Sved A, Abercrombie E. Prior exposure to chronic stress results in enhanced synthesis and release of hippocampal norepinephrine in response to a novel stressor. *J Neurosci* 1991;11: 1478–1484.
59. Pacak K, Armando I, Komoly S, Fukuhara K, Weise

V, Holmes C, Kopin I, Goldstein D. Hypercortisolemia inhibits yohimbine-induced release of norepinephrine in the posterolateral hypothalamus of conscious rats. *Endocrinology* 1992;1369–1376.

60. Pacak K, Kvetnansky R, Palkovits M, Fukuhara K, Yadid G, Kopin I, Goldstein D. Adrenalectomy augments in vivo release of norepinephrine in the paraventricular nucleus during immobilization stress. *Endocrinology* 1993;133:1404–1410.
61. Nisenbaum L, Abercrombie E. Presynaptic alterations associated with enhancement of evoked release and synthesis of norepinephrine in hippocampus of chronically cold-stressed rats. *Brain Res* 1993;608:280–287.
62. Archer T. Serotonin and fear retention in the rat. *J Comp Physiol Psychol* 1982;96:491–516.
63. Azmitia E, McEwen BS. Adrenalcortical influence of rat brain tryptophan hydroxylase activity. *Brain Res* 1974;78:291–302.
64. Neckers L, Sze P. Regulation of 5-hydroxytryptamine metabolism in mouse brain by adrenal glucocorticoids. *Brain Res* 1975;93:123–132.
65. Singh V, Corley K, Phan T-H, Biber M. Increases in the activity of tryptophan hydroxylase from rat cortex and midbrain in response to acute or repeated sound stress are blocked by adrenalectomy and restored by dexamethasone treatment. *Brain Res* 1990;516:66–76.
66. Kuroda Y, Mikuni M, Ogawa T, Takahashi K. Effect of ACTH, adrenalectomy and the combination treatment on the density of 5-HT2 receptor binding sites in neocortex of rat forebrain and 5-HT2 receptor-mediated wet-dog shake behaviors. *Psychopharmacology* 1992;108: 27–32.
67. Kuroda Y, Mikuni M, Nomura N, Takahashi K. Differential effect of subchronic dexamethasone treatment on serotonin-2 and beta-adrenergic receptors in the rat cerebral cortex and hippocampus. *Neurosci Letts* 1993;155: 195–198.
68. McKittrick CR, Blanchard D, Blanchard R, McEwen BS, Sakai R. Serotonin receptor binding in colony model of chronic social stress. *Biol Psych* 1995;37:383–393.
69. Martire M, Navarra P, Pistritto G, Preziosi P. Adrenal steroid-induced chances in serotonin receptors in rat hippocampus and hypothalamus. *Pharm Res Commun* 1988;20:415–416.
70. Mendelson S, McEwen BS. Autoradiographic analyses of the effects of restraint-induced stress on 5-HT1A, 5-HT1C and 5HT2 receptors in the dorsal hippocampus of male and female rats. *Neuroendocrinology* 1991;54: 454–461.
71. Mendelson S, McEwen BS. Autoradiographic analyses of the effects of adrenalectomy and corticosterone on 5-HT1A and 5-HT1B receptors in the dorsal hippocampus and cortex of the rat. *Neuroendocrinology* 1992;55: 444–450.
72. Chalmers D, Kwak S, Mansour A, AKil H, Watson S. Corticosteroids regulate brain hippocampal 5-HT1A receptor mRNA expression. *J Neurosci* 1993;13:914–923.
73. Burnet P, Mefford I, Smith C, Gold P, Sternberg E. Hippocampal 8-[3H]hydroxy-2-(Di-n-propylamino) tetralin binding site densities, serotonin receptor (5-HT1A) messenger ribonucleic acid abundance and serotonin levels parallel the activity of the hypothalamopituitary-adrenal axis in rat. *J Neurochem* 1992;59:1062–1070.
74. Saphier D. Adrenoceptor regulation of paraventricular nucleus neuronal activity as related to hypothalamo-pituitary adrenocortical responses. In Kvetnansky R, McCarty R, Axelrod J, eds. *Stress: neuroendocrine and molecular approaches.* New York: Gordon and Breach 1992;481–488.
75. Al-Damluji S, White A. Central noradrenergic lesion impairs the adrenocorticotrophin response to release of endogenous catecholamines. *J Neuroendo* 1992;4: 319–323.
76. Welch J, Farrar G, Dunn A, Saphier D. Central 5-HT1A receptors inhibit adrenocortical secretion. *Neuroendocrinology* 1993;57:272–281.
77. McEwen BS. Re-examination of the glucocorticoid hypothesis of stress and aging. In Swaab D, Hofman M, Mirmiran M, Ravid R, van Leeuwen F, eds. *Progress in brain research.* 1992;93:365–383.
78. Akana S, Dallman M, Bradbury M, Scribner K, Strack A, Walker C-D. Feedback and facilitation in the adrenocortical system: unmasking facilitation by partial inhibition of the glucocorticoid response to prior stress. *Endocrinology* 1992;131:57–68.
79. Dallman MF, Akana SF, Levin N, Walker CD, Bradbury MJ, Suemaru S, Scribner KS. Corticosteroids and the control of function in the hypothalamus-pituitary-adrenal (HPA) axis. *Ann NY Acad Sci* 1994;746:22–31.
80. Krishnan RR, Reed D, Wilson W, Saunder W, Ritchie J, Nemeroff C, Carroll B. RU486 in depression. *Prog Neuro-Psychopharm Biol Psychiat* 1992;16:913–920.
81. Brindley D, Rolland Y. Possible connections between stress, diabetes, obesity, hypertension and altered lipoprotein metabolism that may result in atherosclerosis. *Clinical Science* 1989;77:453–461.
82. Kaplan JR, Adams MR, Clarkson TB, Manuck SB, Shivery CA. Social behavior and gender in biomedical investigations using monkeys: studies in atherogenesis. *Lab Anim Sci* 1991;41:1–9.
83. Winokur A, Maislin G, Phillips J, Amsterdam J. Insulin resistance after oral glucose tolerance testing in patients with major depression. *Am J Psychiat* 1988;145:325–330.
84. Anda R, Williamson D, Jones D, Macera C, Eaker E, Glassman A, Marks J. Depressed affect, hopelessness and the risk of ischemic heart disease in a cohort of U.S. adults. *Epidemiology* 1993;4:285–294.
85. Szerwinski S, Kurowski T, O'Neill T, Hickson R. Initiating regular exercise protects against muscle atrophy from glucocorticoids. *Am J Physiol* 1987;63:1504–2510.
86. Scheurink AJ, Steffens A, Bourituius H, Dreteler G, Bruntink R, Remie R, Zaagsma J. Sympathoadrenal influence on glucose, FFA and insulin levels in exercising rats. *Am J Physiol* 1989;256:R161–R168.
87. Stallknecht B, Kjaer M, Mikines K, Maroun L, Ploug T, Ohkuwa T, Vinten J, Galbo H. Diminished epinephrine response to hypoglycemia despite enlarged adrenal medulla in trained rats. *Am J Physiol* 1990;259:R998–R1003.
88. Kennett G, Dickinson S, Curzon G. Central serotonergic responses and behavioral adaptation to repeated immobilisation: the effect of the corticosterone synthesis inhibitor metyrapone. *Eur J Pharm* 1985;119:143–152.
89. Anisman H, Zacharko RM. Depression: The predisposing influence of stress. *Behav Brain Sci* 1982;5:89–137.
90. Green BL, Lindy JD, Grace MC, Leonard AC. Chronic posttraumatic stress disorder and diagnostic comorbidity in a disaster sample. *J Nerv Mental Dis* 1992;180:760–766.

PART II

Models of the Impact of Stress on Brain Function

Based on the preclinical investigations reviewed in this volume, it is clear that stress alters the function of numerous neuronal systems in the brain and that these effects can be longlasting. Further, it is likely that stress-induced changes in brain function have clinical consequences. This section of the book contains seven chapters that move beyond identification of which neuronal systems are affected by stress, toward the development of neural models of stress effects that are relevant to what is known about the impact of stress on humans.

The chapter by Gold and associates discusses the mechanism by which stress affects memory. This work provides a useful foundation for the chapter by Krystal and colleagues in which specific memory problems and dissociative phenomena in post-traumatic stress disorder (PTSD) patients are presented.

The chapter by Deutch and Young examines the effects of stress on dopamine and how that relates to subcortical and cortical interactions, and adaptive coping behaviors. Falls and Davis review the neural mechanisms that may mediate fear inhibition, such as extinction and conditioned inhibition. Both of these chapters have obvious clinical relevance. Traumatized individuals frequently develop chronic maladaptive, ineffective coping skills. Persistent fear behaviors continue for many years in patients with PTSD, even in the absence of precipitants, as if the mechanisms of fear inhibition have been disrupted.

The chapter by Post and colleagues emphasizes that sensitization and kindling may be important neural substrates of PTSD. It is possible that these phenomena account for the observation that symptoms resulting from trauma can frequently be longlasting and even worsen with time. Michaelson and associates present a model of the effect of stress on the hypothalamic-pituitary-adrenal (HPA) axis that has implications not only for PTSD but also for the effects of stress on human disease more generally. Charney and colleagues attempt to synthesize data from a variety of perspectives to develop a functional neuroanatomy of PTSD that identifies those brain structures, neurotransmitters, and neural mechanisms that may mediate many of the symptoms associated with PTSD.

Neurobiological and Clinical Consequences of Stress: From Normal Adaptation to PTSD, edited by M. J. Friedman, D. S. Charney, and A. Y. Deutch, Lippincott-Raven Publishers, Philadelphia © 1995.

9

Stress Regulation of Memory Processes

Role of Peripheral Catecholamines and Glucose

Paul E. Gold and Richard C. McCarty

Department of Psychology, University of Virginia, Charlottesville, Virginia 22903

If administered near the time of an experience, a wide range of pharmacological agents and other treatments can enhance or impair the memory of that experience (1–3). Such treatments are effective at regulating memory storage when the treatments are administered very soon after the experience, but are ineffective when administered at later times. Thus, the effects of these treatments on memory decrease with time after the experience, with the temporal properties defining retrograde amnesia and retrograde enhancement gradients. The extensive evidence demonstrating such effects on memory suggest the possibility that endogenous biological responses to experience might similarly regulate memory formation. In particular, the findings raise the possibility that hormonal responses to the experience might regulate the formation of new memories (4). More specifically, according to this view, some hormones released in response to a given training experience should be endogenous modulators of memory storage processes and, by extension, many of the drugs that enhance or impair memory might act through hormonal consequences of the treatments. In animal studies of memory, many of the tasks employ stressful events, leading to a principle interest in the role of stress-related hormones in regulating brain mechanisms responsible for the formation of new memories.

Examples of how stress-related hormonal responses might regulate memory can be found in considering several major public events that seem prominent in most of our memory stores. Depending on the reader's age, these are likely to include memories of: the attack on Pearl Harbor; the John Kennedy, Martin Luther King, and Robert Kennedy assassinations; the Challenger disaster; and the San Francisco earthquake. It is worth noting that these memories are not restricted to the primary events but also include answers to such questions as: Where were you when you heard . . . ?; Who told you about . . . ?; and What were you wearing when you heard . . . ? On the other hand, answers to parallel questions about where you were, who spoke to you, what you wore, etc. for the days before the event are typically not available for recall. Although the veracity of memories of this type is not clear, there is less doubt about the intensity and quantity of detail remembered later, whether correctly or incorrectly (5).

Even at this superficial level it is clear that, if stress-related hormonal responses contribute to memory formation, the hormonal control of memory is especially important for a temporal epoch rather than for specific attributes of memory. Taking this together with evidence for multiple parallel memory systems in the brain (6–10), it is likely that stress-related hormones regulate memory formation in many neural systems, thereby influencing memory for different memory attributes.

Of the hormones tested, the most complete evidence available indicates that release of epinephrine from the adrenal medulla into the circulation is a major contributor to the regulation of memory formation. As will soon be discussed, the mechanism by which epinephrine acts on memory appears to include increases in circulating glucose levels that then act on the central nervous system (CNS). In addition, there is growing evidence that these responses (i.e., epinephrine release with subsequent increases in blood glucose levels) may also be important for memory deficits as seen during aging and cognitive pathologies.

CIRCULATING EPINEPHRINE AND GLUCOSE EFFECTS ON LEARNING AND MEMORY

Demonstrations of the role of epinephrine in regulating memory formation began with tests in rodents trained to avoid punishment in a one-trial inhibitory avoidance task. In this task, a rat (or mouse) is placed in a well-lit start compartment and permitted to cross into a dark compartment. Upon entering the dark compartment, the animal receives a single footshock. If trained with a sufficiently intense shock, the rat exhibits a long latency to reenter the shock compartment on a test trial administered, for example, the next day. If trained with a shock of lower intensity, the rat will cross into the dark compartment more quickly than if trained with a more intense shock. Thus, the latency to enter the shock compartment, used as a measure of memory, increases as the intensity of the shock administered during training is increased; rats will show greater avoidance of a more intense shock than of a less intense shock. In itself, this is a rather unremarkable result, as is the finding that hormonal responses also increase as the intensity of the training experience increases. To identify the separate contribution of epinephrine responses to training to memory in this task, rats are trained with a mild shock, selected to result in low latencies to cross into the shock compartment on a test trial; this mild shock also results in only a very modest and transient release of epinephrine. Immediately after training, the rat is removed from the training apparatus and receives an in-

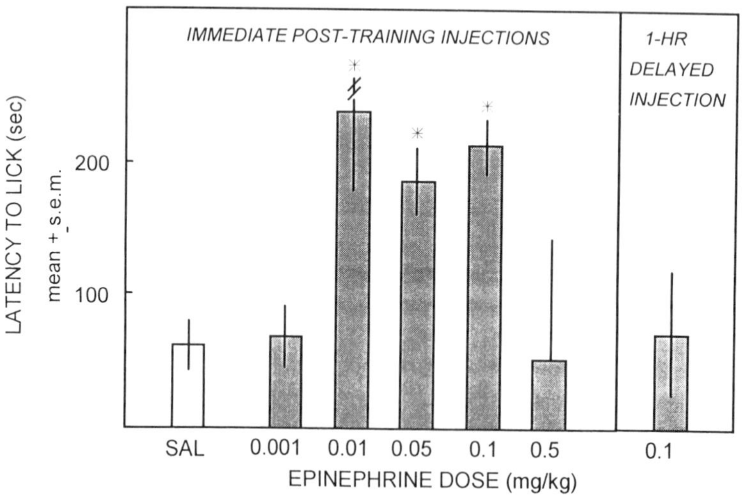

FIG. 1. Epinephrine enhancement of memory for inhibitory avoidance training. Rats received injections (s.c.) of epinephrine immediately or 1 hr after training and were tested for memory 24 hr later. Note that the dose-response curve is an inverted-U function and that injections delayed 1 hr after training did not affect later performance. From Gold and van Buskirk (11).

jection of epinephrine at a dose that matches the circulating epinephrine levels of rats trained with a more intense foot shock. When tested the next day, rats that received exogenous epinephrine avoid the dark compartment for a longer time than do rats that received saline; the longer latencies to enter the dark compartment are taken as evidence of enhanced memory. The findings of one such experiment are illustrated in Fig. 1 (11). Note that the dose-response curve for epinephrine effects on memory is an inverted-U function. While intermediate doses of epinephrine enhance memory, higher doses impair memory. From studies measuring circulating epinephrine levels following an epinephrine injection or foot shock, intermediate doses of epinephrine that enhance memory result in circulating epinephrine levels that are very close to those obtained after training with a foot shock adequate to produce good memory scores on the inhibitory avoidance task (12,13). This is illustrated in Fig. 2 (13).

Enhancement of memory with epinephrine has been seen with a variety of training procedures (2,3,14). In addition, many of the treatments that enhance and impair memory appear to act via modifications of circulating catecholamine levels. One set of evidence for this view is that peripherally administered adrenergic antagonists, such as propranolol or phenoxybenzamine, block the effects on memory of not only epinephrine but many other classes of memory-enhancing or -impairing treatments, including electrical stimulation of the amygdala, electroconvulsive shock, and cycloheximide (15). Although the effects of epinephrine on memory are clear and robust, the neurobiological mechanisms underlying these actions are less well defined. Epinephrine does not enter brain in significant amounts (16), but the hormone might modulate brain functions by brainstem mechanisms [e.g., by activating neurons of the nucleus of the solitary tract (17)]. In addition, epinephrine may regulate memory formation by intermediate mechanisms outside of the CNS. This view is consistent with the findings that peripherally administered adrenergic antagonists block the effects on memory of epinephrine and a broad range of other treatments.

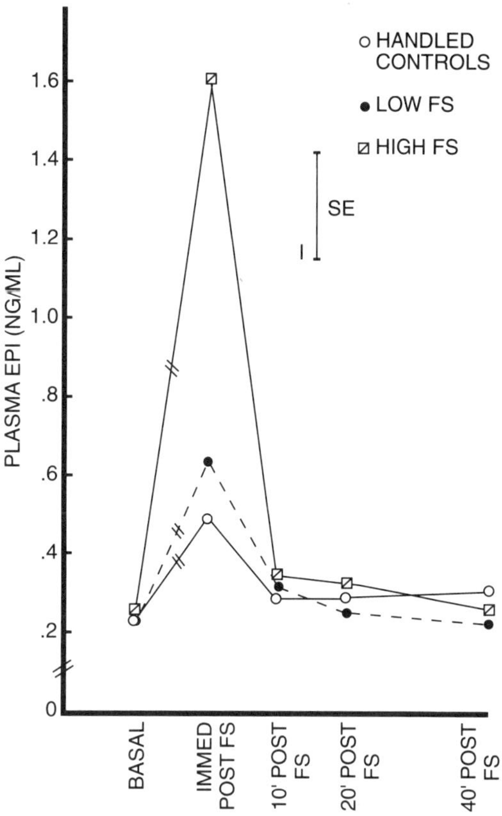

FIG. 2. Mean plasma epinephrine levels before and after footshock as used in inhibitory avoidance tasks. Note the graded epinephrine responses to handling and to foot shock. The peak epinephrine response to foot shock (1.6 ng/ml) is comparable to that obtained after low footshock followed by an epinephrine injection at a dose (0.1 mg/kg) that enhances memory (not shown). Ns = 6–7 / group; the range of standard errors is indicated. From McCarty and Gold (13).

If epinephrine does not itself act directly on central mechanisms, the hormone is likely to act through a peripheral intermediate step. Recent findings suggest that epinephrine-induced increases in circulating glucose levels contribute to the effects of epinephrine on memory. Initial experiments demonstrated that glucose mimics the effects of epinephrine on memory (Fig. 3) (18). This experiment, assessing glucose effects on memory, is analogous to that shown for epinephrine in Fig. 1. As seen with epinephrine, glucose injections administered shortly after

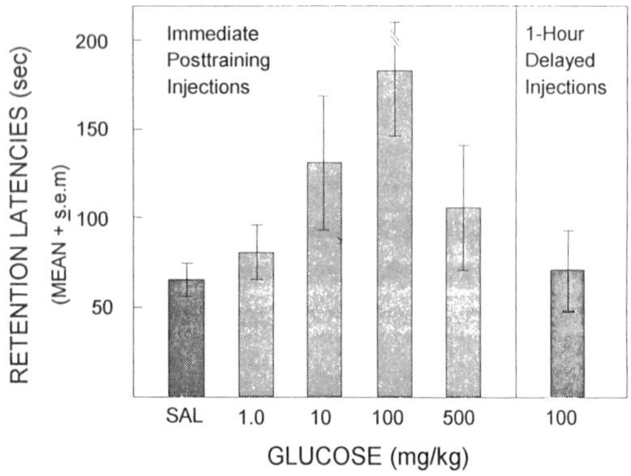

FIG. 3. Glucose enhancement of memory for inhibitory avoidance training. As with epinephrine, glucose injections after training enhanced memory in a dose- and time-dependent manner. From Gold (18).

training resulted in improved learned performance on a test trial the next day. Also comparable to results obtained with epinephrine, the effects of glucose on memory were time dependent and were most effective when glucose was administered immediately after training. For both epinephrine and glucose, the dose-response curves were inverted-U functions; at high doses, epinephrine and glucose produce retrograde amnesia. When blood glucose levels were assessed after injections of epinephrine and glucose, doses optimal for enhancing memory resulted in similar increases in blood glucose levels (19,20). Results just noted indicate that systemically injected adrenergic antagonists block the effects of epinephrine and other treatments on memory. Glucose is an important exception; adrenergic antagonists do not block either enhancement or impairment of memory by glucose (21). These results are consistent with the view that glucose effects on memory are subsequent to epinephrine actions at hepatic adrenergic receptors to liberate glucose stores into the circulation. Using systemic injections, enhancement of memory with glucose has been seen in several laboratories for a variety of tasks (22–29; see reviews 14,15,30). In addition, epinephrine and glucose injections ameliorate age-related deficits in memory in aged rodents, and glucose enhances memory in generally healthy elderly humans (14).

GLUCOSE ACTIONS ON BRAIN SYSTEMS

Analyses of the neuropharmacological bases by which glucose might affect brain processes began with examinations of interactions between glucose and drugs directed at specific neurotransmitter systems. Several experiments tested the efficacy of glucose as a treatment that might attenuate drug-induced impairments in memory. In studies using systemic injections, the findings indicated that glucose ameliorates the effects of many drugs, including reversing impairments in learning and memory produced by opiate and γ-aminobutyric acid (GABA) receptor agonists, and cholinergic (both muscarinic and nicotinic) and glutamatergic (N-methyl-D-aspartate (NMDA)) antagonists (see review 3). An example of one such study is shown in Fig. 4 (data from several experiments) (31). The behavioral task employed in these experiments was a spontaneous alternation task. The apparatus is a Y-shaped maze in which rats are allowed to roam freely for 8 minutes. As the rats move through the maze, they tend to choose the least recently visited arm for successive entries. The memory component in this task is that the rat must remember which arm it has entered most recently in order to select a different arm. Under control conditions, rats alternate on about 70% of their choices. Rats were tested for spontaneous alternation performance after receiving systemic injections of one of several amnestic agents: the opiate receptor agonist, morphine; the GABA receptor agonist, muscimol; the cholinergic antagonists, scopolamine (muscarinic) or mecamylamine (nicotinic); and the NMDA receptor antagonist, NPC 12626. Each drug was injected alone or in combination with glucose. As expected, systemic injections of all amnestic drugs impaired spontaneous alternation performance. Coadministration of glucose attenuated each of these memory impairments.

Thus, in terms of memory measures, glucose interacts with several drugs that influence differ-

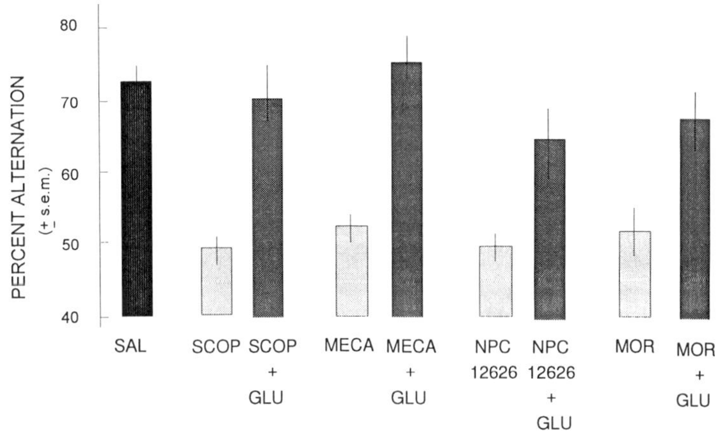

FIG. 4. Attenuation by systemic injections of glucose of impairments in spontaneous alternation performance produced by scopolamine (muscarinic antagonist), mecamylamine (nicotinic antagonist), NPC 12626 (NMDA antagonist), and morphine (opiate agonist). From Walker and Gold (31).

ent neurotransmitter systems. Analogous interactions are also evident when glucose is used as a challenge for the effects of many of these drugs on locomotor activity, sleep, tremors, and brain 2-deoxyglucose uptake (3). Therefore, besides there being broad interactions with several neurotransmitters, the interactions are also evident across a range of neural and behavioral measures.

Although these findings provide strong support for the view that glucose influences brain functions, the varied actions and interactions do not permit definition of mechanisms by which glucose might act. More recent experiments have employed microinjections of glucose and drugs as discussed here into specific brain targets. Using direct brain microinjections, the findings indicate two major differences compared to results seen using systemic injections: 1) glucose interactions with drugs that affect central neurotransmitter systems are more specific than was seen with systemic injections, and 2) the behavioral measure sensitive to drug injections varies with the brain injection site. We have identified two brain injections sites where glucose has effects on learning and memory. One site is the amygdala, a brain area in which lesions and injections of some drugs interfere with memory of emotional events, including both aversive and appetitive training tasks (6,8,32). We found that injections of an opiate agonist, morphine, prior to training in an avoidance task impaired later memory for that training experience; coinjections of glucose, also directly into the amygdala, attenuated the morphine-induced impairment (33) (Fig. 5). Causing results similar to those obtained with systemic injections, injections of drugs directed at different neurotransmitter systems (i.e., propranolol or AP5) also impaired avoidance memory. However, in contrast to the results obtained with systemic injections, glucose effects are, to this point, restricted to interactions with opiate agonists. An additional important difference between the results obtained with systemic and direct amygdala injections is that the memory impairments were seen when using avoidance tasks, but were not evident in the spontaneous alternation task.

We have also investigated the effects of injections of glucose and other drugs directly into the medial septum. The medial septum contains cholinergic cell bodies with axons projecting to the hippocampus (34,35). On the basis of experiments employing brain lesions, direct drug injections, and neurophysiolc gical recordings, the septo-hippocampal system appears to participate in learning and memory for spatial tasks and other situations in which relationships between stimuli are important features of the learned responses (7,8,36). Our findings indicate that mor-

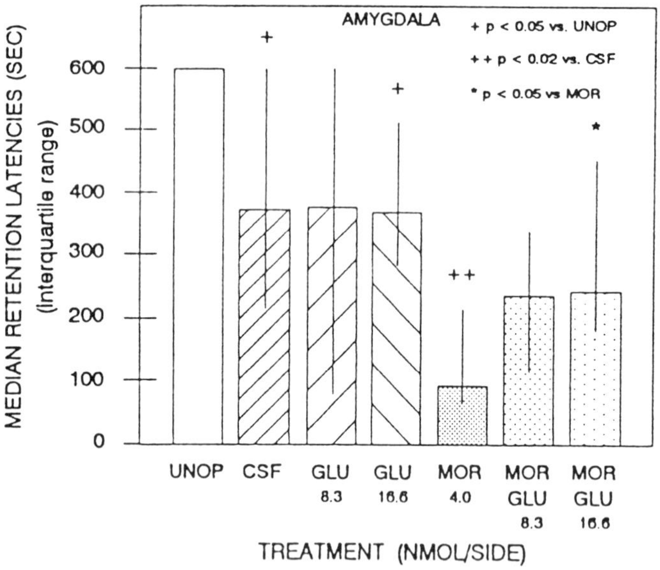

FIG. 5. Effects of intraamygdala morphine and morphine plus glucose on memory for inhibitory avoidance training. Glucose attenuated the memory deficits produced by morphine. From Ragozzino and Gold (33).

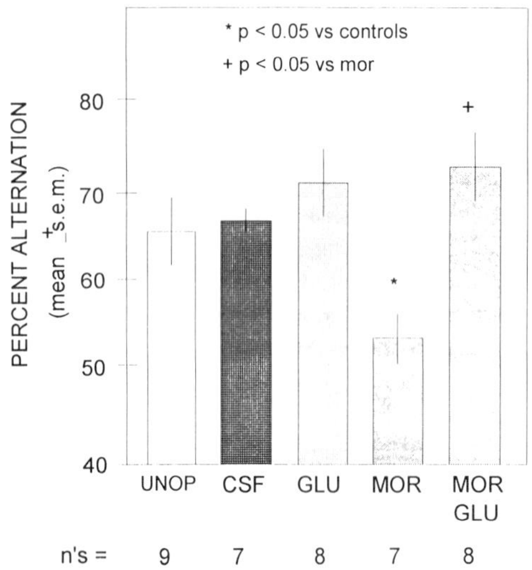

FIG. 6. Effects of intraseptal morphine and morphine plus glucose on spontaneous alternation performance. Coadministration of glucose with morphine blocked the impairment produced by morphine alone. From Ragozzino et al (37).

phine, muscimol, scopolamine, and propranolol all impair spontaneous alternation performance when injected into the medial septum (37). Intraseptal injections of at least some of these drugs also impair memory for inhibitory avoidance training. For those interactions examined thus far, glucose injections into the medial septum attenuate deficits in spontaneous alternation performance after intra-septal injections of morphine, but glucose does not attenuate the effects of either scopolamine or propranolol under the same conditions (Fig. 6) (37). Thus, interactions of glucose with other drugs appear to be more restricted when the drugs are injected into specific brain regions than when delivered via systemic injections. These findings may permit identification of the pharmacological mechanisms (e.g., neuronal phenotype by neurotransmitter class) by which glucose acts.

We have also begun to explore the possibility that glucose effects on learning and memory might be a result of glucose metabolism, either in energy production or in production of one or more of several neurotransmitters synthesized from glucose. Recently, we found that microinjections of pyruvate, like those of glucose, into the medial septum reverse impairments in spontaneous alternation performance produced by morphine (38). These findings, shown in Fig. 7 (38), suggest that glucose metabolism through pyruvate may contribute to the mechanisms by which glucose regulates brain functions.

As just noted, there is a cholinergic projection from the medial septum to the hippocampus. Whatever the direct action of glucose within the medial septum, it is reasonable that glucose actions on memory may be mediated by one or more outputs from the medial septum. Using in vivo microdialysis/HPLC procedures, we therefore examined the effects of systemic and intraseptal injections of morphine and glucose on acetylcholine release in the hippocampus. Previous findings suggested that opiate receptors in the septum inhibit release of acetylcholine in the hippocampus (39,40). When morphine was injected either systemically or into the medial septum, the acetylcholine content of microdialysis samples collected in the hippocampus was significantly reduced. Importantly, coinjections of glucose (either systemic or intraseptal), fully blocked the morphine-induced decrease in release of acetylcholine in the hippocampus (41, 42). Some of the data obtained with intraseptal injections are illustrated in Fig. 8 (42). Several days after assessing hippocampal acetylcholine release in response to intraseptal injections, thc same rats received the same septal injections and were then tested for spontaneous alternation performance. The alternation scores were significantly correlated with levels of acetylcholine release in individual animals; those rats with the greatest decreases in acetylcholine release were the rats with the poorest alternation performance. These findings provide evidence that glucose is exerting actions within the medial septum that regulate the activity of projection neurons, including the cholinergic projections, from septum to hippocampus. Future experiments are needed to determine whether the acetylcholine projection contributes significantly to the effects of glucose on learning and memory.

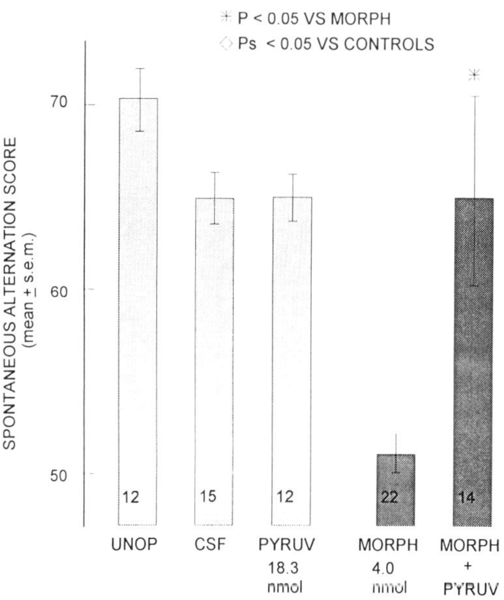

FIG. 7. Effects of intraseptal morphine and morphine plus pyruvate on spontaneous alternation performance. As with glucose, coadministration of pyruvate with morphine blocked the impairment produced by morphine alone, suggesting a contribution of the metabolism of glucose to its effects on memory. From Ragozzino et al (38).

The findings described in this section demonstrate that systemically injected glucose, and by extrapolation circulating glucose, has a wide range of actions on brain functions, including interactions with several neurotransmitter systems on a wide range of behavioral and neural measures. However, the findings obtained with direct brain injections suggest that many of the effects seen with systemic administration are likely to be indirect. Glucose appears to interact directly with at most a limited set of neurotransmitters. Furthermore, the behavioral and neural measures on which glucose acts vary with the site of brain injection.

STRESS-INDUCED RELEASE OF PERIPHERAL CATECHOLAMINES: ROLE OF EXPERIENCE

The studies just summarized point to a role for peripheral catecholamines, especially epinephrine, in memory modulation. Aversive training events (e.g., footshock in inhibitory avoidance paradigms and swimming in cold water for

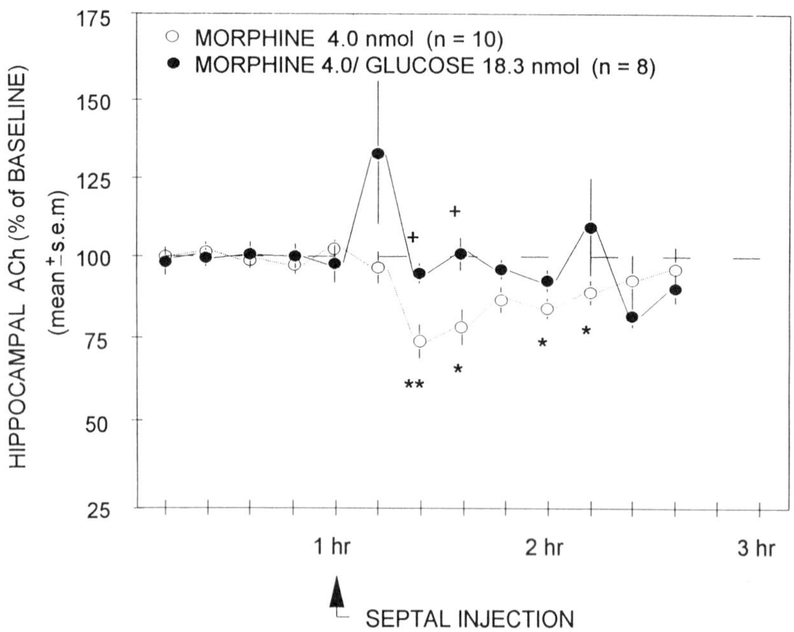

FIG. 8. Hippocampal acetylcholine output assessed in 12-min samples with microdialysis-HPLC procedures after intraseptal injections of morphine or morphine plus glucose. Note that intraseptal morphine, at a dose that impairs spontaneous alternation performance, resulted in decreased acetylcholine output in the hippocampus. As with the behavioral effect, this response to morphine was attenuated by concurrent injection of glucose. From Ragozzino and Gold (42).

the Morris water maze task) lead to release of epinephrine from the adrenal medulla, which stimulates release of glucose from liver, with glucose entering the CNS to affect memory storage processes. An important issue relating to this memory modulatory pathway is the relationship between stressful stimulation and adrenal medullary discharge of epinephrine. In animals with no prior history of experimenter-controlled aversive stimulation, stress-induced increases in plasma levels of norepinephrine and epinephrine vary monotonically with the intensity of the stressor (43,44). In contrast, plasma catecholamine responses of animals previously exposed to chronic intermittent stress differ dramatically from those of unstressed controls.

Several studies have demonstrated in clear fashion that the response of the sympathetic-adrenal medullary system to a given stressor reflects, at least in part, the prior stress history of an animal (45). Given these findings, it is essential that experimental studies of learning and memory take into account prior stress history of subjects. Previously stressed animals may have reduced plasma catecholamine responses to a familiar stressor (habituation) or exaggerated plasma catecholamine responses to a novel stressor (sensitization). These experience-based modifications of plasma epinephrine responses to an aversive training event could enhance or impair memory, based upon the intensity of the aversive stimulus.

To demonstrate habituation of plasma catecholamine responses to chronic intermittent exposure to the same stressful stimulus, Konarska and her coworkers (46) conducted experiments using one of three different stressors: intermittent footshock, restraint, and swim stress. Laboratory rats were exposed to one of the three stressors for 10–30 minutes per day for 26 consecutive days. After exposure to stress on day 26, each rat was provided with an indwelling tail artery catheter for sampling of blood in freely behaving animals. Handled control animals were included in each experiment.

Blood samples were collected from control and chronically stressed rats before, during, and after stressful stimulation on day 28 and were processed for later analysis of norepinephrine and epinephrine. For purposes of analysis, areas under curve (AUCs) were computed for plasma norepinephrine (NE) and plasma epinephrine responses above baseline values for individual animals. In each of three experiments, plasma NE and epinephrine (EPI) AUCs were reduced by 40% to 60% in chronically stressed rats compared to their firsttime stressed controls (Fig. 9). Given the high degree of stressor predictability in these experiments, each rat was capable of activating the sympathetic-adrenal medullary system to the minimum extent necessary for maintaining internal homeostasis (47).

In a separate set of experiments, Konarska and colleagues (48) utilized a crossover design to examine plasma catecholamine responses of chronically stressed laboratory rats to a novel stressor. Animals were again exposed to one of three different stressors each day for 26 consecutive days. Following placement of arterial catheters, previously stressed animals were exposed to a novel stressor on day 28. Handled controls for each of the three experiments were also exposed to the "novel" stressor on day 28. The three groups included: 1) chronic restraint—acute foot shock; 2) chronic swim stress—acute restraint; and 3) chronic foot shock—acute swim stress.

The findings from this crossover design are summarized in Fig. 10. Note that plasma catecholamine AUCs for previously stressed animals were 35% to 155% greater than values for handled controls exposed to the same stressor. In this series of experiments, chronically stressed animals were again presented with a highly predictable schedule of brief daily sessions of stress. Then, on day 28, these same animals were confronted with a dramatic departure from this highly predictable regimen of stress. Their exaggerated plasma catecholamine responses suggest that exposure to a novel stressor is a greater challenge to physiological and behavioral homeostasis than the same stressor is to a control animal with no prior history of experimenter-controlled stressful stimulation (49).

Our findings suggest that nonassociative processes governing adrenal medullary discharge

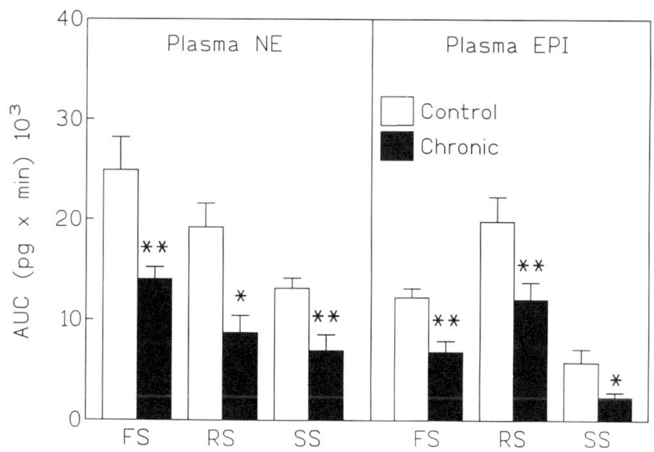

FIG. 9. Area under curve (AUC, pg × min) measures for plasma norepinephrine (NE) and epinephrine (EPI) responses of control, firsttime stressed rats (*light bars*) and rats exposed to chronic intermittent stress (*dark bars*). Stressors included foot-shock stress (FS), restraint stress (RS), and swim stress (SS). Values are presented as means + 1 SEM for 7–8 rats per group. *P<0.05, **P<0.01 (two-tailed t test).

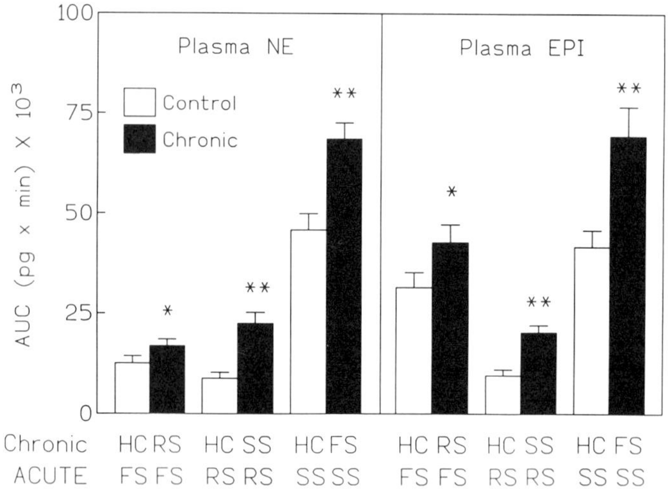

FIG. 10. Area under curve (AUC, pg × min) measures for plasma norepinephrine (NE) and plasma epinephrine (EPI) responses of handled control (HC) rats stressed for the first time (*light bars*) and chronically stressed rats exposed to a novel stressor (*dark bars*). The chronic intermittent stress condition is listed above and the acute stress condition is listed below for each treatment group. Stressors included foot-shock stress (FS), restraint stress (RS), and swim stress (SS). Values are presented as means + 1 SEM for 7–10 rats per group. *$P<0.10$, **$P<0.01$ (two-tailed t test).

of epinephrine following exposure to aversive training events may have a substantial influence on memory for those events. These influences are best viewed within the context of an inverted-U function for memory modulation (14). For example, the plasma epinephrine response to a mild training foot shock in an inhibitory avoidance paradigm would be reduced in animals previously exposed to foot-shock stress, leading to low retention latencies. In contrast, the plasma epinephrine response to a mild training foot shock in an inhibitory avoidance paradigm would be enhanced in previously restrained animals, leading to high retention latencies. In these two hypothetical examples, exposure of two groups of animals to the same training foot shock would be attended by quite different effects on retention latencies, with impairment or enhancement correlated with habituation and sensitization, respectively, of plasma epinephrine responses to training.

CONCLUSIONS

The findings reviewed in this chapter document some of the evidence indicating that stress-related neurobiological responses to an experience can regulate the formation of memory for that experience. The magnitude of the hormonal responses is related to memory by an inverted-U function. Thus, there is a range within which the hormonal responses must fall to have optimal memory. Moderate stress levels enhance memory while high stress levels impair memory. This view is most clearly defined for epinephrine, but applies also to glucose and to many drugs directed at different neurotransmitter systems. The findings described here also point to control of circulating glucose levels, with subsequent actions on the CNS, as an important intermediate step between sympathetic-adrenal medullary functions and actions on cognitive functions. One hypothesis is that glucose acts on the brain to regulate neuronal changes responsible for storing new information. The biological bases are as yet unclear, with possible actions through opiate, cholinergic, and metabolic mechanisms under investigation.

An interesting and less well studied component of this field of study is the degree to which sympathetic-adrenal medullary responses are themselves based on an organism's past experience. Release of catecholamines from the adrenal medulla is a function of not only stress intensity but also stress history. Prior experience can result in either habituation or sensitization of these responses. The implication of this is that there is adaptation imposed on sympathetic functions (i.e., plasticity within a system that regulates further neuronal plasticity). It is easy to imagine that habituation and sensitization of sympathetic responses based on earlier experiences would result in individuals who might respond to similar new experiences with either diminished or exaggerated sympathetic activity (i.e., resulting in either dulled or heightened affect). In extreme forms, such changes may contribute to inappropriate cognitive and emotional responses to new experiences.

ACKNOWLEDGMENTS

Preparation of this article and much of the research from this laboratory described in this article was supported by research grants from the National Institute on Aging (AG 07648), the Office of Naval Research (N0001489-J-1216), and the National Science Foundation (BNS-9012239).

REFERENCES

1. McGaugh JL, Herz MJ. *Memory consolidation.* San Francisco: Albion, 1972.
2. McGaugh JL. Involvement of hormonal and neuromodulatory systems in the regulation of memory storage. *Ann Rev Neurosci* 1989;12:255–287.
3. Gold PE. Modulation of emotional and non-emotional memories: same pharmacological systems, different neuroanatomical systems. In McGaugh JL, Weinberger NM, Lynch G, eds. *Brain and memory: modulation and mediation of neuroplasticity.* New York: Oxford Press, 1995:41–74.
4. Gold PE, McGaugh JL. A single-trace, two process view of memory storage processes. In Deutsch D, Deutsch JA, eds. *Short-term memory.* New York: Academic Press, 1975;355–390.
5. Winograd E, Neisser U, eds. *Affect and accuracy in recall: studies of "flashbulb" memories.* New York: Cambridge University Press, 1992.
6. Kesner RP. Learning and memory in rats with an emphasis on the role of the amygdala. In Aggleton JP, ed. *The amygdala: Neurobiological aspects of emotion, memory, and mental dysfunction.* New York: Wiley-Liss, 1992;379–399.
7. Kesner RP, Bolland BL, Dakis M. Memory for spatial locations, motor responses, and objects: triple dissociation among the hippocampus, caudate nucleus, and extrastriate visual cortex. *Exptl Brain Res* 1993;93:462–470.
8. McDonald RJ, White NM. A triple dissociation of memory systems: hippocampus, amygdala and dorsal striatum. *Behav Neurosci* 1993;107:3–22.
9. Squire LR. *Memory and the brain.* New York: Oxford University Press, 1987.
10. Zola-Morgan S, Squire LR. The components of the medial temporal lobe memory system. In Squire LR, Butters N, eds. *Neuropsychology of memory.* New York: Guilford Press, 1992;325–335.
11. Gold PE, van Buskirk RB. Facilitation of time-dependent memory processes with posttrial epinephrine injections. *Behav Biol* 1975;13:145–153.
12. Gold PE, McCarty R. Plasma catecholamines: changes after footshock and seizure-producing frontal cortex stimulation. *Behav Neural Biol* 1981;31:247–260.
13. McCarty R, Gold PE. Plasma catecholamines: effects of footshock level and hormonal modulators of memory storage. *Horm Behav* 1981;15:168–182.
14. Gold PE. Modulation of memory processing: enhancement of memory in rodents and humans. In Butters N, Squire LR, eds. *Neuropsychology of memory.* New York: Guilford Press, 1992;402–414.
15. Gold PE. An integrated memory regulation system: from blood to brain. In Frederickson RCA, McGaugh JL, Felten DL, eds. *Peripheral signaling of the brain: role in neural-immune interactions, learning and memory.* Toronto: Hogrefe and Huber Publishers, 1991;391–419.
16. Axelrod J, Weil-Malherbe H, Tomchick R. The physiological disposition of ^{3}H-epinephrine and its metabolite metanephrine. *J Pharmacol Exptl Ther* 1959;157:251–256.
17. Williams CL, McGaugh JL. Reversible inactivation of the nucleus of the solitary tract impairs retention performance in an inhibitory avoidance task. *Behav Neural Biol* 1992;58:204–210.
18. Gold PE. Glucose modulation of memory storage processing. *Behav Neural Biol* 1986;45:342–349.
19. Hall JL, Gold PE. The effects of training, epinephrine, and glucose injections on plasma glucose levels in rats. *Behav Neural Biol* 1986;46:156–176.
20. Hall JL, Gold PE. Plasma glucose levels predict the susceptibility of memory enhancement to disruption by adrenergic antagonists. *Eur Journal Pharmacol* 1992; 221:365–370.
21. Gold PE, Vogt J, Hall JL. Posttraining glucose effects on memory: Behavioral and pharmacological characteristics. *Behav Neurol Biol* 1986;46:145–155.
22. Messier C, White NM. Memory improvement by glucose, fructose and two glucose analogs: a possible effect on peripheral glucose transport. *Behav Neural Biol* 1987;48:104–127.
23. Messier C, Destrade C. Improvement of memory for an operant response by post-training glucose in mice. *Behav Brain Res* 1988;31:185–191.
24. Messier C, Durkin T, Mrabet O, Destrade C. Memory-improving action of glucose: indirect evidence for a facilitation of hippocampal acetylcholine synthesis. *Behav Brain Res* 1990;39:135–143.
25. Packard MG, White NM. Effect of posttraining injections of glucose on acquisition of two appetitive learning tasks. *Psychobiology* 1990;18:282–286.
26. Means LW, Fernandez TJ. Daily glucose injections facilitate performance of a win-stay water-escape working

memory task in mice. *Behav Neurosci* 1992;106:345–350.
27. Rodriguez WA, VanAusdle LR, Dhanens K, Mondragon AN. Glucose modulates recently reactivated memories. *Psychobiology* 1993;21:93–100.
28. Ahlers ST, Shurtleff D, Schrot J, Thomas JR, Paul-Emile F. Glucose attenuates cold-induced impairment of delayed matching-to-sample performance in rats. *Psychobiology* 1993;21:87–92.
29. Kopf SR, Opezzo JW, Baratti CM. Glucose effects on memory are not state dependent. *Behav Neural Biol* 1993;60:192–195.
30. White N. Peripheral and central memory-enhancing actions of glucose. In Frederickson RCA, McGaugh JL, Felten DL, eds. *Peripheral signaling of the brain: role in neural-immune interactions and learning and memory.* Toronto: Hogrefe and Huber Publishers, 1991;421–441.
31. Walker DL, Gold PE. Impairment of spontaneous alternation performance by an NMDA antagonist: attenuation with non-NMDA treatments. *Behav Neural Biol* 1992;58:69–71.
32. Davis M. The role of the amygdala in conditioned fear. In Aggleton JP, ed. *The amygdala: neurobiological aspects of emotion, memory, and mental dysfunction.* New York: Wiley-Liss, 1992:255–305.
33. Ragozzino ME, Gold PE. Task-dependent effects of intra-amygdala morphine injections: Attenuation by intra-amygdala glucose injections. *J Neurosci* 1994;14: 7478–7485.
34. Lewis PR, Shute CCD. The cholinergic limbic system: projections to hippocampal formation, medial cortex, nuclei of the ascending cholinergic recticular system, and the subfornicular organ and the supra-optic crest. *Brain* 1967;90:521–540.
35. Meibach R, Seigel A. Efferent connections of the septal area in the cat: an analysis utilizing retrograde and anterograde transport methods. *Brain Res* 1977;119:1–20.
36. Eichenbaum H, Otto T, Cohen NJ. The hippocampus—What does it do? *Behav Neural Biol* 1992;57:2–36.
37. Ragozzino ME, Parker ME, Gold PE. Memory impairments with morphine injections into the medial septum: Attenuation by glucose administration. *Brain Res* 1992; 597:241–249.
38. Ragozzino ME, Hellems K, Lennartz RC, Gold PE. Pyruvate infusions into the septal area attenuate spontaneous alternation impairments induced by intraseptal morphine injections. Submitted.
39. Botticelli LJ, Wurtman RJ. Septohippocampal cholinergic neurons are regulated trans-synaptically by endorphin and corticotropin neuropeptides. *J Neurosci* 1982; 2:1316–1321.
40. Moroni F, Cheney DL, Costa E. Inhibition of acetylcholine turnover in rat hippocampus by intraseptal injections of β-endorphin and morphine. *Naunyn-Schmiedeberg's Arch Pharmacol* 1977;299:149–153.
41. Ragozzino ME, Wenk GL, Gold PE. Glucose attenuates morphine-induced decrease in hippocampal acetylcholine output: an in vivo microdialysis study in rats. *Brain Res* 1994;655:77–82.
42. Ragozzino ME, Gold PE. Glucose injections into the medial septum reverse the effects of intraseptal morphine infusions on hippocampal acetylcholine output and memory. *Neuroscience* 1995; in press.
43. Natelson BH, Tapp WN, Adamus JE, Mittler JC, Levin BE. Humoral indices of stress in rats. *Physiol Behav* 1981;26:1049–1054.
44. Natelson BH, Creighton D, McCarty R, Tapp WN, Pitman D, Ottenweller JE. Adrenal hormonal indices of stress in laboratory rats. *Physiol Behav* 1987;39:117–125.
45. McCarty R, Konarska M, Stewart RE. Adaptation to stress: a learned response? In Kvetnansky R, McCarty R, Axelrod J, eds. *Stress: neuroendocrine and molecular approaches.* New York: Gordon and Breach, 1992; 521–535.
46. Konarska M, Stewart RE, McCarty R. Habituation of sympathetic-adrenal medullary responses following exposure to chronic intermittant stress. *Physiol Behav* 1989;45:255–261.
47. McCarty R, Stone EA. Chronic stress and regulation of the sympathetic nervous system. In Usdin E, Kvetnansky R, Axelrod J, eds. *Stress: the role of catecholamines and other neurotransmitters.* New York: Gordon and Breach, 1984;563–576.
48. Konarska M, Stewart RE, McCarty R. Sensitization of sympathetic-adrenal medullary responses to a novel stressor in chronically stressed laboratory rats. *Physiol Behav* 1989;46:129–135.
49. McCarty R. Regulation of plasma catecholamine responses to stress. *Seminars in the Neurosciences* 1994, 6:197–204.

Neurobiological and Clinical Consequences of Stress: From Normal Adaptation to PTSD, edited by M. J. Friedman, D. S. Charney, and A. Y. Deutch, Lippincott-Raven Publishers, Philadelphia © 1995.

10

A Model of the Stress-Induced Activation of Prefrontal Cortical Dopamine Systems

Coping and the Development of Post-Traumatic Stress Disorder

Ariel Y. Deutch and Cheryl D. Young

Departments of Psychiatry and Pharmacology, Yale University School of Medicine, New Haven, Connecticut 06508; and National Center for Post-Traumatic Stress Disorder, Veterans Affairs Medical Center, West Haven, Connecticut 06516

The physiological and behavioral responses called stress are of critical importance to the survival of the organism. These integrated responses are homeostatic adaptive strategies, ranging from changes in metabolism (e.g., the steroid-induced shift to glycogenolysis) to more complex behaviors. These changes have the common goal of reducing exposure to a stressor or reducing the traumatic effects of a stressor.

Since the stress response is critical to the survival of an organism, it is reasonable to posit that there should be some redundancy in the neurochemical mechanisms and neuroanatomical substrates that comprise the central stress systems. As can be rather easily seen from perusal of the chapters in the first section of this volume, virtually every neurotransmitter can be shown to respond to stress. However, while most transmitters respond to stress, the central sites at which these changes occur are relatively restricted. In particular, the prefrontal cortex (PFC) appears to be a site at which changes in many transmitter systems are observed. Thus, the release and metabolism of specific afferents to the PFC, including dopaminergic, noradrenergic, serotonergic, and cholinergic neurons, are increased by exposure to stressors. The activities of interneurons and projection (pyramidal) neurons in the PFC are in turn altered by the changes in the afferents impinging on the interneurons. Thus, the function and release of γ-aminobutyric acid (GABA) and various neuropeptides (in interneurons) and excitatory amino acids (present in pyramidal cells and afferent fibers) are altered by stress.

The dopaminergic innervation of the PFC is remarkably sensitive to stress. First described a generation ago (1), prefrontal cortical dopamine (DA) afferents sharply increase metabolism and release of DA in response to very mild stressors (1,2). In fact, the PFC is unique in that a large body of literature has revealed that virtually all of the factors thought to be involved in the development of posttraumatic stress disorder (PTSD) can be shown to influence or regulate the response of the PFC dopaminergic neurons. These factors range from genetic and early perinatal events hypothesized to be related to the development of PTSD (see Chapter 23), to the changes in central noradrenergic (see Chapter 3), serotonergic, cholinergic, and peptidergic neurons (see Chapter 6), to the relationship of the cortical dopaminergic system, and to drugs of abuse, including psychostimulants and ethanol (see Chapter 5). Moveover, the PFC has widespread efferents (3), including direct projections to spinal cord vasomotor centers (4), and projections to the hypothalamus and critical brainstem autonomic regions as well as to such regions as the amygdala and hypothalamus (see Chapter 1) that allow the PFC to participate in the coordinated responses to stressors.

In many ways, the cortical DA response to stress can be considered a microcosm of the central changes thought to occur in PTSD. We will briefly review regulation of the DA neurons that innervate the PFC, present a model for the genesis of the stress-induced increase in PFC DA release, and attempt to relate the function of dopaminergic regulation of PFC neurons to a more widespread network of sites (including the amygdala) that are functionally coupled in stress. As befits a topic as broad as stress, there is some redundancy in this volume's coverage of the PFC DA response to stress (see Chapters 4 and 5). We will discuss the PFC in relation to a more widespread stress "system" that involves the amygdala (see Chapters 7 and 11). We will also try to minimize the redundancy with other contributions to this volume, and refer readers to these other chapters for more detailed descriptions of certain aspects of the stress response.

ANATOMY AND PHYSIOLOGY OF THE PREFRONTAL CORTICAL DOPAMINE INNERVATIONS

Anatomical Organization of Cortical Dopamine Afferents

Dopamine afferents to the PFC are present in all mammals. The original observation of a cortical DA innervation was made in the rat (5). In this species, DA fibers discretely innervate the medial and sulcal (suprarhinal) prefrontal cortices, but are not observed in other cortical regions rostral to the genu of the corpus callosum (6,7). There are also relatively minor dopaminergic innervations of the posterior cingulate and suprarhinal cortices, pyriform cortex, and entorhinal cortex (6,7). In contrast to the localized distribution of cortical DA axons in the rat, there is a much more widespread dopaminergic innervation of the cerebral cortex in primate species, with DA projections found in a variety of sites, including association, motor, and sensory cortices (8,9).

In all species examined, the DA innervations of the cortex originate solely from neurons in the ventral mesencephalon. In the rodent, the cortical DA innervations are derived from neurons in the ventral tegmental area of Tsai (VTA; A10 cell group), with a small number of neurons in the substantia nigra (SN; A9 cell group) and retrorubral field (RRF; A8 cell group) projecting to the PFC and to the more lateral (suprarhinal, pyriform, and entorhinal) cortices (8–11). In primate species the DA innervation of the cortex, including the dorsolateral PFC, appears to be derived predominantly from DA neurons in the SN and RRF, with a smaller contribution of VTA DA neurons than would be expected on the basis of the rodent data (12,13).

Certain DA neurons that project to forebrain targets contain, in addition to DA, a number of neuropeptides, including neurotensin and cholecystokinin (see Chapter 6). We have recently reviewed the anatomy and functional attributes of colocalization in DA neurons (14).

The midbrain DA cell body regions receive afferents from a large number of sources (15,16). Particularly prominent are those arising from the PFC, ventral pallidum, lateral hypothalamus, and dorsal raphe region. Since stressors must gain access to the midbrain in order to activate the DA neurons that project to the PFC, it is apparent that afferents to the VTA have been among the most intensely studied of the regulatory features that govern the activity of forebrain DA innervations (17); the reader is referred to reviews by Deutch and Roth (2) and Kalivas (16) for extended discussions of these afferents.

DA axons in the PFC contact both pyramidal cells and interneurons. It was originally thought that the actions of DA in the PFC were paracrine in nature. Although contacts between DA neurons and pyramidal cells in the PFC were evident, these contacts did not display the typical pre- and postsynaptic thickenings that are the electron microscopic hallmark of synapses (18). However, it was subsequently shown in serial reconstructions that DA axons form very small symmetric synapses with PFC pyramidal cells (19–22). Paradoxically, recent ultrastructural data suggest that D_1 DA receptor-like immunoreactivity is not present at the point where DA axons form synapses with pyramidal cells (23), again harkening to a paracrine mode of action (at least with regard to the D_1 receptor-like protein revealed by immunohistochemistry). In addition to DA afferents synapsing onto pyramidal neurons, recent data indicate that DA axons contact and synapse with certain GABAergic interneurons (24–26).

The DA axons that form synapses with pyramidal cells are frequently part of a triadic arrangement, with both an asymmetric (excitatory) input and the inhibitory DA axon contacting the same pyramidal cell (21). The source of the excitatory (presumably glutamatergic) axon is not known. The glutamatergic contribution to the triad could conceivably originate from the medial thalamus (27,28), basolateral amygdala (29), other cortical regions, and possibly the VTA (30); a recent preliminary report indicates that the hippocampal projections to the PFC are not the source of the excitatory part of the triad (31). Since DA appears to have a modulatory role in the PFC (see next section), the source of the excitatory input with which DA shares a pyramidal cell target may be of crucial importance, dictating in part when DA afferents to the cortex are activated.

Physiology of the Cortical Dopaminergic Systems

Dopamine inputs to the PFC inhibit the activity of pyramidal cells in vivo (32–34). The dopaminergic inhibition of cortical output neurons appears to depend upon an atypical D_2-like receptor (33,35). In sharp contrast to a large body of data indicating that DA inhibits PFC pyramidal cells (including an intracellular in vivo study [36]), an in vitro study indicates that DA induces a very small (2–3 mV) depolarization and that high concentrations of DA increase the number of spikes evoked by a single depolarizing pulse (37). These data are most parsimoniously explained by DA increasing the input resistance of the cell, therefore suggesting a modulatory role for DA.

The proposed modulatory role for DA in the PFC offers a reasonable reconciliation between the observed DA-elicited inhibition of pyramidal cells in vivo and the different electrophysiological effects observed in vitro. Under in vitro conditions, pyramidal cells are not spontaneously active, while pyramidal cells in vivo fire vigorously. Stimulation of the mediodoral thalamic glutamatergic projections to the PFC result in an excitatory effect on PFC pyramidal cells that is reduced by DA (38). This suggests that DA exerts a modulatory role over excitatory afferents to the PFC. When considered in light of the fact that pyramidal cells are not spontaneously active in the slice preparation (but are spontaneously active in vivo), it appears likely that a dopaminergic inhibition of PFC pyramidal cells is observed only when pyramidal cells are driven by certain excitatory afferents. In other words, the electrophysiological data suggest that DA operates to modulate excitatory drive over pyramidal cells. While any of the sources of excitatory afferents to the PFC may be important in this regard, the neurons that contribute the excitatory component of the triadic arrangement with DA in the PFC may be particularly important.

In contrast to the inhibitory effect of DA on pyramidal cells, DA increases the firing rate of interneurons and increases the release of GABA from these cells in the PFC (39,40). The actions of DA appear to be mediated predominantly through a D_2 receptor, with a synergistic action upon D_1 stimulation (39). DA agonists elicit an increase in [^{3}H]GABA release (39) and extracellular GABA levels in the PFC (41); DA antagonists decrease extracellular GABA levels (42). Since GABA inhibits pyramidal cell activity, one can conceptualize DA as inhibiting pyramidal cell activity through two mechanisms: 1) directly via contacts with pyramidal cells, and 2) indirectly by enhancing GABA release from interneurons.

STRESS-INDUCED ACTIVATION OF MESOCORTICAL DOPAMINE NEURONS

It has been almost 25 years since the discovery that stress increased the release of DA in the PFC (2). During the time since publication of that paper, the response of mesoprefrontal DA neurons to stress has been extensively investigated (2,43,44). The attention devoted to this topic has in part been due to the unusual sensitivity of the mesoprefrontal cortical DA innervation to stress. Although a large number of different stressors reliably increase DA utilization and release in the PFC of the rat, many stressors also concomitantly increase activation of other cortical neurochemical systems, including noradrenergic and serotonergic innervations. The simultaneous activation of the three monoamine systems in the PFC by stress appears to reflect primarily the intensity or duration of the stressor, since the DA innervation of the PFC can be activated by mild stressors (such as being placed in the environment of stressed conspecifics [45] or exposure to neutral stimulus previously paired with mild footshock [46–48]) that do not increase the activity of the cortical noradrenergic inputs. We will briefly review recent findings concerning the stress-induced activation of mesoprefrontal cortical neurons. Deutch and Roth (2) offer a more extensive review of earlier work on this topic. In addition, the reader is referred to Chapters 4 and 5 for recent developments related to the interactions of amino acid transmitters and dopamine, and the sensitization of forebrain DA systems to stress, respectively.

The Biochemical Responsiveness of the PFC DA Systems to Stress

Activation of the dopaminergic innervation of the PFC by stressors does not appear to reflect spillover of DA and metabolites from noradrenergic neurons (49,50), nor does the cortical DA response depend upon a stress-induced increase in corticosterone (51–53). Mild stressors appear to influence rather selectively the PFC DA innervation relative to other forebrain DA systems. Thus, a very mild footshock (e.g., 200 μA) will increase DA utilization in the PFC, but not in the nucleus accumbens or striatum, while a slightly higher foot shock (230 μA) will activate the nucleus accumbens DA innervation (43,47); there is a similar graded response with duration of restraint stress (43). The DA metabolism in the VTA (site of the DA cell bodies that project to the PFC) is also increased in response to mild stressors (45–47), as is DA metabolism in certain nuclei within the amygdala (54). Interestingly, exposure to a neutral stimulus previously paired with foot shock will activate the PFC but not nucleus accumbens DA systems if the footshock parameters are suitably mild (46,47,51), but exposure to a neutral stimulus previously paired with a higher intensity (current) foot shock will result in the activation of both the DA innervations of the PFC and nucleus accumbens (55,56).

The differential responsiveness of telencephalic DA systems to stress is mirrored by a heterogeneity of the prefrontal cortical DA innervations to stress. As just noted, the medial PFC of the rat is comprised of a number of distinct cytoarchitectonic regions. The PFC efferents are topographically organized both within and across the different cytoarchitectonic regions of the PFC (3). The different projections of PFC efferents subserve different functions. For example, the infralimbic cortex has more prominent projections to the hypothalamus and brainstem autonomic regions than do more dorsal PFC regions (3,57), and is thought to be involved in the early alerting to stress (58). We reasoned that stress may differentially alter DA release from axons innervating the different PFC regions, and found that exposure to a very mild footshock stressor that did not enhance DA utilization in the entire medial PFC resulted in a significant increase in DA utilization in the infralimbic part of the PFC (59,60). The DA innervations of more dorsal PFC regions such as the prelimbic cortex are less sensitive to stress, consistent with the postulated functions of different PFC regions (3).

We also found that the greater sensitivity of

the DA innervation of the infralimbic PFC to stress is mirrored by greater sensitivity of DA neuron cell bodies in VTA areas that project to the infralimbic cortex. Thus, restraint stress increased expression of the immediate-early gene c-fos (see Chapter 1) in a distinct subset of mesoprefrontal cortical DA neurons (61) located in the caudal and medial VTA nuclei that give rise to the DA innervation of the ventral PFC.

There are also differences in the degree to which DA metabolism is activated in the two hemispheres (i.e., there is a lateralization of the prefrontal cortical DA innervation to stress [62,63]). Surprisingly, this lateralized response bias can be detected using a relatively small sample size, and appears to be related to side biases in turning behavior (63). The observation that the lateralization of the PFC DA response to stress is related to side preference in motor behavior suggests that there are individual variations in the degree of responsiveness of cortical DA systems to stress; these individual differences may be determined in part by genetic factors.

The Functional Significance of Enhanced Cortical DA Release in Stress: The Acquisition of Coping

In a study of the effects of conditioned fear on PFC DA utilization (47), we paired footshock delivered through a grid floor (unconditioned stimulus, or UCS) with a neutral auditory conditioned stimulus (CS) for either 1 day or 5 days; the prefrontal cortical DA response to the presentation of the CS was then examined on either day 2 or 6, respectively. We were puzzled to note that in animals that received 5 days of CS–UCS pairings, there was no increase in PFC DA utilization in response to the CS, and a markedly depressed response to the UCS. In contrast, 1 day of CS–USC pairings led to the expected significant increase in cortical DA utilization in response to the CS alone, with a significantly greater response to the UCS (47). One clue pointing to the functional significance of these findings was the observation that rats introduced to the operant chamber on day 4 or day 5 of pairings immediately placed themselves prostrate on the grid floor. We interpreted this unusual behavior to indicate that the animals had developed a coping strategy to the stressor. We surmised that by placing themselves across the grid floor, the rats were able to effectively reduce any high current-density shocks, which are due to an arc of current passing between the floor and a part of the body not quite apposed to the grid floor.

These data led us to suspect that the stress-elicited release of DA in the PFC was involved in the acquisition (but not expression) of a coping response to the stressor (2,64). This idea was buttressed by a series of studies by Scatton and colleagues in inbred strains of rats that exhibit markedly different responses to novel and threatening environments, the Roman Low-Avoidance (RLA) and Roman High-Avoidance (RHA) strains (65). These studies revealed that the ''hypoemotional'' RHA rats exhibited a stressor-induced increase in PFC DA metabolism, but that the ''hyperemotional'' RLA rats did not (64,66,67). Since the rats expressing the ''least emotion'' appeared to exhibit the greatest stressor-induced DA activation, it was concluded that ''the stress-induced increase in cortical DOPAC [3,4-dihydroxyphenyl acetic acid] is unlikely to be a direct reflection of the emotional status of the animals'' (66). Scatton and associates thus proposed that enhanced PFC DA function in response to stress may be reflected in heightened attention or cognitive processes to cope with the stressor (64). This fits well with our inability to detect a stress-induced increase in DA utilization in animals after chronic footshock exposure. The data from the RLA and RHA rat studies also were consistent with our suggestion that release of PFC DA was of importance in the acquisition, rather than performance, of the coping strategy. Finally, the observation that the RHA (hypoemotional) rats no longer exhibit an augmented PFC DA metabolism associated with an active avoidance task once the rats are successful in performing the task (67) is also in accord with our hypothesis.

The hypothesis that DA release in the PFC is involved in the acquisition of a coping response provides a unifying explanation for diverse findings. However, several authors have demonstrated that cortical DA response to stress can be sensitized by prior exposure to stress (see Chapter 5). Although most of these studies examined cross-sensitization of stress and drugs of abuse (68,69), and therefore may shed some light on the choice of the drugs that PTSD patients frequently abuse, there are also reports of sensitization to the stress-induced increase in extracellular PFC DA levels after chronic stress. Rats repeatedly exposed to cold stressor exhibit a significantly greater augmentation of cortical DA levels after an acute tail-shock stressor than do rats exposed to the tail shock without previous cold-stress exposure (70). There are, to the best of our knowledge, no published data indicating sensitization of the cortical DA response to stress that have not used a novel stressor for the final (challenge) day.

We recently examined the effects of daily restraint stress on the ability of an acute restraint stress to evoke an increase in DA metabolism (71). In this study we used 30 minutes of restraint, since this stressor increases both DA and serotonin release in the PFC. Consistent with our previous work, we found that those animals restrained for the first time showed a markedly increased DA and serotonin utilization in the PFC. In contrast, rats restrained every day for 6 days and then restrained on day 7 (test day) did not display an increase in DA utilization (in fact, DA metabolism was nonsignificantly depressed), but still evinced an increase in serotonin utilization. These data fit well with data from Scatton and colleagues, who noted that PFC DA metabolism in both RHA and RLA rats placed into a forced locomotion paradigm was decreased after chronic exposure to the same forced locomotion (66). These observations are consistent with the proposal that the function of prefrontal cortical DA release in response to stress is to participate in the acquisition of a coping response, and show that this condition is neurochemically specific.

Problems Associated with Attributing Internal States to Animals

We have interpreted certain behavioral and neurochemical changes as suggesting that the functional significance of the stress-elicited increase in PFC DA release is to contribute to the development of a coping strategy, the aim of which is to reduce immediate exposure to the stressor or dampen adverse effects upon subsequent exposure to the stressor. The difficulties and appropriateness of applying explanatory emotional concepts to nonhuman animals have long represented problems for neuroscientists in general and psychologists in particular. However, it is now widely accepted that the ability to experience internal states such as anxiety is not likely unique to humans, and thus the inability of nonhuman animals to describe their internal sensations or emotional states should not preclude the formulation of explanatory concepts linking stressors with observable biochemical, physiological, and behavioral responses. The terms "anxiety" and "coping" are a conceptual convenience to provide an explanatory link between observable experimental manipulations (e.g., the administration of a stressor) and the observable consequences or correlates of those manipulations.

The distinction between the definitions of anxiety, fear, and stress is vague. Stress (or a stress response) has been defined, for example, as "the nonspecific response of the body to any demand" (72), the biochemical/physiological responses to a stressor (73), and the "right side" of an inverted U-shaped arousal curve where the left side represents enhanced performance and the right side represents deteriorating performance (74). "Stress response" is also used by many in an interchangeable manner for fear or anxiety.

Fear and anxiety are sometimes distinguished by the presence or absence of an environmental threat. Fear is typically defined as an emotional reaction to a concrete immediate threat, in contrast to the definition of anxiety as an emotional reaction when the threat is not known or not physically present (i.e., the reaction is disproportionate to the actual threat [75,76]). Thus, although animals' behavioral manifestations of fear and anxiety are similar, the two states may be differentiated by the presence or absence of an appropriate direct stimulus (77). This leads one to the distinction that the neural substrates of fear are unconditionally activated, whereas the neural substrates of anxiety are conditionally activated. This simply means that fear and anxiety are states that result from different activating stimuli (78), but does not indicate that different neural systems are necessarily involved in fear and anxiety.

Although it is generally accepted that animals experience fear, there is room to question whether animals experience anxiety, since the experience of anxiety, even in humans, is often inferred rather than observed (79). In humans, anxiety has been defined as an autonomically-derived state of unpleasantness that consists of heightened vigilance coupled with apprehension of impending and inevitable danger over which one will be powerless (80). Empirical support for the view that nonhuman animals experience anxiety is provided by the similar physiological and behavioral states (e.g., motor tension, autonomic hyperactivity, apprehensive expectation, vigilance, scanning) observed in both nonhuman and human animals in circumstances assumed to be anxiety inducing (79). Moreover, anxiolytic drugs such as diazepam inhibit normal stressor-induced behavior in rats and other nonhuman animals (81). While these arguments are somewhat circular, the converging data are consistent with the observations that the physiological sequelae of exposure to a fear-provoking state in man (such as electrical shock) and to conditioned stimuli for the actual event (such as a neutral tone) are comparable in human and nonhuman animals. To those interested in the question of operationally defining anxiety in nonhuman species, it appears that the ability to distinguish man from nonhuman animals may be most suitably made on the basis of assuming that only humans can experience anxiety.

A NEUROBIOLOGICAL MODEL OF THE STRESS-INDUCED ACTIVATION OF MESOPREFRONTAL CORTICAL DOPAMINE NEURONS

We have previously emphasized that the release of DA in the PFC in response to a stressor represents an increase in the firing rate of mesoprefrontal cortical DA neurons (2). This argument was based in part on two different types of evidence: 1) several different pharmacological manipulations of the DA cell bodies in the ventral midbrain attenuate the stress-elicited increase in DA metabolism and release in the PFC (see 2,16) and 2) stress-induced metabolic activation of A10 DA neurons in the VTA is not observed when one transplants VTA neurons to a forebrain site (82). The findings again suggest that the effects of stress are due to afferents impinging on A10 DA neurons that increase the firing rate of these cells, rather than operating through a presynaptic impulse-independent regulation of DA release in the PFC.

A recent study reported that local infusion of an AMPA family excitatory amino acid receptor antagonist into the PFC blocked the stress-elicited increase in cortical extracellular DA levels (83). These data were interpreted to indicate that the stress-elicited increase in cortical DA release was mediated locally through presynaptic influences over DA axons. An alternative explanation for these data, however, focuses on the fact that there are both glutamatergic afferents to the PFC and glutamatergic efferents from the PFC. Thus, blocking an excitatory glutamatergic input onto pyramidal cells prevents the activation of a long-loop excitatory influence of the PFC over mesoprefrontal cortical DA somata in the VTA. Accordingly, if the stress-elicited increase in prefrontal cortical DA release functions in the acquisition of a coping response to stress (2,64), then the enhanced DA tone in the PFC can be assumed to dampen some stress-induced event.

We propose that glutamatergic afferents to the PFC exert an excitatory influence over pyramidal neurons. Some of these pyramidal cells project to the VTA (84), where they drive mesoprefrontal cortical DA neurons. In turn, the release of DA in the PFC dampens the excitatory input onto the pyramidal cell. This scheme takes into account several findings: 1) stress increases glutamate release in the PFC (85,86); 2) both DA axons and nondopaminergic excitatory inputs synapse with PFC pyramidal neurons (the triadic arrangement [21] discussed earlier); and 3) electrophysiological data indicate that DA modulates the excitatory drive exerted upon pyramidal cells by excitatory amino acid inputs (38). Finally, the model that we propose is consistent with a large body of data indicating that the stress-induced increase in DA tone in the PFC depends upon the presence of intact afferents to the somatodendritic areas of the midbrain DA neurons (2,16); among these afferents may be the glutamate-containing projections originating in the PFC.

This model is testable, since it hinges on identification of the excitatory input to the pyramidal cell that is located in proximity to the DA terminal. Thus, one should be able to lesion the relevant excitatory input(s) to the PFC and thus block the stress-elicited activation of DA afferents to the PFC. Although there are several possible sites of origin for these afferents (including certain medial thalamic nuclei, the basolateral amygdala, CA1 neurons of the hippocampus, other cortical sites, the basal forebrain, or even nondopaminergic neurons in the VTA), perhaps the most likely source for this critical excitatory input is the area that contributes to the triadic arrangement with the pyramidal cell and a DA input. A recent report suggests that the hippocampus is not the source of the excitatory input that contributes to the triadic arrangement (31).

Among the most likely candidates for the critical excitatory input to the PFC is the basolateral amygdala. In the rat there are extensive projections from the basolateral nucleus onto the medial PFC (29,87,88); projections to different parts of the PFC are derived from the anterior and posterior basolateral nucleus, with the more posterior division innervating the infralimbic and ventral prelimbic PFC (88). Both direct and indirect evidence are consistent with the speculation that the activation of glutamatergic amygdala inputs to the PFC are the proximate cause of a compensatory increase in cortical DA release. Direct evidence is provided by the recent observation that excitotoxic lesions of the basolateral amygdala prevent the stress-induced increase in PFC DA metabolism and release (55,56). Less direct but consistent with the model is the key role that the amygdala plays in determining the emotional salience of stressors (see Chapters 7 and 11). Finally, lesions of the PFC do not alter the acquisition of a conditioned emotional response (CER) but delay extinction of the CER (89; see Chapter 7).

We propose that the amygdaloid inputs to the PFC may be a critical source of afferent drive to the PFC pyramidal cell that is activated by stress. In turn, excitatory glutamatergic efferents from the PFC to the A10 DA neurons increase the firing rate of mesoprefrontal cortical neurons, which finally dampens the cortical output. Since neurons in the PFC project to the basolateral amygdala as well as VTA (3), it is also possible that the PFC projection neurons contacted by DA afferents directly dampen a positive feedback loop originating in the amygdala.

We have emphasized the amygdala in this loop because of the obvious relevance of the amygdala to fear and stress (see Chapters 7 and 11). However, it is possible, if not likely, that there are multiple sources of excitatory drive to the PFC that are important. Particularly interesting in this regard would be projections to the PFC from the thalamic paraventricular nucleus and the mediodorsal thalamic nucleus (90,91).

Most of the data upon which we have based our model of the stress-elicited activation of the PFC DA innervation was obtained from studies of rats. However, there are key differences between rodent and primate species. Among the anatomical differences is the fact that the amygdala of primates projects to orbital prefrontal and anterior cingulate cortices, but does not innervate the dorsolateral prefrontal cortex (12). This observation suggests that our model (at least with regard to the amygdala component) is consistent with the orbital cortex being the critical cortical site in primate species. The medial PFC of the rat is comprised of several distinct cytoarchitectonic regions that are thought to correspond to spatially-segregated prefrontal cortical regions in primates. As just noted, the ventral PFC (infralimbic cortex) of the rat is the region of the PFC in which the DA response to stress is most vigorous (59) and is the area of the PFC that has direct autonomic projections (57). This ventral portion of the rodent PFC is thought to be homologous to the orbital cortex in primates. Thus, our model is applicable to primates and rodents, and emphasizes in particular those extended circuits formed with the orbital PFC.

RELEVANCE OF THE PREFRONTAL CORTICAL DOPAMINE SYSTEMS TO PTSD

Systematic study of the neurobiology of PTSD is only beginning. We hope that we have provided a useful organizational framework for explaining the activation of the prefrontal cortical DA system that may be of value to scientists studying stress. We have also tried to pose several questions that scientists embarking on clinical studies may find provocative.

We have presented a model to explain the mechanisms that govern stress-induced activation of the prefrontal cortical DA system. It may be useful to conceptualize, at least as a starting point for new investigations, the development of PTSD in terms of the neurobiology and functional significance of the stress-elicited increase in PFC DA release.

We proposed that the functional significance of stress-elicited DA release in the PFC lies in the acquisition of coping strategies. This would suggest that some functional defect in cortical DA systems in PTSD patients would lead to an inability to develop an effective coping strategy for dealing with trauma; this failure would subsequently render the person more susceptible to contextual stimuli associated with the stressor. In the face of repeated exposure to stimuli associated with the original stressor, the inability to develop a coping strategy would result in stimulus generalization, and thus a continued susceptibility to flashbacks.

The hypothesis that there is a functional defect in orbital cortical DA function in PTSD may explain several characteristics of PTSD patients. These patients are hypervigilant and frequently paranoid. Such behaviors, however, suggest a hyperdopaminergic condition in PTSD patients, running counter to the defect in functional cortical DA tone that we propose. Recent theories of the pathogenesis of schizophrenia have had to deal with a similar paradox, and have led to the hypothesis that cortical dopaminergic hypofunction results in a transsynaptic increase in DA tone in subcortical sites and the generation of positive symptoms (92–94). Studies in animals have documented that depletions of prefrontal cortical DA increase subcortical DA tone under challenge conditions (94–96). It is important to note that lesions of the PFC DA innervation do not increase subcortical DA tone under basal conditions (94), but only under challenge conditions, such as stress. In other words, even in the presence of a greater than 80% decrease in cortical DA levels there remains sufficient dopaminergic tone to maintain adequate inhibition of cortical outflow neurons under basal conditions; the defect becomes manifested under conditions of demand, such as stress or associated stimuli. This fits well with current conceptions of PTSD. We hasten to add that we are advancing the possibility of a functional DA defect in the orbital cortex, which may in turn enhance subcortical DA function. However, we do not suggest that PTSD and schizophrenia are related. Although there may be orbital cortical dysfunction in schizophrenia (and thus the emergence of stress-induced exacerbation of the psychotic process), it appears likely that there is a functional alteration in the dopaminergic innervations of other cortical regions, including the dorsolateral PFC and perhaps medial temporal lobe structures.

A functional deficit in cortical DA tone would also explain the apparently poor response of PTSD patients to neuroleptic drugs (see Chapter 26), which are DA receptor antagonists. However, controlled studies of this issue are lacking. Interestingly, clozapine differs from typical antipsychotic drugs by virtue of its actions on the PFC (97–101), including the observation that it sharply increases extracellular DA levels (98,99). It would therefore be interesting to know if clozapine is useful in the treatment of PTSD.

Although one can develop arguments that are consistent with a hypothesis of a defect in cortical DA tone in PTSD, what data can be marshaled in direct support of the hypothesis? Unfortunately, there are no direct data. A few studies have examined peripheral indices of DA function in PTSD, including reports documenting a change in 24-hour urinary homovanillic acid (HVA) and DA levels (102,103). However, peripheral measures of DA function are notoriously poor indices of changes in central DA systems, and as such do not provide any useful test of our hypothesis. Clearly, studies aimed at determining functional alterations in central DA systems would be most useful. In the absence of methods to measure directly cortical DA function in man, useful alternative strategies might include examination of CSF HVA levels in PTSD patients, since studies of nonhuman primates indicate that CSF HVA levels correlate with DA concentration in the dorsolateral PFC but not in any other region (104). At the current time, PET and SPECT in vivo imaging methods do not have the requisite sensitivity to detect either fluoro-DOPA or DA transporter signals from the cortex.

However, in vivo functional studies of cortical regions (including the orbital cortex) are possible with fluoro-deoxyglucose or using methods that detect changes in blood flow. Again, we are aware of no such studies in PTSD. Interestingly, studies of co-morbid psychiatric conditions with PTSD reveals significant association with obsessive-compulsive disorders (OCD) (see Chapter 24); imaging studies in OCD have revealed changes in metabolic activity in the orbital cortex and allied striatal regions (105,106).

We have put forth a hypothesis of the pathophysiology of PTSD that focuses on the prefrontal cortex, and specifically directs attention to the orbital cortex. We have focused on dopamine systems because of the striking effects of stress on central DA systems. We wish to note, however, that other monoamines in the cortex also exert inhibitory effects over cortical projection neurons, either directly or indirectly (40). Cortical norepinephrine release is increased by stress (see chapter 3), as is serotonin release (107–109). Although these monoaminergic transmitters in the PFC are not as sensitive to the effects of stress as the mesoprefrontal cortical DA system, they may function cooperatively in the scheme that we have outlined to influence cortical projections, including those to brainstem monoaminergic cell group regions and the amygdala. Indeed, in view of the myriad changes in transmitter systems that have been observed in PTSD, it seems likely that the central pathology involves more than one transmitter system.

ACKNOWLEDGMENTS

We are indebted to Drs. Matt Friedman and Dennis Charney for helpful discussions, sage advice, and patience. This work was supported in part by the National Center for PTSD, Veteran Affairs Medical Center, West Haven, Connecticut, and by MH-45124.

REFERENCES

1. Thierry AM, Tassin JP, Blanc G, Glowinski J. Selective activation of the mesocortical dopamine system by stress. *Nature* 1976;263:242–244.
2. Deutch AY, Roth RH. The determinants of stress-induced activation of the prefrontal cortical dopamine system. *Prog Brain Res* 1990;85:357–393.
3. Sesack SR, Deutch AY, Roth RH, Bunney BS. The topographical organization of the efferent projections of the medial prefrontal cortex in the rat: an anterograde tract-tracing study using *Phaseolus vulgaris* leucoagglutinin. *J Comp Neurol* 1989;290:213–242.
4. Bacon SJ, Smith AD. A monosynaptic pathway from an identified vasomotor centre in the medial prefrontal cortex to an autonomic area in the thoracic spinal cord. *Neurosci* 1993;54:719–728.
5. Thierry AM, Blanc G, Sorbel A, Stinus L, Glowinski J. Dopaminergic terminals in the rat cortex. *Science* 1973;182:499–501.
6. Hokfelt T, Martensson R, Bjorklund A, Kleinau S, Goldstein M. Distributional maps of tyrosine hydroxylase-immunoreactive neurons in the rat brain. In Bjorklund A, Hokfelt T, eds. *Handbook of chemical neuroanatomy: vol 2: Classical transmitters in the CNS, Part 1.* Amsterdam: Elsevier; 1984.
7. Lindvall O, Bjorklund A. General organization of cortical monoamine systems. In Descarries L, Reader TR, Jasper HH, eds. *Monoamine innervation of cerebral cortex,* New York: Alan R. Liss; 1987.
8. Lewis DA, Campbell MJ, Foote SL, Goldstein M, Morrison JH. The distribution of tyrosine hydroxylase-immunoreactive fibers in primate neocortex is widespread but regionally specific. *J Neurosci* 1987;7: 279–290.
9. Berger B, Gaspar P, Verney C. Dopaminergic innervation of the cerebral cortex: unexpected differences between rodents and primates. *Trends Neurosci* 1991;14: 21–27.
10. Fuxe K, Hokfelt T, Johansson O, Jonsson G, Lidbrink P, Ljungdahl A. The origin of the dopamine nerve terminals in limbic and frontal cortex. Evidence for mesocortico DA neurons. *Brain Res* 1991;82:349–355.
11. Emson PC, Koob GF. The origin and distribution of dopamine-containing afferents to the rat frontal cortex. *Brain Res* 1978;142:249–267.

12. Porrino LJ, Crane AM, Goldman-Rakic PS. Direct and indirect pathways from the amygdala to the frontal lobe in rhesus monkeys. *J Comp Neurol* 1981;198:121–136.
13. Gaspar P, Stepniewska I, Kaas JH. Topography and collateralization of dopaminergic projections to motor and lateral prefrontal cortex in owl monkeys. *J Comp Neurol* 1993;325:1–21.
14. Deutch AY, Bean AJ. Colocalization in dopamine neurons. In Bloom FE, Kupfer DJ, eds. *Psychopharmacology: the fourth generation of progress.* New York: Raven Press; 1994.
15. Phillipson OT. Afferent projections to the ventral tegmental area of Tsai and interfascicular nucleus: a horseradish peroxidase study in the rat. *J Comp Neurol* 1978;187:85.
16. Kalivas PW. Neurotransmitter regulation of dopamine neurons in the ventral tegmental area. *Brain Res Rev* 1993;18:75–113.
17. Roth RH, Wolf ME, Deutch AY. The neurochemistry of midbrain dopamine systems. In Meltzer HY, ed. *Psychopharmacology: the third generation of progress.* New York: Raven Press; 1987.
18. Beaudet A, Descarries L. Fine structure of monoamine axon terminals in cerebral cortex. In Descarries L, Reader TR, Jasper HH, eds. *Monoamine innervation of cerebral cortex.* New York: Alan R. Liss, 1984.
19. van Eden CG, Hoorneman EMD, Buijs RM, Matthijssen MAH, Geffard M, Uylings HBM. Immunocytochemical localization of dopamine in the prefrontal cortex of the rat at the light and electron microscopic level. *Neurosci* 1987;22:849–862.
20. Seguela P, Watkins KC, Descarries L. Ultrastructural features of dopamine axon terminals in the anteromedial and the suprarhinal cortex of adult rat. *Brain Res* 1988;442:11–22.
21. Goldman-Rakic PS, Leranth C, Williams SM, Mons N, Geffard M. Dopamine synaptic complex with pyramidal neurons in primate cerebral cortex. *Proc Natl Acad Sci USA* 1989;86:9015–9019.
22. Smiley JF, Williams SM, Szigeti K, Goldman-Rakic PS. Light and electron microscopic characterization of dopamine-immunoreactive axons in human cerebral cortex. *J Comp Neurol* 1992;321:325–335.
23. Smiley JF, Levey AI, Ciliax BJ, Goldman-Rakic PS. D_1 dopamine receptor immunoreactivity in human and monkey cerebral cortex: predominant and extrasynaptic localization in dendritic spines. *Proc Natl Acad Sci USA* 1994;91:5720–5724.
24. Verney C, Alvarez C, Geffard M, Berger B. Ultrastructural double-labelling study of dopamine terminals and GABA-containing neurons in rat anteromedial cerebral cortex. *Eur J Neurosci* 1990;2:960.
25. Smiley JF, Goldman-Rakic PS. Heterogeneous targets of dopamine synapses in monkey prefrontal cortex demonstrated by serial section electron microscopy: a laminar analysis using the silver-enhanced diaminobenzidine sulfide (SIDS) immunolabeling technique. *Cerebral Cortex* 1993;3:223–238.
26. Sesack SR, Snyder CL, Lewis DA. Axons terminals immunolabeled for dopamine or tyrosine hydroxylase synapse on GABA-immunoreactive dendrites in rat and monkey cortex. *J Comp Neurol* 1993; submitted.
27. Gigg J, Tan A, Finch DM. Glutamatergic excitatory responses of anterior cingulate neurons to stimulation of the mediodorsal thalamus and their regulation by GABA: an in vivo iontophoretic study. *Cerebral Cortex* 1992;2:477–484.
28. Pirot S, Jay TM, Glowinski J, Thierry, A-M. Anatomical and electrophysiological evidence for an excitatory amino acid pathway from the thalamic mediodorsal nucleus to the prefrontal cortex in the rat. *Eur J Neurosci* 1994;6:1225–1234.
29. Krettek JE, Price JL. Projections from the amygdaloid complex to the cerebral cortex and thalamus in the rat and cat. *J Comp Neurol* 1979;172:687–722.
30. Kaneko T, Akiyama H. Nagatsu I, Mizuno N. Immunohistochemical demonstration of glutaminase in catecholaminergic and serotonergic neurons of rat brain. *Brain Res* 1990;507:151–154.
31. Carr DB, Wolf HC, Sesack DR. Hippocampal afferents to the rat prefrontal cortex: synaptic targets and relation to dopamine terminals. *Soc Neurosci Abstr* 1994; 20:225.26.
32. Peterson OT, St Mary JS, Harding NR. Cis-flupenthixol antagonism of the rat prefrontal cortex neuronal response to apomorphine and ventral tegmental area input. *Brain Res Bull* 1987;18:723–729.
33. Sesack SR, Bunney BS. Pharmacological characterization of the receptor mediating electrophysiological responses to dopamine in the rat medial prefrontal cortex: a microiontophoretic study. *J Pharmacol Exp Ther* 1989;248:1323–1333.
34. Thierry A-M, Godbut R, Mantz J, Glowinski J. Influence of the ascending monoaminergic systems on the activity of the rat prefrontal cortex. *Prog Brain Res* 1990;85:366–377.
35. Thierry AM, Le Douarin C, Penit J, Ferron A, Glowinski J. Variation in the ability of neuroleptics to block the inhibitory influence of dopaminergic neurons on the activity of cells in the rat prefrontal cortex. *Brain Res Bull* 1986;16:155–160.
36. Bernardi G, Cherubini E, Marciani MG, Mercuri N, Stanzione P. Responses of intracellularly recorded cortical neurons to the iontophoretic application of dopamine. *Brain Res* 1982;245:267–274.
37. Penit-Soria J, Audinat E, Crepel F. Excitation of rat prefrontal cortical neurons by dopamine: an in vitro eletrophysiological study. *Brain Res* 1987;425: 363–374.
38. Ferron A, Thierry AM, Le Dourin C, Glowinski J. Inhibitory influence of the mesocortical dopaminergic system on spontaneous activity or excitatory response induced from the thalamic mediodorsal nucleus in the rat medial prefrontal cortex. *Brain Res* 1984;302:257.
39. Retaux S, Besson MJ, Penit-Soria J. Synergism between D_1 and D_2 dopamine receptors in the inhibition of the evoked release of [^{3}H] GABA in the rat prefrontal cortex. *Neurosci* 1991;43:323–329.
40. Gellman RL, Aghajanian GK. Pyramidal cells in piriform cortex receive a convergence of inputs from monoamine activated GABAergic interneurons. *Brain Res* 1992;600:63–73.
41. Grobin AC, Deutch AY. Apomorphine increases extracellular GABA levels in the prefrontal cortex of the freely moving, conscious rat. *Soc Neurosci Abstr* 1994;20:1154.
42. Bourdelais AJ, Deutch AY. The effects of haloperidol and clozapine on extracellular GABA levels in the prefrontal cortex of the rat: an in vivo microdialysis study. *Cerebral Cortex* 1994;4:69–77.

43. Roth RH, Tam S-Y, Ida Y, Yang J-X, Deutch AY. Stress and the mesocorticolimbic dopamine systems. *Ann NY Acad Sci* 1988;537:138–147.
44. Dunn AJ. Stress-related activation of central dopaminergic systems. *Ann NY Acad Sci* 1988;537:188–205.
45. Kaneyuki H, Yokoo H, Tsuda A, Yoshida M, Mizuki Y, Yamada M, Tanaka M. Psychological stress increases DA turnover selectively in mesoprefrontal dopamine neurons of rats: reversal by diazepam. *Brain Res* 1991;557:154–161.
46. Herman JP, Guillonneau D, Dantzer R, Scatton B, Semerdjian-Rouquier L, Le Moal M. Differential effects of inescapable footshocks and of stimuli previously paired with inescapable footshocks on dopamine turnover in cortical and limbic areas of the rat. *Life Sci* 1982;30:2207–2214.
47. Deutch AY, Tam S-Y, Roth RH. Footshock and conditioned stress increase 3,4-dihydroxyphenylacetic acid (DOPAC) in the ventral tegmental area but not substantia nigra. *Brain Res* 1985;333:143–146.
48. Ida Y, Suda A, Sueyoshi K, Chair I, Tunica M. Blockade by diazepam of conditioned fear-induced activation of rat mesoprefrontal dopamine neurons. *Pharm Biochem Behav* 1989;33:477–479.
49. Iuvone PM, Dunn AJ. Tyrosine hydroxylase activation in mesocortical 3,4-dihydroxyphenylethylamine neurons following footshock. *J Neurochem* 1986;47: 837–844.
50. Scatton B, Dennis T, Curet O. Increase in dopamine and DOPAC levels in noradrenergic terminals after electrical stimulation of the ascending noradrenergic pathways. *Brain Res* 1984;298:193–196.
51. Claustre Y, Rivy JP, Dennis T, Scatton B. Pharmacological studies on stress-induced increase in frontal cortical dopamine metabolism in the rat. *J Pharmacol Exp Ther* 1986;238:693–700.
52. Dunn AJ. Stress-related changes in cerebral catecholamine and indolamine metabolism: lack of effect of adrenalectomy and corticosterone. *J Neurochem* 1988; 51:406–412.
53. Imperato A, Puglisi-Allegra S, Casolini P, Angelluci L. Changes in brain dopamine and acetylcholine release during and following stress are independent of the pituitary-adrenal axis. *Brain Res* 1991;538:111–117.
54. Coco ML, Kuhn CM, Ely TD, Kilts CD. Selective activation of mesoamygdaloid dopamine neurons by conditioned stress: attenuation by diazepam. *Brain Res* 1992;590:39–47.
55. Goldstein LE, Rasmusson AM, Bunney BS, Roth RH. The NMDA glycine site antagonist (+)-HA-966 selectively regulates conditioned stress-induced metabolic activation of mesoprefrontal cortical dopamine but not serotonin systems: a behavioral, neuroendocrine, and neurochemical study in the rat. *J Neurosci* 1994; 18:4937–4959.
56. Goldstein LE, Rasmusson AM, Bunney BS, Roth RH. Role of the amygdala in the integration and coordination of behavioral, neuroendocrine and neurochemical responses to psychological stress in the rat. *J Neurosci,* 1995; submitted.
57. Hurley KM, Herbert H, Moga MM, Saper CB. Efferent projections of the infralimbic cortex of the rat. *J Comp Neurol* 1991;308:249–276.
58. Frysztak RJ, Neafsy EJ. The effect of medial frontal cortex lesions on respiration, "freezing," and ultrasonic vocalizations during conditioned emotional responses in rats. *Cerebral Cortex* 1991;1:418–425.
59. Deutch AY. Heterogeneity of the prefrontal cortical dopamine system in responsiveness to stress. *Soc Neurosci Abstr* 1991;17:529.
60. Deutch AY, Zahm DS, Bourdelais AJ. The nucleus accumbens core and shell: delineation of corticostriatal circuits and their functional attributes. In Kalivas PW, Barnes CD, eds. *Limbic motor circuits and neuropsychiatry,* Boca Raton, FL: CRC Press; 1993.
61. Deutch AY, Lee MC, Gillham MH, Cameron D, Goldstein M, Iadorola MJ. Stress selectively increases Fos protein in dopamine neurons innervating the prefrontal cortex. *Cerebral Cortex* 1991;1:273–292.
62. Carlson JN, Fitzgerald LW, Keller RW Jr, Glick SD. Side and region dependent changes in dopamine activation with various durations of restraint stress. *Brain Res* 1991;550:313–318.
63. Carlson JN, Fitzgerald LW, Keller RW Jr, Glick SD. Lateralized changes in prefrontal cortical dopamine activity induced by controllable and uncontrollable stress in the rat. *Brain Res* 1993;630:178–187.
64. Bertolucchi-D'Angio M, Serrano A, Scatton. Involvement of mesocorticolimbic dopaminergic systems in emotional states. *Prog Brain Res* 1990;85:405–416.
65. Driscoll P. Genetic models in brain and behavior research, part I. *Experientia* 1988;44:463–464.
66. D'Angio M, Serrano A, Driscoll P, Scatton B. Stressful environmental stimuli increase extracellular DOPAC levels in the prefrontal cortex of hypoemotional (Roman high-avoidance) but not hyperemotional (Roman low-avoidance) rats. An in vivo voltammetric study. *Br Res* 1988;451:237–247.
67. Driscoll P, Claustre Y, Fage D, Scatton B. Recent findings in central dopaminergic and cholinergic neurotransmission of roman high-and low-avoidance (RHA/VERH and RLA/VERH) rats. *Behav Br Res* 1987;26:213.
68. Sorg BA, Kalivas PW. Effects of cocaine and footshock stress on extracellular dopamine levels in the medial prefrontal cortex. *Neurosci* 1993;53:695–703.
69. Hamamura T, Fibiger HC. Enhanced stress-induced dopamine release in the prefrontal cortex of amphetamine-sensitized rats. *Eur J Pharmacol* 1993;237: 65–71.
70. Gresch PJ, Sved AF, Zigmond MJ, Finlay JM. Stress-induced sensitization of dopamine and norepinephrine efflux in medial prefrontal cortex of the rat. *J Neurosci* 1994;63:575–583.
71. Bourdelais AJ, Grobin AC, Deutch AY. Effects of acute and chronic stress on prefrontal cortical dopamine and serotonin systems. *Soc Neurosci Abstr* 1993;19:487.
72. Selye H. *The stress of life.* New York: McGraw-Hill; 1976.
73. Vogel WH. Coping, stress, stressors and health consequences. *Neuropsychobiology* 1985;13:129–135.
74. Koob GF. Arousal, stress, and inverted U-shaped curves: implications for cognitive function. In Lister RG, Weingartner HL, eds. *Perspectives on cognitive neuroscience.* New York: Oxford University Press; 1991.
75. Taylor DP, Riblet LA, Stanton HC, Eison AS, Eison MS, Temple DL. Dopamine and antianxiety activity. *Pharm Biochem Behav* 1982;17[Suppl 1]:25–35.

76. Clare AW. Anxiolytics in society. In Iversen SD, ed. *Psychopharmacology: recent advances and future prospects.* New York: Oxford University Press; 1985; 65–74.
77. Davis M. The role of the amygdala in fear and anxiety. *Ann Rev Neurosci* 1992;15:353–375.
78. Panskepp J. The psychoneurology of fear: evolutionary perspectives and the role of animal models in understanding human anxiety. In Burrows GD, Roth SM, Noyes R, eds. *The neurobiology of anxiety,* New York: Elsevier; 1990.
79. Rowan AN. Animal anxiety and animal suffering. *Appl Anim Behav Sci* 1988;20:135–142.
80. Lal H, Emmet-Oglesby MW. Behavioral analogues of anxiety. *Neuropharmacology* 1983;22:1423–1441.
81. Gray JA. Behavioural and neural-system analyses of the actions of anxiolytic drugs. *Pharm Bioch Beh* 1988;29:767–769.
82. Herman JP, Rivet JM, Abrous N, Le Moal M. Intracerebral dopaminergic transplants are not activated by electrical footshock stress activating in situ mesocorticolimbic neurons. *Neurosci Let* 1988;90:83–88.
83. Jedema HP, Moghaddam B. Glutamatergic control of dopamine release during stress in the rat prefrontal cortex. *J Neurochem* 1994;63:785–788.
84. Sesack SR, Pickel VM. Prefrontal cortical efferents in the rat synapse on unlabeled neuronal targets of catecholamine terminals in the nucleus accumbens septi and on dopamine neurons in the ventral tegmental area. *J Comp Neurol* 1992;320:145–160.
85. Gilad GM, Gilad VH, Wyatt RJ, Tizabi Y. Region selective stress-induced increase of glutamate uptake and release in rat forebrain. *Brain Res* 1990b; 525:335–338.
86. Moghaddam B. Stress preferentially increases extraneuronal levels of excitatory amino acids in the prefrontal cortex: comparison to hippocampus and basal ganglia. *J Neurochem* 1993;60:1650–1657.
87. McDonald AJ. Organization of amygdaloid projections to the prefrontal cortex and associated striatum in the rat. *Neurosci* 1991;44:1–14.
88. Kita H, Kitai ST. Amygdaloid projections to the frontal cortex and the striatum in the rat. *J Comp Neurol* 1990;298:40–49.
89. Morgan MA, Romanski LM, LeDoux JE. Extinction of emotional learning: contribution of medial prefrontal cortex. *Neurosci Let* 1993;163:109–113.
90. Berendse HW, Groenewegen HJ. Restricted cortical termination fields of the midline and intralaminar nuclei in the rat. *Neurosci* 1991;42:73–102.
91. Groenewegen HJ. Organization of the afferent connections of the mediodorsal thalamic nucleus in the rat, related to the mediodorsal-prefrontal topography. *Neurosci* 1988;24:379–431.
92. Weinberger DR. Implications of normal brain development for the pathogenesis of schizophrenia. *Arch Gen Psychiat* 1987;44:660–669.
93. Davis K, Kahn RS, Ko G, Davidson M. Dopamine in schizophrenia: a review and reconceptualization. *Amer J Psychiat* 1992;148:1474–1486.
94. Deutch AY. The regulation of subcortical dopamine systems by the prefrontal cortex: interactions of central dopamine systems and the pathogenesis of schizophrenia. *J Neur Trans* [Suppl] 1992;36:61–89.
95. Deutch AY, Clark WA, Roth RH. Prefrontal cortical dopamine depletion enhances the responsiveness of mesolimbic dopamine neurons to stress. *Brain Res* 1990;521:311–315.
96. Rosin DL, Clark WA, Goldstein M, Roth RH, Deutch AY. Effects of 6-hydroxydopamine lesions of the prefrontal cortex on tyrosine hydroxylase activity in subcortical dopamine systems of the rat. *Neurosci* 1992; 48:831–839.
97. Deutch AY, Duman RS. The effects of antipsychotic drugs on prefrontal cortical Fos expression: cellular localization and pharmacological characterization. *Neurosci,* in Press.
98. Moghaddam B, Bunney BS. Acute effect of typical and atypical antipsychotic drugs on the release of dopamine from the prefrontal cortex, nucleus accumbens, and striatum of the rat: an in vivo microdialysis study. *J Neurochem* 1990;54:1755–1760.
99. Yamamoto BK, Pehek EA, Meltzer HY. Brain regional effects of clozapine on amino acid and monoamine transmission. *J Clin Psychiat* 1994;55 [Suppl B]:9–14.
100. Daly DA, Moghaddam B. Actions of clozapine and haloperidol on the extracellular levels of excitatory amino acids in the prefrontal cortex and striatum of conscious rats. *Neurosci Let* 1993;152:61–64.
101. Yamamoto BK, Cooperman MA. Differential effects of chronic antipsychotic drug treatment on extracellular glutamate and dopamine concentrations. *J Neurosci* 1994;14:4159–4166.
102. Yehuda R, Southwick S, Giller EL, Ma X, Mason JW. Urinary catecholamine excretion and severity of PTSD symptoms in Vietnam combat veterans. *J Nerv Ment Dis* 1992;180:321–325.
103. Yehuda R, Giller EL, Southwick SM, Kahana B, Boisoneau D, Ma X, Mason JW. Relationship between catecholamine excretion and PTSD symptoms in Vietnam combat veterans and holocaust survivors. In Murburg MM, ed. *Catecholamine function in post-traumatic stress disorder: emerging concepts.* Washington, DC: American Psychiatric Press; 1994.
104. Elsworth JD, Leahy DJ, Roth RH, Redmond DE Jr. Homovanillic acid concentrations in brain, CSF, and plasma as indicators of central dopamine function in primates. *J Neur Trans* 1987;68:51–62.
105. Insel TR, Winslow JT. Neurobiology of obsessive compulsive disorder. *Psychiat Clin North America* 1992; 15:813–824.
106. Baxter LR Jr, Schwartz JM, Bergman KS, Szuba MP, Guze BH, Mazziotta JC, Alazraki A, Selin CE, Ferng HK, Munford P, et al. Glucose metabolic rate changes with both drug and behavior therapy for obsessive-compulsive disorder. *Arch Gen Psychiat* 1992;49:681–691.
107. Dunn AJ. Changes in plasma and brain tryptophan and brain serotonin and 5-hydroxyindoleacetic acid after footshock stress. *Life Sci* 1989b;42:1847–1853.
108. Kawahara H, Yoshida M, Yokoo H, Nishi M, Tanaka M. Psychological stress increases serotonin release in the rat amygdala and prefrontal cortex assessed by in vivo microdialysis. *Neurosci Let* 1993;162:81–84.
109. Inoue T, Koyama T, Yamashita I. Effect of conditioned fear stress on serotonin metabolism in the rat brain. *Pharmacol Biochem Behav* 1993;44:371–374.

Neurobiological and Clinical Consequences of Stress: From Normal Adaptation to PTSD, edited by M. J. Friedman, D. S. Charney, and A. Y. Deutch, Lippincott-Raven Publishers, Philadelphia © 1995.

11

Behavioral and Physiological Analysis of Fear Inhibition

Extinction and Conditioned Inhibition

William A. Falls and *Michael Davis

*Department of Psychology, Northern Illinois University, DeKalb, Illinois 60115; and *Department of Psychiatry, Yale University School of Medicine, New Haven, Connecticut 06508*

"I can't get the memories out of my mind! The images come flooding back in vivid detail, triggered by the most inconsequential things, like a door slamming or the smell of stir-fried pork. Last night, I went to bed, was having a good sleep for a change. Then in the early morning a storm front passed through and there was a bolt of crackling thunder. I awoke instantly, frozen in fear. I am right back in Viet Nam, in the middle of the monsoon season at my guard post. I am sure I'll get hit in the next volley and convinced I will die. My hands are freezing, yet sweat pours from my entire body. I feel each hair on the back of my neck standing on end. I can't catch my breath and my heart is pounding. I smell a damp sulfur smell. Suddenly I see what's left of my buddy Troy, his head on a bamboo platter, sent back to our camp by the Viet Cong. Propaganda messages are stuffed between his clenched teeth. The next bolt of lightning and clap of thunder makes me jump so much that I fall to the floor. . . ." (Paraphrased from conversations with Dr. R. L. Gellman, Department of Psychiatry, Yale University School of Medicine).

Perhaps there are no more vivid memories than those which are stored in the brains of soldiers who have experienced excruciatingly horrible combat situations. Witness the account just given of the 50-year-old Viet Nam veteran who cannot hear a clap of thunder, see an Oriental woman, or touch a bamboo placemat without reexperiencing the sight of his decapitated friend. Even though this occurred in a faraway place more than 28 years ago, the memory is still vivid in every detail and continues to produce the same state of hyperarousal and fear as it did on that fateful day.

Once called combat fatigue, war neurosis, or shell shock, it is now clear that intense trauma can produce vivid memories that can last a lifetime. They can be triggered by stimuli that were associated with the original traumatic event, and in some individuals the memories are so intrusive that normal functioning is no longer possible.

Traumatic memories are formed when formerly neutral stimuli (e.g., a bamboo place mat) are paired with aversive events (e.g., the sight of a head without a body). Converging evidence from many different laboratories indicates that an area of the brain called the amygdala is critically involved in both the formation and expression of aversive memories (1–7).

Evidence for involvement of the amygdala in both the formation and retention of conditioned fear has been extensively reviewed (8) and will not be repeated here. Clinically, however, a major problem seen in posttraumatic stress syndrome and certain other types of anxiety disorders is an inability to suppress or inhibit these terrible memories. As the Viet Nam veteran stated, "I can't get the memories out of my mind! The images come flooding back in vivid detail, triggered by the most inconsequential things, like a door slamming or the smell of stir-fried pork." Despite the passage of 28 years and living in an environment very different from Viet Nam, the fear persists. Hence, an important area of inquiry concerns the way that unwanted memories can be inhibited and why this is so difficult following traumatic fear conditioning. This chapter will review selected aspects of the literature on two phenomena, extinction and conditioned inhibition, which deal with fear reduction and have considerable relevance to the problems seen in psychiatric disorders such as post-traumatic stress syndrome. It will begin by reviewing the theories of extinction put forth by Pavlov (9) and later by Konorski (10,11) as these have been influential in modern thinking about extinction (12). It will then discuss at some length whether extinction results from an erasure of the original memory or a buildup of competing inhibition, a critical distinction for an eventual analysis of brain systems involved in extinction. The relationship between extinction and conditioned inhibition and the paradox of avoidance behavior relevant to extinction will be covered next. Clinical implications of data related to extinction and avoidance behavior will then be reviewed. Finally, brain structures and neurotransmitters involved in extinction will be discussed. The review will be restricted to Pavlovian fear conditioning and avoidance conditioning; hence, the extensive literature on extinction related to positive reinforcement (e.g., the partial reinforcement extinction effect [13]) will not be covered.

DEFINITION OF EXTINCTION

When Pavlov paired the sound of a bell (the conditioned stimulus, or CS) with the sight of food (the unconditioned stimulus, or US), his dogs began to salivate (conditioned response, or CR) during presentation of the bell alone. Before such pairings, the animal never salivated to the sound of the bell, but only when given the food. A similar phenomenon can be seen today, when the family cat runs into the kitchen upon hearing the sound of an electric can opener. However, when Pavlov presented the bell over and over again without the food, salivation after the sound of the bell gradually diminished and finally did not occur at all. This decrease in the amplitude or probability of the conditioned response following presentations of the CS in the absence of the US is known as extinction. Importantly, extinction is considered to be an *active process* because it requires that the CS be presented in the absence of the US. In contrast, the term "forgetting" is commonly used to refer to a loss of responding that results merely from the passage of time without any intervening CS presentation.

In fear conditioning, it is more difficult to directly observe the CR. Because of this, it is generally assumed that during the conditioning phase the aversive stimulus (the site of a decapitated head) produces a constellation of emotional behaviors (e.g.: cold, clammy skin; hair on the back of the neck standing straight up; heightened startle) that collectively define a state of fear (14). Neutral stimuli (e.g., bamboo place mat) directly paired with the aversive stimulus or other stimuli that remind one of the general setting in which the traumatic event took place (e.g., the smell of stir-fried pork, the site of an Oriental woman) now can elicit the same state of fear. In this instance, extinction would be defined as a loss of this fear reaction when these conditioned stimuli (e.g., bamboo place mat) were presented over and over again without being paired with the original fear-provoking stimulus (head on the platter).

THEORIES OF EXTINCTION

Extinction as a Result of Active Inhibition

The theoretical mechanism(s) for the loss of responding that occurs in experimental extinction is still a matter of debate. Pavlov (9) and later Konorski (10,11) set the stage for most modern theories of extinction. Interestingly, both sought to explain extinction based on not only behavioral data but also on the current understanding of nervous system function.

Pavlov's Theory of Internal Inhibition

Pavlov (9) was the first to systematically investigate extinction of conditioned responses. Among the classic experiments that led him to formulate his theory were demonstrations of "spontaneous recovery" and "disinhibition." Pavlov (9) observed that an extinguished conditioned response would "spontaneously regenerate" with the passage of time (i.e., an animal previously trained whose response was extinguished would, upon being brought back into the experimental situation, demonstrate the extinguished CR). Similarly, once a CR was extinguished, the simple presentation of a novel stimulus, a disinhibitor, would result in the return of the CR. Because under these circumstances the animal could still perform the conditioned response after extinction, Pavlov (9) argued that extinction could not be regarded as an irreversible destruction of the nervous connections that first allowed the CS to elicit the conditioned response (9). Instead, he believed that under conditions of nonreinforcement the excitatory CS (i.e., a CS capable of eliciting a CR) would become negative or inhibitory.

Pavlov (9) saw learning as taking place entirely in the cortex of the animal. During nonreinforcement, the CS produced an inhibitory effect in the cerebral hemispheres that would compete with the ability of the CS to produce a CR. Internal inhibition, as it was called, was due to a spreading wave of inhibition initiated at the cortical center that corresponded to the cerebral location of the CS. Once inhibition was initiated at the cortical center, it would irradiate over the entire cerebral cortex, affecting a smaller and smaller inhibitory influence with increased distance from the point of initiation. Thus, conditioned responses would be inhibited in proportion to the proximity of their cortical initiation zones to the inhibitory cortical center.

Konorski's Theory of Inhibition Via Associative Connections

Owing mainly to the work of Sherrington, Konorski (10) recast Pavlov's theory of nervous system activity so that it could be more easily integrated into the current understanding of neurophysiology. Like Pavlov (9), Konorski (10) viewed extinction as a consequence of inhibition resulting from nonreinforcement of an excitatory CS. Unlike Pavlov, however, Konorski conceived of neural centers (specific neural structures?) in the brain that are activated by specific stimuli and that over the course of training develop progressively stronger connections with other neural centers. He argued that during acquisition, positive or excitatory connections are strengthened between neural centers corresponding to the CS and the US. Once the connections are strengthened, activation of the CS center alone is capable of initiating activity in the US center, thus producing a CR. During nonreinforced CS presentations, negative or inhibitory connections are strengthened side by side with the excitatory connections. With nonreinforcement, the purely excitatory conditioned response (output of the US center) is transformed into an excitatory-inhibitory CR. But, as nonreinforcement continues, inhibition gradually begins to win out and the conditioned response is reduced. Therefore, extinction of a well-trained CS is due to direct competition of positive and negative connections, respectively, for the activation or inhibition of the US center.

Perhaps the greatest contribution of this theory comes in the statement of how these connections are formed. Excitatory connections are strengthened when activation of the CS center coincides with a rise in activation of the US center. Conversely, inhibitory connections are strengthened when the activation of the CS center coincides with a *fall* in activation of the US center; this occurs when the CS is presented in the absence of the US.

This theory is important because it makes explicit the notion that extinction is the result of an accrual of inhibition, and it makes specific predictions concerning the formation and detection of inhibition resulting from extinction. For example, if the inhibitory connections formed during extinction could be removed temporarily, then the excitation of the US center and therefore the CR would be restored instantaneously. In addition, this theory predicts that the rapidity with which extinction occurs should be directly proportional to the strength of the excitatory connections and, finally, that the development of extinction should follow a negatively accelerating function. Both of these predictions derive from fact that the well-trained CS itself activates the US center; the greater the activation of the US center, the greater the potential fall in activation and the larger the increment in the inhibitory connection with nonreinforcement. Put another way, with a well-trained CS, a US would be expected very much, so that its absence would come as a great surprise and hence be a very salient event. The full significance of this concept was recognized and quantified in the highly influential model of Rescorla and Wagner (15). Finally, the explicit statement of direct competition between excitatory and inhibitory connections provided a clear rationale for the procedures that Pavlov devised to detect inhibition, such as summation and retardation (see next section).

Konorski (10) went beyond Pavlov by providing a somewhat more physiologically accurate theory of inhibition and a more explicit mechanism of the plasticity occurring during extinction. Several important features of Konorski's theory are worth restating: 1) extinction was thought of as an *active associative inhibition* that competed with existing excitation; inhibition was thought to raise the performance threshold for conditioned response elicitation (12). 2) active associative inhibition was the result of strengthening inhibitory CS–US connections; and 3) strengthening these inhibitory connections was a direct result of nonreinforcement of an excitatory CS.

Extinction as a Loss of Excitation

An active inhibitory association can account for the basic finding of extinction, but is the concept of inhibition really necessary in order to explain the loss of conditioned responses resulting from nonreinforcement? Because acquisition of responding is commonly attributed to the strengthening of excitatory connections, extinction might be a symmetrical process (i.e., a weakening of these excitatory connections as a result of nonreinforcement). This "erasure" hypothesis of extinction has had several proponents. The theories of Skinner (16) and Estes (17), derived mainly from instrumental learning experiments, and the model proposed by Rescorla and Wagner (15,18) exemplify this general class of theory. Although they stopped well short of claiming that nonreinforcement produces unlearning of a response, both Skinner (16) and Estes (17), proposed that extinction resulted from learning a new, competing response, one that is more appropriate to nonreinforcement. The new response was associated with the same stimuli that were previously associated with the extinguished response. Therefore, extinction was the result of *rewriting* old associations with new associations. Rescorla and Wagner (15), on the other hand, proposed that extinction resulted from a *weakening* of the previously established connections (see Pearce and Hall (19) for a review of the Rescorla-Wagner model and extinction). Thus for Skinner (16), Estes (17), and Rescorla and Wagner (15), nonreinforcement was assumed to permanently erase part of the original association.

Inhibition or a Loss of Excitation: The Experimental Data

What kinds of experiments can distinguish between extinction being the result of active inhibition versus a loss of excitation? Assuming that extinction is the result of inhibition, a procedure that temporarily removes inhibition in order to reveal excitation would be ideal. However, this is not easily accomplished at the behavioral level. Instead, studies have been directed toward demonstrating that the excitatory conditioned response survives extinction, indicating that CS–US associations have not been erased completely. Demonstrations of spontaneous recovery, rapid reconditioning, and reinstatement all support the idea that excitation survives extinction. However, the data are somewhat less than conclusive and generally fail to support directly an inhibition hypothesis, but more simply the case that extinction *fails* to erase all of the excitatory associations.

At this juncture, it is important to appreciate the distinction between learning and performance. Because associative strength (i.e., learning) cannot be measured directly, it must be inferred from the behavior of the subject (i.e., performance). If a conditioned response is observed, one can conclude that CS–US associations have been formed, given proper controls. However, if a conditioned response does not occur, it cannot necessarily be concluded that CS–US associations did not form. Associations could have formed but are below threshold for detection, or are masked by some inhibitory process. Hence, when considering whether the loss of responding in extinction results from active inhibition or a loss of excitation, it is important to appreciate that even in the absence of conditioned responses, CS–US associations may still exist.

Spontaneous Recovery

Pavlov (9) observed that an extinguished conditioned response would recover to a preextinction baseline with relatively minor disturbances in the experimental protocol. In one experiment conducted by Dr. Eliason and reported by Pavlov (9), three nonreinforced presentations of a well-trained CS at a 10-minute inter-trial interval resulted in complete extinction of a previously established salivation conditioned response (from 8 drops of saliva to 0). After an interval of 20 minutes, the CS was again presented and a complete recovery of the conditioned response was observed (7 drops). This effect was termed spontaneous recovery and taken to imply that the original excitatory conditioning had not been lost but simply overlaid with inhibition. However, as argued by Skinner (16) and Estes (17), extinction may still involve an erasure of many excitatory associations. However, because the conditions of original training are not identical to the conditions during extinction, some connections may still be intact; it is these connections that support spontaneous recovery. In fact, very few theorists now accept spontaneous recovery as evidence for inhibition in extinction. Instead, it is viewed as one example of a class of phenomena that merely indicate that excitation is not completely lost following nonreinforced CS presentations.

Rapid Reacquisition

For quite some time it has been assumed that reacquisition following extinction proceeds more rapidly than the initial acquisition. A series of experiments by Konorski and Szwejkowska employing salivary conditioning in dogs seemed to support this belief. They showed that while extinction of the conditioned response generally proceeded slowly, occurring in 20 to 40 trials, reacquisition to a preextinction baseline occurred in just 2 to 3 trials (11,20,21). Rapid reacquisition appears to be a reproducible phenomenon (e.g., Frey and Butler; Smith and Gormezano) (22,23) but one that is not easily explained by the theories of extinction discussed thus far.

More recently, the phenomenon of rapid reacquisition has been challenged (24,25). Bouton (24) points out that in Konorski's original demonstrations of rapid reacquisition the experiments were designed to compare the rates of reacquisition of the extinguished CS to the previous rate of *extinction* of the same CS, not to acquisition of a novel stimulus at a comparable point in the experiment. If the goal is to assess the effect of extinction upon subsequent reacquisition, then the proper comparison would seem to be the rate of acquisition to a CS that has *not* undergone extinction.

To test this, Bouton (24) used a conditioned emotional response paradigm, devised by Estes and Skinner (26), to measure the acquisition and extinction of conditioned fear. In this test, rats are first trained to press a bar to obtain food. This phase of the experiment has nothing to do with fear, but provides a reliable pattern of behavior that can be used to assess fear. In the fear-conditioning phase, a CS such as a tone is paired with a shock. Later on, presentation of the CS suppresses the rate of bar pressing that is used as the measure of fear. Longer periods of suppression are thought to reflect greater fear. In this experiment, Bouton (24) compared the rates of reacquisition of three groups with different histories of acquisition and extinction. Two groups of rats received tone (CS)–footshock (US) pairings followed by either 24 or 72 nonreinforced tone presentations (groups 24E and 72E, respectively). A control group was given equal context exposure but never received acquisition or extinction. Following these treatments, all three groups were given tone–footshock pairings (i.e., reacquisition for groups 24E and 72E). In contrast to the data on rapid reacquisition, Bouton (24) found that group 24E reacquired the conditioned emotional response at the same rate that the naive control group acquired responding to the novel tone. Even more interesting, while both experimental groups showed equal and complete extinction of the conditioned emotional response, group 72E showed significantly slower reacquisition than group 24E.

Bouton (24) argues that slow reacquisition may be explained best within the context of theories dealing with altered CS processing (c.f. Pearce and Hall [19]; Wagner [27], 1978). However, these data are perfectly congruent with an inhibition hypothesis of extinction. As Konorski (10) proposed, extinction to a zero conditioned response level involves the building up of inhibitory CS–US connections that exactly oppose the excitatory connections acquired during acquisition. From this point, reacquisition would proceed at the same rate as original acquisition since neither excitatory nor inhibitory influences win out. Assuming that, in the Bouton (24) experiment, 24 nonreinforced CS presentations achieved this balance of excitatory and inhibitory tendencies, then the data demonstrating comparable rates of acquisition in a naive control and group 24E are consistent with Konorski's notion. Moreover, if 72 nonreinforced trials tipped the balance in favor of inhibition, reacquisition in group 72E should be slower than acquisition to a novel stimulus. A theory of extinction based solely upon erasure of excitation cannot predict slow reacquisition. In the limit, if all of the excitatory connections are erased, then reacquisition should proceed at a rate similar to initial acquisition but never more slowly. Therefore, of the two alternatives considered, an inhibition hypothesis appears better able to explain slow reacquisition.

Reinstatement

Both Pavlov (9) and Konorski (10) reported that the simple presentation of a US alone was sufficient to renew responding to an extinguished CS, best exemplified by a study of Rescorla and Heth (28). Using a standard conditioned emotional response procedure, groups of rats were first given either explicitly paired or unpaired presentations of a tone and footshock. In a second phase, all of the groups were given nonreinforced presentations of the tone. Next, one half of each of the paired and unpaired groups were given a single reminder footshock identical to the footshock US used in training. Twenty-four hours later, all of the animals were tested for conditioned suppression of bar pressing to the tone. Following nonreinforced tone presentations, both the paired and unpaired groups showed the same lack of conditioned suppression to the tone, (i.e., the paired group was extinguished). However, at testing, the paired group that received the reminder shock 24 hours earlier showed renewed suppression to the tone. The groups that did not receive a reminder shock, as well as the unpaired group that did receive a reminder shock, did not show renewed suppression to the tone. The failure to see a reinstatement-like effect in the unpaired group indicates that reinstatement was not the result of sensitization but instead works through an associative process.

Rescorla and Heth (28) assume that extinction decreases CS–US associations but also decreases how the US is represented. A reminder shock reinstates the full value of the US representation of shock, therefore allowing it to be fully activated by the remaining excitatory CS–US connections. Several experiments support this idea. In contrast, Bouton and Bolles (29) suggest that reinstatement results from the summation of the weakly excitatory-extinguished CS with a context that was made excitatory by the reminder

US. In their experiment, reminder shocks were given either in a context in which they were to be tested or in a context discriminatively different from the test context. Reinstatement occurred only when the rats were tested in the context in which they received the reinstating shocks. Moreover, extinguishing the context in which the reminder shocks were given prior to testing for reinstatement in that context prevented or erased reinstatement (Bouton and Bolles, Experiment 2; c.f. Rescorla and Cunningham) (29,30). Similar results have been obtained more recently by Callen, McAllister, and McAllister (31) using a passive avoidance procedure. It would thus appear that a nonassociative US–memory renewal account of reinstatement is not necessary. Summation of an excitatory context with residual excitation of the extinguished CS may account for the reinstatement of conditioned responses seen following reminder shocks. Regardless of which account one uses, the phenomenon of reinstatement indicates that extinction does not completely erase excitatory CS–US connections, although it does not clearly support an active inhibitory account.

Context Specific Extinction

In an impressive series of experiments, Bouton and colleagues have shown that extinction may be specific to the experimental context in which nonreinforcement occurs. For example, Bouton and Bolles (29) used four groups of rats that were first given tone–foot-shock pairings in a training chamber. After conditioning, two groups received nonreinforced tone presentations in context B and two groups in context C. Two of the four groups were subsequently tested for extinction in the same context in which nonreinforcement occurred, while the other two were tested in the alternate context. While all four groups showed extinction in the contexts in which they were nonreinforced, the rats subsequently tested in a context different from the one in which they were extinguished showed renewed suppression to the tone, a phenomenon Bouton terms *renewal.* Control animals that were tested and extinguished in the same context did not show renewed suppression. The data clearly show that under circumstances in which the rat can discriminate contexts, extinction performance can come under the control of contextual cues. (For a critique of context specific effects, see Lovibond, Preston, and Mackintosh [32].)

The context-specific extinction effect argues strongly against an erasure hypothesis of extinction. The rat was fully capable of performing the conditioned response in a context other than the one in which it was nonreinforced. In order to explain these data, it is necessary to invoke an associative mechanism of extinction. Again, in the language of Konorski, stimuli presented during nonreinforcement acquire inhibitory CS–US connections that summate with existing excitatory CS–US connections. While in the presence of those stimuli, the inhibitory CS–US connections cancel out the excitation, thus producing no net CS–US activation and an extinguished CR. But, if the inhibition is removed, the excitation prevails and the conditioned response is restored. Perhaps Bouton and Bolles (29) succeeded in allowing the context to acquire a great deal of the inhibition during nonreinforcement; removing the inhibition was as easy as changing contexts. Note that this is similar to the conventional conditioned inhibition procedure in which a neutral stimulus is nonreinforced in the presence of an excitatory CS (see information that follows).

To test whether the context in which extinction took place was inhibitory, Bouton and King (33) used a summation test in which rats previously trained to be fearful of a light in another distinctive context were now tested in the context where extinction had taken place. If the extinction context was generally inhibitory, one would expect less fear of the light in this context compared to the context where extinction had not taken place. Surprisingly, Bouton and King (33) failed to detect inhibition by the extinction context with this summation technique. Because of this, these investigators have abandoned a contextual inhibition hypothesis and suggest that some other contextually controlled mechanism is at work (34,35). For example, they surmise that the context may retrieve a specific memory of nonreinforcement, such as, "the tone is not followed by shock in this place." However, this memory is rather specific and does not generalize to: "because the tone is not being reinforced in this place, I guess the light also will not be reinforced."

Taken together, the data reviewed thus far clearly indicate that extinction does not result in a complete erasure of the original CS–US association. Despite the appearance of no further memory toward the end of an extinction session, where no conditioned response whatsoever may occur, evidence of some remaining amount of CS–US association can be seen following a period of time (spontaneous recovery), presentation of a US alone (reinstatement), or testing in a different context (renewal). However, whether extinction results from an active inhibitory mechanism or some other process has not been unambiguously demonstrated using these paradigms. On the other hand, there is wide agreement that a closely related phenomenon, known as conditioned inhibition, does involve active inhibition. In fact, it has been argued that extinction is a special case of conditioned inhibition. We believe that stimuli that should serve as conditioned inhibitors and thereby inhibit fear do not operate properly in certain forms of psychiatric disorders, including post-traumatic stress disorder. Because of this, the phenomenon of conditioned inhibition will be reviewed next.

CONDITIONED INHIBITION

Pavlov (9) observed that if some "indifferent agent having no marked effect on the animal" (see page 208) was added to a well-trained CS and the combination was not followed by the US, the indifferent stimulus would gradually acquire the ability to reduce the reflex produced by the CS, even though the CS continued to produce the reflex by itself. Pavlov argued that this observation indicated that the formerly indifferent stimulus acquired the ability to evoke a process of inhibition that reduced the ability of the CS to produce the CR. He called this process "conditioned inhibition," because it had to be acquired with training. He called the stimulus that produced the reduction in conditioned responding a "conditioned inhibitor."

Because an inhibitory stimulus does not normally evoke a measurable response (but see Batson and Best [36]; Wasserman et al. [37]), special test procedures had to be devised to detect it. These were clearly outlined by Rescorla (38). Based on the assumption that inhibition antagonizes excitation and the two sum algebraically, Rescorla (38) argued that an inhibitory stimulus should 1) reduce excitatory conditioned responses to a CS when the inhibitor and excitor are placed in compound (the summation test) and 2) show slower acquisition of excitatory conditioning if subsequently paired with a US (the retardation test; recall the Bouton 72E group).

The summation test or "combined cue test" is the basic method for observing conditioned inhibition (38,39). In this procedure, the putative conditioned inhibitor is presented in compound with an excitatory CS. If the combination produces a decrease in the conditioned response below the level observed when the CS is presented alone, that stimulus is said to act as a conditioned inhibitor of that excitatory process. Moreover, when the conditioned inhibitor is removed, excitation should return to its original level.

Although summation is often taken as good evidence for conditioned inhibition, care must be taken to rule out alternative explanations for the decrement observed during summation. For example, the combination of two stimuli during summation testing may result in either a generalization decrement around the excitatory stimulus or differential attention to the two stimuli (38). In both instances, responding to the excitatory CS could be decremented even though no active inhibition process was involved. However, with appropriate control procedures at the time of the summation test, generalization decrement and differential attention can be ruled out. Another potential problem occurs specifically with the typical conditioned inhibition procedure that employs nonreinforcement of an inhibitor-excitor compound (i.e., a tone is consistently paired with shock, but a light-tone combination is nonreinforced). In these experiments, summation is most often assessed by observing the decrement in conditioned responding during the same inhibitor-excitor compound used in training (9). However, with this test alone, it may be impossible to dissociate conditioned inhibition from configural learning (40) (i.e., the subject may simply perceive the light-tone compound as a separate unique stimulus that is not appreciate that it is made up of separate tone and light elements). Because the unique compound was consistently nonreinforced, it would be expected to have less excitatory strength than the tone alone, but this might say nothing about what the subject has learned about the separate light element. To control for this, conditioned inhibition must be assessed in summation with some other conditioned excitor not previously presented in compound with the inhibitor, the so called transfer test (9). A configural interpretation can be ruled out if a decrement in conditioned responding to this novel compound occurs, because this particular compound never occurred before and hence the subject could never have learned that it was different from the excitatory CS itself. (However, see Skinner [40] for an attentional account of transfer).

A second procedure routinely used to detect inhibition also follows from the idea that conditioned inhibition competes with conditioned excitation. This idea was recognized early on by both Pavlov (9) and Konorski (10). If a putative conditioned inhibitor is paired with the same US used to develop it, the rate of acquisition of the conditioned response should be slowed. This retardation effect occurs because the inhibition produced by the conditioned inhibitor antagonizes the developing excitatory process that occurs as a result of reinforcement.

Like summation, retardation also must be accompanied by proper controls. Retarded acquisition of a conditioned response may be observed because of differential attention to the putative conditioned inhibitor rather than because of conditioned inhibition (38). For example, in another type of design called latent inhibition, a novel CS is presented in the absence of a US and then a regular conditioning session takes place where the CS is now paired with a US. Prior exposure to the CS slows the rate of subsequent acquisition of conditioned responses to that stimulus, an effect that is operationally identical to retardation (41,42). Because latent inhibition has been attributed to decreased attention to the pre-exposed stimulus rather than active inhibition (c.f. Mackintosh, [43]) retardation could result from a similar process.

Hence, to effectively argue that a stimulus is a conditioned inhibitor, it may be necessary to apply both the summation and retardation tests (c.f. Papini and Bitterman [44]). For example, if decreased attention alone accounts for retardation, then this stimulus should fail to show inhibition in the summation test (45). Likewise, if generalization decrement or increased attention to the putative inhibitor fully explained the decrement in conditioned responding observed in the summation test, then, presumably, this stimulus should not show retardation (38). Implicit in Rescorla's (38) argument is that only by applying both the summation and retardation techniques can one be assured that a stimulus is a true conditioned inhibitor (however, see Williams, Overmier and LoLordo (46); Papini and Bitterman (44) for additional discussion of the necessity of these tests).

Using these criteria, it is now well established that a neutral stimulus can acquire inhibitory properties when placed in compound with an excitatory CS and nonreinforced. Following this procedure, the conditioned inhibitor passes both the summation and retardation tests, and transfers to inhibit another excitatory CS that was never before placed in compound with it, leaving little doubt that it controls a process of inhibition.

Importantly, the conditioned inhibition procedure does not differ substantially from the normal extinction procedure. Both involve nonreinforcement in the presence of an otherwise excitatory CS, and in both a decrement in conditioned responses is observed over repeated nonreinforced trials. Therefore, one may argue whether it is necessary to assume a separate mechanism for extinction and conditioned inhibition.

The reason that the two forms of response decrement are not considered to be equivalent has been the failure to directly detect inhibition by the extinguished CS (12,38,47). However, as

Rescorla (38) has pointed out, there is actually no reason to suspect that an extinguished CS and a conditioned inhibitor would yield the same results on a summation or retardation test. An extinguished CS has both excitatory and inhibitory CS–US connections (c.f., Konorski; Pavlov [9–11]). Inhibition by an extinguished CS can only be detected with the traditional procedures if the net CS–US inhibition is greater than the net CS–US excitation (recall Bouton's 24E group, which did not show a change in the rate of reacquisition, thus demonstrating no retardation). On the other hand, a conditioned inhibitor is a neutral CS at the outset of conditioning. It only acquires inhibition and has no competing excitatory CS–US connections. In this case, inhibition is easily detected.

Given the broad acceptance of an ''inhibition hypothesis'' of conditioned inhibition, one approach has been to demonstrate that the inhibition and extinction procedures are interchangeable. Along these lines, DeVito and Fowler (48) devised an ingenious experiment in which they demonstrated that conditioned inhibition can be enhanced by extinction of the conditioned inhibitor. Using a lick suppression procedure, two groups of rats first received clicker–footshock pairings followed by nonreinforced presentations of the clicker and a tone designed to make the tone CS a conditioned inhibitor (i.e., clicker–US, tone-plus-clicker–no US). Group extensive received a total of 144 tone-plus-clicker trials, while group moderate received 42 tone-plus-clicker trials. In a third phase, both groups received tone-alone extinction trials. A third group, group control, received training identical to group moderate, but the tone-alone phase was omitted. The overall level of inhibition was assessed with a summation procedure. Prior to tone-alone trials, group extensive acquired greater inhibition to the tone than groups moderate and control. Following nonreinforced tone presentations, group moderate showed the same level of inhibition as group extensive, and they both showed reliably more inhibition than group control. Therefore, for group moderate, the intervening nonreinforced tone presentations enhanced an already moderate level of inhibition. Although these data are by no means conclusive, they are consistent with the notion that conditioned inhibition and extinction operate through similar mechanisms (i.e., nonreinforcement in the presence of excitation leads to the accrual of inhibition).

To summarize, there is very little argument over whether or not inhibitory associations mediate conditioned inhibition. This is because the inhibition occasioned by presumptive conditioned inhibitors is easily identified with the traditional summation and retardation procedures. On the other hand, an inhibition hypothesis of extinction has met with skepticism, primarily because of the failure to detect inhibition by an extinguished CS apart from its ability to reduce its own CR. However, the inhibition hypothesis of extinction does not demand that the extinguished CS be a net inhibitor. But perhaps the most compelling argument for the inhibition hypothesis of extinction is that procedures producing conditioned inhibition do not differ substantially from procedures that produce extinction; in fact, there are data to suggest that the two processes are interchangeable. Moreover, it is clear that a theory of extinction owing only to a loss of excitation is inadequate. Coming full circle, both Pavlov and Konorski (9,10), although lacking some of the essential data discussed here, viewed extinction as one manifestation of the nervous system's propensity for inhibition. More work will be needed to better understand whether inhibition is the only mechanism of extinction.

AVOIDANCE BEHAVIOR

All of the experiments discussed thus far have used Pavlovian procedures to address questions on extinction. In these experiments, conditions are arranged to present an exact number of stimuli in precise combinations for specified periods of time. This is why Pavlovian conditioning has been so powerful in elucidating exactly what animals learn and how stimulus events control this learning. Outside the laboratory, however, things are rarely that well controlled. Animals do not allow fear stimuli to be presented for exact periods of time but escape from them whenever possible, a behavior that can interact powerfully with the process of extinction. For this reason, it is appropriate to also consider some of the extensive literature on extinction of active avoidance behavior.

The Paradox of Active Avoidance

Imagine that a dog is placed in a box equipped with a grid floor and a low hurdle separating him from the other side. A tone comes on and 5 seconds later the grid floor is electrified, leading him to escape to the other side by jumping the hurdle. A few minutes later, the same sequence is repeated. In a few trials the animal jumps the hurdle when the tone comes on and thereby avoids the shock altogether. On the next several trials, the latency of this avoidance response becomes even shorter, so that the dog jumps as soon as the tone comes on, continuing to avoid the shock. Depending on various parameters, the dog may make hundreds of these avoidance responses and never get a shock.

At first blush the explanation for this behavior seems straightforward; the dog jumps the hurdle when the tone comes on to avoid getting the shock. Put another way, the shock serves as a negative reinforcer in this situation. But as Schoenfeld (49) pointed out, how can the nonoccurrence of a US (i.e., the shock) act as a reinforcement? Furthermore, why doesn't this behavior extinguish, since we know that repeated presentations of a CS (the tone in this case) in the absence of a US will lead to extinction? Furthermore, what prompted the dog to make the very first avoidance response, because it was never trained that jumping the hurdle early during the CS presentation would lead to avoidance of the shock?

Two-Factor Theory of Avoidance

To resolve this paradox, Mowrer (50) proposed that active avoidance behavior involved two separate mechanisms. First, the animal learned to be afraid of the CS by virtue of Pavlovian fear conditioning, given that the CS was paired with the US. Fear of the CS was then assumed to activate the animal, eventually leading him to escape this aversive CS by making the hurdle response. Thus the so-called avoidance response was viewed by Mowrer as a response of escape from the CS rather than avoidance of the shock. An additional assumption was that when this response occurred, it reduced fear, which itself reinforced the behavior, making a subsequent avoidance response more likely.

This theory resolved several aspects of the avoidance paradox but still appeared not to explain some data. For one thing, dogs well trained in active avoidance paradigms did not seem fearful, but merely jumped the hurdle as soon as the CS came on. In addition, even if the motivation for the hurdle response was actually fear of the CS, it remained unclear why the behavior did not extinguish or at least took a very long time to extinguish in comparison to other measures of Pavlovian fear conditioning.

To explain why little fear was seen when animals were successfully emitting avoidance responses, Solomon and Wynn (51) suggested that, as conditioning proceeds, the animal makes the avoidance response so quickly that there is not enough time for fear to be elicited. Or, as suggested by Levis (52), with continued training a very small amount of fear is all that is required to produce a response, a level of fear that is below threshold for producing other signs of fear (e.g., autonomic measures). In a sense, this is how the avoidance response becomes automatic.

To explain the very slow rates of extinction, Solomon and Wynn (51) introduced their "conservation of anxiety hypothesis," based on a number of observations. First, they noticed that over training the avoidance latencies shortened progressively and that the signs of fear seen early in training were no longer evident when short latency responses were made. However, if the dog happened not to make a short latency response, the continued presence of the CS then elicited a reappearance of fear. Moreover, long latency avoidance responses were typically followed by short latency responses in the next test trial.

Given this behavior, the dog arranged conditions in which the CS was only on for a very short period of time. Solomon and Wynn (51) suggested that this prevented response extinction, a conclusion consistent with later work showing that that rate of extinction is directly related to the total duration of CS exposure (53,54). In fact, in an avoidance study in humans, subjects who failed to extinguish had shorter avoidance latencies than subjects who did extinguish (55).

Eventually, however, according to Solomon and Wynn, the animal will extinguish to the very beginning part of the CS, leading to a longer latency response. However, this now exposes the animal to a part of the CS that has not become extinguished, and leads to a high level of fear that then serves as a reconditioning trial, making the first part of the CS fearful once again (i.e., reinstatement). This then leads to a very short latency avoidance response and once again protects the rest of the CS from being exposed and hence protects it from extinction. In fact, Levi and Boyd (56) have data on rats showing a striking oscillation of avoidance latencies over trials, where a long latency response tends to be followed by a short latency response and vice versa. With extended training, the animal begins to make the avoidance response at a longer latency, thus extending the duration of CS exposure and hastening extinction. Eventually, the latency becomes long enough so that the animal does not make an avoidance response (i.e., extinction has occurred).

These ideas account for a good deal of data within the literature on active avoidance, although they have not gone unchallenged. For example, Seligman and Johnston (57) assert that, although fear is necessary for acquisition of active avoidance behavior, it does not motivate sustained asymptotic avoidance behavior. They pointed out that Solomon's dogs sometimes responded for hundreds of trials, whereas fear measured in a conditioned emotional response paradigm extinguished after only 40 trials (e.g., Annau and Kamin, [58]). Because of this, they concluded that fear would have extinguished long before avoidance behavior and hence fear could not continue to produce the avoidance response. Instead, they proposed a more cognitive account where the avoidance response continued because the animal expected that the absence of a response will lead to a shock.

However, work by Donald Levis indicates that this cognitive account may not be necessary. Levis (52) points out that the total duration of CS exposure in a typical Solomon avoidance study where a dog made 500 avoidance responses without receiving a shock would be about 1,000 seconds, assuming an average response latency of 2 seconds. In contrast, total CS exposure for complete extinction in the Annau and Kamin study quoted by Seligman and Johnston (57) was 2,400 seconds. Furthermore, even when animals are making short latency avoidance responses, they still appear to be fearful of the CS, provided sensitive enough tests are used. For example, Levis and Boyd (56) trained rats to an asymptotic level of active avoidance and showed that the onset of the CS still caused substantial suppression of bar pressing in a conditioned emotional response paradigm.

In summary, the data seem to be consistent with Mowrer's original two-factor theory of active avoidance behavior. Moreover, the slow rates of extinction may be explained by the animal arranging conditions to reduce CS exposure and hence retard extinction.

CLINICAL IMPLICATIONS

Because a good deal of psychotherapy involves procedures to rid patients of unwanted fear memories, a behavioral analysis of extinction has a number of clinical implications, such as those listed by Bouton and Swartzentruber (59). As they point out, "performance after extinction is inherently unstable" (see page 128). Phenomena such as spontaneous recovery and reinstatement may explain why conditioned fears and phobias in humans sometimes seem to return spontaneously without any obvious cause. The renewal effect may explain why fears reduced successfully in the therapist's office reappear when the patient returns to his home or his job. If a drug is used as an adjunct to therapy, renewal of fear could occur when the fearful stimulus is encountered in the absence of the drug. In fact, animal experiments show that when benzodiazepines are given during extinction, fear of the CS returned when testing later occurred in the absence of the drug (60).

As Levis (61) has stressed, reduction of pathological behavior driven by prior aversive conditioning can be approached through basic principles of extinction. Central to this implosive therapy is the need to repeatedly expose the patient to not only "CS patterns directly correlated with symptom onset but also to cues reactivated by the exposure procedure and those hypothesized to be response for symptom developments" (see p. 67 of Levi [61]). Consistent with animal work, extinction is assumed to be a direct function of total exposure. Finally, Levis (61) assumes that extinction will be directly related to the level of emotional response generated during presentation of the CS pattern, similar to animal work on conditioned inhibitors showing that conditioned inhibition is directly related to the excitatory strength of the CS+ with which it is put in compound (c.f. Wagner [62]).

The concept of conditioned inhibition may also have a great deal of clinical relevance. For example, one of our colleagues told us about a particular patient suffering from post-traumatic stress disorder. This particular patient had undergone excruciating combat trauma in Viet Nam and later spent many years in treatment. Eventually, however, he was doing much better and got married. It was a June wedding and he was wearing a white tuxedo with a pink carnation in the lapel. It happened to be a rainy day. As he and his new bride were walking down the steps of the church, a car passed by and backfired. The veteran instantly dove into a nearby patch of muddy bushes, ruining his white tuxedo and pink carnation.

This is a poignant example illustrating a profound failure of conditioned inhibition. Thus, there were many cues that should have served as safety signals on this June morning. The veteran was in the United States, not in Viet Nam. He was wearing a white tuxedo, not battle fatigues. And he was walking out of a church with his new bride. Nevertheless, when he heard the sound of a car backfiring reminiscent of mortar fire, none of the safety signals could prevent the formerly adaptive behavior of diving for cover. Because this inappropriate behavior seems to involve an inability to inhibit traumatic fear memories, understanding the neural basis of extinction and conditioned inhibition should eventually lead to better treatments for these types of anxiety disorders.

STUDIES ON THE PHYSIOLOGICAL BASIS OF EXTINCTION

Anatomy

One way to approach an understanding of the anatomical substrate of extinction and/or conditioned inhibition would be to begin by investigating the contribution of brain structures known to exert inhibitory influences on behavior. Several general inhibitory structures are known to exist in the brain. These include sensory and frontal cortex, hippocampus, and septum.

Sensory Cortex

Some of the earliest experiments on CS-induced inhibition in the brain were conducted by Tchilingaryan (as cited in Rescorla [38]). Using electrical stimulation of the motor cortex to elicit paw flexion, Tchilingaryan found that upon presentation of an extinguished CS, higher levels of cortical stimulation were required to elicit paw flexion. This suggests that there is active inhibition of the cortex or cortical outflow initiated by an extinguished CS. Others have looked at changes in unit recordings in cortex during acquisition and extinction of conditioned responses. Brons and Woody (63) showed that following tone-glabella tap pairings in the cat there was an increase in the excitability of sensorimotor cells (i.e., a decrease in the threshold current necessary to elicit a spike). Following nonreinforced tone presentations, there was little change in the excitability of these same cells despite the fact that they observed complete behavioral extinction. The Brons and Woody (63) study suggests an extrasensorimotor influence upon the behavioral output of the cat's eyeblink response following extinction. In contrast, Gassanov, Merzhanova, and Galashina (64) observed patterned organization of activity in auditory and visual cortices resulting from conditioning of an auditory or visual CS, respectively, with forelimb stimulation in cats. Following behavioral extinction, this patterned activity was disrupted, suggesting that the effect of extinction was either intrinsic to the cortical structures or afferent to them. Similarly, Diamond and Weinberger (65) showed that changes in the frequency of auditory cortex receptive fields as a result of tone-forepaw stimulation pairings were reversed by nonreinforced tone trials. These electrophysiological data suggest that changes are evident in cortex during extinction. However, it is difficult from data such as these to assess the functional role of the cortex in extinction and whether these changes are occurring in the cortex or in structures afferent or efferent to them.

Recently it has been shown that while cortical structures are not involved in the acquisition or performance of conditioned fear responses, they appear to be involved in the extinction of fear. LeDoux, Romanski, and Xagoraris (66) performed bilateral ablations of visual cortex or sham operations on rats prior to giving them repeated light–footshock pairings. Ablated and sham controls acquired conditioned fear response at similar rates as assessed by the light's ability to suppress licking of a water tube. However, with repeated nonreinforced light presentations, visual-cortex–ablated rats failed to show extinction of lick suppression relative to sham controls. In a similar study employing heart-rate conditioning in the rabbit, Teich et al. (67) showed that although bilateral lesions of either auditory or visual cortex did not appear to disrupt acquisition to a tone CS, auditory cortex lesions but not visual cortex lesions blocked extinction of conditioned heart rate. Based on known anatomical connections between sensory cortex and thalamic structures, the authors of both experiments concluded that during extinction the sensory cortices exert a modality-specific inhibition of the thalamic structures that is important for performance of conditioned responses.

From these data it would appear that primary sensory cortex is necessary for the acquisition or induction of extinction, because its removal before extinction trials are initiated blocks extinction. What is not clear from these studies is whether primary sensory cortex is responsible for extinction because it: 1) *encodes* the memory of nonreinforcement; 2) *detects nonreinforcement* and then signals other structures to encode the information; or 3) is a relay structure of sensory information to other structures important for extinction.

The Ledoux et al. (66) and Teich et al. (67) studies looked at the role of sensory cortex in the acquisition of extinction. It is not known from these experiments whether sensory cortex is necessary for extinction performance. In order to assess the role of sensory cortex in performance of extinction, the lesions would have to be made after extinction of the conditioned response. For example, if lesions made following extinction failed to attenuate extinction performance (i.e., extinction remained intact), then the conclusion would be that the memory for extinction is stored somewhere other than the cortex, but that the cortex is important for the acquisition of extinction (e.g., it conveys sensory information or detects that nonreinforcement has occurred). If, on the other hand, lesions made following extinction of the conditioned response restored preextinction performance, then the conclusion would be that the memory for extinction is contained in the cortex or that the cortex conveys information to structures containing the memory. Whatever the outcome, studies assessing both acquisition and performance of extinction would provide better insight into the function of the cortex in extinction.

Using such a design, we failed to find that lesions of the visual cortex made either before or after nonreinforced presentations of a light CS blocked extinction of conditioned fear, using the fear-potentiated startle test (68). Hence, extinction can proceed normally without sensory cortex under some circumstances. Interestingly, in this experiment, excitatory training and extinction took place in different experimental contexts, which was not the case in the LeDoux et al. (66) experiment. Because context can participate importantly in extinction (69), perhaps sensory cortex is required for extinction when cues such as the experimental context cannot be used to signal nonreinforcement. Currently, we are redoing this experiment under conditions where training and extinction take place in the same context.

To explain their results, Ledoux et al. (66) suggested that lesions of visual cortex interfered with extinction by preventing the relay of visual information to other cortical structures that are responsible for feeding forward inhibition to critical thalamic areas. Although their experimental design did not allow them to directly test this hypothesis, the fact that the cortical effect on extinction was modality specific may indicate that there is some sensory convergent zone where the inhibition is initiated; two possible areas are the frontal cortex and the hippocampus.

Frontal Cortex

The role of the frontal cortex in emotional behavior is far from clear. Animals will self stimulate with electrodes in the medial prefrontal cortex and several instances of "pure reward" sites have been found (cf. Miserendino and Coons [70]). Electrical stimulation of the prefrontal cortex can inhibit defensive reactions elicited by hypothalamic stimulation in cats (71), or inhibit motor and autonomic components of the defensive reaction in rats elicited by electrical stimulation of either the hypothalamus or amygdala (72). More recently it was reported that high levels of electrical stimulation of the prefrontal cortex in rats inhibited cardiac responses produced by a CS (73). However, controls for external inhibition, where a novel stimulus is thought to simply divert attention away from the CS, were not included. Hence, it is difficult to assess whether this effect was specific to stimulation of the frontal cortex, especially because it also occurred following stimulation of the nucleus accumbens.

Conversely, compromising normal transmission within the prefrontal cortex can influence the effects of aversive stimulation. For example, depletion of dopamine- or ibotenic-acid–induced lesions of the medial prefrontal cortex can exacerbate the effects of stress on dopamine turnover in other brain areas (74,75).

Concerning extinction, Butter (76) found that frontal cortex lesions attenuated extinction of an operantly conditioned bar press response in monkeys. More recently, Frysztak and Neafsey (77) reported that lesions of the ventral medial prefrontal cortex in rats made before conditioning markedly reduced the amount of freezing and high-frequency vocalizations elicited by a tone CS during extinction testing. These authors suggest that the medial prefrontal cortex appears to be an essential component of the brain's circuitry for emotional responses to stress. In contrast, Morgan et al. (78), reported that lesions of ventral portions of the medial prefrontal cortex retarded, but did not block, extinction of conditioned fear to a tone measured with freezing, also when the lesions were made before fear conditioning and extinction. Using the fear-potentiated startle test, we found a slight increase in the magnitude of condition fear, perhaps due to a retardation of extinction, when lesions of the medial prefrontal cortex were made after conditioning but before extinction testing (79). Clearly, therefore, much more work has to be done to evaluate the role of the frontal cortex in extinction.

Hippocampus

Although a complete review of the hippocampal literature is well beyond the scope of this chapter, this brain area has received a great deal of experimental attention and is widely believed to be involved in extinction. Part of the difficulty associated with determining the mechanism of extinction involves knowing what a system needs to accomplish in order to acquire and produce the inhibition. Earlier, it was suggested that a major problem confronting a theory of inhibition was how to design a system capable of discriminating occasions of reinforcement from nonreinforcement. Recall that Konorski (10) proposed that the strengthening of inhibitory connections would occur when the CS nodal activation coincided with the fall in activation of the US node, which is presumed to occur when the US is omitted. In a subsequent revision of his theory, Konorski (11) proposed that inhibition resulted from the strengthening of excitatory connections from the CS center to a non-US center that in turn had hard-wired inhibitory connections to the US center. In both instances, some mechanism for detecting the absence of the US following CS presentation would be required. In fact, Douglas (80) suggests that the hippocampus is a nonreinforcement detector providing the organism with the means to ''tune out'' information that is of no motivational consequence. It is possible that the hippocampus recognizes that the CS is no longer followed by the US and feeds forward inhibition to relevant sensory or conditioned response production centers. For example, hippocampal lesions prevent discrimination reversals, where a stimulus previously reinforced is now nonreinforced and a stimulus previously nonreinforced is now reinforced (81,82) which would also require the animal to recognize a change in the reinforcement contingency.

As attractive as this hypothesis is, the literature is inconsistent regarding the effects of hippocampal lesions on extinction. A variety of conditioning paradigms have been used to assess this question, including the rabbit nictitating membrane response (81,83,84), conditioned heart rate (82), conditioned suppression (85) and active avoidance (86–88). While some of these experiments have found that hippocampal lesions attenuate extinction (83,86–88) others have found no effect of hippocampal lesions (84) and still another has shown facilitated extinction (85). Each of these studies adds to the confusion over the role of the hippocampus in extinction. One thing to note is that these studies differed on several grounds that may or may not account for the differences observed in extinction.

Several of the studies that have found an attenuation of extinction with hippocampal lesions also report facilitation of acquisition. For example, Schmaltz and Theos (83) report that rabbits with bilateral aspiration of the hippocampus acquired the conditioned nictitating membrane response more rapidly than unoperated or cortical controls and subsequently extinguished less rapidly. Instead of assuming direct involvement of the hippocampus in extinction, results such as these suggest that the attenuation of extinction is somehow related to the rapid acquisition. For one, rapid acquisition may mean more profound acquisition, and it is well known that the deeper the initial learning, the slower the extinction. Related to this, there is also evidence to suggest that hippocampal damage produces an increased level of fear in rats. Antleman and Brown (89) showed that, compared to control rats, hippocampal lesioned rats showed rapid acquisition of active avoidance to an auditory CS that was paired with a shock US. And when these same rats were presented with the same auditory CS in a lick suppression procedure, the hippocampal lesioned rats showed greater lick suppression than sham controls and hippocampal lesioned rats that had received explicitly unpaired CS–US pairings. The results of the Antelman and Brown (89) study suggest that hippocampal rats show greater conditioned fear to the CS. If this is a general finding, then it is not clear how to interpret the effect of hippocampal lesions on extinction.

As mentioned earlier, Solomon (84) found that lesions of the dorsal hippocampus failed to attenuate extinction and conditioned inhibition of the rabbit conditioned nictitating membrane response. Solomon (84) agrees with Douglas (80) that the hippocampus is a detector of nonmotivationally significant events; however, he argues instead that extinction is a motivationally significant event. The fact that reinforcement has stopped is of great motivational significance to the animal. If the hippocampus is responsible for detecting those stimuli that are irrelevant and of no motivational significance, then it should not be active during extinction. Solomon's failure to find that hippocampal lesions block extinction is consistent with his interpretation of the Douglas model. Although Solomon's disagreement with the Douglas model may simply be over semantics, further work on the role of the hippocampus in extinction is clearly warranted.

Septal Area

Consistent with a role in the inhibition of fear and anxiety, destruction of the lateral septal nucleus produces a variety of behavioral effects indicative of hyperemotionality. For example, electrolytic lesions of the lateral septal nucleus produces exaggerated responses to innocuous (e.g., pencil taps and prodding) as well as noxious (i.e., heat, sound and electric shock) stimuli and increases the incidence of mouse killing (an index of hyperaggression) (90–92). Moreover, electrical stimulation of the lateral septal nucleus eliminates the hyperreactivity associated with lesions of the ventromedial hypothalamic nucleus (93). Together, these results have led to the hypothesis that the lateral septal nucleus exerts tonic inhibitory control over systems responsible for emotionality. We have recently reported that lesions of the lateral septal nucleus facilitate the startle reflex (94). Interestingly, the facilitatory effect of lateral septal lesions was blocked by lesions of the amygdala, once again consistent with the hypothesis that the lateral septal nucleus exerts inhibitory control over systems responsible for emotionality (94).

If the lateral septal nucleus is responsible for inhibiting fear, then manipulation of septal function may be expected to alter conditioned fear. Experiments have shown that destruction of the septal nuclei facilitates the acquisition of conditioned fear (95–98). In addition, intralateral septal injections of the serotonin reuptake inhibitor fluoxetine enhances the memory for inhibitory avoidance (99). Because serotonin is known to inhibit cells in the lateral septal nucleus (100), this result is consistent with a disinhibition of conditioned fear. Perhaps the most convincing evidence for the role of the lateral septal nucleus in the inhibition of fear comes from the work of Yadin and Thomas (101). They have shown that unit discharge in the dorsal lateral septal nucleus increases with presentation of a conditioned inhibitor (101) and decreases with presentation of a conditioned excitor (101,102), thus further implicating the lateral septum's role in inhibition of fear and more interestingly, in conditioned inhibition of fear. More recently, Yadin and Thomas (103) reported that stimulation of the same area of dorsolateral septal nucleus inhibits restraint stress-induced ulcers.

Amygdala

As mentioned earlier, there is almost universal agreement that the amygdala is critically involved in the formation and retention of aversive memories (1). Specifically, the amygdaloid nuclei are believed to be convergent zones for CS and US sensory information and are known to provide afferents to nuclei that control many behaviors modulated by aversive CSs. It is important to know whether a site important for acquisition and performance of conditioned fear is also important for extinction of conditioned fear. However, assessing this involves a rather difficult procedural problem. Unlike the visual cortex, which is not involved in acquisition or expression of conditioned fear, the amygdala is essential for both. Therefore, lesioning the amygdala would not allow one to assess its role in extinction, because the fear response itself would be obliterated. A reversible lesion occurring at the time of nonreinforced CS presentations would allow one to assess the role of the amygdala in extinction.

Along these lines, we performed an experiment to assess the role of the amygdala in extinction of conditioned fear, using a compound that reversibly blocks excitatory amino-acid transmission in the amygdala (Falls and Davis, unpublished observations). Rats were implanted with bilateral guide cannula aimed at the basolateral nucleus of the amygdala. Following two sessions of light-shock pairings, the initial level of conditioned fear was measured and rats were matched into two groups with equivalent levels of fear on this test. Twenty-four hours later, the rats were infused with either the nonspecific excitatory amino acid antagonist γ-D-Glutamylglycine or artificial cerebrospinal fluid vehicle immediately before being exposed to a session of nonreinforced light trials. Because excitatory amino acids are known to mediate some forms of transmission in the amygdala, an infusion of the antagonist would functionally serve as a reversible lesion. Twenty-four hours later the rats were again tested for potentiated startle. Rats receiving the artificial cerebrospinal fluid vehicle showed near complete extinction of potentiated startle as compared to their initial levels. However, rats receiving the excitatory amino acid antagonist did not show extinction. These data suggest that activity in the amygdala during nonreinforced light presentations is essential for extinction of light-potentiated startle.

Unlike the sensory cortex and the hippocampus, the amygdala is thought to be critical for the acquisition and performance of potentiated startle. Therefore, there are several very interesting explanations of these data. Perhaps the amygdala detects the occurrence of the CS and provides this information to other structures that are then responsible for feeding forward inhibition in the presence of nonreinforcement, either back to the amygdala or to one of its afferent or efferent structures. Alternatively, because the amygdala is thought to be a convergent zone of CS and US information, it may detect the absence of reinforcement and signal other structures for inhibition to be returned back to the amygdala or to one of its afferent or efferent structures. Lastly, perhaps the amygdala is itself responsible for inhibiting potentiated startle. The fact that activity in the amygdala is critical for extinction may indicate that the plasticity associated with extinction occurs in the amygdala itself. It should be acknowledged, however, that this experiment did not include a control for context-specific extinction produced by a change in drug state from extinction training to extinction testing (e.g., Bouton et al. [60]), that might account for these results. However, if this were true, the ''drug context'' would have to be that produced by local blockade of excitatory amino acid transmission in the amygdala. Whether this would be a sufficient cue to signal nonreinforcement remains to be tested.

Knowing that the sensory cortex and amygdala may both be important for acquisition of extinction provides a starting point for analysis of the structures involves in extinction. Further work combining anatomical and lesion techniques, like those described, will be critical for determining the structures involved in extinction, and how and where they may inhibit responding.

Pharmacology

Peptides: ACTH and Vasopressin

Work by DeWied, Van Wiersima, Izquierdo, and Richardson (104, 107–109, 111), and their co-workers indicates that administration of various peptides such as adrenocorticotropic hormone (ACTH) or vasopressin either before or after extinction training attenuates subsequent extinction performance. In one such experiment, Richardson, Riccio, and Devine (104) first trained rats to actively avoid footshocks in a shuttle box apparatus. Following training, the rats received a session of nonreinforced avoidance trials. Twenty-four hours later, the rats were tested for retention of the avoidance response. Ten minutes prior to the retention test, the rats were injected subcutaneously with either ACTH or vehicle. Although there were no differences in the number of avoidance responses among the groups during nonreinforcement, rats that had received ACTH prior to testing made more avoidance responses than vehicle-injected rats. Thus, ACTH appeared to attenuate extinction performance by reinstating avoidance responding.

The reinstatement of extinguished responding occurs not only with ACTH but also with vasopressin, and both peptides have been shown to attenuate extinction of active avoidance (104–107), passive avoidance (108,109) and conditioned heart rate (110). Moreover, the attenuation of extinction is evidenced both by blocked or slowed extinction when the peptide is administered prior to nonreinforcement (105,110) and by reinstatement of extinguished responding when administered after nonreinforcement but before testing (104,109).

The effect of these peptides on extinction has been equated with the reminder phenomenon reported by Rescorla and Heth (28) in which a single noncontingent foot shock can reinstate extinguished responding (109). Both of these effects have been attributed to their ability to reinstate the cues of original acquisition. More specifically, it is assumed that during original training endogenous ACTH is released (perhaps with the presentation of the aversive US) and thus becomes a component of the original training memory. Exogenous administration of ACTH prior to testing increases the similarity between the internal states at training and testing, and therefore facilitates retrieval of the excitatory association and reinstatement of the extinguished response (111). It is especially interesting in this context that drugs like yohimbine, which activates the release of norepinephrine in the brain, often produce flashbacks (memory of prior trauma) in patients with post-traumatic stress disorder (112). Because elevated levels of brain norepinephrine would have accompanied the traumatic event, this may be a very good human example of reinstatement.

In a direct test of this hypothesis, Ahlers and Richardson (111) sought to reduce the contribution of ACTH to the original training memory. Dexamethasone, a synthetic glucocorticoid that effectively inhibits release of endogenous ACTH, was used to reduce the contribution of endogenous ACTH during acquisition. Pretest administration of ACTH was effective in reinstating pre-nonreinforcement levels of responding only for rats given vehicle during acquisition and nonreinforcement. Administration of ACTH had no effect on extinguished avoidance responding in rats given dexamethasone prior to acquisition. These results suggest that ACTH prior to testing reinstates a portion of the stimulus complex acquired during acquisition, and therefore facilitates retrieval of the excitatory associations.

In the final analysis, this hypothesis does not seem to account for all the data. The reinstatement-like effect of ACTH can occur over very long injection-to-test intervals. ACTH administered 7 days prior to extinction testing (24 hours after extinction) fully reinstates extinguished avoidance responding (113). It seems unlikely that facilitated retrieval of acquisition memory occurs over this very long injection-to-test interval. Instead, effects over very long intervals are often indicative of some form of learning. On the other hand, peptides such as vasopressin can produce long-lasting sensitization-like effects (114,115) and recent data show that a single exposure to a cat can lead to long-lasting changes in the behavior of rats (116). Hence, neural mechanisms involved in sensitization may be a fruitful area of inquiry for explaining some of these phenomena.

Alternatively, because these peptides are known to be critically involved in the organism's response to stress, perhaps they exert their effects by producing a rather general increase in stress-related anxiety or fear. An increased level of fear might be expected to potentiate responding to all aversive CS. However, it is also difficult here to imagine an increased level of fear persisting over a 7-day injection-to-test interval. Moreover, experiments have demonstrated that intracerebroventricular and local microinjection of ACTH and vasopressin, in doses too small to be effective if administered peripherally, have behavioral effects similar to those obtained with systemic injections (106,107,117, 118) but without the concomitant changes in autonomic responses indicative of a stress reaction (106). In addition, ACTH attenuates extinction not only when an aversive US is used, but also in cases where a positive reinforcer is used (119). It would therefore appear that stress or a heightened level of nonspecific fear is not a major contributor to the effect of ACTH and vasopressin on extinction.

Instead, the effect of ACTH may be to somehow strengthen the associative CS–US connections formed during acquisition (120). Regardless of the injection-to-test interval, strengthening of excitatory-associative connections would lead to a decrement in extinction performance (i.e., through summation of excitatory and inhibitory associations). Consistent with this notion, pretraining administration of ACTH and vasopressin facilitates acquisition of active avoidance (121) and post training administration of ACTH or vasopressin facilitates the retention of active avoidance (120). Interestingly, Bohus, Nyakas, and Endroczi (122) found that the facilitatory effect of ACTH did not occur until after a minimum level of conditioning was demonstrated. This indicates that a certain level of conditioning must occur before ACTH can modulate acquisition, again suggesting that ACTH acts to strengthen existing CS–US connections.

It is conceivable that ACTH and vasopressin act directly at the CS–US convergent zones where memory is thought to reside. Recently, vasopressin has been found to improve the length of time that long-term potentiation is maintained in vitro (123). Because long-term potentiation may be a cellular mechanism involved in memory, these data suggest that vasopressin may strengthen certain types of memory. Alternatively, the strengthening of the CS–US connections could occur through modulation of effective intensity of the US at the CS–US convergent site. A stronger US would facilitate acquisition, improve retention, and attenuate extinction, perhaps through a US-specific sensitization-like process. Evidence for this later hypothesis comes from lesion and microinjection studies. For example, the parafasicular nucleus is believed to be important for the processing of noxious information (124). ACTH failed to reinstate extinguished avoidance responding in rats bearing lesions of the parafasicular nucleus. Similarly, vasopressin showed a blunted facilitatory effect on acquisition of active avoidance in parafasicular lesioned animals (124). Moreover, microinjection of ACTH into the area of the parafasicular nucleus attenuated extinction in a manner similar to systemic injections but in a concentration that was two to ten times smaller. Similarly, lesions of the amygdala block the effects of ACTH and vasopressin on extinction of an avoidance response (118).

The significance of these peptides may be twofold. Their release may not only act to facilitate retention of CS–US associations, but may also inhibit the formation of inhibitory CS–US associations. Thus far, the effect of these peptides on extinction has been treated as an effect secondary to their ability to increase acquisition and retention. However, their presence in certain brain areas may actually prevent inhibitory CS–US associations (i.e., extinction). Removal of the posterior pituitary, which decreases levels of ACTH, facilitates extinction while having no effect on acquisition (125). If ACTH influences extinction only through a direct effect on excitatory CS–US connections, then the effects on acquisition and extinction should always be parallel, facilitated acquisition should attenuate extinction, and, likewise, attenuation of acquisition should facilitate extinction. A divergence in this pattern suggests that independent processes are at work. The deWied (125) study is by no means definitive evidence for a direct effect of ACTH on an inhibitory process, and further study of the possible direct effects ACTH on an inhibition is warranted. For example, will ACTH block the establishment of a conditioned inhibitor?

Vasopressin and ACTH may provide the key to how the brain recognizes the absence of an unconditioned stimulus. One of the major problems confronting an inhibition hypothesis of extinction is the necessity of a comparator mechanism that can detect the absence of the US (126) and thus signal for inhibition to accrue (127). Vasopressin and ACTH are released in response to the aversive unconditioned stimulus and are therefore present in higher levels during conditioning than during extinction when the US is omitted. Lower levels of ACTH or vasopressin might perhaps permit inhibition to develop between the relevant neural structures either directly or through the release of some third factor triggered by the absence of increased levels of these peptides. Although it is consistent with the data discussed thus far, this hypothesis is difficult to distinguish from the view that ACTH and vasopressin simply act to strengthen excitatory connections, with the effects on extinction being a consequence of facilitated acquisition.

In summary, the peptides ACTH and vasopressin, which are released in response to noxious stimulation, facilitate acquisition and retention, and attenuate extinction of avoidance responses. The effect on acquisition and retention appears to involve strengthening of the CS–US associations. However, it is unclear whether the attenuation of extinction is the result of this strengthening or a direct result of ACTH and vasopressin preventing the formation of inhibitory connections.

GABA

A number of investigators have demonstrated the involvement of γ-aminobutyric acid (GABA) in conditioning. Administration of GABA agonists (loosely defined as drugs that influence GABA activity by affecting the GABA/benzodiazepine complex) such as muscimol, or of ethanol and the benzodiazepines, disrupts conditioning of active and passive avoidance responses (128,129). Unlike ACTH and vasopressin, which will influence performance when administered both before and some time after acquisition, the effect of the GABA agonists only occurs when they are administered prior to or immediately after acquisition. For example, the benzodiazepine diazepam, which facilitates GABAergic transmission, attenuates performance of active avoidance when tested 24 hours later, if given 30 minutes prior to acquisition of an active avoidance response. However, diazepam given 90 minutes after acquisition has no effect on subsequent performance (129). Therefore, GABA agonists appear to act by influencing the formation (i.e., acquisition) of associative connections rather than by influencing the maintenance or postacquisition strengthening of the connections. Similarly, GABA antagonists facilitate subsequent performance only when they are administered prior to or immediately after acquisition (128).

The influence of GABA antagonists on memory formation is not simply limited to excitatory associations. The GABA agonists and antagonists influence extinction in the same manner as acquisition; agonists attenuate and antagonists facilitate acquisition of extinction (129–132). For example, diazepam administered 30 minutes prior to nonreinforced tone trials attenuates extinction of tone-elicited active avoidance assessed 24 hours later (129), and the GABA antagonist picrotoxin facilitates extinction of a tone-induced suppression of Y-maze exploration (133). Unlike ACTH and vasopressin, which may attenuate extinction indirectly through their facilitation of excitation, GABA agonists and antagonists appear to act directly on the formation of both excitatory and inhibitory associative connections.

These data point to an important feature of extinction that is often ignored. Because GABA agonists and antagonists influence the acquisition of excitation and extinction in the same manner, perhaps the acquisition of excitation and extinction share some underlying process. This process may be the mechanism through which associative connections are formed. Konorski (10) proposed that excitation and extinction involve the strengthening of CS–US connections. The differences in excitation and extinction are in the sign of the connections and the circumstances under which excitation or extinction (i.e., inhibition) occur; that is, excitation is triggered by a coincident rise in activation of both the CS and US centers, whereas inhibition is triggered by a rise in activation of the CS center coincident with a fall in activation of the US center. The strengthening of connections is common to both excitation and extinction. Although the GABA data suggest that strengthening may be common to both excitation and extinction, it is not necessary that the explicit mechanism used by the nervous system to strengthen these different connections be the same.

Excitatory Amino Acids

It is not entirely clear how GABA might influence the strengthening of connections. GABA may be directly involved (e.g., by directly modulating currents necessary for long-term synaptic plasticity), or indirectly involved (e.g., by altering the excitability of cells afferent to those that undergo synaptic plasticity) in modulating synaptic plasticity. Therefore, it is difficult to know whether excitation and inhibition are formed by the same mechanism. However, recent data from our laboratory provides more compelling evidence suggesting that the acquisition of excitation and extinction may share the same mechanism.

The *N*-methyl-D-Aspartate (NMDA) receptors may be critically involved in synaptic plasticity related to learning. Electrophysiological and pharmacological studies have indicated that neuronal activation through the NMDA receptor is critical for a form of synaptic plasticity known as long-term potentiation (134). Consistent with the possible involvement of NMDA receptors in learning, infusion of the competitive NMDA receptor antagonist DL-2-amino-5-phosphonovaleric acid (AP5) prevents acquisition of several different behavioral responses (135,136).

For example, Miserendino et al. (135) demonstrated that AP5 infused directly into the amygdala blocks acquisition of fear-potentiated startle. Because synaptic activity in the amygdala is important for extinction of potentiated startle, we asked whether AP5 infused into the amygdala during nonreinforced CS presentations would prevent extinction of fear-potentiated startle (137).

Rats were implanted with bilateral cannulae aimed at the basolateral nucleus of the amygdala. Following light–footshock pairings, the rats' initial level of potentiated startle was assessed. One day later, the rats were infused with either AP5 or artificial cerebrospinal fluid vehicle immediately before being exposed to a session of nonreinforced light trials. Twenty-four hours later, the rats were tested for the amount of fear to the light remaining. Rats receiving artificial cerebrospinal fluid vehicle showed extinction, whereas rats who received AP5 showed no extinction. Furthermore, the work of Miserendino et al. (135) indicated that the AP5 blockade of acquisition of excitation could not be attributed to reduced salience of the light or to an anxiolytic-like effect of AP5. Therefore, it would appear that not only is the plasticity underlying extinction similar to the plasticity underlying excitation, but that, at least for potentiated startle, both may be occurring in the amygdala.

SUMMARY

At the behavioral level, a great deal is now known about extinction, whereby presentation of a CS in the absence of the US with which it was formerly paired leads to a reduction in the magnitude or probability of a conditioned response. Although a CS may no longer elicit a conditioned response following extinction, it is clear that many, if not most, of the original CS–US associations still exist so that the original memory has not been erased. Phenomena seen after extinction, such as spontaneous recovery (recovery of a conditioned response with the passage of time), reinstatement (recovery of a conditioned response following presentation of a US), or renewal (recovery of a conditioned response when testing takes place in a context different from the one used in extinction), all demonstrate this. Although data such as these can be readily explained by assuming that extinction leads to a separate inhibitory CS–US connection that competes with the original excitatory CS–US connection, this has been very difficult to demonstrate experimentally. On the other hand, a closely related phenomenon, conditioned inhibition, clearly seems to result from an inhibitory process. Conditioned inhibition is typically produced when one stimulus (e.g., light) is paired with shock, whereas another stimulus (e.g., a tone), when placed in compound with the light, is not followed by shock. One can show that the tone in this example will reduce fear in the presence of the light (summation) or a novel fearful stimulus (transfer). Moreover, if a US is then paired with the tone, fear of that tone will develop more slowly following such training compared to a condition in which the tone was not a conditioned inhibitor (retardation).

In contrast to the rich behavioral literature on extinction and conditioned inhibition, much less is known about the neural mechanisms involved in these phenomena. Sensory and frontal cortex, hippocampus, septum, and amygdala have all been implicated in extinction, but thus far a clear picture has not emerged. Peptides such as ACTH and vasopressin clearly retard extinction of active avoidance behavior. Some of these results can be accounted for by reinstatement, whereby exogenous administration of the peptides recreates the internal state experienced in original acquisition. However, other processes such as sensitization or retroactive strengthening of CS–US connections may be required to account for the longlasting effects of these peptides. Like excitatory conditioning, extinction also seems to be degraded by GABA agonists or excitatory amino acid antagonists and improved by GABA antagonists.

When investigating the neural basis of extinction, it is often difficult to determine whether a given treatment alters the magnitude of extinction by affecting extinction itself or the expression of excitatory conditioning. Hence, in extinction one only has one measure, namely the reduction of the conditioned response. In the conditioned inhibition paradigm, however, it is possible to measure the magnitude of the conditioned response by itself and then its reduction in the presence of a conditioned inhibitor. Therefore, we believe that conditioned inhibitor provides a superior experimental paradigm for analyzing brain systems involved in the inhibition of fear. If one could determine which brain area(s) and neurotransmitter systems are activated by a conditioned inhibitor, and where the inhibitor ultimately exerts its inhibitory effect on conditioned fear, important and novel information with considerable clinical relevance would be gained.

REFERENCES

1. Davis M. The role of the amygdala in conditioned fear. In Aggleton JP, ed. *The amygdala: neurobiological aspects of emotion, memory, and mental dysfunction.* New York: Wiley-Liss; 1992;255–305.
2. Gray TS. Autonomic neuropeptide connections of the amygdala. In Tache Y, Morley JE, Brown MR, ed. *Hans Selye Symposia on neuroendocrinology and stress.* New York: Springer Verlag; 1989;92–106.
3. Gloor P. Amygdala. In Field J, ed. *Handbook of physiology: sec. I. neurophysiology.* Washington, DC: American Physiological Society; 1960;1395–1420.
4. Kapp BS, Pascoe JP. Correlation aspects of learning and memory: vertebrate model systems. In Martinez JL, Kesner RP, ed. *Learning and memory: a biological view.* New York: Academic Press; 1986;399–440.
5. Kapp BS, Wilson A, Pascoe JP, Supple WF, Whalen PJ. A neuroanatomical systems analysis of conditioned bradycardia in the rabbit. In Gabriel M, Moore J, ed. *Neurocomputation and learning: foundations of adaptive networks.* New York: Bradford Books; 1990.
6. Ledoux JE. Emotion. In Plum F, ed. *Handbook of physiology, nervous system.* Washington, DC: American Physiological Society; 1987;416–459.
7. Sarter M, Markowitsch HJ. Involvement of the amygdala in learning and memory: a critical review with emphasis on antamocical relations. *Behav Neurosci* 1985;99(2):342–380.
8. Aggleton JP. *The amygdala: neurobiological aspects of emotion, memory, and mental dysfunction.* New York: Wiley-Liss; 1992.
9. Pavlov IP. *Conditioned reflexes.* Oxford: Oxford University Press; 1927.
10. Konorski J. *Conditioned reflexes and neuronal organization.* London: Cambridge University Press; 1948.
11. Konorski J. *Integrative Activity of the Brain.* Chicago: Chicago University Press, 1967.
12. Rescorla RA. Conditioned inhibition and excitation. In Dickinson A, Boakes RA, ed. *Mechanisms of learning and motivation: a memorial volume to Jerzy Konorski.* Hillsdale, NJ: Erlbaum; 1979.
13. Amsel A. Frustrative nonreward in partial reinforcement and discrimination learning: some recent history and a theoretical extension. *Psychol Review* 1962; 69:306–328.
14. McAllister WR, McAllister DE. Behavioral measurements of conditioned fear. In Brush FR, ed. *Aversive conditioning and learning,* New York; Academic Press; 1971;105–179.
15. Rescorla RA, Wagner AR. A theory of Pavlovian conditioning: variations in the effectiveness of reinforcement and nonreinforcement. In Black AH, Prokasy WF, ed. *Classical conditioning II: current research and theory.* New York: Appleton-Century-Crofts; 1972;64–99.
16. Skinner BF. Are theories of learning necessary? *Psychol Review* 1950;57(4):193–216.
17. Estes WK. Statistical theory of spontaneous recovery and regression. *Psychol Review* 1955;62(3):145–154.
18. Wagner AR, Rescorla RA. Inhibition in Pavlovian conditioning. In Boakes RA, Halliday MS, ed. *Inhibition and Learning,* London: Academic Press; 1972.
19. Pearce JM, Hall G. A model of Pavlovian conditioning: variations in the effectiveness of conditioned but not unconditioned stimuli. *Psychol Review* 1980;87:332–352.
20. Konorski J, Szwejkowska G. Chronic extinction and restoration of conditioned reflexes: I. Extinction against the excitatory background. *Acta Bio Exper* 1950;15(12):155–170.
21. Szwejkowska G, Konorski J. Chronic extinction and restoration of conditioned reflexes: IV. The dependence of the course of extinction and restoration of conditioned reflexes on the "history" of the conditioned stimulus. *Acta Bio Exper* 1952;16:95–113.
22. Frey PW, Butler CS. Extinction after aversive conditioning: an associative or nonassociative process? *Learn Motiv* 1977;8:1–17.
23. Smith M, Gormezano I. Effects of alternating classical conditioning and extinction sessions on the conditioned nictitating membrane response in the rabbit. *Psychonomic Sci* 1965;3:91–92.
24. Bouton ME. Slow reacquisition following extinction of conditioned suppression. *Learn Motiv* 1986;17:1–15.
25. Bouton ME, Swartzentruber D. Slow reacquisition following extinction: context, encoding, and retrieval mechanisms. *J Exp Psychol Anim Behav Proc* 1989; 15(1):43–53.
26. Estes WK, Skinner BF. Some quantitative properties of anxiety. *J Exp Pschol* 1941;29:390–400.
27. Wagner AR. Expectancies and the priming of short term memory. In Hulse SH, Fowler H, Honig WK, ed. *Cognitive processes in animal behavior,* Hillsdale, NJ: Erlbaum; 1978;177–209.
28. Rescorla RA, Heth CD. Reinstatement of fear to an extinguished conditioned stimulus. *J Exp Pschol Anim Behav Proc* 1975;1:88–96.
29. Bouton ME, Bolles RC. Role of contextual stimuli in reinstatement of extinguished fear. *J Exp Pschol Anim Behav Proc* 1979;5(4):368–378.
30. Rescorla RA, Holland PC. Associations in Pavlovian conditioned inhibition. *Learn Motiv* 1977;8:429–447.
31. Callen EJ, McAllister WR, McAllister DE. Investigations of the reinstatement of extinguished fear. *Learn Motiv* 1984;15:302–320.
32. Lovibond PF, Preston GC, Mackintosh NJ. Context specificity of conditioning, extinction, and latent inhibition. *J Exp Pschol Anim Behav Proc* 1984;10(3): 360–375.
33. Bouton ME, King DA. Contextual control of conditioned fear: tests for the associative value of the context. *J Exp Pschol Anim Behav Proc* 1983;9(3):248–256.

34. Bouton ME, Bolles RC. Context, event-memories, and extinction. In Balsam PD, Tomie A, ed. *Context and learning,* Hillsdale, NJ: Erlbaum; 1985.
35. Bouton ME, King DA. Effect of context with mixed histories of reinforcement and nonreinforcement. *J Exp Pschol Anim Behav Proc* 1986;12(1):4–15.
36. Batson JD, Best MR. Single-element assessment of conditioned inhibition. *Bull Psychonomic Society* 1981;18:328–330.
37. Wasserman EA, Franklin SR, Hearst E. Pavlovian appetitive contingencies and approach vs. withdrawal to conditioned stimuli in pigeons. *J Comp Physio Psychol* 1974;86:616–627.
38. Rescorla RA. Pavlovian conditioned inhibition. *Psychological Bulletin* 1969;72(2):77–94.
39. Hearst E. Some persistent problems in the analysis of conditioned inhibition. In Boakes RA, Halliday MS, ed. *Inhibition and learning.* New York: Academic Press; 1972;5–39.
40. Skinner BF. *Behavior of Organisms.* New York: Appleton-Century-Crofts, 1938.
41. Lubow RE, Moore A. Latent inhibition: effect of frequency. *J Comp Physiol Psychol* 1959;52:415–419.
42. Lubow RE. Latent inhibition: effects of frequency on nonreinforced exposure of the CS. *J Comp Physiol Psych* 1965;60:454–457.
43. Mackintosh NJ. *The psychology of animal learning.* New York: Academic Press, 1974.
44. Papini MR, Bitterman ME. The two-test strategy in the study of inhibitory conditioning. *J Exper Psychol: An Beh Proc* 1993;19:342–352.
45. Reiss S, Wagner AR. CS habituation produces a "latent inhibition effect" but no active "conditioned inhibition." *Learn Motiv* 1972;3:237–345.
46. Williams DA, Overmier JB, LoLordo VM. A reevaluation of Rescorla's dictums about Pavlovian conditioned inhibition. *Psychological Bulletin* 1992;111(2):275–290.
47. Reberg D. Compound tests for excitation in early acquisition and after prolonged extinction of conditioned suppression. *Learn Motiv* 1972;3:246–258.
48. Devito PL, Fowler H. Enhancement of conditioned inhibition via an extinction treatment. *Animal Learn Behav* 1987;15(4):448–454.
49. Schoenfeld WN. An experimental approach to anxiety, escape and avoidance behavior. In Hoch PH, Zubin J, ed. *Anxiety.* New York: Grune and Stratton; 1950.
50. Mowrer OH. On the dual nature of learning—a reinterpretation of "conditioning" and "problem-solving." *Harvard Educational Review* 1947;17:102–148.
51. Solomon RL, Wynne LD. Traumatic avoidance learning: the principle of anxiety conservation and partial irreversibility. *Psychol Review* 1954;61:353–385.
52. Levis DJ. The case for a return to a two-factor theory of avoidance: the failure of non-fear interpretations. In Klein SB, Mowrer RR, ed. *Contemporary learning theories: Pavlovian conditioning and the status of traditional learning theory.* Hillsdale, NJ: Lawrence Erlbaum; 1989;227–277.
53. Malloy PF. Incubation of human fear effects of UCS intensity, CS exposure, and individual differences. Doctoral Dissertation, Binghamton: SUNY; 1981.
54. Shipley RH. Extinction of conditioned fear in rats as a function of several parameters of CS exposure. *J Comp Physio Psychol* 1974;87:699–707.
55. Williams RW, Levis DJ. A demonstration of persistent human avoidance in extinction. *Bull Psychonomic Society* 1991;29(2):125–127.
56. Levis DJ, Boyd TL. Symptom maintenance: an infrahuman analysis and extension of the conversation of anxiety principle. *J Abnormal Pschol* 1979;88: 107–120.
57. Seligman MEP, Johnston JC. *A cognitive theory of avoidance learning.* Washington, DC: VH Winston; 1973.
58. Annau Z, Kamin LJ. The conditioned emotional response as a function of the intensity of the US. *J Comp Physio Psychol* 1961;54(4):428–432.
59. Bouton ME, Swartzentruber D. Sources of relapse after extinction in Pavlovian instrumental learning. *Clinic l Psychological Review* 1991;11:123–140.
60. Bouton ME, Kenney FA, Rosengard C. State-dependent fear extinction with two benzodiazepine tranquilizers. *Behav Neurosci* 1990;104(1):44–55.
61. Levis DJ. *Implosive theory: a comprehensive extension of conditioning theory of fear/anxiety to psychopathology.* New York: Academic Press, 1985.
62. Wagner AR. Stimulus selection and a "modified continuity theory." In Bower GH, Spence JT, ed. *The psychology of learning and motivation,* New York: Academic Press; 1969.
63. Brons JF, Woody CD. Long-term changes in excitability of cortical neurons after Pavlovian conditioning and extinction. *J Neuro Physio* 1980;44(3):605–615.
64. Gassanov UG, Merzhanova GK, Galashina AG. Interneuronal relations within and between cortical areas during conditioning in cats. *Behav Brain Res* 1985; 15:137–146.
65. Diamond DM, Weinberger NM. Classical conditioning rapidly induces specific changes in frequency receptive fields of single neurons in secondary and ventral ectosylvian auditory cortical fields. *Brain Res* 1986;372: 357–360.
66. LeDoux JE, Romanski L, Xagoraris A. Indelibility of subcortical memories. *J Cognitive Neuroscience* 1989; 1(3):238–243.
67. Teich AH, McCabe PM, Gentile CC, Schneiderman LS, Winters RW, Liskosky DR, Schneiderman N. Auditory cortex lesions prevent the extinction of heart rate conditioning to tonal stimuli in rabbits. *Brain Res* 1989;480:210–218.
68. Falls WA, Davis M. Visual cortex ablations do not prevent extinction of conditioned fear to a visual conditioned stimulus. *Behav Neural Bio* 1993;60:259–270.
69. Bouton ME, Bolles RC. Contextual control of the extinction of conditioned fear. *Learn Motiv* 1979;10: 455–466.
70. Miserendino MJD, Coons EE. Hedonic interactions of medial prefrontal cortex and nucleus reticularis gigantocellularis. *Brain Res* 1989;483:233–250.
71. Siegel A, Edinger H, Lowenthal H. Effects of electrical stimulation of the medial aspects of the prefrontal cortex upon attack behavior in cats. *Brain Res* 1974; 66:467–479.
72. Maskati H, Ali A, Zbrozyna AW. Stimulation of prefrontal cortex area inhibits cardiovascular and motor components of the defense reaction in rats. *J Autonomic Nervous System* 1989;28:117–126.
73. Zbroznya AW, Westwood DM. Stimulation in prefrontal cortex inhibits conditioned increase in blood pressure and avoidance bar pressing in rats. *Physiol Behav* 1991;49:705–708.

74. Deutch AY, Clark WA, Roth RH. Prefrontal cortex dopamine depletion enhances the responsiveness of mesolimbic dopamine neurons to stress. *Brain Res* 1990;521:311–315.
75. Jaskiw GE, Karoum FK, Weinberger DR. Persistent elevations in dopamine and its metabolites in the nucleus accumbens after mild subchronic stress in rats with ibotenic acid lesions of the medial prefrontal cortex. *Brain Res* 1990;534:321–323.
76. Butter CM. Preservation in extinction and in discrimination reversal tasks following selective frontal ablations in macaca mulatta. *Physiol Behav* 1969;4: 163–171.
77. Frysztak RJ, Neafsey EJ. The effect of medial frontal cortex lesions on respiration, "freezing," and ultrasonic vocalizations during conditioned emotional responses in rats. *Cerebral Cortex* 1991;1:418–425.
78. Morgan MA, Coons EE, LeDoux JE. Differential contributions of ventral and dorsal areas of medial prefrontal cortex to fear conditioning. Society for Neuroscience Abstracts. Washington, DC: 1993.
79. Rosen JB, Hitchcock JM, Miserendino MJD, Falls WA, Campeau S, Davis M. Lesions of the perirhinal cortex, but not of the frontal, medial pre-frontal, visual or insular cortex block fear-potentiated startle using a visual conditioned stimulus. *J Neurosci* 1991;12: 4624–4633.
80. Douglas RJ. Pavlovian conditioning in the brain. In Boakes RA, Halliday MS, ed. *Inhibition and learning*. London: Academic Press; 1972.
81. Berger TW, Weikert CL, Basset JL, Orr WB. Lesions of the retrosplinal cortex produce deficits in reversal learning of the rabbit nictitating membrane response: implications for potential interactions between hippocampal and cerebellar brain systems. *Behav Neurosci* 1986;100(6):802–809.
82. Buchanan SL, Powell DA. Divergencies in Pavlovian conditioned heart rate and eyeblink responses produced by hippocampectomy in the rabbit (Oryctolagus cuniculus). *Behav Neural Bio* 1980;30:20–38.
83. Schmaltz LW, Theious J. Acquisition and extinction of a classically conditioned response in hippocampectomized rabbits (oryctolagus cuniculus). *J Comp Physio Psychol* 1972;79(2):328–333.
84. Solomon PR. Role of hippocampus in blocking and conditioned inhibition of the rabbit's nictitating membrane response. *J Comp Physio Psychol* 1976;91: 407–417.
85. Leaton RN, Borszcz GS. Hippocampal lesions and temporally chained conditioned stimuli in a conditioned suppression paradigm. *Psychbiology* 1990;18 (1):81–88.
86. Ackil JE, Mellgren RL, Halgren C, Frommer GP. Effects of preexposures on avoidance learning in rats with hippocampal lesions. *J Comp Physiol Psych* 1969;69(4):739–747.
87. Isaacson RL, Douglas RJ, Moore RY. The effect of radical hippocampal ablation on acquisition of avoidance response. *J Comp Physio Psychol* 1961;54 (6):625–628.
88. Lovely RH, Grossen NE, Moot SA, Bauer RH, Peterson JJ. Hippocampal lesions and inhibition of avoidance behavior. *J Comp Physio Psychol* 1971;77(2): 345–352.
89. Antelman SM, Brown TS. Hippocampal lesions and shuttle box avoidance behavior: a fear hypothesis. *Physiol Behav* 1972;9:15–20.
90. Albert DJ, Richmond SE. Septal hyperreactivity: a comparison of lesions within the septum and adjacent to the septum. *Physiology and Behavior* 1975;15: 339–347.
91. Albert DJ, Brayley KN. Mouse killing and hyperreactivity following lesions of the medial hypothalamus, the lateral septum, the bed nucleus of the stria terminalis, or the region ventral to the anterior septum. *Physiol Behav* 1979;23:439–443.
92. Brown GE, Remley NR. The effects of septal and olfactory bulb lesions on stimulus reactivity. *Physiol Behav* 1971;6:497–501.
93. Albert DJ, Brayley KN, Milner JA. Connections from the lateral septum modulating reactivity in the rat. *Physiol Behav* 1978;21:761–767.
94. Melia KR, Davis M. Effects of septal lesions on fear-potentiated startle, and on the anxiolytic effects of buspirone and diazepam. *Physiol Behav* 1991;49: 603–611.
95. Dalby DA, Shuttlesworth DE. Effect of septal lesions, required response, and shock on the acquisition of a two-way conditioned avoidance response in rats. *Physiol Psychol* 1978;6(1):11–14.
96. Deagle JH, Lubar JF. Effect of septal lesions in two strains of rats on one-way and shuttle avoidance acquisition. *J Comp Physiol Psych* 1971;(77):2.
97. Powell DA, Milligan WL, Mull P. Lateral septal lesions enhance conditioned bradycardia in the rabbit. *J Comp Physio Psychol* 1982;96(5):742–754.
98. Thomas JB, Van Atta L. Hyperirritability, lever-press avoidance, and septal lesions in the albino rat. *Physiol Behav* 1972;8:225–2323.
99. Lee EHY, Lin WR, Chen HY, Shiu WH, Liang KC. Fluoxetine and 8-OH-DPAT in the lateral septum enhances and impairs retention of an inhibitory avoidance response in rats. *Physiol Behav* 1992;51:681–688.
100. Joels M, Gallagher JP. Actions of serotonin recorded intracellularly in the rat dorsolateral septal neurons. *Synapse* 1988;2:45–53.
101. Yadin E, Thomas E. Septal correlates of conditioned inhibition. *J Comp Physio Psychol* 1981;95(2): 331–340.
102. Strickland CE, Thomas E, Yadin E. Effects of chlordiazepoxide on septal single unit activity following Pavlovian conditioning: a regional comparison. Society for Neuroscience Abstracts; Anahiem, CA; 1992.
103. Yadin E, Thomas E. Lateral septal stimulation abolishes cold-immobilization stress-induced ulcers. Society for Neuroscience Abstracts; Anaheim, CA; 1992.
104. Richardson R, Riccio DC, Devine L. ACTH-induced recovery of extinguished avoidance responding. *Physiol Psychol* 1984;12:184–192.
105. Hagan JJ. Effects of lysine vasopressin and response prevention on avoidance responding in extinction. *Behav Neural Bio* 1982;36:204–210.
106. Koob GF, Dantzer R, Bluthe R, Lebrun C, Bloom FE, LeMoal M. Central injections of arginine vasopressin prolong extinction of active avoidance. *Peptides* 1986; 7:213–218.
107. van Wimersma TB, de Weid D. Effects of systemic and intracerebral administration of two opposite acting ACTH-related peptides on extinction of conditioned avoidance behavior. *Neuroendocrinology* 1971;7: 291–301.

108. Ader R, deWied D. Effects of lysine vasopressin on passive avoidance learning. *Psychonomic Science* 1972; 29(1):46–48.
109. Lashley RL, Richardson R, Riccio DC. ACTH-and non contingent footshock-induced recovery of extinguished passive avoidance response. *Physiol Behav* 1987;40:677–680.
110. Hernandez LL, Powell DA. Vasopressin analog delays extinction of classically conditioned bradycardia. Peptides 4 37-41. Vasopressin analog delays extinction of classically conditioned bradycardia. *Peptides* 1983;4: 37–41.
111. Ahlers ST, Richardson R. Administration of dexamethasone prior to training blocks ACTH-induced recovery of an extinguished avoidance response. *Behav Neurosci* 1985;99(4):760–764.
112. Southwick SM, Krystal JH, Morgan CA, Johnson D, Nagy LM, Nicolaou A, Heninger GR, et al. Abnormal noradrenergic function in posttraumatic stress disorder. In press.
113. Ahlers ST, Richardson R, West C, Riccio DC. ACTH produces long lasting recovery following partial extinction of an active avoidance response. *Behav Neural Bio* 1989;51:102–107.
114. Bernard DM, Pitman QJ, Veale WL. Increased motor disturbances in response to arginine vasopressin following hemorrhage or hypotonic saline: evidence for central AVP release in rats. *Brain Res* 1983;273: 59–65.
115. Willcox BJ, Poulin P, Veale WL, Pitman QJ. Vasopressin-induced motor effects: localization of a sensitive site in the amygdala. *Brain Res* 1992;596:58–64.
116. Adamec RE, Shallow T. Lasting effects on rodent anxiety of a single exposure to a cat. *Physiol Behav* 1993;54(1):101–109.
117. Kovacs GL, Bohus B, Versteeg HG, deKleot ER, DeWied D. Effect of oxytocin and vasopressin on memory consolidation: sites of action and catacholaminergic correlates after local microinjection into limbic-midbrain structures. *Brain Res* 1979;175:303–314.
118. van Wimersma Greidanus TB, Croiset G, Bakker E, Bouman H. Amygdaloid lesions block the effect of neuropeptides (vasopressin, ACTH 4-10) on avoidance behavior. *Physiol Behav* 1979;22:291–295.
119. Osborne B, Silverhart T, Markgraf C, Seggie J. Effects of fornix transection and pituitary-adrenal modulation on extinction behavior. *Behav Neurosci* 1987;101(4): 504–512.
120. Izquierdo I, Pereira EM. Post-training memory facilitation blocks extinction but not retroactive interference. Behavioral and Neural Biology 5 108-113. Post-training memory facilitation blocks extinction but not retroactive interference. *Behav Neural Bio* 1989;51:108–113.
121. Bohus B, Endroczi Ex. The influence of pituitary-adrenocortical function on the avoiding conditioned reflex in rats. *Acta Physiol Hungary* 1965;26:183–189.
122. Bohus B, Nyaas C, Endroczi E. Effects of adrenocorticotropic hormone on avoidance behavior of intact and adrenalectomized rats. *Inter J Neuropharm* 1967;7: 307–314.
123. van den Hoof P, Urban IJA, deWied D. Vasopressin maintains long-term potentiation in rat lateral septum slices. *Brain Res* 1989;505:181–186.
124. van Wimersma Greidanus TB, Bohus B, deWied D. Differential localization of the influence of lysine vasopressin and of ACTH 4-10 on avoidance behavior: a study in rats baring lesions in the parafasicular nuclei. *Neuroendocrinology* 1974;14:280–288.
125. deWied D. The influence of the posterior and intermediate lobe of the pituitary and the pituitary peptides on the maintenance of a conditioned avoidance response in rats. *Inter J Neuropharm* 1965;4:157–167.
126. Pfautz PL, Wagner AR. Transient variations in responding to Pavlovian conditioned stimuli have implications for the mechanism of "priming." *Animal Learn Behav* 1976;4(2):107–112.
127. Donegan NE, Gluck MA, Thompson RF. Integrating behavioral and biological models of classical conditioning. In Hawkins RD, Bower GH, ed. *Psychology of learning and motivation: vol. 22. Computational models of learning in simple neural circuits.* New York: Academic Press; 1989.
128. Castellano C, Pavone F. Effects of ethanol on passive avoidance behavior in the mouse: involvement of GABAergic mechanisms. *Pharmacology, Biochemistry and Behavior* 1988;29:321–324.
129. Pereia ME, Rosat R, Huang CH, Godoy MGC, Izquierdo I. Inhibition by diazepam of the effect of additional training and of extinction on the retention of shuttle avoidance behavior in rats. *Behav Neurosci* 1989;103(1):202–205.
130. Goldman MS. Effect of chlorodiazepoxide administered early in extinction on subsequent extinction of a conditioned emotional response in rats: implications for human clinical use. *Psychol Reports* 1977;40: 783–786.
131. Pereia ME, Dalmaz C, Rosat R, Izquierdo I. Diazepam blocks the interfering effects of post-training behavioral manipulations on retention of a shuttle avoidance task. *Psychopharm* 1988;94:402–404.
132. Ziskind D, Amit Z. Resistance to extinction of an avoidance response in rats following the administration of chlorodiazepoxide (Librium) or diazepam (Valium). *Psychopharmacologia* 1974;38:231–238.
133. McGaugh JL, Castellano C, Brioni J. Picrotoxin enhances latent extinction of conditioned fear. *Behav Neurosci* 1990;104(2):264–267.
134. Bliss TV, Collingridge GL. A synaptic model of memory: long-term potentiation in the hippocampus. *Nature* 1993;361:31–39.
135. Miserendino MJD, Sananes CB, Melia KR, Davis M. Blocking of acquisition but not expression of conditioned fear-potentiated startle by NMDA antagonists in the amygdala. *Nature* 1990;345:716–718.
136. Morris RGM, Anderson E, Lynch GS, Baudry M. Selective impairment of learning and blockade of long-term potentiation by an N-methyl-D-aspartate receptor antagonist AP5. *Nature* 1986;319:774–776.
137. Falls WA, Miserendino MJD, Davis M. Extinction of fear-potentiated startle: Blockade by infusion of an excitatory amino acid antagonist into the amygdala. *J Neuroscience* 1992;12(3):854–863.

Neurobiological and Clinical Consequences of Stress: From Normal Adaptation to PTSD, edited by M. J. Friedman, D. S. Charney, and A. Y. Deutch, Lippincott-Raven Publishers, Philadelphia 1995.

12

Sensitization and Kindling

Implications for the Evolving Neural Substrates of Post-Traumatic Stress Disorder

Robert M. Post, Susan R. B. Weiss, and Mark A. Smith

Biological Psychiatry Branch, National Institute of Mental Health, Bethesda, Maryland 20892

Behavioral sensitization and kindling have in common the ability to increase behavioral or physiological responsivity to repeated presentation of the same inducing stimulus. As such, both involve memory-like mechanisms, with the behavioral adaptation of unique interest because the output increases (sensitizes) rather than decreases (tolerates) in magnitude. Both models convey long-lasting consequences for the neurobiology and behavioral reactivity of the organism. In this fashion, each of these two very different physiological processes becomes an interesting model for various and differential aspects of post-traumatic stress disorder (PTSD) and its long-lasting consequences to the organism, including increased responsivity to selective external and internal cues. In both of these models, recent rapid progress has been achieved in understanding at least some of their underlying neurobiological determinants. In particular, an initial glimpse can be provided as to how changes in gene expression might be involved in the long-term alterations resulting from both sensitization and kindling. This may help create a paradigm shift (1), and a focus on the exquisite adaptations involved in the molecular genetics of experience (i.e., normal and pathological mechanisms of learning and memory at the level of genomic regulation), just as earlier data stressed the importance of protein synthesis for long-term memory. The models may also help to redefine PTSD as a synaptic connectivity disease.

While neither model meets most nor all of the requirements of an animal model of PTSD, as discussed by Yehuda and Antelman (2), each may provide a different conceptual framework for consideration of some of the factors and principles that may underlie long-term changes in responsivity to stressors. Specifically, we suggest that some of the principles involved in laying down the differential memory traces of sensitization and kindling may be useful bridging structures for considering parallel processes in different neural substrates that could be occurring in the related processes of stress sensitization and memory "branding" in PTSD. We suggest that these models may provide a primitive template for considering how acute or repeated stressors may not only leave permanent memory traces, but also affect the biochemistry and microstructure of the brain, potentially by impacting immediate-early-gene- and late-effector-gene expression. Stress sensitization and kindling also have their own spatiotemporal unfolding of cascades of neurobiological events that may have important mechanistic and pharmacotherapeutic implications.

Cocaine sensitization also has the positive attribute of a model of PTSD, in that it shows cross-sensitization to a variety of stressors. Numerous investigators have indicated that stress can sensitize to psychomotor stimulant-induced behavioral sensitization and vice versa (3–4). As such, stimulant-induced behavioral sensitization and stress sensitization are not entirely homologous, but may occupy neurobiological substrates sufficiently in common to demonstrate such prominent crossreactivity. Kindling also focuses on the cumulative effects of repeated stimuli, which can progress from behavioral and physiological phenomena that are exogenously triggered to those that are endogenously triggered (i.e., automatic or spontaneous). Both the cocaine-induced behavioral endpoint (motor activity) and the kindled endpoint (electrophysiological seizures) are easily observable and readily quantifiable, and thus are particularly helpful in tracking memory-like alterations in responsivity based on magnitude, frequency, or duration of initial stimuli. Sensitization may be most relevant to the emotional and contextual aspects of memory, while kindling may be most relevant to the idetic and permanent aspect of memory. Elements of both may thus be involved in PTSD.

Thus, these indirect models have a utility when focusing on qualitative and quantitative aspects of the inducing stimuli, as well as variations in the host, as important variables in determining long-term consequences. They highlight the role of both conditioned (contextual) and unconditioned elements in the process, based on long-term changes in physiology, biochemistry, and synaptic and neural anatomy involving complex patterns of changes in gene expression. In this fashion they help one focus on the longitudinal unfolding of the PTSD syndrome, by emphasizing the dynamic evolving nature of the behavioral changes and their underlying neural substrates. The anatomy and pharmacology of both kindling and sensitization vary as a function of time (phase of syndrome evolution). These and related principles can thus form a conceptual and heuristic bridge for further exploring their relevance for PTSD and its more classical behaviorally homologous, stress-induced laboratory models.

COCAINE SENSITIZATION

Dose, Frequency of Repetition, and Context

In cocaine sensitization, prior exposure to cocaine increases behavioral response in a fashion that depends on dose, number of repetitions, chronicity versus intensity of administration, and environmental context, in addition to a wealth of other variables in the host organism, for example, (gender, genetic strain, and developmental phase (5). Following a single large dose of cocaine (40 mg/kg i.p.) on day 1, animals show a greater response to rechallenge on day 2 with a low dose of cocaine (10 mg/kg i.p.) compared with animals pretreated with saline. This behavioral sensitization is context dependent or conditioned, as animals pretreated with the same high dose of cocaine in a different environment show no greater response to the cocaine challenge on day 2 than saline controls (6). If animals receive multiple (three or more) daily high doses of cocaine, a context-independent type of sensitization may also emerge; animals pretreated even in a different environmental context will show increased reactivity (although animals pretreated in the same test environment show a more robust and longer-lasting sensitization). A single low dose of cocaine on day 1 is insufficient to produce sensitization. However, this low dose, received once daily for 1 week to 10 days, is sufficient to produce a robust and long-lasting sensitization. It is, again, highly context-dependent in that the animals treated with the repeated low doses of cocaine in a different environment are no more reactive than saline controls.

Thus, as summarized in Fig. 1, higher doses produce more robust behavioral sensitization than lower doses. However, with repetition, lower doses also come to evoke a highly robust and long-lasting sensitization. With sufficient repetition of high enough doses of cocaine, a context-independent sensitization emerges, suggesting that different mechanisms come into play. With very high doses of cocaine, the local anesthetic components of cocaine become relevant, and the animal may eventually demonstrate seizures in response to a dose of cocaine that

COCAINE		EFFECT									
Number of Injections	Dose mg/kg i.p.	Behavioral Sensitization	Duration	Activity Context Dep.	Activity Context Indep.	Stereotypy Context Dep.	Stereotypy Context Indep.	Saline Conditioning	Sensitization Neuroleptic Independent	Seizure Kindling	Death
↑↑↑	COC_{65}									++	++
x 10 days	COC_{160} subcut. (K. Gale)	++			++		++		++		
↑↑↑↑↑↑↑↑↑↑	COC_{10}	++	months	++	0			++			
↑↑↑	COC_{40}	+		0	++	++	++	±			
↑	COC_{40}	++	days	++	0	0	0	0	0		
↑↑↑	COC_{20}	0	0	0	0	0	+				
↑↑↑	COC_{10}	0	0	0	0	0	0				
↑	COC_{10}	0	0								

FIG. 1. Size and number of arrows indicate dose of cocaine and number of administrations. COC, cocaine; subscript, dose in mg/kg administered once daily i.p. except in the study of Gale (100), when doses were subcutaneous; 0=no effect; ±=equivocal; +=moderate effect; ++=marked or definite effect. Effects of cocaine increase and are more persistent with increases in either dose or number of repetitions. The effects of the highest doses are less dependent on environmental context and conditioning, and are associated with seizures, kindling, and lethality. Thus, effects shift from dopaminergic-mediated behavioral sensitization (motor endpoint) to local anesthetic-mediated kindling (seizure endpoint) as the dose is increased.

was previously subconvulsant, a process known as "pharmacological kindling."

Thus, as in PTSD syndromes, dose and numbers of repetitions are important variables in determining the magnitude, duration, and conditioned components of cocaine-induced behavioral sensitization. In addition, the degree of chronicity versus intermittency of dosage is an important variable. King and associates (7) have demonstrated that continuous rather than intermittent stimulant administration is associated with decreased rather than increased reactivity to the stimulant in some circumstances.

If one examines only the issue of magnitude of induction and number of repetitions, as illustrated in Fig. 1, a provisional template of prediction could be constructed for the putative magnitude and duration of stress sensitization, as illustrated in Fig. 2. That is, similar to a low dose of cocaine, a minor stress may not be asso-

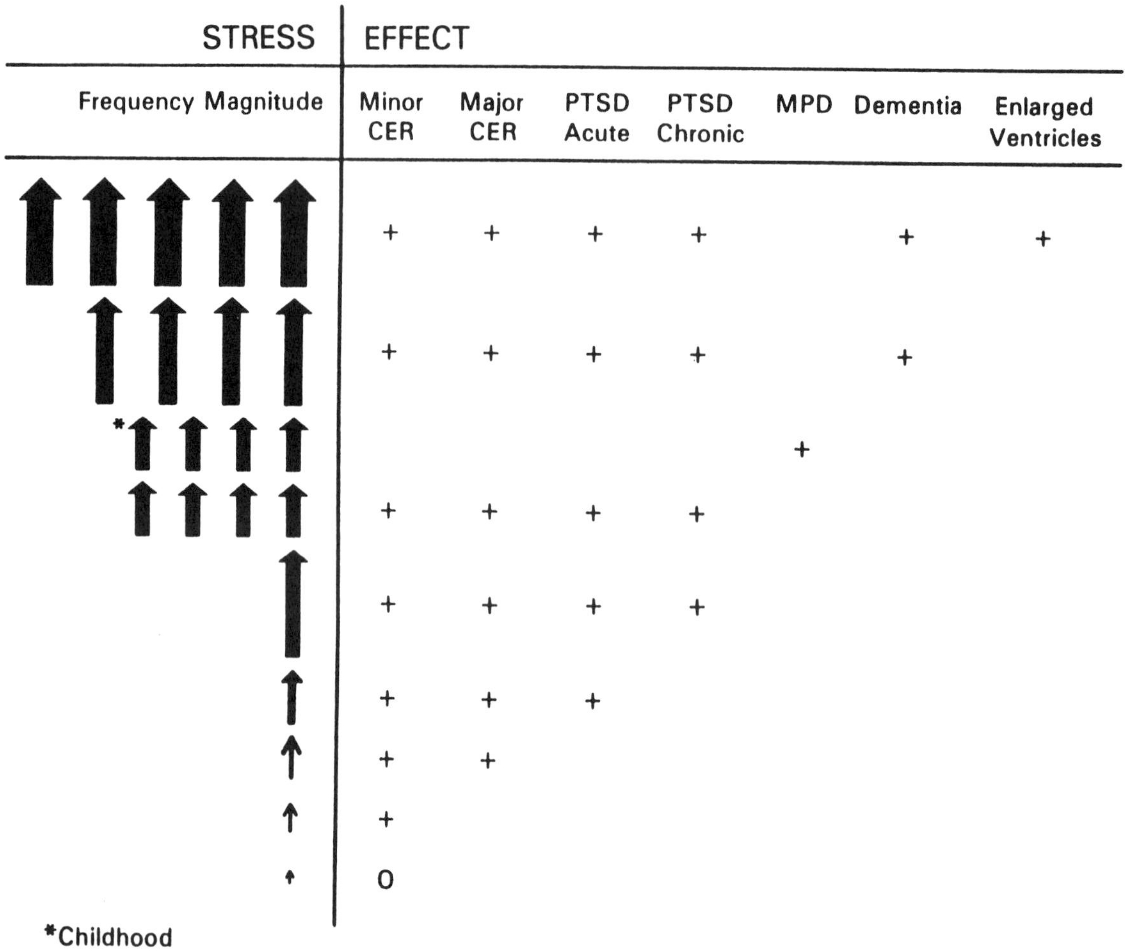

FIG. 2. Hypothetical schema of the relationship between magnitude and number of repeated exposures to stressors and development, severity, and persistence of behavioral pathology. CER, conditioned emotional response; PTSD, post-traumatic stress disorder; MPD, multiple personality disorder. *Developmental phase in which stressors occur may also be a crucial variable. A variety of other variables, including quality, unexpectedness, uncontrollability, and intermittency, in addition to severity and frequency, may markedly influence behavioral outcome.

ciated with any significant sensitization, while a more major or full-blown stressor (equivalent to the high dose challenge with cocaine) might produce robust sensitization. Repeated intermittent stressors may also produce robust sensitization, with a greater conditioned component of the sensitization being predicted. With repeated intermittent extreme stressors, the most profound effects might occur. As postulated by Putnam (8), if repeated intermittent extreme stressors (such as those associated with physical or sexual abuse) were occurring during crucial phases of neural development and the associated evolution of personality, deviations might occur in both of these processes that result in some of the symptomatology of dissociative and multiple personality disorders. Finally, at the most extreme levels of stress, an anatomical change may occur in association with preliminary reports of organic mental syndromes and dementia, particularly in victims of extreme torture. Anecdotal reports support the possibility that increases in cortical atrophy can be observed in these circumstances (9), and a more recent controlled investigation revealed decrements in the size of the hippocampus as measured by MRI.

Conditioned Biochemistry

The neural substrates underlying context-dependent and context-independent behavioral sensitization have begun to be explored, and they reveal interesting potential candidate biochemical and neuroanatomical systems. Fontana and associates (10) have found that the conditioned components of sensitization are associated with an increased dopamine overflow in the nucleus accumbens (but not in striatum or amygdala) in the one-day, high-dose cocaine sensitization paradigm. These data are of interest in suggesting that an essentially psychological effect (conditioning) can be mediated by regionally specific neurochemical correlates (increased dopamine overflow in the nucleus accumbens) in response to a cocaine challenge. Brown, Fibiger, and associates (11) failed to see a similar conditioned increase in dopamine overflow in response to a saline challenge, suggesting the possibility that the cocaine challenge provides a more robust conditioned cue because of the presence of both exteroceptive and interoceptive cues as well as the presence of cocaine providing a reuptake blockade augment subtle alterations in dopamine overflow. The preliminary data with cocaine are consistent with those of Kalivas and associates (12,13) and earlier investigators who have revealed neurochemical alterations with conditioned components of sensitization. These findings become of interest in relation to conditioned cuing in stress sensitization, raising the possibility that conditioned stressors are capable of inducing profound and regionally specific changes in dopamine function.

Evolving Pharmacological and Neuroanatomical Substrates

Both differences in pharmacological responsivity and the effect of lesions suggest that the substrates underlying sensitization achieved by a single cocaine challenge may differ from those involving more repeated cocaine administration. For example, drugs such as neuroleptics, which are effective in the blockade of the development of cocaine-induced behavioral sensitization, are ineffective in blocking the expression of this sensitization (14–15). That is, haloperidol administered prior to a day-1 cocaine challenge prevents behavioral sensitization as revealed by the day-2 dose rechallenge. In contrast, if animals receive high doses of cocaine on day 1 and then are pretreated with haloperidol prior to the day-2 rechallenge, the drug is unable to block the expression of this sensitization. Similar findings have been observed with selective D1 and D2 antagonists, as reviewed by Fontana and associates (16) and Post (17). Taken together, these data suggest that dopaminergic mechanisms are important to the development of behavioral sensitization, but that other mechanisms are induced in the maintenance or expression of this sensitization.

Preliminary studies have suggested that the α-2 agonist clonidine, the benzodiazepine agonist diazepam, and the 5-HT_3 antagonist MDL7002 are all capable of blocking the expression of sensitization suggesting that these neural systems may become involved in this later phase of the process (14,17). These findings, and the differential effects of these substances on the induction versus expression of cocaine-induced behavioral sensitization, are highly reminiscent of those involved in long-term potentiation where NMDA receptors are involved in the induction but not the expression of this phenomenon. In this regard, it is of interest that the allosteric blocker of *N*-methyl-D-aspartate (NMDA) receptor MK801 is capable of blocking cocaine-induced behavioral sensitization as well (18; and Pert et al., unpublished observation). Whether MK801 also blocks expression of behavioral sensitization is currently under investigation.

Given these pharmacological dissociations suggesting that different chemical mechanisms underlie the development versus expression of sensitization, we also asked whether a different neuroanatomical substrate may be involved in the one-day conditioned or three-day conditioned and unconditioned senzitization paradigm. Lesions of the amygdala or nucleus accumbens blocked development of the conditioned of cocaine-induced behavioral sensitization, while lesions of the dorsal and ventral hippocampus, cerebellum, and frontal cortex were without effect (Pert and Weiss, unpublished observation). We have postulated that the nucleus accumbens is required in order to mediate the rewarding aspects of cocaine or its motor program necessary for sensitization, while the amygdala is important for conveying context-dependent cues related to the environment in this 1-day conditioning paradigm.

However, in the repeated high-dose paradigm, cocaine induces sensitization that is no longer blocked by amygdala lesions. This suggests that, with repetition, sensitization becomes amygdala-independent and involves other pathways.

These data are reminiscent of the double dissociation observed with representational versus habit memory in the paradigm of Mishkin and associates (19). In these instances, representational, or one-trial, learning and memory mechanisms have been demonstrated to be based on limbic substrates, while habit memory (often induced by paradigms that require repeated presentations) appears to be independent of the limbic system but may be blocked by lesions of the striatum. (20,21). We are currently investigating whether striatal lesions are sufficient to block the repeated cocaine sensitization paradigm.

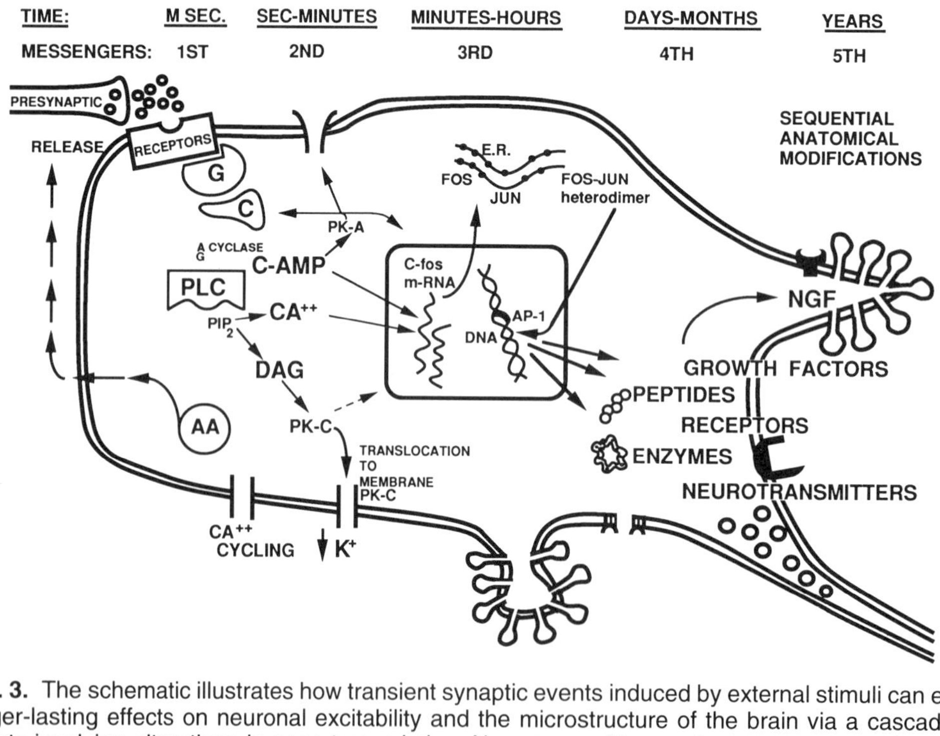

FIG. 3. The schematic illustrates how transient synaptic events induced by external stimuli can exert longer-lasting effects on neuronal excitability and the microstructure of the brain via a cascade of effects involving alterations in gene transcription. Neurotransmitters activate receptors and second messenger systems, which then induce immediate-early genes (IEGs) such as c-fos and c-jun. Fos and Jun proteins are synthesized on the endoplasmic reticulum (ER) and then bind to DNA to further alter transcription of late effector genes (LEGs) and other regulatory factors, the effects of which could last for months or years. PLC, phospholipase C; PIP_2, phosphatidyl inositol 4,5-biphosphate; AA = arachidonic acid; DAG = diacylglycerol; Pk-C = protein kinase C; AP-1 = activator protein 1 (binding site on DNA); ER, endoplasmic reticulum; Pk-A, protein kinase A; NGF, nerve growth factor.

Taken together, these data suggest that different modes of learning and memory, based at least in part on differential numbers of inductions, may occupy different neural substrates, and that these may evolve over time. This theme is further illustrated in the next section on the evolving neural substrates that underlie amygdala kindling, and in the paradigm of Rose (22), addressing more subtle aspects of learning and memory.

Immediate-Early Genes

Recent data suggest that the psychomotor stimulants amphetamine and cocaine produce increases in the immediate-early gene (IEG) c-fos, which is thought to act as a transcription factor helping to choreograph further alterations in gene expression, as outlined in Fig. 3. That is, following a specific extracellular signal that depolarizes or fires the cell in association with second messenger events, Fos, Fos-related antigens (FRAS), Jun, and related transcription factors are induced, which forms heterodimers that bind strongly to an AP-1 DNA binding site in the promoter region of many genes (23). Induction of these transcription factors may influence the expression of intermediate substances or late effector genes that could modulate neurotransmitter, peptide, or receptor levels in a more long-lasting fashion.

Cocaine and amphetamine appear to induce c-fos through a D1 receptor mechanism (24–28). Recent data also indicate that these psychomotor stimulants are capable of activating another IEG, zif268, which has been intimately associated with other processes of learning and memory such as long-term potentiation (29). Both increases and decreases in c-fos and other IEG

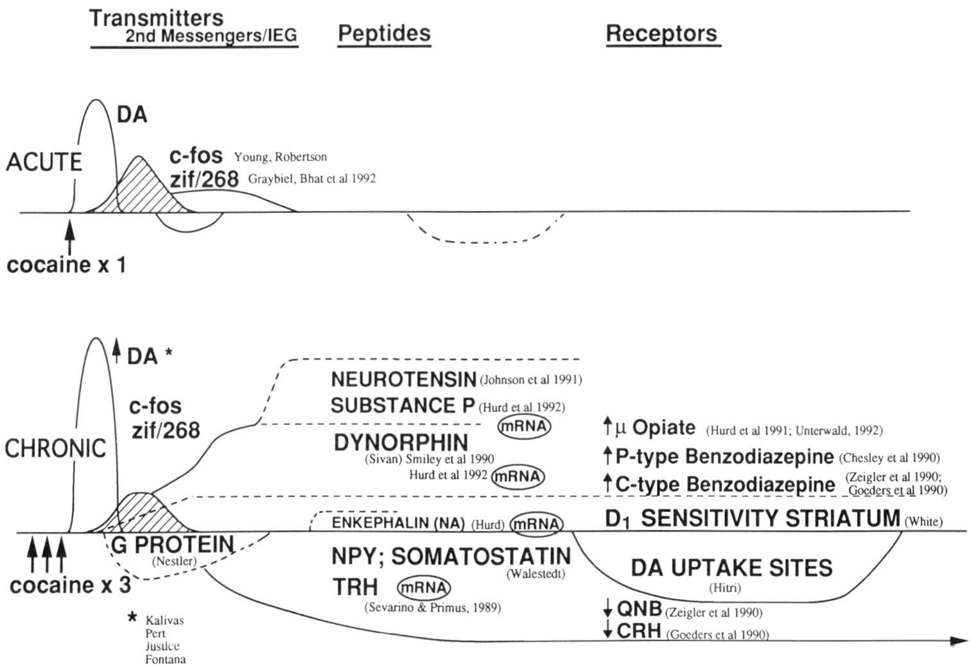

FIG. 4. Evolution of biological changes associated with stimulant-induced behavioral sensitization. Acute cocaine administration (↑) induces immediate-early genes (IEGs) such as c-fos and zif268 while chronic or repeated administration (↑ ↑ ↑) is associated with transient to long-lasting changes in late effector genes (LEGs), neuropeptides, and other adaptations that may provide the framework for longlasting changes in behavioral reactivity.

induction have been observed following different regimens, in part related to stimulant administration dose, numbers of repetitions, and intermittency or interval between injections (30–33).

While induction of a given IEG has not been definitively linked to a specific alteration in neurotransmitters, receptors, or peptides, it is of interest that chronic but not acute administration of cocaine is associated with the upregulation of a variety of neuropeptides and receptors, as illustrated in Fig. 4. These changes thus become putative neural substrates for relatively long-lasting effects of chronic cocaine administration on behavioral responsivity in an unconditioned fashion. In addition, one could postulate a role of the IEGs in the conditioned component of sensitization, as conditioned increases in c-fos have been observed in a variety of circumstances (34–37).

KINDLING: EVOLVING NEURAL SUBSTRATES REVEALED BY IN SITU HYBRIDIZATION AND PHARMACOLOGY

Goddard and associates (38) originally described the phenomenon of kindling as repeated intermittent stimulation of the brain associated with an increase in convulsive responsivity, culminating in the development of full-blown seizures in response to a previously subconvulsive stimulus. With kindling evolution, there is an initial lowering of the threshold for afterdischarges in the amygdala, then subsequent increases and spread of these discharges to other regions of the brain; this results in the emergence of full-blown seizures involving bilateral clonic movements of the head, trunk, and forepaws, (stage 4) with rearing and falling (stage 5) (39). Following sufficient numbers of repetitions of full-blown kindled seizures, animals develop spontaneous epilepsy, where seizures begin to occur in the absence of exogenous stimulation.

Clark and associates (40) have begun to map the potential neural substrates underlying kindling-induced seizure progression, using in situ autoradiography of the mRNA for c-fos. In this instance, a radioactive label is placed on an oligonucleotide probe so that labeling is detected wherever mRNA for Fos is induced. This provides a marker for cellular activation at the level of the nucleus in a fashion analogous to that seen with deoxyglucose autoradiography. However, in this latter instance, deoxyglucose is thought to reflect metabolism largely in nerve terminals as opposed to the nucleus. Nonetheless, Sagar and associates (41) have observed close relationships between deoxyglucose and in situ hybridization of c-fos in a variety of paradigms, suggesting that Fos may indeed be a useful marker for neuronal activation.

Using in situ hybridization, Clark and associates (40) have observed that the initial stage of kindling involving only behavioral arrest (stage 1) is associated with c-fos induction in the brain ipsilateral to the side of the amygdala stimulation in either the piriform cortex or the dentate gyrus, depending on the length of the afterdischarge. With seizure evolution and progression to that involving unilateral convulsions of the forepaw (stage 3), c-fos induction becomes bilateral and, with full-blown stage 4 and 5 seizures, c-fos induction not only involves the piriform and the dentate gyrus bilaterally but also the occipital and cingulate cortices. Clark and associates (1993, unpublished observations) have also assessed the pattern for c-fos induction in one animal to be observed to undergo a spontaneous seizure to assess the possible neural substrates involved in these autonomous as opposed to triggered seizures. In contrast to the triggered seizures, which involve bilateral or unilateral activation of the side of the brain ipsilateral to the electrode, the spontaneous seizure produced unilateral activation of the dentate gyrus and piriform cortex on the side contralateral to the electrode placement. This suggests possible inducement of a mirror focus that becomes autonomous in generating the spontaneous seizure.

The finding of evolving neural substrates involved in amygdala kindling progression, as revealed by c-fos induction, is highly convergent with data assessed with more conventional electrophysiological recording techniques and emerging pharmacology. That is, drugs blocking the initial developmental phases of kindling evolution are not the same as those involved in maintenance of completed kindled seizures (42). The pharmacology of completed and spontaneous seizures also has been found to differ. For example, carbamazepine, which is a highly effective treatment for completed kindled seizures, does not affect their development. Conversely, glutamate antagonists are highly effective in blocking development of amygdala kindling but are less efficacious on the completed phases. Alterations in α-2 noradrenergic mechanisms also have been shown to slow the development of kindling but are largely without effect on completed kindled seizures.

Perhaps the most interesting double dissociation has been revealed in the pharmacological responsivity of completed versus spontaneous seizures in the work of Pinel and associates (43). These investigators observed that the benzodiazepine diazepam was highly effective in the initial development and completed phases of kindling but was without effect on later spontaneous seizures. Conversely, phenytoin, which was largely ineffective in the development and completed phases of kindling, was able to block spontaneous seizures. Taken together, these data suggest that different anatomical and biochemical substrates underlie these differential phases of amygdala kindling evolution, as revealed by marked transitions in pharmacological responsivity. Thus, these data highlight that what appear as a unitary process of repeated amygdala stimulation of the brain, on a once-daily basis induces a very different set of neural processes with kindling evolution.

This spatiotemporal evolution in kindling characteristics is further revealed in the assessment of neuropeptide, neuroreceptor, and micro- and macroneuroanatomical effects that have now been revealed as a function of kindling stage. Again, while the precise consequences of induction of the IEGs such as c-fos, FRAS, Jun, zif268, and the like have not been adequately elucidated, it is presumed that the overall balance of these transcription factors acting on their receptor DNA-binding sites, in conjunction with a variety of other genetic signals, program the cell for the second and third wave of longer-term induction of genes and for longer-term modifications of peptides, proteins, and a variety of substances (e.g., neurotrophic, toxic, and apoptotic factors) that may alter the microanatomical environment in a regionally specific fashion (44,45).

For example, Rosen and associates (32,46) have observed that the mRNA for the peptide thyrotropin-releasing hormone (TRH) is induced in the same regions of brain involved in c-fos induction, first on a unilateral basis and then bilaterally with the transition to major motor seizures. This TRH induction in piriform cortex and dentate gyrus also appears regionally similar to that of c-fos and, in some instances, the two coexist in the same cells (47). This evidence, taken with the finding that the gene for TRH has an AP1-like binding site on it (at which Fos and Jun might bind to initiate TRH transcription), suggests that the initial activation of IEGs is ultimately responsible for the later, more long-lasting, but still transient induction of a variety of proteins and neuropeptides, including TRH. These changes are schematized in Fig. 5, which suggests that initial stimulation of the brain in early phases of kindling may activate IEGs, but that only with sufficient numbers of stimulations are the necessary requirements met for induction of late effector genes such as neuropeptides in sequentially involved areas of brain, first on a unilateral and then bilateral basis as a function of kindling evolution.

During amygdala kindling, the dentate granule cells of the hippocampus and those in the hilar regions (in between the "V" of the dentate gyrus) are highly plastic, both biochemically and neuroanatomically. While TRH is increased in these dentate granule cells, dynorphin is decreased, suggesting transcriptional activation and inhibition of different neuropeptides, even within the same cell (32), that is based on signals from the external environment. These cells are able to reprogram their receptor and peptide machinery transiently to meet altered requirements for responsivity. Smith and associates (48,34) have observed that corticotropin-releasing hormone (CRH) is induced in the dentate hilar region following amygdala-kindled afterdischarges and seizures. The induction of CRH in the den-

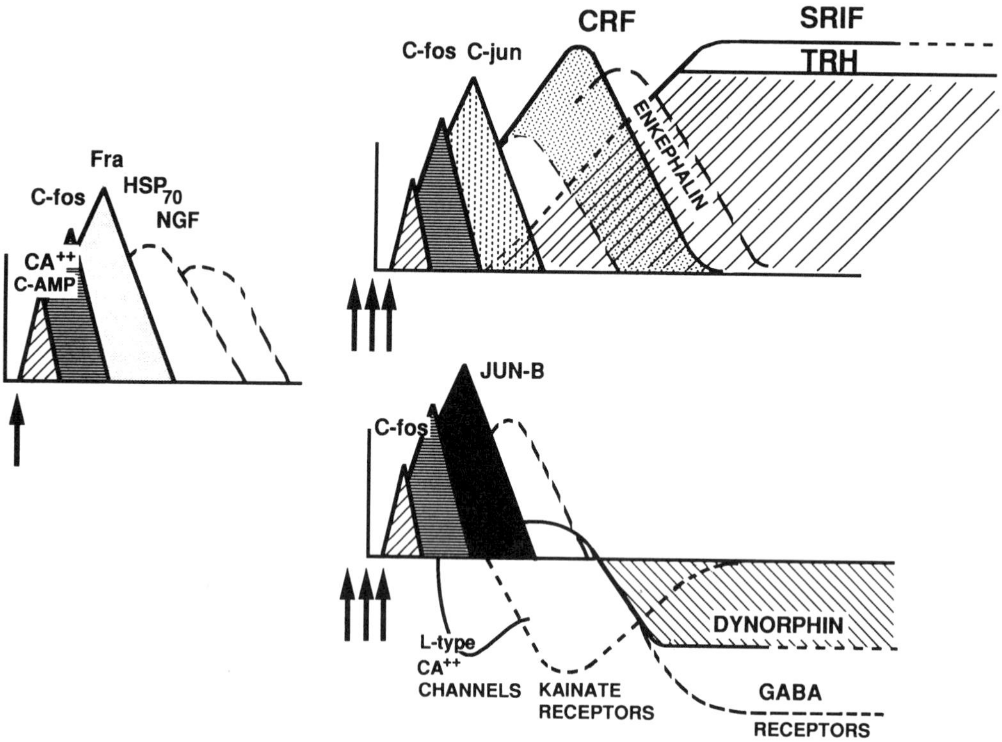

FIG. 5. In kindling, single stimulation (↑, *left side*) may result in activation of second and third messenger systems as well as a variety of immediate early genes (IEGs). However, repeated stimulation (↑ ↑ ↑, *right side*) alters late effector genes (LEGs), resulting in *increases* in peptides (*top*) or *decreases* (*bottom*) in other peptide and receptor systems. These changes may represent either processes related to the primary pathology of kindling (i.e., the long-lasting "memory trace") or shorter-lived secondary compensatory adaptations and anticonvulsant mechanisms. A similar increased induction of IEGs and LEGs may occur in other systems following repeated stressors that could be either adaptive or pathological.

tate hilar area is of particular interest, since this involves production of CRH messenger in cells not ordinarily expressing CRH, also indicating a transcriptional reprogramming of the machinery of these cells based on exogenous stimulation (48,34). These changes are schematically illustrated in Fig. 6. In addition to inductions of c-fos in the granule cells of the dentate gyrus, there are also longer-lasting increases in receptor binding for benzodiazepines, γ-aminobutyric acid ($GABA_A$) receptors (assayed by radioactive muscimol), and t-butylbicyclophosphorothionate (TBPS) binding. Presumably, these increases in $GABA_A$ and benzodiazepine binding mediate endogenous adaptive anticonvulsant mechanisms, tending to make the animal less seizure-prone for a period of days after being given kindled seizures (45). Glucocorticoid and mineralocorticoid receptor mRNAs are also increased in these granule cells, findings which are of particular interest in relationship to the transcriptional effects of glucocorticoid receptors and their ability to impair transcriptional activity of c-fos and vice versa (49,50). Thus, there appears to be "cross-talk" at the level of transcription factors, potentially paving the way to providing the complexity of detailed regulation necessary to up- or down-regulate memory-like events and their behavioral counterparts according to prior experience and the organism's history of deriving "meaning" from these events.

Sutula and co-workers (51) have demonstrated sprouting of the dendrites of these dentate granule cells as well as cell loss in the dentate

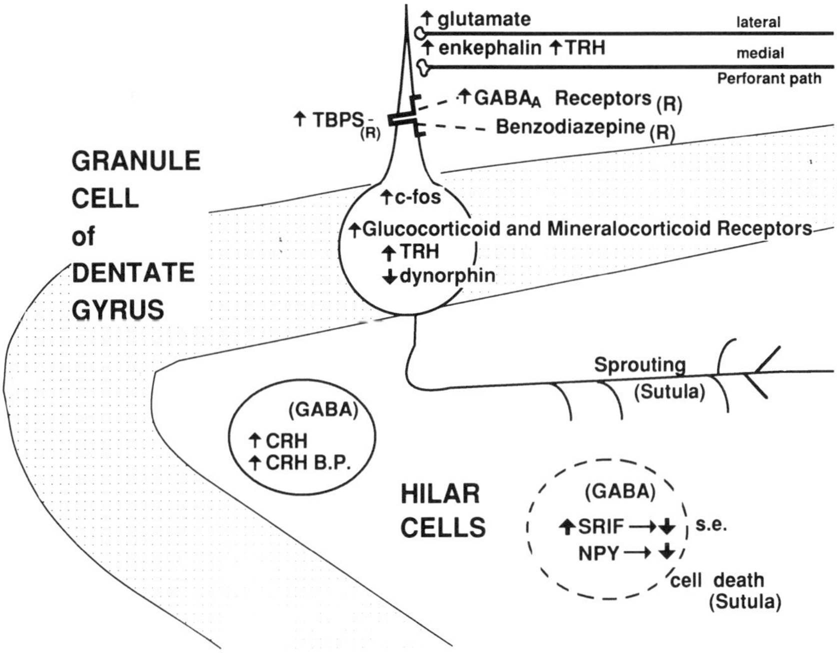

FIG. 6. Alterations in gene expression lead to transient to long-lasting changes in the biochemistry and neuroanatomy (sprouting and cell death) of the hippocampus. Some of these may account for the permanence of the kindled "memory trace."

hilar area (52,53). Kindling is associated with the induction of nerve growth factor and brain-derived neurotrophic factor (BDNF), which could be associated with mossy fiber sprouting. In addition, Smith et al. (54) have recently observed changes in BDNF and in neurotrophin 3 (NT3) expression in the hippocampus following immobilization stress, which are the opposite of the changes seen in kindling. The mechanisms underlying the kindling-induced progressive cell loss in the dentate hilar area described by Sutula (51) are unknown; these may reflect either excitotoxic or apoptotic (preprogrammed cell death) processes (55).

These data highlight that the process of kindling is associated not only with short- and long-term biochemical alterations at the level of neurotransmitters, peptides, and receptors, but also with a resculpting of the synaptic anatomy and microanatomy of the hippocampus, with both neuronal sprouting (mossy fiber system) and cell loss (in the dentate hilar area) occurring. How these anatomical changes in this area of brain are ultimately related to the relative permanence of neuronal hyperexcitability in kindling and other phenomena, including the eventual development of spontaneity, remains to be delineated. Nonetheless, they highlight that brief, but repeated, extracellular signals may come to leave behind an altered biological substrate at the level of both synaptic conductivity and anatomy of the brain in this paradigm, which has been used as a model for neuronal learning and memory (56). Indeed, these findings pose the question as to whether temporally related but different cascades of neurobiological changes underlie the evolution of PTSD.

IMPLICATIONS OF SENSITIZATION AND KINDLING FOR DEVELOPMENT AND EVOLUTION OF PTSD

As just noted, stimulant-induced behavioral sensitization in many instances shows cross-sensitization to that produced by stress, and vice versa (3). This ability to induce bidirectional cross-sensitization suggests that stimulant-induced and repeated stress-induced sensitization occupy common neuronal substrates, at least in part. This is not entirely unexpected, as cocaine not only increases catecholamines and indoleamines often released by stress, but also appears to release CRH and adrenocorticotropic hormone (ACTH) (57,58).

In addition to having the ability to induce progressive degrees of dysphoria with more chronic administration, cocaine is also able to induce panic attacks after repeated administration (59,60). Since many subjects with cocaine-induced panic disorders report: 1) a period of cocaine administration without panic (development phase); 2) a period where cocaine consistently induces panic (completed phase); and then 3) a phase where panic attacks occur spontaneously even after use of cocaine has been discontinued (spontaneity phase), a parallelism exists between kindling to a seizure endpoint and this evolution of a subconvulsant psychiatric syndrome. In addition, the local anesthetic components of cocaine may be sufficient to trigger a literal ''pharmacological kindling process'' where seizures are the endpoint. This reasoning is based on the observation that the pure local anesthetics procaine and lidocaine, which are not psychomotor stimulants, are also capable of engendering seizures in a progressive fashion in response to repeated administration of a previously subconvulsant dose of the drug (14,61). Following sufficient numbers of lidocaine-kindled seizures, spontaneous seizures are also observed (Weiss, 1993, unpublished observations). Thus, in electrical, pharmacological, and possibly behavioral (i.e., as in panic attacks) kindling the syndrome can evolve progressively from one that requires triggering to one that occurs spontaneously. Recent observations that procaine (62) selectively activates limbic structures in man (including amygdala, insula, orbital frontal cortex, and cingulate gyrus), as measured in ^{15}O PET blood flow studies (63), further support the notion that limbic kindling-like processes could underlie the development of local anesthetic-induced panic syndromes and their progression to spontaneity.

Many of the neurohormones and neuropeptides secreted during stress also appear capable of activating either sensitization or kindling pathways. For example, depending on the intervals between drug administrations, opiates can produce either sensitization or tolerance to the locomotor-activating properties (64). Moreover, at high doses the enkephalin-like opiate compounds administered directly into the amygdala on an intermittent basis are also able to produce kindling to a convulsant endpoint (65). Repeated administration of glucocorticoid compounds can also be associated with a sensitization to locomotor activity (see the review of Wolkowitz (66)) and CRH can produce atypical limbic seizures (67,68) in a process that mirrors some aspects of kindling but not others. That is, a single administration of intracerebroventricular CRF can induce the delayed onset of a typical limbic seizure some 4 to 12 hours following administration. Ehlers et al. (67) postulated that repeated CRH-induced excitatory episodes over this interval culminate in a kindling-like process in the amygdala. However, Weiss and associates (68) observed that repeated once-daily administration of CRH results in tolerance to the seizure-inducing processes of CRH rather than a kindling-like progression. Even when this tolerance has become manifest, however, previously CRH-treated animals are still more responsive to electrical amygdala kindling compared with vehicle-treated controls. This suggests that there is a persistent altered neuronal excitability in the amygdala that is evidenced by the increased proclivity to kindle, even though the seizures inducing effects of CRH itself showed tolerance. Perhaps in a similar fashion, the endocrine response to exogenous CRH is reduced in patients with PTSD (69), despite their heightened sensitivity to conditioned stressors. Tolerance to CRH may be a result of supersensitivity to glucocorticoid feedback as demonstrated by Yehuda and colleagues (70), who found an increased number of lymphocyte glucocorticoid receptors and an enhanced sensitivity to dexamethasone suppression of the cortisol response in patients with PTSD (71). It is also of interest that amygdala kindling increases the expression of glucocorticoid and mineralocorticoid mRNAs in the hippocampus (72), which could result in heightened negative feedback of the hypothalamic-pituitary-adrenal axis.

Of particular interest to a discussion of the late effects of stress and PTSD on irritability and aggression are the observations of Weiss et al. (68) of CRH-induced episodes of severe attack behavior against conspecific rats. These episodes can occur intermittently following CRH even in the absence of seizures. As such, this neuropeptide system might be further explored as a candidate mediator for long-lasting effects of stress on irritability and aggression in PTSD.

Intracerebroventricular injection of CRF is also able to induce c-fos in piriform cortex and dentate gyrus in animals with and without associated CRH seizures, respectively (40,73). Many of the other neurotransmitters involved in the stress response are also capable of inducing c-fos and a variety of other transcription factors. The list of substances acting at their respective receptors reported to induce c-fos either in vivo or in vitro include NMDA, adrenergic B_1, α_1, α_2, acetylcholine M_1, μ opiate, dopamine D_1, vasoactive intestinal peptide (VIP), and many others in addition to CRH (see reference [42]). Thus, it would appear that the neurotransmitter pathways directly activating the associated neurohormonal and peptide systems that are also activated in association with stress (depending on its magnitude, frequency, meaning, etc.) are all capable of exerting not only direct and time-limited effects, but may subsequently alter responsivity in a longer-lasting fashion by virtue of their ability to induce IEGs (74–76). This wave of IEG expression elicited by unconditioned or even conditioned stimuli (35–37,48) may then initiate a cascade of subsequent intermediate and late gene expression, leaving behind intermediate and late neurobiological adaptations with both positive and negative consequences to the organism (45). There is increasing evidence that normal mechanisms of learning and memory involve induction of the IEGs, which may be involved in coding longer-lasting changes in synaptic excitability and microstructure (77–80). Even the more fragmentary in vivo and in vitro model of learning and memory, long-term potentiation (LTP), has been shown to be closely linked to the induction of a variety of IEGs, the correlation being closest with the induction of zif268. Recent evidence from Kandel's laboratory (81) additionally implicates alterations in three separate second messenger pathways (protein kinase A (PKA), protein kinase C (PKC), and calcium calmodulin kinase II activity) in association with two retrograde messenger signals (the gases carbon monoxide and nitric oxide). Presumably, each one of these second messenger pathways would also have access to the nucleus by respective effects on induction of associated transcription factors (82). Moreover, the conversion from post-tetanic potentiation (PTP) to long-term potentiation (LTP) requires transcriptional activation.

Normal processes of learning and memory, and LTP that have been documented in vivo in the awake-behaving animal (83) may represent only a fragment of the effects that occur in PTSD. That is, learning and memory taking place under these severe threatening or emotional conditions is, by definition, far different from that in the usual realm. It is taking place under extreme emergency conditions often associated with fright, terror, or the image or imminence of death. Emotional arousal and extreme components of the stress response are thus paradigmatic of PTSD, and unique neural systems may be called into play in these instances. Hokfelt and associates (84) emphasize that neuropeptide cotransmitters are released by a category of events different from their more classical cotransmitters (e.g., norepinephrine, dopamine, or acetylcholine). That is, the neuropeptides are released under more extreme conditions of fast or paroxysmal firing in rhythmic or tetanic bursts. This is very likely to be the situation under extreme emergency stress responses with a threat to life or limb, or other actual impending catastrophe, as definitively implied in the precipitating circumstances of PTSD.

Given this perspective, it is not unreasonable to assume that components of extreme forms of stress sensitization and kindling-like cultivation of indelible memory traces may be generated simultaneously. Since components of both stress sensitization and kindling-like memory traces are likely to occur in parallel, it is postulated that this would involve the transcriptional machinery for each process, potentially involving widespread neural pathways and systems throughout the brain time domains that evolve similarly and differently. In relatively simple models of learning and memory, it has been estimated that 50–100 million neurons are involved in a visual discrimination task in the cat, as evidenced by alterations in metabolic activity (85). One could speculate that alterations in virtually every neuronal system in the brain may be occurring in the extreme situation associated with the induction of PTSD.

Regardless of the precise quantitation of the number of neuronal elements involved, we suggest the conceptual utility in considering complex PTSD resulting from repeated traumatization as a dual process of both sensitization- and kindling-like events, and that assessing the potential implications and predictive validity of the dual modeling may be fruitful. Obviously, this prospect is fraught with severe limitations such as we have already addressed. Cocaine sensitization may not be identical to stress sensitization, yet it may be relevant to the affective and contextual elements of PTSD, with long-lasting but not permanent alterations in behavioral responsivity. Kindling, as suggested by Goddard (38), represents an extreme form of neuronal memory process, with its primary outcome being the manifestation of full-blown seizures (which are not normal accompaniments of PTSD), further limiting the conclusions to be drawn. However, it may be useful conceptually to conjecture that, just as seizures are provoked or spontaneous paroxysmal discharges of motor circuits, flashbacks are provoked or spontaneous paroxysmal discharges of memory circuits. Since both sensitization and kindling do not mimic the inducing conditions of PTSD, their potential utility must reside in other areas. We suggest the utility of mapping PTSD events and organizing them according to some of the principles elucidated in the sensitization and kindling paradigms. In order to accomplish this, it is necessary to generate templates of the crucial elements of PTSD syndrome so that one can assess which components are showing tolerance, which are showing sensitization and persistence, and which are progressing to kindling-like autonomy.

As in behavioral sensitization to a single high dose of cocaine, it is readily apparent that a single catastrophic stressful situation is sufficient to induce PTSD. As illustrated in Fig. 7A,B some of the crucial components of the PTSD syndrome are mapped as a function of amplitude and time, with the initial PTSD-inciting stressor represented by the large arrow with its associated induction of the range of arousal of symptoms (e.g., anxiety, panic, and dissociative symptoms). These extreme arousal symptoms may then gradually be attenuated, depending on the time course of the stressor, and followed by a period of "compensatory" or adaptive changes, with the associated numbing and depression that are observed as later consequences of the inciting event and come to represent the name of the disorder as post-traumatic stress disorder. Coexistent with this new baseline of chronic numbing and withdrawal are more paroxysmal components of the syndrome, where cues in the environment may provide for conditioned triggering of arousal events in the form of hyperstartle, reexperiencing phenomena, and, ultimately, flashbacks.

We suggest that repeated conditioned or triggered flashbacks may, in themselves, acquire a kindling-like capacity to proceed to autonomy and eventually begin to occur in the absence of obvious triggers in the environment. Hyperreactivity to startle may persist as a chronic, basal, altered daytime reactivity as well. This is of interest in light of recent data suggesting that repeated immobilization stress induces change in neurotropin (NT-3) gene expression in the locus coeruleus (54), perhaps providing a neural mechanism for stressors (e.g., long-lasting hyperarousal and hyperstartle). Nightmares and sleep disturbance may represent paroxysmal and more chronic impairments, respectively, emerging during the nighttime hours as memory-like events, and these compensatory mechanisms may gain access to this phase of the circadian cycle.

Given the matrix effect illustrated in Fig. 1 for cocaine sensitization, how might the map of PTSD appear when the inducing circumstances are both extreme and repeated? Extrapolating from the cocaine sensitization model, we predict that the sensitized system would be greater in

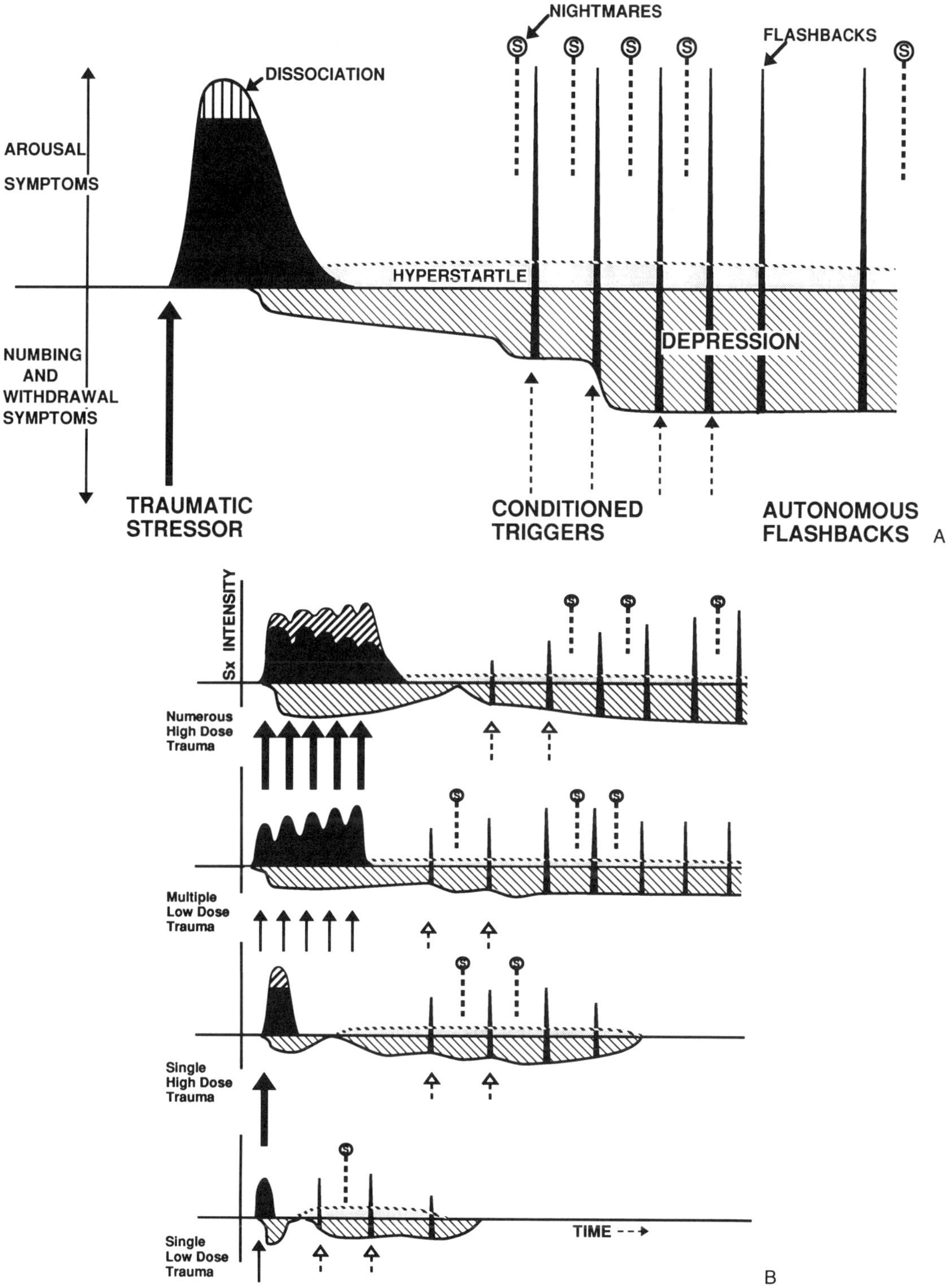

FIG. 7. A. Life chart schema for the evolving symptoms of post-traumatic stress disorder. (See text for description.) B. Impact of trauma frequency and severity on PTSD symptoms. As in cocaine-induced behavioral sensitization (*Fig. 1*), this hypothetical schema portrays the interaction of stressor severity with frequency of repetition in determining the magnitude and persistence of PTSD symptoms.

amplitude, more widespread, and more persistent in duration. The literature on PTSD reviewed by Herman (86) is consistent with this formulation. Just as the magnitude of a dose of cocaine appears to be a crucial variable in inducing sensitization, the magnitude of the stressor as measured by a variety of characteristics, including proximity to a given disaster situation, is a strong correlate of severity and/or duration of a PTSD symptom methodology. The sensitization model would also predict that the number of repetitions of stressors would affect the magnitude and duration of symptomatology as they interacted with severity (Fig. 7b).

One would also postulate, based on the cocaine sensitization model, that the intermittence of a stressor recurrence might, as well, be a crucial variable. Both sensitization and kindling are more robust under intermittent stimulus characteristics as opposed to more chronic administration (61). In fact, with chronic administration of cocaine, behavioral reactivity may even be down regulated (7). Similarly, chronic electrical stimulation of the amygdala is *not* associated with evolution of kindled seizures, and only when greater intermittence of stimulation is achieved (optimally once every 24 hours or more) does kindling to seizure endpoint take place. However, in the PTSD situation, rapidly recurring or superimposed stressors may represent a cumulative process and, over all, increase the initial magnitude of the inducing stimulation.

Just as in behavioral sensitization, severity, frequency, and intermittency of a stressor may be only several of many variables affecting outcome measures. The context of the stressor, the degree of social support provided, genetic and constitutional factors, prior experience with stressors acting as either vulnerability or immunizing factors, gender, and the like may all play a role in affecting outcome. It is in this context that data from the behavioral sensitization model may be particularly useful, as these factors can be isolated and systematically studied for their impact on the magnitude and persistence of sensitization in a way that is often difficult to factor out in clinical studies of post-traumatic stress disorder. In this fashion, the sensitization model can identify potential variables for further study and as they might affect PTSD outcome, and thus be heuristically valuable.

Repeated episodes of trauma may result in more severe forms of PTSD. Terr (87) theorized that "type I" trauma-stress disorders in children characterized by anxiety and intrusive recollections occur as a result of a single traumatic event, while the more severe "type II" disorders characterized by psychic numbing, rage, and dissociation (including multiple personalities) result from repeated traumas (8). Again, in a study that separated childhood PTSD into two groups based on duration of symptoms, acute (<4 months in duration) versus chronic (>8 months in duration), exaggerated startle and generalized anxiety were found in the acute group, while restricted affect and dissociative episodes were more often associated with chronic PTSD (88). Likewise, a history of prior victimization in women who have been raped results in PTSD symptoms of greater severity that take longer to remit, compared with victims of a single rape episode (see review [89]). Recent data of Bremner et al. (90) suggest that prior experience with stressful life events (physical abuse in childhood) may predispose soldiers to develop adult PTSD symptomatology induced by Vietnam War experience. In this way, prior life experiences may have a sensitizing or kindling-like effect in laying down neural substrates and vulnerability factors for subsequent reactivity to stressors.

What are the other predictions derived from the model? As noted in the previous paragraph, in both sensitization and kindling there is evidence from direct mapping with c-fos, as well as with other ligands, that the neural substrates mediating long-term behavioral and physiological consequences show a spatiotemporal evolution. This single postulate is the model's most important implication. To the extent that the *principles* (if not the precise pathways) of sensitization and kindling are applicable to the situation in PTSD, as we have argued, they would suggest that PTSD represents a continuously evolving process with its own series of stages of unfolding exacerbations, permanent residuals, late compensations, and potential healing processes (Figs. 7b, 8). While lidocaine kindling

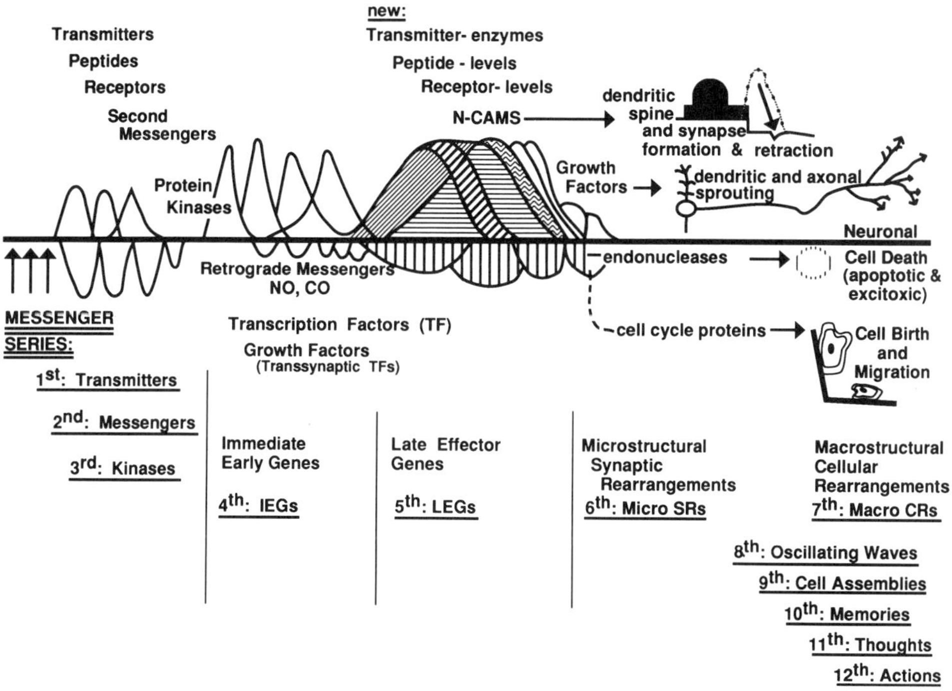

FIG. 8. This is an extension of Figs. 3 and 5, portraying an evolving cascade of messenger systems, each with its own complex regulatory mechanisms and cross-talk with other systems (*not illustrated*). In addition to showing immediate-early genes (IEGs) and late effector genes (LEGs) as illustrated in Figs. 3 and 5, Fig. 8 suggests that environmental stimulation can engage mechanisms that change the connectivity of the brain on a biochemical as well as microstructural basis, including cell sprouting, migration, or even cell death. These synaptic changes may ultimately be reflected in larger functional units (eighth and ninth messengers) that encode thoughts, memories, and preparation for action.

might appear relentlessly progressive, this is not the case, as a subgroup of animals develops sufficient endogenous compensatory mechanisms (45) to inhibit lidocaine-induced seizures (61). Repeated i.c.v. CRH leads to adaptation and tolerance to its convulsant effects (68), even while amygdala reactivity remains increased.

Conceptualization of PTSD as a continuously evolving process changes the investigative and therapeutic strategies applied to the illness. If the neurobiological correlates of PTSD are very much a function of stage of evolution, and if subjects are not assessed accordingly, a variety of variables may be highly confounded. Kindling and behavioral sensitization also suggest a further way that neural systems may be simultaneously and sequentially adapted in their own unique patterns. To the extent that sensitization and kindling represent only fragments of what might be occurring in PTSD, the order of magnitude of potential neural changes in the brain become staggering.

Specifically, if one examines the preliminary evidence illustrated in Fig. 6 of the changes observed in dentate granule and dentate hilar cells during kindling evolution, the potential scope of parallel changes in this and other neural systems during the more complicated and widespread process of PTSD becomes imaginable (Fig. 8). The dentate granule cells have markedly reprogrammed themselves with induction of immediate-early genes, mineralocorticoid and glucocorticoid receptors, inhibitory neurotransmitter receptors in the benzodiazepine-GABA-TBPS-chloride ion channel complex turning on of growth factors, rebalancing of neuropeptides (with increases in TRH and decreases in dynorphin), and even neuronal sprouting. At the same time, peptide events in the hilar cells are being reprogrammed with increased CRH expression, and in these (or nearby) cells a presumptive process of cell death is occurring. Based on the work of Bailey et al. (79), Mayford et al. (78), Grant et al. (91), and Kandel (81) in a more subtle model of learning and memory, massive alterations in a variety of more microscopic synaptic machinery may also be occurring, involving synaptic rearrangement, endocytosis of neural adhesion molecules, and a cascade of alterations in cyclic-AMP, protein kinases, IEGs, and late effector genes (LEGs) (i.e., second, third, and fourth messengers, etc. (Fig. 8)). Thus, as in this area of brain in PTSD, each of these changes would have its own temporal characteristics and changing anatomical representations in the brain. Some of the alterations may, like those in kindling, be permanent. Neuronal sprouting versus cell death based on stress-induced changes in neurotrophic factor gene expression (54), via excitotoxicity or apoptosis, may similarly resculpt the CNS based on experience. We may ultimately find that some elements of traumatic memories are literally branded into the CNS. Presumably in later phases of PTSD evolution, as in other models of learning and memory (20,92,93), the memory trace would become relatively independent of the hippocampus and related medial temporal structures. Like repeated kindling leading to autonomy, repeated reexperience phenomena may come to shift memory from a conscious representational basis to habit memory associated with unconscious processes, possibly mediated by striatal, noncholinergic mechanisms (20,21). Even in a nontraumatic one-trial avoidance learning paradigm with 1-day-old chicks, the cortical areas involved in "holding" the memory trace shift from right to left dorsal areas of brain, then from left to right ventral areas (22,77). This conceptualization could provide one way of considering post-traumatic amnesia, with memories of events being transferred to relatively inaccessible neural substrates. Another view is that some of the adaptive compensatory neural mechanisms are only transiently activated and their dissipation results in predominance of the primary pathological (memory or seizure) process (45). In kindling, endogenous anticonvulsant substances (such as TRH, NPY, and $GABA_A$ receptors) are transiently induced and may prevent occurrence of seizures. In PTSD, stress-induced compensatory mechanism (such as oxytocin and endogenous opiates) with amnesic potential (94,95) via either direct effects or state-dependency (96) may similarly be induced and temporarily block the access of the memory trace to conscious awareness. With the dissipation of these compensatory adaptations (97), the more permanent memory trace may become conscious.

As implied from the evolving neural substrates in sensitization and kindling as a function of time and stage of development, and their unique pharmacological responsivities (15,59), the efficacy of pharmacological interventions in PTSD may vary according to stage of evolution of the process as well. Just as neuroleptics block the development but not the expression of sensitization, we would predict that some agents that would be useful in the blockade of stress sensitization of PTSD, if one were fortunately able to apply them concurrently with the stressor (not to mention in advance of the stressor), would, however, be unable to block various aspects of the expression of the sensitization once the process had been engendered. Since treatment of PTSD would typically be targeted at the intermediate and late aspects (i.e., the expression) of sensitization, studying this phase of sensitization in animal models and clinically testing treatment strategies so derived may be a fruitful approach to uncovering novel therapeutics. It is of interest that two anticonvulsant drugs that block the mid phase of fully developed kindled seizures have shown some promise in alleviating some paroxysmal PTSD symptoms in open clinical trials. Carbamazepine was effective in reducing intrusive recollections, recurrent dreams, and flashbacks, as well as improving mood and sleep (98). Valproate targeted symptoms of hyperarousal such as irritability, anxiety, and increased vigilance (99); it also improved avoidant symptoms such as emotional numbing and social withdrawal. Anticonvulsant benzodiazepines are also widely used in PTSD, but their efficacy leaves much to be desired. To the extent that they were used prophylactically or immediately post-traumatically, they might be more effective, just as they are in the early and mid phase of kindling, but not in the late spontaneous phases (59–61). Pharmacotherapeutic and psychotherapeutic interventions targeted to the immediate post-traumatic phase may be particularly worthy of further study. The success of interventions with glutamate antagonists, even after the occurrence of an ischemic event, in preventing the late effects of hippocampal cell loss in the CA_1 region suggest that ways of preventing the full force of stressor-induced cascade on gene expression may likewise be fruitful.

To the extent that the late phase of autonomous flashbacks has a certain parallelism with spontaneity in the kindled seizure model, one would additionally postulate that these late automatic and autonomous manifestations of PTSD may, as well, have a different anatomy and pharmacology compared with the earlier, more triggered variety. This represents a testable hypothesis that may help organize appropriate experimental designs and pharmacological trials so that the question can be specifically examined.

If evolving neural substrates underlie different stages of PTSD symptom unfolding, one would also postulate that different forms of psychotherapy may be indicated as a function of stage of evolution as well. If a prominent condition triggers hyperreactivity and flashbacks, appropriate flooding and deconditioning techniques may be useful. However, in later autonomous phases, a different approach may be required. Viewed from another perspective, one might raise the question of what manipulations would be most appropriate for restructuring memories and behavioral repertoires ingrained in the "habit" system. In this fashion we hope that longitudinal perspectives offered by the sensitization and kindling models, and their emphasis on the evolving spatiotemporal effects on gene expression (Fig. 8) will have organizing and heuristic value even if they fall short of meeting most of the criteria for a valid model of PTSD (2). Sensitization and kindling do provide memory models with clear-cut and easily measurable behavioral and physiological endpoints; as such, sensitization and kindling may help model different affective/contextual versus idetic/permanent components of PTSD, each with its own engagement of memory systems at the level of gene expression and changes in synaptic connectivity. Studies aimed at behavioral desensitization and unkindling may also ultimately be helpful in conceptualizing novel ways of reversing even some of the most persistent, autonomous, and pernicious elements of PTSD.

REFERENCES

1. Goelet P, Castellucci VF, Schacher S, Kandel ER. The long and the short of long-term memory—a molecular framework. *Nature* 1986;322:419–422.
2. Yehuda R, Antelman SM. Criteria for rationally evaluating animal models of posttraumatic stress disorder. *Biol Psychiatry* 1993;33:479–486.
3. Kalivas PW, Stewart J. Dopamine transmission in the initiation and expression of drug- and stress-induced sensitization of motor activity. *Brain Res Rev* 1991; 16:223–244.
4. Antelman SM. Stressor-induced sensitization to subsequent stress: implications for the development and treatment of clinical disorders. In Kalivas PW, Barnes CD, ed. *Sensitization in the nervous system.* Caldwell NJ: Telford Press, 1988:227–254.
5. Post RM, Contel NR. Human and animal studies of cocaine: implications for development of behavioral pathology. In Creese I, ed. *Stimulants: neurochemical, behavioral, and clinical perspective.* New York: Raven Press, 1983:169–203.
6. Weiss SRB, Post RM, Pert A, Woodward R, Murman D. Context-dependent cocaine sensitization: differential effect of haloperidol on development versus expression. *Pharmacol Biochem Behav* 1989;34:655–661.
7. King GR, Joyner C, Lee T, Kuhn C, Ellinwood EH, Jr. Intermittent and continuous cocaine administration: residual behavioral states during withdrawal. *Pharmacol Biochem Behav* 1992;43:243–248.
8. Putnam FW. Dissociative disorders in children: behavioral profiles and problems. *Child Abuse Negl* 1993; 17:39–45.
9. Jensen TS, Genefke IK, Hyldebrandt N, Pedersen H, Petersen HD, Weile B. Cerebral atrophy in young torture victims [Letter]. *N Engl J Med* 1982;307:1341.
10. Fontana DJ, Post RM, Pert A. Conditioned increase in mesolimbic dopamine overflow by stimuli associated with cocaine. *Brain Res* 1993;629:31–39.
11. Brown EE, Fibiger HC. Cocaine-induced conditioned locomotion: absence of associated increases in dopamine release. *Neuroscience* 1992;48:621–629.
12. Deutch AY, Roth RH. The determinants of stress-induced activation of the prefrontal cortical dopamine system. In Uylings HBM, Van Eden CG, De Bruin JPC, Corner MA, Feenstra MGP, eds. *Progress in brain research.* Amsterdam: Elsevier Science, 1990; 367–403.
13. Deutch AY, Lee MC, Gillham MH, Cameron DA, Goldstein M, Iadarola MJ. Stress selectively increases Fos protein in dopamine neurons innervating the prefrontal cortex. *Cerebral Cortex* 1991;1:273–292.

14. Weiss SRB, Post RM, Szele F, Woodward R, Nierenberg J. Chronic carbamazepine inhibits the development of local anesthetic seizures kindled by cocaine and lidocaine. *Brain Res* 1989;497:72–79.
15. Post RM, Altshuler LL, Ketter TA, Denicoff K, Weiss SRB. Antiepileptic drugs in affective illness: clinical and theoretical implications. In Smith DB, Treiman DM, Trimble MR, eds. *Advances in Neurology, vol. 55, Neurobehavioral problems in epilepsy*. New York: Raven Press, 1991;239–277.
16. Fontana D, Post RM, Weiss SRB, Pert A. The role of D1 and D2 dopamine receptors in the acquisition and expression of cocaine-induced conditioned increases in locomotor behavior. *Behav Pharmacol* 1993;4: 375–387.
17. Post RM, Weiss SRB, Fontana D, Pert A. Conditioned sensitization to the psychomotor stimulant cocaine. *Ann NY Acad Sci* 1992;654:386–399.
18. Karler R, Calder LD, Chaudhry IA, Turkanis SA. Blockade of "reverse tolerance" to cocaine and amphetamine by MK-801. *Life Sci* 1989;45:599–606.
19. Mishkin M, Malamut B, Bachevalier JK. Memories and habits: two neural systems. In Lynch G, McGaugh JL, Weinberger NM, eds. *Neurobiology of learning and memory*. New York: Guilford Press, 1984;67–77.
20. Mishkin M, Appenzeller T. The anatomy of memory. *Scientific Am* 1987;256:80–89.
21. Aigner T, Mishkin M. Scopolamine impairs recall of one-trial stimulus-reward associations. *Behav Brain Res* 1993;54:133–136.
22. Rose SP. How chicks make memories: the cellular cascade from c-fos to dendritic remodelling. *Trends Neurosci* 1991;14:390–397.
23. Morgan JI, Curran T. Stimulus-transcription coupling in the nervous system: involvement of the inducible proto-oncogenes fos and jun. *Annu Rev Neurosci* 1991; 14:421–451.
24. Young ST, Porrino LJ, Iadarola MJ. Cocaine induces striatal c-fos immunoreactive proteins via dopaminergic D1 receptors. *Proc Natl Acad Sci USA* 1991;88: 1291–1295.
25. Cole AJ, Bhat RV, Patt C, Worley PF, Baraban JM. D1 dopamine receptor activation of multiple transcription factor genes in rat striatum. *J Neurochem* 1992;58:1420–1426.
26. Graybiel AM, Moratalla R, Robertson HA. Amphetamine and cocaine induce drug-specific activation of the c-fos gene in striosome-matrix compartments and limbic subdivisions of the striatum. *Proc Natl Acad Sci USA* 1990;87:6912–6916.
27. Moratalla R, Robertson HA, Graybiel AM. Dynamic regulation of NGFI-A (zif268, egr1) gene expression in the striatum. *J Neurosci* 1992;12:2609–2622.
28. Robertson GS, Vincent SR, Fibiger HC. Striatonigral projection neurons contain D1 dopamine receptor-activated c-fos. *Brain Res* 1990;523:288–290.
29. Bhat RV, Cole AJ, Baraban JM. Chronic cocaine treatment suppresses basal expression of zif268 in rat forebrain: in situ hybridization studies. *J Pharmacol Exp Ther* 1992;263:343–349.
30. Hope B, Kosofsky B, Hyman SE, Nestler EJ. Regulation of immediate early gene expression and AP-1 binding in the rat nucleus accumbens by chronic cocaine. *Proc Natl Acad Sci USA* 1992;89:5764–5768.
31. Iadarola MJ, Chuang EJ, Yeung CL, Hoo Y, Silverthorn M, Gu J, Draisci G. Induction and suppression of proto-oncogenes in rat striatum after single or multiple treatments with cocaine or GBR-12909. *NIDA Research Monograph* 1993;125:181–211.
32. Rosen JB, Cain CJ, Weiss SRB, Post RM. Alterations in mRNA of enkephalin, dynorphin and thyrotropin releasing hormone during amygdala kindling: an in situ hybridization study. *Mol Brain Res* 1992;15:247–255.
33. Bhat RV, Worley PF, Cole AJ, Baraban JM. Activation of the zinc finger encoding gene krox-20 in adult rat brain: comparison with zif268. *Mol Brain Res* 1992; 13:263–266.
34. Smith MA, Banerjee S, Gold PW, Glowa J. Induction of c-fos mRNA in rat brain by conditioned and unconditioned stressors. *Brain Res* 1992;578:135–141.
35. Campeau S, Hayward MD, Hope BT, Rosen JB, Nestler EJ, Davis M. Induction of the c-fos proto-oncogene in rat amygdala during unconditioned and conditioned fear. *Brain Res* 1991;565:349–352.
36. Brown EE, Robertson GS, Fibiger HC. Evidence for conditional neuronal activation following exposure to a cocaine-impaired environment: role of forebrain limbic structures. *J Neurosci* 1992;12:4112–4121.
37. Pezzone MA, Lee W-S, Hoffman GE, Rabin BS. Induction of c-fos immunoreactivity in the rat forebrain by conditioned and unconditioned aversive stimuli. *Brain Res* 1992;597:41–50.
38. Goddard GV, McIntyre DC, Leech CK. A permanent change in brain function resulting from daily electrical stimulation. *Exp Neurol* 1969;25:295–330.
39. Racine R. Kindling: the first decade. *Neurosurgery* 1978;3:234–252.
40. Clark M, Weiss SRB, Post RM. Expression of c-fos mRNA in rat brain after intracerebroventricular administration of corticotropin-releasing hormone. *Neurosci Lett* 1991;132:235–238.
41. Sagar SM, Sharp FR, Curran T. Expression of c-fos protein in brain: metabolic mapping at the cellular level. *Science* 1988;240:1328–1331.
42. Post RM, Weiss SRB, Pert A. Animal models of mania. In Willner P, Scheel-Kruger J, eds. *The mesolimbic dopamine system: from motivation to action*. Chichester, England: John Wiley & Sons, 1991;443–472.
43. Pinel JPJ. Effects of diazepam and diphenylhydantoin on elicited and spontaneous seizures in kindled rats: a double dissociation. *Pharmacol Biochem Behav* 1983;18: 61–63.
44. Post RM. Transduction of psychosocial stress into the neurobiology of recurrent affective disorder. *Am J Psychiatry* 1992;149:999–1010.
45. Post RM, Weiss SRB. Endogenous biochemical abnormalities in affective illness: therapeutic versus pathogenic. *Biol Psychiatry* 1992;32:469–484.
46. Rosen JB, Abramowitz J, Post RM. Co-localization of TRH mRNA and Fos-like immunoreactivity in limbic structures following amygdala kindling. *Mol Cell Neurosci* 1993;4:335–342.
47. Rosen JB, Chuang E, Iadarola MJ. Differential induction of fos protein and fos related antigen following acute and repeated cocaine administration. *Brain Res Mol Brain Res* 1994;25(1-2):168–172.
48. Smith M, Weiss SRB, Abedin T, Post RM, Gold P. Effects of amygdala-kindling and electroconvulsive seizures on the expression of corticotropin releasing

hormone (CRH) mRNA in the rat brain. *Mol Cell Neurosci* 1991;2:103–106.
49. Lucibello FC, Slater EP, Jooss KU, Beato M, Muller R. Mutual transrepression of Fos and the glucocorticoid receptor: involvement of a functional domain in Fos which is absent in Fos B. *EMBO J* 1990;9:2827–2834.
50. Yang-Yen HF, Chambard JC, Sun YL, Smeal T, Schmidt TJ, Drouin J, Karin M. Transcriptional interference between c-jun and glucocorticoid receptor: mutual inhibition of DNA binding due to direct protein-protein interaction. *Cell* 1990;62:1205–1215.
51. Sutula T, Xiao-Xian H, Cavazos J, Scott G. Synaptic reorganization in the hippocampus induced by abnormal functional activity. *Science* 1988;239:1147–1150.
52. Cavazos JE, Sutula TP. Progressive neuronal loss induced by kindling: a possible mechanism for mossy fiber synaptic reorganization and hippocampal sclerosis. *Brain Res* 1990;527:1–6.
53. Sutula TP. Reactive changes in epilepsy: cell death and axon sprouting induced by kindling. *Epilepsy Res* 1991;10:62–70.
54. Smith MA, Makino S, Kvetnansky R, Post RM. Stress and glucocorticoids affect the expression of brain derived neurotrophic factor and neurotrophin-3 mRNA in the hippocampus. *J Neurosci* 1995;15(3):1768–1777.
55. Margolis RL, Chuang D-M, Post RM. Programmed cell death: implications for neuropsychiatric disorders. *Biol Psychiatry* 1994;35:946–956.
56. Goddard GV, Douglas RM. Does the engram of kindling model the engram of normal long term memory. *Can J Neurol Sci* 1975;2:385–394.
57. Rivier C, Vale W. Cocaine stimulates adrenocorticotropin (ACTH) secretion through a corticotropin-releasing factor (CRF)-mediated mechanism. *Brain Res* 1987;422:403–406.
58. Calogero AE, Gallucci WT, Kling MA, Chrousos GP, Gold PW. Cocaine stimulates rat hypothalamic corticotropin-releasing hormone secretion in vitro. *Brain Res* 1989;505:7–11.
59. Post RM, Rubinow DR, Ballenger JC. Conditioning and sensitization in the longitudinal course of affective illness. *Br J Psychiatry* 1986;149:191–201.
60. Post RM, Weiss SRB. Behavioral pharmacology of carbamazepine: differential effects on kindling. *Int Clin Psychopharmacol* 1987;2[2]:51–72.
61. Post RM, Weiss SRB, Pert A, Uhde TW. Chronic cocaine administration: sensitization and kindling effects. In Raskin A, Fisher S, eds. *Cocaine: clinical and biobehavioral aspects.* New York: Oxford University Press, 1987;109–173.
62. Kellner CH, Post RM, Putnam F, Cowdry RW, Gardner D, Kling MA, Minichiello MD, Trettau JR, Coppola R. Intravenous procaine as a probe of limbic system activity in psychiatric patients and normal controls. *Biol Psychiatry* 1987;22:1107–1126.
63. Ketter TA, Andreason PJ, George MS, Herscovitch P, Post RM. Paralimbic rCBF increases during procaine-induced psychosensory and emotional experiences. *Soc Biol Psychiatry* 1993;33 (Suppl 6A):66A.
64. Shuster L, Yu G, Bates A. Sensitization to cocaine stimulation in mice. *Psychopharmacology* 1977;52:185–190.
65. Cain DP, Corcoran ME. Intracerebral beta-endorphin, met-enkephalin and morphine: kindling of seizures and handling-induced potentiation of epileptiform effects. *Life Sci* 1984;34:2535–2542.
66. Wolkowitz OM. Prospective controlled studies of the behavioral and biological effects of exogenous corticosteroids. *Psychoneuroendocrinology* 1994;19:233–255.
67. Ehlers CL, Henriksen SJ, Wang M, Rivier J, Vale W, Bloom FE. Corticotropin releasing factor produces increases in brain excitability and convulsive seizures in rats. *Brain Res* 1983;278:332–336.
68. Weiss SR, Post RM, Gold PW, Chrousos G, Sullivan TL, Walker D, Pert A. CRF-induced seizures and behavior: interaction with amygdala kindling. *Brain Res* 1986;372:345–351.
69. Smith MA, Davidson J, Ritchie JC, Kudler H, Lipper S, Chappell P, Nemeroff CB. The corticotropin-releasing hormone test in patients with post-traumatic stress disorder. *Biol Psychiatry* 1989;26:349–355.
70. Yehuda R, Lowy MT, Southwick SM, Shaffer D, Giller EL Jr. Increased number of glucocorticoid receptors in posttraumatic stress disorder. *Am J Psychiatry* 1991; 148:499–504.
71. Yehuda R, Southwick SM, Krystal JH, Bremner D, Charney DS, Mason JW. Enhanced suppression of cortisol following dexamethasone administration in posttraumatic stress disorder. *Am J Psychiatry* 1993;150: 83–86.
72. Clark M, Smith MA, Weiss SRB, Post RM. Modulation of hippocampal glucocorticoid and mineralocorticoid receptor mRNA expression by amygdaloid kindling. *Neuroendocrinology* 1994;59:451–456.
73. Clark M, Post RM, Weiss SRB, Nakajima T. Expression of c-fos mRNA in acute and kindled cocaine seizures in rats. *Brain Res* 1992;592:101–106.
74. Nakajima T, Daval JL, Gleiter CH, Deckert J, Post RM, Marangos PJ. C-fos mRNA expression following electrical-induced seizure and acute nociceptive stress in mouse brain. *Epilepsy Res* 1989;4:156–159.
75. Bullitt E. Induction of c-fos-like protein within the lumbar spinal cord and thalamus of the rat following peripheral stimulation. *Brain Res* 1989;493:391–397.
76. Ceccatelli S, Villar MJ, Goldstein M, Hokfelt T. Expression of c-Fos immunoreactivity in transmitter-characterized neurons after stress. *Proc Natl Acad Sci USA* 1989;86:9569–9573.
77. Anokhin KV, Mileusnic R, Shamakina IY, Rose SP. Effects of early experience on c-fos gene expression in the chick forebrain. *Brain Res* 1991;544:101–107.
78. Mayford M, Barzilai A, Keller F, Schacher S, Kandel ER. Modulation of an NCAM-related adhesion molecule with long-term synaptic plasticity in aplysia. *Science* 1992;256:638–644.
79. Bailey CH, Chen M, Keller F, Kandel ER. Serotonin-mediated endocytosis of apCAM: an early step of learning-related synaptic growth in aplysia. *Science* 1992; 256:645–649.
80. Alkon DL, Nelson TJ: Specificity of molecular changes in neurons involved in memory storage *FASEB J* 1990; 4:1567–1576.
81. Kandel ER. Genes, synapses and declarative forms of memory. *Abstracts of the 146th Annual Meeting, American Psychiatric Association* 1993;50 (Abstract).
82. Bading H, Ginty DD, Greenberg ME. Regulation of gene expression in hippocampal neurons by distinct calcium signaling pathways. *Science* 1993;260:181–186.

83. Granger R, Lynch G. Higher olfactory processes: perceptual learning and memory. *Curr Opin Neurobiol* 1991;1:209–214.
84. Hokfelt T, Johansson O, Ljungdahl A, Lundberg JM, Schultzberg M. Peptidergic neurones. *Nature* 1980; 284:515–521.
85. John ER, Tang Y, Brill AB, Young R, Ono K. Double-labeled metabolic maps of memory. *Science* 1986;233: 1167–1175.
86. Herman JL. *Trauma and recovery.* New York: Basic Books, 1992.
87. Terr LC. Childhood traumas: an outline and overview. *Am J Psychiatry* 1991;148:10–20.
88. Famularo R, Kinscherff R, Fenton T. Symptom differences in acute and chronic presentation of childhood post-traumatic stress disorder. *Child Abuse Negl* 1990; 14:439–444.
89. Foa EB, Riggs DS. Posttraumatic stress disorder and rape. In Oldham JM, Riba MB, Tasman A, eds. *American psychiatry press review of psychiatry, vol. 12.* Washington DC: American Psychiatric Press, 1993; 273–303.
90. Bremner JD, Southwick SM, Johnson DR, Yehuda R, Charney DS. Childhood physical abuse and combat-related posttraumatic stress disorder in Vietnam veterans. *Am J Psychiatry* 1993;150:235–239.
91. Grant SGN, O'Dell TJ, Karl KA, Stein PL, Soriano P, Kandel ER. Impaired long-term potentiation, spatial learning, and hippocampal development in fyn mutant mice. *Science* 1992;258:1903–1908.
92. Squire LR, Zola-Morgan S. The medial temporal lobe memory system. *Science* 1991;253:1380–1386.
93. Kim JJ, Fanselow MS. Modality-specific retrograde amnesia of fear. *Science* 1992;256:675–679.
94. Pitman RK, Orr SP, Lasko NB. Effects of intranasal vasopressin and oxytocin on physiologic responding during personal combat imagery in Vietnam veterans with posttraumatic stress disorder. *Psychiatry Res* 1993;48(2):107–117.
95. Pitman RK, Van der Kolk BA, Orr SP, Greenberg MS. Naloxone-reversible analgesic response to combat-related stimuli posttraumatic stress disorder: A pilot study. *Arch Gen Psychiatry* 1990;47:541–544.
96. Rawlins N. Modes of action of posttrial manipulation of learning: memories lost and memories regained. *Acta Neurol Scand* 1968;74(suppl 109):61–68.
97. Weiss SRB, Clark M, Rosen JB, Smith MA, Post RM. Contingent tolerance to the anticonvulsant affects of carbamazepine: Relationship to the loss of edogenous adaptive mechanism. *Brain Res Rev* 1995: in press.
98. Lipper S, Davidson JRT, Grady TA, Edinger JD, Hammett EB, Mahorney SL, Cavenar JO Jr. Preliminary study of carbamazepine in post-traumatic stress disorder. *Psychosomatics* 1986;27:849–854.
99. Fesler FA. Valproate in combat-related posttraumatic stress disorder. *J Clin Psychiatry* 1991;52:361–364.
100. Gale K. Catecholamine-independent behavioral and neurochemical effects of cocaine in rats. In Sharp CW, ed. *Mechanisms of tolerance and dependence NIDA research monograph series no. 54.* Washington DC: US Government Printing Office, 1984;323–332.

Neurobiological and Clinical Consequences of Stress: From Normal Adaptation to PTSD, edited by M. J. Friedman, D. S. Charney, and A. Y. Deutch, Lippincott-Raven Publishers, Philadelphia 1995.

13

Mediation of the Stress Response by the Hypothalamic-Pituitary-Adrenal Axis

David Michelson, Julio Licinio, and Philip W. Gold

Clinical Neuroendocrinology Branch, National Institute of Mental Health, Bethesda, Maryland 20892

For more than 2,000 years there has been a tradition in Western medicine that has equated health with harmony in physical and emotional life. In this tradition disorders have been understood as disruptions of optimal relationships between different elements, and the restoration of health has been thought to require a restoration of balance. While we now know that this paradigm is oversimplified and for some disorders inaccurate, the concepts of harmony and balance nevertheless form the basis for an increasingly sophisticated model of how the body interacts with and adapts to a changing environment—a model of how what we call stress disturbs the body's normal homeostasis, and how in response the body mobilizes a group of highly specific mechanisms to defend and maintain this balance.

Under normal conditions the living organism is in a state of homeostasis, a complex equilibrium of different systems. Stressors are pressures that threaten this equilibrium, and stress is the state of disequilibrium induced by such pressures. The stress response is a set of adaptive changes that act to restore homeostasis. We and others have suggested that a discrete system has evolved to respond to stress (1), and in what follows we describe this system with particular attention to its component parts and to the way these act together to create the stress response. Phenomenologically, the stress response is made up of both central nervous system and peripheral elements that together allow for cognitive assessment of threats (increased alertness, vigilance, focused attention), suppress behaviors that are nonadaptive in the setting of an acute threat (feeding, reproduction, sleep) and increase physical readiness to act (increased cardiovascular tone, increased availability of energy to the central nervous system and stressed peripheral sites, increased respiration, increased gluconeogenesis). Importantly, the stress response also incorporates mechanisms for its own containment by negative feedback and by inhibition of inflammatory and immune responses. The stress response as we understand it today is largely mediated by two interrelated systems: 1) that controlled by corticotropin-releasing hormone (CRH), and 2) the locus coeruleus–norepinephrine (LC/NE) system. Both of these systems are critical to the stress response, and although the current chapter is primarily concerned with CRH and the hypothalamic-pituitary-adrenal (HPA) axis role in stress response, and it is HPA axis function that is reviewed in detail, it must be remembered that the LC/NE system is equally important.

The HPA axis is a complex system whose components work together under both nonstressed and stressed conditions. As will be discussed, at low levels of activity it acts with a circadian rhythm and is a ''permissive'' system. Under stressful conditions the system is activated and works to maximize the body's ability to respond effectively to disturbances in homeostasis as well as to counterregulate many of the potentially harmful effects of activation of other stress-activated responses. In what follows, we will review the physiology and regulation of the HPA axis, describe how it affects the organism as a whole during the stress response, and consider how alterations in HPA axis function might contribute to the pathophysiology of various disease states.

THE HYPOTHALAMIC PITUITARY ADRENAL AXIS

CRH

Stress responses are mediated through the HPA axis initially by the release of corticotropin-releasing hormone (CRH). The CRH is a 41-amino-acid peptide synthesized in the parvicellular cells of the paraventricular nucleus (PVN) of the hypothalamus. It is released into the hypophyseal portal blood of the median eminence and transported to the anterior lobe of the pituitary. There, in the adrenocorticotropin (ACTH)-containing corticotrophs, through the second messenger system cyclic adenosine monophosphate (AMP), it stimulates cleavage of proopiomelanocortin (POMC), the precursor molecule of ACTH, beta-endorphin, and melanin-stimulating hormone (alpha MSH), as well as release of ACTH. Although the presence of CRH had been hypothesized for many years, it was not until 1981 that Vale et al. elucidated its sequence (2), and over the past decade that the physiology of CRH has become increasingly well understood.

The CRH system is widespread throughout the brain, but is best understood in the paraventricular nucleus of the hypothalamus where it serves as a key component of the HPA axis. The CRH activates the HPA axis, and also increases sympathetic nervous function (3,4). Its central effects are associated with increased arousal, alertness, and other changes characteristic of heightened attention to the environment and increased readiness to respond. The LC-NE system is located in the mid-pons of the brain stem. The LC sends a dense network of terminals throughout the brain, and innervates a majority of neurons in the cerebral cortex. Activation of the LC-NE system is associated with release of norepinephrine from neurons throughout the brain. The two systems appear to be part of a positive feedback loop in which activation of one system is associated with activation of the other. For example, CRH applied to the LC increases LC firing rate (5), while NE is a stimulus to CRH release from the hypothalamus (6). In addition to the fact that the two systems interact to mutually stimulate each other, they also appear to respond similarly to messengers in the neurochemical environment. Thus, both systems are inhibited by opiates and (GABA) γ-aminobutyric acidergic agents, and excited by serotonergic and cholinergic agents (6–11), and both systems are inhibited by glucocorticoids (1).

CRH content is highest in the PVN, but CRH is also found in a variety of extrahypothalamic locations, including Barrington's nucleus in the brain stem (adjacent to the locus coeruleus), the central nucleus of the amygdala (CEA), and the cerebral cortex (12). CRH receptors are located throughout the brain, including areas that lack demonstrable CRH nerve terminals and cell bodies, suggesting a parasynaptic mode of information processing (12,13).

Since the characterization of CRH in 1981, researchers have accumulated a growing body of experimental evidence that CRH not only releases ACTH, but also plays an important role in both the psychological and physical arousal seen in response to stress. The intracerebroventricular (ICV) administration of CRH to the rat produces a series of physiological and behavioral events adaptive during stressful situations that include: 1) activation of the pituitary-adrenal axis and the sympathetic system; 2) facilitation of pathways subserving vigilance, cautious avoidance, and anxiety; and 3) inhibition of pathways subserving vegetative functions such as growth, reproduction, and feeding (14–18). By contrast, ICV administration of the CRH antagonist a-helical CRH (9–41) reduces shock-induced freezing behavior compared to control animals (19). Evidence of the more general relevance of this phenomenon to human behavior comes from work with primates; behavioral effects of CRH and its antagonist on anxiety-related behaviors similar to those just described have been documented in monkeys (20). Direct evidence in humans has been more difficult to obtain, as CRH administered peripherally does not cross the blood-brain barrier, and agonist and antagonist analogues for administration in humans are as yet unavailable.

Although relatively little is known about the determinants of extrahypothalamic CRH secretion, the regulation of CRH secretion by the hypothalamus is becoming increasingly well understood. Preclinical studies in rats have shown that the neurotransmitters norepinephrine, acetylcholine, and serotonin stimulate the release of CRH, as do the cytokines interleukin 1-α (IL-1α) and interleukin 6 (IL-6) (6–9,21–23). On the basis of the available experimental evidence, drugs that dampen anxiety and arousal might be expected to decrease CRH secretion or effects, and indeed benzodiazepines inhibit the release of CRH, as (in humans though not in rats) do opiates (24,25). Chronic—but not acute—administration of a number of antidepressants to rats is similarly associated with marked decreases in CRH messenger RNA (m-RNA) in the PVN, as well as decreases in tyrosine hydroxylase (TH) m-RNA and immunoreactivity in the LC (26–28).

Arginine Vasopressin

The CRH is not the only neuropeptide that stimulates pituitary adrenal activation. The hypothalamus also produces and secretes a second important stimulant of glucocorticoid secretion, the 9-amino-acid peptide arginine vasopressin (AVP). Although most hypothalamic AVP is produced in the magnocellular region and transported to the posterior pituitary to be released as an element active in maintaining fluid and volume balance by conserving free water, AVP is also produced in smaller amounts in the parvicellular region of the hypothalamus and, like CRH, acts at the anterior pituitary to induce the release of ACTH. Compared with CRH, AVP is a relatively weak secretogogue of ACTH; however, in the presence of CRH, AVP has a powerful synergistic effect that increases ACTH release. The AVP-induced ACTH secretion may also differ from CRH-induced ACTH secretion in that it may not be glucocorticoid suppressible (29); as described in the next section, this may allow AVP to play an important role in the maintenance of chronic stress responses.

CRH and ACTH Secretion and Pulsatility

The secretion of HPA axis hormones is not uniform. Variations occur both minute to minute and over the course of the day. Research done on the variability of HPA-axis hormone secretion makes it clear that the secretion of these hormones exhibits both circadian (24-hour) and ultradian (less than 24-hour) rhythmicity.

CRH is secreted by the hypothalamus in minute amounts. It is transported by the portal system directly to the pituitary, where it acts and then circulates systemically. The assessment of hypothalamic CRH rhythmicity by measurement of CRH levels in body fluids such as systemic blood and cerebrospinal fluid is complicated by several factors. Systemic levels of CRH are too low to permit consistent accurate detection, and reflect CRH production not only by the hypothalamus, but by other peripheral organs as well. Similarly, cerebrospinal fluid CRH is thought to reflect the CRH activity not only of the hypothalamus, but also of other areas of the CNS. A precise description of hypothalamic CRH ultradian and circadian rhythms thus remains elusive. Nevertheless, insight into CRH rhythmicity has been gained by the study of ACTH and cortisol levels in different conditions.

To infer information about hypothalamic CRH output from the variations in the peripheral levels of ACTH and cortisol, it is necessary to review some principles of hormone kinetics. The concentration of a hormone in blood is determined by glandular output (secretion rate), volume of distribution, rate of distribution to and from various compartments (e.g., circulation and extracellular fluid (kinetic parameters)), and metabolic clearance rate. Because of individual variations in metabolic clearance, volume of distribution, and kinetic parameters, blood levels may not accurately reflect glandular output. Recently, computerized mathematical models have been developed to use blood hormone concentrations as a starting point for ascertaining actual hormone secretion rates. This technique is called deconvolution, and can be accomplished in two ways: 1) a waveform-defined algorithm is developed, which estimates both hormone half-life and secretory rates in vivo, and 2) a waveform-independent algorithm that requires a priori knowledge of the hormone half-life allows calculation of in vivo secretion rates without assuming any special form for the secretory event (30).

Cortisol and ACTH are secreted in a pulsatile fashion, having both ultradian and circadian rhythms. The circadian rhythms reach their highest levels for both hormones in the early morning and reach a nadir in the late afternoon and evening. Deconvolution analysis has shown that ACTH and cortisol secretion occur in 16–19 rapid bursts of secretion/day. Furthermore, it has been shown that the secretory pulses of cortisol and ACTH are significantly concordant, with cortisol pulses lagging each ACTH secretory burst by about 10 minutes. The integrated plasma levels resulting from those secretory episodes comprise the overall circulating levels of the hormones. Detailed secretory rate studies have shown that ACTH, but not cortisol, secretion is greater in male than in female subjects, suggesting that the adrenal cortex is more responsive to ACTH in females than in males in terms of glucocorticoid production, and that equivalent daily cortisol secretion rates are attained in men and in women at the expense of greater ACTH release in men (31).

Because secretion rates of cortisol are determined by and closely follow the secretory pattern for ACTH, it is tempting to speculate that CRH secretion is similar to and closely precedes ACTH secretion. This assumption is supported by several lines of evidence. Chronic immunization of sheep with anti-AVP antibodies results in no alterations in the basal secretion and diurnal variations of ACTH and cortisol (32), while chronic immunization with anti-CRH antibodies significantly disrupts the mean basal ACTH and cortisol levels, as well as the number and amplitude of pulses of those hormones (33). The ACTH responses to exogenous AVP are highest in the morning and highest to exogenous CRH in the evening. Since by itself AVP is a weak ACTH secretagogue but potentiates the effects of CRH on ACTH secretion, it has been suggested that the diurnal variations in the ability of AVP to stimulate ACTH secretion are the result of variations of ambient endogenous levels of CRH in the hypophyseal-portal system (i.e. higher in the morning and lower in the evening). That would also explain the greater capacity of exogenous CRH to elicit an ACTH response in the evening, because if in the morning endogenous CRH levels are high, exogenous CRH should (and does) produce little further ACTH release from an already highly stimulated pituitary; conversely, in the evening low endogenous levels of CRH would permit exogenous CRH to powerfully induce ACTH secretion (34). Finally, administration of CRH in a pulsatile fashion restores the normal cortisol pattern in patients with adrenal insufficiency (35). These different lines of evidence support the assumption that pulsatile CRH levels modulate ACTH secretory activity.

There is some evidence that factors other than the CRH secretory pulsatility help determine the circadian variation of the HPA axis, since it has been shown that continuous infusion of ovine CRH causes a rise in HPA axis function but does not abolish the circadian rhythmicity of ACTH or cortisol (36). However, the vast majority of available studies suggests that the ambient CRH level is the key determinant of ACTH secretion. It is thus reasonable to propose that the pulsatility of hypothalamic CRH antecedes that of ACTH, and that its architecture is similar to that of ACTH secretion.

ACTH and Cortisol

Pituitary activation results in the release of ACTH, a 39-amino-acid peptide that acts at the adrenal cortex to release cortisol, the primary peripheral stress hormone. Cortisol is derived from cholesterol, which is both synthesized in the adrenal gland and also taken up by the gland in low-density lipoprotein or high-density lipoprotein particles synthesized in the liver. In the adrenal cortex, cholesterol is converted first to pregnenolone, then to 17-a-OH-pregnenolone, a 17-a-OH-progesterone, 11-deoxycortisol, and finally cortisol. ACTH is the primary circulating cortisol releasing factor and stimulates both the release of cortisol and de novo cortisol synthesis. ACTH interacts with adrenal cell membrane receptors, ultimately causing activation of adenylate cyclase, increase in cyclic AMP concentrations, and protein phosphorylation. The major effect of ACTH on steroidogenesis is to stimulate the rate-limiting step (conversion of cholesterol to pregnenolone). In addition to stimulating cortisol synthesis, ACTH is a trophic factor for the adrenal gland (37).

Cortisol has many effects peripherally on both metabolic and immune processes, promoting a redirection of energy and providing feedback to restrain the central effectors of the stress response as well as the peripheral elements of the immune system (38). Glucocorticoid receptors are found in almost every cell throughout the body. Specific effects of glucocorticoids include: 1) raising blood glucose levels through stimulation of gluconeogenesis in the liver, inhibition of insulin secretion, and stimulation of lipolysis; 2) decreasing inflammatory and immune activity; and 3) induction of a number of enzymes and proteins. While these last appear to have a wide variety of functions, one of the more interesting is the induction of the hepatic cytochrome P-450 and the protein metallothionein, since it has been suggested that an important action of glucocorticoids is detoxification (38).

Glucocorticoid Receptors and Feedback Control

The restraining feedback that glucocorticoids provide is a critical function, yet one that remains poorly understood. Evidence from both preclinical and clinical studies is unequivocal that both at the hypothalamus and at the pituitary, elevations in plasma cortisol restrain the activity of the axis. Exactly how this occurs is an area of active investigation, but an important component is receptor-mediated feedback.

Two distinct glucocorticoid receptor types exist in the brain. Type I receptor—also known as mineralocorticoid receptor—is found primarily in the hippocampus, while Type II receptor—the classic glucocorticoid receptor—is ubiquitously present throughout the brain (39). The respective functions of these receptors in mediating glucocorticoid effects, and particularly their roles in regulation of the HPA axis function, are uncertain. Negative feedback control of the HPA axis appears to take place both at the level of the pituitary by cortisol and at the hypothalamus, but there is evidence that the effect at the hypothalamus is an indirect one; pulsatile hypothalamic release of CRH in isolated hypothalamuses is independent of the ambient glucocorticoid environment (40). It has been suggested that Type I receptor, which has a higher affinity for corticosterone and cortisol than Type II receptor, inhibits hypothalamic CRH secretion via hippocampo-hypothalamic neural pathways. It is also possible that Type I receptor regulation is most relevant at diurnal nadirs when cortisol levels are low and Type I receptor is preferentially occupied, while Type II receptor becomes more critically involved as plasma levels rise. Some evidence from studies in rats supports this hypothesis (41,42). Evidence that Type II receptor blockade with the specific Type II receptor antagonist RU 486 suppresses the ACTH response to ovine CRH in humans has been put forward (43), and more recently there was a report of increased serum cortisol levels following administration of a bolus of the Type I antagonist canrenoate (44).

The physiology of receptor-mediated effects has been described for Type II receptor in some detail. Type II receptor is found in the cytosol and associates with glucocorticoids that have crossed cell membranes. It is a member of a superfamily of steroid receptors that includes, among others, the estrogen and progesterone receptors, and the thyroid hormone receptor. Hormone binding and DNA activation are controlled at different sites; mutations leading to loss of portions of the binding site result in a constitutively active molecule, while mutations to the DNA binding region do not prevent hormone binding (45). The Type II receptor is constitutively expressed in all cells (46,47), and its activity is regulated by a variety of modulators, including heat shock proteins of the 90K family (48). This raises the intriguing possibility that the myriad of effects of glucocorticoids in different cells may be explicable on the basis of the expression of different transcription cofactors and regulators in different cell types.

THE HPA AXIS AND STRESS RESPONSE

Recruitment of Stress Responsive Elements

The activation of the HPA axis and the LC-NE system by stress leads to recruitment of central nervous system and peripheral elements that act to counter the perceived threat and to restore homeostasis. These elements include cognitive changes, changes in vegetative functions such as growth and reproduction, metabolic changes, and immune changes. Munck and his coworkers have presented evidence that the HPA axis operates in two different modes (49). At a lower level of activity the HPA axis is associated with normal circadian rhythms, reaching daily morning peaks and evening nadirs in humans as well as with the initial recruitment of stress-responsive elements. In this mode the axis functions in a primarily permissive mode, meaning that at lower levels glucocorticoids are required to sensitize cells that respond to stress (e.g., potentiating pressor effects of catecholamines) and to potentiate the catabolic functions needed to mobilize and redirect energy that will permit an adaptive stress response to take place. At the higher level of activation induced by exposure to stressors, the HPA axis begins to function with different effect. In this mode, rather than potentiating actions, glucocorticoids begin to exert a more generally suppressive effect (immunosuppressive effects are the clearest example of this phenomenon). Munck has suggested that this dual effect could be accounted for by the fact that, for many stress mediators, glucocorticoids decrease production of the substance while increasing production of its receptors. In such a model, increasing concentrations of glucocorticoids would initially increase mediator-receptor complexes (and hence activity), while higher levels of glucocorticoids would, by decreasing the amount of mediator available for binding, decrease the number of receptor-mediator complexes and thus decrease overall activity—effectively becoming a suppressive agent. Whether entirely correct or not, this theory does capture a critical aspect of glucocorticoid function in the stress response—glucocorticoids act to recruit and direct stress responsive elements in the face of a threat to homeostasis, and glucocorticoids also play an important restraining (and hence protective) function by preventing the stress response from continuing uncontained and potentially damaging the host organism.

Among the functions that stress-induced HPA axis activation promotes are behavioral responses, particularly those associated with arousal. These include increased vigilance, focused attention, and alertness, and have been described in the previous paragraph. There are also a number of peripheral responses promoted by HPA axis activation. Many of these are directed at increasing energy available for immediate response where it is most needed (such as flight or tissue repair), and can thus be categorized as being involved in the "adaptive redirection of energy" (1). Such responses include increases in 1) concentration of oxygen and nutrients in the central nervous system and at stressed sites; 2) cardiovascular tone; 3) respiratory rate; 4) gluconeogenesis; and 5) lipolysis.

In a manner that complements its function of providing energy where it is most needed, activation of the HPA axis also tends to suppress the use of energy by systems that would not contribute to enhancing survival and responding to stress effectively, and would thus represent inefficient uses of critical resources in acutely stressful situations. Thus, the systems that promote growth and reproduction are inhibited by activation of the HPA axis by stress (Fig. 1). The CRH suppresses release of leuteinizing hormone-releasing hormone (50–52) by the hypothalamus, while glucocorticoids inhibit LHRH, as well as the pituitary gonadotroph, and de-

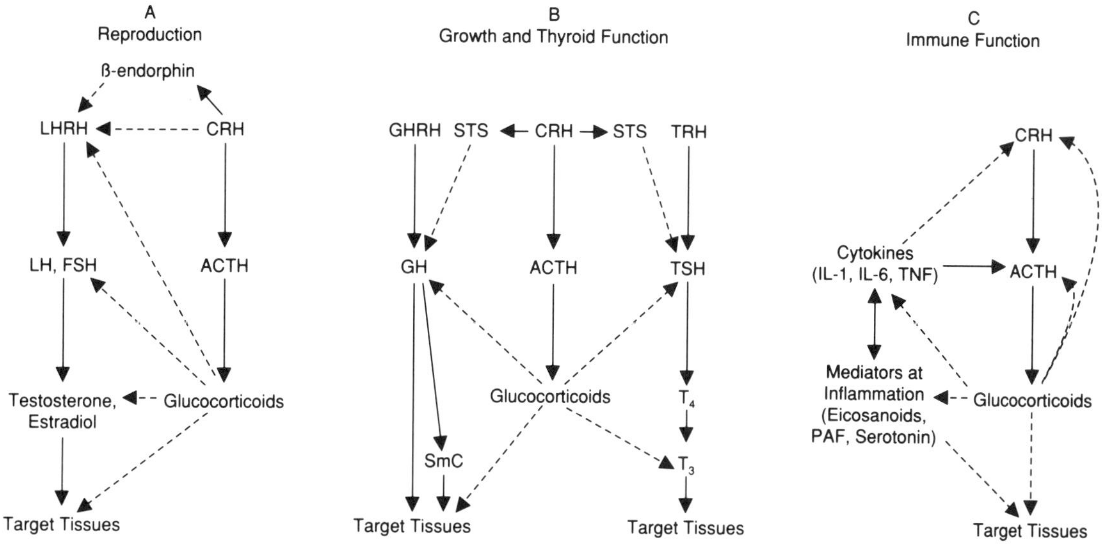

FIG. 1. Interactions between the hypothalamic-pituitary-adrenal (HPA) axis and the reproductive and growth axes, and mediators of immune function. Reproduced from Chrousos GP and Gold PW (1), with permission.

crease the response of target tissues to sex steroids. Analogously, CRH stimulates release of somatostatin and inhibits somatomedin C (53–55). While there is an initial acute increase in growth hormone secretion, more chronically this effect becomes inhibitory. Finally, thyroid axis function is also decreased during stress, although the exact mechanisms through which this is achieved are not known (56,57). There is a decrease in production of thyroid-stimulating hormone, and a decrease in conversion of T4 to the more active T3. This decrease in thyroid activity is thought to be adaptive in reducing diffuse metabolic demands, again allowing a concentration of energy where it will be most needed and effectively utilized.

The effectiveness of the organism in responding to stress is also increased by maximizing the efficiency and appropriateness with which energy is utilized, and activation of the HPA axis plays an important role in regulating the activity of other stress-induced responses. The best characterized of these is the immune system. Unrestrained immune activity in response to a triggering event such as a bacterial pathogen or an allergen can lead to severe tissue damage, as healthy tissue suffers the effects of defenses engendered by affected tissue when a response becomes uncontained. Glucocorticoids decrease production of cytokines and inflammatory mediators, inhibit their effects at sites of inflammation, and alter leukocyte number and function (58).

The effects of stress on nonhypothalamic CRH (e.g., in the central nucleus of the amygdala (CEA)) are less well-described. As just noted, experimental evidence suggests that CRH is distributed throughout the brain and has activating effects consistent with its known role as an effector of the stress response. These effects are regulated differently than is hypothalamic CRH, although the exact mechanism remains uncertain. In the CEA and the bed nucleus of the stria terminalis, persistent elevations in cortisol are associated with increases or no changes in CRH mRNA rather than the consistent decreases seen in the hypothalamus (59–61). Preliminary data in LEW/N rats suggests that after imipramine treatment CRH mRNA in the CEA is elevated (Michelson, D: unpublished data), which is also in contrast to studies showing that CRH mRNA in the PVN is lower after chronic treatment with imipramine and other antidepressants (26, 27).

Chronic Stress and the HPA Axis Function

The role of the HPA axis in chronic stress and its regulation is less well understood than its role acutely. Studies of animals under a variety of chronic stresses show persistent activation of the HPA axis, with preservation of acute reactivity (62–64). Although there is some habituation of the HPA axis after prolonged stress in the sense that serum corticosterone levels normalize after an initial increase, pituitary POMC m-RNA and ACTH content, as well as adrenal weights, remain increased, evidence that the axis remains activated. A recent study showed loss of the diurnal corticosterone nadir in rats chronically stressed with adjuvant-induced arthritis (65). Persistent HPA axis activation as a result of chronic stress, however, appears to be differently regulated compared to acute stress. Induction of arthritis in rats using Freund's adjuvant leads to increases in POMC mRNA and, in some strains, AVP release with either a decrease or no change in CRH mRNA and release, suggesting that although the axis is chronically activated it is not primarily driven by hypersecretion of CRH (62). This is consistent with observations that in repeated stress paradigms AVP may become the primary driver of the HPA axis (66,67). In humans there is relatively little known about the physiology of the HPA axis under chronic stress. Recently we have found that patients with chronic noninflammatory illnesses such as multiple sclerosis (MS) display hypercortisolism relative to healthy controls (68). Unlike patients with major depression, the response to o-CRH in patients with MS is normal, yet the ACTH response to AVP is blunted, suggesting a relative diminution of endogenous CRH secretion consistent with an axis chronically driven by AVP.

HPA AXIS DYSFUNCTION AND DISORDERS OF THE STRESS RESPONSE

The critical involvement of the HPA axis in stress response and its involvement in so many different functions during stress suggest that abnormalities in HPA axis response could be associated with a variety of different pathological states. Such changes in HPA axis function include inappropriate activation, failure of appropriate activation, or altered patterns of activation. Further, since the axis has multiple components and has the capacity to adjust its activity at different levels, the effect of a particular abnormality will vary greatly depending on its character and locus. The best known and best described abnormalities of HPA axis function are those in which the capacity of the adrenal to respond flexibly is lost completely—Cushing's disease (primary and secondary) and Addison's disease. In these illnesses the pathophysiology reflects most dramatically the effects of glucocorticoid excess or deficiency throughout the body—disturbances in normal metabolic functioning, energy utilization, and stress response. With the characterization of CRH, however, and the growing understanding of its central effects, we have begun to learn about another class of HPA axis disorders—those related to central abnormalities.

Major Depression

The best characterized of the states of centrally driven HPA axis hyperactivity is the major depression. Increased production of cortisol measured by 24-hour urinary-free cortisol and a failure to compress cortisol production after treatment with dexamethasone have been shown in as many as 50% of patients with major depression by a number of investigators (69,70). Classically, depression has been thought of as melancholia, and its mood alterations have been associated with melancholic signs and symptoms such as loss of appetite and weight, loss of libido, and the inability to sleep. Internally, melancholic depression is marked by intensely painful arousal and an obsessional preoccupation with personal inadequacy and the inevitability of loss. This aspect of melancholia can be understood as an organized state of anxiety focused on self-representation and experienced as inappropriate guilt and self-contempt (3). These signs and symptoms of increased arousal and decreased vegetative behaviors, considered in the light of the effects of ICV administration of CRH to rats, strongly suggest the possibility that the pathophysiology of melancholic depression is related to a central excess of CRH.

A number of investigators over the past decade have examined this hypothesis, and have found consistent evidence that the hypercortisolism of depressed patients is centrally mediated (i.e., related to the escape of hypothalamic CRH secretion from its usual feedback restraint), and that hypercortisolism in melancholic depression reflects a defect at or above the hypothalamus resulting in the hypersecretion of CRH. Several lines of evidence support this hypothesis. Plasma ACTH responses to synthetic ovine CRH (o-CRH) are attenuated in major depression; since these patients as a group tend to have elevations of circulating glucocorticoids, this blunting is appropriate, reflecting restraint of the pituitary corticotroph cell by the negative feedback (71,72) and suggesting that the pituitary is intact. Resolution of depression after treatment is associated with a restoration of normal glucocorticoid secretion patterns and a normalization of the ACTH response to o-CRH (73). Normal controls given a continuous infusion of o-CRH have a pattern and magnitude of hypercortisolism consistent with that seen in depression (36). The CRH in the cerebrospinal fluid (CSF) of depressed patients correlates positively with indices of pituitary-adrenal activation (74). In addition, preliminary data from the continuous sampling of lumbar CSF for 30 hours indicates that depressed patients secrete an increased number of CRH pulses into the CSF, and that CSF CRH levels range from normal to elevated over the course of a day (Kling MA, unpublished data). Consistent with this data, in single-sample CSF studies we found normal CSF CRH levels in depressed patients, and other groups have reported a range of normal-to-elevated levels of CRH in the CSF of patients with major depression (75). However, while the hypersecretion of CRH in some depressed patients has been documented, it remains unknown whether CRH in other brain regions is also hypersecreted. Preliminary work in rats using lesioning techniques suggests that approximately 50% of CSF CRH originates outside of the PVN (Hong SK, personal communication). Assuming this is also true in humans, it is yet to be determined whether CRH elevations are accounted for by one or both fractions.

Atypical Depression and Fatigue States

Overall, the evidence that overexuberant secretion of CRH is associated with an unrestrained stress response and plays a pathophysiological role in the disordered stress responses of melancholic depression is strong and in keeping with the role of the HPA axis as a mediator of the stress response. But the stress response is not inherently pathological, and the capacity to respond to stress is critical to healthy functioning. Thus it seems likely that hyporesponsive secretion of CRH would create problems related to an inability to adequately respond to stressors or, more simply, the demands of everyday life. A central deficiency of CRH would be expected to lead to an inability to increase attention and a failure of arousal, which could be consistent with lethargy and loss of motivation as well as with a failure of restraint of vegetative behaviors such as appetite and sleep. Descriptively, such syndromes appear to exist as: 1) the chronic fatigue syndrome; 2) atypical depression with its constellation of hyperphagia and hypersomnia; 3) fibromyalgia; and 4) perhaps other conditions as well. Because there have been no good central stimuli of CRH secretion available for clinical investigations in humans, central CRH insufficiency in these illnesses has been difficult to demonstrate definitively, but suggestive indirect evidence of centrally mediated hypocortisolism has been found in chronic fatigue syndrome as well as the seasonal affective disorder (76,77). Conversely, in Cushing's disease, in which there is central hyposecretion of CRH due to unrestrained cortisol secretion driven by the pituitary or the adrenal, patients experience depressions whose features usually include lethargy, fatigue, and hypersomnia, and not the hyperarousal classically associated with melancholia (Kling MA, personal communication).

Post Traumatic Stress Disorder

In post-traumatic stress disorder (PTSD), exposure to severe stressors sets in motion a process in which the stressful experience remains vivid and disturbing, intruding into ongoing psychic life in ways such as flashbacks and nightmares. Descriptions of HPA axis function in this disorder have been conflicting; however, it has been suggested that at least some patients show evidence of a relatively hypoactive axis (78). This finding is intriguing, and its implications uncertain. While it is possible that it represents a chronic adaptation to a persistent state of stress, it may instead be that such patients are susceptible to severe, episodic increases in HPA axis activity due to relatively reduced tonic levels of glucocorticoid feedback, and that such increases are associated with inadequate negative feedback control and sudden, episodic reactivation of the stress response. Such an explanation would also be consistent with the fact that certain patients with major depression and chronic increased HPA axis tone are preoccupied with internally generated anxiety and mood, and are relatively insensitive to the external environment (i.e., the increased tone would be relatively unresponsive to further external stimuli).

Susceptibility to Autoimmune Illness

Although the role of the HPA axis in susceptibility to human autoimmune and inflammatory illness remains uncertain, evidence is accumulating that HPA axis dysfunction could contribute significantly to development of these illnesses. In the acute setting of an inflammatory or immune trigger that acts as a stressor, the integrity of the HPA axis stress response is critical to preventing an unrestrained immune response from overshooting and inflicting damage to the self. The significance of this stress-responsive negative feedback loop between the immune system and the HPA axis has been demonstrated in the LEW/N rat, which is an animal model for a number of autoimmune and inflammatory diseases, including streptococcal cell wall (SCW)-induced arthritis, experimental allergic encephalomyelitis (EAE), and others (79,80). Sternberg showed that increased susceptibility to inflammatory disease (SCW-induced arthritis in the LEW/N rat) was related to a defect in the response of the corticotropin-releasing hormone neuron to inflammatory triggers, resulting in deficient pituitary-adrenal activation and the loss of glucocorticoid-mediated restraint of the immune response. These triggers included not only SCW, but also IL-1-a and serotonin agonists such as quipazine. Sternberg also demonstrated that the relative arthritis resistance of F344/N rats compared to LEW/N rats was related to their intact HPA axis responses to these same inflammatory triggers (79,80). Physiologic replacement doses of corticosteroids (dexamethasone) in LEW/N rats significantly suppressed the severity of the illness in this arthritis-susceptible strain; conversely, suppression of the hypothalamic pituitary adrenal axis in F344/N rats with the corticosteroid receptor antagonist RU486 led to development of systemic inflammation and arthritis in response to SCW in this otherwise inflammatory-disease–resistant strain. Interruption of the F344/N HPA axis at the level of serotonin stimulation with the serotonin antagonist LY53857 was also associated with development of arthritis. Thus, whether present on a genetic basis, as in LEW/N rats, or on a pharmacologic basis, as in RU486-treated F344/N rats, acute interruptions of the HPA axis immune-induced stress response are associated with susceptibility to inflammatory disease.

Potential Mechanisms of HPA Axis Disregulation

While the ways in which HPA axis response varies in different contexts is becoming clearer, the mechanisms leading to appropriate or pathological responses remain uncertain. The LEW/N and F344 strain differences suggest that heredity plays an important role in determining HPA axis reactivity. As just noted, it seems that the presence of a chronic stressor inducing an ongoing pituitary adrenal response is also associated with changes in the function of the HPA axis. There is evidence from both human and animal studies that developmental events are associated with illnesses that are in turn associated with HPA axis abnormalities. The HPA axis response to stress is not—at least in rats—constant through life. Within 2 to 3 days of birth, rats enter a "stress hypo-responsive period" that lasts 2 to 3 weeks and during which stressors that normally produce robust HPA axis responses in adults have little or no effect (81). Several different explanations have been advanced to account for this period, including a relative immaturity of the components of the HPA axis, alterations in receptor modulation of feedback signals, and changes in neural modulation of response. Whatever its basis, however, this period of reduced response is critical to the animal's development. Glucocorticoids are, overall, catabolic, making energy available for immediate use. For a young rat in a period of rapid growth and tissue development, exposure to high levels of glucocorticoids produces lasting deleterious changes. Thus rats that receive glucocorticoids during this period have been found to have reduced brain weights, altered neural morphology, altered social behavior, and altered learning capacities (82), suggesting that the stress-nonresponsive period provides internal protection against the potential harmful effects of stressors. Conversely, early experience appears to have lasting effects on HPA axis function; rats handled regularly neonatally have been found as adults to have higher initial corticosteroid responses to stressors that are also more easily terminated compared to nonhandled animals (67). Similarly, rats separated from their mothers for discrete periods early in life were found to have increased HPA axis response to stressors as adults compared to identical rats that had not undergone maternal separation, and measures of CRH activity in the PVN as well as the CEA and bed nucleus of the stria terminalis were similarly elevated in the separated animals (83).

The effects of development on the organization of the HPA axis, the effects of manipulations of the internal glucocorticoid milieu, and the effects of environmental manipulations on the reactivity of the axis to stress may be relevant to the body of data documenting the importance of early life experience in disorders of stress response such as depression. At the same time, animal models such as the LEW/N and F344/N rats, which exhibit genetically based differences in HPA axis activation in response to stressors, suggest that HPA axis excitability is also governed by inheritance. The stress response is, by definition, an interaction between the organism and its environment. Understanding it more fully will ultimately require an understanding of how genetically embedded tendencies to arousability interact with experience, and how these factors together define patterns of stress response.

REFERENCES

1. Chrousos GP, Gold PW. The concepts of stress and stress system disorders: overview of physical and behavioral homeostasis. *JAMA* 1992;267:1244–1252.
2. Vale W, Spiess S, Rivier C, Rivier J. Characterization of a 41 residue ovine hypothalamic peptide that stimulates secretion of corticotropin and b-endorphin. *Science* 1981;213:1394–1397.
3. Gold PW, Goodwin FK, Chrousos GP. Clinical and biochemical manifestations of depression (first of two parts). *N Eng J Med* 1988;319:348–353.
4. Gold PW, Goodwin FK, Chrousos GP. Clinical and biochemical manifestations of depression (second of two parts). *N Eng J Med* 1988;319:413–420.
5. Valentino RJ, Foote SL, Ashton-Jones G. Corticotropin releasing factor activates noradrenergic neurons of the locus ceruleus. *Brain Res* 1983;270:363–367.
6. Calogero AE, Gallucci WT, Chrousos GP, et al. Catecholamine effects upon rat hypothalamic corticotropin-releasing hormone secretion in vitro. *J Clin Invest* 1988;82:839–846.
7. Calogero AE, Bernardini R., Margioris AN, et al. Effects of serotonergic agonists and antagonists on corticotropin releasing hormone secretion by explanted rat hypothalami. *Peptides* 1989;10:189–200.
8. Calogero AE, Gallucci WT, Bernardini R, et al. Effect of cholinergic agonists and antagonists on rat hypothalamic corticotropin releasing hormone secretion in vitro. *Neuroendocrinology* 1988;47:303–308.
9. Calogero AE, Gallucci WT, Chrousos GP, et al. Interaction between GABAergic neurotransmission and rat hypothalamic corticotropin releasing hormone secretion in vitro. Theoretical and clinical implications. *Brain Res* 1988;463:28–36.
10. Guynet PG, Aghajanian GK. Acetylcholine, substance P and met-enkephalin in the locus coeruleus: pharmacological evidence for independent sites of action. *Eur J Pharmacol* 1979;53;319–328.
11. Pepper CM, Henderson G. Opiates and opiod peptides hyperpolarize locus coeruleus neurons in vitro. *Science* 1980;209:394–396.
12. Swanson LW, Sawchenko PE, Rivier J, et al. Organization of ovine CRF immunoreactive cells and fibers in the rat brain: an immunohistochemical study. *Neuroendocrinology* 1983;36:165–186.

13. Herkenham M. Mismatches between neurotransmitter and receptor localizations in brain: observations and implications. *Neuroscience* 1987;23:1–38.
14. Brown MR, Fisher LA, Spiess J, et al. Corticotropin releasing factor: actions on the sympathetic nervous system and metabolism. *Endocrinology* 1982;111: 928–931.
15. Sutton RE, Koob GF, LeMoal M, et al. Corticotropin releasing factor produces behavioral activation in the rat. *Nature* 1982;297:331–333.
16. Britton DR, Koob G, Rivier J, et al: Intraventricular corticotropin releasing factor enhances behavioral effects of novelty. *Life Sci* 1982;31:363–367.
17. Sirinathsinghji DJS, Rees LH, Rivier J, et al. Corticotropin releasing factor is a potent inhibitor of sexual receptivity in the female rat. *Nature* 1983;305:232–235.
18. Rivier C, Vale W. Influence of corticotropin releasing factor on reproductive functions in the rat. *Endocrinology* 1984;114:914–921.
19. Kalin NH, Sherman JE, Takahashi LK. Antagonism of endogenous CRH systems attenuates stress-induced freezing behavior in rats. *Brain Res* 1988;457:130–135.
20. Kalin NH, Takahashi LK. Animal studies implicating a role of corticotropin-releasing hormone in mediating behavior associated with psychopathology. In Risch SC, ed. *Central nervous system peptide mechanisms in stress and depression.* American Psychiatric Press 1991;53–72.
21. Besedovsky A, Del Rey A, Sorkin E, Dinarello CA. Immunoregulatory feedback between interleukin-1 and glucocorticoid hormones. *Science* 1986;233:652–654.
22. Sapolsky R, Rivier C, Yamamoto G, Plotsky P, Vale W. Interleukin-1 stimulates the secretion of hypothalamic corticotropin-releasing factor. *Science* 1987;238:522–524.
23. Naitoh Y, Fukata J, Toiminaga T, Naki Y, Tami S, Mori K, Imura H. Interleukin-6 stimulates the secretion of adrenocorticotrophic hormone in conscious, free-moving rats. *Biochem Biophy Res Commun* 1988;155: 1459–1463.
24. Kalogeras KT, Calogero AE, Kuribayashi T, et al. In vitro and in vivo effects of the triazolobenzodiazepine alprazolam on hypothalamic-pituitary-adrenal function: pharmacologic and clinical implication. *J Clin Endocrinol Metab* 1990;70:1462–1471.
25. Pechnik RN. Effects of opioids on the hypothalamic-pituitary-adrenal axis. *Annu Rev Pharmacol Toxicol* 1993; 32:353–382.
26. Brady LS, Gold PW, Herkenham M, et al. The antidepressants fluoxetine, idosoxan and phenelzine alter corticotropin releasing hormone and tyrosine hydroxylase mRNA levels in rat brain: therapeutic implications. *Brain Res* 1992;572:117–125.
27. Brady LS, Whitfield HJ, Fox RJ, et al. Long-term antidepressant administration alters corticotropin-releasing hormone, tyrosine hydroxylase and mineralocorticoid receptor gene expression in rat brain: therapeutic implications. *J Clin Invest* 1991;87:831–837.
28. Melia KR, Nestler EJ, Duman RS. Chronic imipramine treatment normalizes levels of tyrosine hydroxylase in the locus coeruleus of chronically stressed rats. *Psychopharmacology* 1992;108:23–26.
29. Bilezikjian VM, Blount AL, Vale WW. The cellular actions of vasopressin in corticotrophs of the anterior pituitary: resistance to glucocorticoid actions. *Mol Endocrinol* 1987;1:451–458.
30. Veldhuis JD, Johnson ML, Bolton WK. Analyzing pulsatile endocrine data in patients with chronic renal failure: a brief review of deconvolution techniques. *Pediatr Nephrol* 1991;5:522–528.
31. Roelfsema F, van den Berg G, Frölich M, Veldhuis JD, van Eijk A, Buurman MM, Etman BH. Sex-dependent alteration in cortisol response to endogenous adrenocorticotropin. *J Clin Endocrinol Metab* 1993;77:234–240.
32. Guillaume V, Conte-Devolx B, Magnan E, et al. Effect of chronic active immunization with antiarginine vasopressin on pituitary-adrenal function in sheep. *Endocrinology* 1992;130:3007–3014.
33. Guillaume V, Conte-Devolx B, Magnan E, et al. Effect of chronic active immunization anticorticotropin-releasing factor on the pituitary-adrenal function in the sheep. *Endocrinology* 1992;130:2291–2298.
34. Salata RA, Jarrett DB, Verbalis JG, Robinson AG. Vasopressin stimulation of adrenocorticotropin hormone (ACTH) in humans. In vivo bioassay of corticotropin-releasing factor (CRF) which provides evidence for CRF mediation of the diurnal rhythm of ACTH. *J Clin Invest* 1988;81:766–774.
35. Avgerinos PC, Schürmeyer TH, Gold PW, et al. Pulsatile administration of human corticotropin-releasing hormone in patients with secondary adrenal insufficiency: restoration of the normal cortisol secretory pattern. *J Clin Endocrinol Metab* 1986;62:816–821.
36. Schulte HM, Chrousos GP, Gold PW, et al. Continuous administration of synthetic ovine corticotropin-releasing factor in man. Physiological and pathophysiological implications. *J Clin Invest* 1985;75:1781–1785.
37. Baxter JD, Tyrrell JB. The adrenal cortex. In Felig P, Baxter JD, Broadus AE, Frohman LA, eds. *Endocrinology and metabolism.* New York: McGraw-Hill, 1987; 511–650.
38. Munck A, Guyre PM, Holbrook NJ. Physiological functions of glucocorticoids in stress and their relation to pharmacological actions. *Endocr Rev* 1984;5:25–44.
39. Reul JMHM, de Kloet ER. Anatomical resolution of two types of corticosterone receptor sites in rat brain with in vitro autoradiography and computerized image analysis. *J Steroid Biochem Mol Biol* 1986;24:269–272.
40. Mershon JL, Sehlhorst CS, Rebar RW, Lie JH. Evidence of corticotropin-releasing hormone pulse generator in the macaque hypothalamus. *Endocrinology* 1992;130: 2991–2996.
41. Ratka A, Sutanto W, Bloemers M, de Kloet ER. On the role of mineralocorticoid (type I) and glucocorticoid (type II) receptors in neuroendocrine regulation. *Neuroendocrinology* 1989;50:117–123.
42. Bradbury MJ, Akana SF, Cascio CS, Levin N, Jacobson L, Dallman MF. Regulation of basal ACTH secretion by corticosterone is mediated by both type I (MR) and type II (GR) receptors in rat brain. *J Steroid Biochem Mol Biol* 1991;40:105–111.
43. Laue L, Chrousos GP, Loriaux DL, et al. 1988 The anti-glucocorticoid and anti-progestin steroid RU 486 suppresses the adrenocorticotropin response to ovine corticotropin releasing hormone in man. *J Clin Endocrinol Metab* 66:290–293.
44. Dodt C, Kern W, Fehm HL, Born J. Antimineralocorticoid canrenoate enhances secretory activity of the hypothalamus-pituitary-adrenocortical (HPA) axis in humans. *Neuroendocrinology* 1993;58:570–574.
45. Evans RM. The steroid and thyroid hormone receptor superfamily. *Science* 1988;240:889–895.
46. Vamvakopoulos NC, Chrousos GP. Hormonal regula-

tion of human corticotropin releasing hormone gene expression: implications for the stress response and immune/inflammatory reaction. *Endocr Rev* 1994;15: 409–420.

47. Kalinyak JE, Griffin CA, Hamilton RW, Bradshaw GJ, Perlman AJ, Hoffman AR. Development and hormonal regulation of glucocorticoid receptor messenger RNA in the rat. *J Clin Invest* 1989;84:1843–1848.
48. Vamvakopoulis N. Tissue-specific expression of heat shock proteins 70 and 90. Potential implications for differential sensitivity of tissues to glucocorticoids. *Mol Cell Endocrinol,* 1993;98:49–54.
49. Munck A, Naray-Fejes-Toth A. The ups and downs of glucocorticoid physiology. Permissive and suppressive effects revisited. *Mol Cell Endocrinol* 1992;90:C1–4.
50. Ono N, Lumpkin MD, Samson WK, McDonald JK, McCann SM. Intrahypothalamic action of corticotropin-releasing factor to inhibit growth hormone and LH release in the rat. *Life Sci* 1984;35:1117–1123.
51. Rivier C, Rivier J, Vale W. Stress-induced inhibition of reproductive function: role of endogenous corticotropin-releasing factor. *Science* 1986;231:607–611.
52. MacAdams MR, White RH, Chipps BE. Reduction in serum testosteronelevles during chronic glucocorticoid therapy. *Ann Intern Med* 1986;140:648–652.
53. Rivier C, Vale W. Involvement of corticotropin-releasing factor and somatostatin in stress-induced inhibition of growth hormone secretion in the rat. *Endocrinology* 1985;117:2478–2482.
54. Unterman TG, Phillips LS. Glucocorticoid effects on somatomedins and somatomedin inhibitors. *J Clin Endocrinol Metab* 1985;61:618–626.
55. Burguera B, Muruais C, Penalva A, Dieguez C, Casanueva F. Dual and selective actions of glucocorticoid upon basal and stimulated growth hormone release in man. *Neuroendocrinology* 1990;51:51–58.
56. Duick DS, Wahner HW. Thyroid axis in patients with Cushing's syndrome. *Arch Intern Med* 1979;139:767–772.
57. Benker G, Raida M, Olbricht T, Wagner R, Reinhardt W, Reinwein D. TSH secretion in Cushing's syndrome: relation to glucocorticoid excess, diabetes, goitre, and the 'sick euthyroid syndrome.' *Clin Endocrinol* (Oxf) 1990;33:777–786.
58. Baxter JD, Tyrell JB. The adrenal cortex. In Felig P, Baxter JD, Broadus AE, Frohman LA, eds. *Endocrinology and metabolism.* New York: McGraw-Hill; 1987; 511–650.
59. Makino S, Gold PW, Schulkin J. Corticosterone effects on corticotropin-releasing hormone mRNA in the central nucleus of the amygdala and the parvocellular region of the paraventricular nucleus of the hypothalamus. *Brain Res* 1994;640:105–112.
60. Swanson LW, Simmons DM. Differential steroid hormone and neural influences on peptide mRNA levels in CRH cells of the paraventricular nucleus: a hybridization histochemical study in the rat. *J Comp Neurol* 1989; 285:413–435.
61. Makino S, Gold PW, Schulkin J. Effects of corticosterone on CRH mRNA and content in the bed nucleus of the stria terminalis; comparison with the effects in the central nucleus of the amygdala and the paraventricular nucleus of the hypothalamus. *Brain Res* 1994;657: 141–149.
62. Harbuz MS, Rees RG, Eckland D, Jessop DS, Brewerton D, Lightman SL. Paradoxical responses of hypothalamic corticotropin releasing factor (CRF) messenger RNA (m-RNA) and CRF-41 peptide and adenohypophysial proopiomelanocortin mRNA during chronic inflammatory stress. *Endocrinology* 1992;130:1394–1400.
63. Kant GS, Leu SR, Anderson SM, Mossgey EH. Effects of chronic stress on plasma corticosterone, ACTH and prolactin. *Physiol Behav* 1987;40:775–779.
64. Dallman MF, Jones MT. Corticosteroid feedback control of ACTH secretion: effect of stress-induced corticosterone secretion on subsequent stress responses in the rat. *Endocrinology* 1973;92:1367–1375.
65. Sarlis MJ, Chowdrey HS, Stephanou A, Lightman SL. Chronic activation of the hypothalom-pituitary-adrenal axis and loss of circadian rhythm during adjuvant-induced arthritis. *Endocrinology* 1992;130:1775–1779.
66. Scaccione S, Muscolo LA, Cigliana G, Navarra D, Nicolai R, Angelucci L. Evidence for a specific role of vasopressin in sustaining pituitary adrenal cortical response in the rat. *Endocrinology* 1991;128:3138–3143.
67. DeGoeij DEC, Kvetnansky R, Whitnall MH, Sezova D, Berkenbosch F, Tildens FSH. Repeated stress-induced activation of corticotropin releasing factor neurons enhances vasopressin stores and colocalization with corticotropin releasing factor in the median eminence of rats. *Neuroendocrinology* 1991;53:150–159.
68. Michelson D, Stone L, Galliven E, et al. Pituitary adrenal function is altered in multiple sclerosis. *J Clin Endocrinol Metab* 1994;79:848–853.
69. Sachar EJ, Hellman L, Fukushima DK, et al. Cortisol production in depressive illness: a clinical and biochemical clarification. *Arch Gen Psychiat* 1970;23:289–298.
70. Carroll BJ, Curtis GC, Davies BM, et al. Urinary free cortisol excretion in depression. *Psychol Med* 1976;6: 43–50.
71. Gold PW, Chrousos GP, Kellner C, et al. Psychiatric implications of basic and clinical studies with corticotropin releasing hormone. *Am J Psychiat* 1984;141:619–627.
72. Gold PW, Loriaux DL, Roy A, et al. Responses to corticotropin releasing hormone in the hypercortisolism of depressin and Cushing's disease: pathophysiologic and diagnostic implications. *N Eng J Med* 1986;314: 1329–1335.
73. Amsterdam JD, Maislin G, Winokur A, Berwish N, Kling M, Gold P. The oCRH stimulation test before and after clinical recovery from depression. *J Affect Disord* 1988;14:213–222.
74. Roy A, Pickar D, Paul SM, et al. CSF corticotropin releasing hormone in depressed patients and normal control subjects. *Am J Psychiatry* 1987;144:641–645.
75. Nemeroff CB, Wilderlov E, Bissette G, et al. CSF concentrations of corticotropin releasing factor-like immunoreactivity in depressed patients. *Science* 1984;226: 1342–1344.
76. Demitrack MA, Dale JK, Gold PW, et al. Evidence for impaired activation of the hypothalamic-pituitary-adrenal axis in patients with chronic fatigue syndrome. *J Clin Endocrinol Metab* 1991;73:1224–1230.
77. Vanderpool J, Rosenthal N, Chrousos GP, et al: Evidence for hypothalamic CRH deficiency in patients with seasonal affective disorder. *J Clin Endocrinol Metab* 1991;72:260–271.
78. Yehuda R, Giller EL, Southwick SM, Lowy MT, Mason JW. Hypothalamic-pituitary-adrenal dysfunction in post-traumatic stress disorder. *Biol Psychiatry* 1991;30: 1031–1048.
79. Sternberg EM, Hill JM, Chrousos GP, et al. Inflammatory mediator-induced hypothalamic-pituitary-adrenal

axis activation is defective in streptococcal cell wall arthritis-susceptible Lewis rats. *Proc Natl Acad Sci USA* 1989;86:2374–2378.

80. Sternberg EM, Young WS Jr., Bernardini R, et al. A central nervous defect in biosynthesis of corticotropin-releasing hormone is associated with susceptibility to streptococcal cell wall induced arthritis in Lewis rats. *Proc Natl Acad Sci USA* 1989;86:4771–4775.
81. Sapolsky RM, Meaney MJ. Maturation of the adrenocortical stress response: neuroendocrine control mechanisms and the stress hyporesponsive period. *Brain Res Rev* 1986;11:65–76.
82. De Kloet ER, Rosenfeld P, Van Eekelen JAM, Sutanto W, Levine S. Stress, glucocorticoids and development. *Prog Brain Res* 1988;73:101–120.
83. Plotsky PM. CRF and AVP in stress: modulation by early experience. Presented at First World Congress on Stress. 1994; Abstract 196:69.

Neurobiological and Clinical Consequences of Stress: From Normal Adaptation to PTSD, edited by M. J. Friedman, D. S. Charney, and A. Y. Deutch, Lippincott-Raven Publishers, Philadelphia © 1995.

14

Toward a Cognitive Neuroscience of Dissociation and Altered Memory Functions in Post-Traumatic Stress Disorder

John H. Krystal, *Alexandre L. Bennett, J. Douglas Bremner, Steven M. Southwick, and Dennis S. Charney

*Department of Psychiatry, Yale University School of Medicine, New Haven, Connecticut 06510, National Center for Post-Traumatic Stress Disorder, Veterans Affairs Medical Center, West Haven, Connecticut 06516; and *Department of Neurology and Neuroscience, Cornell University College of Medicine, New York, New York 10021*

Dissociative states and altered memory functions are integral components of psychological traumatization. The term dissociation was coined by Janet (1) to describe states where the integration of consciousness is disrupted. This concept has been employed to describe a spectrum of subjective experiences in which perceptual, affective, memory, and identity functions are altered. As listed in Table 1, particular symptoms or syndromes associated with dissociative states include distorted sensory perceptions, altered time perception, amnesia, derealization, depersonalization, conversion symptoms, fugue states, and multiple personality (2–6). Dissociation may occur at the time of traumatization (5–12).

Dissociative states and increased vulnerability to hypnotic states may also develop as ongoing sequelae of traumatization (6,12–15). While dissociated, acutely traumatized individuals may appear confused, emotionally dulled, or even catatonic (9,16,17). Decades following traumatization, while recalling their traumatic experiences, individuals may experience time as being slowed, have altered sensory perceptions, and have feelings of unreality (18). Less frequently, adult traumatization may produce fugue states, conversion reactions, or multiple personality as ongoing symptoms of post-traumatic stress disorder (PTSD) (17,19). Childhood psychological traumatization is also associated with dissociative symptoms. In one study, approximately 60% of 450 adults traumatized as children had periods in their lives when they had no memory of their abuse (20). Dissociative symptoms arising from childhood traumatization continue in adulthood (21,22). For example, psychiatric inpatients with histories of childhood trauma have higher levels of dissociative symptoms than nontraumatized inpatients (23).

Flashbacks, perhaps the most distinctive PTSD symptom, link dissociative, memory-related, and arousal regulatory processes pertinent to this disorder. During flashbacks, patients vividly reexperience aspects of the traumatic response while feeling detached from their surrounding environment. Ongoing sensory processing may be altered or disrupted, and patients may report that they are in a fog or that they blacked out (12). Flashbacks involving the recollection of traumatic experiences are frequently associated with intense emotional responses and panic-like states (24). Most flashbacks are brief, lasting only a few minutes. However, some flashbacks

TABLE 1. *Dissociative symptoms associated with PTSD*

Emotion
Withdrawal/detachment
Anxiety/hyperarousal (flashbacks)
Self-Monitoring/Identity
Depersonalization
Derealization
Conversion Reactions
Fugue
Multiple Personality
Memory
Amnesia
Reexperiencing aspects of the trauma (flashbacks)
Shifts in modes of memory encoding (pictorial/iconic vs. linguistic vs. other)
Perceptual
Altered time perception
Altered visual perception (shape, color, size)
Altered contextual perception (proximity, temporal relatedness)
Altered proprioception (body shape, position)
Analgesia
Altered perception in other perceptual domains (olfaction, taste, tactile)
Other Cognitive
Constricted attention
Neglect
Confusion
Altered associative capacities (altered depth of processing?)
Holistic- versus detail-focused information processing style?

may last several hours or several days. Some flashbacks are accurate depictions of a traumatic situation, while others have unreal or distorted qualities, similar to dreams. Out-of-body experiences are one type of perspective distortion frequently associated with dissociative states (5, 25).

Despite progress in identifying, characterizing, and quantitatively assessing dissociative states, there has been surprisingly little study of their neurobiology. Associated with the failure to elucidate a unique neurobiology for dissociative states, there have been few placebo-controlled pharmacotherapy trials for dissociative disorders and no specific antidissociative drugs developed. The absence of antidissociative pharmacotherapies contrasts with the development of anxiolytics, antiobsessionals, antipsychotics, mood-stabilizing agents, and antidepressants. In light of the paucity of research in this area, the commonly held view, that the core features of dissociative disorders are unresponsive to pharmacotherapy, is not surprising (26).

Memory disturbances are also a hallmark of PTSD. The diagnostic criteria for PTSD present a paradox regarding memory functions. Individuals with PTSD experience recurrent and intrusive distressing recollections that are sufficiently vivid to result in a sense of reliving the experience (27). This aspect of the PTSD diagnostic criteria is consistent with the view expressed by Freud and Breuer (2) that post-traumatic hysteria patients "suffer principally from reminiscences" (p. 29). However, PTSD is also defined by an inability to recall important aspects of the trauma (27). For example, one study documented that over 5% of combat veterans in the North African campaign of World War II could not recall important details of a major battle when evaluated shortly after its completion (28). Further, traumatic memories may be distorted versions of the actual events (29,30). In individuals with chronic PTSD, ongoing impairments in short-term memory function have also been described (31).

The past 10 years have marked a revival of interest in the neurobiology of PTSD (32–34). The purpose of this review is to highlight recent progress made in studies of the neurobiology of dissociative states, memory dysfunction, and hyperarousal in PTSD patients. This chapter will review and begin to integrate recent studies of PTSD patients, other patient populations, healthy subjects, and preclinical research. In reviewing this diverse body of literature, we will highlight the potential prognostic and therapeutic significance of recent research findings as a stimulus for future research on the neurobiology of dissociative states and memory disturbance in PTSD.

PTSD AND DISSOCIATIVE STATES

Pharmacologic Studies

Pharmacologic Studies in PTSD Patients

Flashbacks have been precipitated in Vietnam veterans with chronic PTSD following the intravenous administration of sodium lactate (25), yohimbine (35), and m-chlorophenylpiperazine (MCPP) (36). Administration of each of these substances produces panic attacks in a significant proportion of patients with either panic disorder (25,35–39), but not other patients groups. However, PTSD patients are the first group studied to experience flashbacks following administration of these substances.

Rainey and his associates (25) compared the response to intravenous sodium lactate, isoproterenol, and a dextrose placebo in seven Vietnam combat veterans, six of whom also met criteria for panic disorder. All seven patients experienced flashbacks following lactate, two patients also experienced flashbacks after isoproterenol infusion, and one patient experienced a flashback during placebo infusion. The authors described these flashbacks as similar to those occurring naturally as part of PTSD. Dissociative experiences that accompanied flashbacks included depersonalization, derealization, and auditory and visual sensory perceptions. Six of the seven lactate-induced flashbacks, both isoproterenol flashbacks, and the dextrose flashback were followed by panic-like states. However, the absence of reported anxiety ratings makes it impossible to determine whether sub-panic increases in anxiety preceded the flashbacks. The overlap of panic disorder and PTSD in the patients in this study was another limitation of this study, because it raised concerns that lactate-induced flashbacks were a property of panic disorder and not independently associated with PTSD. Little is known about the mechanisms through which lactate produced panic attacks and flashbacks in PTSD patients.

The precipitation of flashbacks and panic attacks in PTSD patients by yohimbine linked noradrenergic systems, implicated in fear and arousal regulation, to the symptoms of PTSD (35). Yohimbine activates central noradrenergic neurons through blockade of α-2 receptors located on noradrenergic neurons. These α-2 receptors mediate, in part, feedback inhibition of noradrenergic neurons (40). Following yohimbine, 40% (8/20) of patients experienced flashbacks and 70% (14/20) experienced panic attacks. No panic attacks and only one flashback emerged following placebo administration. Although 45% of the patients in this study also met DSM-III-R criteria for panic disorder, 43% of the yohimbine-induced panic attacks occurred in individuals without panic disorder. The risk of yohimbine-induced panic attack was increased in patients with panic disorder relative to those without comorbid panic disorder (89% versus 43%). However, history of panic disorder did not appear to influence the likelihood of experiencing a yohimbine-induced flashback. The following vignette illustrates features of a yohimbine-induced flashback:

10:00 A.M.:	[Initiation of yohimbine infusion]
10:05 A.M.:	Subject reports hot and cold flashes, goose bumps, palpitations
10:10 A.M.:	Subject reports clammy hands. He asked the nurse to move away from him, in case he felt like running. "I feel like I'm picking up dead bodies the centrifuge sounds like a helicopter . . . A chopper is shooting at us, we're trying to shoot back at it! One of the guys' head is shot off! Brains are coming at me! I smell burnt flesh . . . I feel scared, I can't hear what's going on. . . ."

The operational definition for flashback employed in this study led to the exclusion of many dissociative states produced by yohimbine in the PTSD patients. The following criteria were employed to define a drug-induced flashback: 1) reexperiencing of a past traumatic event during drug infusion; 2) the reexperiencing must involve one or more sensory modalities; and 3) for patients with a history of flashbacks, the drug-induced state must be similar to naturally occurring flashbacks. Despite the expedient characterization of flashbacks as being present or absent, yohimbine actually produced a continuum of dissociative phenomena. Patients experienced varying degrees of derealization and depersonalization that were often accompanied by other dissociative symptoms. Yohimbine also elicited a range of altered perceptual experiences, some of which were fragmentary or vague. For example, one patient perceived the shadow produced by a sink in the testing facility to be the shadow made by a tank turret. In addition to stimulating flashbacks, yohimbine significantly increased the recall of traumatic memories. Some patients vividly recalled aspects of combat during yohimbine administration, but did not reexperience the actual traumatic events in the form of sensory experiences. Vivid memories were not rated as flashbacks in this study.

The yohimbine study suggested that activation of noradrenergic systems by yohimbine produced panic attacks and flashbacks in a subset of PTSD patients. One question raised by this study was whether the elicitation of flashbacks by yohimbine reflected a specific response to α-2 receptor blockade or whether all anxiogenic drugs produce flashbacks in PTSD patients. In order to investigate this question, yohimbine and MCPP effects were compared in this population (36). This study found that both MCPP and yohimbine produced flashbacks and other dissociative states in veterans with combat-related PTSD. Preliminary analyses indicated that patients tended to experience panic attacks following yohimbine or MCPP, but not both medications. This study also found that drug-induced traumatic memories, autonomic activation, and anxiety states could be associated with the induction of flashbacks, although no single response preceded flashbacks in all cases. This observation raises the possibility that the final common neurobiological pathway mediating the yohimbine and MCPP induction of flashbacks may not be evident in the subjective reports of patients, or that multiple mechanisms might lead to induction of flashbacks.

Case reports suggest that alcohol and opiate withdrawal may increase PTSD symptoms, including flashbacks (41,42; Seibyl, personal communication). Central noradrenergic systems are activated during alcohol and opiate withdrawal, suggesting a possible parallel between yohimbine- and withdrawal-induced flashbacks (41).

Induction of Dissociative States in Healthy Individuals

Three classes of drugs commonly produce dissociative-like states in healthy subjects: 1) antagonists of the *N*-methyl-D-aspartate (NMDA) subtype of glutamate receptor; 2) cannabinoids; and 3) serotonergic hallucinogens. The noncompetitive NMDA receptor antagonist anesthetics, phencyclidine and ketamine, produce a derealized and depersonalized state characterized by marked perceptual alterations and psychosis at subanesthetic doses (43–46).

The capacity of ketamine to produce dissociative-like states in healthy subjects has been rigorously evaluated in a series of studies (47–49). In these studies, dissociative symptoms were rated using the Clinician-Administered Dissociative States Scale (6). At low blood levels, ketamine produced a lightheaded feeling. At higher blood levels, subjects reported the slowing of time and alterations in the vividness, form, and context of sensory experiences. For example, subjects noted that objects appeared brighter or duller than expected, larger or smaller than usual, distorted in shape, or with altered proximity. Also, some subjects had difficulty hearing someone speaking close to them, while reporting that a radio playing quietly in the next room sounded unusually loud. Altered proprioceptive experiences were reported by subjects, who felt that their limbs changed form or were floating in air.

Cognitive effects of ketamine were also prominent. Subjects reported constriction of their field of attention, resulting in the sensation of tunnel vision or the feeling that they were surrounded by fog. As ketamine blood levels rose, the circumference of their field of attention was increasingly constricted. For example, subjects attending to a computer keyboard lost track of events happening on the computer monitor. Ketamine also produced learning and memory impairments. Its effects increased proportionately to the dose administered and the duration of delay between stimulus presentation and testing. Ketamine also interfered with executive functions such as abstraction, assessed by proverb interpretation, and problem solving, evaluated by the Wisconsin Card Sorting Test. Although subjects felt that they had lost control of their thought processes, with effort they could focus on tasks. Ketamine also produced emotional and identity-related responses. At low doses, it had mild anxiolytic properties, while larger doses generally produced euphoria and anxiety. Anxiety stimulated by ketamine tended to follow perceptual alterations and thought disorganization, and to be related to their degree of comfort with the drug-induced disturbances in thought and perception. Some subjects found the perceptual alterations produced by ketamine quite pleasurable, analogous to a ride in an amusement park, while others found ketamine effects frightening.

Ketamine-induced insight impairments may have contributed to the elicitation of anxiety. Following drug infusion, some subjects lost the perspective that their mental status change was produced by ketamine and they became concerned that they had contracted a mental illness. Transient identity alterations were also observed with ketamine. For example, a subject who received ketamine (0.26 mg/kg i.v. bolus followed by 0.65 mg/kg/hr) stated "at first it seemed that I didn't exist, I couldn't process information, after a while, I was convinced that I was an organism, then I realized I was a human being, then after a longer while I remembered that I was a medical student." Ketamine did not lead to the emergence of multiple personalities, flashbacks, or vivid intrusive memories in research subjects. However, symptoms associated with psychosis, including delusions and thought disorder, were observed during ketamine infusion.

Two ongoing studies have attempted to pharmacologically alter ketamine-induced dissociative states by pretreating healthy subjects with lorazepam or haloperidol (48–49). Preliminary data from these studies suggested that 2 mg lorazepam administered orally 2 hours prior to ketamine administration tended to reduce altered environmental perceptions, but had no effects on other dissociative symptoms or psychotic states produced by ketamine. Haloperidol failed to duce dissociative symptoms, vigilance impairments, or amnestic effects produced by ketamine, but reduced ketamine-induced distractibility, abstraction impairments, and bizarreness of thought processes. These data suggested that, at the doses tested, neither lorazepam or haloperidol is a true ketamine antidote. They are also consistent with the literature suggesting that neuroleptics have limited efficacy in treating dissociative symptoms (26).

Dissociative states have also been produced by psychoactive cannabinoids such as tetrahydrocannabinol, the principal psychoactive component of marijuana and hashish. Cannabinoids bind to a specific G-protein–coupled receptor (50) through which they alter cellular functions (e.g., through blockade of N-type calcium channels, inhibition of cyclic adenosine monophosphate (AMP) accumulation, and stimulation of arachidonic acid and intracellular calcium release) (51). Some cannabinoid effects may be mediated by stimulation of glucocorticoid receptors (52) and blockade of NMDA receptors (53). At high doses, cannabinoid intoxication produces depersonalization, derealization, perceptual alterations, and insight impairments (54–56). Cannabis has been reported to produce flashbacks in the drug-free state that resemble cannabis intoxication (57). In one study (58), 3% (1/31) of habitual marijuana users and 1% (3/348) of nonhabitual users reported flashbacks when drug-free, suggesting that flashbacks were not a frequent consequence of cannabis use. However, this study suggested that marijuana use also enhanced the likelihood of experiencing flashbacks following ingestion of the serotonergic hallucinogens.

Serotonergic hallucinogens such as lysergic acid diethylamide (LSD), mescaline, and dimethyltryptamine (DMT) also produce dissociative symptoms. These agents stimulate serotonin-2 (5-HT_2) receptors (59,60) and produce pronounced visual hallucinations, illusions, synesthesia, and expansive or portentious emotional responses (61). Following ingestion of these psychedelics, feelings of derealization or depersonalization may also be prominent. Environmental stimuli may be experienced in a fragmented manner, body image may be distorted, and individuals may feel detached from their surroundings (61–65). Some clinicians also have reported that LSD may facilitate the recall of repressed memories (61), although this capacity has never been rigorously evaluated. Relative to the phencyclidine or ketamine experience, psychedelic hallucinogens tend to produce perceptual effects that predominate over dissociative effects and impairments in higher cognitive functions.

Flashbacks have been reported in healthy individuals following serotonergic hallucinogen use. Freedman (61,66) suggested that LSD intoxication was traumatic for some users because it diminished control over awareness, resulting in intense emotional states experienced as beyond their control. In such cases, LSD flashbacks might have a traumatic etiology. However, some LSD-like experiences, such as synesthesia, may be reexperienced long after drug ingestion by individuals who find such experiences pleasant. These effects do not easily fit a trauma model, suggesting that sensitization, conditioning, or state-dependent learning might also apply (61,66–68). Subject expectancy may also play a role in drug-like flashbacks. One study found that flashbacks may be produced in healthy subjects following placebo administration, if subjects are coached to anticipate that a placebo will produce flashbacks (69). Heaton suggested that the expectancy of flashbacks led subjects to mislabel and selectively attend to aspects of normal experience that are consistent with a flashback-like experience.

Biological Hypotheses

Dissociative states are comprised of a heterogenous spectrum of behavioral states involving multiple brain mechanisms. Dissociative states are associated with alterations in sensory processing and attention, emotion, self-monitoring and identity functions, and learning and memory. Thus, networks within the brain that mediate these functions are involved by implication. In advance of definitive research identifying brain mechanisms associated with dissociative states, it is useful to consider candidate neural networks and neurotransmitter systems that might be involved.

Emerging Circuitry of Altered Attention and Sensory Processing Implicated in Dissociative States

Two domains of attention and sensory processing seem to be simultaneously implicated in the genesis of dissociative states: 1) the capacity to detach attention from or disrupt the processing of particular sensory stimuli, and 2) the capacity to focus attention or become absorbed by other mental functions or sensory processing. During traumatization, withdrawal of attention and disruption of sensory processing serves adaptive functions. For example, the capacity to disrupt the processing of stimuli related to traumatization spares the victim physical and emotional suffering (70). Following the subject's traumatization, PTSD is associated with the capacity to maintain particular trauma-related experiences beyond awareness, only to cause the subject to reexperience them in the form of flashbacks and intrusive thoughts. This reliving involves a shifting of experience from sensory-guided experience to internally-generated states. The dual withdrawal and refocusing of attention is not unique to flashbacks and may be common to trance-like states (71). Dissociative states are frequently associated with a particular type of focused attention that involves the experience of absorption within a memory, thought, or activity (72). During relatively mild states of dissociation, individuals may retain a capacity to self-monitor their relationship to the environment. This capacity appears to be lost in more complete states of dissociation (73). As in the ketamine studies (47,49), dissociative states may be associated with impaired judgement, planning functions, problem solving, and other executive functions. Also, in fugue states or multiple personality, individuals may generate thoughts, plans, and behavior appropriate to their altered sensory input or altered identity functions (27,74). Yet, learning occurring during dissociative states may be highly state-dependent and memories may be irretrievable outside of the particular fugue or identity state. Hilgard (4,74) highlighted the impairments in executive functions found with dissociative and hypnotic states, including reduced self-observation, increased suggestibility, and decreased voluntary control of focused attention. Many of these functions have been attributed to the frontal cortex and cingulate gyrus (75–76). Altered activity of these networks, evoked through behavioral or pharmacologic means, could give rise to altered executive capacities during dissociation.

Dissociative states are not necessarily pathological, and they occur in individuals without dissociative disorders at extremely low or high levels of sensory processing. For example, reduced intensity or variability of sensory stimulation, associated with hypnosis, sleep deprivation, and sensory deprivation, may produce altered states of consciousness with dissociative features (2,10,77–79). As an extreme illustration of this point, sensory polyneuropathies may cause marked depersonalization and derealization associated with feelings of being disembodied (80). Heightened sensory stimulation or arousal may also produce altered sensory processing. Significant levels of arousal and anxiety heighten the salience and vividness of environmental stimuli. Under stress, attention is narrowed to the most salient aspects of the environment, consistent with the need to focus on the danger at hand. Thus, individuals fixate faster and longer on unusual or highly informative objects such as weapons (81), while less critical, but important information about the context of the trauma may not receive much attention (82). At extremely high levels of arousal, coherent integration of sensory information breaks down and dissociative symptoms emerge, even in individuals without dissociative disorders (10,83–87).

The thalamus plays a critical role in modulating responsivity to environmental stimuli associated with sleep and dreaming, and may play a similar role in the genesis of dissociative states. As illustrated in Fig. 1, the thalamus serves as a sensory gate or filter that directly and indirectly modulates the access of sensory information to the cortex, amygdala, and hippocampus (88–91). During slow-wave sleep, for example, thalamic nuclei exhibit slow spindle oscillations that disrupt the transmission of sensory information to cortical and limbic structures (92). During wakefulness, thalamic neurons fire in a relay mode that facilitates transmission of sensory information to cortical regions. The REM sleep, associated with dreaming, is characterized by phasic enhancement of the activity of glutamatergic thalamocortical cells (93,94). In this model, dreams and other sleep-related internally generated experiences may arise as thalamocortical or other direct cortical projections from the amygdala and hippocampus, and bypass the oscillatory thalamic processes that disrupt the flow of sensory information to the cortex (95,96). Thus, like dissociative states, sleep states may neurobiologically preserve associative and mnemonic functions while interrupting sensory processing. Sensory processing alterations associated with dissociative states could indicate the intrusion of sleep-related disturbances in sensory processing into the waking state. If so, then alterations in thalamic activity might link a spectrum of altered states of consciousness such as hypnosis, dreaming, and other conditions in which there is a combination of the features of sleep and waking states (96,97). Dissociative states might also be related to night terrors in which features of waking behavior intrude upon sleep (98–100). Evidence for a thalamic role in maintaining the boundary of sleep-like behavior and wakefulness is provided by patients with paramedian thalamic infarctions. These patients exhibit a profound sense of detachment, reduced responsivity to sensory stimuli, and sleep-like posturing throughout the circadian cycle without the electrophysiologic correlates of non-REM sleep (101).

Sensory distortions associated with stress may develop, in part, as a consequence of the thalamic role in modulating sensory processing. Thalamic nuclei appear to work both in series and in parallel with brain regions involved in arousal, fear, and stress responses to heighten the processing of salient information under conditions of moderate or controllable stress. For example, activation of the central nucleus of the amygdala

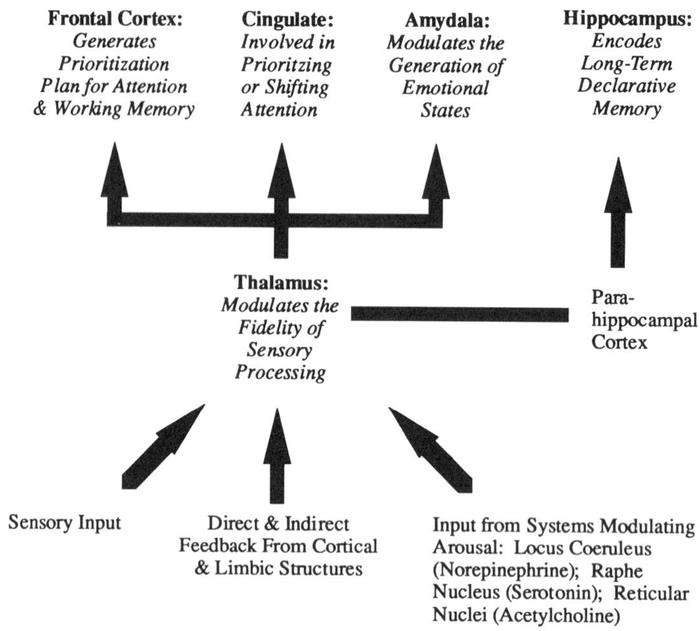

FIG. 1. This schematic illustrates the position of the thalamus within networks that may be involved in the generation of dissociative states. Sensory information reaches the thalamus and is transmitted to limbic and cortical regions responsible for modulating thought, attention, learning and memory, and emotion. The thalamus receives input from limbic regions such as the amygdala, and brainstem regions involved in stress-related arousal. It also receives direct and indirect feedback from cortical regions involved in prioritizing attention. When functioning in relay mode the thalamus facilitates accurate transmittal of sensory information. However, when slow oscillatory firing patterns predominate, the thalamus impedes the flow of sensory information to cortical and limbic regions associated with the predominant focus on internally generated thought processes and sensory experiences associated with dreaming, night terrors, and perhaps dissociation.

during fear states facilitates the thalamic relay of sensory information to cortical and limbic structures (93,102–104). Stress-induced noradrenergic activation would be expected to facilitate thalamic transduction of sensory information by stimulating thalamic α-1 adrenoceptors that increase thalamic activity associated with wakefulness, and to inhibit slow thalamic oscillations (105,106). Postsynaptic α-2 receptors promote thalamic slow oscillations. Thus yohimbine, an α-2 antagonist, could increase thalamic bursting by increasing norepinephrine release and blocking the stimulation of slow oscillations produced by postsynaptic α-2 receptors (105). Serotonergic systems also heighten sensory processing via the 5-HT_2 receptor (106). Both MCPP and the serotonergic hallucinogens stimulate subtypes of this receptor (107). Perceptual alterations associated with extreme or uncontrollable stress suggest that the massive activation of monoamine systems under these conditions may modulate thalamic function in a fashion that results in interference rather than enhancement of the fidelity of sensory transmission.

Alterations in thalamic glutamatergic function could also contribute to sensory gating disturbances. Glutamate is the primary excitatory neurotransmitter within the thalamus (89) and the neurotransmitter involved with thalamic afferents from the amygdala, cerebral cortex, and hippocampus (89,108–111). Cortical regions also modulate thalamic sensory gating functions indirectly through glutamatergic inputs to a modulatory network involving the striatum, globus pallidus, subthalamic nucleus, and thalamus (112). Both NMDA and non-NMDA glutamate receptors are localized to the thalamus, where they have complementary functions (89). Previous reviews have suggested that alterations of the sensory filter function of the thalamus via blockade of NMDA receptors could contribute to the psychotomimetic effects of the NMDA antagonists (112). Given the prominent role of non-NMDA glutamate receptors in cortico-striato-thalamic circuitry, subanesthetic doses of selective NMDA antagonists might be predicted to produce distortions rather than complete blockade of thalamic sensory gating functions. This prediction is consistent with clinical observations suggesting that ketamine produces a state of detachment or withdrawal rather than sleep. Also, ketamine produces sensory distortions and illusions rather than blockade of sensory perceptions or pure hallucinatory experiences (47). The view that NMDA antagonists distort sensory filtering rather than block sensory transmission is consistent with the observation that sensory deprivation reduces, rather than augments, the behavioral effects of phencyclidine (113). Future research is needed to clarify the extent of thalamic contributions to dissociative states.

The thalamus is a heterogenous structure, and component thalamic nuclei have distinctive cortical afferents and efferents, and different patterns of synaptic organization (114). For example, the reticular nuclei of the thalamus function in some ways as an extension of the brainstem and midbrain reticular activating systems (90). However, other thalamic nuclei such as the anteroventral and mediodorsal nuclei appear to be involved in associative processes such as learning (115,116).

Sensory processing alterations and changes in attention may be linked in dissociative states. This connection is suggested by the convergence of corticolimbic networks upon the anterior cingulate gyrus, a brain region implicated in the capacity to shift and focus attention (76,117). These networks are summarized in Fig. 2. Anterior cingulate lesions may produce symptoms reminiscent of thalamic, limbic, and cortical lesions, including confusion, vivid daydreaming, apathy, and impairments in sustained attention and learning (118,119). Direct projections to the anterior cingulate gyrus from midline and intralaminar thalamic nuclei suggest that the cingu-

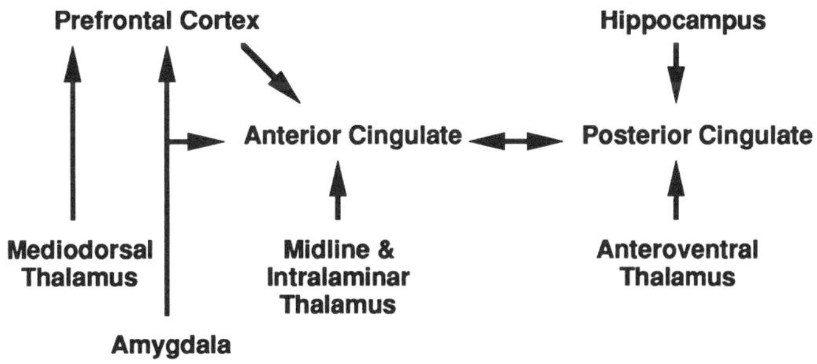

FIG. 2. This schematic illustrates the convergence of corticolimbic networks implicated in the prioritization of attention on the anterior cingulate gyrus. The anterior cingulate gyrus is an area of the brain implicated in emotional expression, the expression of learned behaviors, and the capacity to prioritize and focus attention. It is modulated by thalamic nuclei that influence sensory gating. The anterior cingulate also receives input from prefrontal and posterior cingulate cortices that modulate its attention gating role. The amygdala, implicated in the gating of emotional states, directly and indirectly modulates anterior cingulate activity. Through these mechanisms, the amygdala modulates the prioritization of attention and modulates associative functions of the prefrontal cortex. This schematic oversimplifies the complex and highly redundant interconnectivity of the corticolimbic networks.

late gyrus is responsive to shifts in thalamic sensory processing functions (120). As suggested by the clinical case reports of patients with cingulate lesions, the cingulate gyrus may also be involved in the attribution of salience, and acquisition and retrieval of learned information (121,122). The contributions of the cingulate gyrus to sensory processing, emotional regulation, and learning are facilitated by its connectivity to other brain regions. For example, hippocampal and anteroventral thalamic inputs converge upon the anterior cingulate gyrus via the posterior cingulate gyrus (116,123). Similarly, the anterior cingulate gyrus is an important point of convergence for a network involving the amygdala, prefrontal cortex, and mediodorsal nucleus (108,115,122,124,125).

Cortical Dysconnectivity and Dissociation

Geschwind (126) wrote "there is no evidence for the existence of any all-purpose computer in the brain." Consistent with this view, cortical functions are highly distributed across several cortical regions that require integration in order to generate coherent conscious experience. For example, frontoparietal interactions help to locate memories or mental representations in space, while frontohippocampal interactions appear to contribute contextual information regarding these memories (127). Also, the frontal cortex itself contains many functionally heterogeneous regions. Distinct frontal cortex loci mediate the generation of iconic or working memories for the location and features of environmental stimuli. Brain lesions of one region of the frontal cortex result in memory gaps for spatial features, while lesions of the other region produce an inability to recall faces (127,128). If corticocortical interactions were distributed or disrupted, experiences and cognitive functions dependent upon integrated cortical activity might be distorted. For example, if frontal cortical regions processing features of objects and their spatial attributes were interacting dysfunctionally, one might generate mental representations for stimuli in which features were not correctly matched to their spatial locations (i.e., objects could be experienced out of context or in bizarre or incoherent ways). Disturbances in function arising from abnormal integration of cortical function may be similar, by analogy, to conduction aphasias. In conduction aphasias, both comprehension and fluency are preserved, but speech is paraphasic because information cannot be effectively transmitted from association to motor cortices (129).

Drugs that produce dissociative states disturb cortical integration at several levels. The key output neurons of the cortex are pyramidal neurons that utilize glutamate as their primary neurotransmitter. These neurons are regulated locally by modulatory GABAergic neurons. Pyramidal neurons also receive distant input from subcortical monoaminergic, glutamatergic and peptidergic systems and glutamatergic input from pyramidal neurons in other cortical areas (127,130). In the piriform cortex, serotonergic hallucinogens inhibit pyramidal neuronal activity by stimulating γ-aminobutyric acid (GABA)-ergic interneurons via the $5\text{-HT}_{2A/2B}$ receptors. However, these drugs also activate pyramidal neurons by stimulating 5-HT_{2C} receptors (107). Electrophysiologic data suggest that ketamine distorts the functional connectivity within the cortex by blocking the NMDA receptor-mediated component of glutamatergic corticocortical connectivity. One study, for example, suggested that blockade of NMDA receptors allowed sensory information to reach the cortex, but interfered with the coherent transmission of this information from receptive areas to association cortices (131). Barbiturates, which preferentially block non-NMDA glutamate receptors (132,133), might also interfere with cortical integration, although the human psychopharmacology of non-NMDA glutamate receptors has received little direct study.

Summary: Alteration in Glutamatergic Function—A Final Common Pathway for Dissociation?

Clinical studies related to the neurobiology of dissociative states are summarized in Table 2. Yohimbine and MCPP produce dissociative states only in PTSD patients, who are prone to

TABLE 2. *Summary of pharmacologically facilitated dissociative states*

Substance	Healthy subjects	PTSD Patients
Yohimbine	−	+
MCPP	−	+
Lactate	−	+
Sedative/hypnotics	−	+*
NMDA Antagonists	+	?
Cannabinoids	+	?
Serotonergic hallucinogens	+	?

\+ associated with dissociative state
− not associated with dissociative state
? unclear association with dissociative state
* facilitation of dissociation during guided recollection
? not formally evaluated in patients with PTSD

have these experiences. These drugs primarily produce dissociative states while stimulating anxiety or traumatic recollections. Thus, yohimbine and MCPP may not directly induce dissociation, but rather contribute to the activation of networks, resulting in a dissociative state in vulnerable individuals. In the amygdala, hippocampus, thalamus, and cortex, noradrenergic and serotonergic systems serve to modulate the activity of glutamatergic neurons (127,130). Given the central role of glutamate in corticocortical, thalamocortical, amygdalocortical, and hippocampocortical connectivity, glutamatergic systems in the brain may be considered the framework upon which higher cognitive functions rest. Thus, it may not be surprising that a drug such as ketamine that alters glutamatergic neurotransmission produces dissociative states in healthy individuals. The possibility that glutamate systems might be fundamentally involved in generating dissociative states is consistent with the observation that dissociative states produced by ketamine in healthy people arise as a direct consequence of drug administration and are not dependent on generation of intense emotional responses or memories. The direct evocation of dissociative states by an NMDA antagonist raises the possibility that reductions in NMDA receptor function contribute to dissociative states in humans. If so, then pharmacologic agents that enhance NMDA receptor function might have antidissociative properties (114,134–136).

PTSD AND MEMORY FUNCTION

Clinical Views of Traumatic Memories:

Declarative and Nondeclarative Memory

Post-traumatic memory alterations challenge the explanatory power of current theories of memory function. Traumatic memories are associated with a number of distinctive features, reviewed in Table 3, that are of great interest to clinicians and researchers. A growing literature has identified distinctive subtypes of learning and memory (137). Declarative memory is defined as the conscious recollection of facts and events (138). Encoding in the declarative memory system is dependent on the functioning of hippocampal, and related temporal lobe and diencephalic brain structures. Damage to these areas results in severe impairment in learning of new information. A distinction is made between declarative memory and several other types of learning that are relatively preserved in amnestic patients. These types of learning are referred to as nondeclarative memory and include conditioning, priming, and skill learning. Conditioning is the association of a behavioral or emotional response with a stimulus, the perception of which will reliably elicit the conditioned response. Priming refers to facilitation of the ability to process a stimulus because of prior exposure (139,140). Skill learning involves acquisition of motor, perceptual, or cognitive skills, independent of knowledge about prior experience or the ability to articulate understanding of the acquired skill. These types of learning do not require conscious or volitional recall for expression. Declarative and nondeclarative memory processes participate to varying degrees in the encoding and recall of different types of information.

Declarative memory for material unrelated to the trauma appears to be impaired in PTSD patients, whereas recall of trauma-related material appears to be relatively enhanced. Memory and concentration deficits have been reported in Nazi concentration camp survivors (141) and prisoners of war from World War II and the Korean War (142,143). Also, studies of combat veter-

TABLE 3. *Characteristics of traumatic memories*

	During traumatization
Attention	Focus on salient information, reduced focus on contextual information.
Perception	May be distorted due to dissociation during traumatization
Encoding	Nonverbal modes of memory encoding are prominent, including encoding of visual images, other sensory information, and emotional responses.
	Component processes associated with encoding are distributed, but associative, laying the groundwork for patterns of memory rehearsal and retrieval.
	Salient information readily encoded in long-term memory. Encoded information reflects loss of contextual information due to shifts in attention and perceptual alterations due to dissociation.
	Following traumatization
Attention	Attention bias toward processing cues related to the trauma.
Rehearsal	Memories are shaped in an ongoing fashion. Both the content of the memories and associations to the memories may change with time.
Retrieval	Voluntary access may be reduced (psychogenic amnesia).
	Retrieval experienced as intrusive. However, once recalled, memories may be reexperienced perseveratively.
	Involuntary retrieval mechanisms are prominent, primarily associative, and involve exteroceptive (reminders of the trauma) and interoceptive (emotional and physiologic) cues.
	States associated with reduced levels of arousal (hypnosis, sleep, amytal interview) or impairment of executive functions (amytal interview, ethanol intoxication) may reduce resistance to retrieval.
	Some memories are associated with a loss of perspective (absorption), in that they are reexperienced rather than recalled.
	Once retrieved, memories are accessed in a relatively inflexible manner that may reflect impairments in strategic retrieval processes.
	Retrieval may be associated with intense emotional responses and dissociation.
	Many qualities (key information, vividness, etc.) are highly resistant to decay.

ans from the Vietnam war with chronic PTSD who exhibited immediate and delayed recall impairments (12,144) found that veterans with combat-related PTSD performed worse than veterans without PTSD in learning lists of words unrelated to their traumas. However, PTSD patients exhibited a relative facilitation of recall compared to the nontraumatized veterans, when words were combat-related. Factors involved in post-traumatic learning impairments might include altered allocation of attention, functional alterations in monoamine and neuroendocrine systems that modulate memory (33), and possible long-term neurotoxic effects of severe stress on neurons within the hippocampus implicated in the encoding of memory (145). However, declarative memory impairments in PTSD patients have been identified in groups with significant exposure to drugs of abuse and groups who survived nutritional deprivation and physical torture. Future studies will be needed to determine whether declarative memory deficits in PTSD patients are found in groups without other potential causes of memory impairment.

Traumatization appears to alter encoding patterns of declarative memory, leading to the formation of memories with reduced contextual information. Extreme environmental stimulation has been assumed to affect the ability to process and synthesize information by overloading cognitive capacity and attention resources (146). However, accumulated research suggests that the impact of stress upon memory formation may depend upon the features of the stressful environment and their relationship to the information to be recalled (147). Stress enhances the recall of salient stimuli in the environment, but impairs recall of peripheral details (147–152). This recall bias is not explained by increased novelty or complexity of traumatic situations (153).

Stress and dissociative states may also alter the predominating mode of memory encoding. For example, memories may be encoded using detail-focused strategies in which components of a memory are encoded and rehearsed separately, or holistic strategies where memories are stored and rehearsed as composite images, essentially "snapshots" (154). Normally, detail strategies predominate over holistic strategies. However, during hypnosis, highly hypnotizable individuals tend to utilize holistic rather than detail-focused encoding strategies (154). The photographic or cinematographic nature of traumatic memories suggests that shifts between detail and holistic strategies occur during encoding and rehearsal of traumatic memories. Further, declarative memories may be encoded in multiple modalities, such as verbal, visual, or auditory. Encoding in these domains appears to take place concurrently and to be regulated similarly under most conditions (155,156). However, it is possible that during traumatization, encoding strategies shift away from verbal encoding toward encoding in emotional, pictorial, auditory, and other sensory-based memory systems. This shift would help to explain the posttraumatic development of alexithymia (10). Dissociative states at the time of the trauma may also result in the encoding of bizarre or distorted traumatic memories that accurately depict altered perceptual states occurring at the time that traumatic memories were formed. For example, one veteran in the yohimbine study (36) had vivid recollections of a Cheshire cat-like grin of a dead combat victim, although other features of this victim's face could not be recalled. Alteration in other sensory domains also appears evident in traumatic memories (157). Similarly, critical features of the traumatic context may be disturbed, such as the capacity to place events in time (29) and space (30).

Beyond the selectivity of the information that is encoded at the time of the trauma, repetitive rehearsal of traumatic memories can alter the meaning (16) or the content of encoded memories (30,158). For example, when reviewing stress-related memories, individuals think about emotional events in more personalized and concrete ways relative to neutral events (159). The focus on central details at the expense of contextual information present in the initial encoding process becomes even more pronounced in rehearsal of traumatic memories (147). These distortions in memory introduced during traumatic memory rehearsal may help to explain why people in stressful situations recall the central event well, but confabulate about motives and reactions of participants in the event (153).

Traumatic learning cannot entirely be explained within declarative memory. This point is highlighted by a case in which PTSD developed in the absence of declarative memories of the traumatic event (160). This case involved a patient who developed PTSD following a motor vehicle accident associated with a 3-day loss of consciousness. Although her traumatic brain injury resulted in a permanent loss of conscious recall of the accident, she suffered from frequent, intrusive thoughts about a friend killed in the accident and avoided reminders of the hospital, a rehabilitation center, the grave of her friend, and a second friend disabled in the accident. She showed increased autonomic symptoms, fear of driving, and a feeling that her career potential was foreshortened. Thus, total disruption of encoding of declarative information did not preclude development of PTSD symptoms, suggesting that traumatic learning was independently encoded by nondeclarative mechanisms.

In studies of nondeclarative memory function, PTSD patients showed trauma-related alterations in priming and conditioning. In order to study priming, researchers presented subjects with incomplete words (word stems) that they were required to complete. Combat veterans with PTSD, but not those without PTSD, showed a bias in favor of completing combat-related words (144). This bias was greater for words that they had previously seen compared to completely novel words, suggesting that PTSD patients had an implicit memory facilitation rather than a pure response bias. Since priming generally occurs outside of awareness, this study provided an example of unconscious mechanisms that promoted the recall of traumatic memories in PTSD patients.

Conditioning, another form of nondeclarative learning, is critical to the symptom profile of PTSD. Environmental stimuli associated with the initial traumatic stimulus, the conditioned and unconditioned stimuli, respectively, acquire the capacity to elicit responses that approximate the original traumatic responses, the conditioned and unconditioned responses (161–163). Learned physiologic responses have been clearly demonstrated for PTSD patients. For example, reminders of the trauma increase heart rate, blood pressure, startle responses, plasma epinephrine and norepinephrine, anxiety, tremor, and other indices of central and peripheral adrenergic activity in PTSD patients (164–166). Many forms of nondeclarative and nonprocedural learning fit within the rubric of conditioned responses. As they apply to specific behaviors, learned responses may be traced to the neuronal networks within defined brain regions. For example, in rats, enhancement of the startle response through associative learning results in the long-lasting facilitation of the activity within the central nucleus of the amygdala via an NMDA receptor-mediated process (167). Fear learning in cats also produces conditioned activation of locus coeruleus neurons (59). In humans, exposure to reminders of the trauma may induce recall of traumatic memories and dissociative states (162). These clinical observations suggest that conditioned alterations might occur in the systems regulating memory retrieval and dissociative states in individuals with PTSD.

Post-Traumatic Attention Biases May Promote Cue-Dependent Traumatic Memory Recall

The previous section suggested that traumatization altered the nature of material that is encoded into declarative and nondeclarative memory at the time of the trauma. Further, associative learning might result in the creation of conditioned emotional states. However, a vulnerability toward the involuntary recall of traumatic memories could occur at early levels of information processing, such as focusing attention, rather than at the level of emotional processing or memory retrieval. Recent research suggests that traumatized individuals preferentially process trauma-related cues and are more distracted by these cues than nontraumatized people. For example, patients with PTSD appear to have enhanced ability to identify trauma-related words from lists presented binaurally (168). Also, individuals with rape-related (169) and combat-related (170) traumas are more easily distracted by trauma-associated words than by emotionally charged, positively valenced, or neutral words on modified versions of the Stroop task. The heightened distractibility of PTSD patients by trauma reminders may have a physiological parallel in impairments in prepulse inhibition of the startle response (C.A. Morgan, C. Grillon, and D. Charney, personal communication). Prepulse inhibition deficits, marking a failure in the capacity to suppress the startle response to a loud noise when that noise is preceded by a quiet noise, correlate with increased distractibility in other patient populations (171). The attention bias toward detection and processing of trauma-related environmental cues in PTSD patients may promote retrieval of traumatic memories through associative mechanisms (172,173).

Control of Trauma-Related Reexperiencing

Traumatic memories may intrude into awareness through state- and cue-dependent mechanisms even when they cannot be voluntarily recalled. Traumatic memory retrieval appears to be promoted by environmental cues associated with the trauma (174,175). Also, internal cues, such as emotional state, may provoke retrieval of traumatic memories (176). Traumatized individuals who are not troubled by recollections of their trauma when euthymic appear to more readily recall aspects of their trauma when distressed (9,177). Furthermore, yohimbine and MCPP stimulate autonomic arousal, fear states, and recall of traumatic memories independent of environmental cues associated with the trauma (35,36). Internal or external cues may promote traumatic memory recall in the absence of awareness of these cues (68). The cue dependency of traumatic recollection appears to be a potent and robust phenomenon. The consistency of state-dependent recollections and mood states in PTSD appears to be more robust than many experimental models, where state-dependent learning has been studied and found to produce modest or inconsistent effects (178).

Inaccessibility of traumatic memories also may be state-dependent and associated with emotional numbing in PTSD patients. Traumatization produces a suppression of emotional responsivity in many individuals that is reported as feeling emotionally numb, machine-like, and apathetic (27,179). Reduced emotional responsivity in PTSD patients has been documented in both cognitive and electrophysiological spheres of function. Inpatients with PTSD have difficulty describing their effective response in emotion-laden situations (180). Similarly, using the Rorschach test (157), researchers found that while emotionally numbed, combat veterans with PTSD provided concrete interpretations with a paucity of emotional detail. In contrast, when patients were highly stimulated by Rorschach cards, trauma-related interpretations appeared associated with intense, primitive, and disorganized emotion-laden responses interpreted as "undigested, unsymbolized reliving of traumatic experience" (p. 263). Electrophysiological studies also support the view that many PTSD patients tonically suppress reactivity. Paige et al. (181) found that PTSD patients were in a state of "protective inhibition" of sensory processing relative to healthy subjects on the basis of a study of P200 event-related potentials. Interpreted together, these findings suggest that, when euthymic, patients with PTSD suppress arousal and cognitive associations related to the trauma. Thus, recall impairments in PTSD may arise, in part, from a process that suppresses the retrieval of autobiographical memories. Retrieval disruption does occur in related conditions, such as trance states associated with hypnosis, and may contribute to post-traumatic amnesia in PTSD patients (182,183).

Retrieval impairments may be overcome through associative processes within declarative memory (172) or by inducing particular mood states (184–186). Once trauma-related material intrudes upon consciousness, intense emotions, associations to the trauma, and vivid reexperiencing of the trauma may occur. This relatively dichotomous shift between suppression and reexperiencing has been characterized as the "all-or-none" response pattern in individuals with PTSD (157). Aspects of the shifts between suppression and reexperiencing appear to be beyond conscious modulation (168). However, the common report that individuals with PTSD actively avoid situations in which they might be exposed to reminders of the trauma (27) suggests that voluntary factors also play a modulatory role.

Sedative-hypnotic agents and hypnosis may inhibit mechanisms blocking access to traumatic memories. The pharmacologic facilitation of traumatic memory recall and flashbacks in traumatized individuals began in World War II as part of a therapeutic approach called narcosynthesis. This approach combined barbiturates and guided recollection of traumatic memories (17,187,188). The use of barbiturates to facilitate traumatic memory recall was illustrated by a case reported by Grinker (189):

> That afternoon I gave him 0.25 Gm. of pentothal sodium intravenously. He was then told that he was up in the air on a strafing mission and that the man on his wing was aflame . . . Immediately he [shouted] to his friend . . . pull up and bail out. Why doesn't he pull up and bail out? . . . he went over and over the traumatic situation, crying and sobbing. As this reaction subsided he was allowed to close his eyes and sleep . . . [upon awakening] He stated I must have been asleep. I had a dream about [my friend] . . . (p. 142–143).

Barbiturates and benzodiazepines impair attention, learning, and memory in humans (49,190). Yet, clinical experience suggests that these drugs facilitate recall of previously inaccessible traumatic memories and the production of a dissociative state. Mechanisms through which barbiturates facilitate the recollection of traumatic memories and flashbacks are poorly understood. Barbiturates enhance the actions of GABA at the $GABA_A$ receptor (191) and block the actions of glutamate at non-NMDA receptors (132,133). However, barbiturates are not specific in their capacity to facilitate recollection of traumatic memories in that ether, ethanol, nitrous oxide, and scopolamine-morphine combinations also appear to facilitate the recall of inaccessible memories (192,193). Clinical observations suggest that the sedating and anxiolytic medications employed in narcosynthesis reduce anxiety and thus may lessen the resistance to recalling anxiety-associated memories (17). This view is consistent with a patient who experienced flashbacks during relaxation training (194). Alternatively, these medications, viewed as "truth sera" by the popular press, may impair mechanisms responsible for suppressing access to traumatic memories (195).

Retrieval inhibitions may also be lessened during sleep, perhaps accounting for the intrusion of traumatic experiences into dreams and night terrors. Individuals with PTSD report vivid repetitive nightmares in which traumatic events are reexperienced (196,197). Post-traumatic nightmares occur within REM sleep and are not generally associated with motor behaviors (98,196,197). In contrast, post-traumatic night terrors bear a closer resemblance to flashbacks occurring in the waking state. Night terrors are associated with: 1) confusion upon awakening; 2) reduced responsivity to environmental stimuli; 3) displays of intense emotion; 4) significant autonomic activation; 5) increased sleep motility; 6) complex motor activity; and 7) somnambulism (98,196,198–200). As with flashbacks, recollection of the experiences associated with night terrors may be state-dependent. Despite behavioral evidence that traumatic incidents are being reexperienced during night terrors, such as calls for help and appearing to act out physical struggles, individuals are generally amnestic for the content of their experiences. As with flashbacks and nightmares, night terrors may be precipitated in PTSD patients by reminders of the trauma or environmental stress (9,196). Unlike nightmares, night terrors occur during deep sleep, particularly stage 4, and generally within the first hour after falling asleep (98,99). Nightmares and night terrors may be further distinguished by the effects of paramedian thalamic lesions. These lesions eliminate the stages of sleep responsible for night terrors, but do not alter REM sleep and dreaming (101).

Affective modulation of memory encoding and retrieval appears to be a critical step linking state-dependent recall and suppression of traumatic memories. Simply stated, traumatic memories are suppressed principally because they are associated with significant emotional distress. Retrieval of these memories is experienced as intrusive when elicited involuntarily by internal and external cues, or when intruding upon consciousness under conditions where the capacity to suppress retrieval of traumatic memories is reduced. This combination of mechanisms may help to explain how traumatic memories can be evoked by both the disinhibition associated with alcohol intoxication and the hyperarousal associated with alcohol withdrawal (41). In light of evidence that emotions serve to prioritize access to associative memory networks, it is not surprising that individuals who exhibit disconnection of effective and mnemonic function may have difficulty prioritizing memory access, lack mechanisms for suppressing memory retrieval, and lack a Freudian-style unconscious (201). For example, Temple Grandin, a woman with probable Aspberger's syndrome, noted that her memory retrieval seemed disconnected from emotional functions. She also felt that she lacked the capacity to suppress access to memories: "There are no files in my memory that are repressed . . . There are no secrets . . . nothing is hidden. The amygdala locks the files of the hippocampus. In me the amygdala doesn't generate enough emotion to lock the files of the hippocampus . . . (201)."

There is relatively little known about the neural networks that control memory retrieval as opposed to memory encoding. The hippocampus is a candidate region for this function because of its role in encoding memory. However, humans who have undergone bilateral removal of their hippocampi are able to retrieve established memories, and learn new procedures and cue-dependent priming, despite being unable to form new declarative memories (137). A contribution from the hippocampus should not be ruled out, however, because it contributes importantly to the regulation of thalamocortical activity implicated in memory retrieval (121). The capacity to suppress access to learned information seems phenomenologically related to the ability to prioritize the allocation of attention. For example, in order to generate coherent thought processes, one must continuously facilitate retrieval of relevant information from memory stores while suppressing the retrieval of irrelevant and distracting information. As suggested by psychodynamic theories, the control of memory retrieval is affected by one's emotional state at the time of recall, the nature of the environment at the time of recall, and the nature of and associations to the material that is being recalled (202). Thus, control of memory retrieval appears to involve an executive mechanism, not necessarily under voluntary control, that integrates input from emotional, sensory, mnemonic, and perhaps other systems. From this perspective, post-traumatic amnesia primarily results from the inability to overcome involuntary inhibition of retrieval by voluntary means, rather than an encoding failure. Also, traumatic memories are retrieved though bypassing or overwhelming resistance to recall, altering the meaning or associations to the recalled material, or by impairing the executive functions that inhibit recall.

Frontal cortical networks have been implicated in executive functions related to the control of memory retrieval (203). Frontal lobe lesions, unlike hippocampal lesions, impair retrieval of autobiographical information (204). The frontal cortex has also been implicated in the prioritization of responses, the generation of mental representations within working memory, self-monitoring, and editing of thought (75,127,203). It is nested within networks involving the amygdala, mediodorsal thalamic nucleus, hippocampus, and other regions that provide access to input regarding the nature and meaning of memories that are formed (127). Sedative-hypnotic agents produce impairments in tests sensitive to frontal cortical impairment, as does ketamine (47–49). A frontal cortical perspective may resolve the paradox presented by the narcosynthesis treatment (i.e., that drugs that impair declarative memory functions enhance retrieval of traumatic memories). Barbiturates may facilitate traumatic memories by 1) decreasing the intensity of emotional associations, perhaps by modulating amygdala inputs to the cortex, or 2) directly impairing a frontal cortical capacity to inhibit memory retrieval.

Traumatic Memories and Perseveration

The perseverative nature of traumatic memories adds to their intrusive quality. Once accessible, traumatic memories have been reported to be recurrently intrusive by some patients. Similarly, healthy individuals viewing stressful movies experience repetitive intrusive thoughts of the movie at later times (176,205,206). Pitman (206) highlighted parallels between traumatic memories and a particular type of perseverative thought, the obsessions occurring as a component of obsessive-compulsive disorder (OCD). He noted that traumatic memories were repetitively intrusive, unwanted recollections that were frequently resisted and that disrupted normal activity. One implication of this parallel would be to suggest the pathophysiology and treatment of the intrusive traumatic memories might parallel that of OCD. In particular, OCD is associated with increased metabolic activity in the orbital frontal cortex, anterior cingulate gyrus, and striatum (207,208). Pharmacotherapies and behavioral therapies effective for OCD reduce this increased frontal activity (207,208). Furthermore, blockade of the serotonin transporter by drugs such as fluoxetine, fluvoxamine, and sertraline is more effective in treating OCD than blockade of the norepinephrine transporter by desipramine, a finding not associated with the treatment of depression or panic disorder (209).

However, there are important differences between obsessions and traumatic memories:

1. Traumatic memories are not fully ego-dystonic. Frightening, guilt-laden, or shameful traumatic memories are frequently mood congruent at the time of recall. Also, the Freudian concept of the repetition compulsion suggested that repeated self-exposure to traumatic memories was rewarding in that it prompted their mastery (210). Furthermore, van der Kolk and colleagues (32) pointed out that biological mechanisms associated with traumatic memory recall might induce states of analgesia and euphoria leading to ''addiction to the trauma.'' In addition, some sufferers of PTSD feel that recollection of traumatic memories allows them to effectively memorialize the traumatic event. Thus, as Krystal (10) has stated, traumatic memories may be painful but gratifying to some individuals with PTSD.
2. Traumatic memories are fundamentally reality-based while obsessions are distortions

of reality. The OCD patients may have intrusive thoughts about an event that is unlikely to occur (e.g., accidentally striking a pedestrian while driving a motor vehicle). In contrast, an individual with PTSD may have actually participated in or observed an event of this nature.

3. It is not yet clear whether PTSD patients exhibit increased cortical, cingulate, or striatal metabolism in PET studies (6,211).

Although the parallels between obsessions and traumatic memories may be purely phenomenological at this point, their pathophysiological and pharmacotherapeutic implications warrant further evaluation. For example, a direct comparison of the efficacy of serotonin and norepinephrine transporter inhibitors in PTSD would be a useful addition to the field.

Lessons from Brain Stimulation Studies and Developmental Disorders

The retrieval of memory in the form of reexperiencing rather than recalling is common to posttraumatic flashbacks and flashbacks associated with local activation of cortical and limbic structures. Hughlings Jackson first described the complex polysensory reexperiencing of events that occurred in association with temporal lobe epilepsy as memory flashbacks (212). Patients with clinical and encephalographic evidence of temporal lobe epilepsy exhibited a range of dissociative symptoms that included depersonalization, derealization, auditory and visual hallucinations, and multiple personalities (213). Sacks (80) also described a patient with seizure foci in her medial temporal structures that produced repetitive reexperiencing of Irish folk melodies. Anticonvulsant treatment eliminated the intrusive musical reexperiencing, but also eliminated her ability to recall the melodies.

Penfield and his colleagues elicited dream-like states, memories, and complex experiential phenomena through direct electrical stimulation of structures in the temporal lobe, temporoparietal association areas, hippocampus, and amygdala (214). Temporal lobe stimulation resulted in some individuals reexperiencing frightening events in a polysensory fashion, such as a possible thwarted kidnapping (case 15). However, neutral or pleasant experiences were also produced, such as hearing a choir sing "White Christmas" (case 4).

The amygdala and hippocampus appear to be implicated in the experiential phenomena associated with temporal lobe activation. Gloor and his colleagues (215) found that experiential phenomena were associated with direct stimulation of the amygdala and the hippocampus. Moreover, memories, dream-like states, or other complex experiential phenomena were produced only when temporal cortical stimulation was followed by afterdischarges in the amygdala or hippocampus. Results of this study were consistent with an earlier one that produced complex experiential phenomena through electrical stimulation of the hippocampus and amygdala (216).

The brain stimulation studies suggest that the hippocampus and amygdala control the retrieval of memory in a highly specific manner, much as a program might control access to information stored on a computer. However, this interpretation appears overly simple. Complex experiential phenomena are usually associated with high-intensity stimuli or afterdischarges, suggesting that fairly large cortical areas must be activated (216). Also, stimulation of the same location over several trials does not reliably reproduce experiential phenomena, while stimulation of disparate cortical regions may produce identical experiences (216,217). In addition, surgical excision of an area that produces a memory when directly stimulated does not eliminate the memory (218). A more circumspect interpretation of these data is that memory is stored within distributed networks and that the amygdala and hippocampus stimulations bias the retrieval of memories in a more general fashion, such as facilitating access to an associative network.

One of the striking characteristics of traumatic memories, traumatic dreams, and brain stimulation studies are the inflexible nature of memory retrieval under these conditions. Dreams and memories often replay traumatic scenes in their entirety rather than being retrieved with the cognitive flexibility characteristic of declarative memory. These qualities of memory retrieval are also found in some developmental disorders. Asperger's syndrome, pervasive developmental disorders, and autism are conditions in which memories are retrieved in relatively inflexible patterns. These disorders may be associated with prodigious mnemonic capacities in relatively narrow domains of mental function arising from relatively inflexible preoccupations and thought processes (219). For example, some autistic individuals exhibit savant capacities, such as impressive mental mathematical achievements, even when significantly impaired in many domains of psychosocial function. Temple Grandin, the woman with a developmental disorder referred to earlier, described her memory functions in this fashion: "My mind is like a CD-ROM in a computer . . . But once I get there, I have to play that whole part . . . no fast forward . . . I play the tape . . . it's a slow method of thinking . . . (201, p. 121)." Her thoughts and dreams bear similarities to traumatic memories in that they tend to be concrete, visual, vivid reproductions of actual events (201,220). The neurobiology underlying the loss of retrieval flexibility and efficiency associated with traumatic memory retrieval, limbic stimulation studies, and the developmental disorders is currently unclear. However, reduced mnemonic flexibility has been reported to characterize memory retrieval under conditions where the hippocampus is activated independently of the frontal cortex (221). Memory encoding by the hippocampus is modular, and organizing links between memories arise largely through cue association, as occurs during conditioning (221). Retrieval strategies involving the hippocampus are cue-dependent and not strategic. In other words, the hippocampus cannot efficiently scan stored memories to retrieve a particular memory, even though all declarative memory involves the hippocampus in the encoding process. The organizing and strategizing component of memory retrieval appears to be dependent upon the frontal cortex (221,222). Thus, flashbacks, nightmares, and traumatic recollections may share the qualities of memory retrieval exhibited by individuals during hippocampal stimulation or by individuals with developmental neurologic disorders, because all of these conditions reflect retrieval strategies that involve a relative inhibition, impairment, or bypassing of the frontal component of memory retrieval in the face of relative preservation of the hippocampal component of memory retrieval.

Biological Hypotheses: Traumatic Stress and the Engram

An important unanswered question in the field of traumatic stress is whether, at the cellular level, traumatic memories are encoded and retrieved differently than other memories. Penfield (223) defined the concept of engram or memory trace broadly as "the continuing facilitation of neuron transmission established during mental activity, in related mechanisms in the brain" (p. 108). Clinical research has provided metaphors for explorations of cellular processes that might distinguish the encoding of traumatic and nontraumatic memories. For example, some clinicians have suggested that traumatic memories are somehow engraved into memory stores in a unique fashion that makes decay impossible (224). Alternatively, several formulations have suggested that traumatic memories are preserved as association networks and that the pairing of intense emotional states with the information to be remembered maintains the vividness of traumatic memories (2,168,202,225).

Cellular mechanisms related to long-term potentiation (LTP) may provide clues to cellular processes that may be relevant to the establishment of traumatic memories and long-lasting traumatic sensitization of emotional responses (226). The LTP requires the presence of a presynaptic glutamatergic neuron. Glutamate released by the presynaptic neuron initiates LTP via the NMDA receptor and sustains LTP by stimulating non-NMDA receptors (227). Once bound to the NMDA receptor, glutamate promotes the entry of calcium into the postsynaptic neurons. Calcium entering the cell activates a cascade of modulatory phosphoproteins, producing rapid and long-lasting alterations in gene expression and neuronal structure. Activation of calmodulin also stimulates the synthesis of nitric oxide. Nitric oxide is a gaseous neurotransmitter that diffuses freely out of the neuron and crosses the synapse, where it stimulates presynaptic processes leading to alterations in presynaptic structure and enhanced neurotransmitter release (228,229).

Traumatic learning is largely an associative process. Associative mechanisms are accounted for in LTP models through the capacity of multiple convergent inputs that modulate the induction and maintenance of LTP (230,231). Neurobiological features of the stress response may determine whether LTP is facilitated or prevented. Convergence of glutamatergic, noradrenergic, and peptidergic (vasopressin, corticotropin-releasing factor (CRF)) inputs may be critical for producing stress-induced facilitation or strengthening of LTP (34,226,232). Similarly, thalamic activity in sensory relay modes promotes LTP in the amygdala (103). However, under conditions of extreme stress, several factors may impair LTP, including depletion of noradrenergic neurons, high levels of circulating glucocorticoids, and perhaps disruption of thalamic sensory relay functions (34,226). The LTP can be produced in cortical and limbic networks that have been implicated in distinct aspects of traumatic response. For example, LTP initiation in the hippocampus and associated cortical and subcortical networks has been implicated in transduction of transient memory to longer term forms of memory. The LTP in the amygdala has been implicated in both fear and reward learning. If convergent LTP processes underlie the vividness and emotional valence of traumatic memories, then contributions of the amygdala, hippocampus, thalamus, and cingulate cortex to this process warrant further study.

Pharmacologic studies support the existence of parallels between the neurochemistry of learning and the regulation of LTP. These studies may also provide insights into some of the unique features of traumatic memories. The NMDA antagonists, for example, prevent the initiation of LTP (227), disrupt the acquisition of conditioned fear states in animals (167,233), and produce memory impairments in humans (47). At doses that are too low to disrupt memory encoding, NMDA antagonists make memory retrieval relatively cue- or state-dependent in animals (234,235). Similarly, ketamine makes human thought processes more concrete and impairs abstract problem solving, indicating that it interferes with abstract thought processes generally under voluntary control (48,49). The capacity of ketamine to produce dissociative-like states suggests that dissociation could produce NMDA antagonist-like effects on memory encoding. These hypotheses fit with clinical observations that traumatic memories encoded during dissociative states are relatively concrete depictions of sensory experience, and that retrieval of traumatic memories may be highly dependent upon the induction of intense emotional states or confrontation with trauma-related cues.

A second cellular process, long-term depression (LTD), may be involved in forgetting and extinguishing learned responses (236). Multiple layerings of LTD and LTP may occur within the same synaptic network. These overlapping competing processes could lead to periods of relative facilitation and inhibition of a learned behavioral response. This model helps to explain why learned fear responses may be extinguished and then reinstated by reexposure to fear-related cues or stimuli (34,226). These cellular processes may have important clinical implications. While trauma-related fear learning may be extinguished, one would expect that it would be potentially vulnerable to reawakening through exposure to trauma-related reminders or other stressors (177,237). Thus treatments based on extinguishing learned responses might need to be periodically administered, much as booster antigen presentations are provided to maintain one's immune response to a pathogen, in order to maintain an extinguished response.

IMPLICATIONS FOR TREATMENT OF PTSD

Treatments for dissociative states and traumatic memories cannot delete traumatic memories from the mental record of the victim/survivor. Traumatic experiences are frequently focal experiences shaping the identity of traumatized individuals. Despite their associated distress, traumatic memories guide some traumatized individuals to constructive action, such as working to prevent genocide. Yet patients are frequently concerned that treatment will erase memories. Perhaps some of this concern reflects the degree to which traumatic memories are inaccessible to voluntary recall, so that memories are experienced as either intrusive or nonexistent. Reasonable therapeutic goals include reducing the frequency and severity of dissociative states, enhancing control over the recall of traumatic memories, and reducing distress associated with memory retrieval.

Dissociative phenomena, traumatic memories, and affective regulation are highly interrelated in PTSD patients. As reviewed earlier, traumatic memories and intense emotions may trigger dissociative phenomena in PTSD patients. Similarly, dissociative states and particular emotional states make the recall of traumatic memories more accessible. Completing the triangle, traumatic memories and dissociative phenomena may precipitate strong emotional responses. Thus, reducing the incidence of flashbacks, and the intrusiveness and distress related to traumatic memories must be understood in the context of treating each of the three interactive processes.

The first step in treating dissociative states in traumatized individuals is to alleviate the marked depersonalization, derealization, and extreme emotional arousal. On the World War II battlefield, some psychiatrists employed a technique called narcotherapy where barbiturates were employed to produce a sustained sleep state for up to 3 days (188). This approach is consistent with, but more extreme than, the current use of benzodiazepines to reduce dissociation (26). The long-term benefits of acute anxiolysis are currently unclear. However, Kardiner (16) emphasized the importance of the peri-traumatic period in creating a long-lasting appraisal of traumatic events. Acute anxiolysis may be helpful in reducing negatively-valenced cognitive distortions. Thus, in the context of supportive therapy, benzodiazepine treatment may facilitate development of a more adaptive appraisal of the traumatic stress. Acute anxiolysis might alter the pairing of emotions and memories. Anxiolysis may also reduce or prevent development of dissociative states, facilitating reflective reevaluation of information related to the trauma.

Once hyperarousal is controlled, a second challenge faced by clinicians is to reduce amnesia for traumatic events. Almost every psychotherapeutic strategy for treating acute psychological trauma has as a goal the integration of the traumatic experience within the conscious life of patients (2,10,238). This task is difficult if the patient is amnestic for the trauma. Several guided recollection strategies have been employed to facilitate patients' access to traumatic memories, including relaxation training, free association, dream interpretation, hypnosis, and narcosynthesis (10,188,239,240). Each of these processes takes advantage of an altered state of consciousness associated with increased suggestibility in which there is a reduction in functions usually associated with the frontal cortex, such as reflection, self monitoring, and editing of thought (75).

A potential risk associated with conducting guided recollection in such a state is that ideas introduced by the clinician may be more readily incorporated into the memories of the patient. In such cases, the patient may not be able to distinguish real memories from those created by the therapist, as may occur under hypnosis (241). This concern is particularly relevant to narcosynthesis or other techniques in which the therapist recreates the roles of people within the patient's traumatic memory in order to facilitate memory retrieval (17,239). Furthermore, the use of pharmacologic agents such as amytal in narcosynthesis may produce amnesia for remembered information (242). Because patients may not fully recall information produced during narcosynthesis at later times (189), narcosynthesis may be best viewed as an information-gathering procedure.

Once an individual has access to memories of the trauma, what can the patient-clinician dyad do with them to help the patient? Freud and Breuer (2) initially suggested two strategies for reducing dissociative or conversion symptoms associated with hysteria, abreaction, and the formation of new associations to the traumatic memories. By abreaction, Freud and Breuer meant the discharge, during therapy, of stored feelings that could not be adequately expressed at the time of the trauma. This hydraulic view of emotions has largely been abandoned (70). Alternatively, modern cognitive, behavioral, and insight-oriented therapies focus on altering cognitive, affective, and identity-related associations to the trauma (10,240,243). Psychotherapy may help to reduce the intrusiveness and distress related to traumatic memories by altering associations to the traumatic events, essentially changing the meaning of the trauma to the individual.

Clinicians may also promote the verbal encoding of post-traumatic emotions and memories with the aim of enhancing voluntary control over both domains of symptoms. Traumatic memories may not be encoded or retrieved verbally. Krystal (10) highlighted the difficulty that many individuals with PTSD have in linguistically interpreting their emotional states. He employs cognitive-affective remediation strategies in psychotherapy for PTSD patients with these difficulties. These strategies focus on helping them learn to label their emotional states, a practice he likens to teaching color-blind individuals to interpret shades of gray. Since sensory processing and emotional reactivity are largely involuntary (104) as long as traumatic memories are sensorially encoded and emotionally gated, they will intrude unwanted upon consciousness. Perhaps, by verbally encoding information and emotional responses associated with the trauma, the voluntary control of linguistic function can be brought to bear on previously uncontrollable feelings and memories. It is unclear whether the linguistic functions serve primarily an executive function, similar to an operating system on a computer, or whether parallel linguistic memories are created that provide access to the traumatic information in a form that is less distressing. Both processes could be involved in the therapeutic process. However, the latter view is more consistent with the manner in which some patients have a capacity to comfortably review self-edited versions of their traumatic memories. These same individuals may experience a great deal of distress when exploring their memories in depth.

Managing the recollection of traumatic memories, dissociative states, and intense affects in patients is a tremendous clinical challenge facing the clinician treating PTSD. Guided reexperiencing of the trauma could evoke dissociative states that interfere with associative learning and the generalizing of therapeutic gains beyond the clinical setting. Intense emotions evoked during such recollections could reinforce the association between traumatic memories and intolerable intense emotions, sensitizing individuals to reminders, promoting a sense of helplessness or other negative appraisals of the trauma, and making the individual more reluctant or unable to review traumatic material in subsequent therapy sessions (244). Furthermore, by stimulating intense emotional responses and negative associations in some individuals, flooding may exacerbate depression or provoke impulsive behavior, including substance abuse (244). These potential problems help to explain the need for extensive relaxation training prior to the initiation of guided reexposure therapies such as flooding (240). This step is probably a useful adjunct to all psychotherapies for PTSD patients (245). In addition, one might predict that care must be taken to titrate the level of arousal associated with guided recollection of traumatic memories to the patient's capacity to process information. When, in the course of a therapy session, patients provide clinical data consistent with the induction of dissociative states, further efforts to encourage them to process traumatic material seem unlikely to be fruitful.

The concern that interference with higher cognitive functions limits the clinical utility of altered states of consciousness applies equally to proposed pharmacologic adjuvants to psychotherapy such as the serotonergic hallucinogens (61) and NMDA antagonists (47). Particularly in patients with chronic PTSD who have been in many years of treatment, there seems to be little benefit in guiding them to reexperience the trauma at the expense of repeated dissociative episodes. Carefully conducted flooding therapy, preceded by relaxation training, may reduce intrusive symptoms of PTSD but have no beneficial impact on numbing or avoidance (240).

These authors highlight significant psychological and social deficits that impair treatment response in patients with chronic PTSD. Thus, treatments aimed at reevaluating traumatic memories may have an important, but focused, roles in their therapy. Also, research is needed to characterize optimal strategies for integrating pharmacotherapy approaches with these cognitive and behavioral psychotherapies.

If, analogous to relaxation training, pharmacologic strategies were developed that preserved cognitive functions in the face of strong affects and traumatic memories, the formation of new associations to traumatic memories might proceed more effectively and rapidly. Benzodiazepines, reportedly useful in some PTSD symptoms in patients with dissociative disorders (246) might help by reducing the affective distress, although their amnestic properties might be counterproductive at high doses. Alternatively, one could evaluate pharmacologic approaches to enhancing the cognitive processing. In this regard, drugs that facilitate NMDA receptor function via enhancement of the glycine site, such as cycloserine or milacemide (135,247), should be evaluated for antidissociative and other cognitive-enhancing properties in PTSD patients.

Antidepressants are the most frequently studied pharmacotherapy for PTSD, and research suggests that they provide a moderate degree of relief from flashbacks and intrusive memories. Case reports suggest that tricyclic antidepressants may reduce flashbacks, night terrors, and distress related to intrusive memories of the trauma (248,249). Similar findings have been reported in open label trials of monoamine oxidase inhibitors (250,251) and serotonin reuptake blockers (252–254). Placebo-controlled trials of tricyclic antidepressants and monoamine oxidase inhibitors have supported the findings from the open label trials, although variability between studies and small effect sizes have limited the optimism regarding the efficacy of these agents (251,255–257). The few studies that reported reexperiencing symptoms individually indicated that antidepressants were most effective in reducing the intrusion of traumatic memories and nightmares, and less effective in reducing dissociative phenomena such as flashbacks and amnesia (251,254). The disparity between dissociative and other intrusive symptoms may indicate that antidepressants reduce flashbacks as a secondary consequence of the other effects of these drugs. One mechanism possibly related to the efficacy of these agents is their capacity to prevent or reduce the consequences of noradrenergic hyperreactivity in PTSD patients (33). This hypothesis is consistent with the capacity of fluoxetine treatment to block yohimbine-induced panic attacks in patients with panic disorder (258). The limited efficacy of agents based on monoaminergic transmission suggests that a new direction is needed in the development of pharmacotherapies for dissociative disorders. This new direction may be based on the pharmacology of excitatory amino acid neurotransmission implicated in the genesis of dissociative states.

Carbamazepine has also shown utility in reducing hyperarousal, sleep disturbance, and flashbacks in some PTSD patients (259). Despite limited investigation, carbamazepine has received particular attention due to its capacity to suppress a form of neural sensitization, kindling, that may provide a cellular model for sensitization to repeated stressors (260). Further studies with anticonvulsant agents appear warranted.

Nootrophic pharmacotherapies, such as arginine vasopressin, have been suggested as treatments for memory deficits associated with PTSD (224). Cognitive-enhancing pharmacotherapeutic strategies might be beneficial in reducing encoding deficits associated with short-term memory impairment in PTSD patients. Drugs that might be evaluated for this purpose include ones that facilitate glutamate and acetylcholine function, such as the glycine partial agonist, cycloserine (134), and the cholinesterase inhibitor, tacrine (261). Vasopressin has been used to facilitate memory retrieval in patients with PTSD (262), based on preclinical studies suggesting that vasopressin may enhance memory consolidation and retrieval in animals. However, the preclinical foundation of the vasopressin studies has been questioned because subsequent studies have suggested that the promnestic effects of exogenous administration of this hormone are attributable to its enhancement of arousal (263).

Cognitive-enhancing agents do not appear to be the appropriate pharmacotherapeutic approach for reducing amnesia for the trauma. As noted earlier, post-traumatic amnesia appears to arise from the suppression of retrieval rather than from an ongoing memory encoding deficit. There is no evidence to suggest that tacrine or cycloserine reduce the integrity of memory repression. Agents that do appear to facilitate recollection of traumatic memories, such as lactate, yohimbine, vasopressin, and MCPP, appear to reduce post-traumatic amnesia via facilitating state-dependent retrieval. However, the sedative-hypnotic agents are the most commonly employed agents for this purpose, when combined with some form of guided recollection such as the amytal interview.

SUMMARY AND IMPLICATIONS

This review highlighted the importance of and limited clinical data in neurobiology and pharmacotherapy of dissociative states and traumatic memories in PTSD. In doing so, this chapter attempted to integrate research advances from clinical psychology, cognitive psychology, neuropsychology, functional brain imaging, psychophysiology, psychopharmacology, and a spectrum of preclinical approaches. This chapter's cognitive neuroscience perspective assumed that the features of traumatic memories and dissociative states in PTSD are properties of the underlying neural networks mediating these functions. In beginning to identify and characterize aspects of the regulation of these networks, it is hoped that clinicians and researchers may better understand the features of PTSD and its treatment.

In PTSD, dissociative states, hyperarousal, and traumatic memories are integrally linked. Under normal resting conditions, traumatized individuals may actively and involuntarily suppress emotional arousal and traumatic memories. Yet, traumatic memories may be vividly retrieved under conditions where arousal is heightened, reminders of the trauma are presented, or resistance to recollection is reduced. Functions associated with dissociative states, including attention, identity functions, and control of the fidelity of sensory transmission appear to be centered on networks involving the frontal cortex. In addition to the cortex, this network involves the amygdala, hippocampus, thalamus, and cingulate gyrus. Connectivity between these regions is glutamatergic, but modulated by monoamine systems based in the brainstem and midbrain. A unique feature of glutamatergic neurotransmission is its capacity to generate long-term alterations in neuronal function. Given the implication of glutamatergic systems in dissociative states, learning, and memory, studies of the potential contributions of these systems to the neurobiology and treatment of dissociative states and memory-related PTSD symptoms appear warranted.

ACKNOWLEDGMENTS

This work was supported by funds from the Department of Veterans Affairs to the National Center for Post-Traumatic Stress Disorder, the VA-Yale Alcoholism Research Center, and the Merit Review Grant Program (J.H.K.).

REFERENCES

1. Janet P. *The Major Symptoms of Hysteria.* New York: Macmillen; 1920.
2. Freud S, Breuer J. On the psychical mechanism of hysterical phenomena. In Jones E, ed. *Sigmund Freud, M.D., LL.D. Collected Papers, Vol. 1.* London: Hogarth Press; 1953:24–41.
3. Mayer-Gross W. On depersonalization. *Br J Med Psychol* 1935;15:103–126.
4. Hilgard ER. *Divided Consciousness: Multiple Controls in Human Thought and Action.* New York: Wiley, 1977.
5. Spiegel D, Cardeña E. Disintegrated experience: the dissociative disorders revisited. *J Abnorm Psychol* 1991; 100:366–378.
6. Bremner JD, Southwick S, Brett E, Fontana A, Rosenheck R, Charney DS. Dissociation and posttraumatic stress disorder in Vietnam combat veterans. *Am J Psychiatry* 1992;149:328–332.
7. Janet P. *Automatisme Psychologique.* Paris: Balliere; 1889.
8. Fisher C. Amnesic states in war neurosis: the psychogenesis of fugues. *Psychoanal Q* 1945;14:437–458.

9. Krystal H. *Massive Psychic Trauma.* New York: International Universities, 1968.
10. Krystal H. *Integration & Self-Healing: Affect, Trauma, Alexiihymia.* Hillsdale, NY: Analytic Press; 1988.
11. Carlson EB, Rosser-Hogan R. Trauma experiences, posttraumatic stress, dissociation and depression in Cambodian refugees. *Am J Psychiatry* 1991;148:1548–1552.
12. Bremner JD, Steinberg M, Southwick SM, Johnson DR, Charney DS. Use of the Structured Clinical Interview for DSM-IV Dissociative Disorders for systematic assessment of dissociative symptoms in posttraumatic stress disorder. *Am J Psychiatry* 1993;150:1011–1014.
13. Spiegel D, Hunt T, Dondershine HE. Dissociation and hypnotizability in posttraumatic stress disorder. *Am J Psychiatry* 1988;145:301–305.
14. Bernstein E, Putnam T. Development, reliability, and validity of a dissociation scale. *J Nerv Ment Dis* 1986;174:727–735.
15. Lowenstein RJ, Putnam F. A comparison study of dissociative symptoms in patients with partial-complex seizures, MPD, and PTSD. *Dissociation* 1988;1:17–23.
16. Kardiner A. *The Traumatic Neuroses of War.* Psychosomatic Monograph II-III, Washington, DC: National Research Council, 1941.
17. Grinker RR, Spiegel JP. *War Neuroses.* Philadelphia, PA: Blakiston; 1945.
18. Bremner JD, Mazure CM, Putnam FW, Southwick SM, Marmar C, Hansen C, Lubin H, Roach L, Freeman G, Krystal JH, Charney DS. Measurement of dissociative states with the Clinician-Administered Dissociative States Scale (CADSS) (submitted).
19. McDougle CJ, Southwick SM. Emergence of an alternate personality in combat-related posttraumatic stress disorder. *Hosp Community Psychiatry* 1990;41:554–556.
20. Briere J, Conte J. Self-reported amnesia for abuse in adults molested as children. *J Traum Stress* 1993;6:21–31.
21. Putnam FW, Guroff JJ, Silberman EK, Barban L, Post RM. The clinical phenomenology of multiple personality disorder: review of 100 recent cases. *J Clin Psychiatry* 1986;47:285–293.
22. Hermann JK, Perry J, van der Kolk VA. Childhood trauma in borderline personality disorder. *Am J Psychiatry* 1989;148:490–495.
23. Chu JA, Dill DL. Dissociative symptoms in relation to childhood physical and sexual abuse. *Am J Psychiatry* 1990;147:887–892.
24. Mellman TA, Davis GC. Combat-related flashbacks in post-traumatic stress disorder: phenomenology and similarity to panic attacks. *J Clin Psychiatry* 1985;46:379–382.
25. Rainey JM Jr, Aleem A, Ortiz A, Yeragani V, Pohl R, Berchou R. A laboratory procedure for the induction of flashbacks. *Am J Psychiatry* 1987;144:1317–1319.
26. Kluft RF. An update on multiple personality disorder. *Hosp Community Psychiatry* 1987;38:363–373.
27. American Psychiatric Association, Committee on Nomenclature and Statistics. *Diagnostic and Statistical Manual of Mental Disorders, Revised Third Edition.* Washington, DC: American Psychiatric Association, 1987.
28. Torrie A. Psychosomatic casualties in the Middle East. *Lancet* 1944;29:139–143.
29. Terr LC. Time sense following psychic trauma: a clinical study of ten adults and twenty children. *Am J Orthopsychiatry* 1983;53:244–261.
30. Pynoos RS, Frederick CJ, Nader K, Arroyo W, Steinberg A, Eth S, Nunez F, Fairbanks L. Life threat and posttraumatic stress disorder in school-age children. *Arch Gen Psychiatry* 1987;44:1057–1063.
31. Bremner JD, Scott TM, Delaney RC, Southwick SM, Mason JW, Johnson DR, Innis RB, McCarthy G, Charney DS. Deficits in short-term memory in posttraumatic stress disorder. *Am J Psychiatry* 1993;150:1015–1019.
32. van der Kolk V, Greenberg M, Boyd H, Krystal J. Inescapable shock, neurotransmitters, and addiction to trauma: toward a psychobiology of post traumatic stress. *Biol Psychiatry* 1985;20:314–325.
33. Krystal JH, Kosten TR, Perry BD, Southwick S, Mason JW, Giller EL Jr. Neurobiological aspects of PTSD: Review of clinical and preclinical studies. *Behav Ther* 1989;20:177–198.
34. Charney DS, Deutch AY, Krystal JH, Southwick SM, Davis M. Psychobiologic mechanisms of posttraumatic stress disorder. *Arch Gen Psychiatry* 1993;50:294–305.
35. Southwick SM, Krystal JH, Morgan CA, Johnson DR, Nagy LM, Nicolaou A, Heninger GR, Charney DS. Abnormal noradrenergic function in post traumatic stress disorder. *Arch Gen Psychiatry* 1993;50:266–274.
36. Southwick SM, Krystal JH, Morgan A, Nagy LM, Dan E, Johnson D, Bremner D, Charney DS. Yohimbine and m-chlorophenylpiperazine in PTSD. *1991 New Research Program and Abstracts: American Psychiatric Association, 144th Annual Meeting* 1991;NR348.
37. Pitts FN, McClure JN. Lactate metabolism in anxiety neurosis. *N Engl J Med* 1967;277:1329–1336.
38. Charney DS, Heninger GR, Breier A. Noradrenergic function in panic anxiety: effects of yohimbine in healthy subjects and patients with agoraphobia and panic disorder. *Arch Gen Psychiatry* 1984;41:751–763.
39. Charney DS, Woods SW, Goodman WK, Heninger GR. Serotonin function in anxiety. II. Effects of the serotonin agonist MCPP in panic disorder patients and healthy subjects. *Psychopharmacology* 1987;92:14–24.
40. Starke K, Borowski E, Endo T. Prefrontal blockade of presynaptic *a*-adrenoceptors by yohimbine. *Eur J Pharmacol* 1975;34:385–388.
41. Kosten TR, Krystal JH. Biological mechanisms in post traumatic stress disorder: relevance for substance abuse. In Galanter M, ed. *Recent Developments in Alcoholism.* New York: Plenum Press; 1988;6:49–68.
42. Salloway S, Southwick S, Sadowsky M. Opiate withdrawal presenting as posttraumatic stress disorder: a case of malingering following the L'Ambiance Plaza Building disaster. *Hosp Community Psychiatry* 1990;41:666–667.
43. Luby ED, Cohen BD, Rosenbaum G, Gottlieb JS, Kelley R. Study of a new schizophrenomimetic drug-sernyl. *AMA Arch Neurology Psychiatry* 1959;81:363–369.
44. Domino EF, Chodoff P, Corssen G. Pharmacologic effects of CI-581, a new dissociative anesthetic, in man. *Clin Pharmacol Ther* 1965;6:279–291.
45. Yamakura T, Mori H, Masaki H, Shimoji K, Mishina M. Different sensitivities of NMDA receptor channel

subtypes to non-competitive antagonists. *Neuroreport* 1993;4:687–690.
46. Javitt DC, Zukin SR. Recent advances in the phencyclidine model of schizophrenia. *Am J Psychiatry* 1991; 148:1301–1308.
47. Krystal JH, Karper LP, Seibyl JP, Freeman GK, Delaney R, Bremner JD, Heninger GR, Bowers MB Jr, Charney DS. Subanesthetic effects of the NMDA antagonist, ketamine, in humans: psychotomimetic, perceptual, cognitive, and neuroendocrine effects. *Arch Gen Psychiatry* 1994;51:199–214.
48. Krystal JH, Karper LP, Bennett A, Abi-Dargham A, D'Souza DC, Gil R, Charney DS. Modulation of frontal cortical function by glutamate and dopamine antagonists in healthy subjects and schizophrenic patients: a neuropsychological perspective. *Neuropsychopharmacology* 1994;10(3S):230S.
49. Krystal JH, Karper LP, D'Souza DC, Morrissey K, Bennett A, Abi-Dargham A, Bremner JD, Heninger GR, Bowers MB, Charney DS. Interactive effects of subanesthetic ketamine and lorazepam in humans: psychotomimetic, dissociative, cognitive, and neuroendocrine responses. (submitted)
50. Herkenham M, Lynn AB, Little MD, Johnson MR, Melvin LS, De Costa BR, Rice KC. Cannabinoid receptor localization in brain. *Proc Natl Acad Sci USA* 1990; 87:1932–1936.
51. Felder CC, Briley EM, Axelrod J, Simpson JT, Mackie K, Devane WA. Anandamide, an endogenous cannabimmetic eicosanoid, binds to the cloned human cannabinoid receptor and stimulates receptor-mediated signal transduction. *Proc Natl Acad Sci USA* 1993;90: 7656–7660.
52. Eldridge JC, Landfield PW. Cannabinoid interactions with glucocorticoid receptors in rat hippocampus. *Brain Res* 1990;534:135–141.
53. Feigenbaum JJ, Bergmann F, Richmond SA, Mechoulam R, Nadler V, Kloog Y, Sokolovsky M. Nonpsychotropic cannabinoid acts as a functional N-methyl-D-aspartate receptor blocker. *Proc Natl Acad Sci USA* 1989;86:9584–9587.
54. Bromberg W. Marihuana: a psychiatric study. *JAMA* 1939;113:4–12.
55. Melges FT, Tinklenberg JR, Hollister LE, Gillespie HK. Temporal disintegration and depersonalization during marihuana intoxication. *Arch Gen Psychiatry* 1970; 23:204–210.
56. Dittrich A, Bättig K, von Zeppelin I. Effects of $(-)\Delta^9$-trans-tetrahydrocannabinol (Δ^9-THC) on memory, attention and subjective state: a double blind study. *Psychopharmacologia (Berl.)* 1973;33:369–376.
57. Hollister LE. Health aspects of cannabis. *Pharmacol Rev* 1986;38:2–20.
58. Stanton MD, Mintz J, Franklin RM. Drug flashbacks. II. Some additional findings. *Int J Addict* 1976;11: 53–69.
59. Rasmussen K, Marilak DA, Jacobs BL. Single unit activity of the locus coeruleus in the freely moving cat, II: conditioning and pharmacologic studies. *Brain Res* 1986;371:335–344.
60. Titeler M, Lyon RA, Glennon RA. Radioligand binding evidence implicates the brain 5-HT_2 receptor as a site of action for LSD and phenylisopropylamine hallucinogens. *Psychopharmacology* 1988;94:213–216.
61. Freedman DX. On the use and abuse of LSD. *Arch Gen Psychiatry* 1968;18:330–347.
62. Savage C. Variations in ego feeling induced by D-lysergic acid diethylamide (LSD-25). *Psychoanal Rev* 1955;42:1–16.
63. Liebert RS, Werner H, Wapner S. Studies on the effects of lysergic acid diethylamide (LSD-25). *AMA Arch Neurology Psychiatry* 1958;79:580–584.
64. Klee GD. Lysergic acid diethylamide (LSD-25) and ego functions. *Arch Gen Psychiatry* 1963;8:57–70.
65. Rodin E, Luby E. Effects of LSD-25 on the EEG and photic evoked responses. *Arch Gen Psychiatry* 1966; 14:435–441.
66. Horowitz MJ. Flashbacks: recurrent intrusive images after the use of LSD. *Am J Psychiatry* 1969;126: 565–569.
67. Freedman DX. LSD: the bridge from human to animal. In Jacobs, B.L. ed. *Hallucinogens: neurochemical behavioral, and clinical perspectives* New York: Raven Press; 1984;203–226.
68. McGee R. Flashbacks and memory phenomena: a comment on "flashback phenomena-clinical and diagnostic dilemmas." *J Nerv Ment Dis* 1984;172:273–278.
69. Heaton RK. Subject expectancy and environmental factors as determinants of psychedelic flashback experiences. *J Nerv Ment Dis* 1975;161:157–165.
70. Krystal H. Trauma and affects. *Psychoanal Study Child* 1978;33:81–116.
71. Krippner S, Bindler PR. Hypnosis and attention: a review. *Am J Clin Hypn* 1974;16:166–177.
72. Tellegen A, Atkinson G. Openness to absorbing and self-altering experiences, a trait related to hypnotic susceptibility. *J Abnorm Psychol* 1974;83:268–277.
73. Carlsson EB, Putnam FW. Integrating research on dissociation and hypnotizability: are there two pathways to hypnotizability? *Dissociation* 1989;2:32–38.
74. Bowers KS. Dissociation in hypnosis and multiple personality disorder. *Int J Clin Exp Hypn* 1991;39:155–175.
75. Stuss DT. Biological and psychological development of executive function. *Brain Cog* 1992;20:8–23.
76. Pardo JV, Pardo PJ, Janer KW, Raichle ME. The anterior cingulate cortex mediates processing selection in the Stroop attentional conflict paradigm. *Proc Natl Acad Sci USA* 1990;87:256–259.
77. Bexton WH, Heron W, Scott TH. Effects of decreased variation in the sensory environment. *Can J Psychol* 1954;8:70–76.
78. Lilly JC. Mental effects of reduction of ordinary levels of physical stimuli on intact healthy persons. *Psychiatry Research Report* 1956;5:1–9.
79. Cappon D, Banks R. Studies in perceptual distortion. *AMA Arch Neurology Psychiatry* 1960;10:99–104.
80. Sacks O. *The Man Who Mistook His Wife for a Hat.* New York: Summit Books; 1985.
81. Loftus EF, Loftus GR, Msso J. Some facts about "weapon focus." *Law Hum Behav* 1987;11(1):55–62.
82. Kramer TH, Buckhout R, Eugenio P. Weapon focus, arousal, and eyewitness memory: attention must be paid. *Law Hum Behav* 1990;14:167–184.
83. Oberndorf CP. Role of anxiety in depersonalization. *Internat J Psycho-Analysity* 1950;31:1–5.
84. Cappon D, Banks R. Orientational perception: a review and preliminary study of distortion in orientational perception. *Arch Gen Psychiatry* 1961;5:380–392.

85. Ludwig AM. "Psychedelic" effects produced by sensory overload. *Am J Psychiatry* 1972;128:1294–1297.
86. Krystal JH, Woods SW, Hill CL, Charney DS. Characteristics of panic attack subtypes: assessment of spontaneous panic, situational panic, sleep panic, and limited symptom attacks. *Compr Psychiatry* 1991;32:1–7.
87. Krystal JH, Woods SW, Hill CL, Charney DS. Characteristics of self-defined panic attacks. *1988 New Research Program & Abstracts.* Washington, DC: American Psychiatric Association 1988;NR263.
88. Amaral DG, Cowan WM. Subcortical afferents to the hippocampal formation in the monkey. *J Comp Neurol* 1980;189:573–591.
89. McCormick DA. Neurotransmitter actions in the thalamus and cerebral cortex and their role in the neuromodulation of thalamocortical activity. *Prog Neurobiol* 1992;39:337–388.
90. Steriade M, Llinás RR. The functional states of the thalamus and the associated neuronal interplay. *Physiol Rev* 1988;68:649–741.
91. Turner BH, Herkenham M. Thalamoamygdaloid projections in the rat: a test of the amygdala's role in sensory processing. *J Comp Neurol* 1991;313:295–325.
92. Steriade M, Deschenes M. The thalamus as a neuronal oscillator. *Brain Res Rev* 1984;8:1–63.
93. Steriade M, Datta S, Paré D, Oakson G, Curró Dossi R. Neuronal activities in brain-stem cholinergic nuclei related to tonic activation in thalamocortical systems. *J Neurosci* 1990;19:2541–2559.
94. Steriade M, McCarley RW. *Brainstem control of wakefulness and sleep.* New York: Plenum, 1990.
95. Swanson LW. A direct projection from Ammon's horn to prefrontal cortex in the rat. *Brain Res* 1981;150–154.
96. Llinás RR, Paré D. Of dreaming and wakefulness. *Neuroscience* 1991;44:521–535.
97. Mahowald MW, Schenck CH. Status dissociatus a perspective on states of being. *Sleep* 1991;14:69–79.
98. Fisher C, Kahn E, Edwards A, Davis DM. A psychophysiological study of nightmares and night-terrors. I. Physiological aspects of the stage 4 night terror. *J Nerv Ment Dis* 1973;157:75–98.
99. Kales JD, Kales A, Soldatos CR, Caldwell AB, Charney DS, Martin ED. Night terrors: clinical characteristics and personality patterns. *Arch Gen Psychiatry* 1980;37:1413–1417.
100. Oswald I, Evans J. On serious violence during sleepwalking. *Br J Psychiatry* 1985;147:688–691.
101. Guilleminault C, Quera-Salva M-A, Goldberg MP. Pseudo-hypersomnia and pre-sleep behavior with bilateral paramedian thalamic lesions. *Brain* 1993;116:1549–1563.
102. Davis M. The role of the amygdala in fear-potentiated startle: implications for animal models of anxiety. *Trends Pharmacol Sci* 1992;13:35–41.
103. Clugnet MC, LeDoux J. Synaptic plasticity in fear conditioning circuits: induction of LTP in the lateral nucleus of the amygdala by stimulation of medial geniculate body. *J Neurosci* 1990;10:2818–2824.
104. LeDoux JE. Emotion. In Plum, F. ed. *Handbook of Physiology: The Nervous System V.* Washington, DC: American Physiological Society, 1987;419–459.
105. Buzsáki G, Kennedy B, Solt VB, Ziegler M. Noradrenergic control of thalamic oscillation: the role of $\alpha 2$ receptors. *Eur J Neurosci* 1990;3:222–223.
106. McCormick DA, Wang Z. Serotonin and noradrenaline excite GABAergic neurones of the guinea . . . pig and cat nucleus reticularis thalami. *J Physiol* 1991;442:235–255.
107. Sheldon PW, Aghajanian GK. Excitatory responses to serotonin (5-HT) in neurons of the rat piriform cortex: evidence for mediation by 5-HT_{1C} receptors in pyramidal cells and 5-HT_2 receptors in interneurons. *Synapse* 1991;9:208–218.
108. Aggleton JP, Mishkin M. Projection of the amygdala to the thalamus in the cynomologous monkey. *J Comp Neurol* 1984;222:56–68.
109. Aggleton DP, Desimone R, Mishkin M. The origin, course and termination of the hippocampothalamic projections in the macaque. *J Comp Neurol* 1986;243:409–421.
110. Giguere M, Goldman-Rakic PS. Mediodorsal nucleus: areal, laminar, and tangential distribution of afferents and efferents in the frontal lobe of Rhesus monkeys. *J Comp Neurol* 1988;277:195–213.
111. LeDoux JE, Farb CR. Neurons of the acoustic thalamus that project to the amygdala contain glutamate. *Neurosci Lett* 1991;134:145–149.
112. Carlsson M, Carlsson A. Schizophrenia: a subcortical neurotransmitter imbalance syndrome? *Schizophr Bull* 1990;16:425–432.
113. Cohen BD, Luby ED, Rosenbaum G, Gottlieb JS. Combined sernyl and sensory deprivation. *Compr Psychiatry* 1960;1:345–348.
114. Schwartz ML, Dekker JJ, Goldman-Rakic PS. Dual mode of corticothalamic synaptic termination in the mediodorsal nucleus of the rhesus monkey. *J Comp Neurol* 1991;309:289–304.
115. Orona E, Gabriel M. Multiple-unit activity of the prefrontal cortex and mediodorsal thalamic nucleus during acquisition of discriminative avoidance behavior in rabbits. *Brain Res* 1983;263:295–312.
116. Gabriel M, Sparenborg SP, Stolar N. Hippocampal control of cingulate cortical and anterior thalamic information processing during learning in rabbits. *Exp Brain Res* 1987;67:131–152.
117. Bench CJ, Frith CD, Grasby PM, Friston KJ, Paulesu E, Frackowiak RSJ, Dolan RJ. Investigations of the functional anatomy of attention using the Stroop test. *Neuropsychologia* 1993;31:907–922.
118. Whitty CWM, Lewin W. Vivid day-dreaming an usual form of confusion following anterior cingulectomy. *Brain* 1957;80:72–76.
119. Lapalane D, Degos JD, Baulac M, Gray F. Bilateral infarction of the anterior cingulate gyri and the fornices. Report of a case. *J Neurol Sci* 1981;51:289–300.
120. Vogt BA, Rosene DL, Pandya DN. Thalamic and cortical afferents differentiate anterior from posterior cingulate cortex in the monkey. *Science* 1979;204:205–207.
121. Gabriel M, Vogt BA, Kubota Y, Poremba A, Kang E. Training-stage related neuronal plasticity in limbic thalamus and cingulate cortex during learning: a possible key to mnemonic retrieval. *Behav Brain Res* 1991;46:175–185.
122. Gaffan D, Murray EA, Fabre-Thorpe M. Interaction of the amygdala with the frontal lobe in reward memory. *Eur J Neurosci* 1993;5:968–975.
123. Gabriel M, Sparenborg S. Posterior cingulate cortical lesions eliminate learning-related unit activity in the

anterior cingulate cortex. *Brain Res* 1987;409:151–157.
124. Goldman-Rakic PS, Porrino LJ. The primate mediodorsal (MD) nucleus and its projections to the frontal lobe. *J Comp Neurol* 1985;242:535–560.
125. Gaffan D, Murray EA. Amygdalar interaction with the mediodorsal nucleus of the thalamus and the ventromedial prefrontal cortex in stimulus-reward associative learning in the monkey. *J Neurosci* 1990;10:3479–3493.
126. Geschwind N. Neurological knowledge and complex behaviors. *Cog Sci* 1980;4:185–193.
127. Goldman-Rakic PS. Circuitry of primate prefrontal cortex and regulation of behavior by representational memory. In Plum F, ed. *Handbook of Physiology. Section I: Higher Functions of the Brain.* New York: Oxford University Press, 1987;373–417.
128. Wilson FAW, Scalaidhe SPO, Goldman-Rakic PS. Dissociation of object and spatial processing domains in primate prefrontal cortex. *Science* 1993;260:1955–1958.
129. Geschwind N. The organization of language and the brain. *Science* 1970;170:940–944.
130. Lewis DA, Hayes TL, Lund JS, Oeth KM. Dopamine and the neural circuitry of primate prefrontal cortex: implications for schizophrenia research. *Neuropsychopharmacology* 1992;6:127–134.
131. Corssen G, Domino EF. Dissociative anesthesia: further pharmacologic studies and first clinical experience with phencyclidine derivative CI-581. *Anesth Analg* 1966;45:29–40.
132. Collins GGS, Anson J. Effects of barbiturates on responses evoked by excitatory amino acids in slices of rat olfactory cortex. *Neuropharmacology* 1987;26:161–171.
133. Morgan WW, Bermudez J, Chang X. The relative potency of pentobarbital in suppressing the kainic acid or the N-methyl-D-aspartic acid-induced enhancement of cGMP in cerebellar cells. *Eur J Pharmacol* 1991;204:335–338.
134. Jones RW, Wesnes KA, Kirby J. Effects of NMDA modulation in scopolamine dementia. *Ann N Y Acad Sci* 1991;640:241–244.
135. Saletu B, Grünberger J, Linzmayer L. Acute and subacute CNS effects of miilacemide in elderly people: double-blind placebo-controlled quantitative EEG and psychometric investigations. *Arch Generontol Geriatr* 1986;5:165–181.
136. Nicholls DG. The glutamatergic nerve terminal. *Eur J Biochem* 1993;212:613–631.
137. Squire LR, Zola-Morgan S. The medial temporal lobe memory system. *Science* 1991;253:1380–1386.
138. Cohen NJ, Squire LR. Preserved learning and retention of pattern-analyzing skill in amnesia: dissociation of knowing how and knowing that. *Science* 1980;210:207–209.
139. Squire LR. *Memory and brain.* New York: Oxford University Press, 1987.
140. Tulving E, Schachter DL. Priming and human memory systems. *Science* 1990;247:301–306.
141. Thygesen P, Hermann K, Willanger R. Concentration camp survivors in Denmark: persecution, disease, compensation. *Dan Med Bull* 1970;17:65–108.
142. Goldstein G, van Kammen W, Shelly C, Miller DJ, Van Kammen DP. Survivors of imprisonment in the Pacific theater during World War II. *Am J Psychiatry* 1987;144:1210–1213.
143. Sutker PB, Winstead DK, Galina ZH, Allain AN. Cognitive deficits and psychopathology among former prisoners of war and combat veterans of the Korean conflict. *Am J Psychiatry* 1991;148:67–72.
144. Zeitlin SB, McNally RJ. Implicit and explicit memory bias for threat in post-traumatic stress disorder. *Behav Res Ther* 1991;29:451–457.
145. Sapolsky RM. *Stress, The Aging Brain, and the Mechanisms of Neuron Death.* Cambridge, MA: MIT Press, 1992.
146. Easterbrook JA. The effect of emotion on cue utilization and the organization of behavior. *Psychol Rev* 1959;66:183–201.
147. Christianson SA. Emotional stress and eyewitness memory: a critical review. *Psychol Bull* 1992;112:284–309.
148. Yuille JC, Cutshall JL. Analysis of the statements of victims, witnesses and suspects. In Yuille JC, ed. *Credibility Assessment.* Norwell, MA: Kluwer Academic, 1989.
149. Wagenaar WA, Groeneweg J. The memory of concentration camp survivors. *App Cog Psychology* 1990;4:77–87.
150. Brown R, Kulik J. Flashbulb memories. *Cognition* 1977;5:73–99.
151. Burke A, Heuer F, Reisberg D. Remembering emotional events. *Mem Cogn* 1992;20:277–290.
152. Christianson SA. The relationship between induced emotional arousal and amnesia. *Scand J Psychol* 1984;25:147–160.
153. Christianson SA, Loftus EF. Remembering emotional events: the fate of detailed information. *Cog Emotion* 1991;5:81–108.
154. Crawford HJ, Allen SN. Enhanced visual memory during hypnosis as mediated by hypnotic responsiveness and cognitive strategies. *J Exp Psychol [Gen]* 1983;112:662–685.
155. Paivio A. *Imagery and verbal processes.* New York: Holt, Rinehart, & Winston, 1971.
156. Heil M, Rosler F, Hennighausen E. Dynamics of activation in long-term memory: the retrieval of verbal, pictorial, spatial, and color information. *J Exp Psychol [Learn Mem Cog]* 1994;20:185–200.
157. van der Kolk BA, Ducey CP. The psychological processing of traumatic experience. Rorshach patterns in PTSD. *J Traum Stress* 1989;2:259–274.
158. Craik FIM, Jacoby LL. Elaboration and distinctiveness in episodic memory. In Nilsson LG, ed. *Perspectives on memory research.* Hillsdale, NJ: Erlbaum, 1979.
159. Heuer F, Reisberg D. Emotion, arousal and memory for detail. In Christianson SA, ed. *The handbook of emotion and memory: research and theory,* Hillsdale, NJ: Erlbaum, 1992;151–180.
160. Macmillian TM. Post-traumatic stress disorder and severe head injury. *Br J Psychiatry* 1991;159:431–433.
161. Keane TM, Zimmering RT, Caddell JM. A behavioral formulation of post-traumatic stress disorder in Vietnam veterans. *Behav Therap* 1985;8:9–12.
162. Kolb LC. A neuropsychological hypothesis explaining posttraumatic stress disorders. *Am J Psychiatry* 1987;144:989–995.
163. Shalev AY, Rogel-Fuchs Y. Psychophysiology of the

posttraumatic stress disorder: from sulfur fumes to behavioral genetics. *Psychosom Med* 1993;55:413–423.
164. Blanchard EB, Kolb LC, Pallmeyer TP, Gerardi RJ. A psychophysiological study of post-traumatic stress disorder in Vietnam veterans. *Psychiatr Q* 1982;54: 220–229.
165. Malloy PF, Fairbank JA, Keane TM. Validation of a multimethod assessment of posttraumatic stress disorder in Vietnam veterans. *J Consult Clin Psychol* 1983; 51:488–494.
166. McFall ME, Murburg MM, Ko GN, Veith RC. Autonomic responses to stress in Vietnam combat veterans with posttraumatic stress disorder. *Biol Psychiatry* 1990; 27:1165–1175.
167. Miserendino MJD, Sananes CB, Melia KR, Davis M. Blocking of acquisition but not expression of conditioned fear-potentiated startle by NMDA antagonists in the amygdala. *Nature* 1990;345:716–718.
168. Litz BT, Keane TM. Information processing in anxiety disorders: application to the understanding of posttraumatic stress disorder. *Clin Psychol Rev* 1989;9: 243–257.
169. Foa EB, Feske U, Murdoc TB, Kozak MJ, McCarthy PR. Processing of threat-related information in rape victims. *J Abnorm Psychol* 1991;100:156–162.
170. McNally RJ, Kaspi SP, Riemann BC, Zeitlin SR. Selective processing of threat cues in posttraumatic stress disorder. *J Abnorm Psychol* 1990;99:398–402.
171. Karper LP, Freeman GK, Grillon C, Morgan CA III, Charney DS, Krystal JH. Sensorimotor gating deficits and increased distractibility in schizophrenia. *J Neuropsychiatry Clin Neurosci,* in press.
172. Tulving E, Thomson DM. Encoding specificity and retrieval processes in episodic memory. *Psychol Rev* 1973;80:352–373.
173. Craik RIM, Tulving E. Depth of processing and the retention of words in episodic memory. *J Exp Psychol [Gen]* 1975;104:268–294.
174. Valent P. The Ash Wednesday bushfires in Victoria. *Med J Australia* 1984;141(5):291–300.
175. Kline NA, Rausch JL. Olfactory precipitants of flashbacks in posttraumatic stress disorder: case reports. *J Clin Psychiatry* 1985;46:383–384.
176. Horowitz MJ. Intrusive and repetitive thoughts after experimental stress. *Arch Gen Psychiatry* 1975;32:1457–1463.
177. Solomen Z, Garb K, Bleich A, Grupper D. Reactivation of combat related posttraumatic stress disorder. *Am J Psychiatry* 1987;144:51–55.
178. Eich JE. The cue-dependent nature of state-dependent retrieval. *Mem Cogn* 1980;8:157–173.
179. Litz BT. Emotional numbing in combat-related posttraumatic stress disorder. *Clin Psychol Rev* 1992;12: 417–432.
180. Krystal JH, Giller EL Jr, Cichetti DV. Assessment of alexithymia in post-traumatic stress disorder and somatic illness: introduction of a reliable measure. *Psychosom Med* 1986;48:84–94.
181. Paige ST, Reid GM, Allen MG, Newton JEO. Psychophysiological correlate of posttraumatic stress disorder in Vietnam veterans. *Biol Psychiatry* 1990;27:419–430.
182. Geiselman RE, Bjork RA, Fishman DL. Disrupted retrieval in directed forgetting: a link with posthypnotic amnesia. *J Exp Psychol [Gen]* 1983;112:58–72.
183. Kihlstrom JF. Instructed forgetting: hypnotic and nonhypnotic. *J Exp Psychol [Gen]* 1983;112:73–79.
184. Lang PJ. Imagery in therapy: an information processing analysis of fear. *Behav Therap* 1977;8:862–886.
185. Weingartner H, Miller H, Murphy DL. Mood-state dependent retrieval of verbal associations. *J Abnorm Psychol* 1977;86:276–284.
186. Bower GH. Mood and memory. *Am Psychol* 1981; 36:129–148.
187. Sargent W, Slater E. Acute war neuroses. *Lancet* 1940; 2:1–2.
188. Bartemeier LH, Kubie LS, Menninger KA, Romano J, Whitehorn JC. Combat exhaustion. *J Nerv Ment Dis* 1946;104:489–525.
189. Grinker RR. Treatment of war neuroses. *JAMA* 1944; 126:142–145.
190. Kirk T, Roache JD, Griffiths RR. Dose-response evaluation of the amnestic effects of triazolam and pentobarbital in normal subjects. *J Clin Psychopharmacol* 1990; 10:160–167.
191. Olsen RW. GABA-benzodiazepine-barbiturate receptor interactions. *J Neurochem* 1981;37:1–13.
192. Erickson MH. Hypnotic treatment techniques for the therapy of acute psychiatric disturbances in war. *Am J Psychiatry* 1945;101:668–672.
193. Rosen H, Myers HJ. Abreaction in the military setting. *AMA Arch Neurol Psychiatry* 1947;57:161–172.
194. Fitzgerald SG, Gonzolez E. Dissociative states induced by relaxation training in a PTSD combat veteran: failure to identify trigger mechanisms. *J Traum Stress* 1994;7:111–115.
195. Kardiner A, Spiegel H. *War stress and neurotic illness.* New York: Hoeber, 1947.
196. Fisher C, Byrne J, Edwards A, Kahn E. A psychophysiological study of nightmares. *J Am Psychoanal Assoc* 1970;18:747–782.
197. Greenberg R, Pearlman CA, Bampel D. War neuroses and the adaptive function of REM sleep. *Br J Med Psychol* 1972;45:27–33.
198. Lavie P, Hertz G. Increased sleep motility and respiration rates in combat neurotic patients. *Biol Psychiatry* 1979;14:983–987.
199. Hafez A, Metz L, Lavie P. Long-term effects of extreme situational stress on sleep and dreaming. *Am J Psychiatry* 1987;144:344–347.
200. van der Kolk B, Blitz R, Burr W, Sherry S, Hartmann E. Nightmares and trauma: a comparison of nightmares after combat with lifelong nightmares in veterans. *Am J Psychiatry* 1984;141:187–190.
201. Sacks O. An anthropologist on Mars. *The New Yorker.* December 27, 1993, 106–125.
202. Reiser MF. *Memory in Mind and Brain: What Dream Imagery Reveals.* New York: Basic Books, 1990.
203. Baddeley AD. *Working Memory.* New York: Oxford University, 1986.
204. Baddeley A, Wilson B. Frontal amnesia and the dysexecutive syndrome. *Brain and Cognition* 1988;7:212–230.
205. Horowitz M, Wilner N. Stress films, emotion, and cognitive response. *Arch Gen Psychiatry* 1974;33:1339–1344.
206. Pitman RK. Posttraumatic obsessive-compulsive disorder: a case study. *Compr Psychiatry* 1993;34:102–107.
207. Baster LR, Schwartz JM, Bergman KS, Szuba MP, Buze BH, Mazziotta JC, Alazraki A, Selin CE, Feering

HK, Munford P, Phelps ME. Caudate glucose metabolic rate changes with both drug and behavior therapy for obsessive-compulsive disorder. *Arch Gen Psychiatry* 1992;49:681–689.
208. Swedo SE, Pietrini P, Leonard HL, Schapiro MB, Rettew DC, Goldberger EL, Rapoport SI, Rapoport JL, Grady CL. Cerebral glucose metabolism in childhood-onset obsessive-compulsive disorder: revisualization during pharmacotherapy. *Arch Gen Psychiatry* 1992;49:690–694.
209. Goodman WK, Price LH, Delgado PL, Palumbo J, Krystal JH, Rasmussen SA, Heninger GR, Charney DS. Specificity of serotonin reuptake inhibitors in the treatment of obsessive-compulsive disorder: Comparison of fluvoxamine and desipramine. *Arch Gen Psychiatry* 1990;47:577–585.
210. Fenichel O. *The psychoanalytic theory of the neuroses.* New York: Norton, 1945.
211. Semple WE, Goyer P, McCormick R, Morris E, Compton B, Muswick G, Nelson D, Donovan B, Leisure G, Berridge M, Miraldi F, Schulz SC. Preliminary report: brain blood flow using PET in patients with posttraumatic stress disorder and substance abuse histories. *Biol Psychiatry* 1993;34:115–118.
212. Taylor J. *Selected writings of John Hughlings Jackson on epilepsy and epileptiform convulsions.* London: Hodder and Stroughton, 1931.
213. Mesulam MM. A cortical network for directed attention and unilateral neglect. *Ann Neurol* 1981;10:309–325.
214. Penfield W, Perot P. The brain's record of auditory and visual experience, a final summary and discussion. *Brain* 1963;86:595–696.
215. Gloor P, Olivier A, Quesney LF, Andermann F, Horowitz S. The role of the limbic system in experiential phenomena of temporal lobe epilepsy. *Ann Neurol* 1982; 12:129–144.
216. Halgren E, Walter RD, Cherlow DG, Crandall PH. Mental phenomena evoked by electrical stimulation of the human hippocampal formation and amygdala. *Brain* 1978;101:83–117.
217. Horowitz MJ, Adams JE, Rutkin BB. Visual imagery on brain stimulation. *Arch Gen Psychiatry* 1968;19: 469–486.
218. Baldwin M. Electrical stimulation of the mesial temporal region. In Ramey ER, O'Doherty DS, eds. *Electrical studies on the unanesthetized brain,* New York: Hoeber, 1960;159–176.
219. Szatmari P. Asperger's syndrome: diagnosis, treatment, and outcome. *Psychiatr Clin N Am* 1991;14: 81–93.
220. Sacks O. Letter to the editor. *The New Yorker,* February 14, 1994, pg. 11.
221. Moscovitch M. Memory and working-with-memory: a component process model based on modules and central system. *J Cog Neurosci* 1992;4:257–267.
222. Moscovitch M. Confabulation and the frontal systems: strategic versus associative retrieval in neuropsychological theories of memory. In Roediger HL III, Craik FIM, eds. *Varieties of memory and consciousness: essays in honour of Endel Tulving.* Hillsdale, NJ: Lawrence Erlbaum, 1989;133–160.
223. Penfield W. Memory and perception. In Hamberg DA, Pribram KH, Stunkard AJ, eds. *Perception and its disorders.* Baltimore, MD: Williams & Wilkins, 1970; 108–122.
224. Pitman RK. Post-traumatic stress disorder, hormones, and memory. *Biol Psychiatry* 1989;26:221–223.
225. Pitman RK. Posttraumatic stress disorder, conditioning and network theory. *Psychiatr Annal* 1988;18:182–189.
226. Krystal JH. Animal models for posttraumatic stress disorder. In Giller EL, ed. *Biological assessment and treatment of posttraumatic stress disorder.* Washington, DC: American Psychiatric Press, 1990;3–64.
227. Muller D, Joly M, Lynch G. Contributions of quiqualate and NMDA receptors to the induction and expression of LTP. *Science* 1988;242:1694–1697.
228. Dawson TM, Dawson VL, Snyder SH. A novel neuronal messenger molecule in brain: the free radical, nitric oxide. *Ann Neurol* 1992;32:297–311.
229. Zhuo M, Small SA, Kandel ER, Hawkins RD. Nitric oxide and carbon monoxide produce activity-dependent long-term synaptic enhancement in hippocampus. *Science* 1993;260:1946–1950.
230. Robinson GB. Enhanced long-term potentiation induced in rat dentate gyrus by coactivation of septal and entorhinal inputs: temporal constraints. *Brain Res* 1986;379:56–62.
231. Lynch G, Granger R. Variations in synaptic plasticity and types of memory in corticohippocampal networks. *J Cog Neurosci* 1992;4:189–199.
232. Doyére V, Rédini-Del Negro C, Laroche S. Changes in synaptic function of an LTP form in associative learning and memory. *Neurosci Res Comm* 1993;13 (suppl):S23–S26.
233. Kim JJ, DeCola JP, Landeira-Fernandez J, Fanselow MS. N-methyl-D-aspartate receptor antagonist APV blocks acquisition but not expression of fear conditioning. *Behav Neurosci* 1991;105:126–133.
234. Jackson A, Koek W, Colpaert FC. NMDA antagonists make learning and recall state-dependent. *Behav Pharm* 1992;3:415–421.
235. Shapiro ML, O'Connor C. N-methyl-D-aspartate receptor antagonist MK-801 and spatial memory representation: working memory is impaired in an unfamiliar environment but not in a familiar environment. *Behav Neurosci* 1992;106:604–612.
236. Tsumoto T. Long-term depression in cerebral cortex: a possible substrate of forgetting that should not be forgotten. *Neurosci Res* 1993;16:263–270.
237. Jacobs WJ, Nadel L. Stress-induced recovery of fears and phobias. *Psychol Rev* 1985;92:512–531.
238. Horowitz MJ. *Stress response syndromes.* New York: Aronson, 1976.
239. Grinker RR, Spiegel JP. *War Neuroses in North Africa,* New York: Josiah Macy Jr Foundation, 1943.
240. Keane TM, Fairbank JA, Caddell JM, Zimering RT. Implosive (flooding) therapy reduces symptoms of PTSD in Vietnam combat veterans. *Behav Therap* 1989; 20:245–260.
241. Laurence J-R, Perry C. Hypnotically created memory among highly hypnotizable subjects. *Science* 1983;222: 423–524.
242. Ghoneim MM, Hinrichs JV, Mewaldt SP. Dose-response analysis of the behavioral effects of diazepam: I. learning and memory. *Psychopharmacology* 1984;82: 291–295.

243. Foa EB, Steketee G, Olasov Rothbaum B. Behavioral/cognitive conceptualizations of post-traumatic stress disorder. *Behav Therap* 1989;20:155–176.
244. Pitman RK, Altman B, Greenwald E, Longpre RE, Macklin ML, Poire RE, Steketee GS. Psychiatric complications during flooding therapy for posttraumatic stress disorder. *J Clin Psychiatry* 1991;52:17–20.
245. Hickling EJ, Sison GFP Jr, Vanderploeg RD. Treatment of posttraumatic stress disorder with relaxation and biofeedback training. *Biofeedback Self Regul* 1986; 11:125–134.
246. Loewenstein RJ, Hornstein N, Farber B. Open trial of clonazepam in the treatment of post-traumatic stress symptoms in multiple personality disorder. *Dissociation* 1988;1:3–12.
247. Schwartz BL, Hashtroudi S, Herting RL, Handerson H, Deutsch SI. Glycine prodrug facilitates memory retrieval in humans. *Neurology* 1991;41:1341–1343.
248. Marshall JR. The treatment of night terrors associated with the posttraumatic syndrome. *Am J Psychiatry* 1975;132:293–295.
249. Burnstein AT. Treatment of flashbacks by imipramine. *Am J Psychiatry* 1983;140:509.
250. Hogben GL, Cornfield RB. Treatment of war neurosis with phenelzine. *Arch Gen Psychiatry* 1981;38:440–445.
251. Lerer B, Bleich A, Kotler M, Garb R, Hertzberg M, Levin B. Post-traumatic stress disorder in Israeli combat veterans: effect of phenelzine treatment. *Arch Gen Psychiatry* 1987;44:976–981.
252. McDougle CJ, Southwick SM, Charney DS, St. James RL. An open trial of fluoxetine in the treatment of posttraumatic stress disorder. *J Clin Psychopharmacol* 1991;11:325–327.
253. Davidson J, Roth S, Newman E. Fluoxetine in posttraumatic stress disorder. *J Traum Stress* 1991;4:419–423.
254. Nagy LM, Morgan CA III, Southwick SM, Charney DS. Open prospective trial of fluoxetine for posttraumatic stress disorder. *J Clin Psychopharmacol* 1993;13: 107–113.
255. Kosten TR, Frank JB, Dan E, McDougle CJ, Giller EL Jr. Pharmacotherapy for posttraumatic stress disorder using phenelzine or imipramine. *J Nerv Ment Dis* 1991;179:366–370.
256. Davidson JRT, Kudler HS, Smith RD, Mahoney S, Kipper S, Hammett EB, Saunders WB, Cavenar JO. Treatment of post traumatic stress disorder with amitriptyline and placebo. *Arch Gen Psychiatry* 1990;47: 259–266.
257. Reist C, Kauffmann CD, Haier RJ, Sangdahl C, DeMet EM, Chicz-DeMet A, Nelson JN. A controlled trial of desipramine in 18 men with posttraumatic stress disorder. *Am J Psychiatry* 1989;146:513–516.
258. Goddard A, Charney DS, Heninger GR, Woods SW. Effects of the 5-HT reuptake blocker on anxiety induced by yohimbine. *Soc Neurosci Abstr* 1990;16: #1177.
259. Lipper S, Davidson JRT, Grady TA, Edinger J, Hammett EB, Mahorney SL, Cavenar JO. Preliminary study of carbamazepine in post traumatic stress disorder. *Psychosomatics* 1986;27:849–854.
260. Post RM, Weiss SRB. Sensitization, kindling, and anticonvulsants in mania. *J Clin Psychiatry* 1989;59(suppl.): 23–30.
261. Davis KL, Thal LJ, Gamzu ER, Davis CS, Woolson RF, Gracon SI, Drachman DA, Schneider LS, Whitehouse PJ, Hoover TM. A double-blind, placebo-controlled multicenter study of tacrine for Alzheimer's disease. *New Engl J Med* 1992;327:1253–1259.
262. Pitman RK, Orr SP, Lasko NB. Effects of intranasal vasopressin and oxytocin on physiologic responding during personal combat imagery in Vietnam veterans with posttraumatic stress disorder. *Psychiatry Res* 1993; 48:107–117.
263. Dawson GR, Heyes CM, Iverson SD. Pharmacological mechanisms and animal models of cognition. *Behav Pharm* 1992;3:285–297.

Neurobiological and Clinical Consequences of Stress: From Normal Adaptation to PTSD, edited by M. J. Friedman, D. S. Charney, and A. Y. Deutch, Lippincott-Raven Publishers, Philadelphia © 1995.

15

Neural Circuits and Mechanisms of Post-Traumatic Stress Disorder

*Dennis S. Charney, *†Ariel Y. Deutch, *Steven M. Southwick, and *John H. Krystal

*Departments of *Psychiatry and †Pharmacology, Yale University School of Medicine, New Haven, Connecticut 06510; and National Center for Post-Traumatic Stress Disorder, Veterans Affairs Medical Center; West Haven, Connecticut 06516*

A functional neuroanatomy that can account for the spectrum of symptoms associated with Post-Traumatic Stress Disorder (PTSD) has not been definitively established. The long-held idea that limbic brain structures and their projections constitute a functional system mediating emotional behavior has been questioned (1). However, there is neuroanatomical, neurophysiological, and neurochemical evidence supporting the concept of a neural circuit consisting of brain structures within and outside the limbic forebrain that can mediate the effects of traumatic stress (2).

The brain structures constituting such a neural circuit should have several features. There needs to be sufficient afferent sensory input to permit assessment of the fear- or anxiety-producing nature of the traumatic event. The neuronal interactions among the brain structures must be capable of incorporating an individual's prior experience (memory) into the cognitive appraisal of the event. These interactions are important in the attachment of affective significance to specific stimuli and the mobilization of adaptive behavioral responses. Finally, the efferent projections from the brain structures should be able to mediate an individual's neuroendocrine, autonomic, and skeletal motor responses to original traumas, as well as account for the pathological reactions to subsequent traumatic reminders that result in anxiety-related signs and symptoms.

Preclinical and clinical investigations provide strong evidence for linking several brain structures to the signs and symptoms of anxiety and fear associated with trauma. Chief among these are the amygdala, locus coeruleus, hippocampus, hypothalamus, thalamus, periaqueductal gray, and orbitofrontal cortex. In the sections to follow, the neuroanatomy, neurochemistry, and neural mechanisms relevant to the role these brain structures play in anxiety and fear are reviewed. The implications of these findings for a better understanding of the pathophysiology and treatment of PTSD are also discussed.

KEY BRAIN STRUCTURES MEDIATING ANXIETY AND FEAR BEHAVIORS RESULTING FROM TRAUMATIC STRESS: THE AMYGDALA

Neuroanatomy

Afferent Pathways

The amygdala receives an extraordinary afferent input that forms the basis for the pivotal role of the amygdala in the processing of fear- and anxiety-inducing stimuli, as well as the organization of adaptive behavioral, autonomic, neuroendocrine, and skeletal-motor responses.

Brain stem afferent pathways to the amygdala include those from the parabrachial nucleus (3,4) and locus coeruleus (5,6). Among the subcortical afferent pathways to the amygdala are those from the posterior thalamus and the entorhinal cortex (7,8). The amygdala also receives an extensive innervation from visual, auditory, and somatosensory systems in the cortex (9).

Efferent Pathways

The central nucleus of the amygdala, but not other amygdala nuclei, descend into and through the midbrain, pons, and medulla. These projections innervate a number of brain structures implicated in autonomic control, including the dorsal motor nucleus of the vagus (10) and parabrachial nucleus (3,4), which may mediate fear-induced autonomic and respiratory changes, respectively.

The central and basal amygdaloid nuclei amygdala return the substantial noradrenergic input from the locus coeruleus (LC) (5,6). There are also indirect projections to the LC via the paragigantocellularis nucleus (11) and the ventral tegmental area (6). These projections may form the basis of important functional interactions between the LC and the amygdala, which are discussed in detail in the paragraphs that follow.

The subcortical connections of the amygdala include the amygdala-striatal projections, which are among the most substantial efferents of the amygdala. The region of the striatum that receives amygdaloid projections also receives efferents from the prefrontal cortex and dopamine neurons in the ventral tegmentum, and has been termed the ''limbic striatum'' (12). These interactions may be critical in developing motor responses to threatening environmental stimuli, especially in utilizing past experience (13,14).

There are direct projections from the central nucleus of the amygdala to the lateral hypothalamus (15,16). These projections may be involved in the activation of the sympathetic autonomic nervous system produced by traumatic stress (4). Direct projections from the central nucleus of the amygdala to the paraventricular nucleus of the hypothalamus (17–19), or indirect projections by way of the bed nucleus of the stria terminalis and preoptic area, which receive inputs from the amygdala (7) and project to the paraventricular nucleus (20), may mediate the neuroendocrine responses to stressful or fearful stimuli (19).

There are important anatomical interconnections between the hippocampus and the amygdala (21,22). The amygdala projection to the hippocampus is substantially stronger than hippocampal projections to the amygdala. The predominant amygdala projections to the hippocampus are from the lateral nucleus, magnocellular division of the accessory basal nucleus, and the parvicellular division of the basal nucleus.

The amygdala has an extensive reciprocal innervation with the cortex (23). The amygdala projects to essentially all visually related areas of temporal and occipital cortex. There are also substantial connections between the amygdala and polysensory regions of temporal, insular, cingulate, and frontal lobes (24). In addition to via the extensive, direct interconnections between the amygdala and prefrontal cortex, the amygdala can influence the prefrontal cortex through the thalamus and striatum (13,14).

Behavioral Effects of Stimulation of the Amygdala

In laboratory animals, electrical stimulation of the amygdala produces a complex pattern of behavior consistent with the animal being frightened (25–27). Amygdala kindling increases fear responses without altering spatial memory (28). In addition, stimulation of the amygdala results in changes in autonomic, neuroendocrine, and reflex function that frequently accompany fear and anxiety states. Numerous investigations have demonstrated that respiration, blood pressure, and heart rate are increased by amygdala stimulation (see review [4]). There is also an elicitation of jaw movements (25,26, 29) and activation of facial motor neurons (30), which probably relates to the involvement of the amygdala in facial recognition (31) and the alteration in facial expression that occurs with fear. Numerous brain reflexes are generally enhanced by amygdala stimulation, including the massenteric (32), baroreceptor (33), nictitating membrane, (34) and startle reflexes (4). Finally, electrical stimulation of the amygdala increases plasma levels of corticosterone (35,36) (see review [4]).

Electrical stimulation of the amygdala in humans has been investigated in patients with temporal lobe epilepsy (37–39). A variety of emotions are elicited by amygdala stimulation, including fear, anxiety, pleasure, and anger, with fear reported most frequently. It is noteworthy that the emotional content associated with the affect produced by amygdala stimulation has often been related to both the individual's current psychological concerns and past experience. In this context, many of the experiental phenomena produced have had the quality of flashbacks. These clinical investigations suggest that a primary function of the amygdala may be to bring perceptions and memories stored in other brain structures (cortex) to a conscious level, thereby linking an individual's encoded personal life experience with a current life event, resulting in an individualized motivational and affective state (37). This hypothesis is supported by the extensive reciprocal innervation of the neocortex and amygdala. These neuronal interactions may form the basis of the persistent, intrusive memories associated with PTSD and the distressing effects of reminders of the original trauma.

BEHAVIORAL EFFECTS OF AMYGDALA LESIONS

Lesions of the amygdala in laboratory animals result in a spectrum of changes in behaviors, including reduced aggression, a generalized reduction in emotional responsiveness, increased orality, and altered responses to food (40). Numerous investigators have suggested, based upon lesion studies and the neuroanatomy of the amygdala, that the amygdala is critical to the learning of stimulis-reinforcement associations (41,42). In studies of the effects of amygdala lesions on social behavior in monkeys, a consistent observation in lesioned animals is the loss of the ability to attach meaning to sensory information in order to generate appropriate behavioral responses (43,44). In several different species, lesions of the amygdala have been shown to reduce fear (4). Amygdala lesions will greatly increase the number of contacts a rat will make with a sedated cat (45), and reduce the effect of restraint stress on gastrointestinal ulceration and hypersecretion of corticosterone (46–48).

Studies of lesions of the amygdala suggest an important role for the amygdala in the mechanisms underlying fear conditioning. Lesions attenuate conditioned freezing normally seen in response to a stimulis formerly paired with a shock (45,49,50) or a dominant male rat (51,52). Amygdala lesions antagonize the reduction in bar pressing in the operant conflict test (53) and the fear potentiated startle (54).

Direct Recording of Amygdala Activity During Different Behavioral States

The hypothesis that the amygdala contributes to the assessment of the emotional significance of sensory stimuli is further supported by the results of investigations recording amygdala neural activity (55). While amygdala neurons are responsive to single sensory stimuli (flashing lights, clicks), they are more sensitive to complex sensory events with emotional significance, and rewarding and aversive stimulis properties (55–58). For example, a group of neurons in the amygdala are involved in facial recognition and the resultant emotional and social responses (59–60). These neurons receive input from cortical neurons in the temporal lobes (61) that provide information regarding facial identity and expression (62), which will then predispose the organism toward a specific behavioral response.

Several studies have examined amygdala activity in relation to presentation of conditioned stimuli. It has been demonstrated that a neutral stimulis paired with aversive stimulation will increase neural firing in the amygdala (63–65). In the nonhuman primate, amygdala activity is increased with stimuli of high emotional significance, especially threat. This is true both for auditory and visual stimuli (66). There is preliminary data in humans that amygdala neural activity is increased during highly charged emotional states (67). These data, from amygdala lesion and recording studies, point not only toward a critical role for the amygdala in initial assessment and behavioral response to traumatic events to longer term consequences of trauma.

THE LOCUS COERULEUS

Neuroanatomy of the Locus Coeruleus

Afferent Pathways

The neuroanatomy of the largest central noradrenergic nucleus, the locus coeruleus (LC), is characterized by an extensive efferent projection system and a more restricted afferent input (68–70). A retrograde labeling study of afferent input to the LC using iontophoretic horseradish peroxidase (HRP) injections found that forebrain structures such as the neocortex, amygdala, and hypothalamus project to the LC. The hypothalamic innervation is noteworthy because it provides a potential regulatory role for the LC on autonomic function. Brain stem monoaminergic neurons, such as the raphe nuclei and a variety of sensory relay areas, also project to the LC. The existence of these sensory afferents from the spinal cord and nucleus of the solitary tract provide insight into the mechanisms by which the LC is responsive to noxious stimuli and alterations in cardiovascular function (71). Consistent with these observations, a recent neuroanatomical investigation involving anterograde labeling with PhA-L in the cat and monkey found terminal fibers in the LC following injections in lamina 1 of the spinal or medullary dorsal horn (69). These observations are supported by an electrophysiological mapping study revealing that spinal cord lamina 1 cells terminated in the locus coeruleus (70). Contradicting these observations are reports that direct afferents to the LC almost exclusively come from the paragigantocellularis (PG) and the prepositus hypoglossi (PrH) (72). These studies used very small injections of wheat germ agglutinin conjugated to HRP (WGA-HRP) that were restricted to the LC in the rat. The discrepant observations may be due to the fact that WGA-HRP is not an efficient retrograde marker for lamina 1 neurons (73). The finding of additional afferents to the LC using choleratoxin B subunit as a retrograde tracer instead of WGA-HRP is consistent with this assertion (74). The hypothesis that PGi and PrH are the predominant afferents to the LC is also not supported by the demonstration that destruction of these nuclei fail to block LC responses to somatosensory stimuli (75).

If not the primary LC afferent, the PGI remains an important afferent to the LC and is a key sympathoexcitatory region in brain (68). There are widespread afferents to the PGI from diverse brain areas (76). The PGI may be an integrative pathway for activating the LC by a variety of mechanisms. The PGI is involved in control of arterial blood pressure, cardiopulmonary reflexes, and parasympathetic function. It has been suggested that the peripheral sympathetic nervous system is activated in parallel with the LC by projections to both areas from the PGI (68). The locus coeruleus also appears to be activated by projections responsive to corticotropin-releasing factor (CRF), and considerable evidence suggests that CRF has anxiogenic properties. Intracerebroventricular infusion of CRF increases norepinephrine turnover in several forebrain areas (77). The CRF in a dose-dependent fashion increases the firing rate of LC-norepinephrine neurons (78). Stress, which activates norepinephrine neurons, markedly increases CRF concentrations in the LC (79). Moreover, it has recently been demonstrated that infusion of CRF into the LC produces anxiogenic activity, as well as significant increases in the norepinephrine metabolite 3,4-dihydroxy phenylglycol in forebrain areas such as the amygdala and hypothalamus (80).

Efferent Pathways

The LC has diffuse projections to the entire cortical mantle, as well as thalamus, hypothalamus, amygdala, hippocampus, cerebellum, and spinal cord. Through this broad efferent network the LC can mediate a range of cognitive, neuroendocrine, cardiovascular, and skeletal motor responses that accompany anxiety and fear.

The LC is the primary source of norepinephrine fibers in a number of neocortical regions in the primate. Consistent with postsynaptic inhibitory effects of norepinephrine, activation of the LC neurons produces inhibition of neuronal activity in brain regions receiving projections from the LC, including the cochlear nucleus, cerebral and cerebellar cortices, spinal trigeminal nucleus, hippocampus, caudate, and superior colliculus. Similarly, LC stimulation reduces cerebral metabolism and blood flow in these areas (81,82). Since norepinephrine has greater inhibitory effects on spontaneous than on evoked neuronal activity, it has been suggested that a function of the LC-norepinephrine system is to enhance neuronal responses to inputs of behavioral significance or, put another way, to increase the signal-to-noise ratio (10).

Behavioral Effects of Locus Coeruleus Stimulation

Electrical stimulation of the LC produces a series of behavioral responses similar to those observed in naturally occurring or experimentally-induced fear (83). These behaviors are also elicited by administration of drugs such as yohimbine and piperoxone that activate the LC by blocking alpha-2 adrenergic autoreceptors (84). Drugs that decrease the function of the LC by interacting with inhibitory opiate (morphine), benzodiazepine (diazepam), and alpha-2 (clonidine) receptors on the LC decrease fearful behavior and partially antagonize the effects of electrical stimulation of the LC in the monkey (85). These studies suggest that abnormally high levels of LC activity producing increased release of NE at postsynaptic projection sites throughout the brain may act to augment some forms of fear or pathological anxiety, depending on the environmental conditions.

Studies conducted in healthy human subjects and psychiatric patients are largely consistent with this assertion. Piperoxone and yohimbine cause anxiety in humans. Patients with panic disorder and post-traumatic stress disorder have abnormally increased responses to yohimbine (86–88). A single report of electrical stimulation of the LC region in humans reported that the subjects experienced feelings of fear and imminent death (89). In a subject with chronically implanted stimulation electrodes in the vicinity of the LC, electrical stimulation during sleep was associated with marked insomnia (90).

Behavioral Effects of Locus Coeruleus Lesions

Bilateral lesions of the LC in the monkey decrease the natural occurrence of fearful behavioral responses in a social group situation, and in response to threatening confrontations with humans (83). Agents that decrease the function of the LC also decrease fearful behavior and partially antagonize the effects of electrical stimulation of the LC in the monkey. Lesions of the dorsal bundle that originates in the locus coeruleus and projects mostly to the neocortex, hippocampus, and cerebellar cortex produce anxiolytic type responses in anxiogenic behavioral test situations (91).

Many studies in rodents using the neurotoxin, 6-OHDA, have shown effects on a variety of behavioral tests that have been interpreted as inconsistent with the hypothesis that noradrenergic systems are critically involved in anxiety and fear (92). The validity of these behavioral paradigms to reflect anxiety and fear has been the subject of much debate (83,84,93).

LOCUS COERULEUS ACTIVITY AND BEHAVIORAL STATES ASSOCIATED WITH STRESS AND FEAR

The effect of stressful stimuli on LC activity has been assessed in freely moving cats (94–97). Conditions that are behaviorally activating but not stressful, such as exposure to inaccessible rats or food, do not increase LC firing. In contrast, stressful and fear-inducing stimuli, such as loud white noise, air puff, restraint, and confrontation with a dog, produce a rapid, robust, and sustained increase in LC activity (94–97). Of interest, these increases in LC function are accompanied by sympathetic activation. Generally, the greater the sympathetic activation in response to the stressor, as indicated by heart rate, the greater the correlation observed. Thus, a stimulus intensity threshold for coactivation of central and peripheral NE systems may exist.

A parallel activation of LC neurons and splanchnic sympathetic nerves is produced by noxious stimuli. The LC, like sympathetic, splanchnic activity, is highly responsive to various peripheral cardiovascular events such as alterations in blood volume or blood pressure. Internal events that must be responded to for survival, such as thermoregulatory disturbance, hypoglycemia, blood loss, an increase in pCO2, or a marked reduction in blood pressure, cause robust and long-lasting increases in LC activity (98–101).

There are also peripheral visceral influences on LC activity. In rats, distention of the urinary bladder, distal colon, or rectum activate LC neurons. These findings suggest that traumatic stress-induced changes in autonomic or visceral function may result in specific behavioral responses via the brain LC-norepinephrine system. The LC-norepinephrine network may be an important component of the efferent arm of circuits activated by traumatic stress. On the one hand, the system, when functioning normally, may be critical in facilitating the planning and execution of behaviors important for survival (98). On the other hand, it is possible that severe stress renders the system hyperresponsive subsequent to the stress, resulting in chronic anxiety-related symptoms.

OTHER BRAIN STRUCTURES IN THE NEURAL CIRCUIT OF FEAR

Thalamus

The thalamus is another brain structure important in fear responses. There is evidence from classically conditioned emotional responses that the processing of threatening stimuli involves the relay of sensory signals to limbic forebrain directly from the thalamus and cortex (102,103). Fear responses to acoustic stimuli are disrupted by thalamic lesions, and responses depend on the connection between the thalamus and amygdala (104).

Most importantly, as will soon be described in detail, the thalamus serves as an important link between peripheral sensory inputs and other brain structures that mediate fear responses. The thalamic nuclei relay sensory information from auditory, visual, and somesthetic systems to the cortex and limbic forebrain. Thus, the thalamus is an important neuronal interface between the environment and forebrain structures.

Hippocampus

The hippocampus has been hypothesized to play a key role in anxiety development. Lesions of the hippocampus increase punished responses like anxiolytic drugs (105). Hippocampal lesions have little effects on classically conditioned, modality-specific fear, in contrast to the amygdala (106). Furthermore, there is recent data that damage to the hippocampus formation in monkeys impairs memory, but does not affect emotional behavior. In contrast, damage to the amygdaloid complex alters emotional behavior, but not memory (107,108).

However, there is evidence of a sensory relay role for the hippocampus in fear conditioning. Lesions of the hippocampus interfere with conditioning of fear responses to polymodal contextual stimuli (106,109). It has been speculated that the role of the hippocampus in contextual fear conditioning involves projections from the subiculum to the amygdala (106).

There is strong evidence that the cortical projections relaying sensory information to the hippocampus play a role in many cognitive functions (108,110–112). Projections from the amygdala to the hippocampus could permit the experience of fear to modulate or influence shortterm memory function. The neuronal loop among the cortical structure, hippocampus, and amygdala may serve as part of the neuronal network that attaches cognitive significance to fear-inducing events and facilitates memory traces that enable the individual to rapidly initiate adaptive behavioral responses.

Hypothalamus

The hypothalmic nuclei receive projections from many limbic and brain stem structures (amygdala, septum, piriform cortex, locus coeru leus, dorsal motor nucleus of the vagus), and project to sympathetic regions in the spinal cord and medulla (113). In this context, many of the neuroendocrine and autonomic changes resulting from stress, fear, and anxiety can be understood.

Uncontrollable stress and conditioned fear produces increases in norepinephrine turnover in the hypothalamic nuclei, including the paraventricular nucleus (PVN) (114,115). Further, stimulation of the LC produces increases in NE turnover in the PVN and supraoptic nucleus (SON) (116), indicating an important functional linkage between the LC and the hypothalamus.

The afferent (LC, lateral tegmentum, nucleus of solitary tract cell ventrolateral medulla) and efferent (lower brain stem, intermediolateral cell column of spinal cord, median eminence) projections of the PVN emphasize the PVN's importance in regulation of neuropeptide release and sympathoadrenal outflow associated with anxiety and fear. The PVN is a major contributor to the increased release of CRH, which occurs following stress (117–120). The functional interactions with the LC also apply here, because norepinephrine stimulates CRH release from the PVN. The PVN and SON synthesize the neuropeptides enkephalin, vasopressin, and oxytocin, hormones whose levels are increased by stress (121–124).

The medial basal hypothalamic nuclei (arcuate, ventromedial, periventricular) give rise to fibers that terminate in the median eminance, where a number of hormonal releasing and inhibitory factors are released. There is evidence that stress induces changes in cholecystokinin (CCK) and substance P concentrations in the hypothalamus, suggesting that these peptides may also be involved in a integrated hypothalamic response to anxiety and fear (125).

Orbitofrontal Cortex

The orbitofrontal cortex receives projections from visual, auditory, and somatosensory cortex. It also receives projections from the amygdala, entorhinal cortex, and cingulate gyrus, as well as the mediodorsal thalamic nucleus, which itself receives inputs from the amygdala, septum, and cingulate gyrus (126–129). Efferent pathways from the orbitofrontal region include brain regions that regulate autonomic, neuroendocrine, and skeletal motor function (127,130–132).

An important observation is that a number of sensory responses of orbitofrontal neurons are highly dependent on the meaning of the stimulis. The orbitofrontal cortex may possess information on the reinforcement associations of particular stimuli and may be able to modify this information based upon an individual's recent experience. The orbitofrontal neurons activated after an animal has made responses to specific stimuli may code the characteristics of the experience (rewarding or adverse) and, therefore, be integral to a neural mechanism for rapidly altering the reinforcement associations (133). Thus, the orbitofrontal cortex may be involved in the processing of the effectiveness of the individual's behavior (i.e., the capacity to inhibit and change behavior in the face of changing circumstances such as a threat) (134,135).

Given the reciprocal interactions between subcortical limbic structures and the orbitofrontal region, these two regions may interact in the learning and unlearning of the significance of fear-producing sensory events and in the choice and implementation of behaviors important for survival (136). This hypothesis is consistent with observations that damage to the prefrontal cortex in both experimental animals and humans produces profound changes in mood, affect, and social behavior. Patients with frontal lobe lesions or damage exhibit perservative behavior such that they continue to make the same response to a particular situation, even when the response is no longer appropriate (134–137).

NEURAL MECHANISMS THAT CONTRIBUTE TO PERSISTENT SYMPTOMS FOLLOWING TRAUMATIC STRESS (Table 1)

Fear Conditioning

In many patients with PTSD, vivid memories of a traumatic event, autonomic arousal, and even flashbacks can be elicited by diverse sensory and cognitive stimuli that have been associated with the original trauma (138,139). Consequently, patients begin to avoid these stimuli in their everyday life, or a numbing of general emotional responsiveness occurs. Modality-specific and contextual fear conditioning, which are easily demonstrated in the laboratory, may explain some of these observations. Animals exposed to emotionally neutral, visual or auditory conditioned stimuli (CS) in conjunction with an aversive unconditioned stimulus (UCS) will subsequently exhibit a conditioned emotional response (CER) to the CS in the absence of the UCS. The CERs are also produced when an animal is placed in an environment in which an aversive UCS has previously been experienced. In this circumstance the CER is not elicited by a modality-specific stimulis that was paired with a UCS, but instead by complex, polymodal contextual stimuli that were present in the environment when the UCS originally occurred and are present upon reexposure to the environment (106). The CERs can last for years in laboratory animals (140) and are used to infer that a state of fear has been produced (141). Therefore, a neural analysis of fear conditioning in animals may be useful in identifying the neurochemicals and brain structures involved in learning and remembering associations of stimuli with traumatic events, which may form the basis of many of the symptoms associated with PTSD.

TABLE 1. *Neural mechanisms related to the pathophysiology and treatment of PTSD*

Mechanism	Brain structures	Clinical relevance: Pathophysiology	Clinical relevance: Treatment
Fear Conditioning	Sensory Cortex Amygdala Locus Coeruleus Thalamus Hypothalamus	May account for common clinical observation in PTSD that sensory and cognitive stimuli associated with or resembling the original frightening experience elicit panic attacks, flashbacks, and a variety of autonomic symptoms. Pervasive anxiety may be due to contextual fear conditioning.	Psychotherapies designed to reverse the effects of modality-specific fear conditioning are very efficacious. Development of drugs that act selectively on the sensory pathways afferent to the anxiety circuit may decrease conditioned fear.
Extinction	Sensory Cortext Amygdala	A failure in neural mechanisms underlying extinction may relate to treatment-resistant phobias. In PTSD it may relate to persistence in recalling traumatic memories.	Psychotherapies need to be developed that facilitate extinction through use of conditioned inhibitors and learning of "new memories."
Sensitization	Nucleus Accumbens Amygdala Striatum Hypothalamus	May explain adverse effects of early life trauma on subsequent responses to stressful life events. May play a role in the chronic course of PTSD and, in some cases, the worsening of the illness over time.	Suggests efficacy of treatment may vary according to stage of evolution of the disease process. Emphasizes importance of early treatment intervention.

Fear conditioning to sensory modality-specific stimuli (i.e., auditory, visual) can be mediated by subcortical mechanisms involving sensory pathways that project to the thalamus, amygdala, and modality-specific sensory processing areas of the cortex (8). It has been suggested that emotional memories established via thalamoamygdala pathways may be relatively indelible (142). The fear-potentiated acoustic startle paradigm has been particularly useful for delineation of the mechanisms of fear conditioning, because fear is measured by a change in a simple reflex mediated by a defined neural pathway in the brain stem and spinal cord. This test is sensitive to anxiolytic drugs and is disrupted by anatomical lesions known to affect conditioned fear (143). It may be especially informative regarding PTSD given that many PTSD patients exhibit increased startle responses (144), an abnormality generally not reported in other psychiatric disorders. The central nucleus of the amygdala plays a critical role in the fear-potentiated startle response, because it projects directly to one of the brain stem nuclei necessary for startle (145) and lesions of this pathway block the ability of conditioned or unconditioned fear stimuli to elevate startle (146,147). Neurochemical systems involved in regulation of the fear-potentiated startle response include the noradrenergic, dopaminergic, opiate, NMDA, and corticotropin-releasing systems (4).

Contextual fear conditioning involves more complex stimuli from multiple sensory modalities, and projections to the amygdala from higher order cortical areas that integrate inputs from many sources and the hippocampus may be required (106,109). It is noteworthy that lesions of the hippocampus 1 day after fear conditioning abolish contextual fear. Lesions 7 days or longer after fear conditioning have no effect. These finding suggest that the hippocampus may have a time-limited role in associative fear memories evoked by contextual sensory (polymodal), but not unimodal sensory stimuli (109).

Contextual fear conditioning appears to require exposure to a more aversive UCS than modality-specific fear conditioning. Apparently, as the intensity of the UCS increases, the organism becomes more sensitive to a wider range of stimulus factors in the environment. This finding is consistent with clinical data suggesting that the severity of trauma exposure correlates with the severity of PTSD symptomatology.

Extinction

It is possible that the continued ability of conditioned stimuli to elicit a spectrum of anxiety and fear behaviors results from a deficit in the neural mechanisms involved in response reduction or extinction. Experimental extinction is defined as a loss of a previously learned CER following repeated presentations of a conditioned fear stimulus in the absence of a contiguous traumatic event. Extinction has been explained in terms of either an erasure of the original associations that led to the production of the conditioned response (148) or the acquisition of new associations that compete with or mask the expression of the still intact, response-producing associations (149). The erasure hypothesis predicts that, following nonreinforcement, the response-producing associations no longer exist and, therefore, the conditioned response can no longer be performed. The masking hypothesis predicts that the response-producing associations remain after nonreinforcement and, therefore, if it were possible to temporarily remove the masking associations, the conditioned response could be performed.

Several lines of evidence suggest that the original associations are intact following extinction. Expression of extinction may be specific to the stimulus context in which nonreinforcement occurred (150–152). Representation of the unconditioned stimulus even up to 1 year after extinction is sufficient for reinstating extinguished responding to a preextinction level (153–156). These data indicate the essentially permanent nature of conditioned fear and the apparent fragility of extinction. This phenomenon may help to explain the common clinical observation that traumatic memories may remain dormant for many years, only to be elicited by a subsequent stressor or unexpectedly by a stimulus long ago associated with the original trauma (157,158).

These studies indicate that extinction does not erase the original aversive memory, but instead involves the learning of a new memory that masks or inhibits the original one. It is important to emphasize, however, that although extinction can be overcome, in normal animals extinction does result in a reduction of the conditioned fear response. Using traditional measures of conditioned fear such as freezing, potentiated startle, or autonomic indices, nonreinforcement leads to a reduction in all these measures. In healthy humans, many childhood fears become extinguished and do not intrude daily in adulthood. In contrast, PTSD patients describe persistent traumatic memories that do not extinguish. Thus, it is conceivable that some anxiety disorder patients have deficits in brain systems involved in extinction.

The amygdala is not only involved in the acquisition and expression of conditioned fear responses, but may also be necessary for extinction. The NMDA antagonists infused into the amygdala prevent the extinction of fear-potentiated startle (159). Thus, activity in the amygdala during nonreinforced stimuli presentations may be essential for extinction of conditioned fear stimuli. This may result from processes within the amygdala itself, or via structures that project to the amygdala (e.g., hippocampus, prefrontal cortex, septal area) and have been implicated in extinction in several experimental paradigms. Extinction of conditioned fear responses may represent an active suppression by the cortex of subcortical neural circuits (thalamus, amygdala) that maintain learned associations over long time periods (160).

Behavioral Sensitization

Sensitization generally refers to the increase in behavioral or physiological responsiveness that occurs following repeated exposure to a stimulus. Behavioral sensitization can be generally context dependent or conditioned, such that animals will not demonstrate sensitization if the stimulus is presented in a different environment (161). However, if the intensity of the stimulis or drug dose is high enough, behavioral sensitization will occur even if the environment is changed. It has been suggested that different mechanisms are called into play with environment-independent sensitization (162).

The neurochemical and neuroanatomical systems mediating environment-dependent and environment-independent behavioral sensitization have begun to be investigated. The mechanisms of the development and maintenance of stress-induced sensitization in mammals have been most extensively studied in catecholaminergic systems.

Single or repeated exposure to a stressor potentiates the capacity of a subsequent stressor to increase dopamine function in the forebrain (163,164) without apparently altering basal dopamine turnover (165–167). Recently, it has been shown that the conditioned components of sensitization are related to increased dopamine release in the nucleus accumbens (168). Behavioral sensitization to stress may also involve alterations in noradrenergic function. Animals previously exposed to a stressor exhibit increased norepinephrine release in the hippocampus (169), hypothalamus (169), and prefrontal cortex (170) upon stressor reexposure.

Dopamine D_2 receptor antagonists block the development but not the maintenance of sensitization. Conversely, $alpha_2$ receptor agonists and benzodiazepine agonists block the maintenance but not the development of sensitization (162). In addition, lesions of the amygdala or the nucleus accumbens block the development of cocaine-induced behavioral sensitization. In contrast, lesions of the hippocampus and frontal cortex have no effect (Pert A, Weiss SRB, unpublished observations).

There is recent evidence that both the environment-dependent and independent forms of sensitization may be associated with changes in gene expression. Immediate-early genes (such as c-fos) act as transcription factors influencing the expression of intermediate substances or late effector genes that could modulate neurotransmitter or peptide receptors in a long-lasting fashion. Conditioned increases in c-fos have been shown to occur (171,172).

The characteristics of behavioral sensitization suggest that it may be involved in the signs and symptoms of human anxiety. The PTSD may occur in response to a single precipitating event, or alternatively after repeated traumatic events. Severity of illness is positively correlated with the magnitude of the stressor and history of prior trauma. Finally, the stimuli that evoke intrusive memories, flashbacks, and related symptoms in patients are often difficult to determine, and may bear only a distant association to the initial evoking stimulus.

CONCLUDING COMMENTS

Post-traumatic stress disorder is characterized by intrusive traumatic memories manifested by recurring dreams, flashbacks, and psychological distress following: 1) exposure to events that symbolize or resemble the original trauma; 2) persistent avoidance of stimuli associated with the trauma or a numbing of general responsiveness; and 3) persistent symptoms of increased arousal. The neurobiological basis of some of the symptoms of PTSD can be understood in the context of neural circuits and neural mechanisms of anxiety and fear (173).

Memories and previously learned behaviors influence responses to fear-inducing stimuli via such neural mechanisms as fear conditioning, extinction, and sensitization. Although within the medial temporal lobe memory system, emotional responsiveness (amygdala) and memory (hippocampus) may be separately organized, there is considerable interaction between storage and recall of memory and affect. This is exemplified by the critical role of the amygdala in conditioned fear acquisition, sensitization, extinction, and the attachment of affective significance to neutral stimuli.

The hippocampus and amygdala are sites of convergent reciprocal projections from widespread unimodal and polymodal cortical association areas. It is probably through these interactions, as well as cortical-cortical connections, that memories stored in the cortex, which are continually being reinforced by ongoing experience, are intensified and develop greater coherence (108).

The hippocampal memory system is essential to short-term memory. However, long-term memory storage may be organized such that, as time passes, with subsequent additional retrieval opportunities and the acquisition of related material, the role of the hippocampus diminishes until it may no longer be necessary for memory. The repository of long-term memory may be in the same areas of cortex where the initial sensory impressions take place (174). The shift in memory storage to the cortex may represent a shift from conscious representational memory to unconscious memory processes that indirectly affect behavior.

Therefore, once a fear- or anxiety-inducing sensory stimulis is relayed through the thalamus into neural circuits involving the cortex, hippocampus, and the amygdala, relevant memory traces of past traumatic experiences are stimulated. It is likely that the potency of the cognitive and somatic responses to the stimulis will be strongly correlated with prior experiences due to the strengthening of neural connections within the circuit. These functional neuroanatomical relationships can explain how a single sensory stimulus such as a sight or sound can elicit a specific memory. Moreover, if the sight or sound was associated with a particular traumatic event, a cascade of anxiety- and fear-related symptoms will ensue.

The persistence of intrusive memories may be due to the strength of neuronal interactions between cortical regions where many such memories are stored and subcortical regions, such as the amygdala, which serve to attach affect to the memories. The psychological distress and physiological responses to trauma reminders involve the mechanisms of fear conditioning and extinction. Contextual fear conditioning may be particularly relevant to severe cases of PTSD for whom stimulis generalization is a cardinal feature. The autonomic hyperarousal may be symptoms mediated by brain structures of the efferent arm of the anxiety circuit.

The PTSD is a chronic illness. It has been suggested that the pathophysiology of PTSD should be viewed as a continuously evolving neural process. The disorder is characterized by recurrent panic attacks, flashbacks, conditioned phobic behavior, and illness exacerbation by stressful life events. Through the process of sensitization, these experiences may lead to formation of neural substrates and new vulnerability factors, resulting in autonomous anxiety symptoms that were previously elicited only by environmental triggers (see Chapter 12 and [175]).

The neural circuit and neural mechanisms of anxiety and fear described in this chapter may have implications for the psychotherapy and pharmacotherapy of PTSD. As noted, fear conditioning may contribute to the symptoms of PTSD. Cognitive-behavior therapies have been designed to reverse the impact of fear conditioning. Panic control treatment, which focuses on reversing the effects of fear conditioning on somatic symptoms associated with the efferent arm of the circuit, has been shown to be particularly effective (176,177). Brain imaging studies can now be conducted before and after the therapy to determine which brain regions are involved in its therapeutic effects. Some patients may not respond to such therapies because of an inability to extinguish intrusive memories and maladaptive behaviors. For such patients, therapies need to be developed that are specifically designed to facilitate extinction through the use of conditioned inhibitors and the learning of "new memories."

The concepts discussed in this chapter also suggest potentially fruitful approaches toward development of new therapeutic drugs for PTSD. Given the central role of the amygdala in the neural mechanisms of fear conditioning, extinction, and sensitization, drugs that act on receptors located on amygdaloid neurons and reduce amygdala activity may be effective in reducing the symptoms associated with these mechanisms. Drugs that act selectively on the sensory pathways afferent to the anxiety-fear circuitry (by presynaptically decreasing the release of neurotransmitters or neuropeptides connecting cortical or subcortical afferents) could be effective in blocking conditioned fear. Alternatively, drugs that selectively decrease the function of brain regions in the efferent arm of the circuit may be effective in reducing the signs and symptoms of anxiety.

REFERENCES

1. LeDeux JE. Nervous system V. In Blum F, ed. *Emotion handbook of physiology.* Washington, DC: American Physiological Society; 1987:419–459.
2. Gorman JM, Liebowitz MR, Fyer AJ, Stein J. A neuroanatomical hypothesis for panic disorder. *Am J Psychiatry* 1989;146:148–162.
3. Mraovitch S, Kumada M, Reis DJ. Role of the nucleus parabrachialis in cardiovascular regulation in the cat. *Brain Res* 1982;232:57–75.
4. Davis M. The role of the amygdala in conditioned fear. In Aggleton JP, ed. *The amygdala: neurobiological aspects of emotion, memory and mental dysfunction.* New York: Wiley; 1992:255–305.
5. Cedarbaum JM, Aghajanian GK. Afferent projections to the rat locus coeruleus as determined by a retrograde tracing technique. *J Comp Neurol* 1978;178:1–16.
6. Wallace DM, Magnuson DJ, Gray TS. The amygdalo-brainstem pathway: dopamine, noradrenergic, and adrenergic cells in the rat. *Neurosci Lett* 1989;97:252–258.
7. LeDoux JE, Farb C, Ruggiero DA. Topographic organization of neurons in the acoustic thalamus that project to the amygdala. *J Neurosci* 1990;10:1043–1054.
8. LeDoux JE, Cicchetti P, Xagoraris A, Romanski LM. The lateral amygdaloid nucleus: sensory interface of the amygdala in fear conditioning. *J Neursci* 1990; 10:1062–1069.
9. Aggleton JP, Burton MJ, Passingham RE. Cortical and subcortical afferents to the amygdala of the rhesus monkey (Macaca mulatta). *Brain Res* 1980;190:347–368.
10. Aston-Jones G, Chiang C, Alexinsky T. Discharge of noradrenergic locus coeruleus neurons in behaving rats and monkeys suggests a role in vigilance. In Barnes CD, Pomeiano O, eds. *Progress in brain research.* Amsterdam: Elsevier Science Publishers; 1991:501–519.
11. Aston-Jones G, Ennis M, Pieribone J, Nickel WT, Shipley MT. The brain nucleus locus coeruleus: restricted afferent control over a broad efferent network. *Science* 1986;234:734–737.
12. Kelley AE, Domesick VB, Nauta WJH. The amygdalostriatal projection in the rat—an anatomical study by anterograde and retrograde tracing methods. *Neuroscience* 1982;7:615–630.
13. McDonald AJ. Organization of amygdaloid projections to prefrontal cortex and associated striatum in the rat. *Neuroscience* 1991;44:1–14.
14. McDonald AJ. Topographical organization of amygdaloid projections to the caudatoputamen nucleus accumbens, and related striatal-like areas of the rat brain. *Neuroscience* 1991;44:15–33.
15. Krettek JE, Price JL. Amygdaloid projections to subcortical structures within the basal forebrain and brainstem in the rat and cat. *J Comp Neurol* 1978;178:225–253.
16. Price JL, Amaral DG. An autoradiographic study of the projections of the central nucleus of the monkey amygdala. *J Neurosci* 1981;1:1242–1259.
17. Silverman AJ, Hoffman DL, Zimmerman EA. The descending afferent connections of the paraventricular

nucleus of the hypothalamus (PVN). *Brain Res Bull* 1981;6:47–61.
18. Tribollet E, Dreifuss JJ. Localization of neurones projecting to the hypothalamic paraventricular nucleus of the rat: a horseradish peroxidase study. *Neuroscience* 1981;7:1215–1328.
19. Gray TS, Carney ME, Magnuson DJ. Direct projections from the central amygdaloid nucleus to the hypothalamic paraventricular nucleus: possible role in stress-induced adrenocorticotropin release. *Neuroendocrinology* 1989;50:433–446.
20. Sawchenko PE, Swanson LW. The organization of forebrain afferents to the paraventricular and supraoptic nucleus of the rat. *J Comp Neurol* 1983;218:121–144.
21. Saunders RC, Rosene DL. A comparison of the efferents of the amygdala and the hippocampal formation in the rhesus monkey: I. Convergence in the entorhinal, prorhinal, and perirhinal cortices. *J Comp Neurol* 1988;271:153–184.
22. Saunders RC, Rosene DL, Van Hoesen GW. Comparison of the efferents of the amygdala and the hippocampus formation in the rhesus monkey: II. Reciprocal and nonreciprocal connections. *J Comp Neurol* 1988; 271:185–207.
23. Amaral DG, Price JL. Amygdalo-cortical projections in the monkey (Macaca fascicularis). *J Comp Neurol* 1984;230:465–496.
24. Amaral DG, Price JL, Pitanken A, Carmichael ST. Anatomical organization of the primate amygdala complex. In Aggleton JP, ed. *The amygdala: neurobiological aspects of emotion, memory and mental dysfunction.* New York: Wiley; 1992:1–66.
25. Gloor P. Amygdala. In Field J, ed. *Handbook of physiology: Sec. I. Neurophysiology.* Washington, DC: American Physiological Society; 1960:1395–1420.
26. Applegate CD, Kapp BS, Underwood MD, McNall CL. Autonomic and somatomotor effects of amygdala central n. stimulation in awake rabbits. *Physiol Behav* 1983;31:353–360.
27. Kaada BR. Stimulation and regional ablation of the amygdaloid complex with reference to functional representations. In Eleftheriou BE, ed. *The neurobiology of the amygdala.* New York: Plenum Press; 1972:205–281.
28. Nieminen SA, Sirvio J, Teittinen K, Pitkanen A, Airaksinen MM, Riekkinen P. Amygdala kindling increased fear response, but did not impair spatial memory in rats. *Physiol Behav* 1992;51:845–849.
29. Ohta M. Amygdaloid and cortical facilitation or inhibition of trigeminal motoneurons in the rat. *Brain Res* 1984;291:39–48.
30. Fanardjian VVG, Manvelyan LR. Mechanisms regulating the activity of facial nucleus motoneurons. III. Synaptic influences from the cerebral cortex and subcortical structures. *Neuroscience* 1987;20:835–843.
31. Rolls ET. Neurons in the cortex of the temporal lobe and in the amygdala of the monkey with responses selective for faces. *Hum Neurobiol* 1984;3:209–222.
32. Gary Bobo E, Bonvallet M. Amygdala and masseteric reflex. I. Facilitation, inhibition and diphasic modifications of the reflex, induced by localized amygdaloid stimulation. *Electroencephalogr Clin Neurophysiol* 1975;39:329–339.
33. Pascoe JP, Bradley DJ, Spyer KM. Interactive responses to stimulation of the amygdaloid central nucleus and baroreceptor afferents in the rabbit. *J Auton Nerv Sys* 1989;26:157–167.
34. Whalen PJ, Kapp BS. Contributions of the amygdaloid central nucleus to the modulation of the nictitating membrane reflex in the rabbit. *Behav Neurosci* 1991;105:141–153.
35. Mason JW. Plasma 17-hydroxycorticosteroid levels during electrical stimulation of the amygdaloid complex in conscious monkeys. *Am J Physiol* 1959; 196:44–48.
36. Dunn JD, Whitener J. Plasma corticosterone responses to electrical stimulation of the amygdaloid complex: cytoarchitectural specificity. *Neuroendrocrinology* 1986; 42:211–217.
37. Gloor P. Role of the amygdala in temporal lobe epilepsy. In Aggleton JP, ed. *The Amygdala: Neurobiological Aspects of Emotion, Memory and Mental Dysfunction.* New York: Wiley; 1992:505–538.
38. Chapman WP, Schroeder HR, Geyer G, Brazier MAB, Fager C, Poppen JL, Solomon HC, Yakovlev PI. Physiological evidence concerning importance of amygdaloid nuclear region in the integration of circulatory function and emotion in man. *Science* 1954;120: 949–950.
39. Gloor P, Olivier A, Quesney LF, Andermann F, Horowitz S: The role of the limbic system in experiential phenomena of temporal lobe epilepsy. *Ann Neurol* 1982;12:129–144.
40. Weiskrantz L. Behavioral changes associated with ablation of the amygdaloid complex in monkeys. *J Comp Physiol Psychol* 1956;49:381–391.
41. Jones B, Mishkin M. Limbic lesions and the problem of stimulus-reinforcement associations. *Exp Neurol* 1972;36:362–377.
42. Mishkin M, Aggleton J. Multiple functional contributions of the amygdala in the monkey. In Ben Ari Y, ed. *The amygdaloid complex.* Amsterdam: Elsevier; 1981:409–420.
43. Kling AS, Brothers LA. The amygdala and social behavior. In Aggleton JP, ed. *The amygdala: neurobiological aspects of emotion, memory and mental dysfunction.* New York: Wiley; 1992:353–377.
44. Kling A, Lancaster J, Benitone J. Amygdalextomy in the free-ranging vervet. *J Psychiat Res* 1970;7:191–199.
45. Blanchard DC, Blanchard RJ. Innate and conditioned reactions to threat in rats with amygdaloid lesions. *J Comp Physiol Psychol* 1972;81:281–290.
46. Henke PG. The amygdala and restraint ulcers in rats. *J Comp Physiol Psychol* 1980;94:313–323.
47. Beaulieu S, DiPaolo T, Barden N. Control of ACTH secretion by central nucleus of the amygdala: implication of the serotonergic system and its relevance to the glucocorticoid delayed negative feed-back mechanism. *Neuroendocrinology* 1986;44:247–254.
48. Beaulieu S, DiPaolo T, Cote J, Barden N. Participation of the central amygdaloid nucleus in the response of adrenocorticotropin secretion to immobilization stress: opposing roles of the noradrenergic and dopaminergic systems. *Neuroendocrinology* 1987;45:37–46.
49. LeDoux JE, Iwata J, Cicchetti P, Reis DJ. Different projections of the central amygdaloid nucleus mediate

autonomic and behavioral correlates of conditioned fear. *J Neurosci* 1988;8:2517–2529.
50. LeDoux JE, Cicchetti P, Xagoraris A, Romanski LM. The lateral amygdaloid nucleus: sensory interface of the amygdala in fear conditioning. *J Neurosci* 1990; 10:1062–1069.
51. Boluis JJ, Fitzgerald RE, Dijk DJ, Koolhaas JM. The corticomedial amygdala and learning in an agonistic situation in the rat. *Physiol Behav* 1984;32:575–579.
52. Luiten PGM, Koolhaas JM, deBoer S, Loopmans SJ. The cortico-medial amygdala in the central nervous system organization of agonistic behavior. *Brain Res* 1985;332:283–297.
53. Shibata K, Kataoka Y, Yamashita K, Ueki S. An important role of the central amygdaloid nucleus and mammillary body in the mediation of conflict behavior in rats. *Brain Res* 1986;372:159–162.
54. Sananes CB, Davis M. N-Methyl-D Aspartate lesions of the lateral and basolateral nuclei of the amygdala block fear potentiated startle and shock sensitization of startle. *Behav Neurosci* 1992;106:72–80.
55. LeDoux JE. Emotion and the amygdala. In Aggleton JP, ed. *The amygdala: neurobiological aspects of emotion, memory and mental dysfunction.* New York: Wiley; 1992:339–351.
56. Sanghera MK, Rolls ET, Roper-Hall A. Visual responses of neurons in the dorso-laberal amygdala of the alert monkey. *Exp Neurol* 1979;63:610–626.
57. Wilson FAW, Rolls ET. Neuronal responses related to reinforcement in the primate basal forebrain. *Brain Res* 1990;502:213–231.
58. Wilson FAW, Rolls FT. Learning and memory are reflected in the responses of reinforcement-related neurons in the primate basal forebrain. *J Neurosci* 1990; 10:1254–1267.
59. Rolls ET. Neurophysiology and the functions of the primate amygdala. In Aggleton JP, ed. *The amygdala: neurobiological aspects of emotion, memory and mental dysfunction.* New York: Wiley; 1992:143–165.
60. Leonard CM, Rolls ET, Wilson FAW, Baylis GC. Neurons in the amygdala of the monkey with responses selective for faces. *Behav Brain Res* 1985;15:159–176.
61. Perrett DI, Rolls ET, Caan W. Visual neurons responsive to faces in the monkey temporal cortex. *Exp Brain Res* 1991;47:329–342.
62. Hasselmo ME, Rolls ET, Baylis GC. The role of expression and identity in the face-selective responses of neurons in the temporal visual cortex of the monkey. *Behav Brain Res* 1989;32:203–218.
63. Applegate CD, Frysinger RC, Kapp BS, Gallagher M. Multiple unit activity recorded from amygdala central nucleus during Pavlovian heart rate conditioning in rabbit. *Brain Res* 1982;237:457–562.
64. Henke PG. Unit-activity in the central amygdalar nucleus of rats in response to immobilization-stress. *Brain Res Rev* 1983;10:833–837.
65. Pascoe JP, Kapp BS. Electrophysiological characteristics of amygdaloid central nucleus neurons during Pavlovian fear conditioning in the rabbit. *Behav Brain Res* 1985;16:117–133.
66. Kling AS, Lloyd RL, Perryman KM. Slow wave changes in amygdala to visual, auditory, and social stimuli following lesions of the inferior temporal cortex in squirrel monkey (Saimiri sciureus). *Behav Neural Biol* 1987;47:54–72.
67. Halgren E. Emotional neurophysiology of the amygdala within the context of human cognition. In Aggleton JP, ed. *The amygdala: neurobiological aspects of emotion, memory and mental dysfunction.* New York: Wiley; 1992:191–228.
68. Aston-Jones G, Valentino RJ, Van Bockstaele EJ, Meyerson AT. Locus coeruleus, stress and post traumatic stress disorder: neurobiological and clinical parallels. In Murburg M, ed. *Catecholamine function in post traumatic stress disorder: emerging concepts.* Washington, DC: American Psychiatric Press; in press.
69. Craig AD. Spinal and trigeninal lamina 1 input to the locus coeruleus anterogradely labeled with Phaseolus vulgaris leucoagglutinin (PHA-L) in the cat and monkey. *Brain Res* 1992;584:325–328.
70. McMahon SB, Wall PD. Electrophysiological mapping of brain stem projections of spinal cord lamina 1 cells in the rat. *Brain Res* 1985;338:19–26.
71. Cedarbaum JM, Aghajanian GK. Afferent projections to the rat locus coeruleus as determined by a retrograde tracing technique. *J Comp Neurol* 1978;178:1–16.
72. Aston-Jones G, Ennis M, Pieribone J, Nickel WT, Shipley MT. The brain nucleus locus coeruleus: restricted afferent control over a broad efferent network. *Science* 1986;234:734–737.
73. Craig AD, Linington AJ, Kniftki K-D. Significant differences in retrograde labeling of spinothalamic tract celles by horseradish peroxidase and fluorescent tracers fast blue and diamidino yellow. *Exp Brain Res* 1989; 74:431–436.
74. Luppi PH, Aston-Jones G, Akaoka H, Charlety P, Kovelowski C, Shipley MT, Zhu Y, Ennis M, Fort P, Chouvet G, Touvet M. Afferents to the rat locus coeruleus using choleteratoxin B subunit as a retrograde tracer. *Neurosci Abst* 1991;17:1540.
75. Rasmussen K, Aghajanian GK. Failure to block responses to locus coeruleus neurons to somatosensory stimuli by destruction of two major afferent nucli. *Synapse* 1989;4:162–164.
76. Van Bockstaele EJ, Pieribone VA, Aston-Jones G. Diverse afferents converge on the nucleus paragigantocellularis in the rat ventrolateral medulla: retrograde and anterograde tracing studies. *J Comp Neurol* 1989;290:561–584.
77. Dunn AJ, Berridge CW. Corticotropin-releasing factor administration elicits a stress like activation of cerebral catecholaminergic systems. *Pharmacol Biochem Behav* 1987;27(4):685–691.
78. Valentino RJ, Foote SL. Corticotropin-releasing hormone increases tonic but not sensory-evoked activity of noradrenergic locus coeruleus neurons in unanesthetized rats. *J Neurosci* 1988;8:1016–1025.
79. Chappell PB, Smith MA, Kilts CD, Bissette G, Ritchie J, Anderson C, Nemeroff CB. Alterations in corticotropin-releasing factor-like immunoreactivity in discrete rat brain regions after acute and chronic stress. *J Neurosci* 1990;6:2908–2914.
80. Butler PD, Weiss JM, Stout JC, Nemeroff CB. Corticotropin-releasing factor produces fear-enhancing and behavioral activating effects following infusion into the locus coeruleus. *J Neurosci* 1990;10(1):176–183.
81. Justic A, Feldman SM, Brown LL. The nucleus locus coeruleus modulates local cerebral glucose utilization during noise stress in rats. *Brain Res* 1989;490:73–84.
82. Savaki HE, Kadekaro M, McCulloch J, Sokoloff L.

The central noradrenergic system in the rat: metabolic mapping with alpha adrenergic blocking agents. *Brain Res* 1982;234:65–79.
83. Redmond DE Jr. Studies of the nucleus locus coeruleus in monkeys and hypotheses for neuropsychopharmacology. In Meltzer HY, ed. *Psychopharmacology: the third generation of progress.* New York: Raven Press; 1987:467–974.
84. Redmond DE, Huang YH. New evidence for a locus coeruleus norepinephrine connection with anxiety. *Life Sci* 1979;25:2149–2162.
85. Charney DS, Woods SW, Price LH, Goodman WK, Glazer WM, Heninger GR: Noradrenergic dysregulation in panic disorder. In Ballenger JC, ed. *Neurobiology of panic disorders.* New York: Alan R. Liss; 1990:91–105.
86. Charney DS, Heninger GR, Breier A. Noradrenergic function in panic anxiety: effects of yohimbine in healthy subjects and patients with agoraphobia and panic disorder. *Arch Gen Psychiatry* 1984;41:751–763.
87. Charney DS, Woods SW, Goodman WK, et al. Neurobiological mechanisms of panic anxiety: biochemical and behavioral correlates of yohimbine-induced panic attacks. *Am J Psychiatry* 1987;144:1030–1036.
88. Charney DS, Woods SW, Krystal JH, Nagy LM, Heninger GR. Noradrenergic neuronal dysregulation in panic disorder: the effects of intravenous yohimbine and clonidine in panic disorder patients. *Acta Psychiatrica Scand* 1992;86:273–282.
89. Nashold BS, Wilson WP, Slaughter G. *Advances in Neurology,* vol. 4. New York, Raven Press; 1974.
90. Kaitin KI, Bliwise DL, Gleason C, Nino-Murcia G, Dement WC, Libet B. Sleep disturbance produced by electrical stimulation of the locus coeruleus in a human subject. *Biol Psychiatry* 1986;21:710–716.
91. Verleye M, Bernet F. Behavioral effects of lesions of the central noradrenergic bundles in the rat. *Pharmacol Biochem Behav* 1983;19:407–414.
92. McNaughton N, Mason ST. The neuropsychology and neuropharmacology of the dorsal ascending noradrenergic bundle—a review. *Prog Neurobiology* 1980;14: 157–219.
93. Koob GF, Thatcher-Britton K, Britton DR, Roberts DCS, Bloom FE. Destruction of the locus coeruleus or the dorsal NE bundle does not alter the release of punished responding by ethanol or chlordiazepoxide. *Physiol Behav* 1984;33:479–485.
94. Abercrombie ED, Jacobs BL. Single-unit response of noradrenergic neurons in the locus coeruleus of freely moving cats. I. Acutely presented stressful and nonstressful stimuli. *J Neurosci* 1987;7:2837–2843.
95. Abercrombie ED, Jacobs BL. Single-unit response of noradrenergic neurons in the locus coeruleus of freely moving cats. II. Adaptation to chronically presented stressful stimuli. *J Neurosci* 1987;7:2844–2848.
96. Rasmussen K, Morilak DA, Jacobs BL. Single unit activity of locus coeruleus neurons in the freely moving cat. I. During naturalistic behaviors and in response to simple and complex stimuli. *Brain Res* 1986;371(2): 324–334.
97. Levine ES, Litto WJ, Jacobs BL. Activity of cat locus coeruleus noradrenergic neurons during the defense reaction. *Brain Res* 1990;531:189–195.
98. Svensson TH. Peripheral, autonomic regulation of locus coeruleus noradrenergic neurons in brain: putative implications for psychiatry and psychopharmacology. *Psychopharmacology* 1988;92:1–7.
99. Morilak DA, Fornal CA, Jacobs BL. Effects of physiological manipulations on locus coeruleus neuronal activity in freely moving cats. I. Thermoregulatory challenge. *Brain Res* 1988;422:17–23.
100. Morilak DA, Fornal CA, Jacobs BL. Effects of physiological manipulations on locus coeruleus neuronal activity in freely moving cats. II. Cardiovascular challenge. *Brain Res* 1988;422:24–31.
101. Morilak DA, Fornal CA, Jacobs BL. Effects of physiological manipulations on locus coeruleus neuronal activity in freely moving cats. III. Glucoregulatory challenge. *Brain Res* 1988;422:32–39.
102. LeDoux JE, Iwata J, Pearl D. Disruption of auditory but not visual learning by destruction of intrinsic neurons in the medial geniculate body of the rat. *Brain Res* 1986;371:395–399.
103. LeDoux JE, Sakaguchi A, Iwata J. Interruption of projections from the medial geniculate body to an archineostriatal field disrupts the classical conditioning of emotional responses to acoustic stimuli in the rat. *Neurosci* 1986;17:615–627.
104. Iwata J, LeDoux JE, Meeley MP, Arneric S, Reis DJ. Intrinsic neurons in the amygdaloid field projected to by the medial geniculate body mediate emotional responses conditioned to acoustic stimuli. *Brain Res* 1986;383:195–214.
105. Gray JA. Precis of "The neuropsychology of anxiety: an inquiry into the functions of the septo-hippocampal system." *Behav Brain Sci* 1982;5:469–543.
106. Phillips RG, LeDoux JE. Differential contribution of amygdala and hippocampus to cued and contextual fear conditioning. *Behav Neurosci* 1992;106:274–285.
107. Zola-Morgan S, Squire LR, Alvarez-Royo P, Clower RP. Independence of memory functions and emotional behavior: separate contributions of the hippocampal formation and the amygdala. *Hippocampus* 1991;1: 207–220.
108. Squire LR. Memory and the hippocampus: a synthesis from findings with rats, monkeys, and humans. *Psychol Rev* 1992;99:195–231.
109. Kim J, Fanselow M. Modality-specific retrograde amnesia of fear. *Science* 1992;256:675–677.
110. Eichenbaum H, Otto T, Cohen JN. The hippocampus—What does it do? *Behav Neural Biol* 1992;57: 2–36.
111. Zola-Morgan S, Squire LR, Clower RP, Rempel NL: Damage to the perirhinal cortex exacerbate memory impairment following lesions to the hippocampal formation. *J Neurosci* 1993;13:251–265.
112. Lynch G, Granger R. Variations in synaptic plasticity and types of memory in corticohippocampal networks. *J Cog Neurosci* 1992;4:189–199.
113. Swanson LW. The hypothalamus. In Bjorklund A, Hokfelt T, Swanson LW, eds. *Handbook of chemical neuroanatomy. Integrated systems of the CNS, Part I, hypothalamus, hippocampus, amygdala and retina.* North-Holland, Amsterdam: Elsevier; 1987:1–124.
114. Yokoo H, Tanaka M, Yoshida M, Tsuda A, Tanaka T, Mizoguchi K. Direct evidence of conditioned fear-elicited enhancement of noradrenaline release in the rat hypothalamus assessed by intracranial microdialysis. *Brain Res* 1990;536:305–308.

115. Pacak K, Armando I, Fukuharn K, Kvetnansky R, Palkovitz M, Kopin IJ, Goldstein DS. Noradrenergic activity in the paraventricular nucleus during acute and chronic immobilization stress in rats: an in vivo microdialysis study. *Brain Res* 1993;589(1):91–96.
116. Lookingland KJ, Ireland LM, Gunnet JW, Manzanares J, Tian Y, Moore KE. 3-Methoxy-4-hydroxyphenethylenglycol concentrations in discrete hypothalamic nuclei reflect the activity of noradrenergic neurons. *Brain Res* 1991;559:82–88.
117. Imaki T, Shibasaki T, Hotta M, Demura H. Early induction of c-fos precedes increased expression of corticotropin-releasing factor messenger ribonucleic acid in the paraventricular nucleus after immobilization stress. *Endocrinology* 1992;131:240–246.
118. Gray TS, Carney ME, Magnuson DJ. Direct projections from the central amygdaloid nucleus to the hypothalamic paraventricular nucleus: possible role in stress induced adrenocorticotropin release. *Neuroendocrinology* 1989;50:433–446.
119. Haas DA, George SA. Single or repeated mild stress increases synthesis and release of hypothalamic corticotropin-releasing factor. *Brain Res* 1988;46:230–237.
120. Plotsky PM, Meaney MJ. Early, postnatal experience alters hypothalamic corticotropin releasing factor (CRF) mRNA, median eminence CRF content and stress-induced release in adult rats. *Molec Brain Res* 1993;18:195–200.
121. Miaskowski C, Ong GL, Lukic D, Haldar J. Immobilization stress affects oxylocin and vasopressin levels in hypothalamic and extrahypothalamic sites. *Brain Res* 1988;458:137–141.
122. deGoeij CED, Jezova D, Tilders FJH. Repeated stress enhances vasopressin synthesis in corticotropin releasing factor neurons in the paraventricular nucleus. *Brain Res* 1992;577:165–168.
123. Young WS, Lightman SL. Chronic stress elevates enkephalin expression in the rat paraventricular and synaptic nuclei. *Molec Brain Res* 1992;13:111–117.
124. Bartanusz V, Jezova D, Bertini LT, Tilders FJH, Aubry JM, Kiss JZ. Stress induced release in vasopressin and corticotropin releasing factor expression in hypophysiotropic paraventricular neurons. *Endocrinology* 1993;132:895–902.
125. Siegel RA, Dukes EM, Pahnke U, Wuttke W. Stress induced changes in cholecystokinin and substance P concentrations in discrete regions of rat hypothalamus. *Neuroendocrinology* 1987;46:75–81.
126. Amaral DG, Price JL. Amygdalo-cortical projections in the monkey (Macaca fascicularis). *J Comb Neurol* 1984;230:465–496.
127. Fuster JM. The prefrontal cortex. Anatomy, physiology, and neuropsychology of the frontal lobe. New York: Raven Press; 1989.
128. Goldman-Rakic PS, Porrino LJ. The primate mediodorsal (MD) nucleus and its projection to the frontal lobe. *J Comp Neurol* 1985;242:535–360.
129. Vogt BA, Pandya DN. Cingulate cortex of the rhesus monkey: II. Cortical afferents. *J Comp Neurol* 1987;262:271–289.
130. Aggleton JP, Burton MJ, Passingham RE. Cortical and subcortical afferents to the amygdala of the rhesus monkey (Macaca mulatta). *Brain Res* 1989;190:347–368.
131. Baleydier C, Mauguire F. The duality of the cingulate gyrus in monkey. Neuroanatomical study and functional hypothesis. *Brain* 1980;103:525–554.
132. Goldman-Rakic PS, Selemon LD, Schwartz ML. Dual pathways connecting the dorsolateral prefrontal cortex with the hippocampal formation and parahippocampal cortex in the rhesus monkey. *Neuroscience* 1984;12:719–743.
133. Thorpe SJ, Rolls ET, Maddison S. The orbitofrontal cortex: Neuronal activity in the behaving monkey. *Exp Brain Res* 1983;49:93–115.
134. Milner B, Petrides M. Behavioral effects of frontal lobe lesions in man. *Trends Neurosci* 1984;7:403–407.
135. Petrides M. Frontal lobes and memory. In Bolller F, Grafman J, eds. *Handbook of neuropsychology,* vol. 3. Amsterdam: Elsevier; 1989:75–90.
136. Nauta WJH. The problem of the frontal lobe: a reinterpretation. *J Psychiatr Res* 1971;8:167–187.
137. DeBruin JPC. Social behavior and the prefrontal cortex. In Uylings HPM, Van Eden CG, DeBrun JPC, Corner MA, Feenstra MGP, eds. *Progress in brain research,* vol. 85. Amsterdam: Elsevier; 1990:485–497.
138. Litz BT, Keane TM. Information processing in anxiety disorders: Application to the understanding of posttraumatic stress disorder. *Clin Psychol Rev* 1989;9:243–257.
139. McNally RJ, Luedke DL, Besyner JK, Peterson RA, Bohm K, Lips OJ. Sensitivity to stress relevant stimuli in post traumatic stress disorder. *J Anxiety Disord* 1987;1:105–116.
140. Hoffman HS, Selekman W, Fleishler M. Stimulus aspects of aversive controls: long-terms effects of suppression procedures. *J Exp Anal Behav* 1966;9:659–662.
141. Davis M. Animal models of anxiety based upon classical conditioning: the conditioned emotional response and fear potentiated startle effect. *Pharmacol Ther* 1990;47:147–165.
142. LeDoux JE, Romanski L, Xagoraris A. Indelibility of subcortical emotional memories. *J Cognitive Neurosci* 1989;1:238–243.
143. Davis M. Pharmacological and anatomical analysis of fear conditioning using the fear-potentiated startle paradigm. *Behav Neurosci* 1986;100:814–824.
144. Butler RW, Braff DL, Rausch JL, Jenkins MA, Sprock J, Geyer MA. Physiological evidence of exaggerated startle response in a subgroup of Vietnam veterans with combat-related PTSD. *Am J Psychiatry* 1990;147:1308–1312.
145. Rosen JB, Hitchcock JM, Sananes CB, Miserendino MJD, Davis M. A direct projection from the central nucleus of the amygdala to the acoustic startle pathway: anterograde and retrograde tracing studies. *Behav Neurosci* 1991;105:817–825.
146. Hitchcock JM, Davis M. Efferent pathway of the amygdala involved in conditioned fear as measured with the fear-potentiated startle paradigm. *Behav Neurosci* 1991;105:826–842.
147. Hitchcock JM, Sananes CB, Davis M. Sensitization of the startle reflex by footshock: blockade by lesions of the central nucleus of the amygdala or its efferent pathway to the brainstem. *Behav Neurosci* 1989;103:509–518.

148. Estes WK. Statistical theory of spontaneous recovery and regression. *Psychol Rev* 1955;62:145–154.
149. Konorski J. *Conditioned reflexes and neuronal organization.* London: Cambridge University Press; 1948.
150. Bouton ME, Bulles RC. Contexts, event memories, and extinction. In Balsam PDD, Tomic KA, eds. *Context and Learning.* Hillsdale, NJ: Lawrence Erlbaum Associates; 1985:133–166.
151. Bouton ME, King DA. Contextual control of conditioned fear: tests for the associative value of the context. *J Exp Psychol Anim Behav* 1983;9:248–256.
152. Bouton ME, King DA. Effect of context with mixed histories of reinforcement and nonreinforcement. *J Exp Psychol Anim Behav* 1986;12:4–15.
153. Bouton ME, Bolles RC. Role of contextual stimuli in reinstatement of extinguished fear. *J Exp Psychol Anim Behav* 1979;5:368–378.
154. Pavlov JP. *Conditioned reflexes.* Oxford, England: Oxford University Press; 1927.
155. Rescola RA, Heth CD. Reinstatement of fear to extinguished conditioned stimulis. *J Exp Psychol Anim Behav* 1975;104:88–96.
156. McAllister WR, McAllister DE. Reconditioning of extinguished fear after a one-year delay. *Bull Psychonomic Soc* 1988;26:463–466.
157. VanDyke C, Zilberg NJ, MacKinnon JA. Post traumatic stress disorder: a thirty year delay in a World War II veteran. *Am J Psychiatry* 1985;142:1070–1073.
158. Solomen Z, Garb K, Bleich A, Grupper D. Reactivation of combat related post traumatic stress disorder. *Am J Psychiatry* 1987;144:51–55.
159. Falls WA, Miserendino MJD, Davis M. Excitatory amino acid antagonists infused into the amygdala block extinction of fear-potentiated startle. *Soc Neurosci Abstr* 1990;16:767.
160. Teich AH, McCabe PM, Gentile CC, Schneideman LS, Winters RW, Kikowsky DR, Schneideiman N. Auditory cortex lesions prevent extinction of Pavlovian differential heart rate conditioning to tonal stimuli in rabbits. *Brain Res* 1989;480:210–218.
161. Post RM. Transduction of psychosocial stress into the neurobiology of recurrent affective disorders. *Am J Psychiatry* 1992;149:999–1010.
162. Weiss SRB, Post RM, Pert A, Woodward R, Murman D. Context-dependent cocaine sensitization: differential effect of haloperidol on development versus expression. *Pharmacol Biochem Behav* 1989;34:655–661.
163. Caggiula AR, Antelman SM, Aul E, Knopf S, Edwards DJ. Prior stress attenuates the analgesic response but sensitizes the corticosterone and cortical dopamine responses to stress 10 days later. *Psychopharmacology* 1989;99:233–237.
164. Kalivas PW, Duffy P. Similar effects of daily cocaine and stress on mesocorticolimbic dopamine neurotransmission in the rat. *Biol Psychiatry* 1989;25:913–928.
165. Kalivas PW, Duffy P, Abhold R, Dilts RP. Sensitization of mesolimbic dopamine neurons by neuropeptides and stress. In Kalivas PW, Barnes CD, eds. *Sensitization in the nervous system.* Caldwell, NJ: Telford Press; 1990:119–124.
166. Kalivas PW, Striplin CD, Steketee JD, Klitenick MA, Duffy P. Cellular mechanisms of behavioral sensitization to drugs of abuse. *Ann NY Acad Sci* 1992; 654:128–135.
167. Criswell HE, Mueller RA, Breese GR. Long-term D_1-dopamine receptor sensitization in neonatal 6-OHD-lesioned rats is blocked by an NMDA antagonist. *Brain Res* 1990;512:284–290.
168. Fontana DJ, Post RM, Pert A. Conditioned increase in mesolimbic dopamine overflow by stimuli associated with cocaine. *Brain Res,* in press.
169. Nisenbaum LK, Zigmand MJ, Sved AF, Abercrombie ED. Prior exposure to chronic stress results in enhanced synthesis and release of hippocampal norepinephrine in response to a novel stressor. *J Neurosci* 1991; 11:1478–1484.
170. Finlay JM, Abercrombie ED. Stress induced sensitization of norepinephrine release in the medial prefrontal cortex. *Soc Neurosci Abstr* 1991;17:151.
171. Campean S, Hayward MD, Hope BT, Rosen JB, Nestler EJ, Davis M. Induction of the c-FOS protooncogene in rat amygdala during unconditioned and conditioned fear. *Brain Res* 1991;565:349–352.
172. Smith MA, Banerjee S, Gold PW, Glowa J. Induction of c-fos in mRNA in rat brain by conditioned and unconditioned stressors. *Brain Res* 1992;578:135–141.
173. Charney DS, Deutch AY, Krystal JH, Southwick SM, Davis M. Psychobiological mechanisms of post traumatic stress disorder. *Arch Gen Psychiatry* 1993; 50:294–304.
174. Squire LR, Zola-Morgan S. The medial temporal lobe memory system. *Science* 1991;253:2380–2386.
175. Post RM, Weiss SRB, Uhde TW, Clark M, Rosen JB. Preclinical neuroscience advances pertinent to panic disorder: implications of cocaine kindling, induction of the proto-oncogene c-fos, and contingent tolerance. In Hoehn-Saric R, ed. *Biology of anxiety disorders: recent developments.* Washington, DC: APA Press; 1993:129–187.
176. Craske MG, Cerny JA, Barlow DH. Behavioral treatment of panic disorder. *Behav Res Ther* 1989;20:261–282.
177. Klosko JS, Barlow DH. A comparison of alprazolam and behavioral therapy. *J Consult Clin Psychotherapy* 1990;58:77–84.

PART III

Adaptive and Maladaptive Responses to Stress in Humans

The seven chapters in this section move from laboratory to clinical research and suggest that there are similarities between the animal and human responses to stress with respect to both the catecholaminergic and hypothalamic-pituitary-adrenergic (HPA) systems. Humans exposed to traumatic stress who develop post-traumatic stress disorder (PTSD) exhibit distinct abnormalities in sympathetic nervous system responsivity, as reviewed in Chapter 16 by Prins and associates. They also show alterations in pre- and post-synaptic adrenergic function, as discussed by Southwick and colleagues in Chapter 18. Yehuda and associates (Chapter 19) show that PTSD patients exhibit a robust allostatic equilibrium with regard to HPA function that may be distinct to this disorder. Although there are clear parallels between animal and human research findings regarding catecholaminergic and HPA function, there are also important differences. Such differences may reflect differences between moderate versus traumatic stress and/or between acute versus chronic stress.

Chapter 20 by Mason and associates focuses on profound alterations in thyroid function that appear to be unique to PTSD. Woodward's review of sleep research on PTSD patients in Chapter 17 shows that, although disturbed sleep and dreaming have long been recognized as hallmarks of traumatic stress, it has proven quite difficult to identify specific polysomnographic abnormalities among PTSD patients.

Chapters 21 and 22 focus on important findings in human stress research that have yet to be explored in PTSD patients. Heninger (Chapter 21) reviews the psychoimmunologic literature and shows that a variety of immunologic deficits can be detected following exposure to stress. Williams (Chapter 22) broadens this focus to describe a number of psychological as well as biological abnormalities observed among stressed humans that appear to make them especially vulnerable to somatic problems. If these findings can be replicated in humans exposed to traumatic stress, it would indicate that PTSD patients are especially vulnerable to medical problems as a result of these stress-induced psychobiological abnormalities.

It is important to note that, when comparing animal with human investigations, there are many neurobiological systems known to be altered by stress in animals that have received little if any attention in research on humans with PTSD. We hope that future researchers will investigate these unexplored areas.

Neurobiological and Clinical Consequences of Stress: From Normal Adaptation to PTSD, edited by M. J. Friedman, D. S. Charney, and A. Y. Deutch, Lippincott-Raven Publishers, Philadelphia © 1995.

16

Psychophysiological Evidence for Autonomic Arousal and Startle in Traumatized Adult Populations

*Annabel Prins, †Danny G. Kaloupek, and †Terrence M. Keane

*Department of Psychology, *Smith College, Northampton, Massachusetts 01060; and †Tufts University School of Medicine, National Center for Post-Traumatic Stress Disorder, Boston Veterans Affairs Medical Center, Boston, Massachusetts 02130*

The psychological impact of severe trauma, especially combat trauma, has a long literary, clinical, and scientific history (1,2). Clinically recognized combat syndromes such as irritable heart syndrome, shell shock, operational fatigue, and combat exhaustion have each included changes in physiological functioning as central to the syndrome (2). Similarly, noncombat syndromes like schreckneuroses following natural disasters (3) and railroad spine following life-threatening transportation accidents (4) have also included changes in somatic and autonomic activity as key features. More recently, psychophysiological changes associated with rape-trauma syndrome have been reported (5). Indeed, a close examination of all of these trauma syndromes reveals important similarities with regard to symptoms of autonomic hyperarousal (e.g., sleep disturbance, irritability, hypervigilance, and exaggerated startle responses). Today, these cardinal features of trauma are retained in the D symptom cluster of Post-Traumatic Stress Disorder (PTSD).

Until recently, physiological reactivity to reminders of the trauma was also included in the hyperarousal symptom cluster of PTSD. In DSM IV, this symptom has been moved to the reexperiencing symptom cluster because of its phasic features shared with the symptom of psychological distress following trauma-related cues (6). This change in emphasis is consistent with contemporary views of human stress responses that emphasize the importance of stressor meaning in mediating psychophysiological reactivity (7).

Psychophysiological studies of stress responses in humans, particularly those with PTSD, can be distinguished from other biological studies of stress by their reliance on noninvasive measurement procedures. Typically, bioelectrical signals generated from the body are conducted by surface electrodes and transducers. The signals are then filtered, amplified, and recorded for later interpretation of their biological or psychological significance. Biological interpretations tend to reflect a pathophysiological approach (8) where the underlying biological processes are thought to have direct or primary causal significance for the disorder. Psychological interpretations, on the other hand, tend to focus on the associations between physiological activity and psychological events or processes. This psychophysiological approach (8) is reflected in the majority of studies that have examined physiological changes after trauma-relevant stimuli in individuals with PTSD. Conversely, the pathophysiological approach is more commonly reflected in studies that have examined exaggerated startle responses in individuals with PTSD.

In addition to examining phasic responses to trauma-relevant stimuli and startle probes, psychophysiological studies of PTSD have examined baseline differences between individuals with and without the diagnosis. The measures most frequently examined include: electrodermal activity, especially skin conductance (SC) levels and responses; heart rate (HR); systolic and diastolic blood pressure (BP); electromyographic (EMG) activity of the frontalis, corrugator, zygomaticus or orbicularis oculi muscles; and skin temperature. Several studies have attempted to identify physiological markers that relate to the diagnosis, while others have used psychophysiological measures as dependent variables to document treatment effectiveness (9,10,11,12). Thus, psychophysiological measures can facilitate diagnosis, provide measures of hyperarousal and reactivity that do not rely on self-report, and supply useful information for evaluating treatment effectiveness.

As a caveat, we note that psychophysiological assessment often has an elevated image as an objective means for detecting an individual's true psychological state. This is in part due to the use of polygraphy in forensic settings for lie detection. It is important to counteract this misconception by pointing out that psychophysiological measurement is only one of several methods for studying the nature of disorders such as PTSD. It can provide some unique information, at one level of scientific analysis, but it is not inherently more valid nor objective than other methods. On the other hand, psychophysiological methods are well suited to the investigation of autonomic hyperarousal and exaggerated startle in PTSD.

In this chapter, we will first review laboratory studies that have presented representations of traumatic events to adults with and without PTSD. Neither biological challenge studies nor studies that have examined the psychophysiological impact of talking about a trauma will be covered. However, in keeping with the theme of this book, we will examine potential differences in baseline levels of arousal and psychophysiological responding to other types of laboratory stressors. This broader evidence is important because it offers a perspective on two competing hypotheses. Specifically, it can help differentiate between the hypothesis that PTSD is characterized by biological changes that cause sustained increases in autonomic nervous system (ANS) activity, and the hypothesis that PTSD is characterized by psychological changes that lead to reactive increases in ANS activity only when trauma-relevant stimuli are encountered.

For the purposes of the review, physiological activity will be characterized as autonomic activity even though the effects in question might be due to sympathetic activation, parasympathetic withdrawal, or both. Psychophysiological studies of PTSD have not addressed this distinction and, in fact, most do not even appear to recognize that these two ANS branches have independent and complementary effects, as well as their better-known reciprocal effects (13). Our review will focus on HR and SC because they are the most widely used measures of general ANS activity and appear to be the best discriminators between individuals with and without the PTSD diagnosis (14,15).

Finally, we will consider the literature on exaggerated startle responses in adults with and without PTSD. We will focus on orbicularis oculi EMG activity and eyeblink responses because the startle reflex is principally skeletomuscular in nature. Secondarily, the ANS measures of HR and SC will be examined as adjunctive indices of arousal. Following this review, we will address the need to consider individual differences in reactivity, as well as situational and methodological variables influencing psychophysiological responding in future research on the psychophysiology of PTSD.

EMPIRICAL STUDIES INVOLVING EXPOSURE TO TRAUMA-RELATED AND GENERIC STRESSORS

Research on the psychophysiology of trauma appeared in the literature as early as the 1940s with Kardiner (16), who observed that individuals with "chronic war neurosis" were particularly sensitive to stimuli associated with the original trauma and reacted to these stimuli with distinctive patterns of psychophysiological responding. Gillespie (17) also observed increased tonic levels of muscle tension and palpitations in traumatized war veterans. These clinical impressions were later corroborated by controlled laboratory studies.

Tonic differences in arousal were documented by Wenger (18), who found that 225 patients diagnosed with "operational fatigue" demonstrated higher resting heart rates and SC levels as compared with 98 "psychoneurotics" and 448 control subjects. Dobbs and Wilson (19) were the first to conduct a controlled experiment on psychophysiological responses to combat sounds. Eight "decompensated" WWII veterans were compared with 13 "compensated" WWII veterans (i.e., with no signs of psychiatric impairment) and 10 noncombat control subjects. Consistent with the findings of Wenger, Dobbs and Wilson found higher mean baseline HR in both the compensated (78.3 bpm) and decompensated (79.4 bpm) groups as compared with the control group (66.5 bpm). Although measurement artifact due to behavioral agitation in the "decompensated" subjects prevented meaningful comparisons involving their data, Dobbs and Wilson also found statistically greater increases in HR during exposure to combat sounds for the "compensated" group (M=6.1 bpm) as compared to the non-combat controls (M=2.5 bpm).

To date, the majority of studies on the psychophysiology of trauma have employed Vietnam combat veterans as subjects. This may be due to the unique availability within the Veterans Administration (VA) system of large numbers of individuals who have suffered similar traumas, as well as the need for the VA to provide comprehensive assessments for individuals seeking compensation for psychiatric disability. More recently, similar assessment protocols have been applied to other trauma groups (e.g., sexual assault survivors, motor vehicle accident victims) with comparable success.

Psychophysiological assessments typically begin with a resting baseline period, the duration of which has varied across studies from 30 seconds (following a 3-minute relaxation tape) to 30 minutes. Tonic levels of arousal are usually obtained during this period. Next, the subject is exposed to either neutral (e.g., music) or trauma-related stimuli. Assessment stimuli have been presented in one of three ways[1]. Some researchers have used audiotapes of neutral or trauma-relevant sounds, while others have employed visual images (i.e., slides and film) with an accompanying sound track. Still others have presented subjects with scripts describing neutral and idiographic trauma-related experiences. Most often these scripts are read to subjects who are instructed to imagine the scene as clearly and vividly as possible.

The duration of exposure to neutral and trauma-related stimuli also has varied from one 30-second period to 30 minutes. Although within-subject psychophysiological reactivity is most frequently expressed as a difference score from an immediately preceding baseline, in several studies the psychophysiological reactivity index has been defined more conservatively by determining the difference between presentation periods for the trauma-related stimuli and the neutral stimuli. Exposure to generic (i.e., not related to trauma) stressors such as mental arithmetic or imagery with nontraumatic content has also been included in some protocols as an additional source of within-subject comparison. Most often, the between-group comparison involves psychophysiological responses of PTSD subjects relative to a similar group of trauma survivors without PTSD.

BASELINE LEVELS OF AUTONOMIC AROUSAL

Group comparisons with respect to baseline levels of arousal are presented in Table 1. Significant differences in baseline heart rate were reported in 5 out of the 13 studies summarized. The PTSD group had higher resting heart rate than the non-PTSD group in all five studies. Only one study reported significant differences in baseline SC levels (21), and in this instance

[1]A fourth type of trauma-related stimulus presentation was recently reported by McCaffrey et al. (20). These investigators found significant electroencephalogram (EEG) changes following exposure to trauma-related odors (e.g., odors simulating burning flesh) in five Vietnam veterans with PTSD as compared with five Vietnam veterans without PTSD. Unfortunately, McCaffrey et al. (1993) did not collect data on either HR or SC.

TABLE 1. *Baseline levels of heart rate, skin conductance and subjective distress in PTSD*

Study	Groups	Baseline length	Baseline skin conductance	Baseline heart rate	Power	Baseline subjective distress
Blanchard et al. (23)	1. VN–PTSD (11) 2. Non-vets–NMD (11)	10-minute adaptation + 10-minute baseline	values not reported - ns[a]	1. 77.5 2. 70.2*[b]	.39[c]	not collected
Malloy et al. (26)	1. VN–PTSD (10) 2. VN–NMD (10)	5-minute baseline	SC values not obtained	1. 94.0 (15.3) 2. 84.5 (17.2) ns[d]	.35	not reported
Pallmeyer et al. (37)	1. VN–PTSD (12) 2. VN–NMD (10)	12-minute adaptation + 2-minute baseline	values not reported	1. 81.0 (10.7) 2. 69.5 (10.6)*[d]	.79	not collected
Blanchard et al. (110)	1. VN–PTSD (57) 2. VN–NMD (34)	12-minute adaptation + 2-minute baseline	values not reported	1. 78.8 (14.6) 2. 65.2 (12.7)*[d]	1.00	not collected
Pitman et al. (15)	1. VN–PTSD (18) 2. VN–NMD (15)	3-minute relaxation + 30-second baseline	1. 1.45 (1.0) 2. 1.34 (0.6) ns	1. 75.3 (10.6) 2. 66.3 (8.5)*[d]	.84	not reported
Gerardi et al. (111)	1. VN–PTSD (18) 2. VN–Non-PTSD (18)	≈8-minute adaptation + 10-minute baseline	1. .22 (.12) 2. .15 (.14) ns[d]	1. 76.1 (15.5) 2. 63.0 (13.9)*[d]	.83	not collected
Blanchard et al. (94)	1. VN–PTSD (59) 2. VN–NMD (12)	12-minute adaptation + 2-minute baseline	values not reported	1. 72.9 (12.9) 2. 63.3 (11.3) ns[e]	.79	not collected
McFall et al. (22)	1. VN–PTSD (10) 2. mixed non-PTSD (11)	30-minute rest + 30-minute baseline	not collected	1. 63.2 (2.2) 2. 63.5 (3.5) ns[f]	.08	negative affect: (0–68 scale) 1. 12.1 (3.1) 2. 4.7 (1.1)*[f]
Pitman et al. (24)	1. VN–PTSD (7) 2. VN–other anxiety disorders (7)	3-minute relaxation + 30-second baseline	1. 2.8 (4.1) 2. 2.6 (2.8) ns	1. 70.8 (12.0) 2. 73.6 (7.9) ns	.12	not reported
Orr et al. (30)	1. WWII + Korean veterans - PTSD (8) 2. WWII + Korean veterans–Non-PTSD (12)	3-minute relaxation + 30-second baseline	1. 2.6 (1.9) 2. 3.5 (2.7) ns	1. 71.6 (12.2) 2. 71.8 (11.7) ns	.05	not reported
			Other Trauma			
Shalev et al. (21)	1. non-veteran - PTSD (13) 4♀, 9♂ 2. non-veteran - Non-PTSD (13) 7♀, 6♂	3-minute relaxation? + 30-second baseline	1. 4.7 (2.8) 2. 6.7 (2.0)* (opposite direction)	1. 73.2 (12.9) 2. 77.7 (13.1) ns	.22	not reported
Orr & Pitman†	1. childhood sexual assault (20, ♀) 12 current full PTSD; 8 past full PTSD 2. childhood sexual assault - Non-PTSD (13, ♀)	3-minute relaxation + 30-second baseline	1. 3.0 (2.6) 2. 4.3 (3.5) ns[d]	1. 73.9 (9.3) 2. 71.8 (9.8) ns[d]	.15	not reported
Blanchard et al. (31)	1. MVA PTSD (23) 2. MVA non-PTSD (17)[g]	7-minute adaptation + 5-minute baseline	values not reported	1. 70.2 2. 72.0 ns[c,h]	.12	not reported

Non-PTSD=Comparison group includes some subjects with nonpsychotic Axis I disorders.

[a]There was a nonsignificant group X condition interaction.

[b]Original values and significance reported in Blanchard et al. (110).

[c]The standard deviations used for this power analyses was 12.0. Our one-tail *t* test for independent groups was nonsignificant.

[d]Original values obtained from authors (Special thanks to Robert Gerardi and Scott Orr).

[e]A significance test was not conducted in the original study. Our one-tail *t* test for the independent group was significant.

[f]An average of four samples from baseline period (0, 10, 20, and 30 minutes) was used. The negative affect score was obtained from a composite of 17 adjectives (e.g., disgust) placed on a five point scale.

[g]No information on number of male and female subjects in each group was provided.

[h]No information on standard deviations was provided. Original values were obtained from figures in the text.

*$p<0.05$.

†Unpublished paper presented by Orr and Pitman at the 1993 annual meeting of the International Society for Traumatic Stress Studies, San Antonio, Texas.

MVA, motor vehicle accident; NMD, no mental disorder; PTSD, post-traumatic stress disorder.

the PTSD group had *lower* SC levels than the non-PTSD group. The one study that reported on baseline levels of subjective distress (22) found significant differences in negative affect during a 30-minute baseline period, with the PTSD group reporting more distress than the non-PTSD group.

Because it appeared that significant findings in baseline arousal might be related to sample size, we conducted independent one-tailed *t* tests and power analyses for each study. When exact values were not reported, we relied on figures and graphs to estimate baseline HR and SC values. The results of our analyses are provided in Table 1 along with the significance reported by the original investigators. Our analyses suggest that when the power of a study was relatively high (i.e., greater than .75) reliable differences were evident; when power was low (i.e., less than .40), however, the reported differences were not reliable.

Notably, baseline differences in HR are only evident in early studies on combat-related PTSD. More recent studies on both combat and noncombat-related PTSD do not find sustained differences in arousal. Several explanations can be offered to account for the apparent downward shift in baseline levels of arousal for combat-related PTSD, as well as the nonsignificant findings of more recent nonveteran PTSD studies:

1. Many of the early studies on combat-related PTSD employed comparison groups of non veterans (23) or veterans without a mental disorder. These studies were the most likely to find significant differences in baseline levels of arousal. When trauma survivors with other Axis I psychiatric disorders are used as a comparison group, differences in baseline levels of arousal are noticeably reduced. For example, Pitman and colleagues (24) found that Vietnam veterans with other anxiety disorders had slightly *higher* baseline HR than Vietnam veterans with PTSD. Thus, reported differences in baseline levels may be a function of the comparison group used, with higher tonic levels of arousal perhaps being characteristic of anxiety disorders in general (25).
2. Those studies using shorter baselines tended to report greater HR variability and higher baseline values. For example, the highest HR values for both the PTSD and non-PTSD groups were reported by Malloy et al. (26), who employed a baseline duration of only 5 minutes. On the other hand, the lowest HR values with the least variability was reported by McFall et al. (22), who used a 30-minute baseline period preceded by a 30-minute period of adaptation and rest. Indeed, with the exception of the McFall et al. study (22), the so-called baseline period in all of these studies is less than what would be required to index basal physiological states. Recommendations about baseline length by Hastrup (27) point to a minimum of 15 minutes for studies that examine psychophysiology in relation to acute laboratory stressors. The shorter baseline levels in most existing PTSD studies are, therefore, best understood simply as reference levels of ANS arousal under conditions of relatively diminished ambient stress.
3. Although psychophysiological procedures involving exposure to feared cues has become standard in the assessment of anxiety disorders (25), many of the early PTSD studies employing such procedures were met with skepticism and concern regarding subject welfare. In order to obtain approval for these studies, informed consent and procedural instructions often included demand characteristics that may have contributed to elevated levels of arousal. For example, subjects in the Malloy et al. (26) study were instructed "not to push [themselves] to the limit" and to "stop if the scenes become upsetting." Subjects in the Blanchard et al. (23) study were asked after each phase if they wanted to continue. The possibility of a demand characteristic influencing arousal is partially supported by the high degree of escape behavior observed in early studies on the psychophysiology of PTSD (25) as compared to later studies.
4. Relatedly, baseline differences in arousal may be confounded by anticipatory anxiety, given that these values are typically col-

TABLE 2. *Psychophysinogical reactivity and subjective distress following exposure to trauma-related and trauma-unrelated laboratory stressors*

		Trauma				Other stressors			
Study	Groups	Exposure duration and reactivity defined	Skin conductance reactivity	Heart rate reactivity	Subjective distress	Exposure duration and reactivity defined	Skin conductance reactivity	Heart rate reactivity	Subjective distress
Auditory stimuli									
Blanchard et al. (23)	1. VN PTSD (11) 2. Non-Vets–NMD (11)	≤5 trials: 30 seconds per trial; last combat phase–last music phase	not reported–ns[a]	1. 9.1 2. −1.6*[b]	not collected	not specified: mental arithmetic–preceding baseline	not reported–ns[a]	1. 6.5 2. 6.0 ns[c]	not collected
Pallmeyer et al. (37)	1. VN PTSD (12) 2. VN NMD (10)	≤5 trials: 30 seconds per trial; last combat phase–last music phase	not reported	1. 5.4 2. −0.2*[b]	not collected	not specified: mental arithmetic–preceding baseline	not reported	1. 5.2 2. 7.0 ns[c]	not collected
Blanchard et al. (110)	1. VN PTSD (57) 2. VN NMD (34)	≤5 trials: 30 seconds per trial; last combat phase–last music phase	not reported	1. 10.6 2. 3.5*[c]	not collected	not specified: mental arithmetic–preceding baseline	not reported	1. 6.1 2. 8.2 ns[c]	not collected
Gerardi et al. (111)	1. VN PTSD (18) 2. VN non-PTSD (18)	≤5 trials: 30 seconds per trial; last combat phase–last music phase	not reported–ns[a]	1. 9 2. −1*[c]	not collected	not specified: mental arithmetic–preceding baseline	not reported–ns[a]	1. 8 2. 9 ns[c]	not collected
Blanchard et al. (94)	1. VN PTSD (59) 2. VN NMD (12)	≤5 trials: 30 seconds per trial; last combat phase–last music phase	not reported	1. 3.5 2. 0.25*[c]	not collected	not specified: mental arithmetic–preceding baseline	not reported	1. 6.0 2. 9.0[c,d]	not collected
Audio-Visual stimuli									
Malloy et al. (26)	1. VN PTSD (10) 2. VN NMD (10)	≤9 scenes: 60 seconds each; combat scenes–neutral scenes	SC values not obtained	1. 11 2. −0.5*[c]	fear thermometer (0–11 scale) 1. 4.8 2. 2.2*[c]	not collected	not collected	not collected	not collected

McFall et al. (22)	1. VN PTSD (10) 2. mixed non-PTSD (11)	10-minute VN video; comat–preceding baseline	not collected	1. 6.8 (1.8) 2. 1.1 (1.4)*	Negative Affect: (0–68 scale) 1. 23.7 (3.7) 2. 5.5 (0.3)*	10 minute MVA video: MVA–baseline	not collected	1. 1.6 (1.0) 2. 1.6 (1.3) ns	1. 10.2 (2.1) 2. 3.8 (1.8)*
Idiographic stimuli									
Pitman et al. (15)	1. VN PTSD (18) 2. VN NMD (15)	2 × 30-second trauma imagery: imagery–preceding baseline	1. 0.6 (0.6) 2. 0.05 (0.15)*[c]	1. 5.78 (5.5) 2. 3.0 (4.0) ns[c]	Valence Pleasantness: (0–12 scale) 1. 0.6 (0.9) 2. 1.9 (1.5)* Arousal 1. 10.7 (1.2) 2. 10.0 (1.9) ns	A. 30-second precombat trauma imagery script–preceding baseline B. 30-second public speaking fear script–preceding baseline	A. not reported–ns B. 1. 0.0 2. not reported–ns	A. not reported–ns B. 1. 1.7 2. not reported–ns	A. valence and arousal not reported–ns B. valence and arousal not reported–ns
Pitman et al. (24)	1. VN PTSD (7) 2. VN other anxiety disorders (7)	2 × 30-second trauma imagery: imagery–preceding baseline	1. 0.50 (0.45) 2. 0.10 (0.18)*[c]	1. 5.75 (5.5) 2. 3.0 (2.5) ns[c]	Valence Pleasantness: 1. 0.4 (0.9) 2. 2.7 (2.1)* Arousal: 1. 11.2 (1.2) 2. 7.9 (2.3)*	A. 30-second precombat trauma imagery script–preceding baseline B. 30-second public speaking fear script–preceding baseline	A. not reported–ns B. not reported–ns	A. not reported–ns B. not reported–ns	A. not reported B. not reported
Orr et al. (30)	1. WWII + Korean War PTSD (8) 2. WWII + Korean War non-PTSD (12)	2 × 30-second trauma imagery: imagery–preceding baseline	1. 1.1 (1.1) 2. 0.05 (.25)*[c]	1. 9.9 (8.0) 2. 1.5 (1.5)*[c]	Valence Pleasantness: 1. 0.4 (0.4) 2. 0.6 (1.0) ns Arousal: 1. 10.6 (1.6) 2. 10.3 (2.1) ns	A. 30-second precombat trauma imagery script–preceding baseline B. 30-second public speaking fear script–preceding baseline	A. 1. 0.2 (0.4) 2. 0.1 (0.4) ns B. 1. −0.1 (0.3) 2. 0.0 (0.2) ns	A. 1. 6.4 (7.9) 2. 1.0 (3.1)* B. 1. 2.7 (4.9) 2. 2.0 (3.1) ns	not reported
Shalev et al. (21)	1. non-veteran PTSD (13) 4♀ 9♂ 2. non-veteran non-PTSD (13) 7♀ 6♂	30-second trauma imagery: imagery–preceding baseline	1. 0.9 (1.2) 2. 0.3 (0.6) ns	1. 13.9 (10) 2. 2.0 (7.9)*	not reported	A. 2 × 30-second nontrauma stressors–preceding baseline B. 30-second dental fear script–preceding baseline	A. not reported–ns B. not reported–ns	A. not reported–ns B. not treated–ns	not reported
Orr & Pitman[†]	1. childhood sexual assault PTSD (20, ♀) 2. childhood sexual assault non-PTSD (13, ♀)	2 × 30-second trauma imagery script: imagery–preceding baseline	1. 0.72 (0.8) 2. 0.49 (0.7) ns[f]	1. 1.7 (1.9) 2. 0.6 (1.5)*[f]	not reported	30-second nontrauma stressor–preceding baseline	not reported	not reported	not reported

continued

TABLE 2. *Continued.*

Study	Groups	Trauma				Other stressors			
		Exposure duration and reactivity defined	Skin conductance reactivity	Heart rate reactivity	Subjective distress	Exposure duration and reactivity defined	Skin conductance reactivity	Heart rate reactivity	Subjective distress
Blanchard et al. (31)	1. MVA PTSD (23) 2. MVA non-PTSD (17)[g]	A. 2 × 3-minute idiographic trauma imagery: imagery–preceding baseline B. 3-minute MVA video: video–preceding baseline	not reported	Script #1 1. 5.7 2. 1.2*[c] Script #2 1. 2.2 2. −1.0*[c] MVA video 1. −0.2 2. −1.9 ns[c]	Script #1 1. 55 2. 30*[c] Script #2 1. not reported 2. not reported MVA video 1. not reported 2. not reported	3-minute mental arithmetic–preceding baseline	not reported	1. 6.9 2. 5.6 ns[c]	not reported

Non-PTSD=comparison group includes some subjects with nonpsychotic Axis I disorders.

[a]There was a nonsignificant group X phase interaction.

[b]Mean values were obtained from the Blanchard et al. (110) study. Information on standard deviations was not provided.

[c]Mean values were obtained from figures or tables in the text. Information on standard deviations was not provided.

[d]A significance test was not provided in the original study.

[e]Mean values and standard deviations were obtained from figures in the text.

[f]Values were obtained from authors (Scott Orr). Values are expressed as square root microvolts and square root beats per minute.

[g]No information on the number of male and female subjects in each group was provided.

*$p<0.05$.

†Unpublished paper presented by Orr and Pitman at the 1993 annual meeting of the International Society for Traumatic Stress Studies, San Antonio, Texas.

MVA, motor vehicle accident; NMD, no mental disorder; PTSD, post-traumatic stress disorder.

lected when subjects are anticipating exposure to trauma cues[2]. In a recent study that did not include exposure to trauma cues (28), 11 Vietnam combat veterans with PTSD were compared to 11 asymptomatic controls on several measures of basal sympathoadrenal function, including HR. Two to 4 weeks after participating in diagnostic and psychometric testing, these medication-free subjects were brought into the laboratory and placed in a supine position for 30 minutes prior to a 30-minute data collection period. The results showed that the PTSD subjects were not significantly different from control subjects on any measure. These investigators concluded that tonic sympathetic nervous system activity is not significantly elevated for patients with PTSD. Similar findings were reported by Shalev et al. (29) in their study on exaggerated startle responses in PTSD (see startle section). In this study, subjects with PTSD were no different than trauma survivors without a mental disorder on baseline HR, SC, and obicularis oculi EMG activity. Finally, support for an anticipatory effect confounding baseline differences in arousal comes from the one study that reported on baseline levels of subjective distress. As previously noted, PTSD subjects in the McFall et al. (22) study reported more preexposure subjective distress than the non-PTSD group.

PHASIC CHANGES IN AUTONOMIC AROUSAL

Group differences in reactivity to laboratory stressors which are both related to and unrelated to trauma are reported in Table 2. As with Table 1, when exact values were not reported, we used the figures in the text to estimate HR and SC reactivity. Because standard deviations were rarely provided, we were unable to compute independent *t* tests and power analyses for these studies. Significant differences in HR reactivity to trauma-relevant stimuli were reported in 11 of the 13 studies reviewed. Significant differences in SC levels or SC responses were found in three of the seven studies that provided information on this measure. Subjective ratings of distress are reported as absolute scores in Table 2 because baseline reference values were generally not collected. Of the six studies that reported subjective distress ratings, five found significant differences between the PTSD group and the non-PTSD group, with the PTSD subjects reporting more distress after the trauma-relevant stimuli than did the non-PTSD comparison subjects.

Only one study out of 11 (30) found significant differences in HR reactivity to generic stressors, while no studies found significant differences in SC to these stressors. The one study that included subjective distress ratings for the generic stressor (22) found that the PTSD group reported significantly more subjective distress than the non-PTSD group.

It seems safe to conclude that, at least for combat-related trauma, combat veterans with PTSD show significantly more psychophysiological reactivity to combat stimuli than various comparison groups (i.e., normal controls, nonveterans with other psychiatric diagnoses, combat veterans with no mental disorder). For HR, this reliable and robust finding held for all protocols that utilized standardized audio or audiovisual combat stimuli. Significant differences in SC were more common in assessments using idiographic imagery scripts, although HR differences were in the expected direction for these protocols as well. Despite the different scales used to assess subjective distress, Vietnam veterans with PTSD consistently reported more distress than Vietnam veterans without PTSD. Only the WWII and Korean War veterans with PTSD were not significantly different from WWII and Korean War veterans without PTSD on subjective ratings of distress (30).

[2]In a recently published paper by Gerardi et al. (112), 32 Vietnam veterans with PTSD were compared to 26 age-, race-, and sex-matched Vietnam era veterans with no combat experience on a number of physiological variables, including heart rate, while waiting in the admitting area of a large VA Medical Center. Thus, there was no expectation of exposure to combat cues. These authors found that the subjects with PTSD had higher "basal" heart rate levels than the matched controls.

Although the direction of psychophysiological responding in subjects with noncombat PTSD is similar to that observed in subjects with combat-related PTSD, several important distinctions are apparent:

1. The method of stimulus presentation in these studies has involved exposure to idiographic (trauma) imagery scripts. For combat-related PTSD, the presentation of idiographic scripts most clearly demonstrated differences with respect to SC responding. This does not seem to hold for noncombat PTSD. Blanchard et al. (31) used both standardized and idiographic trauma cues and found HR, rather than SC, to be the most reactive to the idiographic stimuli; the standardized stimuli did not produce significant differences in either HR or SC. Similar findings were reported by Shalev et al. (21), who found significant HR changes due to idiographic scripts but no significant changes in SC.
2. The magnitude and variability of responding to noncombat presentations may bear an inverse relationship to the duration and frequency of exposure to the trauma-relevant stimuli. For example, Shalev et al. (21) exposed subjects to one 30-second trauma-relevant imagery script. Although the HR changes associated with this script were considerably higher than those reported in other studies, the degree of variability was also much greater.
3. Studies on noncombat-related PTSD have included female subjects in contrast to the exclusive use of male subjects in studies of combat veterans. It is too early to tell how psychophysiological responding is influenced by gender. In the one study that directly examined gender differences in civilian trauma (21), female subjects with PTSD demonstrated 33% greater physiological responding to their trauma script than male subjects with PTSD. More specific information about gender differences in HR or SC was not provided, and the small number of subjects in the study made the observed differences nonsignificant.

Consistent with the findings from combat-related PTSD, psychophysiological responding does not appear to generalize to generic laboratory stressors. This interpretation is reinforced by intrasubject comparisons of responses to trauma-related and generic laboratory stressors, which indicate that the psychophysiological reactivity seen in PTSD is specific to the trauma-relevant material. Such response specificity supports a classical conditioning model of PTSD (32,33), but additional research is needed to explain the observed discordance found by Orr (30) between self-reports of distress and physiological responding to trauma-related stimuli. At present this discordance does not appear to be unique to PTSD, because it is also found in relation to generic stressors (22,30).

EXAGGERATED STARTLE RESPONSE

Despite self-reports of exaggerated startle in PTSD (34) and standardized procedures for evoking startle responses in humans (35), only a few studies have examined this symptom of PTSD in the laboratory. The protocols used to assess startle responses in PTSD have typically involved exposure to acoustic startle probes, although these probes have varied in frequency, intensity, duration, and rise and fall times. The aspects of startle most frequently reported in these studies include magnitude of responding to an unwarned signal and trials to nonresponse (i.e., habituation). Table 3 provides a review of the published studies on this topic.

As previously noted, startle is primarily a skeletomuscular response that can be easily indexed by eyeblink responses or orbicularis oculi EMG activity. Although three studies have used either of these measures to examine response magnitude or trials to habituation, only the Butler et al. study (36) found significant differences (in response magnitude) between the PTSD and non-PTSD groups. On the other hand, all three studies that used autonomic measures found significant differences in ANS activity, with the PTSD group showing greater SC and HR responding to the startle probes as compared with the control group.

TABLE 3. *Startle responses in PTSD*

Study	Groups	Startle probe	Startle responses	Measures	Findings
Pallmeyer et al. (37)	1. VN PTSD (12) 2. VN NMD (10)	1 × 80 dB, 2000 msec, burst white noise, ? rise time	A. magnitude	HR; change score from baseline	A. *Magnitude* 1. 1.25 2. −1.5*[a]
Ross et al. (34)	1. VN PTSD (9)[b] 2. Non-Vets NMD (9)	150 × 100 dB [SPL], 1000 hz, 50 msec, <1 msec rise and fall time	A. trials to habituation	eyeblink amplitude	A. *trials to habituation* 1. 15 2. 25 ns
Butler et al. (36)	1. VN PTSD (13) 2. VN NMD (12)	6 × 40 msec noise burst at 85, 90, 95, 100, 110, and 116 dB[A].	A. magnitude	right orbicularis oculi EMG response; transformed digital units (1 unit=15 microvolts)	A. *magnitude*[a] −85 dB 1. 2.5 2. 2.0 ns −90 dB 1. 7.5 2. 2.0 ns −95 dB 1. 15.0 2. 2.0* −100 dB 1. 20.0 2. 5.0* −110 dB 1. 32.0 2. 15.0 ns −116 dB 1. 43.0 2. 25.0 ns
Paige et al. (42)	1. VN PTSD (12) 2. VN NMD (6)	4 × 780 hz tone, 500 msec duration at 74, 84, 94, and 104 dB[SPL] with 25 msec rise and fall times	A. magnitude	EEG (not reviewed here) HR	A. *magnitude*[a,c] −74 dB 1. 2.6 2. 2.5 −84 dB 1. 2.6 2. 1.3 −94 dB 1. 2.9 2. 0.5 −104 dB 1. 3.7 2. 1.7
Shalev et al. (29)	1. mixed trauma PTSD (14) 11♂, 3♀ 2. trauma NMD (15) 10♂, 5♀	15 × 95 dB[SPL], 1000 hz, 500 msec tone, 0 msec rise and fall times	A. magnitude B. trials to habituation	left orbicularis oculi EMG response, SC response, HR response; (square root transformations)	A. *magnitude*[d] *−EMG* *1. 0.9 (0.8)* *2. 0.4 (0.5) ns* *−SC* *1. 0.77 (0.37)* *2. 0.29 (0.16)** *−HR* *1. 2.7 (1.6)* *2. 1.5 (0.9)** *B. trials to habituation* −EMG 1. 9.1 (5.9) 2. 5.9 (5.1) ns −SC 1. 13.1 (3.2) 2. 6.3 (5.0)* −HR 1. not reported 2. not reported

[a]Values were obtained from figures in the text.
[b]Only data from 7 PTSD patients were used in analyses.
[c]Statistical significance for each dB level was not provided. However, HR slope was significant.
[d]EMG, SC, and HR slopes were not significant.
*$p<0.05$.
NMD, no mental disorders; PTSD, post-traumatic stress disorder.

Of this set of studies, the one by Pallmeyer et al. (37) has features that may undercut its value in the context of startle testing. First, it is unclear whether the stimulus was powerful enough to elicit a startle response, because the intensity was near startle threshold (80–85 dB), and the absence of information about stimulus rise or fall times suggests that they were not carefully controlled. This is important because Blumenthal (38) has shown that responses to stimuli near the startle threshold are particularly affected by onset rise time. The possibility of insufficient potency is consistent with the finding that the non-PTSD control group showed a *decrease* in HR responding to the startle stimulus. Second, the startle stimulus was delivered within the context of an assessment that involved exposure to trauma cues, a factor that may have led to startle potentiation. Just as anticipatory anxiety may increase baseline levels of arousal, evidence based on non-PTSD subjects suggests that anticipatory anxiety can potentiate the startle reflex (39). Thus, even though the Pallmeyer et al. study did not conform to the temporal proximity between fear stimulus and startle probe often used in the fear-potentiated startle paradigm (40), the affective context of the study may have had a comparable excitatory effect (41).

The fact that the acoustic startle probes used by both Shalev et al. (29) and Paige et al. (42) would have sounded like gunfire adds to the possibility of inadvertent fear potentiation. These investigators have argued that their findings support the hypothesis that PTSD is associated with abnormalities in unconditioned responding. However, the validity of this assertion rests on the untested assumption that the startle probes did not have conditioned significance to the subjects. Although the nature of the trauma experienced by subjects in the Shalev et al. (29) study was not reported, subjects in the Paige et al. (42) study were Vietnam combat veterans whose trauma experiences probably included some form of exposure to gunfire. Given these considerations, differential responding to the startle probes may have reflected conditioned responding alone, or the impact of conditioned responding as a potentiator of unconditioned responding.

The possibility that loud tones may have conditioned relevance receives support from the Butler et al. study (36). These investigators examined responding to both acoustic and tactile startle probes. Despite significant differences in responses to the acoustic startle probe, PTSD and non-PTSD subjects showed similar responses to the tactile probes. Butler et al. suggest that these findings support a stimulus-specific increase in startle response for individuals with PTSD, and argue that this interpretation has "clinical appeal since many veterans report that auditory stimuli, such as an automobile 'backfire,' will result in an exaggerated startle" (36).

SUMMARY AND CONSIDERATIONS FOR FUTURE RESEARCH

Table 4 offers an interpretive summary of our findings. Our review of the psychophysiological evidence provides strong support for the exis-

TABLE 4. *Interpretive summary of psychophysiological evidence for autonomic arousal and startle responses in PTSD*

	Outcome measure			
Assessment period	Heart rate	Skin conductance	EMG	Subjective distress
Baseline[a]	+/−	−	NR	+
Trauma Reactivity	++	++	NR	++
Generic stressor Reactivity	−	−	NR	−
Acoustic Startle[b]	+	+	−	NR

++ clear positive association
\+ probable positive association
+/− inconsistent findings
− probable negative association
−− clear negative association
EMG, electromyographic
NR, not reviewed

[a]See text for argument that elevated baseline levels of arousal may be a function of anticipatory anxiety related to forthcoming exposure to trauma cues in psychophysiological studies of PTSD.

[b]See text for argument that acoustic startle probes may have conditioned significance for combat veterans with PTSD.

tence of differentially greater ANS reactions when individuals with PTSD are exposed to trauma-related stimuli. Although ANS reactions can also be observed in response to laboratory stressors that are unrelated to trauma, our review suggests that the magnitude of this responding does not differ between individuals with and without the disorder. In addition, our review of potential basal differences in ANS arousal associated with PTSD suggests that observed differences in baseline levels are best understood as ''elevations'' linked to anticipatory ''distress.'' We conclude that it is unlikely that stable elevations in ANS activation are characteristic of PTSD *in the absence of some proximal psychosocial cause.* The clinical reports of sustained elevations may be the result of more frequent triggering of affective states that provoke arousal in individuals who have been traumatized. However, this possibility has not as yet been addressed by research.

Our review of the startle research with PTSD populations suggests continuity with the evidence for psychophysiological reactivity. Differences in laboratory startle responding that are associated with PTSD status appear to be the result of fear potentiation of the response. This distinction is important because it emphasizes the role of the ambient emotional state rather than relying on a stable, biologically altered propensity to startle. We also note that the current set of positive startle findings with PTSD patients are based on the use of acoustic startle probes that may have affective relevance for the subjects as a result of their similarity to the sound of gunfire. For both of these reasons, we believe that the concept of conditioned fear may offer more explanatory power for the body of evidence than does the notion of biological alteration. While better understanding of the biological underpinnings of affective reactions will be a valuable element of our understanding of PTSD, current psychophysiological evidence points to the role of conditioned affective cues as a more critical focus at this time.

Beyond the literature review, the potential impact of anticipatory responding is also suggested by evidence that elevations in baseline arousal are associated with greater subsequent psychophysiological reactivity. Litz et al.[3] created groups of combat veterans on the basis of relative HR reactivity to combat audiovisual presentations and found that they differed rather dramatically in terms of resting HR values. Specifically, the more reactive veterans had resting values approximately 20 beats per minute higher than their less reactive counterparts. Assuming that the difference in resting values reflects anticipatory anxiety as we propose, it may be that this background state potentiates subsequent reactivity in a manner analogous to the effect that baseline arousal appears to have on the likelihood of panic attacks in response to challenge (e.g., lactate infusion (43)). This pattern may be related to a broader psychophysiological phenomenon with respect to HR, for which *between-groups* evidence indicates a consistent positive relationship between baseline values and the magnitude of reactivity (44).

The unique psychophysiological response pattern shown by PTSD subjects following trauma cues has been observed in diverse traumatized populations using a variety of stress-inducing stimuli. In a few instances, the pattern of response has been used to statistically discriminate them from both asymptomatic and psychiatric control subjects. These discriminant function analyses using psychophysiological variables have correctly classified PTSD subjects and their non-PTSD counterparts from 80% to 100% (45). However, these overall hit rates do not reflect the consistent finding that specificity (i.e., percentage of non-PTSD subjects correctly classified) is better than sensitivity (i.e., percentage of PTSD subjects correctly classified). In fact, the sensitivities of discriminant function equations have ranged from 61% to 91%, while specificities have ranged from 86% to 100% (46). These findings indicate that as many as 40% of patients with PTSD do not demonstrate the expected psychophysiological reactivity. Interestingly, Butler et al. (36) reported that a similar percentage of PTSD subjects were not

[3]Unpublished paper presented by Litz, Forsyth, Kaloupek, and Slavkin at the 1991 annual meeting of the Association for the Advancement of Behavior Therapy, New York, New York.

responsive to startle probes. This lack of consistency in responding is an important target for future investigation. What follows is discussion of several factors that may account for the psychophysiological nonresponsiveness of some individuals with PTSD.

Violations of the Boundary Conditions of Assessment

First, the integrity of the assessment procedure must be considered. Valid physiological data can be obtained only if certain boundary conditions are met. As with computerized axial tomography (CAT) scans that require patients to be motionless during the procedure, psychophysiological assessments typically place restrictions on the amount of movement subjects can display while data are collected. They also require compliance with task demands that involve viewing, imagining, and/or listening to material that can evoke strong emotional reactions. Instructions or other constraints may be used to counteract the self-protective or self-regulating tendencies that most individuals are likely to manifest under these circumstances. However, because the emotional reactions are aversive, efforts to limit processing of the stimuli (e.g., turning away from slides or curtailing images) or regulate arousal (e.g., deep breathing) are still common. Unfortunately, when boundary conditions are violated in these ways, physiological systems are subject to influences that can generate or dampen ANS arousal in a manner difficult to quantify (47). Thus, it is understandable that response patterns associated with the PTSD diagnosis are not uniform.

Behavioral agitation often observed during assessment is meaningful as an index of emotional distress and should be given particular consideration in clinical evaluations. However, unless special care is taken in the placement of transducers (e.g., electrodes), movement can easily become so extreme that recordings are infeasible, as was observed in the "decompensated" group in the Dobbs and Wilson (19) study. Moreover, less extreme motoric activity caused by emotional distress may result in elevations of psychophysiological responding, particularly cardiovascular responding. In keeping with our earlier discussion of the autonomic factors that underlie some of the measured physiological reactions, it should be noted that such cardiac-somatic coupling is more likely to reflect the withdrawal of parasympathetic influences rather than the increased activation of sympathetic influences (48). The importance of this distinction for PTSD is not currently clear.

Variables that affect physical state may also challenge the boundary conditions for assessment. Factors such as time of day (i.e., circadian phase) and laboratory temperature or humidity are given some consideration in the assessment of PTSD, but it is still typical for them not to be rigorously controlled. On the other hand, numerous extraneous influences are not easily controlled and can acutely disrupt psychophysiological recording. Among these influences are speaking, coughing, fidgeting, changes in breathing pattern, emotional thoughts unrelated to the procedure (e.g., about a prior argument with a spouse), and physical discomfort from a variety of sources (e.g., uncomfortable seating position, full bladder). Within broad limits, these may not be a major problem *as long as the effect of interest is robust.* In addition, many of these potential problems can be minimized by careful planning and preparation of the physical environment where measurement takes place. Control of the psychological and behavioral influences that might confound the emotional targets of interest also can be managed through the study design and procedures.

More difficult to manage is the uncontrolled use of medications by patients, as well as the consumption of caffeine and nicotine by many research subjects prior to psychophysiological testing for PTSD. The unfortunate fact is that there is not much empirical literature to help disentangle the impact of medications from other influences contributing to the observed response patterns. Psychoactive drugs aimed at anxiety or depression are probably most relevant for PTSD research, and available data indicate that some of these medications reduce both HR and SC activity (49). Drugs with clear autonomic effects (e.g., beta blockers) pose an even greater problem because their impact is likely to be substantial, but there is no available adjustment for psychophysiological data. For example, it is not possible to specify a dose-related formula for adjusting the responding of medicated subjects so that the unmedicated equivalent of their responding can be estimated.

The use of nicotine and caffeine can also have pronounced influence on the physiological systems that are typically monitored in PTSD studies. Unfortunately, their effects are neither uniform across physiological systems nor solely tied to consumption. Nicotine produces cardiovascular effects in the form of increased HR, as well as increased systolic and diastolic blood pressure (BP) (50). The effects of nicotine on HR appear to be additive with the effects of stress (51), and interactive with depression and trait anxiety (52). Furthermore, withdrawal from nicotine decreases HR and SC responses, while increasing subjective anxiety, depression, and irritability that may affect behavior during assessment (53,54). Finally, nicotine may act in combination with oral contraceptives to have an enhanced impact on BP during stress (54), warranting particular attention to these factors when young women are tested.

Caffeine also has cardiovascular effects, but the effect on HR is apparently minimal in comparison to the effect on BP (55,56). This effect may be particularly pronounced for individuals who have a family history that places them at risk for hypertension (57). As with nicotine, caffeine withdrawal is associated with irritability and discomfort (58). However, some evidence indicates that anxiety, depression, and hostility can be increased by moderate-to-high acute doses (59). Given this evidence, the fact that use—even heavy use— of these substances is very common in veterans with PTSD (60) makes their potential impact all the more noteworthy for the existing (and future) literature on the psychophysiological assessment of PTSD.

Limitations of Physiological Measures

There are also some problematic features of the measures used to reflect ANS activity. A basic concern is the differential sensitivity that measures appear to have with respect to psychological or behavioral influences. Theory, empirical evidence, and working experience with these measures all indicate that tasks that evoke states such as attention, vigilance, active avoidance, or inhibition of ongoing behavior are better indexed by some physiological systems than by others—and even by some components of systems rather than by others (61,62,63). As an example, active avoidance appears to have primary impact on the cardiovascular system. Thus, consideration must be given to the features of any psychophysiological assessment task when measures are being selected or, conversely, the task must be constructed so as to capitalize on the strengths of available measures.

Another important limitation is that there are no noninvasive measures that provide a direct, error-free index of ANS activity. One strategy for overcoming this problem is to use convergent data from two or more measures to create an index of ANS reactivity. We are currently investigating the use of concurrent HR and SC increase as such a convergent index for PTSD assessment. We make the assumption that both measures reflect some common aspects of the sympathetic arousal linked to PTSD, as well as their own unique sources of autonomic effects and error. Two indices are created to reflect: (1) from among the sample points during a measurement period that shows HR increase, the proportion that shows concurrent SC increase, and (2) from among the sample points that show SC increase, the proportion that shows concurrent HR increase. In principle, these convergent indices should offer greater sensitivity and validity than single system indices currently used for PTSD assessment (64).

Limitations of Diagnostic Methods

The diagnostic interviews and psychometric instruments typically used as criteria for PTSD diagnosis are subject to several sources of error. At minimum, the current methods are nondiscriminating with regard to factors such as patient motivation (e.g., compensation seeking) that may lead to false positive case identification. Another broad possibility is that these methods—or perhaps even the DSM diagnostic system on which they are based—may be overinclusive (65). One aspect of the problem is that the multiple symptom options available under the diagnosis of PTSD make it possible to obtain a diagnosis of PTSD without the presence of psychophysiological reactivity. The fact that psychophysiological reactivity is not isomorphic with the disorder means that the absence of this symptom may reflect a particular subtype of PTSD. If so, the reactivity could have implications for predicting treatment response to both pharmacological and nonpharmacological interventions (c.f. 66,67) or, in a more proactive sense, for serving as a basis for treatment matching (c.f. 68,69). It could also be the case that the diagnostic system should be less flexible with regard to the necessity of this symptom (70).

Finally, it is important to recognize that the foundation of the PTSD diagnosis is subjective information not necessarily comparable to information recorded directly from physiological systems. Evidence from research on autonomic perception and response covariation (71,72,73) makes it clear that self-reports of psychophysiological reactivity are not interchangeable with observations or recordings of such activity. Thus, reports of physiological reactivity or startle are not likely to show high consistency with more direct measures of these phenomena, or at least are not likely to do so for many individuals. A related point about self-report assessment is that not all formats are equivalent. Evidence indicates that subjective ratings of distress that are in a dimensional format (e.g., those anchored to arousal or valence) correlate more highly with psychophysiological measures than do ratings anchored to affective states such as distress or anxiety (74).

Uncontrolled Variation in Individual Biological Influences

Individual differences that can influence psychophysiological reactivity arise from intrinsic subject variables such as age, sex, race, menstrual cycle, and physical fitness level. These variables are rarely considered in psychophysiological research on PTSD, but their relationship to sustained and reactive features of ANS activity has been established in other studies employing psychophysiological methods.

Epidemiological and laboratory studies have shown that HR both at rest and in response to various laboratory stressors tends to decrease with age, while BP levels and reactivity tend to increase (75,76,77). Recent evidence broadens the implications of these trends by showing that chronic stress and social support are possible moderators of these changes (78).

Differences between races and genders have been found with respect to HR, BP, and SC measures in a number of studies (79,80,81,82). Black subjects have generally demonstrated lower HR and SC levels, and higher BP levels and reactions to stress than white subjects. Furthermore, the two races seem to differ in the behavior of these measures during physical stress (e.g., cold pressor (83)) and in relation to psychosocial variables (e.g., depression and anger (84)), thereby complicating data interpretation.

Gender differences in response to laboratory stressors generally indicate greater BP reactivity for men and greater HR reactivity for women. In addition, there appear to be differences across phases of the menstrual cycle that need to be considered whenever premenopausal women are tested (85,86). Finally, there is a mixture of positive and negative findings concerning reductions in physiological response to stressors as a function of both aerobic fitness levels and acute exercise (87,88,89,90,91). Despite the lack of consistent findings, there is no question that sustained and acute physiological adjustments to exercise can be substantial for both men and women. Accordingly, exercise-related information may be important to the interpretation of psychophysiological data from PTSD assessments.

Individual differences in the primary physiological channels of affective response expression, a phenomenon related to the general psychophysiological principle of individual response stereotypy (92), may also contribute to the impression that reactivity is inconsistent across PTSD subjects. For example, some individuals may respond to emotional stimuli primarily with HR changes, while others may respond mainly with SC changes; still others may express physiological reactivity in channels other than HR or SC. As one means for addressing this issue, Levis and Smith (93) have suggested that an individual's dominant response channel can be determined by an independent "biological" stress test, such as a balloon-burst. In their sample of 39 students, significant response consistency was shown between this procedure and a psychological stressor for subjects who were classified as high responders in a given channel. Of course, this approach is limited by the degree to which the number of different physiological responses being monitored covers the range of primary channels for the individuals being assessed.

Each of the previously noted factors is of sufficient strength to affect the relationship between psychophysiological measures and clinically relevant stimulus conditions such as presentations of trauma cues. However, this evidence applies only to visceral responding. Factors such as race and gender are generally treated as random error rather than being subjected to analysis as independent variables in studies that address the motor response of startle (i.e., eyeblink). As a result, there is little evidence available concerning their potential impact in this context.

Uncontrolled Variation in Individual Psychosocial Influences

Experiential and developmental factors such as degree of trauma exposure, presence of comorbid disorders, and patterns of coping also complicate the measurement and interpretation of psychophysiological data. Only a few studies have examined the relationship between experiential factors such as degree of combat exposure and psychophysiological reactivity to trauma cues, or response to startle probes. Although several investigators have failed to find a relationship between combat exposure and psychophysiological reactivity (15,94), Orr et al. (95) found that group differences in HR response to trauma cues were substantially reduced when anxious subjects were matched for combat exposure with PTSD subjects. Given that magnitude of psychophysiological reactivity may be related to event severity, subjects may need to be matched on this dimension when there is concern about separating the effects of trauma exposure from those of the PTSD syndrome per se. A similar suggestion was made by Butler et al. (36) with regard to relationship between combat exposure and startle.

Other forms of psychopathology often accompany PTSD, particularly depression, anxiety, and substance abuse (96,97). Evidence from other areas suggests that these conditions may have an impact on efforts to distinguish psychophysiological (or startle) responders from nonresponders as a function of changes they cause in HR, SC, and facial EMG activity (98,99,100). The evidence is mixed with regard to PTSD research. Pitman et al. (15) found no difference between PTSD subjects diagnosed as having or not having major depression in terms of their physiological reactivity. However, Orr et al. (95) found that depression at the symptom level as measured by the Beck Depression Inventory (BDI) was positively correlated with reactivity. Conversely, Litz et al.[4] found that HR responders (with increases ≥7 bpm following combat slides/sounds) were less depressed in terms of the Minnesota multiphasic personality inventory depression (MMPI-D) scale than were HR nonresponders. Blanchard et al.(94) found no significant differences between responders and nonresponders on either the BDI or MMPI-D.

Psychophysiological responding will depend in part on the way the assessment is appraised by the subject, learned patterns of coping (e.g., self-regulatory behaviors) they bring to the assessment procedure, and opportunities available for coping before and during the procedure. For example, Lader (101) found dramatic decreases

[4]Unpublished paper presented by Litz, Weathers, Kaloupek, Gerardi and Keane at the 1990 meeting of the Association for the Advancement of Behavior Therapy, San Francisco, California.

in ANS arousal during dissociation in patients with severe anxiety. In a more controlled study, Bloom et al. (47) demonstrated that attention diversion during laboratory stress tasks can reduce physiological reactivity. Such efforts to limit processing of stimulus presentations can be viewed as potentially meaningful indicators of distress, but they are more difficult to manage insofar as their threat to the validity of psychophysiological data. Just as there may be a reciprocal relationship between avoidance and arousal/distress as hallmark symptoms of PTSD, it can be difficult to determine whether the absence of arousal and distress during a psychophysiological assessment is the result of successful efforts to limit processing of the stimulus presentations or simply is due to lack of affective relevance. Part of the complication is that many of the maneuvers that subjects use to distance themselves from evocative stimuli are not readily identified by observation or other external monitoring.

While creative efforts to measure avoidance behavior in the context of assessment are needed, at minimum subjects might be given a retrospective opportunity to report on their behavior during the assessment, with particular emphasis on behaviors likely to alter their physiological reactions to the stimulus presentations. Although there are not as yet any formal means for incorporating this information into the scoring of the psychophysiological data, it does provide a basis for eliminating data segments of questionable validity. Perhaps it will eventually be possible to develop ways to use the information to quantify the relative amount of functional stimulus contact that occurred during the procedure, and then to adjust the psychophysiological reactivity index accordingly. Currently, this goal is far from realization.

ADDITIONAL METHODOLOGICAL CONSIDERATIONS

There are at least three additional features of the methods used for psychophysiological assessment that deserve scrutiny because of their potential to influence the quality and interpretation of the resulting data.

The Functional Impact of Laboratory Stressors

Obrist (48) identified a distinction among laboratory stressors that is potentially important for the design and interpretation of psychophysiological studies of PTSD. He proposed, and subsequent research has tended to support his hypothesis, that tasks involving active coping will engage the cardiovascular system more than tasks involving passive coping. By definition, active coping involves effortful attempts to influence task outcome (102). For example, it is elicited by mental arithmetic, cognitive problem solving, and interpersonal influence tasks. In contrast, passive coping is not effortful or aimed at influencing outcome. It is elicited by tasks such as watching a film or engaging in imagery, which primarily involve passive sensory intake. The majority of studies on PTSD have presented trauma-related stressors in the context of tasks that would engage passive coping although, as indicated earlier, the actual nature of task-related coping may be more complex than expected.

These considerations also apply to the evaluation of response to startle probes. In line with the preceding discussion, these probes may be experienced as aversive and subjects may make efforts to reduce their impact. This may be particularly the case when warning signals are delivered prior to the probes. It may be more difficult for subjects to make acute adjustments to moderate the aversiveness of unsignaled probes, but this difficulty may merely lead to more sustained strategies for disengaging from the task. In any case, the critical point is the importance of recognizing both the formal features of the tasks and their functional impact when the resulting psychophysiological data are being interpreted.

The Content Relevance of Trauma Cue Presentations

The relative strengths and weaknesses of standardized and idiographic approaches to trauma cue selection need to be considered. The primary advantage of a standardized presentation is its uniformity and potential for allowing tight experimental control. The primary disadvantage is a potentially limited range of application. For example, presentations suitable for veterans are not likely to be suitable for victims of motor vehicle accidents or crime. Conversely, a primary advantage of idiographic scripts is the potential for making comparisons across different trauma populations. The key to such comparisons is development of a standardized protocol for selecting the material for presentation. The investigator must be able to match presentations across individuals in terms of important conceptual (as opposed to concrete) features. Examples of such features might include portrayal of physical threat, description of a subjective sense of extreme fear, helplessness, or horror, and reference to bodily reactions triggered by the depicted situation. A related advantage is that idiographic presentations may closely approximate the internal (memory) representations of the traumatic experience and thereby improve the validity of assessment. Potential disadvantages of idiographic scripts are primarily due to the use of imagery as the typical format for their presentation. This approach is hampered by the need to accommodate individual differences in the ability to engage in imagery and difficulties related to monitoring task compliance. Admittedly, this latter consideration is only marginally easier to deal with when the presentations are in an audiovisual format. In general, the selection of format will depend upon the priorities and aims of the professionals who conduct the assessment. Thus far, the most direct comparison of formats to be published (31) suggests that the idiographic approach generates more consistent response differences linked to PTSD.

Precision and Timing of Measurement Collection

The majority of published studies have quantified psychophysiological variables so as to reflect relatively extended (e.g., 30 or 60 seconds) recording intervals. Many studies further aggregate the information by averaging over two or more intervals. These methods are suited to empirical questions that concern relatively stable responding throughout an interval, or responding that is subject to significant error in measurement. However, the benefits of aggregation are realized at the cost of sensitivity and may not be optimal for the purposes of PTSD assessment. Empirical evidence from Orr et al.[5] and Blanchard et al. (31) indicates that averaging across two trauma presentations can result in diluted findings. In addition, earlier findings reported by Malloy et al. (26) indicate that most significant physiological responding can occur during exposure to one particular slide in a series of presentations. Therefore, quantification approaches that aim to identify peak intervals for reactivity may offer greater sensitivity and perhaps even better matches with evidence provided by nonphysiological sources (e.g., interviews or self-ratings).

Similarly, it is important to either measure during temporal intervals that are most likely to show responding, or to use methods that identify peak reactivity within longer intervals. Identification of critical measurement windows is most easily explained in the context of startle studies, because the effects are phasic and short-lived. Roth et al. (103) have provided empirically based guidelines for optimal measurement intervals for startle response—intervals that have not been consistently used in PTSD startle studies. Presentations of trauma relevant cues typically are not subject to the same level of specification regarding optimal responding, at least in part because of the uncontrolled individual differences that affect responding. This problem can be overcome by identifying response peaks within the presentation period. Consider, for example, the impact that coping efforts are likely to have on HR responding during a 1-minute trauma-relevant presentation. Initial reactivity followed by disengagement from the task can produce distinctly high and low values in the same interval, leading to an aggregate value that is potentially indistinguishable from that produced by an interval during which the rate was steady at a moderate level. However, the physiological data will better reflect the psychological character of the intervals if peak response is identified.

[5]Unpublished paper presented by Orr and Pitman at the 1993 annual meeting of the International Society for Traumatic Stress Studies, San Antonio, Texas.

TRENDS AND SUGGESTIONS

Addressing the Emotional Specificity of Psychophysiological Findings

To date, most of the research on the psychophysiology of PTSD has focused on establishing the diagnostic and discriminant validity of the disorder without considering the meaning of the psychophysiological responding. Discriminant validity is an empirically determined quality that is not tied to biological or psychological significance. For example, it may be that psychophysiological responding is closely tied to changes in mood state, but this relationship would be overlooked because almost all of the research has used PTSD diagnosis as the primary independent variable rather than attempting to address the mediating influences more directly.

Several investigators have found that emotions other than fear (e.g., anger) are differentially elicited for PTSD and non-PTSD groups following exposure to trauma cues (15,24). The ANS reactivity can be associated with emotional states other than anxiety, and response differences between different emotions are difficult to detect using peripheral ANS measures such as HR and SC (104). However, there is a growing body of evidence that emotion-specific responding is reflected in patterns of facial muscle contraction (105,106). Future studies using facial EMG may help to elucidate the emotional networks activated during exposure to trauma cues.

Evaluating Recovery of Responding

An approach that has been fruitful with regard to other psychophysiological research, including research on anxiety disorders (107), is examination of response recovery. This approach is based in part on the recognition that responding to stressors has adaptive value and, as appears to be the case for generic stress tasks, may not be differentially affected by diagnostic status. The maladaptive aspects of the responding may be evident in the period after the stressor ends, as indexed by elevations in ANS activity that persist despite the absence of provocation. A number of the important aspects of this approach have been outlined by Haynes et al. (108).

Ambulatory Monitoring

Ambulatory assessment has been conducted with respect to other anxiety disorders, most notably panic disorder, with some success (109). Similar methods may be valuable for PTSD for at least three reasons.

1. Ambulatory monitoring may help resolve the issue of the origins of resting elevations in physiological measures associated with the condition. Orr (personal communication, February 28, 1994) and his colleagues have already used 24-hour ambulatory measurement to test for differences in HR between individuals with and without PTSD. Their initial findings indicate no stable differences, consistent with our view that resting differences prior to assessment are the result of anticipatory distress.
2. In psychometric terms, ambulatory measurement can help to establish the ecological validity of laboratory-based psychophysiological assessments by quantifying the relationship between reactivity in the lab and in the natural environment.
3. More important, ambulatory measurement would allow direct measurement of the magnitude of responding in natural settings where patients are confronted with unexpected reminders of their traumatic events.

As has been shown with panic disorder patients, information gained from ambulatory monitoring can contribute to both clinical care for individuals and increase theoretical understanding of the condition in general.

CONCLUSIONS

Finally, two points in conclusion. First, Anderson and McNeilly (79) have persuasively argued that a contextual approach to psychophysiological research is needed. According to this perspective, psychophysiological responding is a function of the ecological niche that the person inhabits at the time of assessment. This point seems particularly appropriate for psychophysiological testing conducted with individuals who have experienced traumatic stress. Their behavior may be determined by and reflective of heightened sensitivity to physical threat. This outlook can be expected to influence their perceptions of and behavior in the assessment environment.

Second, despite the long list of factors that can complicate the use of psychophysiological measures in PTSD research, some consistent findings do emerge. Our hope is that future refinements in the selection and application of psychophysiological measures will enhance their value to both clinical and research efforts with traumatized populations.

REFERENCES

1. Shay J. Learning about combat stress from Homer's Iliad. *J Traum Stress* 1991;4:561–579.
2. Trimble MR. Post-traumatic stress disorder: history of a concept. In Figley CR, ed. *Trauma and its wake.* New York: Brunner/Mazel, 1985;5–14.
3. Stierlin E. *Uber der medizinischen folgezustande der katastrophe von courrieres.* Berlin, 1901.
4. Veith I. *Hysteria: history of a disease.* Chicago: University of Chicago Press, 1965.
5. Burgess AW, Holmstrom LL. *Rape: victims of crisis.* Bowie, MD: Robert J. Brady, Co., 1974.
6. Davidson JRT, Foa EB. Diagnostic issues in posttraumatic stress disorder: considerations for the DSM-IV. *J Abnorm Psychol* 1991;100:346–355.
7. Krantz DS, Manuck SB, Wing RR. Psychological stressors and task variables as eliciters of reactivity. In Mathews KA, Weiss SM, Detre T, Demobroski TM, Falkner B, Manuck SB, Williams RBJ, eds. *Handbook of stress reactivity and cardiovascular disease.* New York: John Wiley & Sons, 1986;85–107.
8. Shagass C. An electrophysiological view of schizophrenia. *Biol Psychiatry* 1976;11:3–30.
9. Boudewyns PA, Hyer L. Physiological response to combat memories and preliminary treatment outcome in Vietnam veteran PTSD patients treated with direct therapeutic exposure. *Behav Ther* 1990;21:63–87.
10. Keane TM, Fairbank JA, Caddell JM, Zimering RT. Implosive (flooding) therapy reduces symptoms of PTSD in Vietnam combat veterans. *Behav Ther* 1989; 20:245–260.
11. Keane TM, Kaloupek DG. Imaginal flooding in the treatment of post traumatic stress disorder. *J Consult Clin Psychol* 1982;50:138–140.
12. Shalev AY, Orr SP, Pitman RK. Psychophysiologic response during script-driven imagery as an outcome measure in posttraumatic stress disorder. *J Clin Psychiatry* 1992;53:324–326.
13. Berntson GG, Cacioppo JT, Quigley KS. Autonomic determinism: the modes of autonomic control, the doctrine of autonomic space, and the laws of autonomic constraint. *Psychol Rev* 1991;98:459–487.
14. Blanchard EB, Kolb LC, Prins A. Psychophysiological responses in the diagnosis of posttraumatic stress disorder in Vietnam veterans. *J Nerv Ment Dis* 1991;179: 97–101.
15. Pitman R, Orr S, Forgue D, deJong J, Claiborn J. Psychophysiologic assessment of posttraumatic stress disorder imagery in Vietnam combat veterans. *Arch Gen Psychiatry* 1987;44:970–975.
16. Kardiner A. *The traumatic neuroses of war.* New York: Hoeber, 1941.
17. Gillespie RD. *Psychological effects of war on citizen and soldier.* New York: Norton, 1942.
18. Wenger MA. Studies of autonomic balance in Army Air Force personnel. *Comparative Psychology Monographs 19* 1948;101:1–11.
19. Dobbs D, Wilson WP. Observations on persistence of war neurosis. *Dis Nerv Sys* 1960;21:606–691.
20. McCaffrey RJ, Lorig TS, Pendrey DL, McCutcheon NB, Garrett JC. Odor-induced EEG changes in PTSD Vietnam veterans. *J Traum Stress* 1990;6:213–224.
21. Shalev AY, Orr SP, Pitman RK. Psychophysiological assessment of traumatic imagery in Israeli civilian patients with posttraumatic stress disorder. *Am J Psychiatry* 1993;150:620–624.
22. McFall ME, Murburg M, Ko G, Veith R. Autonomic responses to stress in Vietnam combat veterans with posttraumatic stress disorder. *Biol Q* 1990;27:1165–1175.
23. Blanchard EB, Kolb LC, Pallmeyer T, Gerardi R. A psychophysiological study of post traumatic stress disorder in Vietnam veterans. *Psychiatr Q* 1982;54:220–229.
24. Pitman R, Orr S, Forgue D, Altman B, deJong J. Psychophysiologic responses to combat imagery of Vietnam veterans with posttraumatic stress disorder versus other anxiety disorders. *J Abnorm Psychol* 1990;99: 49–54.
25. Barlow DH. *Anxiety and its disorders: the nature and treatment of anxiety and panic.* New York: Guilford Press; 1988.
26. Malloy P, Fairbank J, Keane T. Validation of a multimethod assessment of posttraumatic stress disorders in Vietnam veterans. *J Consult Clin Psychol* 1993;51:488–494.
27. Hastrup JL. Duration of initial heart rate assessment in psychophysiology: current practices and implications. *Psychophysiology* 1986;23:15–18.
28. McFall ME, Veith RC, Murburg MM. Sympathoadrenal function in posttraumatic stress disorder. *Biol Psychiatry* 1992;31:1050–1056.
29. Shalev AY, Orr SP, Peri T, Schreiber S, Pitman RK. Physiologic responses to loud tones in Israeli patients with posttraumatic stress disorder. *Arch Gen Psychol* 1992;49:870–875.
30. Orr SP, Pitman RK, Lasko NB, Herz LR. Psychophysiological assessment of posttraumatic stress disorder imagery in World War II veterans. *J Abnorm Psychol* 1993;102:152–159.
31. Blanchard EB, Hickling EJ, Taylor AE, Loos WR,

Gerardi R. The psychophysiology of motor vehicle accident related post-traumatic stress disorder. *Behav Ther* 1994;25:453–467.

32. Keane TM, Fairbank JA, Caddell JM, Zimering RT, Bender ME. A behavioral approach to assessing and treating post-traumatic stress disorder in Vietnam veterans. In Figley CR, ed. *Trauma and its wake.* New York: Brunner/Mazel; 1985:257–294.
33. Keane TM, Zimering RT, Caddell JM. A behavioral formulation of PTSD in Vietnam veterans. *The Behav Therap* 1985;8:9–12.
34. Ross RJ, Ball WA, Cohen ME, Silver SM, Morrison AR, Dinges DF. Habituation of the startle reflex in posttraumatic stress disorder. *J Neuropsychiatry* 1989; 1:305–307.
35. Lang PJ, Bradley MM, Cuthbert BN. Emotion, attention, and the startle reflex. *Psychol Rev* 1990;97:377–395.
36. Butler RW, Bratf SL, Rausch JL, Jenkins MA, Sprock J, Geyer MA. Physiological evidence of exaggerated startle response in a subgroup of Vietnam veterans with combat-related PTSD. *Am J Psychiatry* 1990; 147:1308–1312.
37. Pallmeyer T, Blanchard E, Kolb L. The psychophysiology of combat induced post-traumatic stress disorder in Vietnam veterans. *Behav Res Thery* 1986;24:645–652.
38. Blumenthal TD. The startle response to acoustic simuli near startle threshold: effects of stimulus rise and fall time, duration, and intensity. *Psychophysiology* 1988; 25:607–611.
39. Grillon C, Ameli R, Merikangas K, Woods SW, Davis M. Measuring the time course of anticipatory anxiety using the fear-potentiated startle reflex. *Psychophysiology* 1993;30:340–346.
40. Bradley MM, Cuthbert BN, Lang PJ. Pictures as prepulse: Attention and emotion in startle modification. *Psychophysiology* 1993;30:541–545.
41. Vrana SR, Lang PJ. Fear imagery and the startle-probe reflex. *J Abnorm Psychol* 1990;99:189–197.
42. Paige SR, Reid GM, Allen MG, Newton JEO. Psychophysiological correlates of posttraumatic stress disorder in Vietnam veterans. *Biol Psychiatry* 1990;27: 419–430.
43. Ehlers A, Margraf J, Roth WT. Experimental induction of panic attacks. In Hand I, Wittchen HU, eds. *Panic and phobias.* New York: Springer-Verlag; 1986:53–66.
44. Scher H, Furedy JJ, Heslegrave RJ. Individual differences in phasic cardiac reactivity to psychological stress and the law of initial value. *Psychophysiology* 1985;22:345–348.
45. Keane TM, Wolfe J, Taylor KL. Post-traumatic stress disorder: evidence for diagnostic validity and method for psychological assessment. *J Clin Psychol* 1987;43: 32–43.
46. McFall ME, Murburg MM, Roszell DK, Veith RC. Psychophysiological and neuroendocrine findings in posttraumatic stress disorder: a review of theory and research. *J Anxiety Disorders* 1989;3:243–257.
47. Bloom LJ, Houston BK, Holmes DS, Burish TG. The effectiveness of attentional diversion and situation redefinition for reducing stress due to a nonambiguous threat. *J Res Pers* 1977;11:83–94.
48. Obrist PA. The cardiovascular-behavioral interaction—as it appears today. *Psychophysiology* 1975;13: 95–107.
49. Zahn TP. Psychophysiological approaches to psychopathology. In Coles MGH, Dorchin E, Porges SW, eds. *Psychophysiology: systems, processes, and applications.* New York: Guilford; 1986:508–610.
50. Perkins KA, Epstein LH, Jennings JR, Stiller R. The cardiovascular effects of nicotine during stress. *Psychopharmacology* 1986;90:373–378.
51. Masson CL, Gilbert DG. Cardiovascular responses to a quantified dose of nicotine as a function of personality and nicotine tolerance. *J Behav Med* 1990;13:505–521.
52. Hughes JR. Possible effects of smoke-free inpatient units on psychiatric diagnosis and treatment. *J Clin Psychiatry* 1993;54:109–114.
53. Lyvers M, Miyata Y. Effects of cigarette smoking on electrodermal orienting reflexes to stimulus change and stimulus significance. *Psychophysiology* 1993;30:231–236.
54. Emmons KM, Weidner G. The effects of cognitive and physical stress on cardiovascular reactivity among smokers and oral contraceptive users. *Psychophysiology* 1988;25:166–171.
55. Lane JD, Williams RB. Caffeine affects cardiovascular response to stress. *Psychophysiology* 1985;22:648–655.
56. Pincomb GA, Lovallo WR, Passey RB, Brackett DJ, Wilson MF. Caffeine enhances the physiological response to occupational stress in medical students. *Health Psychol* 1987;6:101–112.
57. Lovallo WR, Pincomb GA, Sung BH, Everson SA, Passey RB, Wilson MF. Hypertension risk and caffeine's effect on cardiovascular activity during mental stress in young men. *Health Psychol* 1991;10:236–243.
58. Ratliff-Crain J, O'Keeffe MK, Baum A. Cardiovascular reactivity, mood, and task performance in deprived and nondeprived coffee drinkers. *Health Psychol* 1989;8:427–447.
59. Veleber DM, Templer DI. Effects of caffeine on anxiety and depression. *J Abnorm Psychol* 1984;93:120–122.
60. Shalev A, Bleich A, Ursano RJ. Posttraumatic stress disorder: Somatic comorbidity and effort tolerance. *Psychosomatics* 1990;31:197–203.
61. Fowles DC. The three-arousal model: implications of Gray's two-factor learning theory for heart rate, electrodermal activity, and psychopathy. *Psychophysiology* 1980;17:87–104.
62. Lovallo WR, Wilson MF, Pincomb GA, Edwards GL, Tomkins P, Brackett DJ. Activation patterns to aversive stimulation in man: passive exposure versus effort to control. *Psychophysiology* 1985;22:283–291.
63. Sherwood A, Dolan CA, Light KC. Hemodynamics of blood pressure responses during active and passive coping. *Psychophysiology* 1990;27:656–668.
64. Cacioppo JT, Berntson GG, Andersen BL. Psychophysiological approaches to the evaluation of psychotherapeutic process and outcome, 1991: contributions from social psychophysiology. *Psychol Assess* 1991; 3:321–336.
65. Atkinson RM, Spar LF, Sheff AG, White RAF, Fitzsimmons JT. Diagnosis of posttraumatic stress disorder in Vietnam veterans: preliminary findings. *Am J Psychiatry* 1984;141:694–696.

66. Solomon SD, Gerrity ET, Muff AM. Efficacy of treatments for posttraumatic stress disorder: an empirical review. *JAMA* 1992;268:633–638.
67. Turpin G. The psychophysiological assessment of anxiety disorders: three system measurement and beyond. *Psychol Assess: J Consult Clin Psychol* 1991;3: 366–375.
68. Michelson L. Treatment consonance and response profiles in agoraphobia: the role of individual differences in cognitive, behavioral, and physiological treatments. *Behav Res Ther* 1986;24:263–275.
69. Ost LG, Johansson J, Jerremalm A. Individual response patterns and the effects of different behavioral methods in the treatment of claustrophobia. *Behav Res Ther* 1982;20:445–460.
70. Kolb LC, Mutalipassi LR. The conditioned emotional response: a sub-class of the chronic and delayed posttraumatic stress disorder. *Psychiatr Ann* 1982;12:979–987.
71. Eifert GH, Wilson PH. The triple response approach to assessment: A conceptual and methodological reappraisal. *Behav Res Ther* 1991;29:283–292.
72. Spinhoven P, Onstein EJ, Sterk PJ, Le Haen-Versteijnen D. Discordance between symptom and physiological criteria for the hyperventilation syndrome. *J Psychosom Res* 1993;37:281–289.
73. Tyrer P, Lee I, Alexander J. Awareness of cardiac function in anxious, phobic, and hypochondriacal patients. *Psychol Med* 1980;10:171–174.
74. Lang PJ, Greenwald MK, Bradley MM, Hamm AO. Looking at pictures: affective, facial, visceral, and behavioral reactions. *Psychophysiology* 1993;30:261–273.
75. De Meersman RE. Aging as a modulator of respiratory sinus arrhythmia. *J Gerontol: Biol Sc* 1993;48:B74–B78.
76. Faucheux BA, Dupuis C, Baulon A, Lille F, Bourliere F. Heart rate reactivity during minor mental stress in men in their 50's and 70's. *Gerontology* 1983;29:149–160.
77. Matthews KA, Stoney CM. Influence of sex and age on cardiovascular responses during stress. *Psychosom Med* 1988;50:46–56.
78. Uchino BN, Kiecolt-Glaser JK, Cacioppo JT. Age-related changes in cardiovascular response as a function of a chronic stressor and social support. *J Pers Soc Psychol* 1992;63:839–846.
79. Anderson NB, McNeilly M. Age, gender, and ethnicity as variables in psychophysiological assessment: sociodemographics in context. *Psychol Assess: J Consul Clin Psychol* 1991;3:376–384.
80. Baum A, Grunberg NE. Gender, stress, and health. *Health Psychol* 1991;10:80–85.
81. Liu K, Ballew C, Jacobs DR, et al. Ethnic differences in blood pressure, pulse rate, and related characteristics in young adults. *Hypertension* 1989;14:218–226.
82. McAdoo WG, Weinberger MH, Miller JZ, Fineberg NS, Grim CE. Race and gender influence hemodynamic responses to psychological and physical stimuli. *Hypertension* 1990;8:961–967.
83. Ironson GH, Gellman MD, Spitzer SB, Llabre MM, DeCarlo-Pasin R, Weidler DJ, Schneiderman N. Predicting home and work blood pressure measurements from resting baselines and laboratory reactivity in black and white Americans. *Psychophysiology* 1989;26:174–184.
84. Delehanty SG, Dimsdale JE, Mills P. Psychosocial correlates of reactivity in black and white men. *J Psychosom Res* 1991;35:451–460.
85. Kaplan BJ, Whitsett SF, Robinson JW. Menstrual cycle phase is a potential confound in psychophysiology research. *Psychophysiology* 1990;27:445–450.
86. Stoney CM, Langer AW, Gelling PD. The effects of menstrual cycle phase on cardiovascular and pulmonary responses to behavioral and exercise stress. *Psychophysiology* 1986;23:393–402.
87. Czajkowski SM, Hindelang RD, Dembroski TM, Mayerson SE, Parks EB, Holland JC. Aerobic fitness, psychological characteristics, and cardiovascular reactivity to stress. *Health Psychol* 1990;9:676–692.
88. Hobson ML, Rejeski WJ. Does the dose of acute exercise mediate psychophysiological responses to mental stress? *J Sport Exerc* 1993;15L:77–87.
89. Light KC, Obrist PA, James SA, Strogatz DS. Cardiovascular responses to stress: relationships to aerobic exercise patterns. *Psychophysiology* 1987;24:79–86.
90. Rejeski WJ, Thompson A, Brubaker PH, Miller HS. Acute exercise: Buffering psychosocial stress response in women. *Health Psychol* 1992;11:355–362.
91. Steptoe A, Kearsley N, Walters N. Cardiovascular activity during mental stress following vigorous exercise in sportsmen and inactive men. *Psychophysiology* 1993;30:245–252.
92. Lacey JI, Lacey BC. Verification and extension of the principle of autonomic response stereotyphy. *Am J Psychol* 1958;71:50–73.
93. Levis DJ, Smith JE. Getting individual differences in autonomic reactivity to work for instead of against you: Determining the dominant "psychological" stress channel on the basis of a "biological" stress test. *Psychophysiology* 1987;24:346–352.
94. Blanchard EB, Kolb LC, Taylor A, Wittrock D. Cardiac response to relevant stimuli as an adjunct in diagnosing post-traumatic stress disorder: Replication and extension. *Behav Ther* 1989;20:535–543.
95. Orr SP, Claiborn JM, Altman B, Forgue DF, deJong JB, Pitman RK, Hertz LR. Psychometric profile of posttraumatic stress disorder, anxious, and healthy Vietnam veterans: correlations with psychophysiologic responses. *J Consult Clin Psychol* 1990;58:329–335.
96. Davidson JRT, Fairbank JA. The epidemiology of posttraumatic stress disorder. In Davidson JRT, Foa EB, eds. *Posttraumatic stress disorder: DSM IV and beyond.* Washington, DC: American Psychiatric Press; 1993:147–172.
97. Keane TM, Wolfe J. Comorbidity in post-traumatic stress disorder: An analysis of community and clinical studies. *J Appl Psychol* 1990;20:1776–1778.
98. Lahmeyer HW, Bellur SN. Cardiac regulation and depression. *J Psychiatr Res* 1987;21:1–6.
99. Sayette MA, Smith DW, Breiner MJ, Wilson GT. The effect of alcohol on emotional response to a social stressor. *J Stud Alcohol* 1992;53:541–545.
100. Storrie MC, Doerr HO, Johnson MH. Skin conductance characteristics of depressed subjects before and after therapeutic intervention. *J Nerv Ment Dis* 1981;69: 176–179.
101. Lader MH. *The psychophysiology of mental illness.* London: Routledge & Kegan Paul; 1975.

102. Smith TW, Baldwin M, Christensen AJ. Interpersonal influence as active coping: effects of task difficulty on cardiovascular reactivity. *Psychophysiology* 1990;27: 429–437.
103. Roth WT, Dorato KH, Kopell BS. Intensity and task effects on evoked physiological responses to noise bursts. *Psychophysiology* 1984;21:466–481.
104. Stemmler G. The autonomic differentiation of emotion revisited: convergent and discriminant validation. *Psychophysiology* 1989;26:617–632.
105. Cacioppo JT, Petty RE, Losch ME, Kim HS. Electromyographic activity over facial muscle regions can differentiate the valence and intensity of affective reactions. *J Pers Soc Psychol* 1986;50:260–268.
106. Dimberg U. Facial electromyography and emotional reactions. *Psychophysiology* 1990;27:481–494.
107. Rapee RM, Brown TA, Antony MM, Barlow DH. Response to hyperventilation and inhalation of 5.5% carbon dioxide-enriched air across the DSM-III-R anxiety disorders. *J Abnorm Psychol* 1992;101:538–552.
108. Haynes SN, Gannon LR, Orimoto L, O'Brien WH, Brandt M. Psychophysiological assessment of poststress recovery. *Psychol Assess: J Consul Clin Psychol* 1991;3:356–365.
109. Clark DB, Taylor CB, Hayward C, et al. Motor activity and tonic heart rate in panic disorder. *Psychiatry Res* 1990;32:45–53.
110. Blanchard EB, Kolb LC, Gerardi R, Ryan P, Pallmeyer T. Cardiac responses to relevant stimuli as an adjunctive tool for diagnosing post-traumatic stress disorder in Vietnam veterans. *Behav Ther* 1986;17:592–606.
111. Gerardi RG, Blanchard EB, Kolb LC. Ability of Vietnam veterans to dissimulate a psychophysiological assessment for post-traumatic stress disorder. *Behav Ther* 1989;20:229–243.
112. Gerardi RG, Keane TM, Cahoon BJ, Klauminzer GW. An In Vivo Assessment of Physiological Arousal in Posttraumatic Stress Disorder. *Abnorm Psychol* 1994; 103:825–827.

Neurobiological and Clinical Consequences of Stress: From Normal Adaptation to PTSD, edited by M. J. Friedman, D. S. Charney, and A. Y. Deutch, Lippincott-Raven Publishers, Philadelphia © 1995.

17

Neurobiological Perspectives on Sleep in Post-Traumatic Stress Disorder

Steven H. Woodward

Psychology Service, Veterans Affairs Medical Center, Palo Alto, California 94304

It is evident that sleeping and responding to stressors represent opposite poles in the system(s) of behaviors animals have evolved to succeed biologically. It is remarkable, then, to discover the high degree of overlap between sets of brainstem nuclei subserving sleep and those mediating responses to threat and stress. In an effort to understand this convergence, we will first consider some basic sleep biology. We will then turn to the brainstem to review those regions where sleep is controlled and important adaptations to stress occur. Next, we will consider the symptoms of post-traumatic stress disorder (PTSD) and review sleep laboratory data obtained from patients in search of evidence confirming or disconfirming possible brainstem interactions between centers mediating sleep and stress.

FIRST PRINCIPLES

A logical point of departure for an analysis of sleep disturbance in PTSD is an examination of the biological functions of sleep and those of the acute stress response. It is generally assumed that the acute post trauma syndrome (PTS) is an adaptive defense to the possible recurrence of threat, a proposition that has guided cognitive models of PTSD (1) and discussions of its neurobiology (2). Innate defense responses subserved by special neuronal machinery lie at the foundation of all animals' behavioral repertoires. They are apparent also at the highest levels of the mammalian central nervous system (CNS). Neurons preferentially responsive to threatening stimuli have been located in the neocortices of lab-reared primates (3). Normal human subjects preferentially attend to such stimuli as snakes, spiders, and angry faces presented subliminally (4). In similar fashion, defense-related learning modes, ranging from sensitization in aplysia (5) to one-trial taste aversion learning in mammals (6), are both phylogenetically ancient and subserved at the highest levels of the CNS.

How does sleep disturbance fit into this picture of PTS/PTSD as an adaptive response to threat? It seems reasonable that individuals who respond to first instances of acute threat by exhibiting avoidance and hyperarousal will survive recurrences at higher rates than those who do not; however, the situation is not so clear for sleep-related symptoms. Although sleep loss may be conceptualized as an adaptive limitation on the time spent in a state of reduced vigilance, it could also be an epiphenomenal side effect of waking PTS. That is, sleep may simply be incompatible with the waking adaptations evolved to survive conditions of recurrent threat. The point is particularly germane when considering nightmares. Whereas imperative waking memories of the trauma (intrusive thoughts, flashbacks) might be viewed as sustainers of avoidance and hypervigilance, what additional advantage do nightmares confer upon an individual already experiencing sleep limitation?

Uncertainty as to the primary function of sleep ramifies throughout explorations of sleep disturbance in the psychiatric disorders. A currently dominant perspective in the biology of sleep is that sleep solves an energy balance problem posed by endothermy (7,8,9). This problem appears to be a serious one, as total sleep deprivation is fatal in most laboratory species (9). This fact may explain why sleep eventually overtakes individuals regardless of environmental demands for vigilance. Sleep is *protected.* Support for this thermoregulatory perspective derives from correlational studies of interspecies differences in sleep quota, body size, and metabolic rate (8), and from elegant and ongoing sleep deprivation studies performed in the rat by Rechtschaffen and colleagues (9). The data deriving from this framework appear consistent with significant functional overlap for rapid-eye-movement (REM) and non–rapid-eye-movement (NREM) sleep in thermoregulatory support. The REM-deprived and total-sleep–deprived rats demonstrate the same pattern of lesions, metabolic abnormalities, hyperphagia, and preference for warm environments. Further complicating efforts to understand sleep disturbances across different psychiatric disorders, it would appear that humans are not as reliant upon REM sleep, at least, for thermoregulatory support, as are laboratory species (9,10). Pharmacological blockade of REM sleep in humans can be sustained, without apparent thermoregulatory disturbance, well into the range (relative for size) where rats demonstrate severe compromise of energy metabolism and even death (11,12). A number of possible accounts for this observation can be offered. A thermoregulatory subrole apparently specific to REM sleep in rats may be absent in humans, or satisfied by very small amounts of REM (9). Alternatively, this function may have been taken over by NREM sleep or by some nonsleep metabolic or behavioral state (13). In general, these accounts are consistent with a *decoupling* of thermoregulatory support from sleep, as proposed by Horne (10). Studies of shift workers (14) and medical residents (15) have confirmed that humans can survive nearly indefinite partial sleep deprivation, though not without negative effects on mood. Increased longer term flexibility in the coupling between sleep and thermoregulation may have left humans better able to adapt to novel combinations of behavioral and thermoregulatory challenge.

Data and theory concerning chronobiological alterations or dysrhythmias may help bridge the gap between the posited thermoregulatory role for sleep and its disturbance in psychiatric disease (16,17). With regard to PTSD, alterations of circadian cortisol and noradrenaline rhythms have been reported in Vietnam combat veterans with PTSD (18,19). Alterations in REM sleep distribution have been reported in Dutch WWII resistance fighters with PTSD (20). Although consideration of dysrhythmias in this context is probably premature, an understanding of how such findings generally relate to the thermoregulatory dimension of sleep physiology may be auspicious for the study of sleep in PTSD. Acute PTS is associated with sustained sympathetic activation (2), which must, in turn, incur high energy costs and excess thermoregulatory burden. It may be heuristic to consider the symptoms of chronic PTSD as an adaptation to extend the response to acute threat while maintaining thermoregulatory balance (c.f. Shiraki and Yousef (21)).

BRAINSTEM MECHANISMS IN SLEEP AND ADAPTATION TO TRAUMA

We turn next to an examination of where, in the brain, adaptations to stress might contact mechanisms subserving sleep. An obvious point of departure is the brainstem locus coeruleus (LC). Consideration of the central noradrenaline (NA) axis in both sleep and stress underscores the breadth of behavior this system subserves. The LC firing parallels vigilance/external demand from quiescence during REM sleep to extreme firing rates during periods of survival threat (22,23,24). Intermediate levels of vigilance are associated with intermediate levels of LC firing. In addition to following the tonic behavioral level with tonic firing rates, the LC-NA axis may perform a "mode-switching" function via phasic bursts of activity. Aston-Jones et al. (23) have suggested that the LC interrupts behavior with activity bursts that are inversely proportional to the level of current activity and directly proportional to the biological priority of the new stimulus. The most intense bursts of LC activity appear when threat stimuli are presented during periods of low vigilance. The LC supports a large, direct NA innervation of cerebral cortex with the following known effects on target cells: 1) via the α_1 receptor, responses to all excitatory inputs are enhanced, and 2) via the β receptor, spontaneous firing rates are reduced while excitatory and inhibitory responses to afferent input are accentuated (24). This latter effect has been likened to an increase in the signal-to-noise ratio of cortical unit responses.

The role of the LC-NA axis in the modulation of CNS state during acute and chronic stress is covered in detail in other chapters in this volume and will be reviewed only briefly here. In monkeys, activation of LC, either by an α_2 adrenergic antagonist such as yohimbine or by direct electrical stimulation produces a fear-like behavioral state (25,26). The LC lesions block behavioral fear responses (27,28,29). Particularly relevant to PTSD is the finding that an intact LC system appears "necessary for the acquisition of fear-conditioned responses" (see Charney et al., p. 147(30)), and that reexposure to fear-conditioned stimuli elevates NA metabolism, presumably via elevated LC activity (31,32,33). The acquisition of conditioned fear responses and sympathetic activation upon reexposure to fear-conditioned stimuli are central features of human PTSD. Chronic stress also appears to be associated with modification of central NA systems (23,34,35,36,37), as laboratory species subjected to inescapable shock demonstrate adrenoceptor downregulation (38,39,40). The LC may also participate in the coordination of CNS status with the hypothalamic-pituitary-adrenal (HPA) arm of the stress response (30,41). Direct injection of the stress-related transmitter, corticotropin-releasing factor (CRF) elevates LC firing (42,43), and stress-modulation of LC activity is CRF-mediated (44,45), at least in part. Data from human PTSD patients are also suggestive of NA and HPA dysregulation. Urinary NE excretion has been reported to be elevated in combat-related PTSD patients compared with normals (46,47,48), and platelet NA receptor α_2 downregulation has been observed (49,50,51). Recently, Southwick et al. (52) demonstrated that yohimbine potentiates both panic and PTSD symptomology, as well as levels of plasma 3-methoxy-4-hydroxyphenylglycol (MHPG) levels in combat-related PTSD patients but not controls. Although existing data justify a suspicion that chronic stress downregulates NA α_2 receptors in LC, the resulting modification of the LC response profile and impact upon targets remains to be conclusively specified.

THE LOCUS COERULEUS IN SLEEP

The efforts of Hobson, McCarley, Steriade, and many others (see comprehensive review (53)) have indicated the fundamental role played by the LC-NA axis in the control of sleep. This literature is distinguished by its focus on the interactions among the LC and the other brainstem nuclei to which it is connected in supporting sleep cycling. In the current view, the core oscillatory "engine" comprises a monoaminergic complex formed by the LC and dorsal raphe (DR), and a cholinergic complex comprising the pedunculopontine tegmental nucleus (PPT) and the lateral dorsal tegmental nucleus (LDT). Within this framework, the LC/DR complex is labeled *REM-off* because its activity approaches zero during REM sleep and because it appears to exert inhibition over components of REM sleep. The PPT/LDT complex is termed *REM-on* because increased activity in these nuclei is implicated in the triggering of REM sleep phenomena in other CNS structures, particularly medial pontine reticular formation. The LC/DR and PPT/LDT nuclear groups appear to engage in a push-pull or opponent-process dynamic during sleep, with LC/DR inhibition of PPT/LDT playing a prominent role. It is well established that depletion of adrenergic and serotonergic receptors in the pons releases REM sleep signs (54,55). Experimental cooling of the LC is associated with the onset of REM sleep (56), and cooling of the DR is effective in releasing pontogeniculooccipital (PGO) waves (57), large excitatory potentials that propagate upwards from the brainstem through the thalamus into cerebral cortex during REM sleep. It has been recently confirmed that both DR and LC directly innervate cholinergic neurons in PPT/LDT (58), the latter via α_2 adrenergic receptors (59).

A strong implication of current models of the brainstem sleep oscillator is that modified LC-NA axis function could have a direct impact upon sleep cycling. The dynamics of *REM-off* and *REM-on* nuclear groups have been aggressively modeled by McCarley and coworkers (60,61,62). Given appropriate synaptic weightings, initial activation levels, and modulatory recurrent excitation and inhibition, the opponent PPT/LDT–LC/DR interaction creates a cycling of unit activities formally similar to classical predator-prey relations (60). The addition of two classes of nonlinearities, firing rate upper and lower bounds on both cell populations, and firing-rate–dependent gains for recurrent excitation and inhibition, enables the simulation of "limit cycle" behavior in which the PPT/LDT–LC/DR dynamic converges on normal REM cycling from a range of initial conditions. The "limit cycle reciprocal interaction model" (LCRIM (61,62)) successfully simulates REM onset timing at atypical circadian phases, as well as REM latency reduction and first REM period extension in depression. The latter simulation is accomplished by a simple lowering of the initial level of *REM-off* cell activity at sleep onset (60) (also see Steriade and McCarley (53), Chapter 13), a manipulation broadly consistent with models of reduced central NA function in depression. Efforts have been made to merge the LCRIM with a complementary model of slow-wave sleep (SWS) authored by Borbely and colleagues (63,64,65,66).

Of potential interest to researchers examining sleep in PTSD is the expanded role that the LC-NA axis plays in recent mathematical models of the sleep cycle. Both the latest version of the LCRIM (61,62) and the Borbely-Achermann model of SWS (65) now incorporate what are described as noise terms to enable simulation of the shortterm perturbations characteristic of sleep. In the LCRIM, noise is simulated, in part, via short bursts of excitation of the *REM-off* LC/DR complex (62) deriving from diverse forebrain, brainstem, and somatic pathways, and is used, as well, to "wake the system up." This picture is generally consistent with integration of exogenous and endogenous inputs accomplished by LC and with the marked increases observed in LC firing immediately prior to waking (22). Besides responding to exogenous sensory inputs, the LC also responds to endogenous events such as blood volume alterations, sudden pCO_2 variations, and distentions of urinary bladder and distal colon (67,68,69). Such stimuli, though related to internal rather than external phenomena, remain imperative sleep arousal cues. This is true even in species such as predators that demonstrate deep, continuous, REM-rich sleep (70,71). Also of interest is the suggestion of Aston-Jones and colleagues that most sensory input to LC is relayed through the nucleus paragigantocellularis (PGi) (72,73,74). In PGi, known sensory responsivity is dominated by the auditory and somatosensory modalities, and by nociception (75), all critical channels for rapid arousal. In sum, the constellation of endogenous and exogenous inputs eliciting activation in the extended PGi/LC-NA complex, ranging from acute threat stimuli to bladder distention, may all share a status of being *rapid arousal cues.* This observation is consistent with a broad conception of the LC-NA axis, not as a "trauma center," per se, but as a generic mode-switching mechanism that interrupts lower priority behaviors such as sleep and upregulates CNS tone to meet higher priority biological needs (22). If arousal from sleep is only a special case of a generic arousal mechanism, then upregulation of that mechanism by trauma may have unavoidable consequences for sleep. It is not clear, as yet, whether the receptor modifications within LC or its targets are consistent with such a picture. When the stress-related LC response modifications are specified, models such as the LCRIM may allow translation of such data into testable predictions regarding sleep cycling and arousal. For example, certain modifications of intrinsic and spontaneous responding in LC could masquerade as, and be simulated as, increased noise in the context of the LCRIM.

REM SLEEP AND THE ADAPTIVE UPREGULATION OF ADRENOCEPTORS

A perspective on REM sleep neurophysiology that deserves mention here is a class of models in which REM sleep is accorded direct functional relevance to NA receptor regulation. As just noted, activity in the LC virtually ceases during REM sleep. Based upon this observation, Hartmann (75), Stern and Morgane (76), and most recently, Siegel and Rogawski (13) have proposed that REM sleep evolved to enable the upregulation of NA systems in the brain fatigued by the sustained vigilance associated with endothermy. According to this perspective, profound depression of LC activity during REM sleep frees the target cells of the continuous NA input characteristic of all other behavioral states and so provides an opportunity for receptor upregulation. Siegel and Rogawski (13) have marshaled considerable support for this view, which predicts, for example, that acute elevation of NA release should effectively substitute for REM sleep, reduce "REM pressure," and obviate REM rebound. Consistent with that prediction, REM rebound does not follow brief pharmacologic elevations of NA activity (77), yet does follow chronic pharmacologic elevation (78,79). As sustained hypervigilance is a core feature of human PTS/PTSD, and as central NA receptor downregulation is suspected (39,40), human PTSD would seem to provide something of a test of this framework.

STARTLE AND *REM-ON* CELLS

To this point, we have focused on the brainstem LC in examining possible points of contact between sleep mechanisms and neural modifications induced by acute and chronic stress. We now turn to another promising region of overlap between the neurobiologies of trauma and sleep: the intimate relationship between REM sleep and the startle response. Jouvet (54), Morrison (80), and their collaborators first noted and explored the considerable similarity between startle and PGO waves, the "stigmata" of REM sleep (81,82,83,84,85,86,87). Morrison suggested that PGO waves are "endogenous startle events," and described REM sleep as a period of almost continuous startle/orienting to endogenous stimuli (80,87). In an important recent study, Wu, Mallick, and Siegel (88) observed in cat that startle PGO waves (elicited by 115 dB stimuli, measured in lateral geniculate) demonstrated waveforms, state-dependent amplitudes (largest during the transition to REM sleep, smallest during waking), and associations with eye movements highly similar to those of spontaneous PGOs. As well, during waking and NREM sleep (i.e., in the absence of REM atonia), startle PGO and startle electromyogram (EMG) response amplitudes were positively correlated. Finally, both startle PGO waves and startle EMG responses demonstrated prepulse inhibition (PPI), attenuation of the startle response by a lower intensity stimulus (prepulse) presented 50–200 milliseconds earlier. The PPI was apparent during waking and NREM sleep, but disappeared in REM sleep.

It is now known that the "obligatory" startle circuit (89) does not directly overlap the site of PGO generation, which has been localized with some confidence to the *REM-on* PPT/LDT (see review (53)). Instead, Koch et al. (90) have demonstrated that an inhibitory cholinergic pathway from PPT/LDT to the nucleus reticularis pontis caudalis (RPC), the modulatory relay in the acoustic startle reflex arc (89), is probably the final link in the mediation of PPI. In this study, acetylcholine (ACh) inhibited the majority of acoustically responsive cells in RPC while neurotoxic lesions of cholinergic PPT/LDT neurons impaired PPI. In light of the described connectivity, the coincidence of startle PGO wave maxima and PPI blockage suggests the inactivation of a common modulatory mechanism during REM sleep. The PPI of waking startle is reduced in acutely traumatized children (91), although results in adults with chronic PTSD remain equivocal (92).

While the startle response can be attenuated by a prepulse, it can be potentiated by the concurrent presentation of a fear-conditioned stimulus. Extensively studied by Davis and colleagues

(89), ''fear potentiation'' of startle also relies upon inputs to RPC, in this case excitatory inputs from the central nucleus of the amygdala (CNA) (93). It is of special interest here that the PPT/LDT complex shares with RPC descending afferents from the CNA, as well as from the lateral hypothalamus/zona incerta region and central grey (58,90). The CNA is crucial for mediation of fear and aggression (94), while the lateral hypothalamus/zona incerta region and central grey are implicated in coordination of motoric and autonomic functions. Semba and Fibiger (53) have suggested that limbic projections to PPT/LDT may enable emotional/motivational modulation of the vigilance functions that are supported by PPT/LDT projections to thalamocortical circuits and basal forebrain. As these authors point out, the syndrome of narcolepsy, in which emotive stimuli can trigger episodes of REM sleep that interrupt waking (cataplexy), seems to provide evidence of limbic influence on *REM-on* centers. This circuitry, particularly afferents arising in the CNA, raises the possibility of fear potentiation of PPT/LDT function.

A third dimension of the association between the *REM-on* PPT/LDT complex and startle is discernible in the contribution of PPT to brainstem mechanisms underlying motor function. Various modes of locomotion as well as postural and muscle tone adjustments can be induced by stimulation of brainstem motor areas such as the ventral tegmental field, dorsal tegmental field, and mesencephalic motor area, the latter including PPT (95,96). In cat, motor behavior following stimulation in the region of the PPT is ''startle-like'' and ''propulsive'' (96). Mori et al. (96) further reported that the PPT-stimulated cat ''gave the impression . . . it was trying to avoid a noxious stimulus'' (page 69). This PPT-induced behavioral program was also notable for *not* requiring continuing stimulation once triggered. Garcia-Rill and colleagues (95) have reported that, in cat and rat, substance P may be an excitatory neurotransmitter with particularly long-lasting effects in this system, which is also inhibited by cholinergic agonists. These findings appear to implicate PPT in a startle-related, defensive-evasive motor program. The possibility that further components of this program are coded in PPT/LDT is suggested by evidence that some portion of the coupling of respiratory behavior to motoric function is also accomplished there (97). In combat-related PTSD patients, exaggerated startle responses typically include a complex defensive-evasive maneuver described as ''hitting the dirt'' and involve sympathetic arousal. In fact, one prominent coding scheme utilizes precisely these dimensions, the complexity of the response and the severity of sympathetic arousal, to index startle severity (98).

The findings just mentioned indicate that the PPT/LDT complex, in addition to being intimately connected with RPC and modulated by limbic/emotive inputs perhaps including the CNA, is implicated in PPI of startle, in a startle-related defensive/evasive response, and in the PGO wave that may be an endogenous concomitant of startle. The possibility that the responsivity of the PPT/LDT complex is modified in PTSD seems clearly warranted. Glenn (86) has gone so far as to suggest that the ''central issue . . . of REM sleep [can] be approached by determining what it is about the startle response that makes its endogenous activation in sleep so advantageous'' (page 92). Is it an evolutionary accident that REM sleep incorporates repeated activation of components of a defensive/evasive response of which startle may be the leading edge? Could REM sleep provide for the sequestered exercise of this response? The possibility of excess activation of the PPT/LDT axis, whether by upregulated excitatory or downregulated inhibitory modulation, forms the basis of the theory of PTSD nightmares proposed by Ross and colleagues (99) to be reviewed in the next paragraph.

It is valuable to review, before moving on, that in addition to its posited role in sleep cycling, the PPT/LDT complex participates in the cholinergic ''reticular activating system'' that mediates arousal throughout the forebrain (61). The PPT/LDT complex include units that fire at high rates during waking and REM (61) in addition to REM-specific units, and provides the dominant cholinergic projection to virtually all thalamic nuclei (see review (53)). Cholinergic projections to thalamus from brainstem outnumber monoaminergic projections by five to ten times (100). Stimulation of the peribrachial region of the PPT/LDT blocks so-called "spindle" rhythmicity in the nucleus reticularis thalami (NRT) (101) that, in turn, are implicated in "gating" transmission from thalamus to cortex (102). Ascending brainstem cholinergic axis activity relayed through the nucleus basalis is implicated in the suppression of the delta rhythms characteristic of SWS (62,103). Thus, the ascending cholinergic axis shares with the LC-sourced adrenergic axis widespread arousal functions whose upregulation, as may occur in PTSD, could have negative consequences for sleep.

A THEORY OF REM SLEEP NIGHTMARES IN PTSD

Before turning to an examination of nightmare phenomena in PTSD, a few preparatory comments are due. The relation between sleep and nightmare disturbance in PTSD has been difficult to assess for at least three reasons: 1) dream mentation is researchable only in humans; 2) studies of dream mentation generally rely upon arousing subjects to obtain dream reports, impeding study of the relation of dream mentation to baseline sleep; and 3) for reasons not well understood, nightmares are rarely observed in the sleep laboratory (104,105)—so rarely, in fact, that the drawing of affirmative conclusions regarding the stage of sleep from which nightmares emerge, for example, is hazardous, at best. This situation may change in the future with the application of novel methods of longterm, in-home dream recording and sleep data acquisition.

Trauma-related dream anxiety episodes, or nightmares, may be a unique feature of PTSD. Drawing upon the connections between startle and REM phasic phenomena just reviewed, Ross and colleagues (98) reasoned that the nightmares of PTSD might also reflect exaggerated outflow from brainstem nuclei subserving PGO wave generation. A point in favor of the Ross model of nightmares in PTSD is its convergence with the currently dominant "activation-synthesis" model of dreaming (106). The distribution of sensory percepts in dream reports, favoring visual percepts, and secondarily, auditory ones, is in general agreement with the distribution of PGO wave excitation of cerebral cortex in laboratory species (107). This isomorphism suggests that dream mentation represents the cortex "best guess" (synthesis) as to the meaning of phasic brainstem input (activation). On the motor side, the activation-synthesis model proposes that the cortex interprets or "dreams out" corollary discharges from brainstem motor areas. The distribution of movement percepts in dream reports, 78% of which indicated leg movements in one survey (107), is taken to reflect the distribution of brainstem-encoded effector programs. As just reviewed, the PPT encodes a defensive/evasive locomotive program with coupled autonomic support (96,97). The "dreaming out" of corollary discharges from such a program might be predicted to involve threat and defense-related content normatively, as characterize approximately 50% of normals' dream reports (108). If so, exaggerated outflow from the brainstem centers in question might be reasonably predicted to potentiate negative dream content into the nightmare range. As well, the "dreaming out" of corollary discharges from autonomic activation apparently coupled to the defensive/evasive program encoded in PPT might also play a role in the fear percepts central to nightmares.

Ross and colleagues (109) predicted that exaggerated activity in centers of PGO wave generation would be associated with elevated numbers and/or higher amplitudes of surface-recordable REM phasic events. The latter include eye movements, orbicularis oculi EMG bursts, stapedius EMG bursts, leg muscle bursts, and brief disturbances of respiratory and cardiac rhythms (110). In support of this hypothesis, Ross and colleagues (109) have observed that combat-related PTSD patients exhibit higher frequencies of leg muscle twitches and higher REM percentages of sleep than controls. Our laboratory has been engaged in testing a further prediction of the Ross theory, that REM phasic activity will vary systematically across PTSD patients differing in levels of nightmare complaint. We found that REM-sleep EMG bursts were more frequent in combat-related PTSD patients with

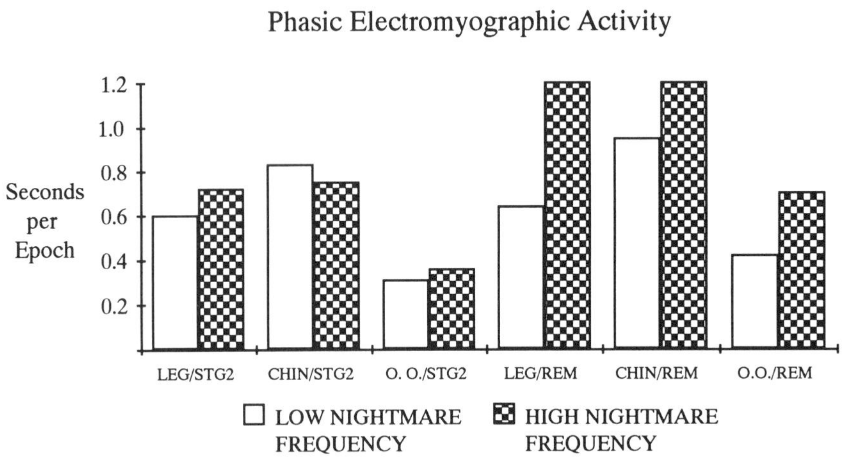

FIG. 1. Plots mean values of phasic electromyogram (EMG) elevation (in seconds per 30-second sleep epoch) by group and site. Sites include leg, chin, and orbicularis oculi (O.O.). Elevated nightmare frequency is associated with elevated EMG bursting during REM sleep, particularly at the leg site. Data were taken from 32 unmedicated combat-related post-traumatic stress disorder (PTSD) patients. EMG data are means of two postadaptational nights.

elevated nightmare complaint as compared with patients denying nightmares (111,112). As Fig. 1 illustrates, though statistically significant, the differences are quite small—an additional 0.5 second of phasic EMG per minute of REM sleep. This effect tended to be stronger at leg (anterior tibialis) than chin sites, a finding in apparent agreement with the priority of locomotive coding in the *REM-on* complex (97). We have found trends towards elevated eye movement activities (numbers and amplitudes of movements; Woodward et al., unpublished data), and higher respiration and heart rates (113) in PTSD patients complaining of more nightmares. Sleep architecture differences were nearly absent between these groups, but involved a surprising reduction in REM percentage of sleep in the high nightmare sample. It is not clear that this apparent contradiction between elevated phasic REM and reduced tonic REM activity might be resolved.

A point deserves mention regarding the possible mechanism of excess activation in the brainstem centers related to PGO waves in PTSD. Close examination of the data of Wu et al. (88) discloses a marked increase in the latency variability of *elicited* PGO waves during REM sleep as compared to waking and NREM. This variability may show evidence of a mechanism operating to maintain *out-of-phase* relations among REM phasic components during sleep. We could speculate that such a mechanism is necessary so that the separate REM phasic activities do not summate and arouse the animal from sleep. The occasional activations and myoclonic jerks that break through REM atonia may reflect occasional *in-phase* activations of REM phasic phenomena. Whatever function is proposed, an implication of the Wu et al. data is that REM phasic phenomena could exert more potent effects, not only by becoming stronger individually but also by increasing their in-phase activation. The latter effect may be detectable in computer-aided examination of REM phasic phenomena at high temporal resolutions, but would not be detectable in studies examining single event types in isolation.

NREM NIGHTMARES IN PTSD?

Although the initial observations of correspondence between REM sleep and dream mentation were made by Aserinsky and Kleitman 40 years ago (114), it has been known for more than 30 years that dream mentation occurs in other sleep stages (115). Foulkes (115) obtained dream reports from 56% of awakenings from stage 1 NREM sleep compared to 83% of awakenings from REM. Although the REM/dreaming paradigm remains dominant, an alternative "continuous state" perspective (116) holds that cortical activation, whether or not due to PGO wave activation, is the key determinant of dreaming. The stages out of which dream mentation is most reliably reported, REM sleep and stage 1, are also those with the highest levels of cortical activation as indicated by low-voltage, desynchronized scalp electroencephalogram (EEG). The volume of REM dream recall, a factor confounded with all measures of dream report, correlates positively with cortical activation indexed by EEG power (117). Although the continuous state perspective invokes no mechanism with obvious relevance to nightmares, it does imply that nightmares should be observable in both REM and NREM sleep given comparable levels of EEG and generalized arousal. The rare PTSD nightmares observed in the laboratory have arisen approximately as often from NREM as from REM sleep, and typically in association with movement (118,119,120,121). In addition, increases in myoclonic activity across REM and NREM sleep in PTSD patients have been observed (109,122). Sleep diary data collected in our laboratory point in the direction of two species of trauma-related nightmares, one emerging early in the night associated with sympathetic activation, and one occurring later in the REM-rich part of the night and not strongly associated with sympathetic activation (123,124). While these data do not confirm a continuous state model of dreaming, they support heterogeneity in the physiological underpinnings of nightmare phenomena in PTSD rather than an exclusive REM-sleep basis.

The activation-synthesis and continuous state frameworks agree that cortical activation is necessary for generation of reportable dream mentation. The continuous state framework does not rule out involvement of phasic brainstem activation. Stage 1 sleep both produces the most frequent dream reports among NREM stages and is characterized by the greatest activity in LC (22). The synthesis of afferent input and "dreaming out" of corollary discharges from brainstem motor areas require activated cortex, not REM sleep, per se. Although PGO waves also occur at low rates outside of REM, they are most common immediately prior to REM onsets and thus are not likely to contribute to stage 1 dream mentation. Could phasic LC-NA axis outflow underlie NREM nightmare reports? Such nightmares should emerge preferentially from stage 1 sleep and so be distinct from night terrors that emerge from SWS, and nocturnal panics that emerge during the transition from stage 2 to SWS (125). This hypothesis might be partially testable using very–low-dose yohimbine provocation following Southwick et al. (52), an approach that might have the great value of allowing simultaneous study of sleep continuity and nightmare disturbance.

TRAUMA-RELATED NIGHTMARE CONTENT

Neither of the reviewed perspectives on dreaming address the question of how specific recurrent dream content might arise; yet, in combat-related PTSD, nightmares "are often repetitive replicas of true combat scenes, appearing more like a memory than a dream . . ." (98) (see page 700). The recurrent trauma-related nightmare, even if accounting for only a minority of dream reports, forces a consideration of the contents of individuals' memories as a factor in nightmare generation. Trauma-related memories are often considered analogous to fear-conditioned responses in animals (30). The contents of human memory in anxiety states have been productively viewed in light of the Lang model of "fear networks" (126) (also see Chemtob et al. (1)). Fear networks are a hypothetical class of brain representations that integrate memories of threatening objects, events, and conditions with appropriate responses such as sympathetic arousal and specific defense-related behaviors and cognitions. Characteristics attributed to fear networks are amplifications of features common to contemporary models of associative memory (c.f. Rumelhart and McClelland (127)). These include intra-associations within a network such that partial information is sufficient to excite the entire representation, and bidirectionality of associations such that efferent limbs may trigger sensory/cognitive limbs. Fear networks thus integrate robust memories having low thresholds and high trace strengths with holistic (i.e., cognitive-motor-autonomic) defensive responses. Fear networks resemble classic brainstem defense responses in that they are rapidly triggered by appropriate stimuli, recruit a holistic defensive program, and are resistant to forgetting. If trauma-related memories in PTSD are viewed as *acquired defense responses,* their activation during REM sleep in the form of nightmares would be reminiscent of the activation of startle-like phenomena during REM. One might again ask whether REM sleep provides a behavioral mode in which the animal, shielded by the skeletal muscle atonia, "exercises" defense-related arousal? At the same time, the activation of conditioned fear responses during REM sleep appears problematic, since the LC-NA system, "necessary for the *acquisition* of fear-conditioned responses," (30) (see page 147) is quiescent at that time.

REM "PRESSURE" IN PTSD

Earlier in this chapter, it was demonstrated that brainstem systems implicated in the adaptation to trauma significantly overlap those involved in the control of sleep. It is natural to ask, then, how sleep cycling might be affected by trauma-related adaptations. An increase in REM "pressure" evidenced by a greater abundance and earlier onset of REM sleep would seem predictable in chronic PTSD on a number of grounds. First, there are the observations made by Ross et al. (109) that are consistent with excess *REM-on* complex activation in PTSD. Second, the Siegel and Rogawski (13) perspective would suggest that the sustained hypervigilance of human PTSD leads to central NA receptor downregulation (as has been observed in chronically stressed rats (38,39,40)), increased NA upregulatory drive, and increased REM pressure. Third, most chronic PTSD patients exhibit features of depression, which has been repeatedly associated with REM latency reduction and increases in REM time and REM density (eye movements per unit time; see review (128)). Given the apparent overdetermination of REM pressure increases in PTSD, it is notable that these increases have often resisted confirmation. The REM latency in PTSD has been observed to be reduced in some studies (129,130), but normal (19) or increased (119,130,131,132) in others. In similar fashion, REM percentage of sleep in PTSD has been variously reported as increased (109,133), normal (131), and decreased (118,119,120,132). A recent study in our laboratory (134) comparing combat-related PTSD patients with and without depression found no differences in REM latency or REM percentage of sleep, though REM latency in the depressed group was markedly variable (26 to 158 minutes). Mean REM latency values for both groups were in the normal range for age of 60 to 70 minutes. When subjects were reclassified according to a median split of depression severity scores, the more depressed PTSD patients exhibited *lower* REM percentages of sleep, a finding not attributable to the circadian phase of morning arousal. The REM percentages of sleep, overall, were higher than published age-matched normative values (23% versus 18%) but in line with those of Ross et al. (109). The only sleep architectural feature differing significantly between groups in the expected direction was SWS, which was more abundant in the nondepressed PTSD patients. These data were interpreted as indicating that PTSD modified the expression of depression-related effects on sleep cycling, or alternatively, that the depressive syndrome associated with PTSD is biologically distinct primary unipolar depression. The latter hypothesis has important support from other quarters (135,136).

While suggesting that comorbid depression is not a key factor in this context, the study described in the previous paragraph did not counter the other two posited determinants of increased REM pressure in PTSD, excess *REM-on* complex activation and adrenoceptor downregulation. Is there an alternative to discarding both of these compelling perspectives on REM sleep in PTSD? It may be possible that the noise construct incorporated into the LCRIM can rationalize these contradictions. In simulations of sleep cycling using the LCRIM, amplification of the noise parameter (modeled as a Poisson shot-noise process with random pulse amplitudes and durations and exponential decay) delays REM onset by partially resetting the REM oscillator (62). Following the LCRIM, increased noise in sleep in PTSD could exert an influence on REM sleep countervailing increased REM pressure. Other mathematical consequences of an increase in the noise parameter in the LCRIM include sleep onset delay, sleep offset advance, sleep fragmentation by brief arousals, SWS reduction, and REM cycle elongation (62) (Massaquoi, personal communication). Thus, an increase in the noise parameter, associated in the LCRIM with external influences on the LC/DR complex but also potentially deriving from modified intrinsic function in this complex, may be able to simulate some of the sleep continuity deficits seen in PTSD (see information that follows). Competition between factors operating to accelerate or delay REM onset might help explain the marked variability in REM latency observed within and across studies. One or the other factor might predominate under different conditions, in different groups, or at different points in time. A clear prediction derivable from the LCRIM, however, is that REM onset delay should be associated with marked sleep continuity disturbance.

A review of existing studies suggests that with increasing time since trauma, there is an increase in observable REM pressure in combat-related PTSD, particularly as evidenced by REM sleep percentage. The single study of very acute, severe combat-related PTSD (119) found that REM sleep was greatly reduced (<10%) and delayed (mean REM latency >200 minutes). These observations were made against a background of severe sleep discontinuity, including multiple awakenings, high EMG activity, bursts of alpha in the EEG, and episodes of tachycardia. At the other end of the time-since-trauma spectrum, Van Kammen et al. (133) observed, in a group of WWII POWs studied 40 years post-traumatization, that sleep continuity disturbance, though still evident, was markedly less severe than in the acute cases reported by Schlosburg, despite the advancing age of the POW sample. The POWs' mean REM percentage of sleep was significantly higher than that of an age-matched control group (21% versus 18%, $p < 0.05$). Studies of samples at intermediate times since trauma generally fall within these extremes. Studies of Lavie et al. (132) and Hefez et al. (118) were performed on samples within a few years of traumatization and reported somewhat ameliorated sleep continuity, with REM latency reduced and REM time increased as compared with the cases of Schlosberg et al. The REM pressure, as indexed by REM latency and percentage of sleep, was still reduced as compared vis a vis controls. The PTSD subjects of Glaubman et al. (131), studied approximately 12 years post-trauma, exhibited moderate sleep discontinuity and significantly increased REM latency (151 minutes versus 103 minutes), but only non-significantly reduced REM time as compared to normals. The Vietnam combat-related subjects of Kramer and Kinney (104), studied 10 to 15 years post-trauma, also demonstrated intermediate severities of sleep discontinuity and REM reduction. In a sample of Vietnam combat-related patients studied approximately 20 years post-trauma and exhibiting sleep continuity not different from controls, Ross et al. (109) observed a trend towards reduced REM latency and significantly increased REM time (26% versus 19%, $p < 0.02$). As noted, our combat-related subjects, also studied approximately 20 years post-trauma, demonstrated near-normal sleep continuity, normal REM latencies (~70 minutes), and REM percentages slightly higher than published age-matched normative values (23% versus 17.5% (137)). Mellman et al. (19) found no reemergence of REM pressure in recently studied Vietnam veterans; however, their subjects exhibited significantly impaired sleep continuity. The study of Greenberg et al. (129), in which substantial sleep discontinuity and REM latency reduction were observed simultaneously, stands as a prominent counter-example to this hypothesis. Since this analysis of the literature is essentially cross-sectional, it is silent on the issue of whether the proposed time-since-trauma factor is actually a function of time, or instead of nonrandom attrition from the underlying populations.

SLEEP CONTINUITY DEFICITS

Complaints of fragmented, nonrestorative sleep are common in PTSD. Mellman et al. (138) surveyed a sample of combat-related PTSD patients and obtained a mean self-reported awakening frequency of 2.5 times/night. This figure agrees closely with sleep diary data we have obtained from combat-related PTSD inpatients (Woodward et al., unpublished data). Many studies of laboratory sleep in PTSD have made observations consistent with disturbed sleep continuity. Such observations have included frequent awakenings (104,118,119,131,138), excess body movement (109,119,131), and increased EMG activity (109,119). In one of the most interesting findings in this area, Mellman et al. (19) observed a negative correlation between baseline corrected nocturnal urinary MHPG and total sleep time. This finding provides specific support for an impact of modified NA system function on sleep in PTSD; moreover, there was a nearly significant group difference in the tendency for arousals to emerge from stage 1 sleep, during which LC is most active. The use of larger samples and of adjunctive psychometric data may help clarify this area. Woodward at al. (139) observed that the amount of awaking during sleep correlated positively with PTSD severity as measured by the Mississippi Scale (MISS) (140). Thus, failure to observe sleep continuity deficits in some studies may be related to low PTSD severity independent of traumatic exposure severity. In an overlapping sample of 50 unmedicated PTSD patients (141), we have observed a robust pattern of inverse correlations between sleep EEG power in the sigma (12–16 Hz) and beta (16–32 Hz) bands, and both the MISS and the schizophrenia scale of the Minnesota multiphasic personality inventory–2 (MMPI-2) (142). The MMPI-2 scale 8 elevation is common in PTSD due to patients' endorsements of social alienation, estrangement, unusual experiences, and perceptual abnormalities. Correlations between MISS and MMPI-scale 8 and EEG were consistently stronger with REM than with NREM sleep EEG power. No correlations were observed between EEG power and depression severity scores, though the latter were as strongly correlated with MISS as were MMPI-2 scale 8 scores. We are far from understanding the physiologic basis of such a finding, though a relationship between PTSD symptom severity and brain mechanisms mediating EEG desynchrony is suggested. Among such systems, the ascending cholinergic system is prominent (53,61).

As noted, traditional analyses of sleep architecture have sometimes failed to confirm sleep continuity deficits in PTSD. Sleep latencies can be low and sleep efficiencies relatively high in the laboratory sleep of PTSD patients (109,134). Although a general lightening of sleep is often indicated by a small shift away from the deeper stages 3 and 4 sleep to stage 1 (119,104,109, 133,143), these shifts are often statistically insignificant. Nonconcordance between sleep architecture and sleep complaint is reminiscent of the syndrome of ideopathic insomnia (143), and so has cast doubt on the existence of PTSD-related sleep disturbances in the minds of some investigators. These doubts have been reinforced by the interesting observation, now reported by two laboratories, that under certain conditions PTSD patients exhibit elevated auditory arousal thresholds compared to normals (143,144). Such findings appear to directly contradict complaints of noise sensitivity and arousability.

At least three sources of error could interfere with the observation of sleep continuity and depth deficits in PTSD patients, when using traditional laboratory-based polysomnographic analyses. First, the standard unit of polysomnographic analysis is the 30-second paper epoch. In standard practice (145), neither awakening nor movement is scored unless either comprises more than half an epoch. Thus, awakenings or movements lasting less than 15 seconds, potentially crucial to understanding sleep continuity

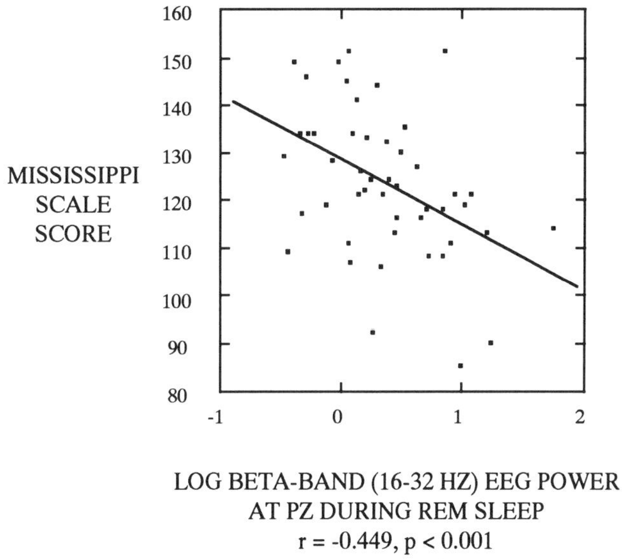

FIG. 2. Plots Mississippi scale scores against log beta (16–31.8 Hz) EEG power-in-band during REM sleep. Data were taken from 50 unmedicated combat-related post-traumatic stress disorder (PTSD) patients. EEG data are means of two postadaptational nights.

deficits in PTSD and in other disorders, are ''filtered out'' of traditional sleep staging. Second, the abundance of SWS is often interpreted as evidence for or against the intactness of NREM sleep; yet, the PTSD sleep literature is characterized by wildly divergent SWS estimates, from a low of 3% (104) to a high of 30% (132). Such differences seem too large to be explained by sample differences in age or cerebral atrophy (c.f. Peters et al. (146)), but instead suggest interlaboratory differences in the scoring of SWS. Traditional SWS scoring involves spectral estimation based on visual analysis of the EEG waveform and is inherently difficult to standardize. Computer-aided spectral estimation of digitized EEG data, though not free of controversy, promises to reduce the error rate in SWS estimation due to visual spectral estimates. Third, the sleep laboratory itself may exert an ameliorative effect on sleep in PTSD. Supportive of such an effect is the paucity of nightmares observed in the laboratory in normal subjects (76). It has been our experience with Vietnam combat-related PTSD inpatients that they often experience an enhanced sense of safety while sleeping in the laboratory. Positive references are frequently made to the all-night presence of sleep technicians. The possibility that a ''guard'' outside one's sleeping quarters could exert an anxiolytic effect in survivors of jungle warfare is not unreasonable. We are now collecting data in an attempt to directly quantify subjective safety in order to determine if it captures variance in laboratory sleep; however, the elimination of laboratory bias, achievable through home sleep recording, may be necessary to answer important questions regarding sleep continuity deficits in PTSD, as well as questions regarding nightmares.

Before leaving this topic, it is interesting to consider how safety cues might influence the individual sleeper. Their entire effect could be exercised during the presleep state, leading to a sleep-promoting physiological/psychological condition enduring throughout the night. Alternatively, safety cues could be reexperienced through the course of sleep. This reexperiencing could be mediated through memory (i.e., dreaming); in normals, up to one third of collected dream reports include content related to the sleep laboratory and staff (147). Or, safety cues could be reexperienced through brief awakenings. The level of engagement of the sleeper with his/her environment may be of particular interest in PTSD. Efforts have been made to assess arousal thresholds and CNS correlates of sensory processing during sleep in normals (148,149,150, 151); however, the meaning of such stimuli has not been varied. Before the elevated arousal thresholds demonstrated by PTSD patients in response to neutral stimuli are viewed as contradicting subjective reports of arousability, the possibility that such stimuli effectively constitute safety cues, reminding subjects that they are in the laboratory, must be eliminated.

SUMMARY

We have reviewed aspects of sleep neurobiology potentially relevant to an understanding of sleep disturbance in PTSD. We have focused on possible overlaps between noradrenergic and cholinergic mechanisms implicated in both sleep and response to stress. Scrutiny of the noradrenergic system suggested that its functions extend beyond threat-related arousal to a more generic arousal function also attuned to interoceptive cues. The possibility fell out from this analysis that arousal from sleep may be a special function of a generic arousal mechanism, with the consequence that stress-related modification of arousal would have unavoidable consequences for sleep. We next examined the intimate anatomic and functional interaction between cholinergic brainstem nuclei subserving both startle-related defense behavior and REM sleep. These nuclei, too, participate in an ascending activating system that modulates cortical arousal in sleep and waking. Ross et al. (98) suggest that dysregulation of components of this system produces REM nightmares in PTSD, a proposal fully compatible with the activation-synthesis model of REM dreaming. The recurrence of trauma-related content in PTSD nightmares, however, points to a role for noradrenergic mechanisms mediating acquisition of conditioned fear responses. An alternative model of LC-NA based NREM stage 1 nightmares compatible with continuous state theory of dreaming was proposed. Review of nightmare phenomena from the perspective of brainstem mechanisms in sleep and trauma underscored the reality that, though the association of nightmares with sleep state has been difficult to establish empirically, such data may yet yield analytic leverage. Because the noradrenergic and cholinergic arousal axes are variably and even reciprocally active during sleep, the stage of emergence of PTSD-related sleep symptoms may be informative in ways that waking symptom expression cannot be. Advanced methods of long-term home sleep recording and dream report acquisition now under development may be useful in answering the still most asked question of sleep in PTSD: from what state do nightmares arise?

We next considered the topic of REM pressure, seemingly overdetermined by putative brainstem and receptor adaptations associated with PTSD, yet inconsistently observed in PTSD patients. Published findings could be loosely rationalized by drawing heavily upon the "limit cycle reciprocal interaction model" of sleep cycling (60,61,62). This approach proposed that factors operating to increase REM pressure were opposed by factors operating to delay REM onset, the latter captured in a construct labeled noise. Amplification of the noise construct also showed promise for the simulation of sleep continuity deficits in PTSD. Noradrenergic effects on sleep continuity were suggested by the observations of Mellman et al. (19) that "baseline-corrected" urinary MHPG was inversely correlated with sleep duration. A number of obstacles to the exploration of sleep continuity disturbance in PTSD were noted, mostly deriving from reliance upon traditional laboratory polysomnography. A deeper understanding of such phenomena may be obtained by using advanced methods of signal analysis and home sleep recording. A plea was also entered for the application of "challenge" studies using both auditory and pharmacological probes (c.f. Sitaram et al. (152)).

To control sleep is to control arousal. Arousal is a cornerstone of the response to threat. These elementary observations recommend examination of apparent commonalities between the mechanisms subserving arousal from sleep and those subserving arousal to the hyperactivated state used to confront threat. Empirical support for a unitary deficit in arousal regulation could take the form of covariances between waking and sleeping symptoms of PTSD. What is the relation between the startle response and both nightmare and sleep continuity disturbance? What is the relation between nightmare disturbance and the strength of conditioned fear responses? These questions remain to be answered. At the same time, the notion of a unitary defect of arousal is itself somewhat contradictory to the paradigm offered by sleep state control—one of cholinergic-aminergic interactions. Waking behavior certainly depends upon some balanced coactivation of these systems, and the evidence suggests that key stress-related adaptations are distributed to both. Both systems possess features such as input from the CNA and modulatory influence over cortex that would seem obligatory in mechanisms regulating an animal's response to threat. Whatever significance is ultimately attached to the alternating quiescences and grossly out-of-balance activations exhibited by these systems during sleep, it is apparent that analysis of one or the other in isolation must be restricted to a limited domain. As a rule, large-scale modulations of arousal involve both. A deeper understanding of the mechanisms of arousal supported by the cholinergic and noradrenergic axes may contribute, in turn, to a deeper understanding of both sleep and waking disturbance in acute and chronic stress.

ACKNOWLEDGMENTS

The author gratefully acknowledges the excellent manuscript assistance provided by Suzette Basore, and the generous advice provided by Richard J. Ross, M. Michelle Murburg, Steve G. Massaquoi, Kathleen L. Benson, and Brett Litz. The author's own research is supported by the Department of Veterans Affairs, the National Center for Post-traumatic Stress Disorder, Clinical Laboratory and Education Division, and the D.V.A.M.C. Medical Center, Palo Alto, California.

REFERENCES

1. Chemtob C, Roitblat HL, Hamada RS, Carlson JG, Twentyman CT. A cognitive action theory of post-traumatic stress disorder. *J Anx Dis* 1988;2:253–275.
2. Krystal JH, Kosten TR, Southwick S, Mason JW, Perry BD, Giller EL. Neurobiological aspects of PTSD: review of clinical and preclinical studies. *Beh Ther* 1989;20:177–198.
3. Hasselmo ME, Rolls ET, Bayliss GC. The role of expression and identity in the face-selective responses of the neurons in the temporal visual cortex of the monkey. *Beh Brain Res* 1989;32:203–218.
4. Ohman A. Face the beast and fear the face: animal and social fears as prototypes for evolutionary analyses of emotion. *Psychophys* 1986;23:123–145.
5. Goelet P, Kandel ER. Tracking the flow of learned information from membrane receptors to genome. *Trends Neurosci* 1986;9:492–499.
6. Garcia J, McGowan BK, Green KF. Biological constraints on conditioning. In Black AH, Prokasy WF, eds. *Classical conditioning II: current research and theory.* New York: Appleton-Century-Crofts; 1972.
7. Berger RJ. Bioenergetic functions of sleep and activity rhythms and their possible relevance to aging. *Fed Proc* 1975;34:97–102.
8. Zepelin H, Rechtschaffen A. Mammalian sleep, longevity, and energy metabolism. *Brain Behav Evol* 1974;10:425–470.
9. Rechtschaffen A, Bergmann BM, Everson CA, Kushida CA, Gilliland MA. Sleep deprivation in the rat: X. Integration and discussion of the findings. *Sleep* 1989;12:68–87.
10. Horne J. *Why we sleep; the functions of sleep in humans and other mammals.* New York: Oxford University Press; 1988.
11. Wyatt RJ, Fram DH, Buchbinder R, Snyder F. Treatment of intractable narcolepsy with a monoamine oxidase inhibitor. *N Engl J Med* 1971;285:987–991.
12. Akindele MO, Evans JI, Oswald I. Mono-amine oxidase inhibitors, sleep and mood. *Electroencephalogr Clin Neurophysiol* 1970;29:27–56.
13. Siegel JM, Rogawski MA. A function for REM sleep: regulation of noradrenergic receptor sensitivity. *Brain Res Rev* 1988;13:213–233.
14. Naitoh P, Kelly TL, Eglund C. Health effects of sleep deprivation. *Occup Med* 1990;5:209–237.
15. Samkoff JS, Jacques CHM. A review of studies concerning effects of sleep deprivation and fatigue on residents' performance. *Acad Med* 1991;66:687–693.
16. Rosenthal NE, Levendosky AA, Skwerer RG, et al. Effects of light treatment on core body temperature in seasonal affective disorder. *Biol Psychiat* 1990;27:39–50.
17. Lesch KP, Mayer S, Disselkamp-Tietze J, Hoh A, Schoellnhammer G, Schulte HM. Subsensitivity of the 5-hydroxytryptamine1A (5-HT1A) receptor-mediated hypothermic response to ipsapirone in unipolar depression. *Life Sci* 1990;46:1271–1277.
18. Yehuda R, Levengood RA, Teicher M, Binder-Brynes K, Trestman RL, Gerber D, Siever L. Circadian secretion of cortisol in PTSD. Presented to the International Society for Traumatic Stress Studies, San Antonio, TX; 1993.
19. Mellman TA, Kumar A, Kulick-Bell R, Kumar M, Nolan B. Nocturnal/daytime urine noradrenergic measures and sleep in combat-related PTSD, in press.
20. de Groen JHM, op den Velde W, van Duijn H, Falger PRJ, Hovens JEJM, Flos MFR. Posttraumatic night-

mares and timing of REM sleep. *J Interdiscipl Cycle Res* 1990;21:192–193.

21. Shiraki K, Yousef MK, eds. *Man in stressful environments: thermal and work physiology.* Springfield, IL: Charles C. Thomas; 1987.
22. Foote SL, Bloom FE, Aston-Jones C. Nucleus locus coeruleus: new evidence of anatomical and physiological specificity. *Physiol Rev* 1983;63:844–914.
23. Aston-Jones G, Valentino RJ, Van Bockstaele EJ, Meyerson AT. Locus coeruleus, stress, and PTSD: neurobiological and clinical parallels. In Murburg MM, ed. *Catecholamine function in post-traumatic stress disorder.* Washington, DC: American Psychiatric Press; 1994, pp. 17–62.
24. Aston-Jones G, Bloom FE. Norepinephrine-containing locus coeruleus neurons in behaving rats exhibit pronounced responses to non-noxious environmental stimuli. *J Neurosci* 1981;1:887–900.
25. Grant SJ, Huang YH, Redmond DE Jr. Behavior of monkeys during opiate withdrawal and locus coeruleus stimulation. *Pharmacol Biochem Behav* 1988;30:13–19.
26. Redmond DE Jr, Krystal JH. Multiple mechanisms of withdrawal from opioid drugs. *Ann Rev Neurosci* 1984;7:443–478.
27. Redmond DE Jr, Huang YH, Snyder DR, et al. Behavioral changes following lesions of the locus coeruleus in *M. arctoides. Neurosci Abs* 1976;1:472.
28. Redmond DE Jr. New and old evidence for the involvement of a brain noradrenergic system in anxiety. In Fann WE, Karacan I, Pakorney AD, Williams RL, eds. *Phenomenology and treatment of anxiety.* New York: Spectrum; 1979;153–203.
29. Svensson TH. Peripheral, autonomic regulation of the locus coeruleus noradrenergic neurons in the brain: putative implications for psychiatry and psychopharmacology. *Psychopharmacology* 1987;92:1–7.
30. Charney DS, Southwick SM, Krystal JH, Deutch AY, Murburg MM, Davis M. Neurobiological mechanisms of PTSD. In Murburg MM, ed. *Catecholamine function in post-traumatic stress disorder.* Washington, DC: American Psychiatric Press, 1994, pp. 131–160.
31. Cassens G, Kuruc A, Roffman M, et al. Alterations in brain norepinephrine metabolism and behavior induced by environmental stimuli previously paired with inescapable shock. *Behav Brain Res* 1981;2:387–407.
32. Rassmussen K, Morilak DA, Jacobs BL. Single unit activity of the locus coeruleus in the freely moving cat, I: During naturalistic behaviors and in response to simple and complex stimuli. *Brain Res* 1986;371: 324–334.
33. Tsaltas E, Gray JA, Fillenz M. Alleviation of response suppression to conditioned aversive stimuli by lesions of the dorsal noradrenergic bundle. *Behav Brain Res* 1987;13:115–127.
34. Glavin GB. Stress and brain noradrenaline: a review. *Behav Neurosci Rev* 1985;9:233–243.
35. Tsuda A, Tanaka M. Differential changes in noradrenaline turnover in specific regions of rat brain produced by controllable and uncontrollable shocks. *Behav Neurosci* 1985;99:802–817.
36. Kosten TR, Krystal JH. Biological mechanisms in posttraumatic stress disorder: Relevance for substance abuse. In Galanter M, ed. *Recent advances in alcoholism. vol. 6.* New York: Plenum; 1988:49–68.
37. Murburg MM, McFall ME, Veith RC. Catecholamines, stress, and posttraumatic stress disorder. In Giller EL Jr, ed. *Biological assessment and treatment of posttraumatic stress disorder.* Washington, DC: American Psychiatric Press; 1990:27–64.
38. Simson PE, Weiss JM. Altered activity of the locus coeruleus in an animal model of depression. *Neuropsychopharmacology* 1988a;1:287–295.
39. Simson PE, Weiss JM. Responsiveness of locus coeruleus neurons to excitatory stimulation is uniquely regulated by alpha-2 receptors. *Psychopharmacol Bull* 1988b;24:349–354.
40. Simson PE, Weiss JM. Altered electrophysiology of the locus coeruleus following uncontrollable stress: relationship to anxiety and anxiolytic action. In Murburg MM, ed. *Catecholamine function in post-traumatic stress disorder.* Washington, DC: American Psychiatric Press, 1994, pp. 63–86.
41. Yehuda R, Southwick SM, Perry BD, Mason JW, Giller EL Jr. Interactions of the hypothalamic-pituitary-adrenal axis and the catecholaminergic system in posttraumatic stress disorder. In Giller EL Jr, ed. *Biological assessment and treatment of posttraumatic stress disorder.* Washington, DC: American Psychiatric Press; 1990:115–134.
42. Valentino RJ, Foote SL, Aston-Jones G. Corticotropin-releasing factor activates noradrenergic neurons of the locus coeruleus. *Brain Res* 1983;270:363–367.
43. Avanzino GL, Ermirio R, Cogo CE, et al. Effects of corticosterone on neurons of the rat locus coeruleus. *Neurosci Lett* 1987;80:85–88.
44. Valentino RJ. Corticotropin-releasing factor: putative neurotransmitter in the noradrenergic nucleus locus coeruleus. *Psychopharmac Bull* 1989;25:306–311.
45. Valentino RJ, Bockstaele EJV, Aston-Jones G. Corticotropin-releasing factor immunoreactive (CRF-IR) neurons are localized in nuclei which project to the locus coeruleus. *Soc Neurosci Abstracts* 1990;16:519.
46. Mason JW, Giller EL, Kosten TR, Yehuda R. Psychoendocrine approaches to the diagnosis and pathogenesis of posttraumatic stress disorder. In Giller EL Jr, ed. *Biological assessment and treatment of PTSD.* Washington, DC: American Psychiatric Press, 1990b.
47. Mason JW, Giller EL, Kosten TR, Ostroff RB, Podd L. Urinary free-cortisol levels in posttraumatic stress disorder patients. *J Nerv Men Dis* 1986;174:14–149.
48. Kosten TR, Mason JW, Giller EL. Sustained urinary norepinephrine and epinephrine elevation in posttraumatic stress disorder. *Psychoneuroendocrinol* 1987; 12:13–20.
49. Perry BD. Placental and blood element neurotransmitter receptor regulation in humans: potential models for studying neurochemical mechanisms underlying behavioral teratology. In Boer GJ, Fenstra MGP, Mirmiran M, eds. *Biochemical basis of functional neuroteratology: permanent effects of chemicals on the developing brain.* Amsterdam: Elsevier, 1988;189–207.
50. Perry BD, Southwick SM, Giller EL Jr. Altered platelet alpha2-adrenergic receptor affinity states in post-traumatic stress disorder. *Am J Psychiat* 1987;144:1511–1512.
51. Perry BD, Southwick SM, Yehuda R, Giller EL Jr. Adrenergic receptor regulation in posttraumatic stress disorder. In Giller EL Jr, ed. *Biological assessment*

and treatment of posttraumatic stress disorder. Washington, DC: American Psychiatric Press, 1990;87–114.

52. Southwick SM, Krystal JH, Morgan CA, et al. Abnormal noradrenergic function in posttraumatic stress disorder. *Arch Gen Psychiat* 1993;50:266–274.
53. Steriade M, McCarley RW. *Brainstem control of wakefulness and sleep.* New York: Plenum Press; 1990.
54. Jouvet M. The function of dreaming: A neurophysiologist's point of view. In Gazzaniga M, Blakemore C, eds. *Handbook of psychobiology.* New York: Academic Press; 1975.
55. Hobson JA, Steriade M. Neuronal basis of behavioral state control. In Mountcastle VB, Bloom FE, eds. *Handbook of physiology.* Bethesda: American Physiological Society; 1986.
56. Cespuglio R, Gomez ME, Faradji H, Jouvet M. Alterations in the sleep-waking cycle induced by cooling of the locus coeruleus area. *Electroencephalogr Clin Neurophysiol* 1982;54:570–578.
57. Cespuglio R, Gomez ME, Walker E, Jouvet M. Effets du refroidissement et de la stimulation des noyaux du systeme du raphe sur les etats de vigilance chez le chat. *Electroencephalogr Clin Neurophysiol* 1979;47: 289–308.
58. Semba K, Fibiger HC. Afferent connections of the laterodorsal and the pedunculopontine tegmental nuclei in the rat: A retro- and antero-grade transport and immunohistochemical study. *J Comp Neurol* 1992;323: 387–410.
59. Williams JA, Reiner PB. Noradrenaline hyperpolarizes identified rat mesopontine cholinergic neurons in vitro. *J Neurosci* 1993;13:3878–3883.
60. McCarley RW, Massaquoi SG. A limit cycle mathematical model of the REM sleep oscillator system. *Am J Physiol* 1986;251:R1011–R1029.
61. McCarley RW, Massaquoi SG. Neurobiological structure of the revised limit cycle reciprocal interaction model of REM cycle control. *J Sleep Res* 1992;1: 132–137.
62. Massaquoi SG, McCarley RW. Extension of the limit cycle reciprocal interaction model of REM cycle control: an integrated sleep control model. *J Sleep Res* 1992;1:138–143.
63. Borbely AA. A two-process model of sleep regulation. *Human Neurobiol* 1982;1:155–165.
64. Achermann P, Borbely AA. Simulation of human sleep: ultradian dynamics of electroencephalographic slow-wave activity. *J Biol Rhythms* 1990;5:141–157.
65. Achermann P. Modeling human slow-wave activity. Presented to the Sleep Research Society, Los Angeles, CA; 1993.
66. Achermann P, Borbely AA. Combining different models of sleep regulation. *J Sleep Res* 1992;1:144–147.
67. Morilak DA, Fornal CA, Jacobs BL. Effects of physiological manipulations on locus coeruleus neuronal activity in freely moving cats. I. Thermoregulatory challenge. *Brain Res* 1987a;422:17–23.
68. Morilak DA, Fornal CA, Jacobs BL. Effects of physiological manipulations on locus coeruleus neuronal activity in freely moving cats. II. Cardiovascular challenge. *Brain Res* 1987b;422:24–31.
69. Morilak DA, Fornal CA, Jacobs BL. Effects of physiological manipulations on locus coeruleus neuronal activity in freely moving cats. III. Glucoregulatory challenge. *Brain Res* 1987c;422:32–39.
70. Allison T, Van Twyver H. The evolution of sleep. *Nat Hist* 1970;79:56–65.
71. Allison T, Cicchetti DV. Sleep in mammals: ecological and constitutional correlates. *Science* 1976;194:732–734.
72. Aston-Jones G, Ennis M, Peribone VA, et al. The brain nucleus locus coeruleus: Restricted afferent control of a broad efferent network. *Science* 1986;234:734–737.
73. Aston-Jones G, Shipley MT, Ennis M, et al. Restricted afferent control of locus coeruleus neurons revealed by anatomic, physiologic and pharmacologic studies. In Marsden CA, Heal DJ, eds. *The pharmacology of noradrenaline in the central nervous system.* Oxford: Oxford University Press; 1990:187–247.
74. Aston-Jones G, Shipley MT, Chouvet G, et al. Afferent regulation of locus coeruleus neurons: anatomy, physiology and pharmacology. *Prog Brain Res* 1991;85: 47–75.
75. Van Bockstaele EJ, Akaoka H, Aston-Jones G. Brainstem afferents to the rostral (juxtafacial) nucleus paragigantocellularis: integration of exteroceptive and interoceptive sensory inputs in the ventral tegmentum. *Brain Res* 1993;603:1–18.
76. Hartmann E. The D-state and norepinephrine-dependent systems. *Int Psychiat Clin* 1970;7:308–328.
77. Stern WC, Morgane PJ. Theoretical view of REM sleep function: Maintenance of catecholamine systems in the central nervous system. *Behav Biol* 1974;11:1–32.
78. Drucker-Colin R, Benitez J. REM sleep rebound during withdrawal from chronic amphetamine administration is blocked by chloraphenicol. *Neurosci Lett* 1977;6: 267–271.
79. Feinberg I, Hibi S, Braun M, Cavness C, Westerman G, Small A. Sleep amphetamine effects in MBDS and normal subjects. *Arch Gen Psychiat* 1974;31:723–731.
80. Morrison AR, Reiner P. A dissection of paradoxical sleep. In McGinty D, Drucker-Colin R, Morrison A, Parmeggiani PL, eds. *Brain mechanisms of sleep.* New York: Raven Press; 1985:97–110.
81. Kaufman LS, Morrison AR. Spontaneous and elicited PGO spikes in rats. *Brain Res* 1981;214:61–72.
82. Kaufman LS. PGO waves in rats in the non-paradoxical sleep states. *Brain Res* 1983;276:73–80.
83. Pivik T, Dement WC. Phasic changes in muscular and reflex activity during non-REM sleep. *Exp Neurol* 1970;27:115–124.
84. Pivik T, Metz J. Phasic EMG inhibition and spinal reflex modulation during synchronized sleep in the cat. *Exp Neurol* 1975;48:493–501.
85. Glenn LL, Dement WC. Motoneuron properties during electromyogram pauses in sleep. *Brain Res* 1982;243: 11–23.
86. Glenn LL. Brainstem and spinal control of lower limb motoneurons with special reference to phasic events and startle reflexes. In McGinty DJ, et al, eds. *Brain mechanisms of sleep.* New York: Raven Press; 1985.
87. Bowker RM, Morrison AR. The startle reflex and PGO spikes. *Brain Res* 1976;102:185–190.
88. Wu M-F, Mallick BN, Siegel JM. Lateral geniculate spikes, muscle atonia and startle responses elicited by auditory stimuli as a function of stimulus parameters and arousal state. *Brain Res* 1989;499:7–17.
89. Davis M, Falls WA, Campeau S, Kim M. Fear-potentiated startle: a neural and pharmacological analysis. *Behav Brain Res* 1993;58:175–198.

90. Koch M, Kungel M, Herbert H. Cholinergic neurons in the pedunculopontine tegmental nucleus are involved in the mediation of prepulse inhibition of the acoustic startle response in the rat. *Exp Brain Res* 1993;97:71–82.
91. Ornitz EM, Pynoos RS. Startle modulation in children with posttraumatic stress disorder. *Am J Psychiat* 1989;146:866–870.
92. Butler RW, Braff DL, Rausch JL, Jenkins MA, Sprock J, Geyer MA. Physiological evidence of exaggerated startle response in a subgroup of Vietnam veterans with combat-related PTSD. *Am J Psychiat* 1990; 147:1308–1312.
93. Rosen JB, Hitchcock JM, Sananes CB, Miserendino MJ, Davis M. A direct projection from the central nucleus of the amygdala to the acoustic startle pathway: Anterograde and retrograde tracing studies. *Behav Neurosci* 1991;105:817–825.
94. Davis M. The role of the amygdala in conditioned fear. In Aggleton JP, ed. *The amygdala: neurobiological aspects of emotion, memory and mental dysfunction.* New York: Wiley-Liss; 1992:255–305.
95. Garcia-Rill E, Kinjo N, Atsuta Y, Ishikawa Y, Webber M, Skinner RD. Posterior midbrain-induced locomotion. *Brain Res Bull* 1990;24:499–508.
96. Mori S, Sakamoto T, Ohta Y, Takakusaki K, Matsuyama K. Site-specific postural and locomotor changes evoked in awake, freely moving intact cats by stimulating the brainstem. *Brain Res* 1989;505:66–74.
97. Garcia-Rill E. The pedunculopontine nucleus. *Prog Neurobiol* 1991;36:363–389.
98. Blake DD, Weathers F, Nagy LM, Kaloupek DG, Klauminzer G, Charney DS, Keane TM. A clinician rating scale for assessing current and lifetime PTSD: The CAPS-1. *The Behav Therap* 1991;14:187–188.
99. Ross RJ, Ball WA, Sullivan KA, Caroff SN. Sleep disturbance as the hallmark of posttraumatic stress disorder. *Am J Psychiat* 1989;146:697–707.
100. DeLima AD, Singer W. The brainstem projection to the lateral geniculate nucleus in the cat: identification of cholinergic and monoaminergic elements. *J Comp Neurol* 1987;259:92–121.
101. Hu B, Steriade M, Deschenes M. The effects of brainstem peribrachial stimulation on reticular thalamic neurons: the blockage of spindle waves. *Neurosci* 1989; 31:1–12.
102. Yingling CD, Skinner JE. Gating of thalamic input to cerebral cortex by nucleus reticularis thalami. *Prog Clin Neurophysiol* 1977;1:70–96.
103. Buzsaki G, Bickford RG, Ponomareff G, Thal LJ, Mandel R, Gage FH. Nucleus basalis and thalamic control of neocortical activity in the freely moving rat. *J Neurosci* 1988;8:4007–4026.
104. Kramer M, Kinney L. Sleep patterns in trauma victims with disturbed dreaming. *Psychiatr J Univ Ottawa* 1988;13:12–16.
105. Hartmann E. *The nightmare.* New York: Basic Books; 1984.
106. Hobson JA, McCarley RW. The brain as a dream-state generator: an activation synthesis hypothesis of the dream-process. *Am J Psychiat* 1977;134:1335–1348.
107. McCarley RW, Hoffman EA. REM sleep dreams and the activation-synthesis hypothesis. *Am J Psychiat* 1981;138:904–912.
108. Trenholme I, Cartwright R, Greenberg G. Dream dimension differences during a life change. *Psychiat Res* 1984;12:36–45.
109. Ross RJ, Ball WA, Dinges DF, Kribbs NB, Morrison AR, Silver SM, Mulvaney FD. Rapid eye movement sleep disturbance in posttraumatic stress disorder. *Biol Psychiat* 1994;35:195–202.
110. Rechtschaffen A. The psychophysiology of mental activity during sleep. In McGuigan FJ, Schoonover RS, eds. *The psychophysiology of thinking.* New York: Academic Press; 1973.
111. Woodward SH, Arsenault NJ, Basore SM. REM phasic EMG activity is elevated in frequent trauma-related nightmare sufferers. *Sleep Res* 1994, Abst. 209.
112. Woodward SH, Friedman MJ, Bliwise DL, Gusman DF, Arsenault NJ, Basore S, Murray C. Polysomnographic correlates of nightmare frequency in PTSD (in preparation).
113. Woodward SH, Arsenault NJ, Basore SM. Heart and respiration rates during sleep in trauma-related nightmare sufferers. *Sleep Res* 1994, Abst. 210.
114. Aserinsky E, Kleitman N. Regularly occurring periods of ocular motility and concomitant phenomena during sleep. *Science* 1953;118:361–375.
115. Foulkes D. Dream reports from different stages of sleep. *J Abnorm Psychol* 1962;65:14–101.
116. Antrobus J. The neurocognition of sleep mentation: rapid eye movements, visual imagery, and dreaming. In Bootzin RR, Kihlstrom JF, Schacter DL, eds. *Sleep and cognition.* Washington, DC: American Psychological Association; 1990.
117. Meier B. Physiological correlates of REM dream recall. *Sleep Res* 1988;17:149.
118. Hefez A, Metz L, Lavie P. Long-term effects of extreme situational stress on sleep and dreaming. *Am J Psychiat* 1987;144:344–347.
119. Schlosberg A, Benjamin M. Sleep patterns in three acute combat fatigue cases. *J Clin Psychiat* 1978; 39:546–549.
120. Kramer M., Shoen LS, Kinney, L. Nightmares in Vietnam veterans. *J Am Acad of Psychoanal* 1987;15: 67–81.
121. van der Kolk BA, Blitz R, Burr WA, et al. Nightmares and trauma. *Am J Psychiat* 1984;141:187–190.
122. Engdahl B. Chronic posttraumatic stress disorder and sleep disturbances among former prisoners of war and combat veterans. Presented to the American Psychological Association, Toronto; 1993.
123. Woodward SH, Arsenault EJ, Bliwise DL, Gusman DF. The temporal distribution of combat nightmares in Vietnam combat veterans. *Sleep Res* 1991a;20: 152.
124. Woodward SH, Arsenault EJ, Bliwise DL, Gusman DF. Physical symptoms accompanying dream reports in combat veterans. *Sleep Res* 1991b;20:153.
125. Hauri P, Friedman MJ, Ravaris CL. Sleep in patients with spontaneous panic attacks. *Sleep* 1989;12: 323–337.
126. Lang PJ. The cognitive psychophysiology of emotion: fear and anxiety. In Tuma AH, Maser JD, eds. *Anxiety and the anxiety disorder.* Hillsdale, NJ: Lawrence Erlbaum, 1985.
127. Rumelhart DE, McClelland JL. *Parallel distributed processing: explorations in the microstructure of cognition. vol. 1: Foundations.* Cambridge, MA: MIT Press; 1987.

128. Reynolds CF, Kupfer DJ. Sleep in depression. In Williams RL, Karacan I, Moore CA, eds. *Sleep disorders: diagnosis and treatment.* 2nd ed. New York: Wiley; 1988.
129. Greenberg R, Pearlman CA, Gampel D. War neuroses and the adaptive function of REM sleep. *Br J Med Psych* 1972;45:27–33.
130. Kauffman CD, Reist C, Djenderedjian A, et al. Biological markers of affective disorders and post-traumatic stress disorder: a pilot study with desipramine. *J Clin Psychiat* 1987;48:366–367.
131. Glaubman H, Mikulincer M, Porat A, Wasserman O, Birger M. Sleep of chronic post-traumatic patients. *J Traum Stress* 1990;3:255–264.
132. Lavie P, Hefez A, Halperin G, et al. Long-term effects of traumatic war-related events. *Am J Psychiat* 1979;136:175–178.
133. van Kammen WB, Christiansen C, van Kammen DP, Reynolds CF. Sleep and the prisoner-of-war experience—40 years later. In Giller EL Jr, ed. *Biological assessment and treatment of posttraumatic stress disorder.* Washington, DC: American Psychiatric Press; 1990;159–172.
134. Woodward SH, Friedman MJ, Bliwise DL. Sleep and depression in combat-related PTSD inpatients. *Biological Psychiatry,* in press.
135. Yehuda R, Southwick SM, Krystal JH, Bremner D, Charney DS, Mason JW. Enhanced suppression of cortisol following dexamethasone administration in posttraumatic stress disorder. *Am J Psychiat* 1993a;150: 83–86.
136. Yehuda R, Southwick SM, Nussbaum G, Wahby V, Giller E, Mason JW. Low urinary cortisol excretion in patients with posttraumatic stress disorder. *J Nerv Men Dis* 1990;178:366–369.
137. Williams RL, Karacan I, Hursch CJ. *Electroencephalography (EEG) of human sleep: clinical applications.* New York: Wiley & Sons, 1974.
138. Mellman TA, Kulick-Bell R, Ashlock LE, Nolan B. Sleep events in combat-related post-traumatic stress disorder. Presented to the American Psychiatric Association, Washington, DC: 1992.
139. Woodward SH, Bliwise DL, Friedman MJ, Gusman DF. The relation of sleep architecture and psychometric findings in veterans hospitalized for PTSD (submitted).
140. Keane TM, Cadell JM, Taylor KL. Mississippi scale for combat-related PTSD: three studies on reliability and validity. *J Consult Clin Psychol* 1988;56:85–90.
141. Woodward S, Bliwise D, Arsenault N, Murray C. PTSD-related hyperarousal assessed during sleep. (submitted).
142. Butcher JM, Dahlstrom WG, Graham JR, Tellegen AM, Kaemmer B. *MMPI-2: Manual for administration and scoring.* Minneapolis: University of Minneapolis Press, 1989.
143. Dagan Y, Lavie P, Bleich A. Elevated awakening thresholds in sleep stage 3–4 in war-related post-traumatic stress disorder. *Biol Psychiat* 1991;30:618–622.
144. Schoen L, Kramer M, Kinney L. Auditory thresholds in the dream disturbed. *Sleep Res* 1984;13:102.
145. Rechtschaffen A, Kales A. *A manual of standardized terminology, techniques and scoring system for sleep stages of human subjects.* Los Angeles, CA: Brain Information Service/Brain Research Institute, University of California, 1968.
146. Peters J, van Kammen DP, van Kammen WB, Neylan T. Sleep disturbance and computerized axial tomographic scan findings in former prisoners of war. *Comp Psychiat* 1990;31:535–539.
147. Piccione P, Thomas S, Roth T, Kramer M. Dreaming. *Sleep Res* 1976;5:120.
148. Price LJ, Kremen I. Variations in behavioral response threshold within the REM period of human sleep. *Psychophys* 1980;17:133–140.
149. Harsh J, Badia P, O'Rourke D, Burton S, Revis C, Magee J. Factors related to behavioral control by stimuli presented during sleep. *Psychophys* 1987;24:535–541.
150. Campbell K, Bell I, Bastien C. Evoked potential measures of information processing during natural sleep. In Broughton R, Ogilvie R, eds. *Sleep, arousal and performance.* Boston: Birkhauser; 1991:88–116.
151. Kutas M. Event-related brain potential (ERP) studies of cognition during sleep: is it more than a dream? In Bootzin RR, Kihlstrom JF, Schacter DL, eds. *Sleep and cognition.* Washington, DC: American Psychological Association; 1990:43–57.
152. Sitaram N, Dube S, Jones D, Pohl R, Gershon S. Acetypecholine and α_1-adrenergic sensitivity in the separation of depression and anxiety. *Psychopath* 1984;17: 24–39.

Neurobiological and Clinical Consequences of Stress: From Normal Adaptation to PTSD, edited by M. J. Friedman, D. S. Charney, and A. Y. Deutch, Lippincott-Raven Publishers, Philadelphia © 1995.

18

Clinical Studies of Neurotransmitter Alterations in Post-Traumatic Stress Disorder

Steven M. Southwick, *Rachel Yehuda, and C. A. Morgan III

*Department of Psychiatry, Yale University School of Medicine, New Haven, Connecticut 06510, National Center for Post-Traumatic Stress Disorder, Veterans Affairs Medical Center, West Haven, Connecticut 06516; and *Department of Psychiatry, Mount Sinai School of Medicine, Bronx Veterans Affairs Medical Center, Bronx, New York 10468*

Many of the earliest terms for post-traumatic stress disorder (PTSD), such as "shell shock" (1), reflected the belief that post-traumatic clinical symptomatology was linked to an underlying disturbance of the nervous system. Some believed that the physical shock from exploding shells actually caused permanent alterations in brain structure, leaving the shell-shocked veteran in a chronic state of hypersensitivity and hyperarousal. To distinguish this condition from other neuroses and to emphasize its physiological underpinnings, Kardiner coined the term "physioneurosis" in 1941 (2). In addition to hypervigilance, insomnia, irritability, and hyperarousal, veterans with shell shock or physioneurosis suffered from symptoms of "reexperiencing" and "avoidance."

Although preliminary neurobiologic investigations of combat trauma began as early as 1918, most biologic studies of trauma have been conducted since 1980 when PTSD was added to the formal nosology of the DSM-III. As a result, unlike preclinical stress data, the number of neurobiologic investigations directly related to human trauma is relatively small. Nevertheless, over the past decade, a series of psychophysiologic, hormonal, neurotransmitter, receptor binding, electrophysiologic and brain imaging studies have begun to characterize the biological nature of this disorder.

In this chapter we will describe and summarize clinical studies of neurotransmitter alterations in PTSD. While preclinical studies of stress have repeatedly demonstrated alterations in multiple neurotransmitter systems, studies in traumatized humans have largely been limited to the sympathetic nervous system and the hypothalamic-pituitary-adrenal (HPA) axis. The HPA axis studies are summarized in Chapter 19. In this chapter we will focus primarily on sympathetic nervous system neurotransmitters (norepinephrine, epinephrine, and dopamine), although mention will be made of opiate and serotonin systems.

It is important to note that other neurotransmitter systems are also likely involved in the fight-flight response. Furthermore, because neurotransmitter systems are closely interrelated, it is difficult, if not impossible, to determine accurately the specific behavioral effects of a single neurotransmitter. For example, the HPA axis and sympathetic nervous system appear to modulate one another during acute stress (3).

In humans, most biological studies of PTSD have involved combat veterans. As a result, much of what is discussed in this chapter draws heavily on findings from this patient population. It is important to note, however, that PTSD can develop in response to a wide variety of traumatic stressors, including natural disasters, violent crimes, accidents, child abuse, and war. To the extent that neurobiologic alterations are related to the core symptoms of PTSD, it would be anticipated that findings from combat veterans will be replicated in civilian populations, since the core symptoms of PTSD are similar across all traumas.

Although to date numerous techniques and strategies have been used to study neuroendocrine alterations in PTSD, it is useful to classify these studies into two major categories: 1) naturalistic or baseline observations (Table 1), and 2) challenge strategies (Table 2). In naturalistic studies, components of a particular neurobiologic system are investigated under normal resting conditions. The goal of these studies is to examine the biological systems of interest as they function under normative conditions, without external provocation. Clearly, however, the process of informing an individual that he or she is being evaluated, reviewing potential hazards of the study, obtaining informed consent, attaching electrodes, drawing blood, etc., may serve as environmental influences that impact on the study measures. Challenge strategies, on the other hand, are designed to evaluate biological systems under controlled conditions that intentionally provoke the system. Behavioral, physiologic, and neuroendocrine responses are typically measured after provocation with external stimuli such as trauma scripts or exogenously administered biological substances. When baseline and challenge strategies are used in tandem, it is possible to obtain more complete information about the nature and origin of biological alterations.

SYMPATHETIC NERVOUS SYSTEM

In order to maximize both mobilization and utilization of energy under conditions of extreme stress, the sympathetic nervous system tends to discharge as a unit. Coordinated sympathetic discharge will: 1) shunt blood from temporarily unnecessary splanchnic and renal regions to active muscle groups; 2) rapidly increase energy supply to skeletal musculature by mobilizing blood glucose; 3) accelerate heart rate and increase blood pressure, allowing for greater perfusion of muscles and vital organs; 4) dilate pupils so that more light enters the eye; and 5) constrict skin vasculature in order to limit blood loss should injury occur (4,5). Widespread emergency-stimulated sympathetic nervous system discharge prepares the organism for what Cannon termed the "flight or fight" response (6).

While stress-induced sympathetic nervous system activation serves a protective role in the short run, it appears that long-term negative sequelae may ensue for some individuals. That is, the sympathetic nervous system appears to become hyperresponsive to a host of trauma-related stimuli in many individuals who develop PTSD. The nature and extent of this hyperresponsiveness have been the subject of intense recent study and now will be summarized.

HISTORICAL FRAMEWORK FOR STUDIES OF NERVOUS SYSTEM ALTERATIONS IN PTSD

In 1918, two studies became the first laboratory investigations of sympathetic nervous system activity among traumatized humans. First, Meakins and Wilson exposed combat veterans with shell shock to sounds of gunfire and the smell of sulfuric flames, and found that as a group they had greater increases in heart rate and respiratory rate than healthy controls (1). Second, Fraser and Wilson demonstrated exaggerated psychophysiologic arousal with marked increases in subjective anxiety, heart rate, and blood pressure among war veterans compared to healthy controls in response to intravenous epinephrine (EPI) administration (7).

During World War II, Grinker and Spiegal described combat soldiers who appeared to suffer from a chronic stimulation of the sympathetic nervous system. "They perspire freely, are tremulous, restless, irritable, sleep poorly, and look very sick. At times these symptoms suddenly increase, especially in response to mild auditory and verbal stimuli, and the patients react as if they have received an injection of adrenalin" (8). This belief that altered catecholamine function played a critical role in combat neurosis led some clinicians and researchers to advocate bilateral denervation of the adrenal glands as a form of treatment for highly symptomatic war veterans (9).

The first contemporary psychophysiological study in combat veterans was performed by Dobbs and Wilson (10), who demonstrated that combat sounds led "decompensated" World War II soldiers to become more agitated compared with World War II soldiers who were not described as decompensated. Since 1980, there has been a series of psychophysiology studies measuring heart rate, blood pressure, and galvanic skin response at baseline and in response to trauma-related cues (11–20). The data have shown marked elevations in psychophysiologic parameters during provocation but few, if any, differences at baseline between combat veterans and normal controls. This hyperreactiveness has not been observed in combat veterans without PTSD (16) or in combat veterans with anxiety disorders other than PTSD (21), suggesting that neither combat alone nor the presence of anxiety disorders other than PTSD is sufficient to explain postwar physiologic reactivity. Furthermore, hyperreactive psychophysiologic responses to reminders of trauma also have been observed in civilians with PTSD. For example, Blanchard et al., while studying victims of motor vehicle accidents, found heart rate, systolic blood pressure, and electrodermal activity to be significantly elevated in response to personal trauma-related cues among subjects who developed PTSD compared to those who did not develop PTSD (22). Similarly, Shalev et al. reported significant responses to traumatic imagery in Israeli civilians with post-traumatic stress disorder (20).

Indeed, consistent alterations in psychophysiologic responses to stress-related stimuli in humans prompted researchers to investigate the biochemical underpinnings of sympathetic nervous system activation in traumatized populations. These studies have primarily focused on norepinephrine (NE), EPI, and dopamine (DA), the three key sympathetic nervous system neurotransmitters.

CATECHOLAMINE ALTERATIONS IN PTSD

An enormous body of preclinical literature has been devoted to the relationship between stress and catecholamines. For example, it has been shown repetitively that stressful or fearful stimuli of many types produce marked increases in central and peripheral NE. Noradrenergic

TABLE 1. *Baseline catecholamine studies in PTSD*

	Compared to controls	References
24-hour urine		
Norepinephrine	Increased	Kosten et al. (31), Yehuda et al. (30) Davidson and Baum (33)[a]
	No difference	Pitman and Orr (32)[b]
Epinephrine	Increased	Kosten et al. (31), Yehuda et al. (30)
	No difference	Pitman et al. (32)[b]
Dopamine	Increased	Yehuda et al. (30)
Plasma		
Norepinephrine	No difference	McFall et al. (34), Blanchard et al. (13) Hamner et al. (35), Southwick et al. (36)
Epinephrine	No difference	McFall et al. (34)
MAO activity	Decreased	Davidson and Baum (33)
Receptors		
$Alpha_2$ receptors	Decreased	Perry et al. (40)
cAMP	Decreased	Lerer et al. (48,49) 1987, 1990

[a] Eight-hour urine collection
[b] Compared to combat controls
MAO, monoamine oxidase; PTSD, post-traumatic stress disorder.

TABLE 2. *Catecholamine challenge studies in PTSD*

	Compared to controls	Reference
In Vitro Studies		
Epinephrine/alpha$_2$ receptors	Increased responsivity	Perry et al. (39)
Isoproterenol/cAMP	Decreased responsivity	Lerer et al. (48)
Forskolin/cAMP	Decreased responsivity	Lerer et al. (48) (49)
Non Trauma-related Stressors		
Exercise stress/NE	Increased responsivity	Hamner et al. (35)
Noncombat film/NE	No difference	McFall et al. (34)
Simulated Traumatic Stressor		
Combat film/E	Increased responsivity	McFall et al. (34)
Combat sounds/NE	Increased responsivity	Blanchard et al. (13)
Neuroendocrine Challenge		
Desipramine/growth hormone	No difference	Dinan et al. (60)
Clonidine/growth hormone	Decreased responsivity	Hansenne et al. (61)
Lactate	Increased responsivity	Rainey et al. (63)
Yohimbine/MHPG	Increased responsivity	Southwick et al. (36)
Yohimbine/Startle	Increased responsivity	Morgan et al. (57)
Yohimbine/Cerebral metabolism	Decreased metabolism	Bremner et al. (69)

brain systems appear to play a critical role in orientation to novel stimuli, vigilance, selective attention, and cardiovascular responses to life-threatening situations (23). Although acute stress-related elevations of NE appear to serve a protective role for the organism, accumulating evidence suggests that certain types of stress, especially uncontrollable stress, can contribute to chronic maladaptive alterations in catecholaminergic systems (24). Thus, for example, preclinical investigations have shown that uncontrollable stress can cause chronic increased responsivity of locus coeruleus neurons to excitatory stimulation (25,26), as well as an associated subsensitivity of alpha-2-adrenergic receptors and a decreased number of postsynaptic beta-adrenergic receptors (27,28). (For a review of stress and catecholamines, see Murburg (29) and Chapter 9 of this volume.)

Baseline Studies (Table 1) NE Studies

Mean 24-hour urinary excretion of NE has been found to be elevated in combat veteran inpatients with PTSD compared to normal controls (30), and compared to psychiatric patients with major depression and schizophrenia (31). In combat veterans with PTSD, Kosten and colleagues reported a mean NE level of 76±10 μg/day (31) while Yehuda et al. (30) and Pitman and Orr (32) reported mean NE levels of 56±14 μg/day and 60.8±26.1 μg/day respectively. Pitman and Orr (32) additionally reported that combat veterans without PTSD had a mean NE level of 58.0±35.0 μg/day, which was not statistically different from combat veterans with PTSD. In contrast, the mean NE levels reported for normals has been 32±10 μg/day (30). Although Pitman and Orr concluded that similar NE values for combat veterans with and without PTSD indicated a failure to replicate the Kosten et al. (31) study, their findings may indicate that combat veterans without PTSD, like combat veterans with PTSD, exhibit an elevation of 24-hour urinary catecholamines. Thus, alterations in NE excretion may be a function of trauma, rather than the disorder of PTSD per se. Further studies using all three groups (i.e., combat veterans with PTSD, combat veterans without PTSD, and nontraumatized individuals without psychiatric disorder) are necessary to clarify this issue.

In a study of civilians, Davidson and Baum (33) measured urinary catecholamine levels in a community sample of individuals living within 5 miles of the Three Mile Island nuclear power station 5 years after the accident. The authors found higher urinary NE levels in residents living closer to Three Mile Island compared those living 80 miles away. These results implied increased "stress" levels in individuals with greater exposure to the accident based on proximity. However, Davidson and Baum did not assess subjects for the presence or absence of PTSD.

Unlike findings from 24-hour urine studies, baseline plasma NE investigations have generally reported no significant differences between PTSD patients and normal controls (13,34). McFall et al. (34) and Blanchard et al. (13) in separate studies measuring NE response to war-related laboratory stimuli both reported resting NE values that were similar in PTSD patients and controls. Hamner et al. (35) similarly found no baseline differences between PTSD patients and controls. As part of a yohimbine infusion study, Southwick et al. (36) reported comparable baseline plasma levels of the major norepinephrine metabolite, 3-methoxy-4-hydroxyphenylglycol (MHPG), in PTSD patients and controls. It should be noted, however, that these studies typically use NE values obtained by single-stick venipuncture, which may not provide an accurate baseline catecholamine measure due to the stress of the venipuncture itself (37).

EPI and DA Studies

Twenty-four-hour urinary DA and EPI excretion also have been reported as elevated in combat veterans with PTSD (30,31). The urinary DA finding is consistent with Hamner et al.'s report of significant elevations in plasma DA among 12 Vietnam combat veterans with PTSD compared with major depressed patients and healthy controls (38). Studies of urinary EPI, on the other hand, have been less consistent. While Kosten et al. (31) and Yehuda et al. (30) both reported increased levels of 24-hour urinary EPI excretion among inpatients with PTSD, Pitman and Orr (32) did not find an elevation. In a study measuring baseline plasma EPI in PTSD and non-PTSD subjects, McFall et al. (34) found no differences between groups. Finally, in the just mentioned Yehuda et al. study (30), urinary NE and DA, but not EPI excretion, correlated with severity of PTSD symptomatology, particularly intrusive symptoms.

Adrenergic Receptor Function

Alpha-2-adrenergic receptors play an important role in translating the neurochemical message of NE and EPI, and as such may provide information about the longterm effects of alterations in catecholamine neurotransmission. It has further been suggested that alpha-2-adrenergic receptor activity may reflect overall sympathetic "tone" (39). Perry et al., in two separate radioligand bindings studies—one involving combat veterans (40) and the other investigating traumatized children (41)—found fewer total alpha-2-adrenergic receptor binding sites per platelet in subjects with PTSD compared to controls. Similar findings have been noted in medical illnesses characterized by excessive catecholamine hypersecretion, such as congestive heart failure (42) and hypertension (43). Interestingly, decreased alpha-2 adrenergic receptor number also has been demonstrated in patients with borderline personality disorder (44,45). It is possible that borderline personality disorder and PTSD share a common neurobiologic alteration of the noradrenergic system. Indeed, it has been recently demonstrated that treatment-seeking combat veterans with PTSD show an unusually high incidence of borderline personality disorder (46). As discussed by Perry, chronic elevation of circulating catecholamines may explain this decrease in receptor number (39). Of note, in a separate study, comparison groups of patients with major depression and generalized anxiety disorder both had significantly greater numbers of platelet alpha-2-adrenergic receptors than PTSD patients, suggesting that PTSD differs from these two disorders with regard to regulation of catecholamine metabolism (47).

Finally, Lerer et al. examined adenylate cyclase activation in platelet membranes, as well as adenosine 3′,5′-cyclic phosphate (cAMP) signal tranduction in both platelet membranes and intact lymphocytes of PTSD patients. In separate studies, basal cAMP (48) and basal adenylate cyclase (49) levels were noted to be lower in PTSD patients compared to controls.

Challenge Studies (Table 2)

To date, four different challenge strategies have been applied to the study of the neurobiology of PTSD. These include in vitro paradigms, provocation with nontrauma-related stressors, provocation with trauma-related stressors, and neuroendocrine challenges. Each of these strategies will now be described.

In Vitro Paradigms

An in vitro challenge refers to an experimental paradigm in which the biological system of interest, (e.g., a platelet preparation containing alpha-2 receptors) is isolated from the organism, maintained in a ''live'' state, and then subjected to exogenous stimulation. Such a paradigm allows investigators to directly control a host of variables that would be difficult if not impossible to account for in the living organism. To investigate dynamic functioning and regulation of the alpha-2-adrenergic receptor in patients with PTSD, Perry exposed intact platelets to high concentrations of EPI. Typically, in response to agonist binding (as is the case with EPI), a percentage of receptor recognition proteins are physically taken into the cell in a process called internalization. While internalized, the receptor site is no longer available for radioligand binding. Compared to controls, EPI-incubated platelets of PTSD patients showed a more rapid and extensive loss of receptor protein from the platelet membranes. These findings appear consistent with a receptor-effector system that has been overtaxed and is now easily ''fatigued'' and more sensitive to downregulation (39).

A second series of in vitro investigations has involved forskolin stimulation of lymphocyte and platelet preparations, as well as isoproterenal stimulation of lymphocyte preparations. In lymphocyte preparations, lower responsiveness to isoproterenol and forskolin stimulation was noted in PTSD patients compared to controls (48,49). Similarly, in platelet membrane preparations the patient group was found to have lower responsiveness to forskolin stimulation. These findings appear to reflect a diminished responsivity of the receptor adenylate-cyclase complex in patients with PTSD. Together with Perry et al.'s finding (39–41), these reports point to potential abnormalities at both the adrenergic receptor level and at sites distal to the receptor.

Provocation with Nontrauma Related Stressors

In PTSD, very few reports have examined relative differences between individuals with PTSD and individuals without PTSD in basic stress parameters. Therefore, to date, information concerning the basic stress response is limited. In one study using an exercise treadmill, Hamner et al. (35) investigated the effects of physical stress on plasma catecholamines in 12 combat veterans with PTSD and in 8 normal controls. The two groups did not differ significantly in maximal blood pressure and heart rate response to exercise, pre- and post-test NE levels, or percent increase in mean plasma norepinephrine. However, plasma MHPG levels increased significantly in the PTSD patients but not in the controls. The authors note that an exaggerated MHPG response suggests increased sensitivity to noradrenergic activation in the PTSD group after exposure to a physical stressor unrelated to combat trauma. In contrast to these findings, McFall et al. (34) measured psychophysiological reactivity, circulating plasma catecholamines, and subjective distress in 10 combat veterans with PTSD and 11 controls in response to a film depicting an automobile accident. Although combat veterans with PTSD reported greater subjective distress, they did not show any differences compared to normals on any measure of autonomic arousal, including heart rate, blood pressure, NE, and EPI. It should be stated that the sample size in both of these studies is probably too small to allow sufficient power for detecting differences. Therefore, negative results may be attributed to an insufficient subject number, whereas a positive result in such studies may be the result of chance. Clearly, further studies are needed to clarify these interesting questions.

Another line of neurobiological investigation with potential relevance to neurotransmitter alterations in PTSD involves the acoustic startle response. Startle is a ubiquitous cross species, mammalian CNS response to a sudden exteroceptive stimulus (50). The organism may startle in response to visual, tactile, and auditory stimuli. The startle response is measured as a whole-body movement in the rat and as the magnitude of the eyeblink response in humans. Because startle is a brainstem-mediated response, it may afford a means more direct than cardiac reactivity for investigating the CNS in individuals who have been traumatized. When the organism is presented with a stimulus that is conditioned to fear, startle increases (51). Study of the startle reflex, especially fear-potentiated startle, may have relevance to PTSD since accentuated startle is a cardinal symptom of PTSD in humans, and since preclinical studies have shown that magnitude of startle is sensitive to manipulations of several neurotransmitter systems hypothesized to be involved in the pathophysiology of PTSD (52). For example, drugs that increase NE transmission and that have been associated with increased fear also increase fear-potentiated startle. Thus, in the rat, NE and fear-potentiated startle both increase after yohimbine administration and decrease after infusion of clonidine (53). To date, studies in traumatized humans have largely shown normal acoustic startle in the absence of stressful test conditions (54–56). However, recent data suggest that startle may be more pronounced in subjects with PTSD than in controls under conditions of fear and during hyperadrenergic states (57–59).

Provocations with Trauma-related Stressors

In addition to the psychophysiological studies just reviewed and those summarized in Chapter 16, a number of investigations have measured stress-induced psychophysiologic and catecholamine responses in parallel. McFall et al. (34) found higher levels and a parallel rise in subjective distress, blood pressure, heart rate, and plasma EPI during and after a combat film among combat veterans with PTSD compared to controls. This parallel rise suggested that heightened physiological reactivity was indeed related to circulating catecholamines, more specifically to EPI. Because the increase in EPI and other indices of sympathetic nervous system activity were more pronounced in response to combat-related stimuli than noncombat related stimuli, the authors interpreted these findings as being consistent with the idea that biological responses may be conditioned in PTSD. In a similarly designed study involving combat veterans with and without PTSD, Blanchard et al. measured plasma NE and heart rate before and after exposure to auditory stimuli reminiscent of combat. Veterans with PTSD had significantly greater increases in both NE and heart rate than controls (13).

Neuroendocrine Challenge Studies

The desipramine growth hormone challenge has been used as a probe of postsynaptic alpha-2-adrenergic receptor function in a study of eight traumatized women. Dinan et al. found no difference between traumatized subjects and controls in desipramine-stimulated growth hormone levels (60). Hansenne et al. reported a blunted growth hormone response to intravenous clonidine in a 20-year-old car accident victim with PTSD (61). Since the growth hormone response to clonidine is generally thought to be an index of noradrenergic function, blunting suggested heightened noradrenergic sensitivity with possible downregulation of noradrenergic receptors. After successful treatment, the authors reported a normal growth hormone response to clonidine that they interpreted as further evidence for a relationship between noradrenergic dysregulation and PTSD-specific symptoms (61).

Intravenous lactate infusion has been shown to cause panic attacks in patients with panic disorder (62). Although the precise mechanism of lactate-induced anxiety and panic is not known, central noradrenergic stimulation has been suggested. In a study of seven Vietnam veterans with PTSD, Rainey et al. reported panic attacks in six of seven subjects and flashbacks in all seven (63). However, because the six subjects who had lactate-induced panic attacks also met criteria for comorbid panic disorder, it was not possible to determine whether the responses were secondary to panic disorder, PTSD, or both.

Yohimbine, a more direct probe of noradrenergic activity, has been tested in 20 Vietnam combat veterans with PTSD compared to 18 controls. Yohimbine is an alpha-2-adrenergic receptor antagonist that activates noradrenergic neurons by blocking the alpha-2-autoreceptor, thereby increasing presynaptic noradrenergic activity. Although yohimbine acts on multiple neurotransmitter systems, at the dose employed in this study, its primary effect is on the noradrenergic system (64). Yohimbine produced panic attacks in 70% and flashbacks in 40% of patients with PTSD (36). There were no yohimbine-induced panic attacks or flashbacks among the control group. A more than twofold greater elevation of plasma MHPG following yohimbine administration suggested abnormal presynaptic noradrenergic reactivity in the PTSD patients.

With regard to specificity, yohimbine has not produced similar effects in MDD, schizophrenia, obsessive-compulsive disorder, or even generalized anxiety disorder (65). However, it has resulted in comparable behavioral and cardiovascular responses in panic disordered patients (66), suggesting that PTSD and panic disorder share a common neurobiological abnormality related to altered sensitivity of the noradrenergic system. Importantly, because 43% of the patients with PTSD who had yohimbine-induced panic attacks in this study did not meet comorbid criteria for panic disorder, the yohimbine-induced panic attacks could not be explained solely by the presence of panic disorder. On the other hand, patients meeting criteria for both panic disorder and PTSD had an 89% incidence of yohimbine-induced panic attacks. It may be that comorbid panic disorder in patients with PTSD simply reflects a more pronounced abnormality of the noradrenergic system. This notion is supported by recent evidence suggesting that comorbid panic disorder in some patients with PTSD is most likely trauma-induced rather than familial in origin (67).

In both normal subjects and in combat veterans with PTSD, yohimbine also has been used as a biological probe to study the effects of catecholamine transmission on the acoustic startle reflex (57,68). As predicted by preclinical studies, Morgan et al. found that yohimbine significantly increased both startle reflex and serum MHPG in normal subjects (68). These increases were significantly correlated in time, suggesting that magnitude of startle was related to catecholamine transmission. In a separate study, yohimbine infusion caused significantly greater increases in acoustic startle among PTSD patients compared to controls (57). Although MHPG from this study has not yet been analyzed, in prior published work yohimbine has been shown to produce greater increases in MHPG among combat veterans with PTSD compared to normal subjects (36). Taken together, these studies suggest that exaggerated startle in patients with PTSD may indeed be related to increased noradrenergic transmission. Of note, however, because preclinical data has clearly implicated other neurotransmitters in exaggerated startle, it is likely that exaggerated startle seen in humans with PTSD is modulated by multiple transmitters.

To study the effects of yohimbine on brain metabolism, a single bolus of [F-18]2-fluoro-2-deoxyglucose (FDG) was administered to 10 Vietnam combat veterans with PTSD and 10 healthy age-matched controls immediately following either yohimbine or placebo infusion. Subjects were then scanned for 60 minutes, and a PET image was reconstructed at 30–50 minutes postinjection in order to determine brain tissue activity (69). Consistent with this reported yohimbine study, 6 of 10 PTSD patients had a panic attack and 3 of 10 a flashback in response to i.v. yohimbine but not placebo. A significantly different metabolic response to yohimbine in neocortical brain regions (prefrontal, temporal parietal, orbitofrontal cortices) was noted in patients compared to controls, with controls showing a tendency toward increased and PTSD patients a tendency toward decreased metabolism.

The authors note that these findings agree with preclinical data that supports a dose response effect of NE on brain metabolism where lower levels of NE result in increased metabolism and higher levels in decreased metabolism. Thus, it is hypothesized that PTSD patients, being unusually sensitive to yohimbine, release more NE in response to yohimbine, with a resultant decrease in brain metabolism. This decrease in brain metabolism may cause an increase in signal-to-noise ratio of neuronal activity, with an increase in selective attention or neuronal responsivity to relevant stimuli and a decrease or suppression of background neuronal activity. It is possible that chronic hyperarousal and hypervigilance in individuals with PTSD are related to the effects of increased responsivity of brain catecholamine systems.

SEROTONERGIC ALTERATIONS IN PTSD

Although only two studies have directly examined the 5-HT system in PTSD, a large body of indirect evidence suggests that this neurotransmitter may be important in the pathophysiology of trauma-related symptomatology. In humans, low serotonin (5-HT) functioning has been associated with aggression (70), impulsivity (70), and suicidal behavior (71). Patients with PTSD are frequently described as aggressive or impulsive, and often suffer from depression and suicidal tendencies. Further evidence comes from the observation that serotonin reuptake inhibitors have been found to be partially effective in treating PTSD symptoms such as intrusive memories and avoidance symptoms (72–75).

Baseline Studies

The first report of serotonergic function in PTSD was a study examining paroxetine binding in blood platelets of 20 combat veterans with PTSD under baseline conditions (76). In this study, platelet 5-HT uptake was significantly decreased in PTSD patients compared with normals, and in PTSD patients meeting criteria for comorbid major depressive disorder. Because decreased platelet 5-HT uptake also has been reported in patients with depression and alcoholism, the specificity of these findings has yet to be determined.

Provocation Studies

More recently, the behavioral effects of m-chloro-phenyl-piperazine (MCPP) have been examined in a preliminary study of 14 combat veterans with PTSD (77). MCPP is a 5-HT agonist with predominant effects on 5-HT2 and 5-HT1c receptors. Five of the 14 patients with PTSD had a panic attack and 4 had a flashback following MCPP administration. In contrast, no patient had a panic attack and one patient experienced a flashback following infusion of placebo saline. Thus, a subgroup of patients with PTSD exhibited a marked behavioral sensitivity to serotonergic provocation, raising the possibility of pathophysiologic subtypes among traumatized combat veterans.

Clearly, further studies are needed to delineate possible serotonergic alterations in PTSD. For example, in mood and anxiety disorders, baseline studies of serotonergic function have included measures such as cerebrospinal concentrations of 5-hydroxyindoleacetic acid and platelet 3H-imipramine binding as well as estimates of 5-HT and 5-HIAA content in blood and urine. More recently, a variety of challenge strategies has been used to study the function and sensitivity of the 5-HT system in psychiatric disorders. In addition to MCPP, investigators have studied behavioral and neuroendocrine responses to fenfluramine, tryptophan infusion, and tryptophan depletion. Future studies using these paradigms would help to clarify the possible role of serotonergic systems in PTSD.

OPIATES

Stress causes a release of endogenous opiates, increased opiate peptide levels, and diminished pain sensitivity (78–80). The fact that stress-induced analgesia can be blocked by administration of an opiate antagonist, naltrexone hydrochloride, supports the notion that stress, endogenous opiates, and analgesia are related (80–82). Additionally, evidence suggests that analgesia can accompany neutral stimuli previously paired with aversive stimuli (83). Furthermore, it appears that the endogenous opiate system can become sensitized so that reexposure to lower levels of uncontrollable shock results in the same degree of analgesia previously induced by significantly greater degrees of shock (82). Given these facts, it is reasonable to study opiate systems in individuals who have been severely traumatized.

Baseline Studies

Only two laboratory reports have described baseline opiate metabolism in traumatized humans. Hoffman et al. (84) reported significantly lower a.m. and p.m. plasma b-endorphin levels in 21 PTSD patients compared to 20 controls. The results were viewed as support for van der Kolk's hypothesis (85) that patients with PTSD have a chronic depletion of endogenous opioids and that hyperresponsiveness is related to endogenous opiate withdrawal. According to this hypothesis, patients with PTSD seek out or provoke recurrent stressors and traumas in order to increase opiate release, and hence decrease endogenous opiate withdrawal.

The second baseline study (86) measured circulating levels of methionine-enkephalin and its in vitro plasma degradation half-life in 13 Vietnam combat veterans with PTSD compared to controls. While plasma methionine-enkephalin levels were similar in PTSD patients and controls, degradation half-life was significantly higher in the PTSD group. The authors suggest the possible existence of a circulating endogenous inhibitor of methionine-enkephalon degradation. However, such a substance has never been isolated. The authors further speculate that decreased degradation in the face of normal plasma methionine-enkephalin levels may suggest decreased production and release of this peptide. Although it is suggested that these findings are consistent with those of Pitman et al. (80) (summarized in the next section), it is more likely that the findings are compatible with the notion of a chronic baseline opiate depletion. Because the Hoffman et al. and Wolfe et al. studies measured different opiates, a direct comparison of the results is not feasible.

Provocation Studies

To date, only one challenge study has focused on the opiate system in PTSD. Pitman et al. (80) exposed eight combat veterans with PTSD and eight combat controls to the stress of a combat film. Following the film, subjects were exposed to a pain sensitivity test. Veterans with PTSD showed reduced pain sensitivity compared to veterans without PTSD. This effect was reversible with the opiate antagonist nalaxone. These findings suggest that stress-induced analgesia is, at least in part, mediated by endogenous opiates and that patients with PTSD show an enhanced release of endogenous opiates following exposure to stress. The results of the Pitman et al. study differ from the results of baseline studies. In baseline studies, it appears that opiate levels are either normal or reduced. However, the Pitman et al. study suggests enhanced opiate release in response to a stressful stimulus. The Pitman et al. study fails to support van der Kolk's hypothesis of a biologically-mediated "addiction to trauma" in PTSD, as combat films in veterans with PTSD evoked numbing and blunting of emotional responses as opposed to euphoria or emotional feelings of calm and control. Studies of self-mutilation in traumatized psychiatric patients further support a relationship between psychic numbing and opiate-mediated, stress-induced analgesia (87).

Whether alterations in endogenous opiates contribute to the core symptoms seen in PTSD is not clear. However, based on the studies just discussed, it has been hypothesized that symptoms of avoidance and numbing are related to a dysregulation of opioid systems in PTSD (88). Furthermore, it has been suggested that the use of opiates in chronic PTSD may represent a form of self-medication (88,89). Animal studies have shown that opiates are powerful suppressants of central and peripheral noradrenergic activity (90). If, as suggested earlier in this chapter, some PTSD symptoms are mediated by noradrenergic hyperactivity, then opiates may serve to "treat" or dampen that hypersensitivity and the accompanying symptoms. On the other hand, during opiate withdrawal when opiates are decreased and noradrenergic activity increased, PTSD symptoms may become acutely exacerbated. In fact, many symptoms of PTSD have been compared to symptoms experienced during opiate withdrawal (91).

SUMMARY AND CONCLUSION

In summary, baseline or resting studies generally have found no differences in plasma catecholamine levels between combat veterans with PTSD and normal controls. In contrast, most 24-hour urine studies have reported increased excretion of catecholamines and most challenge studies have found evidence for hyperresponsivity of catecholaminergic systems. It is possible that the discrepancy between single plasma samples versus 24-hour urine values is explained by phasic increases in catecholamine reactivity that are detected and accounted for through the use of 24-hour but not single plasma samples. That is, catecholamine levels in 24-hour urine samples reflect the summation of both phasic physiologic changes in response to meaningful stimuli (as seen in challenge studies) and tonic resting levels of catecholamines, while single plasma samples reflect only tonic activity.

In the aggregate, the aforementioned investigations point to an increased responsivity of the sympathetic nervous system detectable under conditions of "stress" in severely traumatized individuals with PTSD. These findings are consistent with a large body of psychophysiologic data recently reviewed by Orr (92) that shows consistent increases in psychophysiologic reactivity to trauma-related cues in combat veterans with PTSD. Consistent psychophysiologic evidence of resting or baseline elevations has not been found across studies (92). Because far less is known about the serotonin and opiate systems in PTSD, no real conclusions can be drawn except that alterations in these systems also appear to be most easily detected under conditions of "stress" or provocation.

The findings just discussed are consistent with a behavioral sensitization model of PTSD. Behavioral sensitization refers to an increased magnitude of response following repeated presentations of a particular stimulus. Following a stressful event, biochemical, physiologic, and behavioral responses to subsequent stressors increase over time. For example, when animals are exposed to repeated shock, dopamine beta-hydroxylase activity, tyrosine hydroxylase, and synaptic levels of NE all increase (93–95). When exposed to a limited shock, these repeatedly shocked animals respond as if the shock were much greater and release an amount of NE that is appropriate for a much larger stressor. Thus, a compensatory increase in NE synthesis and subsequent release appears to occur over time.

There are a number of parallels between PTSD and laboratory-induced behavioral sensitization. First, prior exposure to traumatic stressors can increase subsequent responses to stressors, depending on initial dose of exposure as well as frequency and intermittency of exposure (96-98). For example, animals that receive a large initial dose of cocaine show greater behavioral responses on reexposure than animals that receive a low dose of cocaine initially (99). Similarly, in traumatized humans severity of PTSD symptomatology is positively correlated with magnitude of traumatic exposure (100,101) and prior exposure to childhood trauma (102). In a study of Israeli combat soldiers who fought in two successive wars, Soloman et al. reported that soldiers were more likely to develop symptoms during the second war if they had suffered acute combat stress during the first one (103).

Second, for many patients with PTSD, symptoms increase in magnitude over time. In a study of WWII veterans 20 years after the war, Archibald and Tuddenham reported an increase in symptoms among patients with PTSD over time (104). In a retrospective study of treatment-seeking Vietnam combat veterans with PTSD, Bremner et al. found that symptomatology typically increased over the first few years after the war and then plateaued, becoming chronic and unremitting (105). Similarly, animal studies of behavioral sensitization, and what Antelman has termed "time dependent change" (106) have shown that stressors can cause longlasting and gradually increasing changes in behavioral and physiological responses to subsequent stressors.

Increased sensitivity and sensitization of the noradrenergic system, as detected by the provocation studies mentioned in the previous paragraph, may leave the individual in a hyperaroused, vigilant, sleep-deprived, and, at times, explosive state that worsens over time. Being irritable and on edge makes it difficult to interact with family members, friends, coworkers, and employers. To quiet these symptoms of hyperarousal, PTSD patients often withdraw and become avoidant. Many also resort to substances, particularly central nervous system depressants, that suppress peripheral and central catecholamine activity. Alterations in other neurobiologic systems may further contribute to multiple symptoms such as intrusive memories, dissociative phenomena, and even numbing. Clearly, prospective biological studies in recently traumatized humans are needed to better assess behavioral sensitization as a model for the natural evolution of PTSD-related symptomatology.

Other models that may be relevant to aspects of PTSD include fear conditioning (51–53), overconsolidation of memory (17,107,108), and failure of extinction (88). Although these models have been offered as paradigms of the neurobiology of PTSD, there is not yet enough empirical neurobiologic data in humans to adequately support any one model. Nonetheless, these models may serve as useful heuristic tools for understanding discrete aspects of the human response to trauma.

Advances in the neurobiologic characterization of PTSD have largely relied on available medical technology. Much of what has been discussed in this chapter has grown out of recent advances in psychophysiologic, hormone, and receptor assay methodology. As technology in areas such as brain scanning becomes increasingly refined, it soon will be possible to more accurately delineate acute and longterm stress-induced changes in central and peripheral nervous system functioning. With a clearer understanding of biologic pathophysiology, it should be possible to develop more specific and effective treatments for the often devastating disorder of PTSD.

REFERENCES

1. Meakins JC, Wilson RM. The effect of certain sensory stimulation on the respiratory rate in cases of so-called "irritable heart." *Heart* 1918;7:17–22.
2. Kardiner A. The traumatic neuroses of war. *Psychosomatic medicine monograph* (I-II). Washington, DC: National Research Council; 1941.
3. Yehuda R, Southwick SM, Perry BD, Mason JW, Giller EL. Interactions of the hypothalamic-pituitary-adrenal and catecholaminergic system in posttraumatic stress disorder. In Earl L. Giller, ed. *Biological Assessment and Treatment of PTSD,* Progress in Psychiatry. Washington, DC: APA Press; 1990:115–134.
4. Mountcastle ZB. *Medical physiology,* 13th Ed. St. Louis, MO: Mosbey Publishing, 1973.
5. Gagnon WF. *The nervous system.* Los Altos, CA: Lange Publishing; 1977.
6. Cannon WB. Emergency function of adrenal medulla in pain and the major emotions. *Am J Physiol* 1914; 3:356–372.
7. Fraser F, Wilson EM. The sympathetic nervous system and the "irritable heart of soldiers." *Br Med J* 1918; 2:27–29.
8. Grinker RR, Spiegel JP. *Men under stress.* Philadelphia: Blakiston; 1945.
9. Crille G. Results of 152 denervations of the adrenal glands in the treatment of neurocirculatory asthenia. *Military Surgeon* 1970;87:509–513.
10. Dobbs D, Wilson WP. Observations on the persistence of neurosis. *Dis Nerv Syst* 1960;21:40–46.
11. Brende JO. Electrodermal responses in post-traumatic syndromes. *J Nerv Ment Dis* 1982;170:352–361.
12. Blanchard EB, Kolb LC, Pallmeyer TP, Gerardi RJ. A psychophysiological study of post traumatic stress disorder in Vietnam veterans. *Behav Res Ther* 1986;24: 645–652.
13. Blanchard EB, Kolb LC, Prins A, et al. Changes in plasma norepinephrine to combat-related stimuli among Vietnam veterans with post traumatic stress disorder. *J Nerv Ment Dis* 1991;179:371–373.
14. Malloy PF, Fairbank JA, Keane TM. Validation of a multimethod assessment of posttraumatic stress disorders in Vietnam veterans. *J Consult Clin Psychol* 1983; 51:488–494.
15. Pallmeyer TP, Blanchard EB, Kolb LC. The psychophysiology of combat-induced posttraumatic stress disorder in Vietnam veterans. *Behav Res Ther* 1986;24: 645–652.
16. Pitman RK, Orr SP, Forgue DF, DeJong JB, Clairborn JM. Psychophysiologic assessment of posttraumatic stress disorder imagery in Vietnam combat veterans. *Arch Gen Psychiatry* 1987;44:970–975.
17. Pitman RK. Posttraumatic stress disorder, hormone, and memory (editorial). *Biol Psychiatry* 1989;26:221–223.

18. Orr SP. Psychophysiologic studies of post traumatic stress disorder. In Giller EL, ed. *Biological assessment and treatment of post traumatic stress disorder.* Washington, DC: American Psychiatric Press; 1990:135.
19. Orr SP, Pitman RK, Lasko NB, Herz LR. Psychophysiological assessment of posttraumatic stress disorder imagery in World War II and Korean combat veterans. *J Abnorm Psychol* 1993;102:152–159.
20. Shalev AY, Orr FP, Pitman RK. Psychophysiologic assessment of traumatic imagery in Israeli civilian patients with posttraumatic stress disorders. *Am J Psychiatry* 1993;150:620–624.
21. Pitman RK, Orr SP, Forgue DF, Altman B, deJong JB, Herz LR. Psychophysiologic responses to combat imagery of Vietnam veterans with posttraumatic stress disorder versus other anxiety disorders. *J Abnorm Psychol* 1990a;99:49–54.
22. Blanchard EB, Hickling EJ, Taylor AC. The psychophysiology of motor vehicle accident related posttraumatic stress disorder. *Biofeedback Self Regul* 1991; 16:116.
23. Aston-Jones G, Valentino R, Van Bockstaele. Locus coeruleus, stress, and PTSD: neurobiological and clinical parallels. In Murburg, M, ed. *Catecholamine function in post-traumatic stress disorder: emerging concepts.* Washington, DC: APA Press; 1994, pp. 17–62.
24. Southwick SM, Bremner D, Krystal JH, Charney DS. Psychobiologic research in posttraumatic stress disorder. *Psychiatr Clin North Am* 1994;17.
25. Simpson PE, Weiss JM. Altered activity of the locus coeruleus in an animal model of depression. *Neuropsychopharmacology* 1988;1:287–295.
26. Simpson PE, Weiss JM. Responsiveness of locus coeruleus neurons to excitatory stimulation is uniquely regulated by alpha-2 receptors. *Psychopharmacol Bull* 1948; 224:349–345.
27. U'Prichard DC, Kvetnansky T. Central and peripheral adrenergic receptors in acute and repeated immobilization stress. In Usdin E, Kvetnansky R, Kopin IJ, eds. *Catecholamines and stress: recent advances.* Amsterdam: Elsevier Science Publishers; 1980.
28. Yocca FD, Friedman E. Effect of immobilization stress on rat pineal beta-adrenergic receptor-mediated function. *J Neurochem* 1984;42:1427–1432.
29. Murburg MM. *Catecholamine function in posttraumatic stress disorder: emerging concepts.* Washington, DC: APA Press; 1994.
30. Yehuda R, Southwick SM, Giller EL, et al. Urinary catecholamine excretion and severity of PTSD symptoms in Vietnam combat veterans. *J Nerv Ment Dis* 1992;180:321–325.
31. Kosten TR, Mason JW, Giller EL, et al. Sustained urinary norepinephrine and epinephrine elevation in post-traumatic stress disorder. *Psychoneuroendocrinology* 1987;12:13–20.
32. Pitman R, Orr S. Twenty-four hour urinary cortisol and catecholamine excretion in combat-related posttraumatic stress disorder. *Biol Psychiatry* 1990;27: 245–247.
33. Davidson LM, Baum A. Chronic stress and posttraumatic stress disorder. *J Consult Clin Psychol* 1986;54: 303–308.
34. McFall M, Murburg M, Ko G, et al. Autonomic response to stress in Vietnam combat veterans with posttraumatic stress disorder. *Biol Psychiatry* 1990;27: 1165–1175.
35. Hamner MB, Diamond BI, Hitri A. Plasma norepinephrine and MHPG responses to exercise stress in PTSD. In Murburg MM. *Catecholamine function in post-traumatic stress disorder: emerging concepts.* Washington, DC: APA Press, 1994, pp. 221–232.
36. Southwick SM, Krystal JH, Morgan AC, et al. Abnormal noradrenergic function in post-traumatic stress disorder. *Arch Gen Psychiatry* 1993a;50:266–274.
37. Veith RC, Murburg MM. Assessment of sympathetic nervous system function in PTSD: a critique of methodology. In Murburg, MM, ed. *Catecholamine function in posttraumatic stress disorder: emerging concepts,* Progress in Psychiatry. Washington, DC: APA Press; 1994:221–232.
38. Hamner MB, Diamond BI, Hitri A. Plasma dopamine and prolactin levels in PTSD. *Biol Psychiatry* 1990;27: 72A.
39. Perry BD, Southwick SM, Yehuda R, Giller EL. Adrenergic receptor regulation in post-traumatic stress disorder. In EL Giller, ed. *Biological assessment and treatment of post-traumatic stress disorder.* Washington, DC: APA Press; 1990:87–114.
40. Perry BD, Giller EL, Southwick SM. Altered platelet alpha2 adrenergic binding sites in post-traumatic stress disorder. *Am J Psychiatry* 1987;144:1511–1512.
41. Perry BD. Neurobiological sequelae of childhood trauma: PTSD in children. In Murburg M, ed. *Catecholamine function in post-traumatic stress disorder: emerging concepts.* Washington, DC: APA Press; 199#:131–158.
42. Weiss RT, Tobes M, Wertz CE, et al. Platelet alpha-2 adrenoceptors in chronic congestive heart failure. *Am J Cardiol* 1983;52:101–105.
43. Hollister AS, Onrot J, Lonce S, et al. Plasma catecholamine modulation of adrenoreceptor agonist affinity and sensitivity in normotensive and hypertensive human platelets. *J Clin Invest* 1986;77:1416–1421.
44. Southwick SM, Yehuda R, Giller EL, Perry BD. Altered platelet $alpha_2$-adrenergic binding sites in borderline personality disorder. *Am J Psychiatry* 1990;147: 1014–1017.
45. Southwick SM, Yehuda R, Giller EL, Perry BD. $Alpha_2$-adrenergic-receptor regulation in major depression and borderline personality disorder. *Psychiatry Res* 1991;34:143–203.
46. Southwick SM, Yehuda R, Giller EL. Personality disorders in war veterans with chronic posttraumatic stress disorder. *Am J Psychiatry* 1993;150:1020–1023.
47. Yehuda R, Perry BD, Southwick SM, Giller EL. Platelet $alpha_2$-adrenergic binding in PTSD, generalized anxiety disorder and major depressive disorder. New Research Abstracts, 143. New York: American Psychiatric Association; May 1990.
48. Lerer B, Ebstein RP, Shestatsky M, et al. Cyclic AMP signal transduction in post-traumatic stress disorder. *Am J Psychiatry* 1987;144:1324–1327.
49. Lerer B, Bleich A, Bennett ER, Ebstein RP, Balkin J. Platelet adenylate cyclase and phospholipase C activity in posttraumatic stress disorder. *Biol Psychiat* 1990;27: 735–744.
50. Davis M, Gendelman D, Tischler M, et al. A primary acoustic startle circuit: lesion and stimulation studies. *J Neuroscience* 1982;91:805.

51. Brown JS, Kalish HI, Farber IE. Conditioned fear as revealed by magnitude of startle response to auditory stimulus. *J Exp Psychol* 1951;41:317–327.
52. Davis M. The role of amygdala in fear potentiated startle: implications for animal models of anxiety. *Trends Pharmacol Sci* 1992;13:35–40.
53. Davis M, Redmond E, Baraban JM. Noradrenergic agonists and antagonists: effects on conditioned fear as measured by the potentiated startle paradigm. *Psychopharmacology* 1979;65:1111–1118.
54. Ornitz EM, Pynoos RS. Startle modulation in children with posttraumatic stress disorder. *Am J Psychiatry* 1989;146:866–870.
55. Shalev AY, Orr SP, Peri T, Schreiber S, Pitman RK. Physiologic responses to loud tones in Israeli patients with posttraumatic stress disorder. *Arch Gen Psychiatry* 1992;49:870–875.
56. Ross JR, Ball W, Cohen M, Silver SM, Morrison AR, Dinges DF. Habituation of the startle reflex in Posttraumatic Stress Disorder. *J Neuropsychiatry* 1989;3:305–307.
57. Morgan CA, Grillon C, Southwick SM, Nagy LM, Krystal JH, Davis M, Charney DSL. Yohimbine facilitated acoustic startle in combat veterans with PTSD. *Psychopharmacology,* 1995;117:466–471.
58. Morgan III CA, Southwick SM, Grillon C, Davis M, Charney DS. Fear potentiated startle in PTSD, submitted.
59. Morgan III CA, Southwick SM, Grillon C, Charney DS. Exaggerated acoustic startle in Desert Storm Veterans with PTSD. *Society for Neurosciences Abstracts,* 1994.
60. Dinan TG, Barry S, Yatham LN, et al. A pilot study of neuroendocrine test battery in post traumatic stress. *Biol Psychiatry* 1990;28:665–672.
61. Hansenne M, Pitchot W, Anseau M. The clonidine test in post traumatic stress disorder. *Am J Psychiatry* 1991; 148:810–811.
62. Balon R, Pohl R, Yeragani VK. The provocation of anxiety states in humans and its possible significance for the pathogenesis of these disorders. In den Boer JA, Ad Sitsen JM, eds. *Handbook of depression and anxiety.* New York: Marcel Dekker; 1994.
63. Rainey JM, Aleem A, Ortiz A, et al. A laboratory procedure for the induction of flashbacks. *Am J Psychiatry* 1987;144:1317–1319.
64. Scatton B, Zirkivic B, Devek J. Antidopamine properties of yohomine. *J Pharmacol Exp Ther* 1980;90: 215:494–499.
65. Charney DS, Woods SW, Price LH. Noradrenergic dysregulation in panic disorder. In Ballinger JC, ed. *Neurobiology of panic disorder.* New York: Wiley Liss; 1990;91–105.
66. Charney D, Woods SW, Goodman WK, et al. Neurobiological mechanisms of panic anxiety: biomechanical and behavioral correlates of yohimbine-induced panic attacks. *Am J Psychiatry* 1987;144:1030–1036.
67. Nagy LM, Merikangas KR, Morgan CA III, et al. Genetic epidemiology of panic attacks and noradrenergic response in PTSD: a family history study. Submitted.
68. Morgan III CA, Southwick SM, Grillon C, Davis M, Krytal JH, Charney DS. Yohimbine facilitated acoustic startle reflex in humans. *Psychopharmacology* 1993; 110:342–346.
69. Bremner JD, Ny CK, Staib L, et al. PET measurement of cerebral metabolic correlates of yohimbine administration in combat-related posttraumatic stress disorder. Submitted.
70. Brown GL, Linnoila MI. CSF serotonin metabolite (5-HIAA) studies in depression, impulsivity, and violence. *J Clin Psychiatry* 1990;51:31–41.
71. Stanley M, Stanley B. Postmortem evidence for serotonin's role in suicide. *J Clin Psychiatry* 1990;51:22–28.
72. McDougle C, Southwick SM, Charney DS, et al. An open trial of fluoxetine in the treatment of post traumatic stress disorder. *J Clin Psychopharmacol* 1991;1: 325–327.
73. Nagy LM, Morgan CA, Southwick SM, et al. Open prospective trial of fluoxetine for post traumatic stress disorder. *J Clin Psychopharmacol* 1993;13:107–113.
74. Davidson JRT, Ross S, Newman E. Fluoxetine in posttraumatic stress disorder. *J Traumat Stress* 1991;4: 419–423.
75. Shay J. Fluoxetine reduces explosiveness and elevates mood in Vietnam combat vets with PTSD. *J Traumat Stress* 1992;5:97–102.
76. Arora RC, Fitchner CG, O'Connor F. Paroxetine binding in the blood platelets of posttraumatic stress disordered patients. *Life Sc* 1993;53:919–928.
77. Southwick SM, Krystal JH, Morgan CA, et al. Yohimbine and m-chloro-Phenyl-Piperazine in PTSD (/Abstract NR 3481) New Research Abstracts of the American Psychiatric Association, 144th meeting, 991.
78. Amir S, Brown ZA, Arnit A. The role of endorphins in stress: evidence and speculations. *Neurosci Biobehav Res* 1986;4:77–86.
79. Hemingway RB, Reigle TG. The involvement of endogenous opiate systems in learned helplessness and stress-induced analgesia. *Psychopharmacology* 1987;3: 353–357.
80. Pitman RK, van der Kolk BA, Orr SP, et al. Naloxone-reversible analgesic response to combat-related stimuli in post traumatic stress disorder. *Arch Gen Psychiatry* 1990;47:541–544.
81. Jackson RL, Maier SF, Coon DI. Long term analgesic effects of inescapable shock and learned helplessness. *Science* 1979;206:91–93.
82. Maier SF. Stressor controllability and stress induced analgesia. *Ann NY Acad Sci* 1986;467:55–72.
83. Fanselow MS. Conditioned fear-induced opiate analgesia: a competing motivational state theory of stress analgesia. *Ann NY Acad Sci* 1986;467:40–54.
84. Hoffman L, Watgson PD, Wilson G, et al. Low plasma beta-endorphin in post-traumatic stress disorder. *Aust NZJ Psychiatry* 1989;23:268–273.
85. van der Kolk BA, Greenberg MS, Orr SP, et al. Endogenous opioids, stress induced analgesia and post-traumatic stress disorder. *Psychopharmacol Bull* 1981;25: 417–421.
86. Wolfe M. Plasma Methionine Enkephalin in PTSD. *Biol Psychiat* 1991;29:295–308.
87. Richardson JS, Zaleski WA. Naloxone and self mutilation. *Biol Psychiatry* 1983;18:99–101.
88. Charney D, Deutch A, Krystal J, et al. Psychobiologic mechanisms of posttraumatic stress disorder. *Arch Gen Psychiatry* 1993;50:294–305.
89. Southwick SM, Krystal JH, Johnson DR, et al. Neurobiology of posttraumatic stress disorder. In Tasman A,

Riba MG, eds. *Review of psychiatry.* Washington, DC: APA Press; 1992.
90. Kosten T, Krystal JH. Biological mechanisms in posttraumatic stress disorder: relevance for substance abuse. In Galanter, ed. *Recent advances in alcoholism* 6, New York: Plenum Press; 1988.
91. Salloway S, Southwick S, Sadowsky M. Opiate withdrawal presenting as posttraumatic stress disorder. *Hosp Comm Psychiatry* 1990;41:666–667.
92. Orr SP. An overview of psychophysiological studies of PTSD. *PTSD Research Quarterly* 1994;5(1):1–7.
93. Irwin J, Ahluwalia P, Anismar H. Sensitization of norepinephrine activity following acute and chronic footshock. *Brain Res* 1986;379:98–103.
94. Karmarcy NR, Delaney RL, Dunn AL. Footshock treatment activates catecholamine synthesis in slices of mouse brain regions. *Brain Res* 1984;290:311–319.
95. Melia KR, Nestler EJ, Haycock J, et al. Regulation of tyrosine hydroxylase (TH) in the locus coeruleus (LC) by corticotropin-releasing factor (CRF): relation to stress and depression. *Neurosci Abstr* 1991;16:444.
96. Antelman SM. Time dependent sensitization as the cornerstone for a new approach to pharmacotherapy: drugs as foreign or stressful stimuli. *Drug Dev Res* 1988;4:1–30.
97. Post RM, Contel NR. Human and animal studies of cocaine: implications for development of behavioral pathology. In Creese I, ed. *Stimulants: neurochemical, behavioral, and clinical perspectives.* New York: Raven Press; 1983:169–203.
98. Post RM. Transduction of psychosocial stress into the neurobiology of recurrent affect disorder. *Am J Psychiatry* 1992;149:999–1010.
99. Weiss SRB, Post RM, Pert A, et al. Context dependent cocaine sensitization: differential effect of haloperidol on development versus expression. *Pharmacol Biochem Behav* 1989;34:655–661.
100. Kulka RA, Schlenger WE, Fairbanks JA, et al. The National Vietnam veterans readjustment study: table of findings and appendices Trauma and the Vietnam war generation. New York: Brunner/Mazel, 1990.
101. Southwick SM, Morgan CA, Nagy LM, et al. Trauma related symptomatology in Desert Storm veterans: a preliminary report. *Am J Psychiatry* 1993b;150:1524–1528.
102. Bremner JD, Southwick SM, Johnson DR, et al. Childhood abuse in combat-related post traumatic stress disorder. *Am J Psychiatry* 1993;150:235 239.
103. Solomon Z, Mikulincer M, Jakob BR. Exposure to recurrent combat stress: combat stress reactions among Israeli soldiers in the Lebanon War. *Psychol Med* 1987; 17:433–440.
104. Archibald HC, Tuddenham RD. Persistent stress reaction after combat. *Arch Gen Psychiatry* 1965;12:475–481.
105. Bremner DJ, Darnell A, Charney DS, et al. The longitudinal course of chronic posttraumatic stress disorder in Vietnam veterans: development, course of illness, and relationship to alcohol and substance abuse. Submitted.
106. Antelman SM, Yehuda R. Time dependent change following acute stress: relevance to the chronic and delayed aspects of PTSD. In Murburg MM, ed. *Catecholamine function in PTSD: emerging concepts.* Washington, DC: American Psychiatric Association Press, 1994.
107. LeDoux JE, Romanski L, Xagoraris A. Indelibility of subcortical emotional networks. *J Cog Neurosi* 1989;1: 238–243.
108. McGaugh JL. Significance and remembrance: the role of neuromodulatory systems. *Psychol Sci* 1990;1:15–25.

Neurobiological and Clinical Consequences of Stress: From Normal Adaptation to PTSD, edited by M. J. Friedman, D. S. Charney, and A. Y. Deutch, Lippincott-Raven Publishers, Philadelphia © 1995.

19

Hypothalamic-Pituitary-Adrenal Functioning in Post-Traumatic Stress Disorder

Expanding the Concept of the Stress Response Spectrum

Rachel Yehuda, *Earl L. Giller, Jr., Robert A. Levengood, †Steven M. Southwick, and Larry J. Siever

*Department of Psychiatry, Mount Sinai School of Medicine, Bronx Veterans Affairs Medical Center, Bronx, New York 10468; *CNS Clinical Research, Pfizer, Inc., Groton, Connecticut 06340; and †Department of Psychiatry, Yale University School of Medicine, New Haven, Connecticut 06510, National Center for Post-Traumatic Stress Disorder, Veterans Affairs Medical Center, West Haven, Connecticut 06516*

This chapter will review the literature on hypothalamic-pituitary-adrenal (HPA) axis functioning in post-traumatic stress disorder (PTSD). The conclusion that will emerge from a review of these findings is that the alterations that have been noted in PTSD differ dramatically from those that have been described in most studies of the normative stress response. Furthermore, the HPA axis findings in PTSD are distinct from those reported in other psychiatric conditions, even conditions with similar symptoms to PTSD, such as major depressive disorder. A model of HPA axis functioning that has been put forth to explain these alterations will be described.

The distinctness of the HPA axis findings in PTSD raises two basic conceptual issues that will also be the focus of discussion. The first issue bears on the formulation of PTSD as a normative consequence of stress. Clearly, the unusual nature of the alterations observed in this stress-responsive system challenges us to reexamine the clinical phenomenology of PTSD in order to evaluate the extent to which this disorder may also represent an atypical reaction to extreme stress. The second issue raised concerns the narrowness of current formulations of the normative stress response. The HPA axis findings in PTSD challenge us to widen our perspective on the stress response spectrum.

CURRENT CONCEPTUALIZATIONS OF PTSD AND THEIR INFLUENCE ON BIOLOGICAL HYPOTHESES

The inclusion of PTSD as a psychiatric disorder in the DSM-III (1) represented a rediscovery in how the field of psychiatry viewed the mental health consequences following exposure to trauma. In essence, the diagnosis of PTSD acknowledged that exposure to trauma precipitated serious and long-lasting symptoms in normal individuals. Prior to this formulation, there was no clear distinction between exposure to a stressor versus exposure to a traumatic event. Thus, the focus was on describing a set of mental health consequences that could occur independently of the nature and severity of the stressor. Regardless, the presence of longlasting psychiatric symptoms following either type of event was seen as developing only in vulnerable individuals, perhaps as secondary to preexisting personality traits or other constitutional risk factors. On the other hand, descriptions such as "gross stress reaction" in the DSM-I (2) and terms such as "war neurosis" (3) acknowledged that psychiatric symptoms could develop in response to trauma. However, these responses were seen as anomalous maladaptations.

The DSM-II contained descriptions of "transient situational disturbance" (4), which was the diagnostic forerunner of what are now called "adjustment disorders." The underlying assumption of these disorders was that normative reactions to stress would be characterized by quick recovery. On the other hand, prolonged responses to trauma were viewed as anxiety or depressive neuroses, and were conceptualized as having roots that predated the occurrence of the trauma (4). Thus, PTSD as described in DSM-III, and more recently in DSM-IV (5), represented an important departure from the "stress disorders" described in prior diagnostic schemata in viewing chronic and debilitating illness following trauma as a consequence of the nature of the traumatic event, even in individuals without constitutional vulnerability.

Importantly, because the symptoms of PTSD were essentially conceptualized as describing a normative response to extreme stress, many biological studies of PTSD—certainly earlier investigations—had as their major impetus and rationale the hypothesis that alterations observed in symptomatic individuals would be analogous to those observed in animal and human studies of stress (6–9).

THE PARADIGM OF THE BIOLOGICAL STRESS RESPONSE

Selye's initial characterization of the biological stress response in 1936 formed the basis of an influential paradigm in the field of science that continues to pervade modern conceptualizations of the changes that occur during and after exposure to a stressor (10). The original observations, which have withstood the test of time, were that provocation by a wide range of "stressors" would result in a complex psychoneuroendocrine response, part of which was an increased release of pituitary-adrenocortical hormones. Selye also described the fact that stress left indelible marks on the organism, and that cumulative stress effects would have a substantial impact on the reservoir of "adaptational energy" that influenced biological adaptation (11).

An important component of the Selye model of stress was that any potential stressor could invoke the stress response. In fact, the presence and magnitude of a stressor could be inferred by the presence and magnitude of the pituitary-adrenocortical response. Therefore, Selye's focus was not on the objective nature of any particular stressor per se but rather on the universality of the stress response. In this context it is interesting that Selye's formulation was quite different from the original focus of the DSM-III on the nature of the traumatic event (i.e., as needing to fulfill the magnitude and severity implicit in "Criterion A"). The inclusion of a caveat to Criterion A in the DSM-IV regarding the fact that a stressor must have elicited a behavioral response of fear and helplessness is more compatible with Selye's notion of the importance of subjective responses to stress.

Selye's observations sparked more than half century of research that aimed to further elucidate the basic neurophysiology of the stress response. What emerged from decades of research has been information regarding: 1) neuromodulators that stimulate and attenuate the release of stress hormones; 2) how the release of HPA axis hormones modifies other biochemical reactions in response to stress; 3) how characteristics of stressors, such as their intensity and frequency, influence the stress response; 4) factors that can influence resilience or vulnerability to stressors; and 5) medical and psychiatric consequences of alterations in normal HPA axis activity. More recently, knowledge has accrued regarding the myriad cellular and molecular changes induced by stress. Many of these topics are reviewed in other chapters in this volume.

Despite the addition of this knowledge, the basic stress response is still defined by the strength of pituitary-adrenocortical activity. It has now become clear, however, that a complex interplay of neuromodulators influences this response. For example, stress-activated neurotransmitter systems in the brain release corticotropin-releasing factor (CRF) from the hypothalamus and other areas of the brain, presumably in a dose-dependent fashion (see review [12]). Release of adrenocorticotropin hormone (ACTH) from the portal system of the anterior pituitary and cortisol from the adrenals is stimulated by CRF and cosecretagogues such as arginine vasopressin (see review [13]). Cortisol and other glucocorticoids then initiate the suppression of other immune, metabolic, and neural defensive reactions that occur in response to stress, and also act via a negative feedback loop to the hippocampus, hypothalamus, and pituitary to regulate subsequent hormone release (12). A variety of neuromodulators and neuropeptides additionally exert negative feedback and regulatory effects on the target organs of the HPA axis (see review [13]). The immune, metabolic, and neural defensive biological responses that occur in response to stress are important for the short-term response to stress, but produce long-term damage to the organism if they are not eventually terminated (14). For example, prolonged, elevated concentrations of glucocorticoids can be toxic, and are associated with several other detrimental consequences such as neuronal death (15), medical diseases (16), and impaired affect and cognitive alterations (17).

It has also become clear in recent years that several factors such as the objective nature of the stressor, the individual's prior stress history, and genetic and environmental modifiers can determine or modify the pituitary-adrenocortical response to stress (18,19). These findings have challenged the unidirectional, dose-dependent increases in pituitary-adrenocortical hormones that currently define the basic stress response (20). In this chapter, we summarize the literature on the HPA axis in PTSD, and consider the extent to which current models of stress responsivity can be informative in understanding the particular type of stress response present in individuals with this diagnosis.

HPA AXIS ALTERATIONS IN PTSD

Consistent with initial hypotheses, most studies to date have demonstrated significant differences in HPA axis parameters in PTSD sufferers compared to normal controls. However, the direction in which the system has been found to be altered is contrary to early predictions; the findings have demonstrated substantial differences in HPA axis parameters in PTSD compared to those described in preclinical and clinical studies of stress, and psychiatric disorders such as major depression. Many of the results are summarized in Tables 1 and 2.

Studies of Urinary Cortisol

Four of five studies performed to date have demonstrated that individuals with PTSD show evidence of low cortisol levels. The first series of investigations described differences in mean 24-hour urinary cortisol excretion between patients with PTSD compared to patients with

TABLE 1. *A summary of findings of HPA axis alterations in PTSD compared to normal comparison group*

Baseline measures	
24-hr urinary cortisol excretion	↓
Plasma cortisol levels	↓
Cortisol regulation ("signal-to-noise" ratio)	↑
Lymphocyte glucocorticoid receptor number	↑
Response to neuroendocrine challenge	
Dexamethasone	
1.00 mg	same
0.50 mg	↑
0.25 mg	↑
Response to CRF	↓
Response to metyrapone stimulation	↑

CRF, corticotropin-releasing factor; HPA, hypothalamic-pituitary-adrenal; PTSD, post-traumatic stress disorder.

TABLE 2. *A summary of findings of HPA axis alterations in PTSD compared to trauma-exposed comparison group*

Baseline measures	
24-hr urinary cortisol excretion	↓, ↑
Lymphocyte glucocorticoid receptor number	↑
Response to dexamethasone challenge	
0.50 mg	↑
0.25 mg	↑

HPA, hypothalamic-pituitary-adrenal; PTSD, posttraumatic stress disorder.

other psychiatric diagnoses and to normal controls. In an initial investigation, lower mean 24-hour urinary cortisol excretion was observed in nine inpatient combat veterans with PTSD compared to patients with major depressive disorder, bipolar mania, paranoid schizophrenia, and undifferentiated schizophrenia (21). These findings were replicated and extended in two other studies examining urinary cortisol excretion in a similar population (i.e., treatment-seeking Vietnam war veterans with PTSD). In one study, urinary cortisol excretion was found to be significantly lower in combat-veteran inpatients with PTSD compared to inpatients with major depression, bipolar mania, schizophrenia, and panic disorder (22). Another study comparing urinary cortisol excretion in treatment-seeking combat veterans with PTSD to nonpsychiatric healthy controls found urinary cortisol excretion to be lower in the PTSD group (23). Importantly, this study also showed no differences in urinary cortisol excretion between outpatients and inpatients, or between PTSD with and without MDD.

In contrast to the findings just cited, another group found that urinary cortisol excretion was significantly higher in combat veterans with PTSD compared to combat veterans without PTSD (6). The mean cortisol excretion reported was 107 μg/day for the PTSD group and 80.5 μg/day for the combat controls. These values are at the very high end of the normal 24-hour cortisol excretion (i.e., normal range of cortisol is considered to be 20–90 μg/day) and between two- and threefold higher than those reported for both PTSD and normals in the two other published studies (i.e., in Mason et al.'s study, the mean urinary cortisol was 33 μg/day, and in the Yehuda et al. studies the means were 42 μg/day and 39 μg/day for the PTSD group). The relatively high cortisol values in this study may have been due to the fact that urine specimens were collected in acid preservative (which may promote hydrolysis of the unconjugated cortisol and/or interfere with the antibody-antigen reaction in the radioimmunoassay procedure, thus yielding artificially high cortisol values (24)). Another important difference between this and the aforementioned studies of urinary cortisol excretion was in the use of a trauma-exposed, non-PTSD comparison group. The inclusion of a trauma-exposed non-PTSD group is essential to biological studies because it allows a differentiation between hormonal differences relating to exposure to trauma and hormonal differences specifically related to PTSD. Given the increasing recognition that most individuals who experience trauma do not develop PTSD, it is reasonable to expect differences in stress-responsive biological differences between these two groups. In the Pitman and Orr study (6), the direction of cortisol secretion in the PTSD group was different when a trauma-comparison group was used instead of a psychiatric or normal control group. Unfortunately, because the authors did not include a nontrauma-exposed control group, it was not possible to determine the extent to which the combat-exposed groups differed from nonexposed subjects.

More recently, 24-hour urinary cortisol excretion was measured in a group of nontreatment-seeking Holocaust survivors with PTSD, a group of Holocaust survivors without PTSD, and a demographically-matched nonpsychiatric control group (25). Because both trauma-exposed individuals without PTSD and nonexposed individuals were studied and compared to normals, it was possible to evaluate the extent to which cortisol levels were related to trauma exposure versus the presence of PTSD symptoms. The results demonstrated that Holocaust survivors with PTSD had a significantly lower mean urinary cortisol excretion as compared with the other two groups. The survivor groups without PTSD showed levels that were statistically similar to those of nonexposed controls.

To determine whether cortisol levels in PTSD represented "trait-related" or "state-related" variables, additional analyses were performed. In the first analysis, the Holocaust survivor group without PTSD was further subdivided into those meeting and those not meeting the criteria for past PTSD. There was no significant difference in mean 24-hour cortisol excretion between these two groups. A second set of analyses using multiple correlation revealed a significant relationship between cortisol excretion and severity of PTSD, as determined by the Clinicians Administered PTSD Scale, that was due to a substantial association with scores on the avoidance subscale. It was concluded that low urinary cortisol levels do not appear to be a function of exposure to trauma per se but rather are related to the current severity of PTSD.

The study of Holocaust survivors also provided information about several other aspects of urinary cortisol excretion in PTSD that had not been determined in previous studies of Vietnam veterans.

1. Because both men and women were studied, potential gender differences in cortisol levels in women with PTSD were examined. Results revealed no significant gender effects on cortisol in any of the groups studied.
2. It was possible to address the issue of the persistence of the low cortisol levels in PTSD. Holocaust survivors with PTSD were studied 50 years after their exposure to the Holocaust, compared to Vietnam veterans who were studied 20 to 25 years following exposure to combat. Because low cortisol levels were present in individuals who sustained a focal trauma over 50 years prior to the biological assessment, it was concluded that low cortisol levels may persist for decades in symptomatic individuals.
3. Substance abuse and the presence of other psychiatric diagnoses could have possibly contributed to the biological alterations in combat veterans with PTSD. The study of community-dwelling Holocaust survivors provided the opportunity to explore whether cortisol levels would be low in a sample of individuals with PTSD who were not treatment seekers, who did not have other psychiatric conditions requiring inpatient or outpatient mental health interventions, who had no past or current substance abuse behavior at all, and who showed good vocational, occupational, and social functioning. The fact that urinary cortisol excretion was lower in this type of sample with PTSD mitigates the numerous criticisms that have often been applied to biological studies of combat veterans (e.g., that biological alterations observed may be artifacts of psychiatric, occupational, and social problems). The presence of these potentially confounding variables in combat veterans do not seem to be related to the low cortisol levels in PTSD.

Studies of nonveteran trauma survivors are particularly critical to a fuller understanding of the HPA axis in PTSD because they allow a consideration of whether biological changes reflect differences in the type of traumatic experiences sustained or the characteristics of the sample being studied. Indeed, in the stress literature numerous studies have showed differences in biological parameters as a function of the specific nature of the stress paradigm. To the extent that biological alterations are related to the presence of a stress disorder and not exposure, differences in HPA axis parameters should not be present as a function of the type of traumatic event that has given rise to the disorder. Therefore, it will be important in future studies to systematically explore potential abnormalities in cortisol secretion as a function of type of trauma sustained.

Studies of Plasma Cortisol and ACTH

Determinations of plasma cortisol and ACTH levels using a single sample obtained by routine venipuncture are not considered to accurately reflect hormone concentration, because the stress of the blood withdrawal procedure itself can cause fluctuations in hormone levels. Not surprisingly, studies using this method of assessment have observed increases (26), decreases (27), or no differences (28–29) in cortisol levels in PTSD subjects compared to control groups.

Two studies reported ACTH levels in PTSD patients that were comparable to those of a normal comparison group (8,26). However, in addition to the caveat in the previous paragraph regarding the use of a single venipuncture sample, it is difficult to accurately assess the pituitary activity based on blood concentrations of ACTH due to negative feedback influences on the pituitary. Because the pituitary gland mediates between hypothalamic stimulation of the pituitary by CRF and inhibition of ACTH resulting from the negative feedback of adrenal corticosteroids, baseline ACTH levels may indeed appear to be "normal," even though the pituitary gland is receiving excessive stimulation from CRF (30). Therefore, plasma ACTH levels do not really provide much information about the regulation or activity of the HPA axis, especially if they appear to be "normal" in the face of other alterations.

More recently, the circadian release of cortisol over the 24-hour diurnal cycle was examined in outpatients with combat-related PTSD, normal controls, and outpatients with major depression. In this study, following an overnight stay in the clinical research center, an i.v. was inserted and blood samples were withdrawn every 30 minutes for a 24-hour period while the subject remained at bed rest and in a fasted state (until 6:00 PM). Under these conditions, basal plasma cortisol release was found to be significantly lower, primarily in the late evening and very early morning hours in the PTSD group (31). Chronobiological analysis of the raw cortisol levels using multioscillator cosinor modeling revealed a greater degree of circadian rhythm and a higher signal-to-noise ratio of cortisol release in subjects with PTSD. That is, relative to lower cortisol excretion, PTSD patients tended to show cortisol fluctuations that were high. These data were interpreted as reflecting a more dynamic HPA axis in PTSD in that this neuroendocrine system may now be more sensitive to environmental stimuli. Future studies exploring the circadian release of other neuropeptides would increase our understanding of this axis and its regulatory influences.

Lymphocyte Glucocorticoid Receptors

Steroid-receptor-binding parameters are important in interpreting studies examining basal hormone secretion, because hormones cannot exert their genomic effects unless they are bound to steroid receptors (32,33). Because lymphocytes and brain glucocorticoid receptors share similar regulatory and binding characteristics, it has been suggested that lymphocyte glucocorticoid receptor function reflects aspects of both peripheral and central cortisol regulation (34, 35).

Results from three studies have now demonstrated a significantly larger number of lymphocyte glucocorticoid receptors in combat veterans with PTSD compared to nonpsychiatric and psychiatric comparison groups with major depression, panic disorder, bipolar mania, and schizophrenia (22,27,36). The finding of a larger number of glucocorticoid receptors is consistent with observations of low cortisol in PTSD, in that low circulating levels of a hormone or neurotransmitter are usually associated with an upregulation or increased number of receptors. However, in the case of PTSD, the direction of causality of this relationship is not known. In more classic models of receptor-ligand interactions, changes in glucocorticoid receptor number are conceptualized as reflecting compensatory responses to the concentration of ligand (37–40). According to these models, increased number of glucocorticoid receptors is a secondary consequence of low cortisol. It is alternatively possible, however, that glucocorticoid receptors actually serve to regulate hormonal release by modifying the strength of negative feedback. In this case, the low cortisol levels would be conceptualized as secondary consequences of increased glucocorticoid receptor activity. Knowledge of causality of the relationship between the glucocorticoid receptors and low cortisol is imperative to a deeper understanding of HPA axis alterations in PTSD. If increased glucocorticoid number is conceptualized as a primary alteration in PTSD, this would imply alterations in parameters that are directly related to the activity of glucocorticoid such as negative feedback inhibition. On the other hand, if increased glucocorticoid number in PTSD is secondary to low cortisol levels, then it would be likely that HPA axis alterations in PTSD would be a result of a decreased stimulation of the adrenals and a resultant decrease in the production of cortisol.

The animal stress literature provides only one known model in which glucocorticoid receptors are altered in the direction observed in PTSD. This model is the "early handling" paradigm, in which neonatal animals are exposed to the stress of handling daily for several weeks. Early handling results in a permanent upregulation of hippocampal glucocorticoid receptors (41–44), and a resultant decrease in cortisol following exposure to subsequent stress (45–48). The upregulation of glucocorticoid receptors is rather unusual, because typically glucocorticoid receptors are thought to downregulate following stress (37). The primary alteration in glucocorticoid receptor number is thought to underlie the attenuated cortisol response to stress observed in adult animals who have received neonatal handling (41–44). The observations of altered HPA axis parameters in early handled rats nicely parallels many of the findings observed in PTSD (i.e., not only in regard to the HPA axis but also with respect to alterations in thyroid hormones; early handled animals show increases in T3 and T4 (49) similar to those recently described in PTSD (49a). Thus, the model of early handling provides an interesting basis for further hypothesis testing regarding the true etiology of HPA axis alterations, and in particular glucocorticoid receptor abnormalities, in PTSD. For example, it is currently believed that the early handling manipulation provides a paradigm for how environmental events shape the sensitivity of stress-responsive neuroendocrine systems (50). Implicit in this model is the idea of a developmentally sensitive period in which exposure to environmental events leads to the establishment of a different type of stress response. This notion would be consistent with the observation that combat veterans with PTSD have a higher exposure to stressful events during childhood (51), and raises the possibility that certain features of the HPA axis alterations observed in PTSD (i.e., in particular, alterations in glucocorticoid receptor number) may be present in individuals prior to trauma exposure in adulthood.

It has not been possible to date to examine whether lymphocyte glucocorticoid receptor number is altered in individuals who develop PTSD prior to their exposure to a traumatic event that will induce this disorder. It is also not definitively known whether changes in lymphocyte glucocorticoid receptor number observed in combat veterans with PTSD reflect the fact that these individuals were exposed to stress or, rather, are related to their manifesting a stress disorder. In a preliminary report, it was observed that glucocorticoid receptor number was significantly higher in combat veterans without PTSD compared to normal controls (27). This finding suggests that trauma exposure per se may result in long-lasting changes in glucocorticoid receptors. On the other hand, the number of glucocorticoid receptors in combat veterans who do not meet current diagnostic criteria for PTSD is found to be smaller than that observed in combat veterans with PTSD. Part of the difficulty in reaching an unequivocal conclusion regarding the nature of the alteration in the non-PTSD veteran group is that some of the subjects met the criteria for past PTSD and had some PTSD symptoms. Future studies exploring associations between dose of trauma exposure, PTSD symptoms, and glucocorticoid receptor number are necessary in order to definitively address the cause of glucocorticoid receptor alterations in victims of trauma.

Neuroendocrine Challenge Studies

Three HPA axis challenge paradigms have been used to study PTSD: the dexamethasone suppression test (DST), the corticotropin releasing factor (CRF) test, and the metyrapone stimulation test.

Cortisol Response to Dexamethasone:

Studies examining the cortisol response to dexamethasone in psychiatric disorders, most notably major depressive disorder, have repeatedly shown a "nonsuppression" of cortisol (i.e.,

8:00 AM postdexamethasone cortisol levels at or above 5.0 μg/100 dl) in about 40% to 60% of depressed patients (see review [52]). A nonsuppression of cortisol following dexamethasone is thought to result from a reduced ability of glucocorticoids to suppress the release of CRF and ACTH, and/or adrenal cortisol hypersecretion. The cortisol response to 1 mg dexamethasone has been investigated in five studies of PTSD. These studies have all reported that PTSD patients without major depression show a "normal" suppression to dexamethasone (7,9,53–55). However, closer examination reveals that PTSD patients as a group show an exaggerated response to dexamethasone. A recent meta-analysis of the DST literature in PTSD (24) showed that averaging the mean cortisol data across all published studies revealed a cortisol value in nondepressed PTSD subjects of 1.74 μg/dl, a value well below the established cutoff of 5.0 μg/dl.

The findings of cortisol suppression following dexamethasone in PTSD patients with major depression are less clear. In an initial study, Kudler et al. (9) reported that PTSD patients with major depressive disorder showed a rate of nonsuppression comparable with what has been observed in major depressive disorder, whereas Halbreich et al. (7) and Kosten et al. (54) found normal responses to dexamethasone even in depressed combat veterans with PTSD. Olivera and Fero (53) reported a 32% rate of nonsuppression in 65 combat veterans with PTSD who met comorbid criteria for major depressive disorder. However, these individuals showed normal suppression after their major depression had remitted. A study examining the cortisol response to dexamethasone in eight civilian women with PTSD (55) also showed normal responses to dexamethasone. Thus, in averaging the 4:00 PM postdexamethasone cortisol values of the 83 PTSD patients with comorbid depression, the overall mean was somewhat higher than that reported for PTSD patients without major depression, but still well below the cutoff of the 5.0 μg/dl used as a threshold for major depression.

Most of the DST studies in PTSD were conducted before it was appreciated that cortisol levels in PTSD patients tended to be low and the number of glucocorticoid receptors larger than in normals. Therefore, these earlier studies were designed to test for nonsuppression in PTSD, and did not focus on the possibility of a hypersuppression to dexamethasone. However, in the discussion of one of these reports, the possibility of a hypersuppression to following dexamethasone was considered (7). Failure to observe the classic nonsuppression response to cortisol coupled with reported HPA axis alterations that appeared distinct from those in depression led to studies designed to detect a potential enhanced suppression of the cortisol response to dexamethasone.

Recently, the question of enhanced cortisol suppression to dexamethasone was explored using 0.50 mg (27–28) and 0.25 mg (27) dexamethasone. A hyperresponsiveness to low doses of dexamethasone as reflected by significantly lower postdexamethasone cortisol levels was observed in PTSD patients compared to normals. This effect was not due to individual differences in dexamethasone bioavailability. Nor was it due to differences in free versus bound cortisol as determined by measurements of corticosteroid-binding globulin (Yehuda et al., unpublished observations). The enhanced suppression of cortisol was accompanied by a concurrent decline in the number of cytosolic lymphocyte glucocorticoid receptors (27). Interestingly, the hyperresponsiveness to dexamethasone was also present in combat veterans with PTSD who met the diagnostic criteria for major depressive disorder (28) and, importantly, was not present in combat veterans without PTSD (27). Furthermore, the enhanced suppression of cortisol to low doses of dexamethasone has not been described in other psychiatric disorders and has the potential for use as a relatively specific marker for PTSD.

Cortisol suppression to dexamethasone in PTSD patients also appeared to be unrelated to exposure to trauma. Combat veterans without PTSD were indistinguishable from nontraumatized normal males in their responses to both doses of dexamethasone. The failure to observe cortisol hypersuppression in traumatized combat veterans without PTSD indicates that this phenomenon is likely related to the pathophysiology of the disorder and not directly to the severity of the stressor that may have precipitated this condition.

ACTH Response to CRF:

A challenge strategy is necessary to explore the pituitary release of ACTH because, as just described, given the highly regulated nature of the HPA axis, baseline ACTH levels may appear "normal" even in cases where the pituitary gland is being overstimulated by CRF. Use of challenge strategies allows direct assessment of pituitary sensitivity, and then an indirect determination of the regulatory influences that gave rise to this sensitivity. Finally, it is possible from such studies to evaluate the potential consequences of pituitary sensitivity for adrenal release of cortisol.

The CRF challenge test measures the pituitary ACTH and adrenal cortisol response to exogenous infusion of the neuropeptide CRF. In several reports, the ACTH responses to CRF has been reported to be "blunted" in major depression (56–58). It has been suggested that decreased ACTH response to CRF reflects a downregulation of pituitary CRF receptors caused by hypothalamic CRF hypersecretion, or occurs as a result of the increased negative feedback inhibition of the pituitary secondary to high cortisol levels (56–59).

A single study of eight PTSD subjects suggests that the ACTH response to CRF is also blunted (8). Given that several different biological routes can result in ACTH blunting to CRF (59), and considering the already specified differences between PTSD and major depression, it is quite possible that the mechanisms underlying blunted ACTH response in PTSD are different from the mechanism just described for major depression. For example, because the attenuated ACTH response in PTSD patients occurred in the presence of normal, nonelevated evening plasma cortisol levels, it may be that ACTH blunting in response to CRF occurred as a result of a hyperresponsivity of the pituitary gland to cortisol resulting directly from an increased number of glucocorticoid receptors (GR) on the pituitary gland (60).

β-Endorphin and ACTH Response to Metyrapone:

Metyrapone administration represents another way to examine pituitary sensitivity. Metyrapone is a drug that blocks synthesis of cortisol from its immediate precursor, causing a drastic reduction in cortisol levels for several hours. Because of the substantial drop in cortisol levels, negative feedback influences of cortisol on the pituitary are dramatically suppressed. The abrupt disruption of cortisol synthesis results in a sudden and significant increase in release of β-endorphin and ACTH. The normal response to metyrapone results in a two- to fourfold increase in plasma levels of β-endorphin and ACTH within 2 to 8 hours of metyrapone administration. Under these conditions, it is possible to use the concentration of β-endorphin and ACTH as indices of hypothalamic CRF activity (61). If levels of β-endorphin or ACTH are substantially higher (i.e., greater than fourfold increase in response to metyrapone), this is evidence that the pituitary is receiving a greater stimulation by CRF and is now hyperresponsive. If levels of these peptides are substantially lower (i.e., less than twofold increase), this suggests a decreased CRF stimulation of the pituitary, and a resultant pituitary hyporesponsivity.

In a pilot study, metyrapone was administered to 8 combat veterans with PTSD and 10 normal controls. Only 1/10 (10%) normal volunteers showed a hyperresponsivity to metyrapone, whereas 6/8 (7.5%) PTSD patients were hypersensitive (62). Although it is important to replicate these findings, these data are consistent with the results of the CRF challenge test in suggesting an increased stimulation of the pituitary gland that may be secondary to increased CRF stimulation in PTSD. The enhanced negative feedback provided by the increased number of glucocorticoid receptors may then serve a compensatory role in regulating ACTH and cortisol levels.

ENHANCED NEGATIVE FEEDBACK IN PTSD: A PSYCHONEUROENDOCRINE MODEL OF SENSITIZATION

The findings observed in PTSD represent an interesting constellation of abnormalities that are consistent with several models of HPA axis functioning. The first possibility is a tonic suppression of the HPA axis. However, if the HPA axis were under a tonic suppression, it would be difficult to explain the hyperresponsivity of the pituitary to metyrapone administration, unless this effect was an artifact of the consequences of metyrapone administration on some other biological system. The results of the metyrapone study also make it unlikely that the results are due to a primary pituitary insufficiency.

The second potential model would be that of insufficient cortisol production or release from the adrenal glands (i.e., partial adrenal insufficiency). In this model, the low cortisol levels in PTSD would be due to a failure at the level of the adrenal, and increases in lymphocyte glucocorticoid receptors would reflect a compensatory change in response to low cortisol levels. Implicit in this possibility is that stress-activated neuromodulators may stimulate hypothalamic release of CRF, as well as subsequent release of ACTH from the pituitary; however, because of reduced activity of the adrenal, cortisol would be low. It would be predicted based on this model that cortisol levels would be attenuated in response to stress or provocation. To date, there is insufficient information to determine whether individuals with PTSD show an attenuated cortisol response to either traumatic or nontraumatic stressors. In fact, preliminary observations suggest that combat veterans with PTSD are capable of showing higher levels of cortisol at times when they are experiencing emotional stress (Mason, personal communication). If individuals with PTSD show the characteristic increases in cortisol following exposure to stressors, this would tend to mitigate against the possibility of a partial adrenal insufficiency. Future studies are needed to address this important issue.

A third model that explains the HPA axis findings in PTSD is the model of an enhanced negative feedback inhibition (63). This model is graphed in Fig. 1. The presence of an enhanced negative feedback inhibition along one or more of the target tissues of the axis accounts for all of the current data of HPA abnormalities observed to date, and is compatible with the findings showing evidence of CRF hypersecretion as well as findings of low cortisol and augmented cortisol suppression to dexamethasone. In this model, PTSD sufferers would be expected to show chronic increases in the release of hypothalamic CRF, likely resulting from differences in neuropeptide modulation. The high levels of CRF would lead to an altered responsivity of the pituitary, as evidenced by the blunted ACTH response to CRF challenge and increased β-endorphin levels following metyrapone in PTSD. However, because of a primary alteration in glucocorticoid receptor responsivity, there would be a stronger negative feedback inhibition resulting in attenuated baseline ACTH and cortisol levels, as well as an enhanced responsivity to dexamethasone. The enhanced negative feedback model is also consistent with the recent findings of a stronger circadian rhythm of cortisol as well as the stronger signal-to-noise ratio. The stronger signal-to-noise ratio of cortisol can be seen as reflecting a system that is maximally responsive to stress, rather than a system that has adapted or habituated to stress.

The model of enhanced negative feedback is fundamentally consistent with the status of PTSD patients as being unusually responsive to stress (as opposed to being less responsive to stress, as would be indicated by models of hormone insufficiency). The PTSD patients often show exaggerated behavioral as well as biological responses to environmental challenge. The maximally low background (i.e., baseline cortisol levels) and an ability to hyperrespond to the environment as necessary (e.g., by showing exaggerated responses to neuroendocrine chal-

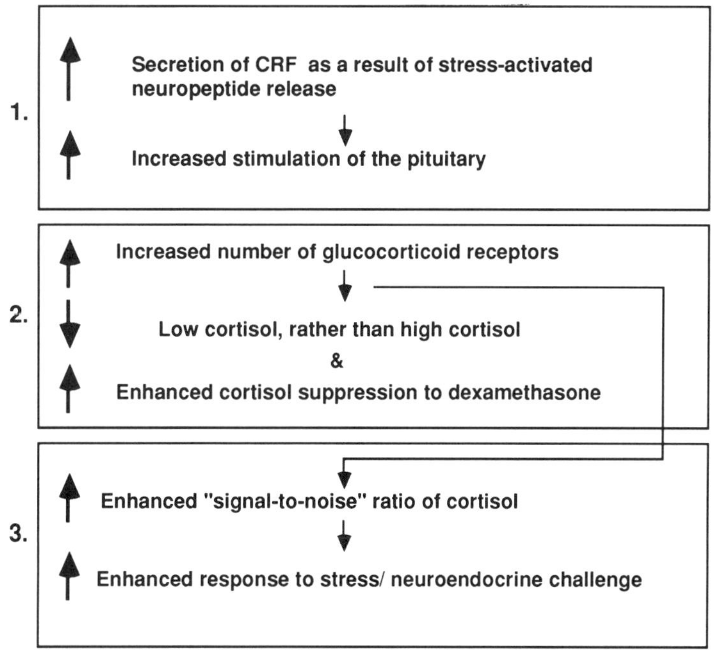

FIG. 1. Enhanced negative feedback of cortisol. A model of HPA axis alteration in PTSD.

lenge) may indeed reflect a type of adaptation to stress that is similar to the symptoms of hyperarousal that describe PTSD. Furthermore, the pattern of hyperresponsiveness may present not only in the HPA axis but also in the catecholamine system (64,65). (See Chapter 19 for review.)

Given the numerous modulatory influences on the HPA axis at the level of the hippocampus, hypothalamus, and pituitary, biological events that lead to the observed HPA axis alterations are likely more complex than is suggested by the model just discussed. For example, given the synergistic effects of vasopressin and CRF, it is likely that altered modulation by arginine-vasopressin may be a critical factor in the expression of HPA axis regulation (66). Furthermore, catecholamines also regulate (and are regulated by) the hormones of the HPA axis (see review (60)). Given the strong evidence for catecholamine alterations (see Chapter xx), it is likely that these neuromodulators also contribute to the alterations observed.

EXPANDING THE STRESS RESPONSE DISORDER SPECTRUM

The HPA axis alterations in PTSD raise some interesting questions about how best to conceptualize PTSD as a stress disorder. The fact that exposure to trauma is necessary for induction of this disorder makes it reasonable to conceptualize PTSD as a "stress response." Furthermore, numerous studies have shown a relationship between the severity of trauma and the presence and severity of post-traumatic symptoms (67–70). The paradox that has arisen from the empirical data is that although PTSD is associated with stress, it does not appear to be a continuation of the typical acute response to stress (as has been suggested, perhaps erroneously, in DSM-IV). Indeed, PTSD may represent an additional type of stress response that occurs in some individuals who are exposed to trauma.

This conclusion is based on the biological data presented in this chapter, but it is also supported by recent epidemiologic studies. As reviewed here, the initial conception of PTSD was that it was a normal occurrence following exposure to trauma. However, at the time that the diagnosis was established, there was little information available concerning the actual prevalence of this disorder among traumatized individuals. Thus, although it was postulated that any individual could become afflicted with PTSD following exposure to extreme stress, it was unknown how many individuals did in fact develop this disorder in response to trauma exposure. It is now well established that the prevalence of PTSD is far more rare than the prevalence of trauma, with prevalence rates of PTSD ranging from 3% to 58% (67–69,71,72). The observation that only a small proportion of individuals develop prolonged symptoms following trauma supports the idea that although PTSD is a stress response, it is not inevitable.

If PTSD is not an inevitable consequence of stress, then it is not that surprising that the biological alterations observed are different from those that have been classically associated with stress responses. It might even be hypothesized that the "classical" Selye stress response profile would be more typical of individuals who have undergone trauma and do not have symptoms of PTSD. The study of the biology of PTSD, then, appears to be the study of a response observed in a subset of individuals who are exposed to trauma. For this reason, it has been suggested that most animal models of stress are likely to have very little heuristic value in promoting understanding of the pathophysiology of PTSD unless they directly model the atypical and characteristic features of the disorder (73). This point cannot be emphasized enough, because the erroneous pairing of animal models of stress with PTSD may lead to a confusion between the normative neurobiological consequences of stress and atypical neurobiological sequelae that constitute the maladaptive and rarer stress response known as PTSD.

Although the specific biological findings summarized in this chapter do not tend to support PTSD as a typical stress response, they do confirm PTSD as a distinct disorder that can be differentiated from other psychiatric conditions. Thus, the very presence of a distinctive set of abnormalities provides biological validation of PTSD as a discrete diagnosis. In this vein it should be noted that studies of other biological systems have revealed alterations in PTSD that are also distinct from those observed in most other psychiatric conditions and differentiable from normals. In particular, psychophysiological, electrophysiological, and neurochemical investigations have provided support for an abnormal responsivity of the sympathetic nervous system and, likely, other neuromodulatory systems to traumatic (64,74–76) and nontraumatic (77,78) stimuli. These findings are reviewed in Chapter 25, and imply that there are biological changes following exposure to a trauma that are particularly associated with the symptoms of PTSD, and that can be differentiated from those observed simply as a result of stress exposure.

IMPLICATIONS AND CONCLUSIONS

The study of PTSD, whose definition rests on being the sequelae of stress, represents an opportunity to express the effects of extreme stress on neuroendocrine functioning from a unique perspective. The findings suggest that, rather than showing a pattern of increased adrenocortical activity and resultant dysregulation of this system, individuals who suffer from PTSD show evidence of a highly sensitized HPA axis characterized by decreased basal cortisol levels, increased number of lymphocyte glucocorticoid receptors, a greater suppression of cortisol to dexamethasone, and a more sensitized pituitary gland compared to individuals without PTSD. Thus, in addition to the classic pattern of increased cortisol levels in response to stress, there may be a contrasting paradigm of cortisol abnormalities following stress, characterized by diminished cortisol levels as a result of a stronger negative feedback inhibition. This paradigm compels us to expand the stress response spectrum.

To the extent that PTSD is conceptualized as a stress disorder, the findings challenge us to regard the stress response as diverse and varied, rather than as conforming to a simple, unidirectional pattern. The findings from epidemiological and longitudinal studies suggest that the search for risk factors other than exposure to trauma that may predispose or increase the likelihood of developing PTSD in individuals who sustain extreme stress may provide an important clue in defining biological heterogeneity in stress response. The neurobiology of PTSD may ultimately be related to risk factors for this disorder such as potentiality, pretrauma history, nonspecific genetic factors, and premorbid personality. These are the questions that must be addressed in the next generation of biological investigations of PTSD.

REFERENCES

1. American Psychiatric Association. *Diagnostic and statistical manual of mental disorders,* 3rd ed. Washington, DC; 1980.
2. American Psychiatric Association. *Diagnostic and statistical manual of mental disorders,* 1st ed. Washington, DC; 1952.
3. Kardiner A. The traumatic neuroses of war. *Psychosomatic Medicine Monograph (I-II).* Washington, DC: National Research Council; 1941.
4. American Psychiatric Association. *Diagnostic and statistical manual of mental disorders,* 2nd ed. Washington, DC; 1968.
5. American Psychiatric Association. *Diagnostic and statistical manual of mental disorders,* 4th ed. Washington, DC; 1994.
6. Pitman R, Orr S. Twenty-four hour urinary cortisol and catecholamine excretion in combat-related posttraumatic stress disorder. *Biol Psychiatr* 1990;27:245–247.
7. Halbreich U, Olympia J, Carson S, et al. Hypothalamic-pituitary-adrenal activity in endogenously depressed post-traumatic stress disorder patients. *Psychoneuroendocrinology* 1989;14:365–370.
8. Smith MA, Davidson J, Ritchie JC, Kudler H, Lipper S, Chappell P, Nemeroff CB. The corticotropin releasing hormone test in patients with posttraumatic stress disorder. *Biol Psychiatr* 1989;26:349–355.
9. Kudler H, Davidson J, Meador K, et al. The DST and post-traumatic stress disorder. *Am J Psychiatry* 1987; 144:1068–1071.
10. Selye H. Thymus and adrenals in the response of the organisms to injuries and intoxications. *Br J Exp Pathol* 1936;17:234–246.
11. Selye H. *The stress of life.* New York: McGraw Hill Book Company; 1956.
12. Stokes PE, Sikes CR. Hypothalamic-pituitary-adrenal axis in affective disorders. In Meltzer HY, ed. *Psychopharmacology: the third generation of progress.* New York: Raven Press, 1987:589–607.
13. Rivier CL, Plotsky PM. Mediation by corticotropin releasing factor (CRF) of adenohypophysial hormone secretion. *Ann Rev Physiol* 1986;48:475–494.
14. Munck A, Guyre PM, Holbrook NJ. Physiological functions of glucocorticoids in stress and their relation to pharmacological actions. *Endocr Rev* 1984;93:9779–9783.
15. Sapolsky RM, Krey LC, McEwen BS. Prolonged glucocorticoid exposure reduces hippocampal neuron number. Implications for aging. *J Neurosci* 1985;5:1221–1226.
16. Ling MHM, Perry PJ, Tsuang MT. Side effects of corticosteroid therapy. *Arch Gen Psychiatry* 1981;38: 471–477.
17. Wolkowitz OM, Reus VI, Weingartner H, et al. Cognitive effects of corticosteroids. *Am J Psychiatry* 1990; 147:1297–1303.
18. Mason JW. A review of psychoendocrine research on the sympathetic-adrenal-medullary system. *Psychosom Med* 1968;30:631–653.
19. Mason JW, Maher JT, Hartley LH, Mougey EH, Perlow MJ, Jones LG. Selectivity of corticosteroid and catecholamine responses to various natural stimuli. In Serban G, ed. *Psychopathology of human adaptation.* New York: Plenum, 1976:147–171.
20. Mason JW. A historical view of the stress field, part I. *J Human Stress* 1975;1:6–12.
21. Mason JW, Giller EL, Kosten TR, Ostroff RB, Podd L. Urinary-free cortisol levels in post-traumatic stress disorder patients. *J Nerv Ment Dis* 1986;174:145–159.
22. Yehuda R, Boisoneau D, Mason JW, Giller EL. Relationship between lymphocyte glucocorticoid receptor number and urinary-free cortisol excretion in mood, anxiety, and psychotic disorder. *Biol Psychiatr* 1993;34: 18–25.
23. Yehuda R, Southwick SM, Nussbaum G, Giller EL, Mason JW. Low urinary cortisol excretion in PTSD. *J Nerv Ment Dis* 1991;178:366–369.
24. Yehuda R, Giller EL, Southwick SM, Lowy MT, Mason JW. Hypothalamic-pituitary-adrenal dysfunction in posttraumatic stress disorder. *Biol Psychiatr* 1991;30: 1031–1048.
25. Yehuda R, Kahana B, Binder-Brynes K, Southwick S, Zemelman S, Mason JW, Giller EL. Low urinary cortisol excretion in Holocaust survivors with posttraumatic stress disorder. *Am J Psychiatry* 1995;152:982–986.
26. Hoffman L, Watson PB, Wilson G, Montgomery J. Low plasma b-endorphin in posttraumatic stress disorder. *Aust N Z J Psychiatry* 1989;23:269–273.
27. Yehuda R, Boisoneau D, Lowy MT, Giller EL. Dose-response changes in plasma cortisol and lymphocyte glucocorticoid receptors following dexamethasone administration in combat veterans with and without posttraumatic stress disorder. *Archives of General Psychiatry,* 1995, in press.
28. Yehuda R, Southwick SM, Krystal JH, Bremner D, Charney DS, Mason JW. Enhanced suppression of cortisol following a low dose of dexamethasone in combat veterans with posttraumatic stress disorder. *Am J Psychiatry* 1993;150(1):83–86.
29. Southwick SM, Krystal JH, Morgan CA, Johnson D, Nagy LM, Nicolaou A, Heninger GR, Charney DS. Abnormal noradrenergic function in posttraumatic stress disorder. *Arch Gen Psychiatry* 1993;50:266–274.
30. Lowy MT, Gormley GJ, Reder AT, et al. Immune function, glucocorticoid receptor regulation and depression. In Miller AH, ed. *Depressive disorders and immunity.* Washington DC: APA Press, 1989;105–134.

31. Yehuda R, Teicher MH, Levengood R, Trestman R, Siever LJ. Circadian rhythm of basal cortisol levels in PTSD. *Ann NY Academy Sci* 1994;746:378–380.
32. DeKloet R, Joels M, Oitzl M, Sutanto W. Implication of brain corticosteroid receptor diversity for the adaptation syndrome concept. *Methods Achiev Exp Pathol* 1991;14: 104–132.
33. Svec F. Glucocorticoid receptor regulation. *Life Sci* 1985;35:2359–2366.
34. Lowy MT. Reserpine-induced decrease in type I and II corticosteroid receptors in neuronal and lymphoid tissue of adrenalectomized rats. *Neuroendocrinology* 1990;51: 190–196.
35. Lowy MT. Quantification of type I and II adrenal steroid receptors in neuronal, lymphoid, and pituitary tissues. *Brain Res* 1989;503:191–197.
36. Yehuda R, Lowy MT, Southwick SM, Shaffer S, Giller EL. Increased number of glucocorticoid receptor number in post-traumatic stress disorder. *Am J Psychiatry* 1991;149:499–504.
37. Saplosky RM, Krey LC, McEwen BS. Stress down-regulates corticosterone receptors in a site specific manner in the brain. *Endocrinology* 1984;114:287–292.
38. McEwen BS, DeKloet ER, Rostene WH. Adrenal steroid receptors and actions in the nervous system. *Physiol Reviews* 1986;66:1121–1188.
39. Schlecte JA, Ginsburg BH, Sherman BML. Regulation of the glucocorticoid receptor in human lymphocytes. *J Steroid Biochem* 1982;16:69–74.
40. Tornello S, Orti E, DeNicola AF, Rainbow TC, McEwen BS. Regulation of glucocorticoid receptor in brain corticterone treatment of adrenalectomized rats. *Neuroendocrinology* 1982;35:411–417.
41. Meany MJ, Aitken DH, Bodnoff SR, Iny LJ, Sapolsky RM. The effects of postnatal handling on the development of the glucocorticoid receptor systems and stress recovery in the rat. *Prog Neuropsychopharmacol Biol Psychiatry* 1985;9:731–734.
42. Meaney MJ, Aitken DH, Bodnoff SR, Iny LJ, Taterewicz JE, Sapolsky RM. Early postnatal handling alters glucocorticoid receptor concentrations in selected brain regions. *Behav Neurosci* 1985;99:765–77.
43. Meaney MJ, Aitken DH, Viau V, Sharma S, Sarieau A. Neonatal handling alters adrenocortical negative feedback sensitivity and hippocampal type II glucocorticoid binding in the rat. *Neuroendocrinology* 1989;50:597–604.
44. Meaney MJ, Aitken DH, van Berkel C, Bhatnagar S, Sapolsky RM. Effect of neonatal handling on age-related impairments associated with the hippocampus. *Science* 1988;239:766–768.
45. Hess JL, Denenberg VH, Zarrow X, Pfeifer WD. Modification of the corticosterone response curve as a function of handling in infancy. *Physiol Behav* 1969;4:109–111.
46. Levine S, Haltmeyer CG, Karas GG, Denenberg VH. Physiological and behavioral effects of infantile stimulation. *Physiol Behav* 1967;2:55–59.
47. Haltmeyer GC, Denenberg VH, Zarrow MX. Modification of the plasma corticosterone response as a function of infantile stimulation and electric shock parameters. *Physiol Behav* 1967;2:61–63.
48. Ader R. The effects of early experience on the adrenocortical response to different magnitudes of stimulation. *Physiol Behav* 1970;5:837–839.
49. Meaney MJ, Aitken DH, Sapolsky RM. Thyroid hormones influence the development of hippocampal glucocorticoid receptors in the rat: a mechanism for the effects of postnatal handling on the development of the adrenocortical stress response. *Neuroendocrinology* 1987;45: 278–283.

49a. Mason JW, Southwick S, Yehuda R, Wang S, Rhey S, Bremner D, Johnson D, Lubin H, Blake D, Zhoi G, Guzman F, Charney DS. Elevation of serum-free T4, total T4, TBG, and total T4 levels in control-related post-traumatic stress disorder. *Arch Gen Psychiatry* 1994;51:629–641.

50. Meaney MJ, Viau V, Bhatnagar S, Betito K, Inv LJ, O'Donnell D, Mitchell JB. Cellular mechanisms underlying the development and expression of individual differences in the hypothalamic-pituitary-adrenal stress response. *J Steroid Biochem Molec Biol* 1991;3:265–274.
51. Bremner JD, Southwick SM, Johnson DR, Yehuda R, Charney DS. Childhood abuse in combat-related post-traumatic stress disorder. *Am J Psychiatry* 1993;150: 235–239.
52. APA task force on laboratory tests in psychiatry. The dexamethasone suppression test: an overview of its current status in psychiatry. *Am J Psychiatry* 1987;144:1253–1262.
53. Olivera AA, Fero D. Affective disorders, DST, and treatment in PTSD patients: clinical observations. *J Traumat Stress* 1990;3:407–414.
54. Kosten TR, Wahby V, Giller E, Mason J. The dexamethasone suppression test and thyrotropin-releasing hormone stimulation test in post-traumatic stress disorder. *Biol Psychiatr* 1990;28:657.
55. Dinan TG, Barry S, Yatham LN, Mobayed M, Brown I. A pilot study of a neuroendocrine test battery in post-traumatic stress disorder. *Biol Psychiatr* 1990;28:665–672.
56. Gold PW, Chrousos G, Kellner C, et al. Psychiatric implications of basic and clinical studies with corticotropin-releasing factor. *Am J Psychiatry* 1984;141:619.
57. Holsboer F, von Bardeleben U, Gerken A, et al. Blunted corticotropin and normal cortisol responses to human corticotropin-releasing factor in depression. *N Engl J Med* 1984;311:1127.
58. Krishnan JRR, Ritchie JC, Reed D, et al. CRF stimulation test results before and after dexamethasone in depressed patients and normal controls. *J Neuropsychiatry Clin Neurosci* 1990;2(1):34–43.
59. Yehuda R, Nemeroff CB. Neuropeptide alterations in affective and anxiety disorders. In DenBoer JA, Sitsen A, eds. *Handbook on depression and anxiety: a biological approach.* New York: Marcel Deckker, Inc., 1994: 543–571.
60. Yehuda R, Southwick SM, Mason JW, Giller EL. Interactions of the hypothalamic-pituitary-adrenal axis and the catecholaminergic system in posttraumatic stress disorder. In Giller EL, ed. *Biological Assessment and Treatment of PTSD.* Washington, DC: American Psychiatric Press, 1990:117–134.
61. *Physicians' Desk Reference,* 47th ed. Montvale, NJ: Medical Economics Data, 1993:905.
62. Pinto S, Yehuda R, Giller E, Kreek MJ. Abnormal natural killer cell activity in PTSD subjects. *National Institute Drug Alcohol Monograph Series.* CPDD Meeting, Florida, June 1992.
63. Yehuda R, Resnick H, Kahana B, Giller EL. Persistent

hormonal alterations following extreme stress in humans: adaptive or maladaptive? *Psychosom Med* 1993; 55:287–297.

64. Southwick SM, Krystal JH, Morgan AC, et al. Abnormal noradrenergic function in post traumatic stress disorder. *Arch Gen Psychiatry* 1993;50:266–274.
65. Murburg M, ed. *Catecholamine function in posttraumatic stress disorder: emerging concepts.* Washington, DC: American Psychiatric Press; 1994.
66. Hedge A, Huffman LJ. Vasopressin and endocrine function. In Gash DM, Doer GJ, eds. *Vasopressin, principles and properties.* New York: Plenum Press; 1987:435–476.
67. Resnick HS, Kilpatrick DG, Danski BD, Saunders BE, Best CL: Prevalence of civilian trauma and posttraumatic stress disorder in a representative national sample of woman. *J Consult Clin Psychol* 1993;61:984–991.
68. Southwick SM, Morgan CA, Nagy LM, et al. Trauma related symptomatology in Desert Storm veterans: a preliminary report. *Am J Psychiatry* 1993;150:1524–1528.
69. Kulka RA, Schlenger WE, Fairbank JA, Hough RL, Jordan BK, Marmor CR, Weiss DS. *Trauma and the Vietnam war generation.* New York: Brunner/Mazel; 1990.
70. Green BL. Psychosocial research in traumatic stress: an update. *J Traum Stress* 1993;7:341–362.
71. Breslau N, Davis GC, Andreski P. Traumatic events and posttraumatic stress disorder in an urban population of young adults. *Arch Gen Psychiatry* 1991;48:216–222.
72. Helzer JE, Robins LN, McEvoy L. Post-traumatic stress disorder in the general population: findings of the epidemiological catchment area survey. *N Engl J Med* 1987; 317:1630–1634.
73. Yehuda R, Antelman S. Criteria for rationally evaluating animal models of posttraumatic stress disorder. *Biol Psychiatr* 1993;33:479–486.
74. Pitman RK, Orr SP, Forgue DF, DeJong JB, Clairborn JM. Psychophysiologic assessment of posttraumatic stress disorder imagery in Vietnam combat veterans. *Arch Gen Psychiatry* 1987;44.970–975.
75. McFall M, Murburg M, Ko G, et al. Autonomic response to stress in Vietnam combat veterans with post-traumatic stress disorder. *Biol Psychiatry* 1990;27:1165–1175.
76. Shalev AY, Orr FP, Pitman RK. Psychophysiologic assessment of traumatic imagery in Israeli civilian patients with posttraumatic stress disorders. *Am J Psychiatry* 1993;150:620–624.
77. Paige S, Reid G, Allen M. Psychophysiological correlates of post traumatic stress disorders. *Biol Psychiatry* 1990;27:419–430.
78. McFarlane AC, Weber DL, Clark R: Abnormal stimulus processing in posttraumatic stress disorder. *Biol Psychiatry* 1993;34:311–320.

Neurobiological and Clinical Consequences of Stress: From Normal Adaptation to PTSD, edited by M. J. Friedman, D. S. Charney, and A. Y. Deutch, Lippincott-Raven Publishers, Philadelphia © 1995.

20

Some Approaches to the Study of the Clinical Implications of Thyroid Alterations in Post-Traumatic Stress Disorder

John W. Mason, Sheila Wang, *Rachel Yehuda, J. Douglas Bremner, †Sherry J. Riney, Hadar Lubin, David R. Johnson, Steven M. Southwick, and Dennis S. Charney

*Department of Psychiatry, Yale University School of Medicine, New Haven, Connecticut 06510, National Center for Post-Traumatic Stress Disorder, Veterans Affairs Medical Center, West Haven, Connecticut 06516; *Department of Psychiatry, Mount Sinai School of Medicine, Bronx Veterans Affairs Medical Center, Bronx, New York 10468; and †National Center for Post-Traumatic Stress Disorder, Veterans Affairs Medical Center, Palo Alto, California 94304*

BACKGROUND: HORMONAL STUDIES IN PTSD

Post-traumatic Stress Disorder (PTSD) in male Vietnam combat veterans has been found to be associated with multiple hormonal alterations involving the cortisol, norepinephrine, epinephrine, testosterone, and thyroid systems (1–8). Early psychoendocrine studies revealed preliminary evidence of an unusual hormonal profile in PTSD, with the overall balance shifted towards low cortisol levels (1,2), high norepinephrine and epinephrine levels (3,4), a high norepinephrine/cortisol ratio (5), high testosterone levels (6), and high total thyroxine (T4) levels in the face of lower free T4 levels (7).

We have recently reported a follow-up study involving a much larger sample of PTSD patients and a broader profile of thyroid assays that has uncovered the striking new finding that many PTSD patients show persistent and disproportionate elevations in both total and free triiodothyronine (T3) in relation to free thyroxine (T4); this elevates the T3/T4 ratio, suggesting that there may be increased peripheral conversion of T4 and T3 in this disorder (8). The study sample included 96 male Vietnam combat veterans with PTSD (4 subsamples of 24 patients each) and 24 male control subjects.

Figure 1 presents a summary of the main thyroid findings, with the PTSD patients showing highly significant elevations in total T3, free T3, and total T4 in the presence of no significant change in free T4, so that the total T3/free-T4 ratio is consequently elevated. It should be noted that the graph presents standardized scores around an arbitrary mean of 10 for each hormone, so that the height of the bars does not indicate absolute hormonal levels but rather the *relative* levels of the patient group versus the control group, showing the marked elevations in the PTSD group in all measures except the free T4 levels. Although a detailed interpretation of this profile is complicated somewhat by the additional presence of a concurrent serum thyroxine-binding globulin (TBG) elevation in the PTSD patients, as reported elsewhere (8), the presence of a marked elevation in *free* T3 levels

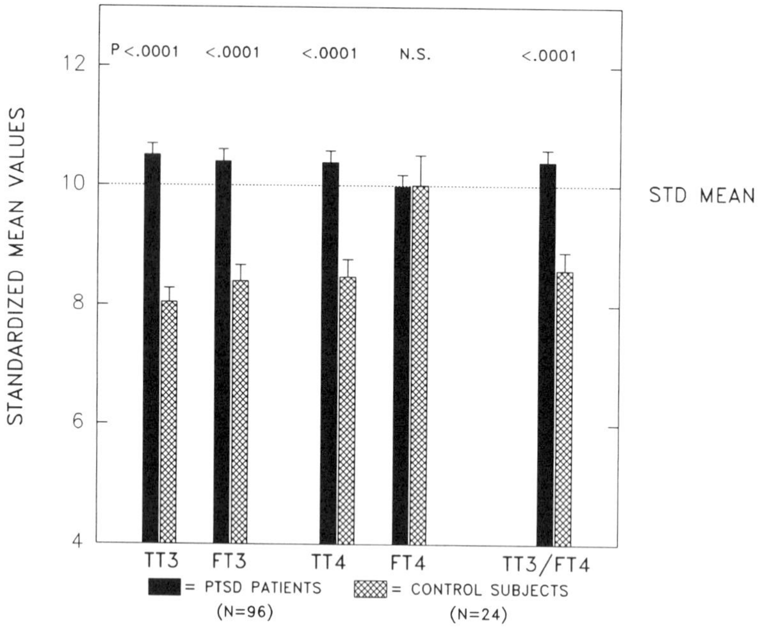

FIG. 1. Summary of thyroid hormone alterations in PTSD.

and in the T3/T4 ratio lends strong support for the hypothesis that the overall profile reflects not only TBG elevation but also increased production of T3 from T4.

As shown in Fig. 2, which is a simplified diagram of the hormonal production pathways of the hypothalamic-pituitary-thyroid axis, the centrally initiated influence of the hypothalamic thyrotropin-releasing hormone (TRH), acting through the anterior pituitary thyroid-stimulating hormone (TSH), is primarily involved in stimulating the secretion of T4 and probably only accounts for about 20% or less of the total T3 production in the body. It is important to realize that by far the greatest source of T3 production, probably at least 80%, is from peripheral enzymatic monodeiodination of T4 in extrathyroidal tissues such as liver and muscle (9,10). It is also noteworthy that one of the factors that appears to promote increased T3 conversion from T4 is elevation of peripheral catecholamine levels (11,12), a condition known to be prominent among the sustained hormonal alterations in PTSD (3,4).

In viewing the PTSD thyroid profile summarized in Fig. 1, then, the pattern of elevated T3 levels in the face of relatively lower free T4 levels might well be explained if the latter hormone were being "used up" excessively to produce increased amounts of T3. As detailed elsewhere, total T4 elevation probably is secondary mainly to an increase in TBG levels and possibly some central stimulation, although TSH levels are not significantly elevated in PTSD patients (8). A sense of the considerable extent to which PTSD patients tend towards elevation of the total T3/free-T4 ratio can be gained from Fig. 3, which presents all the raw data from the individual control subjects and patients, showing that 85% of the patients have values above the mean level for the control group. The first sample of 24 patients (PTSD-1) was from the Menlo Park VA Medical Center and the remaining three samples were from the West Haven VA Medical Center.

It should also be pointed out that our group of control subjects was composed mostly of veterans, some of whom had combat experience,

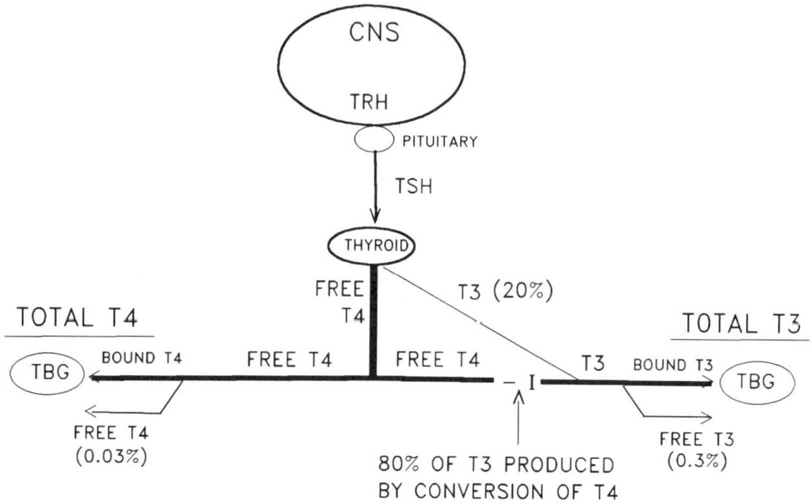

FIG. 2. Organization of the hypothalamic-pituitary-thyroid system.

with Mississippi PTSD Scale scores ranging up to 81 that raise the possibility that they may represent a subsample with partial or mild PTSD. Figure 4 shows that this subsample with combat exposure has a mean total T3 level intermediate between the patients meeting DSM-III-R criteria for PTSD and the control subjects without combat experience. The full extent of the difference between PTSD patients and normal control subjects without exposure to combat,

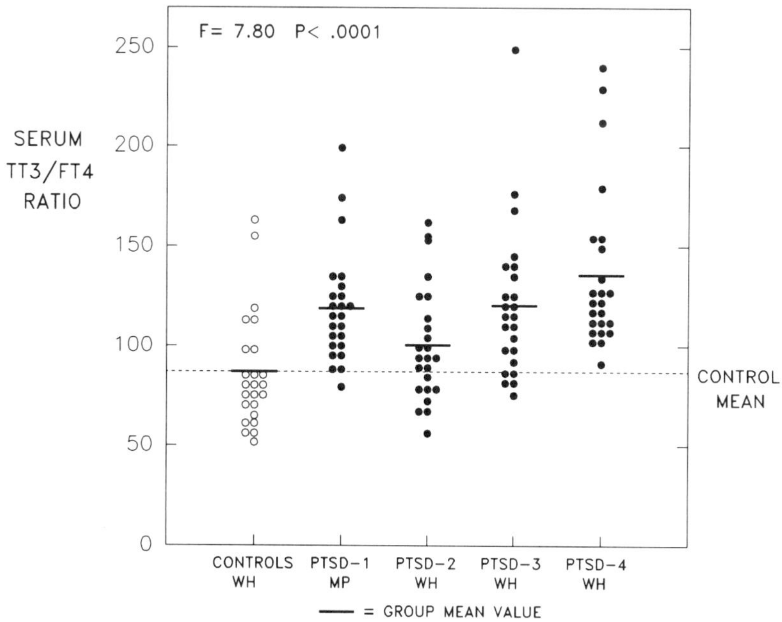

FIG. 3. TT3/FT4 ratio elevations in multiple samples of PTSD patients.

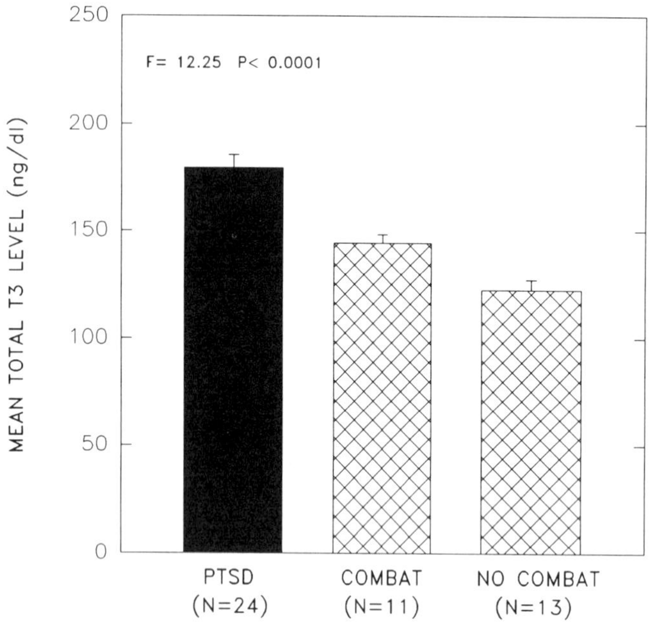

FIG. 4. Total T3 levels in PTSD versus combat and noncombat controls.

therefore, may be somewhat underestimated in our study.

SOME HISTORICAL CONSIDERATIONS

It has long been known that a history of traumatic stress, especially life-threatening crises such as combat experiences, fires, earthquakes, shipwrecks, narrow escapes from accidents, etc., is a common antecedent to the onset of thyrotoxicosis (13,14). Many preclinical studies in both normal humans and animals have shown alterations in thyroid hormone secretion in response to psychologically stressful situations (15). Many clinical studies have provided evidence that thyroid hormones may have a significant role in a variety of psychiatric illnesses, including both affective and schizophrenic disorders (16,17,18). With such strong support for considering the thyroid hormones as "stress hormones," it should perhaps not be surprising to find thyroid alterations in PTSD, by definition a stress-related disorder.

Yet the nature of the observed changes in PTSD are not those of classical or other established forms of clinical hyperthyroidism, but rather of a more subtle, unusual, and perhaps even distinctive type of moderate hyperactivity featuring total and free T3 elevations that are not detected by conventional routine clinical thyroid screening tests and have therefore been overlooked until very recently. The fact that T3 is metabolically two to four times more potent than T4, along with the observation that 64% of our large sample of PTSD patients had total T3 levels above the upper range limit of the control subjects (8), clearly stimulate interest in exploring the possibility that T3 elevations may have a significant clinical role in PTSD, either in reflecting or contributing to pathogenetic factors.

It is, of course, well known that the clinical picture of classical hyperthyroidism includes such psychiatric symptoms as sleep disturbances, restlessness, anxiety, irritability, explosive anger, jumpiness or increased startle, and difficulty in concentrating—in other words, both cognitive and affective disturbances commonly observed in PTSD patients (19). It should also be kept in mind that many previous psychoendocrine studies, particularly those of the cortisol system, have clearly demonstrated that both state or phase factors, such as emotional arousal, as well as trait or characterological factors, such as coping or defensive styles, are reflected in hormonal levels (20). In beginning to pursue an understanding of the clinical significance of the thyroid alterations in PTSD, then, it appears well advised to consider and search for both enduring and transitory clinical or psychological correlates of the elevated T3 levels in PTSD, using both cross-sectional and longitudinal approaches. As the present study unfolded and we became increasingly aware of the unusual thyroid findings, we began in an exploratory way to utilize our core battery of PTSD assessment scales, along with the psychometric measurements available from other concurrent clinical studies involving our patient sample and, as possible, finally began to add psychometric measurements that would permit some screening for personality or trait factors as well as state and symptom factors. The purpose of this chapter is to review some of these preliminary efforts to look for relationships between clinical or psychological factors and the T3 elevations in PTSD, and to consider some important methodological and research strategy issues emerging in our experience so far that appear likely to have important implications for future work in this field.

LONGITUDINAL COURSE OF T3 LEVELS IN PTSD INPATIENTS

In the West Haven PTSD inpatient sample, we were able to obtain hormonal samples at three points during hospitalization—at admission, midcourse, and discharge—on a subsample of 47 patients. Figure 5 shows that the mean total T3 level for this PTSD patient group is similar at the admission and middle points, but drops significantly at the discharge point (ANOVA for repeated measures, $F=7.77$, $p<0.0008$), so that it clearly appears likely that T3 levels may covary with changes in clinical state or phase of illness on a longitudinal basis. It is also of interest, however, that at all three points the PTSD patients sustain T3 levels significantly above the control mean value ($F=53.62$, $p<0.0001$), which appears to support the possibility that there are also important relationships

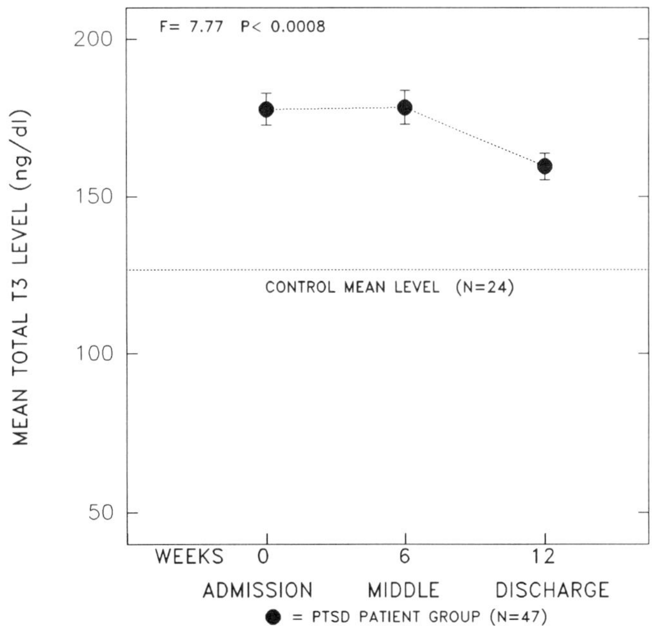

FIG. 5. Longitudinal course of total T3 levels in PTSD inpatients.

between T3 levels and relatively enduring psychological traits or clinical characteristics in PTSD patients.

As a further way of examining state or phase fluctuations over time, the changes in T3 levels in individual patients between the period of hospital admission and the period of hospital discharge were analyzed in our West Haven sample of patients. These patients were admitted in 6 successive cohorts, ranging from 8 to 13 men each, for a 12-week inpatient treatment program involving an intensive schedule of about 32 hours a week of individual and group therapy. The change or difference in total T3 levels between the admission and discharge periods was calculated as a "delta" value, and Fig. 6 presents all the raw data for individuals within the six cohorts. Notice that about 70% of the patients show changes greater than 20 ng%, either increases or decreases, between these two points in time. There are clearly great individual differences between subjects, with the range extending from +74 to −105 ng%. While about 30% of the patients show increases and 70% show decreases, there appears to be some interesting variation between the group mean values, indicating not only an individual but also a cohort or group tendency towards either an increase or a decrease over the course of hospitalization.

An anecdotal example of longitudinal T3 changes in a single cohort of PTSD patients is presented in Fig. 7, showing an apparent relationship of hormonal changes to the phase of treatment. The marked elevation in total T3 levels in six out of seven patients during the middle phase of the treatment program appeared coincident with the initiation of traumatic memory sessions, in which each man recounted in a group setting the most distressing memory of his combat experience. The degree, however, to which patients earnestly participated in these disturbing sessions appeared to vary considerably from cohort to cohort, as well as from individual to individual, and was often difficult to assess, so that the situational criterion of a traumatic memory session alone was not a sufficient basis for developing a "challenge" model for psychoendocrine studies.

The findings presented in both Figs. 6 and 7, demonstrating such examples as the same patient having a total T3 value of 245 ng% at one point

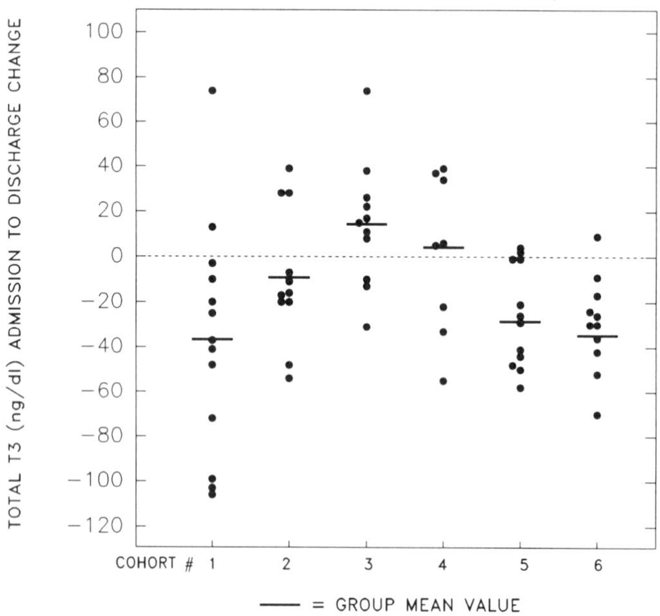

FIG. 6. Admission to discharge change in total T3 levels in PTSD cohorts.

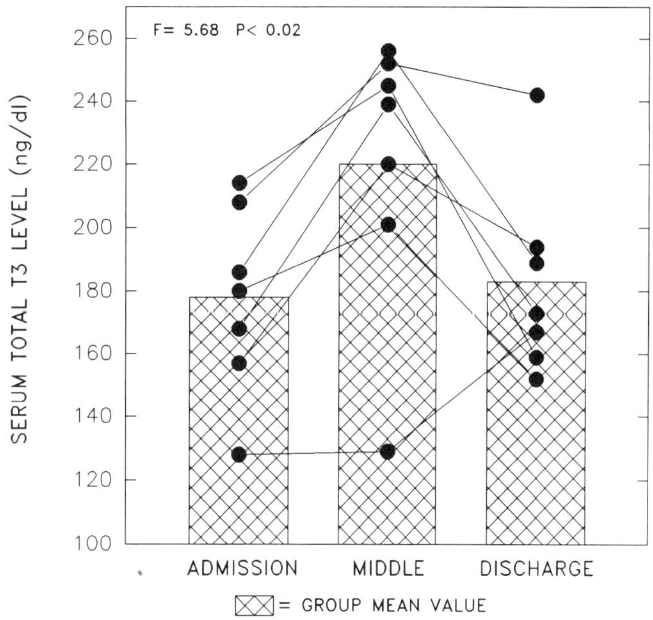

FIG. 7. Total T3 change in relation to treatment phase in PTSD cohort.

in time and 160 ng% at another point, raise the question as to whether such T3 changes might be related simply to concurrent changes in the severity of specific PTSD symptoms or perhaps to other clinical variables (e.g., phasic shifts in symptom patterns, defensive organization, or different stages of decompensation or recompensation in response to treatment or to life events or changing conditions over time).

SOME PRELIMINARY LEADS RELATING T3 AND PSYCHOLOGICAL MEASURES

Core PTSD Symptoms

In a subgroup of 65 of the West Haven cohort patients, we were able to obtain ratings of PTSD symptoms together with hormonal measurements using the Clinician Administered PTSD Scale (CAPS-2), which provides three subscales measuring reexperiencing, avoidance, and hyperarousal symptoms on a state basis (21). The principal findings were that the CAPS total PTSD symptom score correlated significantly with total T3 ($r=.376$, $p<0.002$), with free T3 ($r=.309$, $p<0.01$) and total T4 ($r=.363$, $p<0.003$). Figure 8 presents a scatterplot of the total T3 and CAPS-2 total score data. With regard to the subscale data, the strongest correlations involving T3 were between the hyperarousal subscale score and levels of total T3 ($r=.407$, $p<0.0008$), and free T3 ($r=.352$, $p<0.004$). As might be expected, free T4 did not correlate significantly with any of the PTSD symptom measures. It is also interesting that total T3 was correlated significantly ($r=.291$, $p<0.02$) with the avoidance subscale score, while free T3 was significantly related ($r=.282$, $p<0.03$) to the reexperiencing subscale score (21). With regard to the 30 individual CAPS items, those showing the strongest correlations with TT3 and FT3 included feelings of being overwhelmed, increased startle, difficulty concentrating, hypervigilance, feeling as if a traumatic event were occurring, efforts to avoid activities or situations, and efforts to avoid thoughts or feelings. The main findings of significant relationships between the CAPS total and hyperarousal scores and both TT3 and FT3 were relatively strong, and were similarly found in both the first and sec-

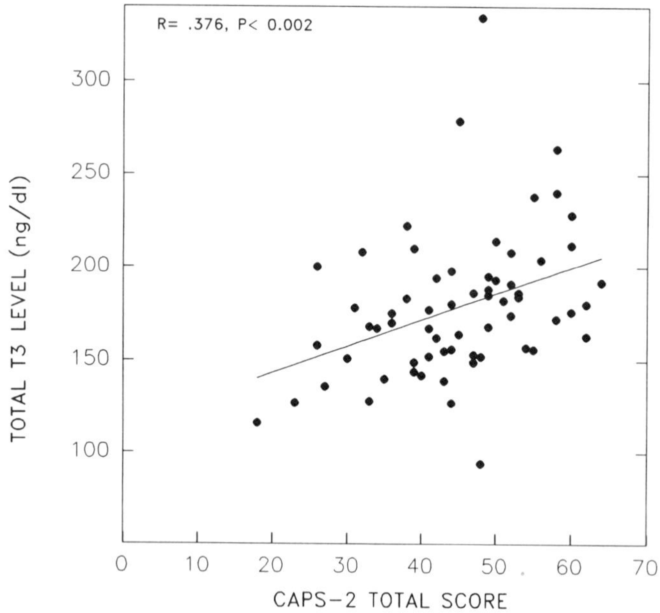

FIG. 8. Correlation between total T3 and CAPS-2 total score in PTSD.

ond halves of our West Haven sample of 65 patients.

It should be added, however, that the relationships between thyroidal and CAPS scores just cited did not reach significance in a smaller sample of PTSD patients (n=24) in the Menlo Park study. A search for possible explanations for the difference in findings between the two studies led to our awareness of some subtle but perhaps significant differences between the West Haven and Menlo Park programs in two areas, one involving clinical criteria for patient selection and the other involving clinical control of milieu conditions. Differences in the selection process of the Menlo Park program included exclusion of more severely ill patients meeting criteria for borderline personality disorder and those with prominent dissociative symptoms. With regard to ward milieu differences in the Menlo Park setting as an important element of treatment approach during the program, all patients were instructed *not* to discuss the war or traumatic memories with each other except in special treatment sessions for that purpose. As might be expected from these factors, the Menlo Park (MP) sample had significantly lower mean levels of CAPS scores than the West Haven (WH) sample, including the CAPS total score (WH=44.8±1.3 versus MP=32.9±2.5, t=4.60, p<0.00001), and the hyperarousal score (WH=17.0±0.4 versus MP=12.4±1.1, t=4.68, p<0.00001). As reported previously (8), there was also a tendency for lower T3 levels in the Menlo Park sample, so that lower mean levels and the associated compression of the range of both symptom and hormonal values in the Menlo Park sample may have accounted for the differences in correlational findings between the two studies.

This comparative study of two independent inpatient samples was one of the first of an increasing series of experiences to call our attention to the importance of detailed sample description—including careful definition of selection factors, severity of illness, clinical and study conditions affecting ward milieu, and other methodological details—in reporting and interpreting findings involving hormonal levels and their relationships with clinical or psychological measures in any given PTSD patient sample.

Other Clinical Features

From our early unpublished efforts to explore other possible relationships between thyroid hormones and prominent clinical features in our PTSD samples, it may be useful to cite two examples that provide further support for the need to attend carefully to the methodological concerns expressed in the previous section—involving details of sample description and study conditions in psychoendocrine studies of combat-related PTSD patients.

In an early subsample of 36 West Haven PTSD inpatients, some pilot work on dissociation by a member of our group (Bremner) led to the observation of a significant positive relationship between levels of total and free T3 and levels of dissociation, measured psychometrically. These relationships were, however, not significant in the Menlo Park sample of 24 patients, which showed a lower mean level and more narrow range of dissociation scores (a finding that may well be related to the policy of excluding patients with prominent dissociative symptoms in the Menlo Park program) and might logically be expected to reduce the likelihood of detecting significant relationships in the correlational analyses. Another possibly important methodological difference in the two studies was that the dissociation test was personally introduced and explained to the West Haven patients by a professional investigator with special interest in dissociation and expert knowledge of the psychometric instrument, whereas it was administered more impersonally as part of a larger package of assorted self-report psychometric tests in the Menlo Park study. In any event, this experience again appeared to dictate caution in approaching and interpreting "replication" psychoendocrine studies in PTSD without careful examination and comparison of patient sample and methodological factors that may be crucial to demonstrating relationships between particular clinical and hormonal features in this disorder.

A second example emerged from an interest of a member of our group (Southwick) in the possibility of novelty-seeking characteristics as a discriminating feature for a subtype of PTSD patient. This led to the observation in the Menlo Park sample ($n=24$) of significant positive correlations between both total and free T3 levels and a psychometric test score for novelty-seeking behavior, which appeared not simply to be related to severity of PTSD illness. These relationships, however, were not found when the same measurements were obtained in a subsequent subsample of West Haven patients, so that the question again was raised about whether or not differences between the patient samples or study conditions might possibly account for the difference in our correlational findings in the two studies. It may be, for example, that relationships between hormonal levels and clinical features such as novelty-seeking, which are generally thought of as trait or personality characteristics, are best demonstrated under relatively basal conditions, when state or stage of illness fluctuations (e.g., such as the elevated hyperarousal in the West Haven sample) are not superimposed in such a way as to obscure the more subtle long-term basal relationships. Another methodological difference in the two studies was that the psychometric and hormonal measures were obtained at the same point in time in the Menlo Park study, while there was considerable variability between the time of hormonal versus psychometric measurements in the West Haven sample; this perhaps raises the possibility of some state fluctuation in novelty-seeking behavior, even though it is generally viewed as an enduring characterological feature. Again, only further studies with careful attention to details of sample characterization, study conditions, and methodological factors will provide a more conclusive understanding of early leads such as these about possible relationships between clinical and hormonal aspects of PTSD.

GENERAL COMMENTS AND CONCLUSIONS

The finding of sustained elevations in levels of the metabolically potent hormone triiodothyronine (T3) in many PTSD patients presents an interesting challenge in the investigation of the clinical significance of this unusual hormonal alteration, not only in terms of its possible linkage to PTSD symptoms, but also in relation to underlying psychological mechanisms that may have an important pathogenetic or clinical role in PTSD. Our initial efforts in this direction reviewed in the previous section must certainly be regarded as very preliminary, but they do provide encouragement and support for further work along these lines and present some leads suggesting potentially fruitful areas for further study. Our experience so far, however, has been perhaps most significant in drawing our attention repeatedly and forcefully to important issues concerning methodological rigor, quality control, and research strategy; these issues appear to be foundational for developing practical guidelines for the design and interpretation of future studies exploring the relationships between hormonal and clinical or psychological factors in PTSD patients.

Importance of Selection Factors in Patient Recruitment

There is a general tendency in research in biological psychiatry to define patient samples solely or largely by DSM-III-R diagnostic criteria and by ratings of severity of core symptoms of the primary illness. Our experience with combat-related PTSD patients within the VA hospital system, however, indicates that a number of additional clinical factors may require consideration in the screening process in order to characterize patient samples in sufficient detail to: 1) help **minimize heterogeneity** within a given patient sample, and 2) permit valid replication studies between different research centers. Some selection limitations are unavoidable, of course, such as our inability to include patients who seek isolation and avoid and refuse any hospital contact. Some patients may seek hospital treatment in a period of special crisis, but would not seek out or volunteer for any elective treatment program. Some will seek outpatient treatment help but decline participation in inpatient treatment programs. The inducements presented, restrictions imposed, or other factors associated with an inpatient treatment program that might influence the inclination of an individual patient to participate may further introduce selection bias in defining a particular patient sample. While patients in all the various categories just described may similarly meet DSM-III-R criteria for the diagnosis of PTSD and may even have comparable Mississippi scores, it appears inadvisable at this early stage of work in the field to assume that some may not represent biologically or clinically different subtypes of PTSD patients.

In recruiting patients for group-oriented longterm inpatient treatment programs, it is commonly a practical policy to exclude, for example, patients with a history of especially violent or asocial tendencies, or other behavior likely to cause ward or group management problems. Patients who are judged unlikely to be able to tolerate the demands and restrictions of a prolonged inpatient program also tend to be excluded. The issue of some VA patients viewing their acceptance into PTSD research or treatment programs as relevant to disability claims or economic advantage is widely recognized as another possible confounding element and one not always easy to assess in the screening process. Another variable that may naturally develop in research centers involved in longterm PTSD programs is the gradual modification of inclusion or exclusion criteria as the supply of new patients in a given region dwindles, in an effort to continue or expand the program. Our experience with psychoendocrine research at this preliminary stage suggests strongly that secondary selection factors such as those just mentioned should not be dismissed as minor, but should be included in the description of patient samples and considered in the interpretation of research findings until a sufficient database has been developed to establish the extent to which such variables may prove to be important to understanding relationships between hormonal and clinical factors in PTSD.

Importance of Clinical Subtyping and Comorbidity Assessment

Closely linked to selection factors described in the previous section is the issue of the desirability of defining and obtaining patient samples with as low a degree of heterogeneity as possible, for both clinical and research purposes. While many clinicians having extensive contact with PTSD patients have impressions about discriminating possible diagnostic subtypes, there appears to be no general agreement concerning the specific criteria that might best define or characterize such subtypes. There is a pressing need at this early stage in the field to generate and test hypotheses with both clinical and biological criteria that may identify clinically significant subtypes of PTSD patient samples. Such approaches could include not only the obvious criteria provided by comorbidity assessments, but also might be based upon clinical impressions involving specific psychological characteristics such as, for example, our pilot work on the novelty-seeking personality dimension as a possible basis for subtype discrimination. While this area is admittedly still in a preliminary exploratory stage, it is one that could have an extremely important strategic influence in facilitating future progress in this field and deserves considerable attention in our present efforts. From a tactical standpoint, however, it appears that any efforts to define patient subtypes must include careful evaluation of whether clinical or biological differences between patients represent enduring background or baseline differences versus episodic state, or phase differences at a particular point in their course of illness, as discussed in the next section.

Importance of Baseline Versus Phase Relationships

One very basic research strategy particularly highlighted by our preliminary work involves the need to consider both **phase** (acute, state, stage) and ongoing **basal** (chronic, trait, character) variables from a clinical and psychological standpoint. Hyperarousal, for example, may primarily represent the phase category, while novelty-seeking may be an example predominantly reflecting the basal category. It is very evident, however, that the two categories are not mutually exclusive, and that some variables showing appreciable acute or episodic fluctuations may also show enduring baseline alterations and, conversely, that some variables that appear predominantly to represent characterologic features may also show phase changes at times. Perhaps the generally intractable clinical course of PTSD favors the possibility that an understanding of the chronic baseline influences may yield the most insight into the pathogenesis of this disorder, and perhaps the most leverage in defining diagnostic subtypes. However, there is clearly a need to establish first the extent, magnitude, and relative importance of phase fluctuations in PTSD patients.

It appears, therefore, important to move in the direction of **longitudinal studies** with repeated hormonal and psychometric measurements over periods of many weeks or months in order to discriminate between phase and baseline relationships, as well as to learn more about the time relationships and the direction of the interactions between hormonal and psychological mechanisms. At this stage in the development of the field, it also appears very advisable to obtain the psychometric assessments, whether viewed as primarily phase or basal measures, as closely as possible to the time of the collection of hormonal samples, preferably on the same day or the following day.

Importance of Milieu Factors

Our experience in the VA hospital system with the study of PTSD inpatients in a treatment program organized within a group or cohort framework has impressed us with the importance of taking social factors carefully into account in the interpretation and reporting of psychoendocrine findings in PTSD patients. It is our impression, for example, that PTSD patients participating in intensive treatment and research programs, which involve living in a 12-patient cohort with constant group interaction over a 12-week period, tend to sustain higher urinary cortisol levels than PTSD patients studied on an individual basis in a mixed-diagnosis ward or on an outpatient basis. We have not as yet been able to look closely at this factor in relation to the thyroid system, but it appears very important to compare the thyroid profile in PTSD patients under highly protective, sanctuary-like conditions—in contrast to heavy-demand, group-oriented conditions—cross-sectionally but especially longitudinally, studying the same individual patient in such contrasting social and treatment settings.

Validity Questions Concerning Psychometric Assessments in PTSD

Another important issue in our experience relates to methodological questions concerning the administration and interpretation of psychometric tests in PTSD studies. In beginning a survey of the results obtained so far with a battery of about 20 psychometric methods in our PTSD patient cohorts, questions have repeatedly arisen concerning the accuracy of self-report test data or even clinician-administered test data depending largely on the verbal response of patients to questions. The problem may even begin with the conditions under which psychometric tests are administered. Variables such as the authority status, style, or attitude of the person presenting and explaining the tests, the attitude of the patients toward that person, the number of tests given consecutively and the time duration of the test session, and the nature of group interaction before and during the session might potentially influence test responses.

It is widely recognized that PTSD patients, particularly in the VA hospital setting, tend to exaggerate or over-endorse their symptomatology in response to direct self-report questionnaires. In looking at MMPI-2 validity measures in our patient sample, mean scores on the "F" validity scale and the Back-page Infrequency Scale, for example, are particularly high—around 100. The "F" scale was originally designed so that high scores would detect deviant or atypical ways of responding to test items, but such ways can range from indicating random responding to: 1) faking bad symptoms; 2) exaggerating symptoms as a plea for help or for self gain; 3) defiant resisting of the testing procedure; 4) short attention span; 5) withdrawal; 6) delusions or other psychiatric symptoms; or 7) perhaps responding simply from a deep need to maintain an acceptable self image. The Back-page Infrequency Scale high score is somewhat similar to the "F" scale score in that it may reflect such deviant responding as faking bad symptoms or over-endorsing items, but it also can indicate that the patient began to lose attention and concentration as the test progressed, and shifted to a largely random pattern of responding. There is a need for keen and clinically sensitive judgment in determining how to conceptualize and deal with these validity questions in evaluating not only MMPI data, but psychometric data in general that is obtained from PTSD patients. Clinical insight may even be useful in guiding the modification of certain standard psychometric instruments in accordance with idiosyncratic features in the perspective or style of many PTSD patients.

Until more guidelines are developed in this area, it may be a valuable strategy in the assessment of specific features in PTSD patients to use whenever possible an array of psychometric methods, including a self-report, and a clinician-administered, projective, and global clinical measurement of the feature in question. The use of projective tests and global clinical assessments, though generally considered less objective and quantitative than the more structured and operationally-defined procedures, may in fact deserve much wider consideration for trial in PTSD studies as instruments that may help in evaluation of the accuracy of self-report problems in this disorder. Other more standard tactics, such as eliminating patients from the data analysis who are extreme outliers on the validity measures of the MMPI or stratifying patients on the basis of validity or personality measures, may also be useful in teasing out the true relationships between psychological and hormonal factors.

Finally, while our preliminary findings suggest that such particular psychological or clinical features as hyperarousal, dissociation, or novelty-seeking behavior may have special relationships to the T3 elevations in PTSD, these observations clearly only represent a beginning in a broader survey that remains to be done, especially perhaps with further assessments in two areas. The first area involves a need to find or develop instruments that will provide valid and reliable measurement of PTSD symptoms that are closely similar to those reported in clinical hyperthyroidism, including sleep disturbances, startle, irritability, explosive anger, difficulty in concentrating, and restlessness. Such measures would permit the study of the possible role of T3 elevations in relation to these important symptoms and impairing factors in PTSD, and perhaps contribute to ideas concerning possible treatment studies linked to thyroid alterations in PTSD. The second area involves the need for assessment of personality or character disorder factors, perhaps including coping and defensive styles, which may be especially likely to be linked with hormonal measures (20) and are a prominent part of the clinical picture in PTSD patients.

ACKNOWLEDGMENTS

Support for this work was provided in part by National Institute of Mental Health grants, a Research Scientist Award K05-MH00346, and a RO1-MH41125 award to J.W.M., as well as National Center for PTSD awards to the West Haven and Menlo Park VA Medical Centers by the Department of Veterans Affairs.

REFERENCES

1. Mason JW, Giller EL, Kosten TR, Ostroff RB, Podd L. Urinary free cortisol levels in post-traumatic stress disorder patients. *J Nerv Ment Dis* 1986;174:145–149.
2. Yehuda R, Southwick SM, Nussbaum G, Wahby V, Giller E, Mason J. Low urinary cortisol excretion in PTSD. *J Nerv Ment Dis* 1990;178:366–369.
3. Kosten TR, Mason JW, Giller EL, Ostroff RB, Harkness L. Sustained urinary norepinephrine and epinephrine elevation in post-traumatic stress disorder. *Psychoneuroendocrinology* 1987;12:13–20.
4. Yehuda R, Southwick S, Giller EL, Ma X, Mason JW. Urinary catecholamine excretion and severity of PTSD symptoms in Vietnam combat veterans. *J Nerv Ment Dis* 1992;180:321–325.
5. Mason JW, Giller EL, Kosten TR, Harkness L. Elevation of urinary norepinephrine/cortisol ratio in post-traumatic stress disorder. *J Nerv Ment Dis* 1988;176:498–502.
6. Mason JW, Giller EL, Kosten TR, Wahby VS. Serum testosterone levels in post-traumatic stress disorder inpatients. *J Traum Stress* 1990;3:449–457.
7. Mason JW, Giller EL, Kosten TR, Southwick SM. The use of psychoendocrine strategies in post-traumatic stress disorder. *J Applied Soc Psychol* 1990;20:1822–1846.
8. Mason J, Southwick S, Yehuda R, et al. Elevation of serum free T3, total T3, TBG and total T4 levels in combat-related posttraumatic stress disorder. *Arch Gen Psychiat* 1994;51:629–641.
9. Borst CC, Eil C, Burman KD. Euthyroid hyperthyroxinemia. *Ann Int Med* 1983;98:366–378.
10. Chopra JJ. Nature, sources, and relative biological significance of thyroid circulating hormones. In Ingbar SH, Braverman LE, eds. *The thyroid.* Philadelphia: JB Lippincott, 1986:136–153.
11. Galton VA. Thyroid hormone-catecholamine interrelationships. *Endocrinol* 1965;77:278–284.
12. Lumholtz IB, Sierbaik K, Nielsen K, Faber J, Kirkegaard C, Friis TH. Effects of propranolol on extrathyroidal metabolism of thyroxine and 3,3′,5-triiodothyronine evaluated by noncompartmental kinetics. *J Clin Endocrin Metab* 1978;47:587–589.
13. Bram I. Psychic trauma and pathogenesis of exophthalmic goiter. *Endocrinol* 1927;11:106–116.
14. Leclerc J, Germain M, Weryla G, Duquenne M, Hartemann P. Role of stressful life events in the onset of Grave's disease. 10th International Thyroid Conference, The Hague; 1991; Abstract.
15. Mason JW. A review of psychoendocrine research on the pituitary-thyroid system. *Psychosom Med* 1968;30: 666–681.
16. Whybrow P, Ferrell R. Thyroid state and human behavior: contributions from a clinical perspective. In Prange AJ, ed. *The thyroid axis, drugs, and behavior.* New York: Raven Press; 1974:5–28.
17. Mason JW, Kennedy JL, Kosten TR, Giller EL. Serum thyroxine levels in schizophrenic and affective disorder diagnostic subgroups. *J Nerv Ment Dis* 1989;177:351–358.
18. Nemeroff CB. Clinical significance of psychoneuroendocrinology in psychiatry: focus on the thyroid and adrenal. *J Clin Psychiat* 1989;50:13–22.
19. Whybrow PC. Behavioral and psychiatric aspects of thyrotoxicosis. In Utiger RD, Braverman LE, eds. *The thyroid.* Philadelphia: JB Lippincott, 1991;863–870.
20. Mason JW. Clinical psychophysiology: psychoendocrine mechanisms. In *American handbook of psychiatry,* 2nd ed., vol 4. New York: Basic Books, Inc., 1975; 553–582.
21. Wang S, Mason J, Southwick S, Johnson D, Lubin H, Charney D. Relationships between thyroid hormones and posttraumatic stress disorder symptoms. *Psychosom Med,* in press.

Neurobiological and Clinical Consequences of Stress: From Normal Adaptation to PTSD, edited by M. J. Friedman, D. S. Charney, and A. Y. Deutch, Lippincott-Raven Publishers, Philadelphia © 1995.

21

Neuroimmunology of Stress

George R. Heninger

Department of Psychiatry, Yale University School of Medicine, New Haven, Connecticut 06508

A major role for both the central nervous system (CNS) and immune system is the generation of adaptive responses that correct for deviations in the homeostasis of the organism. In this context, stress can be defined as that level of altered homostasis to which the organism has problems adjusting. Since the initial stress may involve either the CNS or the immune system, there may be more or less involvement of the other system, depending on the nature of the stressor and the ability of the first system to successfully adjust to it internally. For example, a cognitive stress that is successively dealt with may not activate stress-related neurotransmission and neuroendocrine hormones that will affect the immune system, and, reciprocally, an effective adaptation to immune threat may minimize immune activation and consequent CNS involvement (1).

During the past 40 years, there has been a continually enlarging and detailed body of information on the biology and function of both the CNS and immune systems. The immune system is second only to the CNS in the complexity and specificity of its actions. By design, the immune system is primarily autoregulated so that most of its function in defining "self from nonself" and mounting an effective response to "nonself" invasion does not require major input from the CNS. However, the CNS does have an important modulatory role in immune function, and abnormalities in CNS function can produce measurable changes in the immune system that have clinically significant health consequences (1,2). Reciprocally, primary abnormalities in immune function can cause CNS disease (e.g., myasthenia gravis, multiple sclerosis), and even changes in immune function within the physiologic range can produce CNS symptoms (e.g., the fever, somnolence, malaise, and behavioral changes following immunization) (3). What is increasingly clear is that many different types and clinically important interactions between the CNS and the immune system relate to disease processes. In recent years, use of vastly improved biologic methodologies have allowed interactions between the CNS and immune system to be more clearly described and specified.

This chapter will review: 1) the current state of knowledge about the mechanisms that subserve CNS–immune-system interactions during stress; 2) the evidence that psychologic and behavioral stress can alter and impair immune function; and 3) recent data on the mechanisms by which immune and activation may influence CNS activity.

CENTRAL NERVOUS SYSTEM—IMMUNE SYSTEM INTERACTIONS

Pathways Serving Central Nervous System Effects on Immune Function

Figure 1 illustrates a rough schematic of some of the known relationships between the central nervous system and the immune system. At high

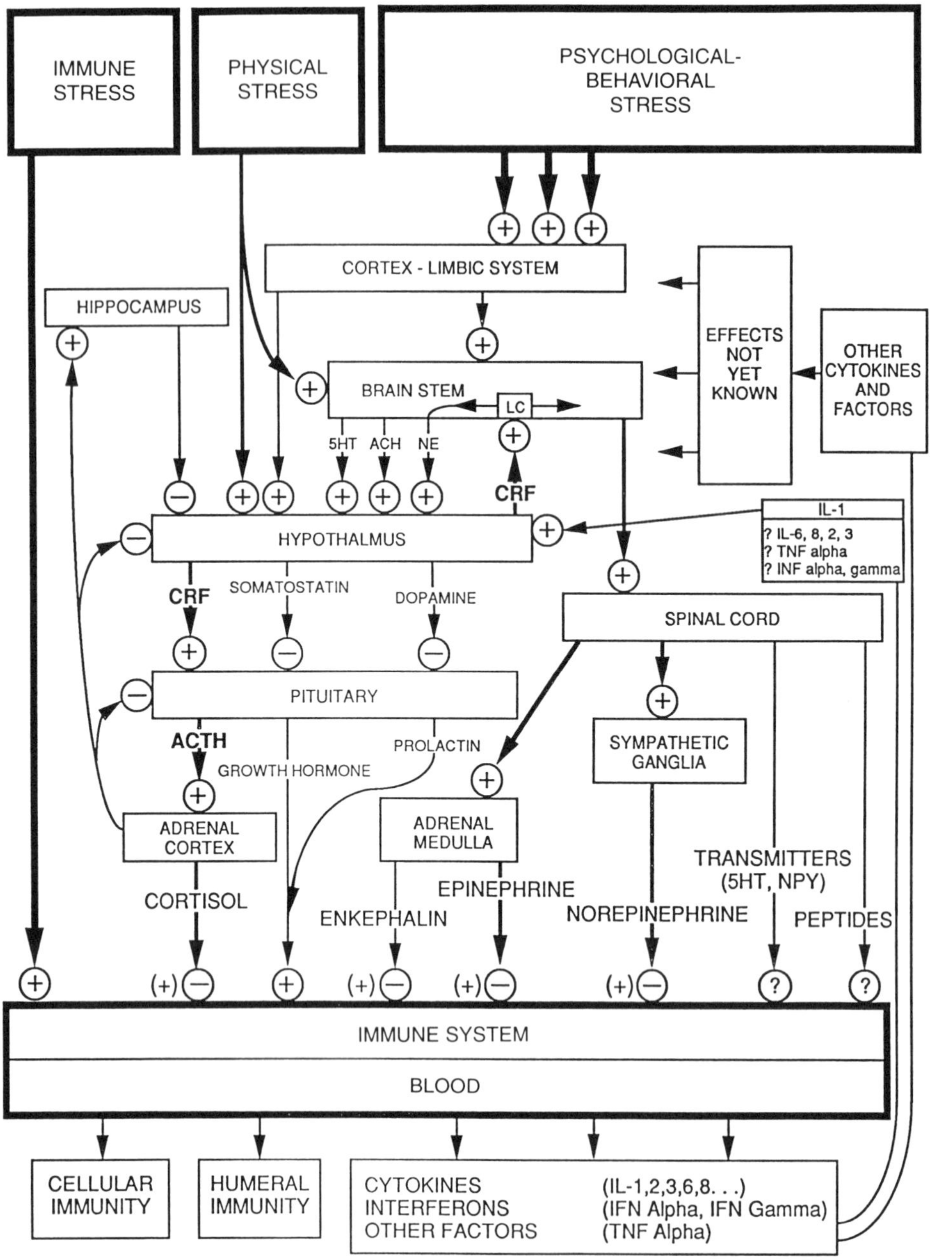

FIG. 1. Schematic representation of the pathways involved in the neuroendocrine response to stress and their effects on the immune system. The anatomic and functional interrelationships (stimulatory or inhibitory) of the different neuroendocrine components involved in the stress response are illustrated. (⊕), stimulation; (−), inhibition; (+), mild stimulation; CRF, corticotropin-releasing factor; NE, norepinephrine; 5-HT, serotonin; ACH, acetycholine; ACTH, adrenocorticotropic hormone; LC, locus coeruleus; IL, interleukin; IFN, interferon; TNF, tumor necrosis factor. (See text for description.)

concentrations, the adrenal steroids have a profound suppressant effect on the immune system (although at lower levels a small stimulatory effect has been observed). Thus, any stress that markedly alters the hypothalamic pituitary adrenal (HPA) axis can have considerable influence on immune system function (4). The detailed neurobiology of the stress response of the HPA axis is more fully described in Chapters 13 and 19. Figure 1 simply illustrates that physical stress (pain, trauma, metabolic disturbance, etc.) and psychological-behavioral stress (via cortical-limbic systems) activate the hypothalamic system, the net result being that corticotropin-releasing factor (CRF) is released from hypothalamic nuclei. There is considerable evidence indicating that CRF is one of the major transmitters coordinating the stress response (4,5). The largest concentration of CRF-containing neurons is found in the paraventricular nucleus (PVN) of the hypothalamus, although other neurons throughout the brain also contain CRF. There is a diverse peripheral and central input to the paraventricular nucleus. With stressful stimulation, there is increased CRF messenger RNA in the PVN resulting in CRF being secreted into the capillaries of the hypophyseal portal venus plexus and consequent stimulation of adrenocorticotropic hormone (ACTH) release from the pituitary. The ACTH is the primary tropic hormone for stimulating corticol release from the adrenal cortex. As plasma cortisol levels rise, there is increased inhibition and suppression of a large number of immune functions (5,6). In addition, there is direct inhibitory feedback of adrenal steroids at the pituitary and hypothalamic level to reduce CRF and ACTH secretion. A second inhibitory feedback loop involves adrenal steroid stimulation of hippocampal cells, which in turn have an inhibitory effect on CRF release from the hypothalamus (5,6).

The CRF also has many other effects related to the stress response (see Chapter 13). Most relevant to the immune system is the stimulation of norepinephrine (NE) release via CRF receptors on the locus coeruleus (LC) (Fig. 1). This activates the central neuroadrenergic system as well as the peripheral adrenergic system, involving both epinephrine (E) release via the adrenal medulla and NE release from peripheral sites, and including direct NE enervation of organs of the immune system (5,6). An additional effect of CRF is its inhibitory role in reducing growth hormone and prolactin secretion. Although prolactin and growth hormone may be increased early in some more intense forms of the stress response, they are generally decreased later. Thus, CRF has direct immunosuppressive activity via adrenal steroids, and an indirect inhibitory effect from impairing the prolactin and growth hormone immune-activating effects (6,7).

The sympathetic nervous system (SNS) is designed to function in parallel to the HPA axis. A wide variety of psychological-behavioral stressors (via cortex-limbic-brainstem-spinal and peripheral pathways) and physical stressors (via brainstem–spinal-cord pathways) result in SNS activation. Depending on the type of stressor, this can result in primary stimulation of the adrenal medulla with E release or stimulation of peripheral noradrenergic terminals in many organs, with NE release or both. Both E and NE have significant immune suppressive effects (although some immune-enhancing effects are seen in low-dose paradigms with acute application). Thus, the SNS via E and NE release provides a second major CNS inhibitory system of immune function that parallels the HPA axis (5,6).

Other CNS effects on immune modulation can be demonstrated, in addition to those from the HPA axis and the SNS. Stress-related immunosuppression can be observed in the absence of the adrenal glands (5). In addition, a large number of transmitter systems have been shown to have terminals within organs of the immune system, including serotonin (5-HT) and a number of peptides such as neuropeptide Y, substance P, calcitonin gene-related protein, vasoactive intestinal peptide, cholecystokinin, and others (8). Oxytocin and vasopressin are located in a nonneural thymic compartment. The magnitude of the immune modulatory effects of these other neurotransmitter systems and their significance following a variety of types of stress is not yet currently known. Enkephalins can be released both with HPA and with SNS stimulation, and their effects, although primarily inhibitory on immune function, are complex and not fully characterized at present (9,10).

It can be seen in Fig. 1 that there is a wide variety of mechanisms by which the CNS modulates immune function. The immune system is extremely complex, and there are many subcomponents that could be separately influenced by the many separate types of CNS effects. In general, cell-mediated immune reactions are apparently more sensitive to stress than antibody production, although stress effects on both have been clearly demonstrated.

Pathways Serving Effects of Immune Activation on the Central Nervous System

It has clearly been demonstrated that cytokines such as interleukin-1 (IL-1) and IL-6, and tumor necrosis factor alpha (TNF-alpha) that are released from immune cells during immune activation can affect CNS function (11). Although the blood-brain barrier may prevent many of the cytokines from entering the brain directly, some brain areas have a relatively weak blood-brain barrier (e.g., the preoptic nucleus of the hypothalamus). Receptors for IL-1, IL-6, and TNF alpha have been observed in this area, and provide a mechanism by which cytokines can bind and therefore transmit their messenger from the circulation to the CNS (3). In addition, a saturable transport system for IL-1 has been demonstrated in the mouse (12). In general, the data suggests that the cytokines complete a negative feedback loop (see Fig. 1) such that increased cytokine levels (IL-1) stimulate CRF release and consequent cortisol release, which in turn reduces immune activation and production of the increased cytokine levels (11,13). This is one of the major regulatory mechanisms that can keep the CNS immune system in balance.

EFFECTS OF STRESS ON IMMUNE FUNCTION

Stress and Viral Infections in Humans

Stress has long been associated with the occurrence of viral infections (the term "catch a cold" is of ancient origin). One of the better models of the effect of stress on immunity to viral infection involves herpes simplex virus (HSV). Following initial infection, HSV remains latent in the neurons of infected individuals and is susceptible to reactivation by a variety of factors, including immune suppression, fever, hormonal changes, and physical and emotional stress. In Table 1, five studies are listed that examine the relationship between reported stress and recurrence of HSV. In four of the five studies, there was an association between increased stress and increased symptoms (14–17). In the one negative study, self-reported stress from daily diaries did not increase over the 6 days prior to recurrence (18). It may be that stress effects have a different time course than this method measured, or that a different pattern or type of stress would relate to reoccurrences. In general, there is good evidence that subjectively perceived stress does relate to increased vulnerability to HSV recurrence.

In Table 1, the relationship between reported stress and alteration in immunological findings relative to seven viruses is tabulated. When interpreting this data it is important to recognize that lower cellular immune control of virus usually corresponds to higher antibody levels in serum. Thus, in Table 1, the increased antibody to HSV, Epstein-Barr virus, and cytomegalovirus would be interpreted as decreased cellular immune control of viral activity. It can be seen in Table 1 that, across all seven viruses, adverse life events interpreted as stressors are associated with an impairment of immune function (14–32). In human immunodeficiency virus (HIV-1) studies, the relationship between life events and indexes of immune function do not appear as strong (31,32). This may relate to this specific type of viral infection and the known devastating effects of HIV-1 on immune function. The magnitude of the destruction of immune function resulting from HIV-1 may be so large that the more subtle relationships between stress and immune function seen in other conditions are obscured by the massive immunologic attack by HIV.

The main findings from one of the more robust studies in Table 1 are illustrated in Fig. 2. In volunteer research subjects under high stress the prior year, an increased rate of respiratory infection was found following administration of nasal drops of five respiratory viruses. The increase in clinical colds was primarily due to the increase

TABLE 1. *Stress decreases antiviral immunity in humans*

Virus / Type of stressor	Decreased immunity	Immunological findings	Subjects	Reference
Herpes Simplex Virus				
Life Events–Social Support	Yes	↑ Recurrences	Pts. with Genital Herpes	14
Life Events–Mood	Yes	↑ Recurrences	" " "	15
Life Events	Yes	↑ Duration	" " "	16
Psychiatric Illness	Yes	↑ Recurrences	" " "	17
Daily Stress–6 days Prior to Recurrence	No	No Change	" " "	18
Living Near Nuclear Accident	Yes	↑ Ab to HSV	Residents near Accident	19
Divorce/Separation	Yes	↑ Ab to HSV	Men	20
Epstein-Barr Virus				
Divorce/Separation	Yes	↑ Ab to EBV	Men	20
Giving Care to Alzheimer's Disease	Yes	↑ Ab to EBV	Caregivers	21 22
Examinations	Yes	↑ Ab to EBV	Medical Students	23
Rhino Virus				
Life Events–Personality Type	Yes	↑ Frequency Infection	Research Volunteers	24
Life Events–Personality Type	Yes	↑ Nasal Secretion and Virus Shed	Research Volunteers	25
Influenza Virus				
Family Dysfunction	Yes	↑ Frequency Infection	Family Members	26
Life Events–Personality Type	Yes	↑ Nasal Secretion and Virus Shed	Research Volunteers	25
Hepatitis B Virus				
Examinations–Social Support	Yes	↓ and Delayed Ab Response to Vaccine	Medical Students	27
Life Events	No	↑ Antibody Response to Vaccine	Medical Students	28
Cytomegalovirus				
Living Near Nuclear Accident	Yes	↑ Ab to CMV	Residents	19
Human Immunodeficiency Virus				
Life Events	Yes	↓ Total Lymphocytes	(HIV+) Male Homosexuals	29
Serostatus Notification	Yes	↓ Natural Killer cells	Male Homosexuals	30
Life Events–Mood	No	No Findings	(HIV+) Patients	31
Life Events	No	No Findings	(HIV+) Male Homosexuals	32

HSV, herpes simplex virus; EBV, Epstein-Barr virus; CMV, cytomegalovirus; HIV, human immunodeficiency virus; Ab, antibody; ↑, increase; ↓, decrease

in infection in the subjects under high stress (24). In this well-designed study, the exposure to virus, and the evaluation of infection and colds were very well controlled and objectively evaluated. The evaluation of stress, however, was only by retrospective self-report. Thus, it is not possible to evaluate the immune effects of different types of stress (e.g., acute versus chronic, or magnitude of stress precipitant versus coping strength).

In human studies, it is not possible to selectively administer different levels of stress to control for the effects of stress magnitude on immunologic variables. This would be necessary in order to demonstrate the actual biopathogenicity of viral infection. However, in animal studies it is possible to control experimental parameters, and when this has been done a clear relationship between stress and viral pathogenicity has been demonstrated. In Fig. 3, the effect of 8 days of foot-shock stress on HSV-specific cellular and hormonal immunity, and the eventual outcome of pathogenicity is illustrated. It can be seen in Fig. 3 that cellular immunity was decreased in

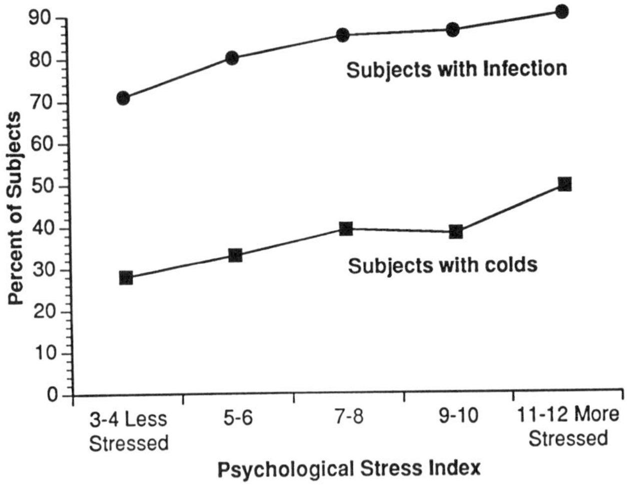

FIG. 2. Subjects completed 3 scales: 1) number of negative major stressful life events in the past year; 2) degree that current demands exceed ability to cope; and 3) current negative affect. The scales were combined into one index that was related to the number of subjects with infection and the number with colds following the controlled administration of nasal drops containing one of five viruses. The percentage of subjects with infection and colds increased in a dose-response pattern with increasing stress index. $P<.002$. From Cohen et al. (24).

the shock group compared to the control groups. In addition, inhibition of the antibody response to HSV was also seen, since the shock group had a smaller antibody response to HSV than the control groups. Most importantly, the shock-stress group had higher levels of HSV plaque-forming units at the site of inoculation than did the control groups (33). This illustrates that stress-induced impairment of specific cellular and humoral immunity are related to overall pathogenicity of HSV viral illness in mice. Thus, these types of animal studies provide strong presumptive evidence that the relationship between stress and HSV illness listed in Table 1 are actually mediated by stress-induced alterations in cellular and humeral immunity to HSV in humans.

Role of the HPA Axis and SNS in Stress-Induced Modulation of Viral Pathogenesis

A review of the data on CNS modulation of immune function (illustrated in Fig. 1) suggests that the HPA axis and SNS are the two primary effectors of CNS actions on immune function. It would be important to know what the relative contribution of the APA axis and SNS are to each of the many separate immune effects of stress. Typical results from an initial study on this question are illustrated in Fig. 4. Ten to eleven days of restraint of mice reduced cytotoxicity of T lymphocytes to less than 12% of home cage controls. This effect was blocked by only 50% by the peripheral adrenergic beta receptor antagonist, nadolol, and not blocked at all by the corticosteroid receptor antagonist R0486. In contrast, the combination of nadolol plus R0486 produced a complete blockade of the effects of stress in reducing cytotoxic T lymphocyte activity. Another aspect of immune function, lymph node cellularity, which is impaired by stress, was completely blocked by R0486 but not by nadolol (34). This demonstrates that both the APA axis and the SNS are contributing to the stress-induced suppression of antiviral cellular immunity.

Genetic Vulnerability to the Effects of Stress on Immunity

Some of the genetic factors contributing to variation in the susceptibility to different infectious diseases are known, but the role of genetics

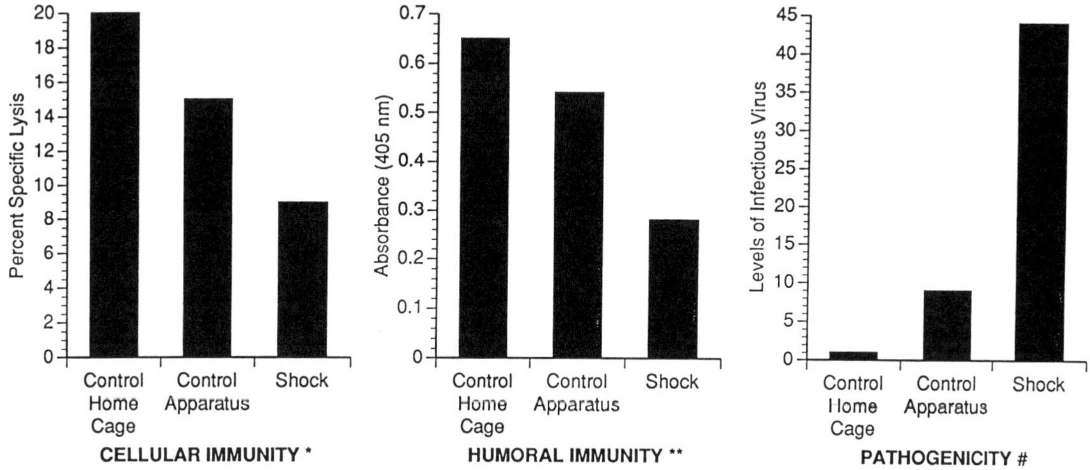

FIG. 3. Female, 6–8-week-old C57BL/6 mice received 10 seconds of 0.6mA light-signaled foot shock every 30 minutes during the 12-hour dark cycle for 8 days. The apparatus control group did not receive shock or the light stimulus as a signal at any time. The home cage control mice remained in their cage throughout the study. HSV-1 strain Patton virus was applied to the scarified rear foot pad after 1 day of shock. After 7 more days of shock: *Cellular immunity was assessed by a cytotoxicity assay for HSV-specific cytotoxic T lymphocytes. Data are the mean of the 5/1 and 10/1 effector to target cell ratios for spleen-combined experiments 1 and 2, $P<.05$.
**Humoral immunity was assessed by an ELISA assay for serum anti-HSV antibody. Data are for mean serum dilutions 1/25-, 1/75-, 1/225-second experiment, $P<.05$.
Pathogenicity was assessed by a plaque-forming assay for titration of infectious virus in both rear foot pads, experiment 2, $P<.05$. From Kusnecov et al. (33).

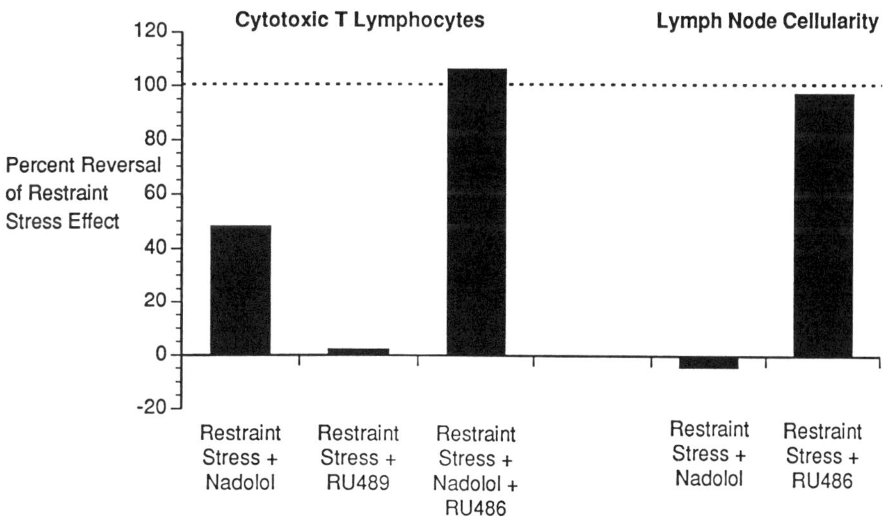

FIG. 4. Male C57BL/6 mice were restrained for 16 hours during the dark cycle 5–6 days before and 5 days after infection with HSV-1 strain Patton. Nadolol, R0486, or both were implanted in slow-release pellets for 7 days prior to HSV-1 infection of the hind foot pad. Cytotoxic T lymphocyte activity of lymphocytes from popliteal nodes utilized a 51 Cr release assay to assess cytotoxicity. Cellularity counts utilized a Coulter counter. Data plotted are from Experiment 2, 25/1 effector target cell ratio, drug treatments versus home cage control. Restraint decreased percentage-specific release in each experiment to 12%, 0% and 5% of home cage control for nadolol, R0486, and both, respectively. Restraint decreased cellularity to 40% and 21% of home cage control for nadolol and R0486, respectively. The percent reversal by each treatment of the restraint stress effect in decreasing cytotoxic T lymphocyte activity and lymph node cellularity is illustrated. From Dobbs et al. (34).

in stress-induced vulnerability to infectious illness in humans is just starting to be explored. Since studies in laboratory animals have demonstrated that there are genetic factors that increase the stress-induced vulnerability to infective illness, this would be an important area of study in humans. The recent increase in tuberculosis during the past few years has especially been observed in patients with immune deficiency who are infected with HIV-1. This has become a major public health problem. Some epidemologic studies have indicated that certain racial groups (blacks) may be more susceptible than other groups to infection with tuberculosis (35). It is of interest that the macrophages from blacks are more permissive to the growth in tuberculosis (36). Whether stress-induced vulnerability contributes to this increased incidence would be of considerable interest.

A study on this point has been conducted in mice and is illustrated in Fig. 5. Cell-mediated immune response is important in the control of the tuberculosis bacilli. Some strains of mice are resistant but others are vulnerable to the growth of mycobacteria. The gene that controls resistance to mycobacteria growth in mice, termed Bcg, has been matched to a gene chromosome 1. Studies have shown different growth patterns of the mycobacteria in different mouse strains. This is in part due to mature tissue macrophages where macrophages from Bcg-resistant mice control the growth of mycobacteria, but the macrophages from Bcg-susceptible mice do not. Figure 5 illustrates that, before restraint, Bcg-resistant mice show less growth of mycobacteria avium within macrophages while the Bcg-susceptible mice have a greater rate of growth. Following restraint stress, the macrophages from Bcg-susceptible mice supported a larger increase in mycobacteria avium growth than the resistant mice. Similarly, in the spleen the Bcg susceptible mice had greater rates of bacterial growth than

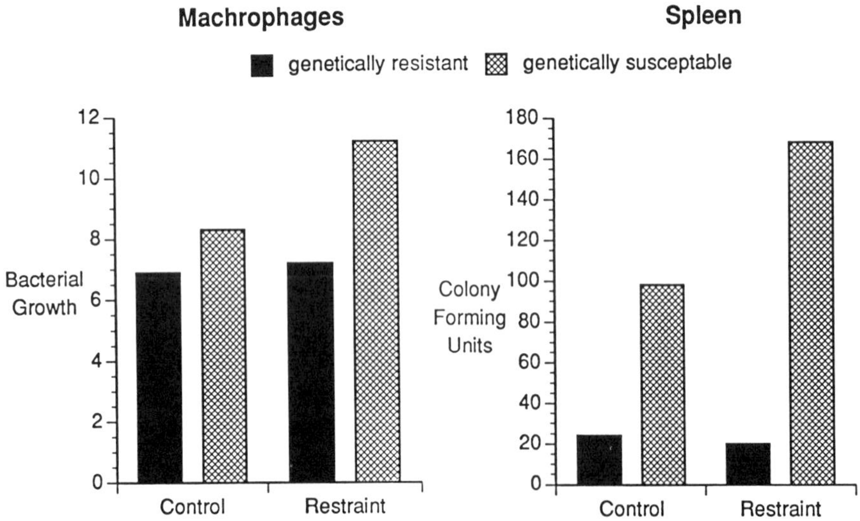

FIG. 5. The effect of restraint stress on antimicrobacterial activity of macrophages (*left*) and spleen (*right*). In resistant (BALB/c Bcg mice) and sensitive (BALB/c Bcg mice), restraint was given 18 hours/day for 10 successive days. Macrophages from spleen were isolated and infected with M.avium. Growth of the microbacteria in macrophages was determined following lysis of the macrophages and pulsing with radioactive urasil after 5 days of in vitro culture. The changes illustrated represent growth of microbacteria rather than phagocytosis, since release of tracer immediately after phagocytosis did not differ between the groups. For the spleen assay, mice were injected with M.avium and stressed for 18 hours a day for 10 days. The numbers of colony-forming units in the spleens were determined 12 days after infection. Both the differences between the resistant and susceptible mice before and after restraint were significant ($P<.001$). From Brown et al. (37).

the resistant mice in the control situation before restraint, but bacterial growth was clearly increased following the restraint stress only in the susceptible mice. These data indicate that stress potentiates the genetic vulnerability of susceptible mice and that this is operant even at the level of the macrophage (37).

The Effects of Exercise in Stress on Natural Killer Cells

Natural killer cells (NK) are efficient, nonspecific effector cells that are not restricted by the major histocompatibility complex, and are rapidly activated and capable of protecting the host until more specific immune-competent cells are activated. The NK cells are an important line of defense in acute and chronic viral infections, and the prevention of the development and progression of cancer, and are involved in certain other immune diseases. These cells are highly influenced by a variety of different stresses that appear to influence the NK cell system in a similar manner (38). The NK cell activity is generally enhanced during severe physical stress but is suppressed afterwards. The manner in which physical and mental stress modulates NK activity can have important implications for many aspects of health.

Physical exercise increases the absolute concentration and also the relative fraction among blood mononuclear cells that express characteristic NK cell markers such as CD16, and CD56 or CD57. The NK cell functional activity also increases during physical exercise. Figure 6 illustrates the effects of bicycle exercise for 1 hour at 75% of the maximal exertion. Blood was collected before, at the end of 1 hour of exercise, and 2 hours after exercise. It can be seen that the concentration of the NK cells was markedly increased with exercise but that it is also suppressed 2 hours later (39). Although several other stress-induced changes were seen in blood mononuclear cell populations, the largest change was in the NK cell concentrations.

The effect of exercise in acutely increasing NK cell concentrations and activity may explain the protective effect of regular exercise on viral infections. The hypothesis has been put forward that during both moderate and severe exercise the immune system is enhanced, but that following severe exercise immunity is depressed. Dur-

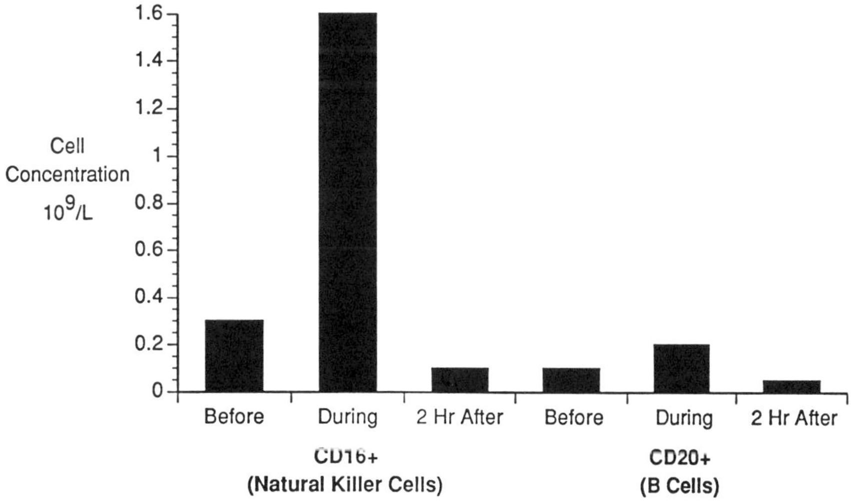

FIG. 6. An illustrative example of the effect of 60 minutes of 75% maximal bicycle exercise. Blood sampled before, during, and 2 hours after exercise bout. Although exercise affects many subsites of lymphocytes, the most reliable effect of exercise is the increase in natural killer cells. From Pedersen and Ullum (39).

ing this period of immune depression, infectious agents, especially viruses, may invade the host and establish infections. In contrast, regular moderate exercise produces the desired increase in NK activity and improved defense against infection, and is not followed but the postexercise suppression that allows vulnerability (39, 40).

Evidence in support of this hypothesis comes from studies of the rate of upper respiratory infections in various exercise groups. In Fig. 7, data from two studies supporting this hypothesis are illustrated. The data illustrated on the left side of Fig. 7 indicate that moderate physical activity may reduce upper respiratory infection (URI) symptomatology. In this study, 36 women were randomized to 30 minutes of walking 5 days per week. In comparison to their more sedentary controls, they experienced less than one half the URI symptoms during the 15-week pe-

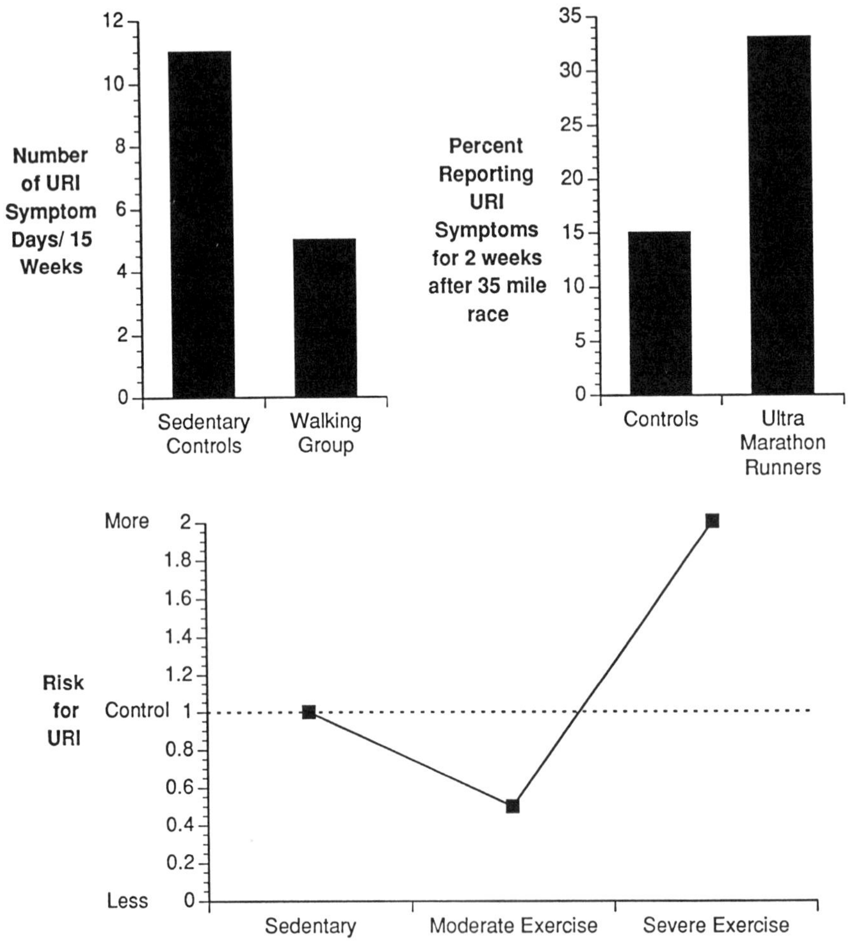

FIG. 7. *Top left:* Rate of URI symptomatology during 15 weeks of exercise or no exercise in 36 women (mean age, 35 years) randomly assigned to moderate exercise (n=18) (45 minutes of walking, 5 days per week) or no exercise (n=18). *Top right:* Rate of URI during 2 weeks after the race in 150 South African runners who participated in a 35-mile (56 km) ultra marathon in comparison to controls. *Bottom:* Theoretical relationship of exercise to risk of URI. The definition of the degree of exercise is currently very crude and needs to be more accurately standardized. Data plotted are from left and right above. From Nieman (40).

riod of the study. In contrast, the data on the right of Fig. 7 illustrate that 150 South African runners who took part in a 35-mile marathon race had more URI symptoms 2 weeks after the race than controls who did not run. The bottom of Fig. 7 illustrates the hypothesized relationship between severity of exercise and vulnerability to URI infections. At low levels of moderate exercise, there is a protective effect while following strenuous exercise, which could be termed "stressful," there is an increased vulnerability to URIs. Given the large number of individuals performing moderate to strenuous exercise, it would be important to determine the nature and amount of exercise and/or the immunologic factors that determine the boundary between the beneficial effects of exercise and the detrimental effects of severe exercise. Building on the concept that moderate exercise is immunologically beneficial, a number of studies have been carried out in HIV-infected individuals. In general, moderate exercise by HIV-infected individuals participating in athletic programs is associated with improved psychological measures, as well as demonstrates some evidence of improved immunologic parameters (41–44).

Chronic Stress and Natural Killer Cell Activity

In contrast to acute stress, which often increases plasma E and NE levels that have been shown to increase in NK cell activity, the effects of chronic stress are much less clearly defined. In normal subjects at rest, a negative correlation between E levels and percentage of NK cells in blood has been observed (45). This suggests the hypothesis that while acute increases in E and NE produce an increase in NK numbers and activity, chronic stress with chronically elevated E and NE levels may actually reduce NK cell numbers.

One of the more destructive and stressful life events is spousal bereavement. Although most widows and widowers adjust to the loss, a substantial number demonstrate serious psychological and/or medical complications. It is possible that the chronic stress suffered from bereavement could alter NK cell activity. Prior studies found that NK cell activity was lower in bereaved women than in nonbereaved controls, and that the bereaved women were more depressed and that NK cell activity was more impaired in women with more dysphoric mood (46). The NK cell function was also impaired in women whose husbands died from lung cancer, and the impaired NK cell activity was most apparent in those with a debilitating increase of severity of depressive symptoms. Thus, this data suggest the hypothesis that it is not bereavement per se, but the adaptation to the event that relates to NK cell activity. Individuals with depressive symptoms show a breakdown in adaptive mechanisms and a resultant impairment of NK cell activity.

In Fig. 8, the data from a recent study of bereavement, depression, and immune function are illustrated (47). Middle-aged widows who had lost their spouses 2 months before the first data were obtained and demographically matched married women were evaluated at 6-month intervals for 13 months. There were no significant differences in a variety of immune measures between the widows and control subjects as a group. However, the subset of 6 widows who met DSM3-R criteria for major depressive syndrome 2 months after their husbands died demonstrated significantly lower NK cell activity than the 15 nondepressed widows. It is of interest that the data in Fig. 8 illustrate that the depressed widows had reduced NK activity compared to controls and nondepressed widows, but that the nondepressed widows had somewhat increased NK activity compared to the controls. Whether successful adaptation to bereavement (with the avoidance of the depressive symptoms) increases NK cell activity remains to be more fully investigated.

Chronic Stress Alters the Response to Acute Stress

As discussed in the previous paragraph, acute stress has been found to increase several subsets of lymphocytes, with particular increases in NK cells being observed. In contrast, chronic stress

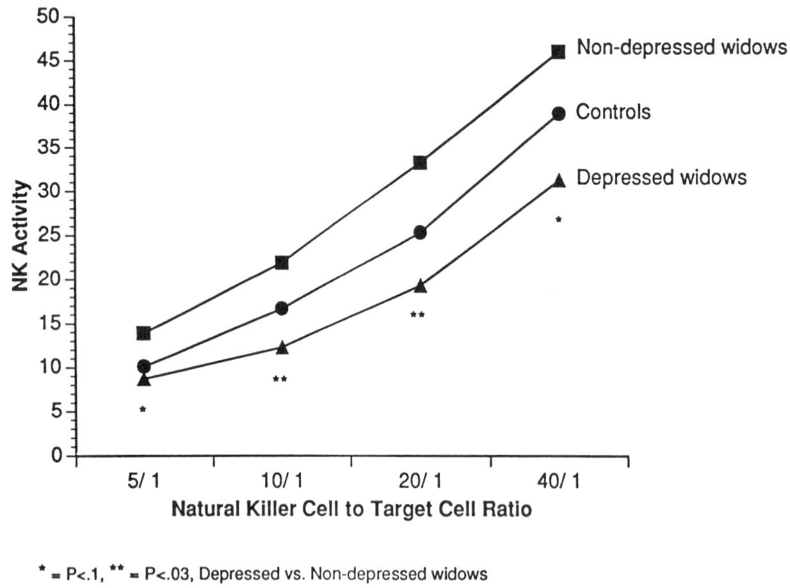

FIG. 8. Twenty-one middle-aged widows and 21 demographically matched married controls were evaluated with diagnostic interviews and immune measures 2 months after bereavement. Six widows met DSM-III R criteria for depression, and they had lower natural killer (NK) cell function, lower T8 concentrations, and a trend toward lower lymphocyte stimulation responses than nondepressed widows. Zisook et al. (47).

has been associated with a decrease in mitogen response in lymphocytes and lower NK cells numbers. In the real world of clinical problems, most problematic stressors are of a chronic nature. To date, there have been very few studies of the interaction between chronic stress and its influence on the acute stress response. Given that chronic stress produces effects opposite to that of acute stress, it would be predicted that chronic stress would impair the acute effects of stress on the immune response.

In Fig. 9, a study investigating the effects of chronic daily stress upon the immunologic response to an acute psychological stressor in high school teachers is illustrated (48). It can be seen that the teachers with low chronic stress responded to the acute stress situation with an increase in their NK cells and CD3 cells. In contrast, the teachers with high chronic stress had a decrease in NK cells and CD3 cells during and following this stressor. The high chronic stress and low chronic stress subjects did not differ when there was no acute stress given. These data demonstrate that chronic stress actually reverses the increased NK cell activity following an acute stressor. This suggests that the duration of stress can have a major influence on immune reactivity. In addition, in this study there were no differential effects of high chronic stress and low chronic stress on the cardiovascular response to the acute stressor. This suggests that the effects of chronic stress differ across different SNS dimensions, and that chronic stress effects on immune activity can be quite different than chronic stress effects on cardiovascular reactivity.

The neuroendocrine mechanisms involved in the effects of acute mental stress on cardiovascular and immune changes are not fully understood. It is generally thought that activation of the SNS and release of E and NE into the circulation are involved in the increase in NK cells. Figure 10 illustrates a study where volunteers performed mental tasks after ingesting the beta adrenergic antagonist, propranolol, or placebo (49). The significant stress-induced increase in NK cell activity seen in the placebo group was blocked by propranolol. Propranolol demon-

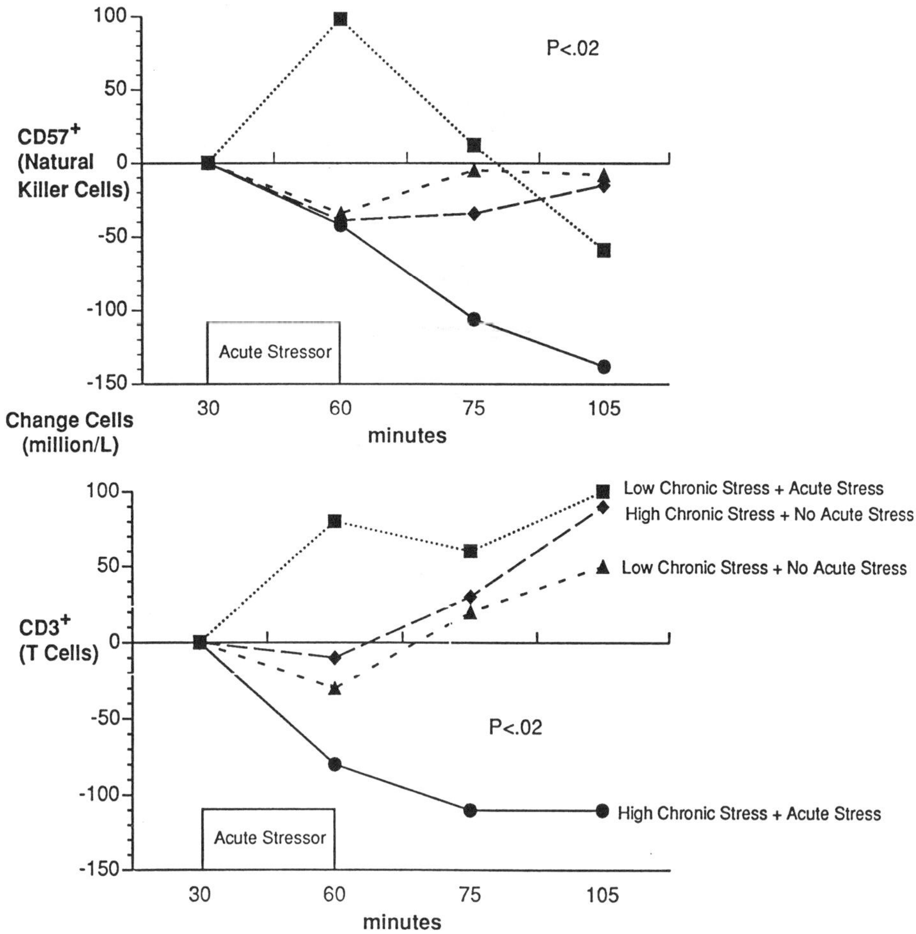

FIG. 9. From 92 male high school teachers (24–55 years of age), those reporting the highest number of everyday problems (high hassles group; top 25% = high chronic stress) and lowest (low hassles group; bottom 25% = low chronic stress) were chosen for study. Ten from the high and 10 from the low chronic stress groups did not perform a task, and 13 from the high and 14 from the low groups did perform a "stressful" task. Blood was sampled as indicated in the Fig., and pulse and blood pressure were sampled before and during the task. The high and low chronic stress groups had a differential response to the acute stress in CD3+, CD8+, and natural killer (NK) cells, but not CD4+, or B cells, or cardiovascular measures, or cortisol, prolactin, beta-endorphin, or adrenocorticotropic hormone (ACTH). From Benschop et al. (48).

strated a marked blockade of the heart rate response to mental stress but did not alter the blood pressure response. Taken together, the data indicate that the increase in NK cells following acute mental stress is dependent on a beta adrenergic mechanism presumably derived from SNS activity. Whether the effects of chronic stress are also mediated through a beta adrenergic receptor system remains to be elucidated.

Complex Relationships of Stress-CNS Effects on Immune Function

It is becoming increasingly evident that different types of stress can produce different patterns of CNS-neuroendocrine activation, and that this could be the basis for numerous differential effects of stress on the many separate aspects of immune system function. Evidence for separate

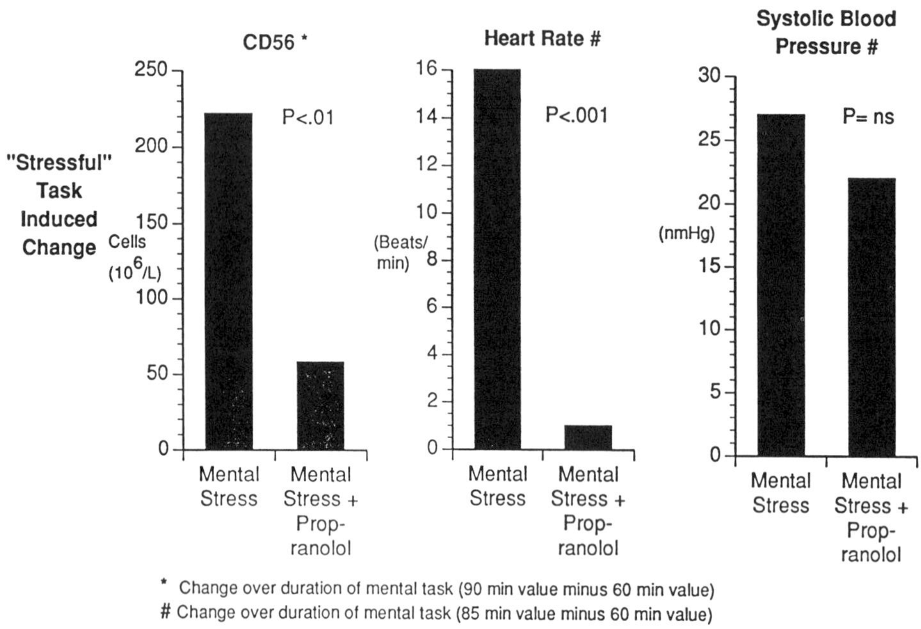

FIG. 10. Healthy male volunteers received either 40 mg propranolol orally or placebo 60 minutes before a 30-minute "active coping" stressful task. The task increased natural killer (NK) cell concentrations and other lymphocytes subsets. The stressful task also altered cardiovascular variables, including increased pulse rate and blood pressure and decreased EKG preinjection period (PEP). Propranolol blocked the stress-induced changes in NK cells, heart rate, and PEP, but stress-induced changes in blood pressure were not blocked by propranolol. From Benschop et al. (49).

types of stress-neuroendocrine patterns and complex CNS immune relationships include: 1) the effect of shock, but not swim stress, on suppressing NK cell inhibition of tumor metastasis is reversed by the opiate antagonist naltexone (50); 2) naltrexone reverses the shock-induced decrease in NK cell function but does not reduce the shock-induced reduction in mitogenic responsiveness (51); 3) lithium chloride treatment blocks the immunosuppressive effect of ''cooling stress'' but does not alter the effects of ''darkness stress'' (52); 4) benzodiazepine agonists block central CRF-induced suppression of NK cell activity (53), but a benzodiazepine inverse agonist at lower dose decreased and a higher dose increased postimmunization serum antibody levels (54); and 5) many forms of stress increase CNS dopamine turnover and inhibit immune function, but dopamine antagonists potentiate and agonists block the effects of restraint stress (55).

Complexity at the immune cellular level is also apparent as evidenced by the findings that, in vitro, isolated NK cell cytotoxicity is increased by 5HT (apparently via a 5HT-1A receptor mechanism) (56), but in whole blood 5HT reduces NK cell cytotoxicity (apparently in part via a 5HT-2 receptor) (57). What is becoming more clear from these types of studies is that the variable effects of stress on immune function can result from differential patterns of neurotransmitter-neuroendocrine activation in the CNS interacting with differential receptor-specific activation of separate components of the immune system. Drugs altering stress effects on neurotransmission can act at both the CNS level and the immune cellular level, and the two effects are not necessarily the same (58).

EFFECTS OF IMMUNE ACTIVATION ON CNS STRESS-RESPONSIVE SYSTEMS

Immune Abnormalities Can Produce Secondary Stress Response in the CNS

Many forms of stress (including physical, metabolic, infectious, and immunologic stress) can have their primary effect on the immune system. The CNS stress response in these instances would be viewed as secondary to the primary abnormality in the immune system. Over the past few years there has been considerable new research on the mechanisms by which a primary immune abnormality could produce a stress response in the CNS (59,60). Some of the evidence that a primary immune event can influence CNS function (see review (8)) includes:

1. Changes in immune function following immunization have been associated with changes in neuron firing rates recorded from discreet regions of the hypothalamus, and increases and decreases in firing rate have been demonstrated in difficult nuclei with different time courses following immunization. Changes are generally less during the secondary immune responses compared to the primary immune response.
2. Changes in CNS monoamines have been found in the hypothalamus and other sites within the CNS following immunization. The changes are differentially seen amongst different nuclei, and like the changes in firing rates they have individual relationships to the time course of changes in immune function. Since hypothalamic, limbic, and brain stem sites are intimately involved in the CNS stress response, an immune-induced alteration in their function could produce a secondary CNS stress response following a primary immune abnormality. Alteration of brain stem and hypothalamic neural activity by an immune response will alter the CRF-ACTH glucocorticoid axis and SNS activity, and this will in turn have a direct effect on immune function. This effect provides a mechanism for bidirectional communication between the brain and the immune system (60).

IL-1 Stimulates the HPA Axis and Other Brain Areas

The IL-1 is released with immune activation, and it has been known for some time that blood levels of glucocorticoids increase at the time of peak immune response to many antigens (61). The IL-1 has been shown to stimulate release of CRF from the hypothalamus, as well as release ACTH from pituitary cells in vitro (62,63). The mRNA for IL-1 has been identified in many brain areas, and IL-1 can be induced in different brain regions following peripheral immune stimulation. In Fig. 11, evidence for IL-1 induction from different brain areas is presented. It can be seen that 2 hours after the lipopolysaccharide injection there is no IL-1 activity in either cell lysates or extracellular fluid, but that there is activity in plasma. Four hours after injection when the plasma does not demonstrate IL-1 activity, there is evidence of IL-1 activity in cell lysates in two areas and in extracellular fluid in all brain regions (64). The data from this study suggest that the lipopolysacharide stimulates IL-1 production via active IL-1 release into extracellular fluid, since the IL-1–like activity in cortex and cerebellum cell lysates was not blocked by an IL-1–specific antibody. The mechanisms by which plasma IL-1 might alter CNS function, as well as possibly induce IL-1 production, remain to be clarified. It is known that many circumventricular areas have a markedly reduced blood-brain barrier that could allow access of IL-1 and other cytokines to provide direct CNS stimulation. It is of considerable interest that a saturable transport mechanism for IL-1 has been described in many brain areas; this could account for a much wider effect of IL-1 on CNS function than that found with only stimulation via the circumventricular areas (12).

Brain Areas	Brain-Stem	Diencephalon	Cortex	Cerebellum	Hippocampus	Plasma
Cell Lysate	—	—	—	—	—	+
Extra Cellular Fluid	—	—	—	—	—	

2 Hr after Injection

Brain Areas	Brain-Stem	Diencephalon	Cortex	Cerebellum	Hippocampus	Plasma
Cell Lysate	+	+	—	—	+	—
Extra Cellular Fluid	+	+	+	+		

4 Hr after Injection

FIG. 11. Lipopolysaccharide (1.0 mg/kg) was given 2 or 6 hours prior to sacrifice. Extracellular fluid and cell lysate were fractionated by gel exclusion chromatography, and fractions were tested for thymocyte stimulation (specificity was confirmed by blockade of stimulation by a monoclonal antibody to interleukin (IL-1) receptor). When lipopolysaccharide was injected IP, IL-1 was detected in plasma 2 hours after injection but not in brain. Six hours after injection there was no IL-1 in plasma, but IL-1 activity was found in all five extracellular fluid areas, and in two of the areas within cell lysates. However, the cell lysate bioactivity was not blocked by the IL-1 receptor antibody, indicating that the thymocyte stimulation was not due to IL-1. The results support the idea that IL-1 is secreted by cells in the brain and that it is found in extracellular fluid of many regions following an immune stimulus in the periphery. From Rivier (62).

Cytokines Can Affect CNS Stress-Responsive Systems

The monoamines, including serotonin, norepinephrine, and dopamine, are intimately involved as part of the CNS stress response system. Almost all of the current pharmacologic treatments for stress disorders in humans involve drugs that alter these monoamine neuromodulatory systems. In understanding the effects of primary immune abnormalities on the CNS stress response, it would be important to know the effect of the various cytokines and other immune factors on these monoamine systems. In Fig. 12, data illustrating differential effects of three cytokines on monoamine turnover are illustrated. It can be seen that interleukin-1 produced changes in corticosteroid as well as increased turnover ratios for the noradrenergic, serotonergic, and dopaminergic systems. In contrast, IL-2 did not alter corticosteroid levels but increased noradrenergic and dopaminergic turnover, while IL-6, which also did not alter corticosteroid levels, increased turnover in the serotonergic and dopamine systems without altering the noradrenergic system. Thus, this data indicate that, depending on the brain area studied, differential effects on monoamine function can be observed following peripheral administration of cytokines known to be released in high levels following immunologic activation (65).

Additional evidence of interactivation CNS stress modulatory system is presented in Fig. 13. Since the peripheral injection of cytokines may produce concentrations outside the physiologic range, it would be important to study a more physiologic immune stressor. In Fig. 13 it can be seen that, during the primary immune response to T-cell–dependent antigen in rats, there is an increase in frontal cortical release of 5HT and that this is prevented by immune suppression. This type of data indicates that the primary alterations in immune system function can directly effect the CNS systems subserving the stress response (66).

The most clear functional loop where the neuroimmune system affects the CNS involves IL-1, which has been shown to release CRF in the CNS. This would have two consequences: 1) increasing corticosteroid secretion via the HPA axis, and 2) increasing SNS output by the stimulation of noradrenergic input to the autonomic

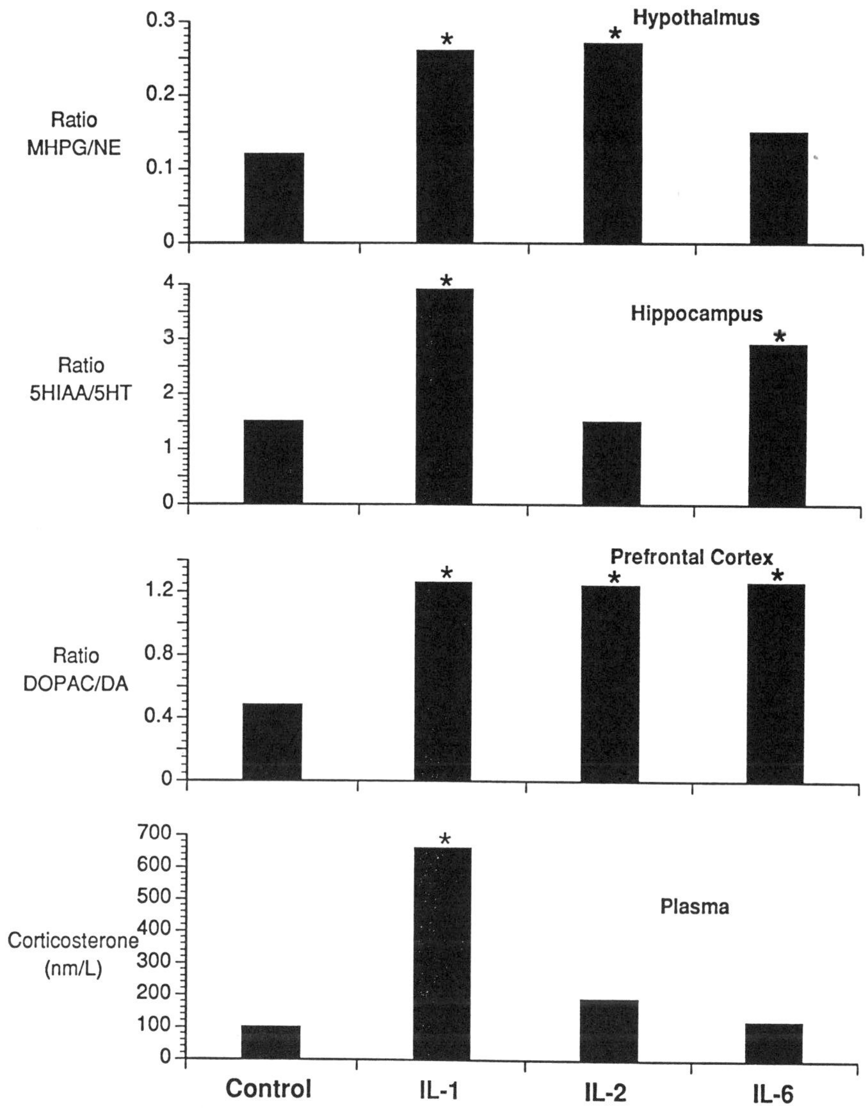

FIG. 12. Recombinant interleukin (IL-1B, IL-2, and IL-6) were administered (200 ng IP) to male resistant (BALB/c) mice. Brain monoamine and monoamine metabolite levels were determined 2 hours later in hypothalamus, hippocampus, and prefrontal cortex. (*=P>.05 compared to control.) The three different cytokines produced differential effects on monoamine turnover in different brain regions that mediate neural-immune interactions and that are also involved in the response to physical and psychological stressors. Of the three cytokines, IL-1 was the only one that increased cortisterone. From Ericsson et al. (63).

nervous system. Both of these effects would increase catecholamine and cortisol levels, which in turn would act to inhibit immune activation, thus reducing IL-1 levels. Thus, excessive levels of IL-1 would tend to inhibit its own production. There are, however, many other effects of immune activation on this CNS in addition to IL-1 that may not fit this simple negative feedback mechanism. For instance, IL-1 decreases acetylcholine release in hippocampus (67), and it is also involved in sleep regulation (68), which is of interest given the relationship of increased

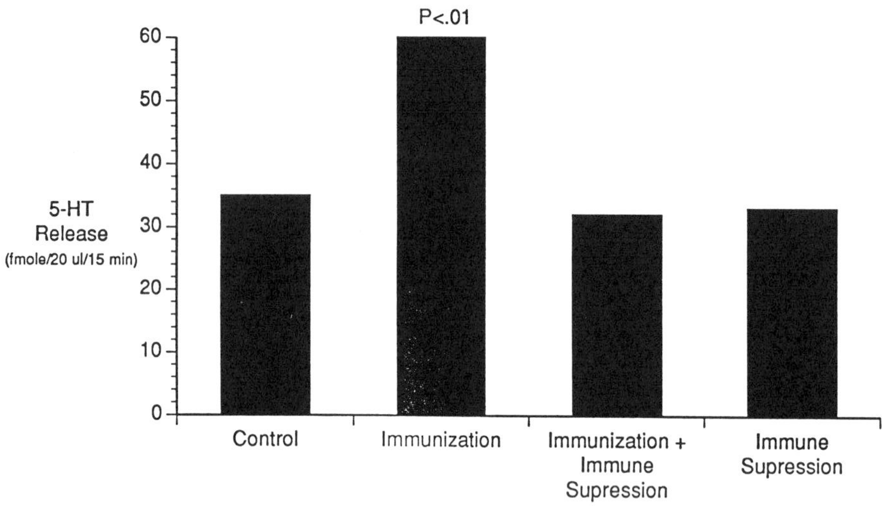

FIG. 13. Male Fisher 344 rats (weight=180–200 g) were injected with phosphate buffered saline (control) or sheep red blood cell immunization intravenously. Immune suppression was achieved by administering cyclosporin-A 12.5 mg/kg by gavage for 7 days before sacrifice. Serotonin (5-HT) release was assessed 3 days after sheep red blood cell immunization. The immune response was readily detectable from days 3–5 days after immunization, with the peak on the 4th day. 5-HT release following depolarization with 120 mm/KCl did not differ between the control, immunization, and immunization plus suppression groups. The observed immune-induced increase to 170% of control for 5HT release was accompanied by an 8% decrease in 5HT concentrations in cortex at the same time. From Quan et al. (64).

NK cell activity with increased sleep seen in both depressed patients and controls (69).

In addition, in vitro studies have demonstrated that leukocytes can produce proopiomelanocortin and endorphins. Thus, it is possible that leukocyte-derived ACTH and other molecules could have direct peripheral as well as central effects (60), although it remains to be seen whether the magnitude of the leukocyte-generated molecules are high enough to produce CNS effects. The administration of high doses of IL-2 and lymphokine-activated killer cells to patients with cancer did produce cognitive decline in over 60% of the 44 patients treated (70), but the dose of IL-2 and activated killer cells may be much higher than the levels generated following most types of immune stress.

Health Consequences of Stress-Induced Alterations of Immune Function

Animal and human data indicates that, at low levels, stress may actually improve some aspects of immune function (40,71). However, most forms of stress have been shown to impair the function of the immune system. How the magnitude of the different immune changes measured translates into increased susceptibility to different illness is not yet clear. Both laboratory animal and human studies indicate that, in general, the immune defense against viral illness is impaired by stress, but the role of stress in altering immunity to other types of infection is less clear (72).

Learning, as exemplified by classical conditioning, can either enhance or depress immune responses, and there is a large literature describing classical conditioning and immune modulation (73). Although there is clear evidence for direct communication and reciprocal feedback loops between the CNS and immune systems, the complexity of the interrelationships and the state-related variability in the minute by minute function of both the CNS and immune systems makes the prediction of outcome in each specific instance extremely difficult. The effect of stress on the immune system can vary greatly depending on the duration, intensity, and timing of stress administration, and the length of time between immunological evaluation and application of the stressor. Obviously, the age, immune status, and "coping strength" of the stressed animal will also influence the outcome.

The complex relationship between different types of stress and variable effects on immune function is further illustrated by the findings in a review of studies in laboratory animals where stress reduced NK cell activity and increased tumor growth. However, this stress effect was not always the case and often depended on the type of stress utilized (50). In a recent study of psychiatric intervention in the treatment of malignant melanoma, it was found that intervention reduced the rate of death and that lower initial NK cell activity related to increased rates of recurrences (74). Even though there are many similar studies suggesting that stress and reduced immune function increase cancer susceptibility, the magnitude of the effect is often small. Much more research needs to be completed in order to more clearly delineate the relationships between types of stress, immune changes, and cancer vulnerability.

SUMMARY

The effects of stress on either the CNS or the immune system are manifest and modified through interactive CNS-immune feedback loops. There are several ways that the CNS stress response can modify immune function, including activation of the HPA axis, and of the SNS adrenergic and opiate systems. Similarly, release of cytokines (particularly IL-1) and other factors during immune activation can also reciprocally modify the function of CNS stress regulatory systems.

One negative feedback loop that has been identified involves immune activation releasing IL-1 into plasma, which in turn stimulates CRF release with a consequent increase in adrenal steroids, which in turn inhibits immune activation; thus, the initial immune activation is kept in control.

Most types of severe or chronic stress impair immune function (although some types of mild acute stress appear to improve it).

There is reliable data that, in general, chronic stress decreases NK cell activity and increases the rate of viral infections. The role of stress-induced impairments of NK activity and immunity in other infections, and in the initiation and progression of cancer are less clear.

Overall, the data suggest that many of the adverse health consequences of stress may be mediated through the immune system.

ACKNOWLEDGMENTS

Supported by USPHS Grants MH25642, MH36229, and MH30929, and a grant from the Stanley Foundation.

REFERENCES

1. McEwen BS, Stellar E. Stress and the individual. Mechanisms leading to disease [Review]. *Arch Int Med* 1993; 153:2093–2101.
2. Plotnikoff N, Murgo A, Faith R, Wybran J. *Stress and immunity*. Boca Raton: CRC Press, 1991.
3. Saper CB, Breder CD. The neurologic basis of fever. *N Engl J Med* 1994;330(26):1880–1886.
4. Reichlin S. Neuroendocrine-immune interactions. [Review]. *N Engl J Med* 1993;329:1246–1253.
5. Irwin M. Stress-induced immune suppression—role of brain corticotropin releasing hormone and autonomic nervous system mechanisms. *Adv Neuroimmunol* 1994; 4:29–47.
6. Black PH. Central nervous system-immune system interactions—psychoneuroendocrinology of stress and its immune consequences. *Antimicrob Agents Chemother* 1994;38:1–6.
7. Kelley KW, Dantzer R. Growth hormone and prolactin as natural antagonists of glucocorticoids in immunoregulation. In Plotnikoff N, Murgo A, Faith R, Wybran J, eds. *Stress and immunity*. Boca Raton: CRC Press, 1991: 433–452.
8. Felten DL, Cohen N, Ader R, Felten SY, Carlson SL, Roszman TL. Central neural circuits involved in neural-immune interactions. In Ader R, Felten DL, Cohen N. eds. *Psychoneuroimmunology,* 2nd ed. New York: Academic Press, Inc., 1991;3–25.
9. Flores LR, Hernandez MC, Bayer BM. Acute immunosuppressive effects of morphine—lack of involvement of pituitary and adrenal factors. *J Pharmacol Exp Ther* 1994;268:1129–1134.
10. Bhargava HN, Thomas PT, Thorat S, House RV. Effects of morphine tolerance and abstinence on cellular immune function. *Brain Res* 1994;642:1–10.
11. Black PH. Immune system-central nervous system interactions—effect and immunomodulatory consequences of immune system mediators on the brain. *Antimicrob Agents Chemother* 1994;38:7–12.

12. Banks WA, Kastin AJ, Gutierrez EG. Interleukin-1 alpha in blood has direct access to cortical brain cells. *Neurosci Lett* 1993;163:41–44.
13. Dekloet ER, Oitzl MS, Schobitz B. Cytokines and the brain corticosteroid receptor balance—relevance to pathophysiology of neuroendocrine-immune communication. *Psychoneuroendocrinology* 1994;19:121–134.
14. VanderPlate C, Aral SO, Magder L. The relationship among genital herpes simplex virus, stress, and social support. *Health Psychol* 1988;7:159–168.
15. Kemeny ME, Cohen F, Zegans LS, Conant MA. Psychological and immunological predictors of genital herpes recurrence. *Psychosom Med* 1989;51:195–208.
16. Silver PS, Auerbach SM, Vishniavsky N, Kaplowitz LG. Psychological factors in recurrent genital herpes infection: stress, coping style, social support, emotional dysfunction, and symptom recurrence. *J Psychosom Res* 1986;30:163–171.
17. Goldmeier D, Johnson A. Does psychiatric illness affect the recurrence rate of genital herpes? *Br J Vener Dis* 1982;58:40–43.
18. Rand KH, Hoon EF, Massey JK, Johnson JH. Daily stress and recurrence of genital herpes simplex. *Arch Int Med* 1990;150:1889–1893.
19. McKinnon W, Weisse CS, Reynolds CP, Bowles CA, Baum A. Chronic stress, leukocyte subpopulations, and humoral response to latent viruses. *Health Psychol* 1989; 8:389–402.
20. Keicolt-Glaser JK, Fisher LD, Ogrocki P, Stout JC, Speicher CE, Glaser R. Marital quality, marital disruption, and immune function. *Psychosom Med* 1987;49: 13–34.
21. Keicolt-Glaser JK, Glaser R, Shuttleworth EC, Dyer CS, Ogrocki P, Speicher CE. Chronic stress and immunity in family caregivers of Alzheimer's disease victims. *Psychosom Med* 1987;49:523–535.
22. Keicolt-Glaser JK, Dura JR, Speicher CE, Trask OJ, Glaser R. Spousal caregivers of dementia victims: longitudinal changes in immunity and health. *Psychosom Med* 1991;53:345–362.
23. Glaser R, Keicolt-Glaser JK, Speicher CE, Holliday JE. Stress, loneliness, and changes in herpesvirus latency. *J Behav Med* 1985;8:249–260.
24. Cohen S, Tyrrell DA, Smith AP. Psychological stress and susceptibility to the common cold [see comments]. *N Engl J Med* 1991;325:606–612.
25. Broadbent DE, Broadbent MH, Phillpotts RJ, Wallace J. Some further studies on the prediction of experimental colds in volunteers by psychological factors. *J Psychosom Res* 1984;28:511–523.
26. Clover RD, Abell T, Becker LA, Crawford S, Ramsey CN, Jr. Family functioning and stress as predictors of influenza B infection [see comments]. *J Fam Pract* 1989;28:535–539.
27. Petry LJ, Weems LB, Livingstone JN. Relationship of stress, distress, and the immunologic response to a recombinant hepatitis B vaccine. *J Fam Pract* 1991;32: 481–486.
28. Glaser R, Kiecolt-Glaser JK, Bonneau RH, Malarkey W, Kennedy S, Hughes J. Stress-induced modulation of the immune response to recombinant hepatitis B vaccine. *Psychosom Med* 1992;54:22–29.
29. Goodkin K, Fuchs I, Feaster D, Leeka J, Rishel DD. Life stressors and coping style are associated with immune measures in HIV-1 infection—a preliminary report. *Int J Psychiatry Med* 1992;22:155–172.
30. Ironson G, LaPerriere A, Antoni M, et al. Changes in immune and psychological measures as a function of anticipation and reaction to news of HIV-1 antibody status. *Psychosom Med* 1990;52:247–270.
31. Perry S, Fishman B, Jacobsberg L, Frances A. Relationships over 1 year between lymphocyte subsets and psychosocial variables among adults with infection by human immunodeficiency virus [see comments]. *Arch Gen Psychiatry* 1992;49:396–401.
32. Kessler RC, Foster C, Joseph J, et al. Stressful life events and symptom onset in HIV infection [see comments]. *Am J Psychiatry* 1991;148:733–738.
33. Kusnecov AV, Grota LJ, Schmidt SG, et al. Decreased herpes simplex viral immunity and enhanced pathogenesis following stressor administration in mice. *J Neuroimmunol* 1992;38:129–137.
34. Dobbs CM, Vasquez M, Glaser R, Sheridan JF. Mechanisms of stress-induced modulation of viral pathogenesis and immunity. *J Neuroimmunol* 1993;48:151–160.
35. Stead WW, Senner JW, Reddick WT, Lofgren JP. Racial differences in susceptibility to infection by Mycobacterium tuberculosis [see comments]. *N Engl J Med* 1990; 322:422–427.
36. Crowle AJ, Elkins N. Relative permissiveness of macrophages from black and white people for virulent tubercle bacilli. *Infection & Immunity* 1990;58:632–638.
37. Brown DH, Sheridan J, Pearl D, Zwilling BS. Regulation of mycobacterial growth by the hypothalamus-pituitary-adrenal axis: differential responses of Mycobacterium bovis BCG-resistant and -susceptible mice. *Infect Immun* 1993;61:4793–4800.
38. Hickson RC, Boone JB, Jr. Physical exercise and immunity. In Plotnikoff N, Murgo A, Faith R, Wybran J. eds. *Stress and immunity.* Boca Raton: CRC Press, 1991; 211–234.
39. Pedersen BK, Ullum H. NK cell response to physical activity—possible mechanisms of action. *Med Sci Sports Exerc* 1994;26:140–146.
40. Nieman DC. Exercise, upper respiratory tract infection, and the immune system. *Med Sci Sports Exerc* 1994;26:128–139.
41. Schlenzig C, Jager H, Rieder H. Supervised physical exercise leads to psychological and immunological improvement in pre-AIDS patients. *Proceedings of the 5th International AIDS Conference,* 1989:337.
42. Florijin Y, Geiger A. Community based physical activity program for HIV-1 infected persons. *Proceedings of the Biological Aspects of HIV Infection Conference,* 1991.
43. Esterling BA, Antoni MH, Schneiderman N, et al. Psychosocial modulation of antibody to Epstein-Barr viral capsid antigen and human herpesvirus type-6 in HIV-1-infected and at-risk gay men. *Psychosom Med* 1992; 54:354–371.
44. Rigsby LW, Dishman RK, Jackson AW, Maclean GS, Raven PB. Effects of exercise training on men seropositive for the human immunodeficiency virus-1 [see comments]. *Med Sci Sports Exerc* 1992;24:6–12.
45. Knudsen JH, Kjaersgaard E, Jensen EW, Christensen NJ. Percentage of NK-cells in peripheral blood in resting normal subjects is negatively correlated to plasma adrenaline. *Scand J Clin Lab Invest* 1994;54:221–225.
46. Irwin M, Weiner H. Depressive symptoms and immune function during bereavement. In Zisook S, ed. *Biopsychosocial aspects of grief.* Washington, DC: American Psychiatric Press, 1987:159–174.
47. Zisook S, Shuchter SR, Irwin M, Darko DF, Sledge

P, Resovsky K. Bereavement, depression, and immune function. *Psychiatry Res* 1994;52:1–10.
48. Benschop RJ, Brosschot JF, Godaert GLR, et al. Chronic stress affects immunologic but not cardiovascular responsiveness to acute psychological stress in humans. *Am J Physiol* 1994;266:R75–R80.
49. Benschop RJ, Nieuwenhuis EES, Tromp EAM, Godaert GLR, Ballieux RE, Vandoornen LJP. Effects of beta-adrenergic blockade on immunologic and cardiovascular changes induced by mental stress. *Circulation* 1994;89: 762–769.
50. Yirmiya R, Ben-Eliyahu S, Liebeskind JC, et al. Modulation of immunity and neoplasia by neuropeptides released by stressors. In McCubbin JA, Kaufmann PG, Nemeroff CB. eds. *Stress, neuropeptides, and systemic disease.* New York: Academic Press, Inc., 1991:261–285.
51. Cunnick JE, Lysle DT, Armfield A, Rabin BS. Shock induced immunosuppression: naltrexone sensitive and insensitive parameters. In Breznitz S, Zinder O, eds. *Molecular biology of stress.* New York: Alan R. Liss, Inc., 1989;205–214.
52. Bubaksatora M, Skowroncendrzak A, Kubera M, Holan V. Protective effect of lithium on the stress-induced depression of cell-mediated immunity in mice. *Int J Immunopharmacol* 1994;16:233–237.
53. Irwin M, Hauger RL, Britton K. Benzodiazepines antagonize central corticotropin releasing hormone-induced suppression of natural killer cell activity. *Brain Res* 1993;631:114–118.
54. Fleshner M, Brohm MM, Laudenslager ML, Watkins LR, Maier SF. Modulation of the in vivo antibody response by a benzodiazepine inverse agonist (DMCM) administered centrally or peripherally. *Physiol Behav* 1993;54:1149–1154.
55. Puri S, Ray A, Chakravarti AK, Sen P. Role of dopaminergic mechanisms in the regulation of stress responses in experimental animals. *Pharmacol Biochem Behav* 1994;48:53–56.
56. Hellstrand K, Hermodsson S. Enhancement of human natural killer cell cytotoxicity by serotonin: role of non-T/CD16+ NK cells, accessory monocytes, and 5-HT1A receptors. *Cell Immunol* 1990;127:199–214.
57. Garssadi SI, Mandi Y, Regely K, Tarodi B, Beladi I. The inhibitory effect of interferon-alpha on the serotonin-induced impairment of human NK cell activity in whole blood. *Brain, Behav Immun* 1993;7:164–175.
58. Boranic M, Pericic D, Poljak-Blazi M, et al. Immune response of stressed rats treated with drugs affecting serotoninergic and adrenergic transmission. *Biomed Pharmacother* 1990;44:381–387.
59. Benveniste EN. Inflammatory cytokines within the central nervous system: sources, function, and mechanism of action. [Review]. *Am J Physiol* 1992;263:C1–16.
60. Blalock JE. A molecular basis for bidirectional communication between the immune and neuroendocrine systems. [Review]. *Physiol Rev* 1989;69:1–32.
61. Dinarello CA, Wolff SM. The role of interleukin-1 in disease [published erratum appears in N Engl J Med 1993 Mar 11;328(10):744]. [Review]. *N Engl J Med* 1993;328:106–113.
62. Rivier C. Stimulatory effect of interleukin-1 beta on the hypothalamic-pituitary-adrenal axis of the rat—influence of age, gender and circulating sex steroids. *J Endocrinol* 1994;140:365–372.
63. Ericsson A, Kovacs KJ, Sawchenko PE. A functional anatomical analysis of central pathways subserving the effects of interleukin-1 on stress-related neuroendocrine neurons. *J Neurosci* 1994;14:897–913.
64. Quan N, Sundar SK, Weiss JM. Induction of interleukin-1 in various brain regions after peripheral and central injections of lipopolysaccharide. *J Neuroimmunol* 1994; 49:125–134.
65. Zalcman S, Greenjohnson JM, Murray L, et al. Cytokine-specific central monoamine alterations induced Bp interleukin-1, -2 and -6. *Brain Res* 1994;643:40–49.
66. Gardier AM, Kachaner S, Shaghaghi EK, et al. Effects of a primary immune response to T-cell dependent antigen on serotonin metabolism in frontal cortex: in vivo microdialysis study in freely moving Fischer 344 rat. *Brain Res* 1994;645:150–156.
67. Rada P, Mark GP, Vitek MP, et al. Interleukin-1 beta decreases acetylcholine measured by microdialysis in the hippocampus of freely moving rats. *Brain Res* 1991; 550:287–290.
68. Opp MR, Krueger JM. Interleukin-1 is involved in responses to sleep deprivation in the rabbit. *Brain Res* 1994;639:57–65.
69. Irwin M, Smith TL, Gillin JC. Electroencephalographic sleep and natural killer activity in depressed patients and control subjects. *Psychosom Med* 1992;54:10–21.
70. Denicoff KD, Rubinow DR, Papa MZ, et al. The neuropsychiatric effects of treatment with interleukin-2 and lymphokine-activated killer cells. *Ann Int Med* 1987;107: 293–300.
71. Weiss JM, Sundar SK, Becker KJ. Stress-induced immunosuppression and immunoenhancement: cellular immune changes and mechanisms. In Goetzl ED, Spector NH. eds. *Neuroimmune networks: physiology and diseases.* New York: Alan R. Liss, Inc., 1989;193–206.
72. Sheridan JF, Dobbs C, Brown D, Zwilling B. Psychoneuroimmunology—stress effects on pathogenesis and immunity during infection. *Clin Microbiol Rev* 1994;7: 200–212.
73. Ader R. Psychoneuroimmunology—conditioning and stress (Vol 44, Pg 77, Yr 1993). *Annu Rev Psychol* 1994;45:U8.
74. Fawzy FI, Fawzy NW, Hyun CS, et al. Malignant melanoma. Effects of an early structured psychiatric intervention, coping, and affective state on recurrence and survival 6 years later. *Arch Gen Psychiatry* 1993;50:681–689.

Neurobiological and Clinical Consequences of Stress: From Normal Adaptation to PTSD, edited by M. J. Friedman, D. S. Charney, and A. Y. Deutch, Lippincott-Raven Publishers, Philadelphia © 1995.

22

Somatic Consequences of Stress

Redford B. Williams

Department of Psychiatry, Duke University Medical Center, Durham, North Carolina 27710

Stress *does* have somatic consequences. This has been widely known for at least 2,000 years. One early recorded instance of stress's somatic consequences can be found in the "Acts of the Apostles" in the New Testament of the *Bible.* There the story is told of Ananias and his wife Saphira, who sold their lands and supposedly donated all the proceeds of the sale to Peter and the Apostles to support the work of the early Christian church. But they did not deserve all the approbation they received; they had held back some of the proceeds from the sale for their own use.

When Peter learned of their deception, he called Ananias in and told him, "You have sinned not only against me but against God, and you shall pay the ultimate price." At this, Ananias fell down dead. After his body had been carried out, Peter called Saphira in and chastised her similarly, whereupon she fell dead also. Here we have what to my knowledge is the first recorded instance of sudden cardiac death brought on by acute stress.

While we have "known" for over 2,000 years of the harm stress can wreak on our bodies, it has only been in the last 40–50 years that scientific research has provided more systematic evidence to confirm the health-damaging somatic consequences of stress. It should come as no surprise that the scientific literature describing this research draws upon many disciplines and is very broad, complex, and of varying quality.

In this chapter I shall confine my attention to those areas where I consider the available evidence regarding somatic consequences of stress to be of very good quality. While this approach may result in some worthy topics being omitted due to oversight or my own blind spots, I am confident that the areas covered merit serious consideration not only to increase understanding of the impact of stress on health and disease, but also to guide the development of interventions to ameliorate that impact.

Stress is not a unitary concept. In Fig. 1 I present a conceptual model that will be helpful in understanding the domains that are critical in determining stress and its somatic consequences, as well as help to organize the broad and complex literature to be reviewed in this chapter.

As presented in this model, stress is a process involving an *interaction* between the person and the environment. The nature of this interaction depends upon both the person's psychological characteristics and the characteristics of the environmental situation. The psychological characteristics are determined by gene-environment interactions over the prior lifespan of the individual, and can be either enduring and stable traits (e.g., certain personality types) or more transient states (e.g., periods of emotional upheaval or clinical conditions). Environmental characteristics impacting health can be either physical or social. As determined by the psychological and environmental characteristics active in any given situation, the person-environment interaction is transduced—presumably in the service of adaptation and coping—by the brain into "motor" messages to the body's tissues and organs. The

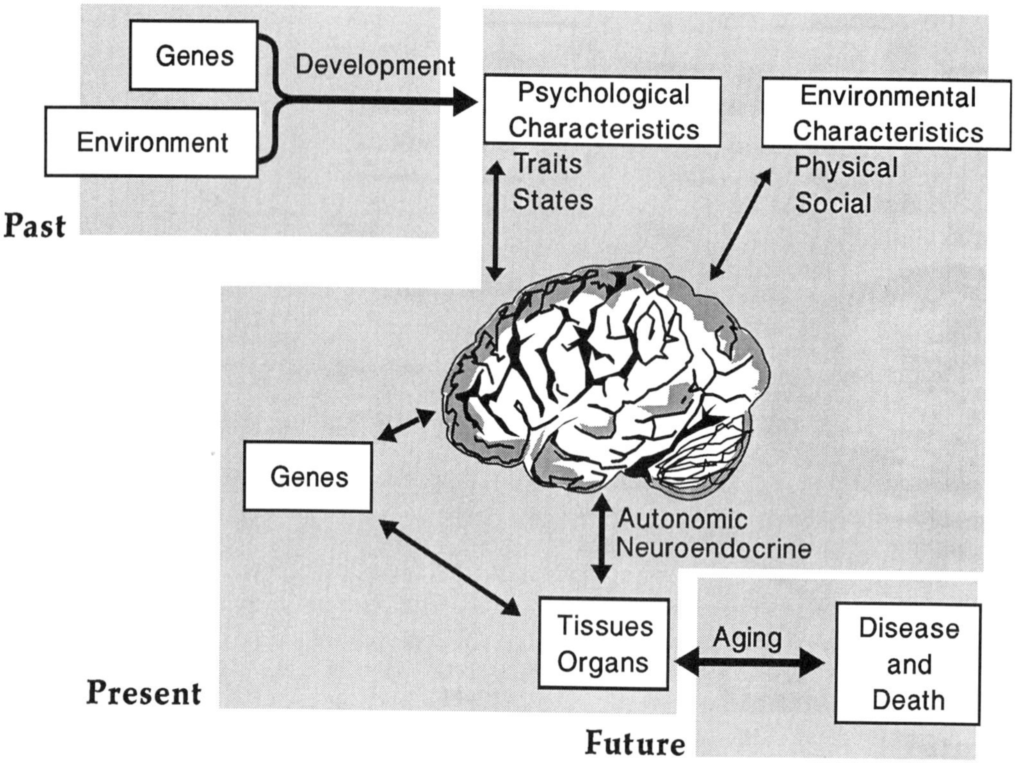

FIG. 1. Conceptual model showing those domains that are critical sources of stress and the pathways whereby this stress is transduced into biobehavioral responses that lead to disease and death over the life cycle.

resulting "behaviors"—which can be both biological in nature at the level of specific tissues and organs (e.g., the cardiovascular system, the immune system, etc.) or more grossly behavioral at the level of the whole organism (e.g., smoking, eating, drinking)—contribute over time during the aging process to the development of certain diseases and, ultimately, death.

I stated earlier that I would review only those areas where I considered the evidence regarding the somatic consequences of stress to be "of very good quality." In making this determination, I began with an *epidemiological* criterion: to be considered a source of stress with significant somatic consequences, a given psychological or environmental characteristic must have been shown in *prospective longitudinal studies* to predict increased risk of major medical disorder or poorer prognosis once such a disorder is present. I believe there are two psychological and two environmental characteristics that meet this criterion: 1) the psychological trait of a hostile personality; 2) the psychological state of depression; 3) physical and social environmental situations that impose high demands for coping upon the individual in combination with low levels of control over how those demands are met; and 4) social environments that provide low levels of social support.

In this chapter, therefore, I shall review first for each of these sources of stress the epidemiological evidence for their somatic consequences in terms of the most profound longterm consequences, disease and death. Second, I shall consider for each source of stress the more proximate somatic consequences—biological and behavioral effects—that are likely responsible for the long-term effects on health. Third, I shall include in a more speculative mode a consideration of the possible neurobiological mechanisms responsible for the stress effects on biobehavior leading to disease and death. And, finally, I shall include some thoughts regarding how what we know about the somatic consequences of stress might be applied to develop interventions that will prevent or ameliorate those consequences.

EPIDEMIOLOGICAL EVIDENCE

Psychological Traits: Hostile Personality

The first systematic research documenting somatic health consequences of any stress-inducing psychological characteristic emerged from the pioneering studies of two San Francisco cardiologists, Meyer Friedman and Ray Rosenman. Based on their prospective studies showing that men with the Type A behavior pattern—impatient, hostile, and ambitiously competitive—were about twice as likely to develop coronary heart disease over an 8.5-year follow-up period, a review panel assembled by the National Heart, Lung and Blood Institute and the American Heart Association in the late 1970s was ready to declare Type A behavior an established risk factor (1). The final anointing was put on hold, however, by the appearance of a number of papers that failed to find Type A behavior associated with or predictive of increased coronary risk. Most compelling among these was a paper by Shekelle and coworkers (2) that failed to find Type A (assessed by the "gold standard" structured interview) predicting coronary events in over 3,000 middle-aged men followed up over a 7-year period.

Rather than discard entirely the core hypothesis that there is a personality type that predisposes to the development of coronary heart disease, many investigators responded to these negative Type A studies by refocusing their attention on the specific components making up the global Type A pattern. What has emerged over the ensuing decade of research (see Williams (3) and Smith (4) for detailed reviews) is a growing consensus that it is only the *hostility* and *anger* component of the Type A behavior pattern that is coronary-prone. Using both interview- and questionnaire-based assessments of hostility in both cross-sectional and prospective designs, and despite some negative studies, the consistent finding has been that persons with high levels of hostility are at higher risk of developing coronary heart disease.

Since one of the widely used measures of hostility, the Cook and Medley "Ho" scale, is derived from the Minnesota multiphasic personality inventory (MMPI), it has been possible to rescore archival MMPI data from older studies and show that hostility predicts not only increased coronary risk but increased risk of dying from all causes, including cancer (5–7). An item analysis of the Ho scale items (6) suggested that it is a cynical, mistrusting attitude, the frequent experience of angry feelings, and the overt aggressive expression of those feelings that account for the scale's success in predicting somatic health problems.

Psychological States: Depression

While the hostile personality trait falls within the realm of "normal" psychological functioning, the presence of an abnormal state of significant clinical depression has also recently been shown to have important somatic health consequences. This adverse impact has been most clearly demonstrated in studies of depression and prognosis once coronary disease is present. In a follow-up study of over 200 patients following a myocardial infarction, the 16% who exhibited signs of major depression experienced a 6-month mortality that was fivefold higher than that seen in the nondepressed patients, even after controlling for the severity of the infarct (8).

Recently emerging evidence also implicates depression as a risk factor for the development of major medical disorder in healthy individuals. Thus, Anda et al. (9) found that depression as indexed by high scores on the depression subscale of the General Well-Being Schedule predicted increased risk of developing coronary heart disease in a cohort of 2,832 participants in the National Health Examination Follow-up Study.

While the evidence implicating depression in the development and course of major medical disorder is less extensive than that implicating the hostile personality trait, I am now convinced that further research will show depression to have an equally strong negative impact on health.

The Physical and Social Environment: High Demands, Low Control

Thanks in large part to the work of Karasek and coworkers (see review (10)), the *specific* environmental situation consisting of a job that places high demands on the worker while allowing little decision latitude in how those demands are met creates a situation of high *job strain.* A typical high-strain job would be an assembly-line position, where the requirement to screw in a widget every 15 seconds places high demands for physical output on the worker, while the inability to stop or slow the line leaves the worker with very little control over how he or she meets the demands. Employment in high-strain jobs has been found in prospective studies (10) to predict increased risk of developing coronary heart disease and other serious illnesses. In cross-sectional studies, workers in high-strain jobs have also been found more likely to suffer from high blood pressure, and to have increased ambulatory blood pressure levels and increased left ventricular mass (11).

In addition to the specific example of job strain, the concept of demand/control imbalance is also useful in understanding the *general case* whereby environmental factors can be stressful in ways that have somatic consequences. As developed by Holmes and Rahe (12), the concept of ''life events'' stress has been related to increased risk of medical disorders. Just as certain types of jobs place high demands upon the individual, with little control over how those demands are met, certain types of life events (e.g., getting married, birth of a child, loss of a job, etc.) create an analogous demand/control imbalance. It appears reasonable to conceptualize other physical/social environmental constructs (e.g., examination stress in medical students, caregiver stress in relatives of Alzheimer's patients, low socioeconomic status) that have been shown in other lines of research to have adverse impacts on biological functioning or health in similar terms; they impose high physical and/or psychological demands upon the individual, while at the same time permitting little control over how those demands are met.

The Social Environment: Low Social Support

As reviewed by House (13), an extensive body of research has shown that among persons who are healthy at study inception, those who are socially isolated (i.e., who have low levels of social support) are at higher risk of developing coronary disease, as well as other major, life-threatening illnesses.

The impact of low social support, or social isolation, is even more dramatic among patients who already have coronary disease: even after controlling for severity of myocardial damage, those patients who are socially isolated show death rates that are two- to fivefold higher over follow-ups of up to 5 years than those observed in nonisolated patients (14–16).

In summary, the epidemiological evidence points to a hostile personality trait, a depressed psychological state, physical/social environments that impose high demands with low control, and social environments that are lacking in support as sources of stress that have somatic consequences severe enough to increase risk of death and disease. While I have considered each of these constructs separately, it should be noted that they undoubtedly interact with each other in ways that magnify the effects. A health illustrative example: Orth-Gomer has found that men who are both Type A *and* socially isolated are at higher risk of developing coronary heart disease than men with either risk characteristic alone (17).

While the epidemiological evidence reviewed thus far makes a strong case for the health-damaging somatic consequences of the sources of stress included in this review, the case of the *biological plausibility* of their impact on disease processes is strengthened still further by a consideration of their biological and behavioral effects.

BIOBEHAVIORAL MECHANISMS

Hostile Personality

Increased sympathetic nervous system reactivity to stress is a biologically plausible mechanism whereby psychosocial stress factors contribute to pathogenesis of somatic diseases. It is now well established that, relative to their nonhostile counterparts, hostile persons exhibit enhanced sympathetically mediated cardiovascular responses when angered by various experimental manipulations in laboratory studies (18, 19). Preliminary evidence from our laboratories at Duke (unpublished data) indicates that this enhanced sympathetic reactivity also occurs among hostile persons in real life, as indicated by increased daytime urinary catecholamine excretion and downregulated lymphocyte beta adrenergic receptors in hostile as contrasted to nonhostile subjects.

In addition to their excessive sympathetic reactivity to stress, hostile persons also show reduced parasympathetic nervous system antagonism of sympathetic effects on the heart (20). Ambulatory studies have also confirmed a reduced level of parasympathetic nervous system activity among hostile persons in naturalistic environments (21).

Besides having these biological characteristics persons with hostile personality engage in behaviors that also would be expected to increase their risk of developing major medical disorders, according to recent research. In one large-scale follow-up study of over 4,000 persons on whom hostility scores from college years were available (22), those with high MMPI Ho scores at age 19 were more likely than their low Ho-scoring counterparts to show the following at age 42: 1) increased body mass index; 2) increased cholesterol/HDL ratio; 3) increased alcohol consumption; 4) increased smoking; and 5) increased caffeine consumption. An earlier study (2) had found an association between high Ho scores and increased smoking and alcohol use. In a large-scale cross-sectional study of 6,000 young adults (23), those with Ho scores in the highest quartile were found more likely to be smokers and to consume on average 600 more calories per day than those with Ho scores in the lowest quartile.

Depression

There is a striking overlap between the potentially pathogenic biobehavioral characteristics found in persons with hostile personalities and those found in persons suffering from major depression. Using sophisticated techniques to characterize sympathetic nervous system (SNS) function, Vieth and coworkers (24) have documented increased SNS outflow in patients with major depression. Other studies have shown depressed patients to have reduced levels of parasympathetic function as well (25). Both increased sympathetic and decreased parasympathetic function have been implicated in the increased mortality found among depressed myocardial infarction victims (8). Another biological pathway, likely associated with the known SNS and cortisol excesses, whereby depression may be contributing to increased risk of medical disorders, is via suppression of immune function (26).

In addition to a change in autonomic balance favoring the SNS over the parasympathetic that is similar to that found in hostile persons, depression has also been found associated with a similar profile of health-damaging behaviors. Both smoking and alcohol consumption have been found increased among depressed patients as well as persons with a history of depression (27–29).

High-Demand/Low-Control Environments

As just noted, jobs that impose high demands with low levels of control over how those demands are met have been found associated with increased ambulatory blood pressure levels, as well as increased left ventricular mass. While direct evidence may be lacking at present, it seems highly likely that both these effects will eventually be found linked to increased SNS function among persons working in high-strain jobs.

With respect to the more general case in which environmental stress is conceptualized as any situation that creates a demand/control imbalance, there is extensive evidence, reviewed by Kiecolt-Glaser and Glaser (30), that such environmental situations (e.g., examination stress in medical students, marital strife, caregiver stress in relatives of Alzheimer's patients) are associated with decreases in a wide array of immune system functions. That such stress-associated downregulation of immune function has the potential to increase risk of disease was documented in a study by Cohen and coworkers showing increased susceptibility to the common cold among those inoculated with cold virus who also reported a high level of recent stress (31).

It is well known that high-demand/low-control environments increase the use of a number of substances (e.g., cigarettes, alcohol, caffeine-containing beverages) that could also add to the risk of disease.

Low Social Support

Several studies have identified biological correlates of low social support that would be expected to contribute to pathogenic mechanisms of disease. For example, increased sympathetic function has been observed among persons reporting low levels of social support (32). Such increased SNS function might be one mechanism contributing to the decreased immune function (30) that has been found in persons who are socially isolated or who report low social support.

Social isolation also appears to have an adverse impact on behaviors related to somatic health and disease. Persons who are socially isolated have been reported to be less likely to stop smoking or to adhere to a prescribed medical regimen (33,34).

In summary, the psychological and environmental characteristics that have been found in epidemiological research to predict increased risk of major medical disorders and death have also been found in other research—both epidemiological and laboratory-based—to be associated with biological (increased SNS and decreased parasympathetic function, immunosuppression) and behavioral (increased smoking, excessive alcohol use, over- or undereating, nonadherence to medical regimens) characteristics that are *biologically plausible* contributors to pathogenic processes.

Moving closer to the tissue and organ levels (see Fig. 1) where the pathogenesis of diseases such as coronary atherosclerosis and cancer is actually occurring, it appears likely that our growing understanding of the cell and molecular biology of disease processes can be applied to identify the specific mechanisms whereby these biobehavioral concomitants of stress cause disease. What follows is a somewhat speculative projection of how this work might proceed.

A Potential Cellular and Molecular Mechanism Whereby Somatic Consequences of Psychosocial Factors Affect Health

As reviewed by Adams (35), monocytes/macrophages are the critical starting point for many of the functions of the immune system that have been implicated in stress-related disease processes. One such function relates to a scavenger receptor that, when ligated by oxidized-LDL, causes expression of a phenotype that would be expected to facilitate macrophage involvement in atherogenesis and retard macrophage involvement in tumor-cell killing and rejection. In recent pilot studies (unpublished), we have been able to show that ligation of macrophage beta adrenergic receptors causes expression of a phenotype that is similar in many respects to that activated by oxidized LDL cholesterol.

The biobehavioral characteristics of persons with the psychological traits and/or states, and/or exposure to stressful environments associated with increased risk of medical disorder and death (as described earlier in this chapter) would be expected to activate this same macrophage phenotype. Thus, high cholesterol levels and increased smoking would lead to increased levels of oxidized LDL in hostile or depressed persons, which could combine with their increased catecholamines to produce additive or even synergistic increases in the macrophage phenotype favoring both atherogenesis and tumorogenesis. Future research can evaluate the degree to which the pathogenic phenotype is expressed in monocytes (the precursor of the tissue macrophage) obtained from hostile and depressed persons (as well as persons with low social support and/or who work in high-strain jobs). If the expected increased expression of this phenotype is found in such persons, it would constitute evidence at the cell and molecular levels of a pathogenic mechanism activated by the biobehavioral concomitants of psychosocial stress factors.

Shifting the focus from the peripheral cell and molecular mechanisms of pathogenesis, we shall now consider the neurobiological mechanisms that may be responsible for transducing psychosocial stress factors into biobehavioral characteristics that accelerate the pathogenesis of somatic disease.

NEUROBIOLOGICAL MECHANISMS

As reviewed in the previous section, persons with the hostile personality trait, as well as persons who are in a depressed state (and possibly those exposed to high-demand/low-control and/or low social support environments) exhibit a cluster of biobehavioral characteristics that, taken either singly or in combination, could account for some of the excess somatic disease and consequently higher death rates to which they are prone. These characteristics include: 1) depressed mood; 2) increased anger/irritation; 3) increased sympathetic nervous system reactivity; 4) decreased parasympathetic nervous system function; 5) increased smoking; 6) increased eating; and 7) increased alcohol consumption.

I have proposed elsewhere (36) the heuristic hypothesis that this clustering of biobehavioral characteristics in persons who score high on the Ho scale is the result of a single underlying neurobiological condition: reduced central nervous system (CNS) serotonergic function. This hypothesis is based on a considerable body of research from both human and animal studies suggesting that reduced CNS serotonin function is capable of mediating all of the biobehavioral characteristics found in persons with high hostility levels. Representative studies from this literature are noted in the next paragraph.

First of all, many of the basic behavioral manifestations of the hostile personality may themselves result from reduced CNS serotonin function. Men with a history of aggressive behavior have been found to have reduced cerebrospinal fluid (CSF) levels of the serotonin metabolite 5-HIAA (37). Low CSF 5-HIAA levels predict increased aggressive behavior over a 2-year follow-up period in children and adolescents (38). Prolactin response to fenfluramine correlates negatively with hostility scale scores (39), suggesting reduced CNS serotonin function in hostile persons. Thus, abundant evidence links the aggressivity component of the hostile personality with reduced CNS serotonin function.

Agonists for the serotonin ($5HT_{1A}$) receptor in the CNS have been found to reduce sympathetic outflow and increase parasympathetic outflow (40), and tryptophan loading reduces cardiac sympathetic nerve firing in cats (41). Thus, low CNS serotonin function is capable of causing the same pattern of altered autonomic balance—increased sympathetic and decreased parasympathetic function—found in hostile persons.

With respect to the risky health behaviors found in hostile persons, the reward property of nicotine appears to be mediated by the $5HT_3$ receptor (42), and the serotonin agonist buspirone may be useful in reducing smoking in humans (43). It is well known that depletion of CNS serotonin leads to increased food consumption and adiposity in animal models (44), while enhancement of CNS serotonin function with selective serotonin reuptake inhibitors (SSRIs) reduces food intake and body weight (45). And finally, there is a well-established inverse association between CNS serotonin function and alcohol use, with alcoholics having lower CSF 5-

HIAA levels (46) and SSRIs appearing useful in reducing alcohol intake (47).

In contrast to this new speculation that low CNS serotonergic function might be responsible for the clustering of biobehavioral characteristics found in hostile persons, this low function has long been suspected to play a role in depression (48).

While making an eminently plausible case that reduced CNS serotonin function *could* cause all of the biobehavioral characteristics that epidemiological and biological research has shown to cluster in hostile persons, the case is at present purely circumstantial. In addition, I recognize that this serotonergic hypothesis is simplistic in focusing on a single neurotransmitter, serotonin, that is known to interact with other neurotransmitters and to exert its effects via an incompletely understood family of receptors, some of which could have opposite effects on the biobehavioral characteristics of hostile and depressed persons. Despite these obstacles, it will be possible to test this hypothesis using research tools that can provide direct evidence of reduced CNS serotonin function in hostile as compared to nonhostile people. These tools have already been applied (48) to document reduced brain serotonin function in depression and other psychiatric disorders.

Very preliminary findings from research currently in progress in my laboratory are providing evidence that reduced CNS serotonin function in hostile persons can be adduced. Even though sample sizes are quite small still, we find that fenfluramine pretreatment produces a significantly greater enhancement (over that seen with placebo pretreatment) of the cortisol response to mental challenge in hostile subjects. This finding supports the hypothesis of reduced serotonin function in hostile persons, in that the enhanced cortisol response by fenfluramine in hostile persons suggests the presence in them of serotonin receptor supersensitivity, itself a possible result of reduced serotonin turnover.

Thus far I have said nothing about the origins of the reduced serotonin function that I hypothesize to underlie the clustering of health-damaging biobehavioral characteristics in hostile persons. The work of Higley and coworkers (49) has documented genetic contributions to serotonin turnover in rhesus monkeys, as well as the effects of altering the social environment.

Indeed, it is possible that major environmental stressors leading to disorders such as post-traumatic stress disorder (PTSD) might affect a wide array of somatic consequences (see Chapter 29) via effects on CNS serotonin function. My colleague at Duke, Dr. Jean Beckham, has found (unpublished data) that veterans with PTSD exhibit the highest hostility levels, using our behavioral assessment, of any group we have yet studied. She hypothesizes that this increased hostility in PTSD patients will be associated with physiologic hyperreactivity that may place them at higher risk of developing cardiovascular disease.

I wonder whether both their hostility and any associated physiological hyperreactivity, as well as other biobehavioral effects (see Chapter 29), may stem from their original trauma having caused a depletion of their CNS serotonin stores. If so, then PTSD might be considered an extreme, stress-induced case of serotonin deficiency leading to *both* increased hostility and health-damaging biobehavioral accompaniments.

IMPLICATIONS FOR PREVENTION AND TREATMENT

Epidemiologic research clearly documents the serious somatic consequences, in the form of death-causing medical illnesses, of psychosocial stress factors (e.g., hostile personality, social isolation, job strain, depression) in humans. Further research has identified both behavioral and biological correlates of these psychosocial stress factors that could be responsible for their health-damaging effects. An heuristic hypothesis regarding CNS mechanisms responsible for clustering of these biobehavioral characteristics in humans with psychosocial risk factors is that they all result from reduced CNS serotonergic function.

If confirmed in further research, this serotonin-deficiency hypothesis would have important implications for *pharmacologic* approaches to prevention and treatment of the somatic consequences of psychosomatic stress factors. For one thing, the SSRIs would deserve serious consideration for at least the prevention of stress effects on prognosis once coronary disease is present. Ultimately, such agents might even deserve consideration in primary prevention programs among high-risk populations (e.g., hostile or depressed persons who smoke).

In the meanwhile, however, there should be no question about the importance of evaluating *behavioral* approaches that could prove effective in preventing the impact of stress on prognosis once disease is present. I say this because there are already published studies showing: 1) reduced recurrence rates in post–myocardial infarction (MI) patients randomized to behavioral training to reduce Type A behavior (50); 2) reduced mortality in post-MI patients randomized to a social support intervention (51); 3) longer survival in women with metastatic breast cancer randomized to a stress-management/support-group intervention (52); and 4) reduced recurrence rates in patients with malignant melanoma randomized to a stress-management/support-group intervention (53).

SUMMARY AND CONCLUSIONS

There are psychological characteristics (hostile personality trait, depressed state) and environmental characteristics (high demand/low control, low social support) that have been shown to have somatic consequences severe enough to have significant adverse impact on disease and death. These characteristics are associated with biobehavioral characteristics that are biologically plausible mediators of the adverse health effects. Further research has the potential to identify both peripheral molecular and CNS neurobiological substrates (e.g., reduced CNS serotonergic function) that mediate the health-damaging effects of these psychological and environmental characteristics. Ultimately, our knowledge of these health-damaging characteristics, their pathogenic mechanisms, and their neurobiological substrates should lead to more effective means of prevention and treatment (54).

REFERENCES

1. Review Panel. Coronary-prone behavior and coronary heart disease: a critical review. *Circulation* 1981;63: 1199–1215.
2. Shekelle RB, Hulley SB, Neaton JD, et al. The MRFIT behavior pattern study. II. Type A behavior and incidence of coronary heart disease. *Am J Epidemiol* 1985; 122:559–570.
3. Williams RB. A relook at personality types and coronary heart disease. *Prog Cardiology* 1991;4:91–97.
4. Smith TW. Hostility and health: current status of a psychosomatic hypothesis. *Health Psychol* 1992;11:139–150.
5. Barefoot JC, Dahlstrom WG, Williams RB. Hostility, CHD incidence and total mortality: a 25-year follow-up study of 255 physicians. *Psychosom Med* 1983;45: 59–63.
6. Barefoot JC, Dodge KA, Peterson BL, Dahlstrom WG, Williams RB. The Cook-Medley hostility scale: item content and ability to predict survival. *Psychosom Med* 1989;51:46–57.
7. Shekelle RB, Gale M, Ostfeld AM, Paul O. Hostility, risk of coronary heart disease, and mortality. *Psychosom Med* 1983;45:109–114.
8. Frasure-Smith N, Lesperance F, Talajic M. Depression following myocardial infarction: impact on 6-month survival. *JAMA* 1993;270:1819–1825.
9. Anda R, Williamson D, Jones D, et al. Depressed affect, hopelessness, and the risk of ischemic heart disease in a cohort of U.S. adults. *Epidemiology* 1993;4:285–294.
10. Karasek RA, Theorell T. *Healthy work: stress, productivity, and the reconstruction of working life.* New York: Basic Books, 1990.
11. Schnall P, Pieper C, Schwartz JE, et al. The relationship between job strain, workplace diastolic blood pressure, and left ventricular mass index: results of a case-control study. *JAMA* 1990;263:1929–1935.
12. Holmes T, Rahe R. The social readjustment rating scale. *J Psychosom Res* 1967;11:213–218.
13. House JS, Landis KR, Umberson D. Social relationships and health. *Science* 1988;241:540–545.
14. Ruberman W, Weinblatt E, Goldberg JD, Chaudhary BS. Psychosocial influences on mortality after myocardial infarction. *N Engl J Med* 1984;311:552–559.
15. Case RB, Moss AJ, Case N, et al. Living alone after myocardial infarction: impact on prognosis. *JAMA* 1992; 267:515–519.
16. Williams RB, Barefoot JC, Califf RM, et al. Prognostic importance of social and economic resources among medically treated patients with angiographically documented coronary artery disease. *JAMA* 1992;267:520–524.
17. Orth-Gomer K, Rosengren A, Wilhelmsen L. Lack of social support and incidence of coronary heart disease in middle-aged men. *Psychosom Med* 1993;55:37–43.
18. Smith TW, Allred KD. Blood pressure reactivity during social interaction in high and low cynical hostile men. *J Behav Med* 1989;11:135–143.
19. Suarez EC, Williams RB. Situational determinants of cardiovascular and emotional reactivity in high and low hostile men. *Psychosom Med* 1989;51:404–418.
20. Fukudo S, Lane JD, Anderson NB, et al. Accentuated vagal antagonism of beta-adrenergic effects on ventricu-

lar repolarization: differential responses between Type A and Type B men. *Circulation* 1992;85:2045–2053.
21. Sloan RP, Shapiro PA, Bagiella E, Bigger JT, Steinman RC, Gorman JM. Cardiovascular autonomic control and hostility. *Am J Cardiol* 1994;74:298–300.
22. Siegler IC, Peterson BL, Barefoot JC, Williams RB. Hostility in late adolescence predicts coronary risk factors at mid-life. *Am J Epidemiol* 1992;136:146–156.
23. Scherwitz LW, Perkins LL, Chesney MA, Hughes GH, Sidney S, Manolis TA. Hostility and health behaviors in young adults: the CARDIA study. Coronary Artery Risk Development in Young Adults Study. *Am J Epidemiol* 1992;136:136–145.
24. Vieth RC, Lewis N, Linares OA, et al. Sympathetic nervous system activity in major depression: basal and desipramine-induced alterations in plasma norepinephrine kinetics. *Arch Gen Psychiat* 1994;51:411–422.
25. Carney RM, Rich M, TeVelde A, et al. The relationship between heart rate, heart rate variability and depression in patients with coronary heart disease. *J Psychosom Res* 1988;32:159–164.
26. Evans DL, Folds JD, Pettito JM, et al. Circulating natural killer cell phenotypes in men and women with major depression: relation to cytotoxic activity and severity of depression. *Arch Gen Psychiat* 1992;49:388–395.
27. Anda RF, Williamson DF, Escobedo LG, Mast EE, Giovino GA, Remington PL. Depression and the dynamics of smoking: a national perspective. *JAMA* 1990;264:1541–1545.
28. Glassman AH, Helzer JE, Covey LS, et al. Smoking, smoking cessation, and major depression. *JAMA* 1990;264:1546–1549.
29. Hartka E, Johnstone B, Leino EV, Motoyoshi M, Temple MT, Fillmore KM. A meta-analysis of depressive symptomatology and alcohol consumption over time. *Br J Addict* 1991;86:1283–1298.
30. Kiecolt-Glaser JK, Glaser R. Psychoneuroimmunology: can psychological interventions modulate immunity? *J Consult Clin Psychol* 1992;60:1–6.
31. Cohen S, Tyrrell DA, Smith AP. Psychological stress in humans and susceptibility to the common cold. *N Engl J Med* 1991;325:606–612.
32. Fleming R, Baum A, Gisriel MM, Gatchel RJ. Mediating influences of social support on stress at Three Mile Island. *J Human Stress* 1982;8:14–22.
33. Mermelstein R, Cohen S, Lichtenstein E, Kamarck T, Baer JS. Social support and smoking cessation and maintenance. *Consult Clin Psychol* 1986;54:447–453.
34. Williams CA, Beresford SA, James SA, et al. The Edgecombe County High Blood Pressure Control Program, III: social support, social stressors, and treatment dropout. *Am J Public Health* 1985;75:483–486.
35. Adams DO. Molecular biology of macrophage activation: a final, common pathway whereby psychosocial factors can potentially affect health. *Psychosom Med* 1994;56:316–327.
36. Williams RB. Neurobiology, cellular and molecular biology, and psychosomatic medicine. *Psychosom Med* 1994;56:308–315.
37. Brown GL, Ebert MH, Goyer DC, Jimerson DC, Klein W, Bunney WE, Goodwin FK. Aggression, suicide and serotonin: relationships to CSF amine metabolites. *Am J Psychiatry* 1982;139:741–746.
38. Kruesi MJP, Hibbs ED, Zahn TP, et al. A 2-year prospective follow-up study of children and adolescents with disruptive behavior disorders: prediction by cerebrospinal fluid 5-hydroxyindoleacetic acid, homovanillic acid, and autonomic measures? *Arch Gen Psychiatry* 1992;49:429–435.
39. Coccaro EF, Siever LJ, Klar HM, et al. Serotonergic studies in patients with effective and personality disorders. *Arch Gen Psychiatry* 1989;46:587–599.
40. Saxena PR, Villalon CM. Cardiovascular effects of serotonin agonists and antagonists. *J Cardiovasc Pharmacol* 1990;7:S17–S34.
41. Verrier RL. Neurochemical approaches to the prevention of ventricular fibrillation. *Fed Proc* 1986;45:2191–2196.
42. Carboni E, Acquas E, Leone P, DeChiara G. $5HT_3$ receptor antagonists block morphine- and nicotine- but not amphetamine-induced reward. *Psychopharmacology* 1989;97:175–178.
43. Gawin FH, Compton M, Dych P. Buspirone reduces smoking. *Arch Gen Psychiatry* 1989;46:288–289.
44. Waldbillig RJ, Bortness TJ, Stanley BG. Increased food intake, body weight, and adiposity following regional neurochemical depletion of serotonin. *J Comparat Physiol Psychol* 1981;95:391–405.
45. Levine LR, Enas GG, Thompson WL, et al. Use of fluoxetine, a selective serotonin uptake inhibitor, in the treatment of obesity: a dose-response study. *Int J Obes* 1989;13:635–645.
46. Ballenger J, Goodwin FK, Major LF, Brown GL. Alcohol and central serotonin metabolism in man. *Arch Gen Psychiatry* 1979;36:224–227.
47. Seller EM, Naranjo CA. Therapeutic use of serotonergic drugs in alcohol abuse. *Clin Neuropharmacol* 1986;9:60–62.
48. Meltzer HY, Lowy MT. The serotonin hypothesis of depression. In Meltzer HY, ed. *Psychopharmacology: the third generation of progress.* New York: Raven Press; 1987:513–526.
49. Higley JD, Thompson WW, Champoux M, et al. Paternal and maternal genetic and environmental contributions to cerebrospinal fluid monoamine metabolites in Rhesus monkeys *(Macaca mulatta). Arch Gen Psychiatry* 1993;50:615–623.
50. Friedman M, Thoresen CE, Gill JJ, et al. Alteration of type A behavior and its effect on cardiac recurrences in post myocardial infarction patients; summary results of the Recurrent Coronary Prevention Project. *Am Heart J* 1986;112:653–665.
51. Frasure-Smith N, Prince R. Long-term follow-up of the Ischemic Heart Disease Life Stress Monitoring Program. *Psychosom Med* 1989;51:485–513.
52. Spiegel D, Bloom JR, Kraemer HC, Gottheil E. Effect of psychosocial treatment on survival of patients with metastatic breast cancer. *Lancet* 1989;2:888–890.
53. Fawzy FI, Fawzy NW, Hyun CS, et al. Malignant Melanoma: effects of an early structured psychiatric intervention, coping, and affective state on recurrence and survival 6 years later. *Arch Gen Psychiat* 1993;50:681–689.
54. Chesney MA, Williams RB. Psychosocial factors and prognosis in established coronary artery disease. The need for research on interventions. *JAMA* 1993;270:1860–1861.

PART IV

Clinical Issues

The seven chapters in this final section attempt to synthesize available information from basic animal or human research, and focus on their relevance to the detection, diagnosis, and treatment of post-traumatic stress disorder (PTSD).

Chapter 23 reviews the epidemiology of PTSD in the general adult population, among children, and among selected cohorts such as military veterans and disaster survivors. Their discussion of risk factors suggests that biological variables such as genetic inheritance and substance abuse may increase vulnerability to PTSD.

A common diagnostic problem results from the fact that the vast majority of PTSD patients have at least one additional comorbid psychiatric disorder. Chapter 24 suggests that psychobiological assessment strategies may improve our capacity to distinguish PTSD from other DSM-IV diagnoses. The chapter identified those approaches currently used in PTSD research that may have potential applicability as diagnostic tools in a clinical setting.

Four chapters focus on treatment. Chapter 25 reviews the complex neurobiological, diagnostic, and therapeutic issues encountered by clinicians when PTSD is associated with alcoholism or substance abuse. Chapter 26 discusses current research on the clinical pharmacology of PTSD from a neurobiological perspective. Chapter 27 reviews cognitive behavioral therapy of PTSD. This approach, which has enjoyed considerable empirical success, can be understood in terms of classic psychological theory regarding fear conditioning and resistance to extinction. Chapter 28 provides a comprehensive review of dynamic psychotherapy for PTSD. It is clear that the complexity, severity, and clinical priorities for traumatized patients may vary considerably and require different therapeutic strategies for different patients. It is less clear how therapeutic options offered to PTSD patients might be altered after neurobiological knowledge is translated into effective treatment approaches. Finally, Chapter 29 shows that the preponderance of evidence indicates that traumatized individuals, especially those who develop PTSD, appear to be at increased risk for medical illness. These findings are consistent with what would be predicted from basic research showing excessive autonomic reactivity, decreased immunologic function, and increased somatic consequences following exposure to stress.

What is notable about this final section of our book is how many post-traumatic neurobiologic abnormalities detected by basic researchers have not been investigated with respect to clinical consequences. We hope that such gaps in the literature will be addressed as soon as possible.

Neurobiological and Clinical Consequences of Stress: From Normal Adaptation to PTSD, edited by M. J. Friedman, D. S. Charney, and A. Y. Deutch, Lippincott-Raven Publishers, Philadelphia © 1995.

23

An Epidemiologic Profile of Post-Traumatic Stress Disorder

Prevalence, Comorbidity, and Risk Factors

John A. Fairbank, William E. Schlenger, *Philip A. Saigh, and †Jonathan R. T. Davidson

*Health and Social Policy Division, Research Triangle Institute, Research Triangle Park, North Carolina 27709; *Ph.D. Program in Educational Psychology, City University of New York, New York, New York 10036; and †Department of Psychiatry and Behavioral Medicine, Duke University Medical Center, Durham, North Carolina 27710*

Epidemiology is the study of health and illness in human populations (1,2). Historically, epidemiology has focused on the frequency of illness in groups of people and the factors that affect distribution of diseases in populations (3). This includes studies of the prevalence of disorders in populations (i.e., how many people have the disorder), how a particular disorder relates to other disorders, and what factors confer increased risk for developing the disorder (4,5). In this chapter, we will examine the current epidemiologic knowledge concerning post-traumatic stress disorder (PTSD), focusing on the results of investigations conducted since 1980, because these studies largely conform to contemporary standards of PTSD diagnosis (6–8). We will successively discuss research findings on the prevalence of PTSD and exposure to extreme events, the comorbidity of PTSD with other psychiatric disorders, and risk factors for PTSD.

PREVALENCE OF PTSD

PTSD Prevalence in Adults

In epidemiologic research, the measurement of the number of cases of a given disease or other condition in a given population at a designated time is called prevalence (9). The prevalence of PTSD has been studied in the general population using nationally representative samples (10,11) and samples drawn from specific geographic areas (12,13). Prevalence of PTSD has also been ascertained in samples drawn from sectors of the health service system (14). Perhaps most frequently, rates of PTSD have been studied in samples of persons exposed to various and specific types of extreme events, such as natural disasters, war, rape, sexual assault, and other forms of criminal activity.

As shown in Table 1, estimates of the lifetime prevalence of PTSD from surveys of the general adult population range from 1.0% to 12.3%. Using data from the Epidemiologic Catchment Area (ECA) program, a National Institute of Mental Health (NIMH)-sponsored multisite collaborative research project on the epidemiology of mental disorders among adults in five metropolitan areas, Helzer et al. (12) reported a lifetime PTSD prevalence of 1.0% among adults in the St. Louis metropolitan area, and Davidson et al. (13) estimated that 1.3% of adults in the Piedmont Region of North Carolina met diagnostic criteria for PTSD. More recently, the

TABLE 1. *Selected studies of lifetime and current PTSD prevalence: General population and service-sector samples*

Investigators	Population	PTSD prevalence Lifetime[a]	Current[b]
Helzer et al. (12)	General adult population: St. Louis metropolitan area	1.0%	–
Davidson et al. (13)	General adult population: Piedmont region of North Carolina	1.3%	–
Breslau et al. (14)	Young adults from an HMO in the Detroit metropolitan area	9.2%	–
Norris (15)	Adult residents of four southern cities: Charlotte, NC, Charleston, SC, Greenville, SC, Savannah, GA.	–	5.1%–6.2%
Resnick et al. (10)	General population of adult women: United States	12.3%	4.6%
Kessler et al. (11)	General population of persons aged 15–54: United States	10.4% (women) 5.0% (men)	3.0% (women) 1.5% (men)

[a]Lifetime Prevalence: Percentage who met diagnostic criteria for PTSD at some time in their lifetime.
[b]Current Prevalence: Percentage who met diagnostic criteria for PTSD within a designated period of time before assessment (e.g., within 30 days of assessment in Norris (15); within 6 months of interview in Resnick et al. (10).
HMO, health maintenance organization; PTSD, post-traumatic stress disorder.

prevalence of PTSD was examined in a nationally representative sample of persons aged 15–54 in the National Comorbidity Survey (NCS) (11). Lifetime prevalence of PTSD in the NCS was 7.8% and current (30-day) prevalence was 2.3%. As shown in Table 1, women were more likely than men to meet diagnostic criteria for both lifetime and current PTSD. In a study of young adults randomly selected from a health maintenance organization (HMO) in the Detroit metropolitan area, Breslau, et al. (14) found a lifetime prevalence rate of 9.2% for PTSD.

Resnick, et al. (10) examined PTSD prevalence rates associated with different types of extreme events in a nationally representative community sample of women. Exposure to crime events, other trauma, and PTSD symptomatology were assessed by trained telephone interviewers using computer-assisted structured survey interviews. The overall prevalence of PTSD was 12.3% lifetime and 4.6% within the past 6 months (i.e., current). Prevalence of lifetime and current PTSD varied by type of traumatic event. Women who reported a crime event were more likely to meet criteria for lifetime (25.8%) and current (9.4%) PTSD than women who reported a noncrime event (9.7% lifetime; 3.4% current). Norris (15) examined rates of PTSD in a community sample of 1,000 men and women residing in four metropolitan areas in the southeastern United States. Respondents were asked if they had ever experienced an event in any of eight categories: 1) robbery; 2) physical assault; 3) sexual assault; 4) tragic death; 5) motor vehicle crash; 6) combat; 7) fire; or 8) other disaster. Current (past year) PTSD prevalence rates by type of exposure were estimated using an index developed from the survey instrument. Rates of current PTSD varied by type of exposure, and were highest for sexual assault (14%), physical assault (13%), and motor vehicle accident (12%), and lowest for combat (2%).

Epidemiologic studies of PTSD have most often focused on specific types of extreme events and examined the prevalence of PTSD among those subjected to that trauma, such as: 1) combat veterans (16,17,18); 2) civilian survivors of or refugees from war (19); 3) survivors of various types of natural and technological disasters (20), victims of a range of crime events, including sexual trauma (21), terrorism and torture (22), and random violence (23); and 4) firefighters and rescue and recovery workers who respond to specific disaster events (24,25). In addition, the prevalence of exposure to trauma and PTSD has been frequently examined in clinical and treatment-seeking samples (26), and special populations such as police officers (27), homeless men and women (28), incarcerated women felons (29), and male juvenile offenders (30).

The prevalence of PTSD has been more widely studied among U.S. Vietnam veterans than among any other at-risk group (see Table 2). Estimates of the prevalence of PTSD among Vietnam veterans include findings from the National Vietnam Veterans Readjustment Study (NVVRS; 18), the Centers for Disease Control (CDC) Vietnam Experience Study (VES; 16), a Department of Veterans Affairs (VA) twin study (17), the St. Louis site of the NIMH-ECA program (12), an American Legion study (31), and a study by the Traumatic Stress Study Center at the University of Cincinnati (32).

To date, the most comprehensive examination of the prevalence of PTSD among Vietnam veterans comes from the NVVRS (18), a congressionally mandated investigation of the prevalence of postwar psychological problems among Vietnam veterans (n=1,632), other veterans of the same era (n=716), and nonveterans (n=668) matched to the theater veterans on age, sex, and race/ethnicity. Cases of PTSD were identified on the basis of comprehensive multimeasure assessment that included both self-report scales and structured clinical interviews. The NVVRS findings indicated that 15.2% of men and 8.5% of women who served in Vietnam were assigned diagnoses of current PTSD (i.e., met the DSM-III-R criteria in the 6 months prior to the interview) at the time the study was conducted. By contrast, the current prevalence among comparable-era veterans was 2.5% among men and 1.1% among women; for nonveterans, prevalence was 1.2% among men and 0.3% among women. Lifetime prevalence rates among Vietnam veterans were 30.9% for men and 26.9% for women.

Other epidemiologic studies of PTSD among Vietnam veterans have focused on smaller subsets of the population. For example, in the CDC Vietnam Experience Study (16), which was one component of a comprehensive, congressionally mandated study of the health effects of Vietnam service, researchers studied a random sample (n=2,490) of Army enlisted men who had served one tour of duty in Vietnam. Using a survey interview as the basis for identifying cases of PTSD, the CDC study team estimated a lifetime prevalence of combat-related PTSD of 14.7% and a current (past month) prevalence of 2.2%. Similarly, in the VA twin study (17), a sample of 2,092 male, monozygotic twin pairs drawn from the Vietnam Era Twin Registry was surveyed via mail and telephone interviews. The current prevalence of PTSD among the twins who served in Vietnam was 16.8%, compared to 5.0% among those who served elsewhere.

In the St. Louis site of the NIMH-ECA study, Helzer et al. (12) studied 64 self-identified Vietnam veterans from a general population sample. The PTSD diagnoses were based on survey interviews. Findings suggested a lifetime prevalence of combat-related PTSD of 6.3%. Keane and Wolfe (26) have noted that Helzer later reported that the weighted prevalence rate for PTSD in Vietnam veterans in the St. Louis ECA survey was 11.8%. In the American Legion study (31), a sample (n=2,858) of American Legion members in six states was surveyed by mail. Results suggested a current PTSD prevalence ranging from 1.8% to 15%, depending on how exposure to combat was operationalized. In the Cincinnati study (32), a sample of Vietnam veterans (n=200) recruited from a variety of sources was assessed for PTSD by trained clinicians using a structured clinical interview. Current PTSD prevalence among those veterans who were not seeking treatment was estimated to be 17%.

In earlier reviews it has been recognized that, in spite of large variations in methodology between the major epidemiological studies of PTSD among Vietnam veterans, the findings are surprisingly consistent (4,33,34). The estimates of the majority of studies lie within the 95% confidence interval of the NVVRS estimates (13.0%–17.4%).

Several recent studies have examined PTSD prevalence among other populations of survivors of war in Southeast Asia. O'Toole and colleagues (35) estimated the prevalence of PTSD

TABLE 2. *Selected studies of PTSD prevalence: Groups of at-risk adults*

Investigators	Group	PTSD prevalence	
		Lifetime	Current
Helzer et al. (12)	Vietnam veterans	6.3%	–
Kilpatrick et al. (21)	Crime victims	–	7.5%
CDC (16)	Vietnam veterans	14.7%	2.2%
Shore et al. (42)	Residents of a community exposed to volcanic eruption	3.6%	–
Goldberg et al. (17)	Vietnam veterans	–	16.8%
Green et al. (32)	Vietnam veterans	–	17.0%
Green et al. (32)	Flood survivors	–	28.0%
Kulka et al. (33)	Vietnam veterans	30.9% (men) 26.9% (women)	15.2% (men) 8.5% (women)
Winfield et al. (38)	Sexual assault survivors	–	3.7%
McFarlane and Papay (25)	Volunteer firefighters exposed to a natural disaster–Australia	–	18.0%
Norris (15)	Survivors of crime, hazards, accidents, and combat	–	Crime–8.5% Hazard–5.8% Accident–8.6% Combat–2.2%
North et al. (23)	Survivors of a mass shooting	–	20.0% (men) 36.0% (women) (one month after the event)
O'Toole et al. (35)	Australian Vietnam veterans	20.9%	11.6%

PTSD, post-traumatic stress disorder.

in a national sample (n=641) of Australian veterans of the Vietnam war. Findings indicated a lifetime PTSD prevalence ranging from 11.7% to 20.9%, depending upon how exposure to trauma and PTSD was assessed. The current (1-month) estimate of PTSD prevalence among Australian Vietnam veterans was 11.6%, as determined by structured clinical interviews. Hinton et al. (19) assessed 201 newly arrived refugees from Vietnam for PTSD and other psychiatric disorders using structured clinical interviews that were translated from English into Vietnamese. These investigators reported that 3.5% of the sample met DSM-III-R criteria for PTSD, with rates higher among ethnic Vietnamese (7.2%) than among ethnic Chinese (0.9%). Prevalence of psychiatric disorder was highest among former soldiers in the Army of the Republic of Vietnam and reeducation camp survivors. Findings of studies on the prevalence of PTSD among young Cambodian refugees are presented in a section of this chapter titled "PTSD Prevalence in Children and Adolescents."

Epidemiologic studies of PTSD that result from exposure to extreme events other than war have also been conducted. These include studies of persons who have been exposed to various forms of criminal victimization (10). An influential study in this area is Burgess and Holstrom's (36) seminal study of the sequelae of rape. Based on interviews with 146 rape victims, Burgess and Holstrom described in detail "rape trauma syndrome," a precursor to PTSD. This included identification and description of the acute phase and a chronic phase of the syndrome.

Following along these lines, Kilpatrick and colleagues have conducted a series of studies of the prevalence and consequences of criminal victimization. For example, Kilpatrick and colleagues (37) studied symptoms in samples of rape victims and controls. Kilpatrick and associates (21) reported findings of a community survey of victimization and symptomatology among women. Findings indicated that about three quarters of the women had been victims, and that 7.5% were characterized by current

PTSD. Winfield et al. (38), using data collected at the North Carolina site of the NIMH-ECA study, found that about 7% of the women aged 18–44 reported a history of sexual assault. A positive history of sexual assault was found to be a risk factor for a variety of psychiatric disorders, including PTSD. Current PTSD prevalence among those with a positive history was estimated to be 3.7%, compared to less than 1% for those with negative history.

Epidemiologic studies have also been conducted in the aftermath of a number of disasters. For example, Green and her colleagues (39,40) have studied survivors of the Buffalo Creek dam collapse. Most recently, this team reported that, 14 years after the event, 28% of the sample followed longitudinally met the criteria for current PTSD. This represented a decrease from the 44% current prevalence estimated at 2 years postexposure. In a similar vein, Smith et al. (41) recently reported findings concerning the sequelae of an event in which a plane crashed into a hotel. From a sample of 46 hotel employees, only some (n=17) of whom were at the site at the time of the crash, it was estimated that 29% of those on-site and 17% of those off-site developed PTSD following the event.

A report by Shore and colleagues (42) addressed PTSD in the population exposed to the Mount St. Helens (MSH) volcanic eruption, as well as a control group that was unexposed to the effects of this natural disaster. Base lifetime prevalence rates were 3.6% in the MSH group and 2.6% in the control group, with a 3.1% overall lifetime prevalence. A study by McFarlane (43) assessed PTSD among a sample of 315 fire fighters of an Australian brushfire. Prevalence rates of 32%, 27%, and 30% were obtained at 4, 11, and 29 months postexposure, respectively.

The examples just mentioned are from a rapidly growing literature that includes studies of the psychological impact of the release of radiation at Three-Mile Island (44), the Beverly Hills Supper Club fire (45), the Lockerbie (Scotland) aircraft disaster (46), and the relationship between PTSD and chronic exposure to community violence (47), to name a few. Clearly, the existing literature demonstrates that a variety of extreme events precipitate PTSD reactions in at least a portion of exposed individuals.

PTSD Prevalence in Children and Adolescents

Few epidemiological studies have specifically examined the general population prevalence of PTSD in children and adolescents (48). However, as shown in Table 3, a number of studies have examined the frequency of PTSD in groups of children and adolescents exposed to specific traumatic events. These studies generally involve youth exposed to war stress, criminal victimization, and natural disasters.

Saigh and his colleagues (49,50) have examined the prevalence of PTSD among children and adolescents exposed to war. In one study (49), 840 children (aged 8–12 years) who were referred for psychological evaluation were assessed for PTSD at approximately 1–2 years following exposure to war in Lebanon. Nearly one third (32.5%) of the children met diagnostic criteria for PTSD. Approximately 5 years after the war, Saigh and Mrouegh (50) assessed 95 adolescents from the Beirut metropolitan area and found that 14.7% of these youths met diagnostic criteria for current PTSD.

Sack and colleagues (51) interviewed over 200 Cambodian youths who had lived through the genocide perpetrated by the notorious Pol Pot regime, fled Cambodia, and subsequently resided in the United States. They reported a current PTSD prevalence rate of 18.2% and a lifetime rate of 21.5% for these youths. In an earlier study, Kinzie et al. (52) observed that 50% of a sample of 40 Cambodian refugee youths met criteria for PTSD approximately 2.5 years after emigration to the United States. Savin et al. (53) have also reported high rates of current (26.3%) and lifetime (31.3%) PTSD among Cambodian youths resettled in refugee camps in Thailand.

Research following the 1991 war in the Persian Gulf area has also focused on child and adolescent PTSD. Five months after Iraq's military forces left Kuwait, Nader and associates (54) assessed 51 Kuwaiti children, adolescents,

TABLE 3. *Selected studies of PTSD prevalence: Groups of at-risk children and adolescents*

Investigators	Sample	PTSD prevalence	
		Lifetime	Current
Kinzie et al. (52)	Cambodian refugee children	–	50.0%
Pynoos et al. (56)	Children exposed to sniper attack on schoolyard	–	58.4%
McLeer et al. (60)	Sexually abused children	–	48.0%
Saigh (49)	Children exposed to war in Lebanon	–	32.5%
Schwartz & Kowalski (57)	Children exposed to a school shooting	–	27.0%
Garrison et al. (58)	Adolescents exposed to Hurricane Hugo	–	1.5%–6.2%
Saigh and Mrouegh (50)	Adolescents exposed to war in Lebanon	–	14.7%
Savin et al. (53)	Cambodian refugees in camps in Thailand	31.3%	26.3%
Weisenberg et al. (55)	Israeli children exposed to missile attacks	–	26.6%
Green et al. (59)	17-yr follow-up of children who survived major flood	32.0%	6.2%
Sack et al. (51)	Cambodian refugees resettled in United States	21.5%	18.2%

PTSD, post-traumatic stress disorder.

and young adults (ages 8–21 years) and reported that 70% of them met criteria for PTSD. Also within the context of the Gulf War, Weisenberg et al. (55) evaluated 492 Israeli children enrolled in grades 5 through 10 for symptoms of PTSD. The children in this study had remained with their families in hermetically sealed rooms and worn gas masks during missile attacks. Based on findings from assessments conducted 3 weeks after the war, 26.6% of the sample met criteria for PTSD.

In a study of children exposed to a sniper attack, Pynoos and associates (56) observed a 58.4% prevalence of acute PTSD 1 month after the attack. Their sample of 159 children was obtained at random from a total school population of about 1,000. Schwarz and Kowalski (57) assessed 64 children exposed to a school shooting and found that 27% of the children met DSM-III-R criteria for PTSD 6 to 14 months after this event, when a moderate level of symptom intensity was required for making the diagnosis.

Garrison et al. (58) have recently reported findings from a school-based study of the frequency of PTSD among adolescents (aged 11–17 years) 1 year after exposure to Hurricane Hugo. Rates of PTSD were estimated by race (white and black) and gender. The PTSD rates were highest for white females (6.2%), followed by black females (4.7%) and white males (3.8%). Rates were lowest for black males (1.5%). A 17-year follow-up study of children and adolescents who survived the floods associated with the collapse of a dam at Buffalo Creek, West Virginia, found that 6.2% met criteria for current PTSD and 32% met criteria for lifetime PTSD (59).

The prevalence of PTSD has also been studied in treatment samples of children who were physically or sexually abused. McLeer et al. (60) reported that 48% of a sample of sexually abused children who were receiving treatment in an outpatient psychiatric clinic met criteria for PTSD. Adams et al. (61) found that PTSD prevalence varied by type of abuse. Among children (aged 4–12 years) receiving treatment in an outpatient clinic, approximately 20% of those who were physically abused, and 43% of those who were sexually abused, met criteria for PTSD. The PTSD prevalence among children who were both physically and sexually abused was 29%.

Prevalence of Exposure to Extreme Events

In addition to studying the prevalence of disorder among those exposed to trauma, the prevalence of exposure itself is an important issue. In a comprehensive review of the disaster literature, Green (20) cited a report by the International Federation of Red Cross and Red Crescent Societies that attests to the startling frequency, mortality, and morbidity of natural and technological disasters worldwide. From 1967 to 1991, a total of 7,766 disasters were reported throughout the world, killing over 7 million people and adversely affecting many times more (20). In the NCS, 60.7% of men and 51.2% of women reported exposure to one or more traumatic events, with 19.7% of the men and 11.4% of the women reporting three or more traumatic events (11).

In their national survey of exposure to crime among adult women in the United States, Resnick et al. (10) found that 12.7% reported a completed rape, 14.3% reported molestation or attempted sexual assaults, 10.3% had been physically assaulted, 13.4% reported the death of a close friend or relative by homicide, and 35.6% reported that at some time in their life they had been the victim of any type of crime. Of those exposed to some type of crime, more than one half (51.8%) of the sample reported that they had experienced more than one type of crime or multiple episodes of the same type of crime.

Orner (62) reviewed the history of armed conflicts involving European countries since 1918 and listed over 60 instances of wars between countries, civil wars, episodes of terrorism, and military interventions. Orner (62) notes that although there exist no reliable, empirically based estimates on the prevalence of exposure to war trauma among veterans and civilians in these countries, the prevalence of exposure to war trauma in Europe alone is likely to be extremely high. With respect to exposure to war trauma among American veterans of the Vietnam War, Kulka et al. (18) found that 64.2% of the men who served in Vietnam had been exposed during their lifetime to one or more probable or definite traumatic events, compared to 47.8% of men who served in the military but not in Vietnam and 44.5% of the men who did not serve in the military. Estimates of lifetime exposure among women were 70.8% of those who served in Vietnam, 49.5% of other women veterans, and 37.2% of the nonveteran women.

Breslau et al. (14) found that 39.1% of a sample of young adult (aged 21–30) members of a health maintenance organization in Detroit, Michigan, had been exposed to traumatic events during their lifetimes. Chu and Dill (63) found that 66% of a sample of female psychiatric inpatients had a history of childhood sexual and/or physical abuse. Similarly, Bryer and colleagues (64) found that three quarters of a sample of female psychiatric inpatients had a lifetime history of physical and/or sexual abuse.

Additional study of community and clinical populations is needed to document exposure levels and improve our understanding of why only some of those exposed to serious trauma actually develop PTSD or other stress-related symptoms. What is clear, however, is that exposure to a vast array of traumatic events is more prevalent than was perhaps once thought to be the case.

COMORBIDITY OF PTSD AND OTHER DISORDERS

In psychiatric epidemiology, comorbidity generally refers to the simultaneous occurrence of two or more mental disorders in one patient (65). As shown in Table 4, both clinical and community studies have demonstrated that people who have PTSD frequently meet the criteria for other psychiatric disorders as well. Sierles and his colleagues at the North Chicago Veterans Administration (VA) Medical Center studied comorbidities in small samples of Vietnam veterans with PTSD being treated as inpatients (66) and outpatients (67). In both groups, more than 80% of patients were found to have one or more comorbid psychiatric disorders. Similarly, Keane and Wolfe (26) studied patients in the Boston VA Medical Center's PTSD program, and found high rates of comorbid substance abuse, major depression, dysthymic disorder, and antisocial personality disorder. An analysis of national hospital discharge data for 539,557 veterans treated in medical, surgical, psychiatric, and substance abuse treatment wards in Department of Veterans Affairs (DVA) Medical Centers during 1991 indicates that 6.2% were identified as having co-occurring PTSD and substance abuse or de pendence (68).

Community studies have also documented high rates of comorbid psychiatric disorder. In the NCA, Kessler et al. (11) reported that 79% of the women and 88.3% of the men with PTSD

TABLE 4. *Selected studies on the comorbidity of PTSD with other disorders*

Investigators	Study groups	Findings
Sierles et al. (66)	Inpatient sample of Vietnam veterans	In both studies, more than 80% of veterans with PTSD had one or more comorbid psychiatric disorder
Sierles et al. (67)	Outpatient sample of Vietnam veterans	
Helzer et al. (12)	Sample of general adult population in St. Louis metropolitan area	Nearly 80% of interviewees with PTSD met criteria for another current or lifetime psychiatric disorder
Kulka et al. (33)	National sample of Vietnam veterans	Most frequent comorbid psychiatric disorders were; for men—alcohol abuse or dependence, generalized anxiety disorder, and major depression; for women—major depression, generalized anxiety disorder, and dysthymia
Keane and Wolfe (26)	Treatment-seeking veterans at Boston VA Medical Center	High rates of comorbid substance abuse (84%), major depression (68%), dysthymia (34%), and antisocial personality disorder (26%)
Davidson et al. (13)	Sample of general adult population in Piedmont Region of North Carolina	62% of individuals with PTSD had a lifetime comorbidity of another psychiatric disorder
Breslau et al. (14)	Young adults enrolled in an HMO in the Detroit metropolitan area	83% of individuals with PTSD met criteria for another psychiatric disorder
McFarland and Papay (25)	Sample of firefighters who had been exposed to a natural disaster in Australia	77% of individuals with PTSD met criteria for another psychiatric disorder, with major depression the most common (51%)
Kessler et al. (11)	National sample of persons aged 15–54	88.3% of men and 79% of women with lifetime PTSD have a history of another psychiatric disorder

PTSD, post-traumatic stress disorder.

also met criteria for another lifetime psychiatric disorder. In contrast, 46.2% of women and 54.8% of men with disorders other than PTSD were found to have comorbid psychiatric disorders. The disorders most prevalent for men with PTSD were alcohol abuse or dependence (51.9%), major depressive episode (47.9%), conduct disorder (43.3%), and drug abuse and dependence (34.5%). The disorders most comorbid with PTSD among women were major depressive episode (48.5%), simple phobia (29.0%), social phobia (28.4%), and alcohol abuse/dependence (27.9%).

In the St. Louis ECA study (12), almost 80% of full PTSD cases had a previous or concurrent psychiatric disorder, compared to about one third of respondents with no post-traumatic stress symptoms. In the North Carolina ECA study (13), 62% of individuals with PTSD had a lifetime comorbidity of a psychiatric illness as compared to only 15% of the control group. Ninety-two percent of the Mount St. Helens PTSD group had co-occurring disorders. McFarlane and Papay (25) investigated the comorbidity of PTSD, other anxiety disorders, and affective disorders in a community sample of fire fighters, 42 months after exposure to a natural disaster. Co-occurring major depressive disorder was assessed in 51%, generalized anxiety disorder in 39%, panic disorder in 33%, obsessive-compulsive disorder in 13%, and manic episode in 8% of firefighters with PTSD, respectively.

The prevalence of co-occurring PTSD and substance use disorders has also been examined in community populations using St. Louis ECA data (69). The ECA respondents were hierarchically classified into one of four substance-use categories: 1) cocaine/opiates (n=144); 2) pills/hallucinogens (n=163); 3) marijuana only (n=308); and 4) heavy alcohol use (n=388). Rates of exposure to traumatic events were found to vary by drug of choice, with 43% of the cocaine/opiate users reporting exposure to a traumatic event, compared to 23% of the pill/hallucinogen users, 18% of the marijuana users, 16% of the problem drinkers, and 13% of a group of nonuser comparison subjects. Nearly one fifth of the cocaine/opiate users also met criteria for PTSD, and the overall rate of PTSD was higher among substance users than comparison subjects. These findings are clearly consistent with earlier analyses of ECA data indicating that people with alcohol and drug use problems are at very high risk for co-occurring psychological disorders (70).

In the NVVRS, Kulka et al. (18) found that virtually all Vietnam veterans with PTSD met the criteria for one or more other psychiatric disorders (including substance abuse; three quarters if alcohol disorders are excluded) at some time during their lives, and half were characterized by a current comorbid disorder. In men, the most prevalent comorbid disorders were alcohol abuse or dependence (75% lifetime, 20% current), generalized anxiety disorder (44% lifetime, 20% current), and major depression (20% lifetime, 16% current). Among women, the most frequently occurring were major depression (42% lifetime, 23% current), generalized anxiety disorder (38% lifetime, 20% current), and dysthymic disorder (33% lifetime).

In addition to psychiatric comorbidity, physical health comorbidities and co-occurring life adjustment problems have also been studied. The CDC (16) studied the physical health of Vietnam theater and Vietnam-era veterans as part of its Vietnam Experience Study (VES). Although Vietnam veterans were found to rate their health more poorly and to report more physical health problems in an interview, physical examination findings showed few health differences. In the NVVRS, Kulka et al. (18) found no differences between Vietnam theater veterans and Vietnam-era veterans in self-evaluations of health status or number of reported health problems, but did find that Vietnam veterans with PTSD reported more health problems and rated their health more poorly than did Vietnam-era veteran or nonveteran controls. Shalev and colleagues (71) similarly found that Israeli veterans with PTSD reported more somatic symptoms than did matched controls, but that the groups did not differ on laboratory test findings. This study also showed a greater frequency of adverse health practices (e.g., smoking, alcohol use) among veterans with PTSD.

The NVVRS (18,34,72) provided a comprehensive examination of the association of PTSD with other life adjustment problems. Although Vietnam theater veterans as a group were found not to differ substantially from Vietnam-era veterans or nonveteran controls in current life adjustment, Vietnam veterans with PTSD were found to have profound and pervasive problems in their daily lives. This included problems in family and other interpersonal relationships, and employment, as well as involvement with the criminal justice system. Thus, PTSD is associated with high levels of psychological distress and with pervasive disruptions in life functioning.

Findings from the North Carolina ECA study show an association between lifetime PTSD and increased rates of attempted suicide, which is not explained by comorbid depression (13). The same study also showed an increased lifetime prevalence of hypertension, bronchial asthma, and peptic ulcer in association with PTSD. It is not unreasonable to speculate that individuals with PTSD may make increased use of health-care facilities over and above their use of counseling, psychotherapeutic, and psychopharmacologic services.

RISK FACTORS FOR PTSD

Epidemiologic studies are empirical sources of information about risk factors for PTSD: variables that are believed to be related to the probability of an individual developing this disorder. A number of studies have examined risk factors and predictors of PTSD in at-risk groups and community samples. Some studies have found evidence of sociodemographic and pretrauma psychosocial risk factors in the development of PTSD. Such factors include: 1) familial psychiatric illness; 2) parental poverty; 3) childhood trauma (e.g., sexual assault, separated or divorced parents before age 10) (13); 4) childhood behavior disorder (12); 5) neuroticism; 6) introversion; 7) prior psychiatric disorder; 8) adverse life events before and after the trauma (43); 9) being female and between ages 36 and 50; 10) being concerned about finances; and 11) having prior physical health problems (73). For example, in their national general population study of persons in the United States aged 15–54, Kessler and colleagues (11) found that the lifetime prevalence of PTSD is highest among women, the previously married, and persons of lower socioeconomic status.

As was the case with prevalence, Vietnam veterans have been scrutinized most closely for PTSD risk factors. The contributions of a comprehensive set of war trauma and prewar characteristics to the prevalence of PTSD were examined in detail in the NVVRS (18). Findings of multivariate analyses of a set of over 80 preexposure characteristics that might have rendered a person more vulnerable to the development of PTSD indicated a relatively small but statistically significant role for such characteristics, particularly: 1) those related to the person's socioeconomic circumstances while growing up; 2) the existence of psychiatric symptoms prior to exposure; and 3) reported childhood abuse. Additional analyses, however, indicated a much larger role for characteristics of the stressor. These NVVRS findings complement and are consistent with the findings of the VA twin study (17), in which potential genetic contributions were controlled by comparing the prevalence of PTSD in male monozygotic twins who served in Vietnam with their identical siblings who served in the military but not in Vietnam. Both studies showed a strong dose-response relationship between exposure to war zone stress and current prevalence of PTSD.

In the North Carolina ECA study (38), the prevalence of PTSD in victims of sexual assault who were physically injured during the assault (14.1%) was 22 times higher than the prevalence of PTSD in noninjured victims of sexual assault (0.64%). In studying those exposed to the Mount St. Helens (MSH) eruption, Shore et al. (74) found a strong dose-response relationship between proximity to the mountain and the contribution of exposure. The MSH study showed that the three MSH stress-related disorders (of which PTSD was one) increased in frequency with the victim's proximity to the eruption. First year postdisaster onset rates of MSH stress-related disorders for males were 0.9%, 2.5%, and 11.1% in nonexposed controls, low-exposure populations, and high-exposure populations, respectively. Onset rates for females in these groups were 1.9%, 5.6%, and 20.9%, respectively. Unfortunately, data were unavailable for PTSD alone. However, in his study of Australian fire fighters, McFarlane (43,75) found that the relative contribution of background versus exposure factors changed over time, with the contributions of the stressor seeming to diminish with time.

In a report by Saigh (76), the descriptive traumatic experiences of 230 children (aged 9 years–12 years, 11 months) who met criteria for PTSD were classified along a dimension of exposure to war in Lebanon. Of these PTSD cases, 58 children were found to have been directly exposed to war trauma, 128 youths observed others in life-threatening situations, 13 heard about the fates of relatives or friends who were casualties of war, and 31 children described experiences that were combinations of these types of events.

Characteristics of the postexposure environment have also been shown to be related to postexposure adjustment. Several studies have described low levels of social support and dysfunctional patterns of social interaction in Vietnam veterans with PTSD relative to comparison groups of veterans without PTSD (72,77). Similarly, studies of rape survivors have demonstrated a positive relationship between postexposure social support and subsequent adjustment. A consistent finding in the literature concerning predictions of postrape functioning has been the positive association between social support and adjustment, in terms of both immediate impact and latency of recovery (78–80).

Although the studies to date only begin to answer the many etiologic questions of interest, the bulk of the evidence is consistent with an etiologic model of PTSD that posits a role for individual vulnerability (potentially including biological, psychological, and sociocultural factors), a role for characteristics of the stressor, and a role for postexposure variables (81, 82). It remains for further research to specify the relative contributions of these three major factors, as well as ways in which they may interact.

SUMMARY

The relevance of epidemiologic studies lies largely in providing data for treatment and public policy planning. The studies discussed in this chapter suggest that PTSD is an important and continuing public health problem, especially among at-risk groups such as survivors of war and criminal activity. Epidemiologic studies have demonstrated that although many kinds of extreme events can cause PTSD (e.g., exposure to combat, criminal victimization, natural disasters), not all of those exposed will develop the disorder. Evidence suggests that exposure to extreme events may be more common than once thought, and that risk factors include characteristics of the person at the time of exposure to the extreme event, characteristics of the event itself, and characteristics of the post-exposure environment.

As noted by Costello and Angold (83), one of the practical goals of epidemiologic research is to "calculate the importance of a risk factor in contributing to the prevalence of a disease" (page 22). In principle, each risk factor can be ranked in terms of its relative importance to PTSD prevalence in a given population (e.g., Vietnam veterans, survivors of natural disaster). To achieve this goal, however, studies are needed that focus on determining the direct and indirect contributions of trauma exposure to rates of PTSD in specific populations through the psychological, social/environmental, and neurobiologic mechanisms comprehensively discussed in the other chapters of this book. In the current debate on health care reform, more detailed studies of risk factors for exposure and subsequent development of PTSD are needed to inform health policy planners and treatment providers about potential strategies for cost-effective prevention and treatment.

REFERENCES

1. Friedman GD. *Primer of epidemiology,* 3rd ed. New York: McGraw-Hill; 1987.
2. Kleinbaum DG, Kupper LL, Morgenstein H. *Epidemiologic research: principles and quantitative methods.* New York: Van Nostrand Reinhold; 1982.
3. Mausner JS, Kramer S. *Mausner & Bahn: Epidemiology—an introductory text.* Philadelphia: Saunders; 1985.
4. Davidson JRT, Fairbank JA. The epidemiology of posttraumatic stress disorder: In Davidson JRT, Foa EB, eds. *Posttraumatic stress disorder. DSM-IV and beyond.* Washington, DC: American Psychiatric Press; 1993.
5. Fairbank JA, Schlenger WE, Caddell JM, Woods GM. Post-traumatic stress disorder. In Sutker PB, Adams HE, eds. *Comprehensive handbook of psychopathology,* 2nd ed. New York: Plenum Press, 1993:145–165.
6. American Psychiatric Association. *Diagnostic and statistical manual of mental disorders,* 3rd ed. Washington, DC: Author, 1980.
7. American Psychiatric Association. *Diagnostic and statistical manual of mental disorders,* 3rd ed., rev. Washington, DC: Author, 1987.
8. American Psychiatric Association. *Diagnostic and statistical manual of mental disorders,* Fourth ed. Washington, DC: Author, 1994.
9. Last JM. *A dictionary of epidemiology.* New York: Oxford University Press; 1983.
10. Resnick HS, Kilpatrick DG, Dansky BS, Saunders BE, Best CL. Prevalence of civilian trauma and posttraumatic stress disorder in a representative national sample of women. *J Consult Clin Psychol* 1993;61:984–991.
11. Kessler RC, Sonnega A, Bromet E, Hughes M, Nelson CB. Posttraumatic stress disorder in the National Comorbidity Survey. *Arch Gen Psy,* in press.
12. Helzer JE, Robins LN, McEvoy L. Posttraumatic stress disorder in the general population: findings of the Epidemiologic Catchment Area survey. *N Engl J Med* 1987; 317:1630–1634.
13. Davidson JRT, Hughes D, Blazer DG, George LK. Posttraumatic stress disorder in the community: an epidemiological study. *Psychol Med* 1991;21:713–721.
14. Breslau N, Davis GC, Andreski P, Peterson E. Traumatic events and posttraumatic stress disorder in an urban population of young adults. *Arch Gen Psy* 1991;48: 216–222.
15. Norris FH. Epidemiology of trauma: frequency and impact of different potentially traumatic events on different demographic groups. *J Consult Clin Psychol* 1992;60: 409–418.
16. Centers for Disease Control Vietnam Experience Study. Health status of Vietnam veterans: I. Psychosocial characteristics. *JAMA* 1988;259:2701–2707.
17. Goldberg J, True WR, Eisen SA, Henderson WC. A twin study of the effects of the Vietnam War on posttraumatic stress disorder. *JAMA* 1990;263:1227–1232.
18. Kulka RA, Schlenger WE, Fairbank JA, Hough RL, Jordan BK, Marmar CR, Weiss D. *Trauma and the Vietnam War generation: report of findings from the National Vietnam Veterans Readjustment Study.* New York: Brunner/Mazel; 1990.
19. Hinton WL, Chen YJ, Du N, Tran CG, Lu FG, Miranda J. DSM-III-R disorders in Vietnamese refugees: preva-

lence and correlates. *J Nerv Ment Dis* 1993;181:113–122.

20. Green, BL. Traumatic stress and disaster: mental health effects and factors influencing adaptation. In Davidson JRT, McFarlane AC, eds. *International Review of Psychiatry* vol. 2, in press.
21. Kilpatrick DG, Saunders BE, Veronen LJ, Best CL, Von JM. Criminal victimization: lifetime prevalence, reporting to police, and psychological impact. *Crime & Delinquency* 1987;33:479–489.
22. Abenhaim L, Dab W, Salmi L. Study of civilian victims of terrorist attacks. *J Clin Epidemiol* 1992;45:103–109.
23. North CS, Smith EM, Spitznagel EL. Posttraumatic stress disorder in survivors of a mass shooting. *Am J Psychiatry* 1994;151:82–88.
24. Caddell J, DeRosa R. An examination of the etiology and development of PTSD in disaster survivors. Presented at International Society for Traumatic Stress Studies, 9th Annual Meeting, October 1993.
25. McFarlane AC, Papay P. Multiple diagnoses in posttraumatic stress disorder in the victims of a natural disaster. *J Nerv Ment Dis* 1992;180:498–504.
26. Keane TM, Wolfe J. Comorbidity in post-traumatic stress disorder: an analysis of community and clinical studies. *J Appl Psychol* 1990;20:1776–1788.
27. Martin CA, McKean HE, Veltkamp LJ. Post-traumatic stress disorder in police and working with victims: a pilot study. *J Police Sci Admin* 1986;14:98–101.
28. North CS, Smith EM. Posttraumatic stress disorder among homeless men and women. *Hosp Community Psychiatry* 1992;43:1010–1016.
29. Jordan BK, Schlenger WE, Fairbank JA, Caddell JM, Kristiansen P, Stutts E. The Women Inmate's Health Study: An Overview. Presented at American Public Health Association annual meeting, November 1992.
30. Burton D, Foy D, Bwanausi C, Johnson J, Moore L. The relationship between traumatic exposure, family dysfunction, and post-traumatic stress symptoms in male juvenile offenders. *J Traumatic Stress* 1994;7:83–93.
31. Snow BR, Stellman JM, Stellman SD, Sommer JF. Posttraumatic stress disorder among American Legionnaires in relation to combat experience in Vietnam: Associated and contributing factors. *Environ Res* 1988;47:175–192.
32. Green BL, Grace MC, Lindy JD, Gleser GC, Leonard A. Risk factors for PTSD and other diagnoses in a general sample of Vietnam veterans. *Am J Psychiatry* 1990;147:729–733.
33. Kulka RA, Schlenger WE, Fairbank JA, Jordan BK, Hough RL, Marmar CR, et al. Assessment of PTSD in the community: prospects and pitfalls from recent studies of Vietnam veterans. *Psychol Assessment* 1991;3:547–560.
34. Schlenger WE, Kulka RA, Fairbank JA, Hough RL, Jordan BK, Marmar CR, Weiss DS. The prevalence of post-traumatic stress disorder in the Vietnam generation: a multimethod, multisource assessment of psychiatric disorder. *J Traumatic Stress* 1992;5:333–363.
35. O'Toole BI, Marshall RP, Grayson DA, Schureck RJ, Dobson M, Ffrench M, Pulvertaft B, Meldrum L, Bolton J, Vennard J. *The Australian Vietnam Veterans Health Study. III. Psychological Health of Australian Vietnam Veterans and the Relationship to Combat.* 1994; unpublished manuscript.
36. Burgess AW, Holmstrom LL. *Rape: victims of crisis.* Bowie, MD: Robert J. Brady; 1974.
37. Kilpatrick DG, Veronen LJ, Resick PA. Assessment of the aftermath of rape: changing patterns of fear. *J Behav Assessment* 1979;1:133–148.
38. Winfield I, George LK, Swartz M, Blazer DG. Sexual assault and psychiatric disorders among a community sample of women. *Am J Psychiatry* 1990;147:335–341.
39. Green BL, Lindy JD, Grace MC, Gleser GL, Leonard AC, Korol M, et al. Buffalo Creek survivors in the second decade: stability of stress symptoms. *Am J Orthopsy* 1990;60:43–54.
40. Green BL, Grace MC, Lindy JD, Gleser GC, Leonard AC, Kramer TL. Buffalo Creek survivors in the second decade: comparison with unexposed and nonlitigant groups. *J Appl Soc Psy* 1990;20:1033–1050.
41. Smith EM, North CS, McCool RE, Shea JM. Acute postdisaster psychiatric disorders: identification of persons at risk. *Am J Psychiatry* 1990;147:202–206.
42. Shore JH, Vollmer WM, Tatum EL. Community patterns of post traumatic stress disorders. *J Nerv Ment Dis* 1989;177:681–685.
43. McFarlane AC. The aetiology of post-traumatic morbidity: predisposing, precipitating and perpetuating factors. *British J Psy* 1989;154:221–228.
44. Bromet EJ, Parkinson DK, Schulberg HC, Dunn LO, Gondek PC. Mental health of residents near the Three Mile Island reactor: a comparative study of selected groups. *J Prev Psy* 1982;1:225–276.
45. Green BL, Grace MC, Lindy JD, Titchener JL, Lindy JG. Levels of functional impairment following a civilian disaster: the Beverly Hills supper club fire. *J Consult Clin Psychol* 1983;51:573–580.
46. Brooks N, McKinlay W. Mental health consequences of the Lockerbie disaster. *J Traumatic Stress* 1992;5:527–543.
47. Fitzpatrick KM, Boldizar JP. The prevalence and consequences of exposure to violence among African-American youth. *J Am Acad Child Adolesc Psychiatry* 1993;32:424–430.
48. March JS, Amaya-Jackson L. Post-traumatic stress disorder in children and adolescents. *PTSD Research Quarterly* 1993;4:1–8.
49. Saigh PA. The validity of the DSM-III posttraumatic stress disorder classification as applied to children. *J Abnorm Psychol* 1989;97:338–342.
50. Saigh PA, Mrouegh A. Scholastic impairments among traumatized adolescents. Presented at the Lake George Research Conference on PTSD. Bolton Landing, NY; January 1993.
51. Sack WH, McSharry S, Kinney R, Seely J, Lewinsohn P. The Khmer Adolescent Project: I. Epidemiologic findings in two generations of Cambodian refugees. *J Nerv Ment Dis* 1994;182:387–395.
52. Kinzie JD, Sack WH, Angell RH, Mason SM. The psychiatric effects of massive trauma on Cambodian children. *J Am Acad Child Adolesce Psychiatry* 1986;25:370–376.
53. Savin D, Sack W, Clarke G, Nee M, Richert I. The Khmer Adolescent Project: II. A study of trauma from Thailand's Site Two refugee camp. 1993; unpublished manuscript.
54. Nader K, Pynoos R, Fairbanks L, Frederick C, Al-Ajeel M, Al-Alsfour A. A preliminary study of PTSD and grief among the children of Kuwait following the Gulf crisis. *British J Clin Psychol,* 1993;32:407–416.
55. Weisenberg M, Schwarzwald J, Waysman M, Solomon

Z, Klingman A. Coping of school-age children in the sealed room during SCUD missile bombardment and postwar stress reactions. *J Consult Clin Psychol* 1993; 61:462–467.
56. Pynoos RS, Frederick C, Nader K, Arroyo W, Steinberg A, Eth S, et al. Life threat and posttraumatic stress in school children. *Arch Gen Psy* 1987;44:1057–1063.
57. Schwarz ED, Kowalski JM. Malignant memories: PTSD in children and adults after a school shooting. *J Am Acad Child Adolesc Psychiatry* 1991;30:936–944.
58. Garrison CZ, Weinrich MW, Hardin SB, Weinrich S, Wang L. Post-traumatic stress disorder in adolescents after a hurricane. *Am J of Epidemiology* 1993;138:522–530.
59. Green BL, Grace MC, Vary MG, Kramer TL, Gleser GC, Leonard AC. Children of disaster in the second decade: a 17-year follow-up of Buffalo Creek survivors. *J Am Acad Child Adolesc Psychiatry* 1994;33:71–79.
60. McLeer SV, Deblinger E, Henry D, Orvaschel H. Sexually abused children at high risk of post-traumatic stress disorder. *J Am Acad Child Adolesc Psychiatry* 1992; 31:875–879.
61. Adams BS, Everett BL, O'Neal E. PTSD in physically and sexually abused psychiatrically hospitalized children. *Child Psychiatry Hum Dev* 1992;23:3–8.
62. Orner RJ. Post-traumatic stress disorders and European war veterans. *Br J Clin Psychol* 1992;31:387–403.
63. Chu JA, Dill DL. Dissociative symptoms in relation to childhood physical and sexual abuse. *Am J Psychiatry* 1990;147:887–892.
64. Breyer JB, Nelson BA, Miller JB, Krol PA. Childhood sexual and physical abuse as factors in adult psychiatric illness. *Am J Psychiatry* 1987;144:1426–1430.
65. Michels R, Marzuk PM. Progress in psychiatry. *N Engl J Med* 1993;552–638.
66. Sierles FS, Chen JJ, McFarland RE, Taylor MA. Posttraumatic stress disorder and concurrent psychiatric illness: a preliminary report. *Am J Psychiatry* 1983;140: 1177–1179.
67. Sierles FS, Chen JJ, Messing ML, Besyner JK, Taylor MA. Concurrent psychiatric illness in non-Hispanic outpatients diagnosed as having posttraumatic stress disorder. *J Nerv Ment Dis* 1986;174:171–173.
68. Walker RD, Howard MO, Lambert MD, Suchinsky R. Psychiatric and medical comorbidities of veterans with substance use disorders. *Hosp Community Psychiatry* 1994;45:232–237.
69. Cottler LB, Compton WM III, Mager D, Spitznagel EL, Janca A. Posttraumatic stress disorder among substance users from the general population. *Am J Psychiatry* 1992;149:664–670.
70. Regier DA, Farmer ME, Rae DS, Locke BZ, Keith SJ, Judd LL, et al. Comorbidity of mental disorders with alcohol and other drug abuse: results from the epidemiologic catchment area (ECA) study. *JAMA* 1990;264: 2511–2518.
71. Shalev A, Bleich A, Ursano RJ. Posttraumatic stress disorder: somatic comorbidity and effort tolerance. *Psychosomatics* 1990;31:197–203.
72. Jordan BK, Schlenger WE, Hough R, et al. Lifetime and current prevalence of specific psychiatric disorders among Vietnam veterans and controls. *Arch Gen Psychiatry* 1991;48:205–215.
73. Shore JH, Tatum EL, Vollmer WM. Evaluation of mental health effects of disaster, Mount St. Helens eruption. *Am J Public Health* 1986;76:76–83.
74. Shore JH, Tatum E, Vollmer WM. Psychiatric reactions to disaster: the Mt. St. Helen's experience. *Am J Psychiatry* 1986;143:590–595.
75. McFarlane AC. The longitudinal course of post traumatic morbidity. *J Nerv Ment Dis* 1988;176:30–39.
76. Saigh PA. The development of posttraumatic stress disorder following four different types of traumatization. *Behav Res Ther* 1991;29:213–216.
77. Keane TM, Scott WO, Chavoya GA, Lamparski DM, Fairbank JA. Social support in Vietnam veterans with post traumatic stress disorder: a comparative analysis. *J Consult Clin Psychol* 1985;53:95–102.
78. Atkeson BM, Calhoun KS, Resick PA, Ellis EM. Victims of rape: repeated assessment of depressive symptoms. *J Consult Clin Psychol* 1982;50:96–102.
79. Burgess AW, Holmstrom LL. Adaptive strategies and recovery from rape. *Am J Psychiatry* 1979;136:1278–1282.
80. Ergood J, Frank E, Holroyd K, Anderson B, Hughes C, West D, et al. *Coping and post-rape symptomatology.* Paper presented at the meeting of the Association for Advancement of Behavior Therapy, 1985, November.
81. Foy DW, Osato SS, Houskamp BM, Neumann DA. Etiology of posttraumatic stress disorder. In PA Saigh, ed. *Post-traumatic stress disorder: a behavioral approach to assessment and treatment.* New York: Allyn & Bacon; 1992.
82. Jones JC, Barlow DH. The etiology of posttraumatic stress disorder. *Clin Psychol Rev* 1990;10:299–328.
83. Costello EJ, Angold A. Developmental epidemiology. In Cicchetti D, Cohen D, eds. *Developmental psychopathology Vol 1: Theory and methods.* New York: Wiley; 1995:23–56.

Neurobiological and Clinical Consequences of Stress: From Normal Adaptation to PTSD, edited by M. J. Friedman, D. S. Charney, and A. Y. Deutch, Lippincott-Raven Publishers, Philadelphia © 1995.

24

Post-Traumatic Stress Disorder and Comorbidity

Psychobiological Approaches to Differential Diagnosis

Matthew J. Friedman and *Rachel Yehuda

*Departments of Psychiatry and Pharmacology, Dartmouth Medical School, Hanover, New Hampshire 03755, and National Center for Post-Traumatic Stress Disorder, Veterans Administration Medical Center, White River Junction, Vermont 05009; and *Department of Psychiatry, Mount Sinai School of Medicine, Bronx Veterans Affairs Medical Center, Bronx, New York 10468*

A striking feature of the psychobiological findings in post-traumatic stress disorder (PTSD), as is reviewed extensively in this volume, is that the nature and direction of neurobiological alterations in PTSD appear to be quite distinct from what has been observed in other psychiatric disorders—even in psychiatric disorders with overlapping symptomatology to PTSD. As such, in addition to providing potential information about pathophysiology, these psychobiological findings have provided some measure of validation for PTSD as a distinct diagnostic entity.

In contrast to the relatively consistent literature supporting PTSD as a biologically identifiable syndrome, prevalence studies have suggested that PTSD as a clinical entity rarely occurs in a discrete or ''pure'' form. As this chapter will further review later, many studies have now documented that PTSD often co-occurs with other psychiatric disorders. Indeed, if an individual meets diagnostic criteria for PTSD, it is likely that he or she will also meet diagnostic criteria for a major mood or anxiety disorder, alcoholism, substance abuse, or a personality disorder. The uniqueness of PTSD from a biological perspective, then, appears somewhat paradoxical in the face of a clinical phenomenology in which most PTSD symptoms can be found in other psychiatric disorders.

The purpose of this chapter is to consider whether, and to what extent, issues of diagnostic comorbidity or differential diagnosis can be better understood—conceptually and practically—in the context of psychobiological findings of PTSD. Our strategy will be to review the evidence for psychiatric comorbidity and explore the phenomenologic similarities and differences between PTSD and three other disorders—major depressive disorder (MDD), panic disorder (PD) and generalized anxiety disorder (GAD)—that are often associated with PTSD. In this context, we will present and clarify the issues that must be addressed by clinicians performing a diagnostic assessment for PTSD and other possible comorbid diagnoses. Next, we will discuss the potential utility of biological findings to clarify some of these issues, and review laboratory-based abnormalities associated with PTSD for the purpose of identifying those procedures that may be most useful and clinically applicable in distinguishing PTSD from other disorders. Finally, we will discuss new research findings that may offer new diagnostic approaches.

PSYCHIATRIC COMORBIDITY IN PTSD

PTSD often co-occurs with other psychiatric disorders. The National Vietnam Veterans Readjustment Study (NVVRS), a national epidemiologic survey of PTSD among Vietnam veterans (1,2), (see Chapter 23) found that 50% of veterans with a current diagnosis of PTSD had at least one other psychiatric disorder. According to NVVRS, veterans with PTSD had the following lifetime prevalence rates: MDD, 20%; alcoholism, 75%; drug abuse, 23%; and personality disorder, 20%. Nontreatment-seeking civilians with PTSD have also been found to have high rates of psychiatric comorbidity (3).

Rates of psychiatric comorbidity have been found to be even higher among treatment seekers with PTSD. A 4-week survey of over 100,000 veterans seeking psychiatric treatment at Department of Veterans Affairs (VA) Hospitals across the U.S. in 1986 found that 80% of veterans with PTSD had at least one additional diagnosis, including MDD, anxiety disorders, alcoholism, drug abuse, and personality disorders (4). Reports of lifetime prevalence rates of MDD among treatment-seeking combat veterans with PTSD ranged from 26%–65% (5). Lifetime rates of alcoholism and drug abuse among treatment-seeking PTSD veterans have been estimated at about 60%–80% (6), and lifetime rates of personality disorder ranged from 40%–60% (5).

In considering the reasons for the unusually high incidence of comorbidity between PTSD and other psychiatric disorders, it is important to point out that there is considerable overlap in DSM-III-R (7) diagnoses in general, especially in the anxiety and mood disorders, that likely reflect the considerable overlap of symptoms between anxiety and mood disorders as illustrated in Fig. 1. As can be clearly seen in the Venn diagram, PTSD shares a number of symptoms with MDD, and the anxiety disorders PD and GAD.

With regard to overlap between PTSD and MDD, both syndromes exhibit impaired concentration, diminished interest, and insomnia. Guilt, though no longer an approved symptom of PTSD (it was included in DSM-III [8] but dropped in DSM-III-R), is still considered important, especially by those who believe that pathological grief is also a post-traumatic syndrome (9). In clinical practice, there is often confusion between certain symptoms of PTSD and depression. For example, emotional detachment and restricted range of affect are often confused with depressed mood. Emotional detachment and/or impaired concentration may be confused with psychomotor retardation. Autonomic hyperarousal may be confused with psychomotor agitation. Finally, suicidal ideation is often present in PTSD patients even though it lacks the diagnostic specificity to have been included in the DSM-III-R. A similar analysis could be made of the overlap between PTSD and PD or GAD.

Given this kaleidoscopic shifting of overlapping symptom clusters, it is perhaps not surprising that if an individual meets diagnostic criteria for PTSD, it is likely that he or she will meet DSM-III-R criteria for one or more additional diagnoses. The important clinical issue is, however, whether or not the presence of a comorbid diagnosis constitutes a true concomitant condition (i.e., with a separate underlying cause and pathophysiology), or an apparent one (i.e., due to the artifact of overlapping symptoms). Indeed, it has been argued that when individuals with PTSD also meet diagnostic criteria for MDD, this may actually signal the presence of a single disorder (e.g., a depressive subtype of PTSD) rather than two separate syndromes (5,10). Similarly, association of PTSD with personality disorder (especially borderline or multiple personality disorder) may actually reflect a unitary disorder beyond the detection of current nosological formulations (i.e., as suggested by Herman's (11) conception of "complex PTSD"). The presence of comorbidity, then, may indicate that current definitions of post-traumatic syndromes are too limiting.

Thus, the critical question is whether PTSD is related to other psychiatric disorders in more intrinsic and fundamental ways than simply as a result of nosological artifacts of symptom overlap. Other disorders may indicate comorbidity with the presence of a more expansive response to trauma. Thus, it is certainly possible, that in

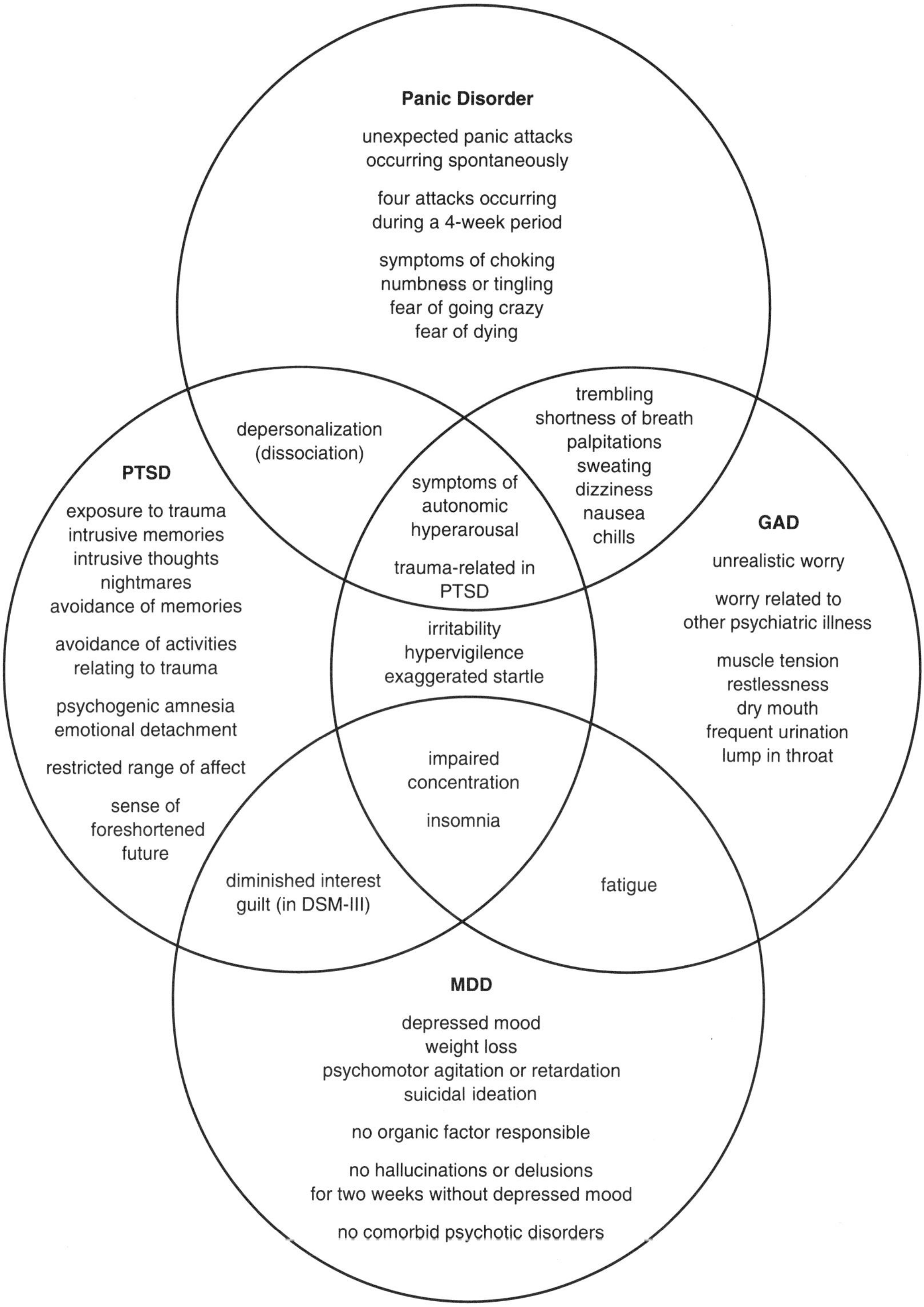

FIG. 1. Overlap of symptoms between post-traumatic stress disorder (PTSD), panic disorder, generalized anxiety disorder (GAD), and major depressive disorder (MDD).

addition to suggesting the presence of subtypes or variants of PTSD, the high comorbidity between PTSD and other psychiatric disorders may reflect different phenomenological expressions of the same underlying pathophysiology (5). Indeed, it has been suggested that stress and trauma are etiologic agents for psychiatric disorders other than PTSD. Furthermore, it can be argued that the concept that stress influences the development and expression of virtually all psychiatric syndromes is an important cornerstone of psychiatry and mental health (Yehuda and McFarlane, in preparation). Thus, in addition to producing a post-traumatic syndrome, a traumatic event may induce or exacerbate a diathesis for other psychiatric disorders. Here the question becomes whether any type of syndrome induced by a trauma, regardless of the specific symptomatology, would be best conceptualized and treated as a post-traumatic adaptation.

In all instances where the occurrence of a comorbid psychiatric illness does not predate a traumatic event, and cannot therefore be definitively identified as a separate (pre-traumatic) entity, it is virtually impossible to determine whether such illnesses constitute distinct comorbid disorders or extensions and variants of the post-traumatic syndrome. However, to the extent that PTSD is associated with a distinct biological profile, which in many instances can be clearly distinguished from alterations occurring in related psychiatric disorders, it is theoretically possible that the interrelationship between these clinical phenomena can be specified more precisely through the use of biological strategies.

Several investigators have provided empirical support for the utility of using psychobiological approaches to complement the DSM-III-R phenomenological assessment of PTSD (12–14). In the next section, we will identify biological and laboratory techniques that may have utility and clinical applicability to the differential diagnosis of PTSD, and critically evaluate their promise and limitations. In doing so, however, it is important to have a conceptual framework that allows for such an evaluation. Table 1 lists five criteria that should be met for a psychobiological assessment procedure to have practical usefulness in the clinical diagnostic work-up of a patient suspected of having PTSD. It will be useful to keep these criteria in mind when evaluating the adequacy of each laboratory test.

TABLE 1. *Criteria for clinical usefulness of a psychobiological assessment procedure*

1. The biological finding obtained through the testing procedure should be highly replicable within and across different laboratories. Furthermore, to have true clinical utility, the testing paradigm should be relatively easy to set up and the information obtained be subject to a relatively unambiguous interpretation by a trained professional.
2. The biological alteration should be reliably present in individuals with PTSD regardless of the type of trauma that has been sustained, and should be differentiable from normals.
3. The biological alteration should not be present in individuals who do not meet criteria for PTSD, even though they may have been exposed to trauma.
4. The biological abnormality should be relatively specific to PTSD and, as such, afford the opportunity of distinguishing PTSD from other diagnostic possibilities.
5. The biological finding, ideally, should be present in individuals with PTSD even if they meet the comorbid criteria for other psychiatric disorders.

PTSD, post-traumatic stress disorder.

LABORATORY ABNORMALITIES IN PTSD

The purpose of this section is not to repeat the review of material presented earlier in this book, but rather to synthesize and evaluate these findings from a clinical perspective. We will also discuss how the identification of psychobiological assessment strategies that have acceptable sensitivity and specificity for PTSD, and that can easily be transferred from the research laboratory to the clinic, ultimately enable us to determine whether comorbid diagnoses assessed by the DSM-III-R are truly distinct disorders or are actually subtypes of PTSD.

Table 2 summarizes the major areas of research on biological abnormalities associated with PTSD. Although it may seem that several types of techniques and methodologies have been used in biological studies of PTSD, it is useful to conceptualize these studies as falling into two major categories: 1) naturalistic, or

TABLE 2. *Laboratory abnormalities in PTSD*

Physiological assessment techniques
Psychophysiological Reactivity
Startle Response
EEG/Sleep Physiology
Event-Related Brain Potentials
Odor-induced EEG
Baseline Neurohormone levels
Catecholamines
Cortisol
Testosterone
Thyroid
ACTH
Endorphins
Urinary Neurohormone Profile
Baseline receptor levels
Lymphocyte Glucocorticoid
Platelet Alpha-2 Adrenergic
Lymphocyte Beta Adrenergic
Challenge Tests[a]
DST
Yohimbine
Stress-Induced Analgesia (With/Without Naloxone)
Clonidine (Growth Hormone)
L-DOPA (Growth Hormone)
CRF (ACTH)
TRH (TSH)

[a]Most physiological assessment approaches are also challenge tests.

PTSD, post-traumatic stress disorder; EEG, electroencephalogram; ACTH, adrenocorticotropic hormone; DST, dexamethasone suppression test; L-DOPA, levodopa; CRF, corticotropin-releasing factor; TRH, thyroid-releasing hormone; TSH, thyroid-stimulating hormone.

baseline observations, and 2) challenge strategies. Naturalistic studies involve measuring components of the particular system of interest under normal, baseline conditions. Ideally, in these studies one examines the biological systems of interest as they function under normative conditions, without external provocation. However, it must be kept in mind that the process of informing an individual that he or she is being evaluated, reviewing potential hazards of the study and obtaining informed consent, attaching electrodes, drawing blood, and other environmental influences may affect the desired measures. Challenge strategies permit evaluation of biological systems under more controlled conditions because they focus on the impact of an external influence (e.g., exposure to a traumatic script or biological substance). When baseline and challenge studies are used in tandem, it is possible to obtain careful information about the nature and origin of biological defects.

Physiological Assessment Techniques

Psychophysiological Reactivity

The earliest attempts to explore the biological basis of PTSD utilized psychophysiological strategies to explore differences in sympathetic nervous system activity under baseline conditions and in response to experimental challenge (e.g., audiotaped sounds of combat, videotaped combat scenes, and, more recently, individualized autobiographical traumatic scripts). In almost all earlier studies, combat veterans with PTSD were found to have higher baseline heart rate, systolic blood pressures, and electromyographical responses compared with normal and combat controls (14). However, later studies did not tend to reveal significant differences between PTSD patients and other groups in baseline measures (15,16). One explanation for the discrepancy between earlier and later studies might be that, as these studies evolved, there was a greater emphasis on methodological details such as psychiatric and medical inclusion/exclusion criteria, and a greater refinement of the psychophysiological protocol in general. (See Chapter 16 for a thorough review of this research.)

In contrast to the lack of consensus concerning the question of baseline psychophysiological differences in PTSD patients compared with other groups, the finding that combat veterans with PTSD show a more dramatic increase in psychophysiological response to combat-related stimuli is unquestionably the most replicable biological finding in the PTSD literature. When exposed to such a stimulus, PTSD subjects exhibit greater cardiovascular arousal as well as increased responsiveness of other autonomic measures such as electrodermal and electromyographic activity. A variety of trauma-related stimuli can reliably evoke this response, including auditory, visual, and olfactory cues. In the case of Vietnam combat veterans, psychophysiological reactivity has been elicited in response to audiotapes of combat sounds (17), videotapes of war zone scenes (18), and a script-driven imagery technique in which the veteran is asked to recollect his own combat experiences (19). This approach has also been used successfully with PTSD patients who have survived motor vehicle accidents (20), Israeli survivors of non-combat trauma (21), and female childhood sexual abuse survivors with PTSD (22). Using the script-driven imagery technique, PTSD subjects could be successfully distinguished from non-PTSD controls with a specificity of 61%–88% and sensitivity of 100% (19,23). In two studies, Vietnam veterans without PTSD were asked to fake their psychophysiological responses to appear more like PTSD veterans. In both cases, it was possible to distinguish the PTSD patients from the non-PTSD comparison subjects (24, 25). Finally, in an innovative extrapolation from this approach, McCaffery et al. (26) detected changes in the electroencephalogram (EEG) in response to trauma-related odors that distinguished PTSD from non-PTSD patients.

Because autonomic hyperarousal is also listed as a symptom of PD and GAD (see Fig. 1), an important question that arises from these studies is the extent to which patients with other anxiety disorders would also show autonomic hyperreactivity. Pitman et al. (27) compared PTSD subjects with a non-PTSD anxiety disorder group (consisting of subjects with PD, GAD, and obsessive-compulsive disorder (OCD). They found that the psychophysiological responses of PTSD subjects were significantly higher than the non-PTSD anxiety group when groups were compared on responsivity to trauma-related scripts. In fact, none of the non-PTSD subjects (including anxious subjects) responded to the script-driven imagery procedure. However, it is important to note that there is no specific reason to assume that patients with PD or GAD would have an intense psychophysiological reaction to a script of a nontraumatic event or traumatic event to which they were not exposed. Perhaps the most stringent test of this paradigm would be to compare traumatized patients with PD or GAD who do not have PTSD with traumatized patients who have PTSD. A comparison of psychophysiological reactivity of the PD, GAD, and PTSD groups to a standard script-driven traumatic imagery process would indicate the specificity of this technique for PTSD assessment.

In this regard, it is also of interest that, although autonomic reactivity to trauma-related stimuli seems to be the most robust psychophysiological assessment technique for PTSD, an exaggerated cardiovascular response can also be elicited in response to a neutral stimulus such as a burst of white noise (28). It is not clear, however, whether this approach has the specificity and sensitivity achieved with trauma-related stimuli.

In assessing the potential utility of psychophysiological reactivity as a laboratory measure to aid in diagnostic assessment, it can be concluded that reactivity as assessed by the cardiovascular response to trauma-related stimuli stands out as one of the best-tested and most feasible psychodiagnostic procedures at this time. It has high specificity and sensitivity, and is difficult to fake. There is sufficient collective experience with this approach so that optimal instrumentation for a psychophysiological laboratory has been determined and an optimal psychophysiological protocol has been established. Furthermore, the script-driven traumatic imagery technique developed by Pitman, Orr, and associates (19) can be used with any trauma survivor. The autobiographical scripts can be carefully crafted to reflect the uniqueness of both the traumatic event and the immediate post-traumatic reaction.

An endorsement of this approach must be tempered, however, by the fact that this method has not been adequately tested on nonveteran cohorts. Therefore, despite its remarkable performance with PTSD patients exposed to war-zone stress, there are only a few preliminary studies in which nonveteran cohorts (such as victims of sexual assault or survivors of motor vehicle accidents) have received psychophysiological assessment. Furthermore, as just described, it is unclear how well this test can distinguish PTSD from other anxiety disorders (e.g., PD, GAD, OCD).

Startle Response

A second indicator of physiological reactivity characteristic of PTSD is the startle response. Indeed, an exaggerated startle response is mentioned as a diagnostic symptom of PTSD. As shown in Fig. 1, however, it is also listed as a symptom of GAD. Startle is usually measured as the latency and amplitude of the eyeblink response to a burst of white noise. Since it has also been shown that this response can be modulated (facilitated or inhibited by nonstartling acoustic prestimulation), another approach is to monitor the degree of facilitation or inhibition rather than the eyeblink reflex itself. Evidence suggests that the startle response in PTSD is an abnormal reflex differing from that seen in normal subjects or patients with panic disorder (29,30). In comparison with non-PTSD subjects, PTSD subjects exhibit shorter latency and increased amplitude of the acoustic-eyeblink reflex (30–32). Ornitz and Pynoos (33) monitored modulation of the startle response in children with PTSD and found that these subjects exhibited a significant loss of the normal inhibitory modulation of startle, suggesting that the traumatic experience had induced a longlasting brainstem dysfunction. Ornitz and Pynoos' study is also unique because it is one (of only two) studies in which a psychobiological approach was tested in children with PTSD. This technique has the distinct advantage over structured interviews and other traditional assessment approaches in that it can be used with preschool children who have neither the linguistic nor conceptual capacity to provide verbal responses or understand abstract ideas.

There have been a number of studies of the startle response in other anxiety disorders. In two studies by Shalev and associates, the acoustic-eyeblink reflex in PTSD has been compared with the startle response in other anxiety disorders such as PD, GAD, and OCD. In the first study, PTSD patients compared to anxiety disorder (PD, GAD, OCD, etc.) patients exhibited a startle response that was significantly greater (30). In the second investigation, it was found that the benzodiazepine, alprazolam, could block the startle response in PD but not in PTSD patients, suggesting a qualitative difference between the startle response in these two disorders (34). Another difference between PTSD startle and that seen in other anxiety disorders is the lack of habituation of this response in PTSD, in contrast to habituation seen in PD, GAD, and OCD (31, 34). Furthermore, alprazolam accelerates startle habituation in PD but not in PTSD (34). A great deal of additional information is needed, however, before concluding that the startle response can be used as an aid in clinical assessment. Since increased startle has been found in other anxiety disorders (29), the extent to which this test can actually differentiate PTSD from other disorders is uncertain. Unless the magnitude of the typical startle response seen in PTSD is so much greater than the startle response seen in GAD, PD, OCD, etc., it may not have much clinical applicability in the differential diagnosis of PTSD.

EEG/Sleep Physiology

Another important use of psychophysiology is in the study of sleep. It is well known that most PTSD patients experience insomnia, and that traumatic nightmares are unique events that differ from classic nightmare/night terror Stage 4 episodes as well as dream anxiety attacks associated with REM sleep (35–37). What is less clear, however, is whether there are characteristic alterations in sleep architecture associated with PTSD (see Chapter 17). It is unclear whether the disagreements found in the published literature to date can be explained by: 1) methodological differences between laboratories; 2) diagnostic imprecision with respect to accurate identification of PTSD patients; 3) interpretive problems because of the frequent association of PTSD with MDD (since MDD does produce its own unique abnormalities in sleep architecture); 4) complicated abnormalities that will not be detected by traditional techniques for monitoring the sleep cycle; or 5) all four of the factors just mentioned. At this point in time, it can be stated that the sleep abnormalities regularly found in

PTSD (increased sleep latency, decreased time asleep, increased awakenings, and increased nocturnal movements) are either too nonspecific or that changes in sleep architecture reported by different investigators are too controversial for the sleep EEG to be useful as a psychobiological assessment strategy.

Event-Related Brain Potentials

The fourth psychobiological abnormality found in PTSD has been reported in two studies. Paige et al. (28) reported a robust difference in the event-related brain potentials (ERPs) recorded in PTSD as compared to non-PTSD patients. In response to increasing intensity of auditory stimulation, the PTSD patients showed a reduced amplitude of the P2 component of the ERP. Using a different procedure with low-intensity auditory stimuli, McFarlane et al. (38) observed delayed N2 and attenuated P3 components of the ERP response among PTSD patients in contrast to normal controls. Despite their intrinsic interest, these findings will have to be replicated before ERP recordings can be considered a potentially useful diagnostic strategy.

Baseline Neurohormone Levels

Catecholamines

Mean 24-hour concentrations of urinary norepinephrine (NE) have been reported elevated in combat veterans with PTSD compared with patients in several other diagnostic categories, such as major depression and bipolar mania (39), and with normal controls (40). Additionally, combat veterans with PTSD also showed a significant increase in the urinary excretion of dopamine (DA) compared to normals. Excretion of NE and DA, but not epinephrine, (EPI) was found to correlate with severity of PTSD symptoms, particularly intrusive symptoms (41). Pitman and Orr (42) failed to observe significant differences in catecholamines between 20 patients with PTSD and 5 combat controls. However, the mean urinary NE excretion observed in the combat controls (58.0 μg/day) was substantially higher than what has previously been observed for normals.

It is important to note, however, that despite the evidence for increased 24-hour urinary catecholamine excretion, evidence of baseline catecholamine activation has not been observed in studies using biological determinations made over short periods of time. For example, as just reviewed, the majority of studies has not found increases in autonomic measures at baseline, nor have baseline changes in plasma catecholamines or catecholamine metabolism been noted (15,16, 43). It is possible that part of the discrepancy between the findings obtained using a 24-hour urine collection versus a single sample plasma determination is related to the fact that the former method allows an estimate of catecholamine activity over a longer time frame, whereas blood sampling (or even resting psychophysiological assessments) relies on immediate autonomic activity over relatively short periods of time. Differences in hospitalization status, severity of PTSD, comorbidity, current or past substance abuse, or other issues may also account for the discrepancy between the findings obtained from urine versus plasma and psychophysiology studies. Future studies are clearly needed to more carefully address these issues. A more thorough review of this topic can be found elsewhere (44–46).

Cortisol

The finding of low urinary cortisol excretion in PTSD compared to other psychiatric groups and normal controls has now been replicated in several studies. In an initial study, lower mean 24-hour cortisol excretion was observed in nine PTSD patients compared to patients in four other diagnostic groups. This finding has been replicated in both inpatient and outpatient combat veterans with PTSD compared to nonpsychiatric, healthy controls, as well as in nontreatment-seeking civilians with chronic PTSD. The only other study examining 24-hour urinary cortisol excretion in PTSD reported an increased urinary cortisol in PTSD (42). This study, however, was methodologically different from the others in method of urine collection, radioimmunoassay, and other variables (45). In a recent attempt to replicate their initial findings, Pitman et al. did not observe a significant elevation in cortisol in subjects with PTSD. It should also be mentioned that a recent study examining the circadian release of cortisol over the 24-hour diurnal cycle has confirmed that the basal plasma release of cortisol (i.e., determined by averaging estimates of cortisol from samples obtained every 30 minutes for a 24-hour period) was found to be significantly lower in patients with PTSD compared to patients with major depression and normal controls (45).

Although the results just cited are promising in suggesting that PTSD patients can be differentiated from normals, and from patients with mood and psychotic disorders, it is unclear whether cortisol is low in patients with PTSD who also have current comorbid major depression. In one study (47), mean urinary-free cortisol excretion from PTSD patients with MDD was not significantly different from the mean urinary-free cortisol excretion of PTSD patients without MDD. However, additional studies with larger samples are needed before drawing this conclusion definitively.

Testosterone

Serum testosterone concentrations were found to be substantially higher in patients with PTSD compared with patients with major depressive disorder, bipolar mania patients, and normals, but were comparable to schizophrenics. While the clinical characteristics of these findings still need to be explored, the data lend further support to the neuroendocrine distinctness of PTSD and major depressive disorder (48).

Thyroid

Thyroid function tests (TFTs) are well-established clinical laboratory procedures that offer a relatively straightforward technique for distinguishing between PTSD and MDD. Mason and associates (see Chapter 20) have shown that thyroid function is elevated among male Vietnam combat veterans with PTSD. In contrast to non-PTSD control groups, such patients exhibit elevated mean serum total thyroxine (T_4), thyroid-binding globulin, total and free triiodothyronine (T_3), and T_3/T_4 ratios. In fact, Mason has observed that, for some PTSD patients, thyroid indices are at or near the thyrotoxic range. In contrast, when abnormal TFTs are found in depressed patients, such individuals tend to exhibit hypothyroidism (49). Most published reports suggest that when there is a question of comorbidity between PTSD and MDD, elevated TFTs are more likely among PTSD patients and the reverse appears to be the case for MDD. The major limitation of this approach is that most patients with either PTSD or MDD will probably have TFTs within the normal clinical range. Therefore, measurement of TFTs cannot be recommended as a first-line psychobiological assessment strategy for detecting PTSD. Furthermore, it is unclear whether TFTs have any usefulness in the detection of other anxiety disorders.

Plasma ACTH and Beta-Endorphin

Hoffman et al. (50) and Smith et al. (51) have both reported normal adrenocorticotropic hormone (ACTH) levels in PTSD patients compared to normals. The former group, however, reported that beta-endorphin levels were lower at both 9:00 AM and 4:00 PM. Certainly, further studies are needed, since a single stick sample of ACTH or beta-endorphin is virtually uninterpretable in the context of a baseline evaluation.

Urinary Neurohormone Profile

In summarizing whether an assessment of neurohormone levels may have diagnostic utility, as has been previously suggested, (13,44), several points should be considered. As just reviewed, the reproducibility of these findings, particularly with regard to cortisol, has now been documented in several studies. One of the major difficulties with using this approach for diagnostic purposes is that both the lowered cortisol levels and elevated catecholamine levels observed in PTSD are within the normal clinical (endocrinological) range. Although several psychoendocrine studies have clearly pointed to the fact that hormonal levels need not be outside the medically defined "normal limits" (which reflect glandular disorders) to have clinical significance or practical value in the assessment and management of psychiatric illness, practically speaking, it is currently difficult to determine whether any specific value of one hormone is associated with the presence of a psychiatric condition. It is perhaps for this reason that Mason et al. have suggested that the concurrent assessment of several hormonal systems provides an opportunity to explore differences in hormonal patterns in various psychiatric disorders. Indeed, PTSD appears to be characterized by a specific profile of hormonal changes that is distinct from that of many diagnostic groups and normal controls (although, as with the other biological tests just mentioned, the relationship between PTSD and other anxiety disorders has not been fully evaluated).

Baseline Receptor Levels

Lymphocyte Glucocorticoid Receptors

Steroid receptor-binding parameters are critical to a proper interpretation of studies examining basal hormone secretion, because hormones cannot exert their genomic effects unless they are bound to steroid receptors. Because of similarities in the regulation and binding characteristics of lymphocyte glucocorticoid receptors (GRs) and those in brain, this measure may reflect aspects of central as well as peripheral cortisol regulation. Results from three studies have now demonstrated an increased lymphocyte GR number in combat veterans with PTSD compared with normal individuals, and patients with other psychiatric disorders such as depression, schizophrenia, bipolar mania, and panic disorder (45, 52).

In evaluating the utility of this measure as a potential laboratory assessment, it should be noted that the only major limitation of this test is that, to date, only combat veterans have been studied. If other traumatized groups are found to have similar alterations in GR number, this measure can be of potential usefulness as a laboratory test, as it seems to offer some measure of discrimination between PTSD and other psychiatric disorders, and to a lesser extent between combat veterans with and without PTSD.

Platelet Alpha-2 Adrenergic Receptors

In an initial study, Perry et al. (53) reported a decreased number of platelet alpha-2 adrenergic receptors in 12 combat veterans with PTSD compared to normals. The finding of decreased alpha-2 adrenergic receptors was replicated in a sample of 40 combat veterans with PTSD compared to normals (54). Also observed was that patients with MDD and GAD had significantly higher numbers of platelet alpha-2 adrenergic receptors compared to normal controls and patients with PTSD (54). Platelet alpha-2 adrenergic receptors of PTSD patients showed a greater degree of downregulation or receptor uncoupling (i.e., decreased number of receptors) following in vitro exposure to epinephrine compared with platelets from healthy control subjects. These data support the notion of a chronic catecholamine hypersecretion as reflected by 24-hour urinary catecholamine studies.

Lymphocyte Beta Adrenergic Receptors

Lerer et al. (55) have shown significantly lower basal and forskolin-stimulated adenylate cyclase activity in Israeli combat veterans compared with normals. Adenylate cyclase activity in response to stimulation of nucleotide binding protein and via the receptor linked to PGE1 and phospholipase C activity were normal in the PTSD group. The examination of these second messenger systems has suggested a reduced signal transduction of the beta adrenergic receptor in PTSD. Interestingly, one might hypothesize that a decreased number of beta adrenergic-binding sites due to a loss of beta high-affinity sites (i.e., sites coupled to the nucleotide-binding protein of the receptor-adenylate cyclase complex) would reduce the activity of the inhibitory nucleotide-binding protein to which this receptor is linked, and would tend to be consistent with an increased (not decreased) catalytic subunit activity in response to forskolin stimulation. Further studies exploring homeostatic regulation of second messenger systems in response to receptor alterations are clearly needed.

In evaluating the utility of measuring adrenergic receptors for diagnostic purposes, it should be noted that no studies have examined any of these receptor parameters in combat veterans without PTSD. Furthermore, although the findings of platelet alpha-2 adrenergic and lymphocyte beta receptor number are relatively consistent among individuals with PTSD, a reduced number of platelet and lymphocyte adrenergic receptors has also been observed in several other medical and psychiatric conditions, (54,55) and appears to reflect a chronic increase in sympathetic nervous system activity, which would by no means be specific to PTSD.

Challenge Tests

Dexamethasone Suppression Test

Studies using the dexamethasone suppression test have continued to show that PTSD patients who are not depressed do not show the classic nonsuppression commonly observed in major depressive disorder. The issue regarding cortisol suppression following dexamethasone in PTSD patients with major depression is less clear. In an initial study, Kudler et al. (56) reported that PTSD patients with major depressive disorder show a comparable rate of nonsuppression to that seen in patients with major depressive disorder, whereas Halbreich et al. (57) and Kosten et al. (58) showed that even depressed combat veterans with PTSD show normal responses to dexamethasone. Olivera and Fero (59) showed a 32% incidence of nonsuppression in 65 combat veterans with PTSD who had a comorbid major depressive disorder. However, these individuals showed normal suppression after their major depression had gone into remission. A study examining the cortisol response to dexamethasone in eight civilian women with PTSD (60) also showed normal responses to this neuroendocrine challenge.

Recently, in order to explore the possibility of enhanced cortisol suppression to dexamethasone (52), Yehuda et al. challenged PTSD patients with 0.50 (10) and 0.25 mg of dexamethasone. A hyperresponsivity to low doses of dexamethasone as reflected by both significantly lower cortisol responses and enhanced translocation of cytosolic lymphocyte glucocorticoid receptors into the nucleus was observed in PTSD patients as compared with normals. Importantly, the hyperresponsivity to dexamethasone was also present in combat veterans with PTSD who met the diagnostic criteria for major depressive disorder. However, in that study patients were not sorted with respect to primary versus secondary MDD, a distinction that may prove important in future studies.

The dexamethasone suppression test (DST) may be a second potential psychobiological assessment approach with sufficient sensitivity, specificity, and feasibility to be considered useful in a clinical context. Here too, however, there must be a conditional endorsement presupposing that hypersensitivity to dexamethasone will be found in PTSD patients who have been exposed to traumatic events other than combat. Furthermore, it will be necessary to show that supersensitivity to dexamethasone among PTSD patients can be reliably demonstrated in several independent laboratories.

Should the hypersuppression to DST be found in other traumatized groups with PTSD, this method would have a major advantage since it is relatively simple to administer. Indeed, the simplicity of this test was one of the major reasons for previous widespread interest in this technique as a potential method for diagnosing MDD. Currently, it does appear that DST can distinguish PTSD from MDD, since the direction of change on the DST is virtually opposite in these two disorders. The MDD patients tend to exhibit nonsuppression in contrast to the hypersuppression seen in PTSD. There is little data to date, however, regarding the DST in

PD, GAD, OCD, or other anxiety or personality disorders. It would be necessary to acquire such information in order to determine the specificity and sensitivity of low-dose DST for PTSD.

What is perhaps most compelling about low-dose DST for PTSD is that the hypersuppression of cortisol observed following dexamethasone administration is compatible with conceptual ideas of PTSD as an atypical stress response (45). Furthermore, the biological abnormality observed on this test represents a descriptor of PTSD that appears to be separate and independent from the psychophysiological assessment procedures described earlier.

Behavioral and Biochemical Response to Yohimbine

Biochemical challenge studies have shown that the selective noradrenergic alpha-2 antagonist, yohimbine, elicits sympathetic arousal as evidenced by increased MHPG levels. Infusion of this drug also elicits panic attacks and Vietnam-related flashbacks in 60% of war veterans tested (61). This response is consistent with the increased peripheral catecholamine excretion and downregulation of platelet alpha-2 adrenergic receptors observed in other studies. It should be mentioned, however, that MHPG increases in response to yohimbine have also been observed in panic disorder.

Stress-Induced Analgesia (With/Without Naloxone)

One of the more provocative neuroendocrine findings in PTSD has been that of a naloxone reversible stress-induced analgesia comparable to that seen in laboratory animals exposed to inescapable stress paradigms (62). In this study, eight PTSD patients reported less pain in response to heat stimulation after viewing a traumatic combat film than eight combat veterans without PTSD. This analgesic response was reversed by the narcotic antagonist, naloxone. Although the data should be considered preliminary due to the small sample size, the findings suggest dysregulation of the endogenous opioid system in PTSD.

Growth Hormone Response to Clonidine and L-DOPA

Jensen et al. (63) found differences in the growth hormone response to clonidine and levodopa (L-DOPA), respectively, in sexually and physically abused boys. In contrast to normals and psychiatric controls who exhibited higher growth hormone levels following challenge with both drugs, sexually abused boys responded only to clonidine, while physically abused boys responded primarily to L-DOPA. The major criticism of the study is that PTSD was not assessed in any of the subjects. Therefore, it is impossible to know whether this approach might be useful in the psychobiological assessment of PTSD.

ACTH Response to CRF

To date, the corticotropin-releasing factor (CRF) challenge test has been utilized to evaluate both hypothalamic CRF secretion and pituitary ACTH activity in psychiatric illness. In several reports, the ACTH response to CRF has been reported to be "blunted" in MDD and anxiety disorders (64–67). A single study of eight PTSD subjects demonstrated that the ACTH response to CRF is blunted as well (51). Therefore, it would appear initially that PTSD patients show the same abnormality as other psychiatric patients in regard to this test.

In interpreting this finding, however, it is important to remember that there are no currently definitive theories as to why ACTH blunting occurs in MDD and anxiety disorders. The fact is, there are several possible explanations for ACTH blunting to CRF and, as previously suggested, given the many different biological routes that can result in ACTH blunting to CRF, and given the already specified differences between PTSD and MDD in other parameters of the HPA axis, it is quite possible that the mechanisms underlying the blunted ACTH response in PTSD are different from those relevant to the blunting observed in MDD (10). Practically speaking, however, although there may be biochemical reasons for the blunted ACTH response to CRF in PTSD that signify a different pattern of HPA responses in this disorder, as a laboratory test to aid in differential diagnoses this test would not be useful, since the same results are found for PTSD and MDD.

Thyroid-Stimulating Hormone Response to Thyroid-Releasing Hormone

Kosten et al. (58) explored the thyroid-stimulating hormone (TSH) response to thyroid-releasing hormone (TRH) in 11 PTSD patients compared with 28 depressed patients. Rather than the classic blunted thyroid-stimulating response observed in major depressed patients, 4 out of 11 patients showed an augmented response to TRH. This study represents the first attempt to utilize challenge strategies to elucidate hypothalamic-pituitary-thyroid axis dysfunction in PTSD.

FEASIBLE PSYCHOBIOLOGICAL APPROACHES TO DIFFERENTIAL DIAGNOSIS

We believe that only six of the laboratory abnormalities listed in Table 2 can even be considered to have potential clinical applicability at this time. Some tests that are very interesting with regard to theoretical implications and potential practical applications need to be replicated in more than one study before they can be considered for clinical use. These include: 1) measurement of ERPs; 2) odor-induced EEG; 3) baseline ACTH and beta-endorphin levels; 4) stress-induced analgesia; 5) growth hormone response to clonidine and L-DOPA; and 5) challenge tests with CRF and TRH. Some tests would be difficult to set up and standardize in a clinical laboratory. These include measurement of baseline adrenergic and glucocorticoid receptor levels. And some tests are just too complex, ambiguous, and controversial when it comes to interpreting the data (particularly regarding research with the sleep EEG).

In order to be considered useful, a psychobiological assessment procedure should ideally meet the feasibility criteria listed in Table 1. Table 3 shows the six laboratory tests that presently show promise as clinical tools for psychobiological assessment. The table also shows how well each test meets the five criteria in Table 1.

The most promising test at this time is measurement of psychophysiological (especially cardiovascular) reactivity. It has good specificity and sensitivity, is hard to fake, and certainly differentiates PTSD from MDD. It is currently in use in many VA hospitals to complement the clinical and psychometric assessment of patients

TABLE 3. *Feasible psychobiological approaches to differential diagnosis*

Test	Feasibility criteria				
	1	2	3	4	5
Psychophysiological Reactivity	X	X	X	?3	X
DST	X	?1,2	X	?3	X
Startle	X	X	U	?1,3	?4
Thyroid Function Tests	X	U	U	No	U
Yohimbine	X	?1,2	U	?1,2,3	X
Urinary Neurohormone Profile	X	?1,2	U	X	U

X Definitely meets criterion.
? Probably meets criteria but hasn't been adequately tested.
?1: Only tested in one laboratory.
?2: Only tested on Vietnam veterans.
?3: Not adequately tested in other anxiety disorders.
?4: Not adequately tested in PTSD patients with comorbid diagnosis.
U Unknown.
No Does not meet criterion.
DST, dexamethasone suppression test; PTSD, post-traumatic stress disorder.

with war-related PTSD. It is also used informally in VA compensation and disability evaluations, as well as for forensic applications (68). In addition, this technique has been successfully used to assess patients who have developed PTSD following the trauma of sexual assault or motor vehicle accidents. With regard to our feasibility criteria (Table 1), Table 3 shows that specifications for a standardized laboratory procedure and standard assessment procedure have been established (criterion 1), psychophysiological reactivity distinguishes PTSD patients from normal subjects (criterion 2), it's not present in traumatized subjects without PTSD (criterion 3), and it is present in patients who have PTSD plus a comorbid disorder (criterion 5). Some question remains about the specificity of this procedure when it comes to distinguishing PTSD from other anxiety disorders, but preliminary results suggest that it also performs well in that context. In the near future, psychophysiological assessment may become a bona fide part of the routine clinical assessment of PTSD.

The DST, while extremely promising, has only been demonstrated by one group of investigators on one traumatized group (Vietnam veterans), although there have been several replications of DST supersuppression in that laboratory setting. It is a well-standardized clinical laboratory procedure that provides good discrimination between PTSD, MDD, and normals. Future studies must monitor the DST response in PD, GAD, and other anxiety disorders. To date, DST supersuppression has not been observed in non-PTSD subjects. Assuming that Yehuda's work can be replicated in other laboratories and with other than Vietnam veteran PTSD cohorts, DST will meet four of the feasibility criteria in Table 1.

Startle is the only other test listed in Table 3 that may someday stand alone as an assessment technique for PTSD. What is particularly lacking at this time, however, is sufficient research demonstrating that the startle response in PTSD is qualitatively and/or quantitatively different from that seen in other anxiety disorders, especially PD. If Shalev's (34) finding can be replicated, that the PD but not the PTSD startle response can be abolished by alprazolam, this technique may prove very useful as a psychobiological assessment for PTSD.

The measurement of thyroid function tests may complement other approaches but cannot stand alone as a diagnostic strategy. This is primarily because thyroid hormone levels in PTSD are generally within the normal range (although at the high end of the range). It is the pattern of such a hormone profile that has been particularly important in research comparing results in PTSD patients with those of other patients. For example, high normal thyroid indices would be consistent with other tests indicating PTSD but would not be conclusive in their own right. Furthermore, it is unclear whether TFTs have any usefulness in differentiating other anxiety disorders from PTSD. Measurement of thyroid function has become fairly routine from one laboratory to another; the problem with this test is interpretive, not technical.

There seems little potential for clinical application of yohimbine challenge as a diagnostic tool. Although it will distinguish PTSD from MDD, it may not distinguish it from PD. On the other hand, yohimbine challenge may prove to be an excellent technique for monitoring the efficacy of treatment in PTSD patients. As with PD, PTSD patients who exhibited a pretreatment, yohimbine response might be expected to be unresponsive to such a challenge following successful treatment.

MULTIDIMENSIONAL APPROACHES TO DIAGNOSIS

Mason et al. (69) were the first to suggest a multidimensional approach to the evaluation of biological disturbance of PTSD. This suggestion originally stemmed from observations showing that multiple neuroendocrine abnormalities were present in this disorder. The initial multivariate strategies of Mason et al. involved performing baseline assessments of multiple hormonal systems, and using appropriate statistical approaches (e.g., multidimensional scaling) to combine the discriminating powers of each of these hormones to achieve an even greater precision in differentiating PTSD from other psychiatric disorders. Results from this initial study suggested that using cortisol and NE levels together provided a sensitivity and specificity for the diagnoses of chronic PTSD that were substantially greater than those achieved by other biological markers for psychiatric diagnoses (13,44,69). Furthermore, a multidimensional analysis using a profile of hormones including cortisol, NE, EPI, testosterone, and thyroxine provided classification accuracy approaching 100% in the differential diagnosis of combat PTSD compared to endogenous MDD (13,44). More recently, Yehuda et al. (45) have suggested that, as different neurobiological measures are added to the initial battery of hormones reported in earlier studies (i.e., receptor measures and endocrine responses to challenge strategies), there is a greater potential to enrich and expand the concept of multidimensional analysis in PTSD. For example, utilizing both hormone concentration and receptor number, as well as neuroendocrine responses to challenge, may provide a clearer picture of biological alteration than the assessment of any single parameter individually. What may be helpful, ultimately, is the putting together of a battery that helps address some of the weaknesses and limitations of each tool individually.

From a practical point of view, this elegant approach could potentially be put into practice in most clinical laboratories. Since most hormones measured in PTSD patients are within the normal range, it is essential that biochemical assay procedures be rigorously standardized from one clinical laboratory to the next so that neurohormonal profiles can be calibrated and interpreted consistently wherever they are measured.

SUMMARY

The purpose of this chapter has been to synthesize old rather than to present new information. The goal has been to develop a diagnostic strategy that will improve our ability to diagnose PTSD. We believe that psychobiologic diagnostic strategies will have an important role in future PTSD treatment and research for three reasons: 1) there is a great deal of symptom overlap between PTSD and other psychiatric disorders; 2) PTSD rarely occurs in a "pure" form, but is most often associated with at least one other DSM-III-R diagnosis; and 3) we do not believe that a diagnostic approach such as the DSM-III-R, that relies entirely on clinical phenomenology, can ever achieve the diagnostic precision needed for treatment and research.

Our strategy has been to review the various psychobiological laboratory findings that have been published regarding PTSD, and to evaluate their potential applicability as clinical tools for distinguishing PTSD from other disorders. We have not considered laboratory procedures that have not been replicated, that would be difficult to set up in a clinical laboratory, or that currently appear too ambiguous to interpret.

Our final list is short. We review six potential psychobiologic diagnostic approaches from the perspective of five feasibility criteria. Several strategies appear quite promising, especially psychophysiological reactivity, the dexamethasone suppression test, probably the startle response, and possibly the 24-hour urinary neurohormone profile. We expect that this list will grow and that new clinical applications will emerge as biological research on PTSD is extended to survivors of different traumatic events (besides the Vietnam War), and as PTSD patients are systematically compared with individuals who suffer from other (and comorbid) affective and anxiety disorders.

REFERENCES

1. Kulka RA, Schlenger WE, Fairbank JA, Hough RP, Jordan BK, Marmar CR, Weiss DS. *Trauma and the Vietnam War generation.* New York: Brunner/Mazel; 1990.
2. Jordan K, Schlenger W, Hough R, Kulka RA, Weiss D, Fairbank JA, Marmar CR. Lifetime and current prevalence of specific psychiatric disorders among Vietnam veterans and controls. *Arch Gen Psychiatry* 1991;48: 207–215.
3. Helzer JE, Robins LN, McEvoy L. Post-traumatic stress disorder in the general population: findings of the epidemiologic catchment area survey. *N Engl J Med* 1987; 317:1630–1634.
4. Friedman MJ, Kolb L, Arnold A, et al. *Chief Medical Director's Special Committee on PTSD: Third Annual Report.* Washington, DC: Department of Veterans Affairs; 1987:17–119.
5. Friedman MJ. Interrelationships between biological mechanisms and pharmacotherapy of posttraumatic stress disorder. In Wolfe ME, Mosnaim AD, eds. *Post-traumatic stress disorder: etiology, phenomenology, and treatment.* Washington, DC: American Psychiatric Press, 1990: 204–225.
6. Kofoed L, Friedman MJ, Peck R. Alcoholism and drug

abuse in patients with PTSD. *Psychiatr Q* 1993;64: 151–171.
7. American Psychiatric Association Committee on Nomenclature and Statistics. *Diagnosis and Statistical Manual of Mental Disorders,* Third Edition–Revised. Washington, DC: American Psychiatric Association 1987.
8. American Psychiatric Association Committee on Nomenclature and Statistics. *Diagnosis and Statistical Manual of Mental Disorders,* Third Edition. Washington, DC: American Psychiatric Association 1980.
9. Horowitz MJ. Stress-response syndromes: a review of post-traumatic stress and adjustment disorders. In Wilson JP, Raphael B, eds. *International handbook of traumatic stress syndromes.* New York: Plenum; 1993:49–60.
10. Yehuda R, Southwick SM, Krystal JH, Bremner JD, Charney DS, Mason JW. Enhanced suppression of cortisol following dexamethasone administration in posttraumatic stress disorder. *Am J Psychiatry* 1993;150(1): 83–86.
11. Herman JL. *Trauma and Recovery.* New York: Basic Books; 1992.
12. McFall ME, Murburg MM. Psychophysiological studies of combat-related PTSD. An integrative review. In Murburg MM, ed. *Catecholamine function in post-traumatic stress disorder: emerging concepts.* Washington, DC: American Psychiatric Press; 1994:161–174.
13. Mason JW, Giller EL, Kosten TR, Yehuda R. Psychoendocrine approaches to the diagnosis and pathogenesis of posttraumatic stress disorder. In Giller EL, ed. *Biological assessment and treatment of post-traumatic stress disorder;* Washington, DC: American Psychiatric Press; 1990:65–86.
14. Blanchard EB. Elevated basal levels of cardiovascular response in Vietnam veterans with PTSD: a health problem in the making? *J Anx Disorders* 1990;4(3):233–237.
15. McFall ME, Veith RC, Murburg MM. Basal sympathoadrenal function in posttraumatic distress disorder. *Biol Psychiatry* 1992;31(10):1050–1056.
16. Murburg MM, McFall ME, Veith RC. Basal sympathoadrenal function in patients with PTSD and depression. In Murburg MM, ed. *Catecholamine function in post-traumatic stress disorder: emerging concepts.* Washington, DC: American Psychiatric Press; 1994:174–188.
17. Blanchard EB, Kolb LC, Pallmeyer BA, Gerardi RJ. A psychophysiological study of post-traumatic stress disorder in Vietnam veterans. *Psychiatr Q* 1982;54: 220–229.
18. Malloy PF, Fairbank JA, Keane TM. Validation of a multimethod assessment of post-traumatic stress disorder in Vietnam veterans. *J Consult Clin Psychol* 1983; 51:488–494.
19. Pitman RK, Orr SP, Forgue DF, de Jong JB, Clairborn JM. Psychophysiologic assessment of post-traumatic stress disorder imagery in Vietnam combat veterans. *Arch Gen Psychiatry* 1987;44:970–975.
20. Blanchard EB, Hickling EJK, Taylor AE. The psychophysiology of motor vehicle accident related posttraumatic stress disorder. *Biofeedback Self Regul* 1991;16 (4):449–458.
21. Shalev AY, Orr SP, Pitman RK. Psychophysiologic assessment of traumatic imagery in Israeli civilian patients with posttraumatic stress disorder. *Am J Psychiatry* 1993;150:620–624.
22. Orr SP, Ahern C, Reaves M, Pitman RK. Psychophysiology of childhood sexual abuse imagery in adults. Presented at *Lake George Research Conference on Post-Traumatic Stress Disorder.* Bolton Landing, NY; January 1993.
23. Orr SP, Pitman RK, Lasko NB, Herz LR. Psychophysiological assessment of posttraumatic stress disorder imagery in World War II and Korean combat veterans. *J Abnorm Psychol* 1993;102(1):152–159.
24. Gerardi RJ, Blanchard EB, Kolb LC. Ability of Vietnam veterans to dissimulate a psychophysiological assessment for post-traumatic stress disorder. *Behav Ther* 1989;20(2):229–243.
25. Orr SP, Pitman RK. Psychophysiologic assessment of attempts to simulate posttraumatic stress disorder. *Biol Psychiatry* 1993;33(2):127–129.
26. McCaffrey RJ, Lorig TS, Pendrey DL, McCutcheon NB, Garrett JC. Odor-induced EEG changes in PTSD Vietnam era veterans. *J Traum Stress* 1993;6(2):213–224.
27. Pitman RK, Orr SP, Forgue DF, et al. Psychophysiologic responses to combat imagery of Vietnam veterans with posttraumatic stress disorder versus other anxiety disorders. *J Abnorm Psychol* 1990;99(1):49–54.
28. Paige S, Reid G, Allen M, Newton J. Psychophysiological correlates of PTSD. *Biol Psychiatry* 1990;27(4): 410–430.
29. Howard R, Ford R. From the jumping Frenchmen of Maine to post-traumatic stress disorder: the startle response in neuropsychiatry. *Psychol Med* 1992;22: 695–707.
30. Shalev AY, Orr SP, Peri T, Schreiber S, Pitman RK. Physiologic responses to loud tones in Israeli patients with posttraumatic stress disorder. *Arch Gen Psychiatry* 1992;49(11):870–875.
31. Butler RW, Braff DL, Rausch JL, Jenkins MA, Sprock J, Geyer MA. Physiological evidence of exaggerated startle response in a subgroup of Vietnam veterans with combat-related PTSD. *Am J Psychiatry* 1990;147(10): 1308–1312.
32. Kolb LC. PTSD: psychopathology and the startle response. *Psychiatr Q* 1991;62(3):233–250.
33. Ornitz EM, Pynoos RS. Startle modulation in children with post-traumatic stress disorder. *Am J Psychiatry* 1989;146:866–870.
34. Shalev AY, Peri T, Bloch M. Effect of alprazolam on the audity startle response in PTSD and Panic Disorder. *New Research Abstracts,* American Psychiatric Association, 146th Annual Meeting, Philadelphia, PA; May 21–26, 1994.
35. Friedman MJ. Post-Vietnam syndrome: recognition and management. *Psychosomatics* 1981;22:931–943.
36. Ross RJ, Ball WA, Sullivan KA, Caroff SN. Sleep disturbance as the hallmark of post-traumatic stress disorder. *Am J Psychiatry* 1989;146:697–707.
37. Ross RJ, Ball WA, Dinges DF, Kribbs NB, Morrison AR, Silver SM, Mulvanly FD. Rapid eye movement sleep disturbance in posttraumatic stress disorder. *Biol Psychiatry* 1994;35:195–202.
38. McFarlane AC, Weber DL, Clark CR. Abnormal stimulus processing in posttraumatic stress disorder. *Biol Psychiatry* 1993;34:311–320.
39. Kosten TR, Mason JW, Giller EL, Harkness L. Sustained urinary norepinephrine and epinephrine elevation in post-traumatic stress disorder. *Psychoneuroendocrinology* 1987;12:13–20.

40. Yehuda R, Southwick SM, Ma X, Giller El, Mason JW. Urinary catecholamine excretion and severity of symptoms in PTSD. *J Nerv Ment Dis* 1992;180: 321–324.
41. Yehuda R, Giller EL, Southwick SM, Kahana B, Boisoneau D, Ma X, Mason JW. Relationship between catecholamine excretion and PTSD symptoms in Vietnam combat veterans and holocaust survivors. In Murburg MM, ed. *Catecholamine function in post-traumatic stress disorder: emerging concepts.* Washington, DC: American Psychiatric Press 1994;203–220.
42. Pitman R, Orr S. Twenty-four hour urinary cortisol and catecholamine excretion in combat related posttraumatic stress disorder. *Biol Psychiatry* 1990;27:234–247.
43. Southwick SM, Krystal JH, Johnson DR, Charney DS. Neurobiology of post-traumatic stress disorder. In Tasman A, ed. *Annual review of Psychiatry* vol. 11. Washington, DC: American Psychiatric Press 1992:347–370.
44. Giller EL, Perry BD, Southwick SM, Yehuda R, Wahby VS, Kosten TR, Mason JW. Psychoendocrinology of posttraumatic stress disorder. In Wolf ME, Monsnaim AD, eds. *Post-traumatic stress disorder: etiology, phenomenology, and treatment* 1st ed. Washington, DC: American Psychiatric Press, 1990;158–167.
45. Yehuda R, Giller EL, Mason JW. Psychoneuroendocrine assessment of posttraumatic stress disorder: Current progress and new directions. *Prog Neuropsychopharmacol Biol Psychiatry* 1993;17:541–550.
46. Murburg MM, ed. *Catecholamine function in post-traumatic stress disorder: emerging concepts.* Washington, DC: American Psychiatric Press, 1994.
47. Yehuda R, Southwick SM, Nussbaum G, Wahby VS, Giller EL, Mason JW. Low urinary cortisol excretion in patients with posttraumatic stress disorder. *J Nerv Ment Disease* 1990;178(6):366–369.
48. Mason JW, Giller EL, Kosten TR, Wahby VS. Serum testosterone levels in posttraumatic stress disorder patients. *J Traum Stress* 1990;3:449–457.
49. Rubin RT, Poland RE, Lesser IM, Martin DJ. Neuroendocrine aspects of primary endogenous depression IV pituitary-thyroid axis activity in patients and matched control subjects. *Psychoneuroendocrinology,* 1987;12: 333–347.
50. Hoffman L, Watson PD, Wilson G, Montgomery J. Low plasma beta-endorphin in post-traumatic stress disorder. *Aust N Z J Psychiatry* 1989;23:269–273.
51. Smith MA, Davidson J, Ritchie JC, Kudler H, Lipper S, Chapell P, Nemeroff CB. The corticotrophin-releasing hormone test in patients with post-traumatic stress disorder. *Biol Psychiatry* 1989;26:349–355.
52. Yehuda R, Boisneau D, Lowy MT, Giller EL. Dose-response changes in plasma cortisol and lymphocyte glucocorticoid receptors following dexamethasone administration in combat veterans with and without posttraumatic stress disorder. *Arch Gen Psychiatry;* (in press).
53. Perry BD, Southwick SM, Yehuda R, Giller EL. Adrenergic receptor regulation in posttraumatic stress disorder. In Giller EL, ed. *Biological assessment and treatment of posttraumatic stress disorder,* Washington, DC: APA Press, 1990.
54. Yehuda R, Perry BD, Southwick SM, Giller EL. Platelet alpha2-adrenergic binding in PTSD, generalized anxiety disorder and major depressive disorder. *American Psychiatric Association New Research Abstracts* 143:NY, 1990.
55. Lerer B, Bleich A, Bennett ER, Ebstein RP, Balkin J. Platelet adenylate cyclase and phospholipase C activity in post-traumatic stress disorder. In Wolfe ME, Mosnaim AD, eds. *Post-traumatic stress disorder: etiology, phenomenology, and treatment.* Washington, DC: American Psychiatric Press, 1990;148–156.
56. Kudler H, Davidson J, Meador K. The DST and post-traumatic stress disorder. *Am J Psychiatry* 1987;144: 1068–1071.
57. Halbreich U, Olympia J, Glogowski J, Carson S, Axelrod S, Yeh CM. The importance of past psychological trauma and pathophysiologic process as determinants of current biologic abnormalities. *Arch Gen Psychiatry* 1988;45:293–294.
58. Kosten TR, Wahby V, Giller E, et al. The dexamethasone suppression test and TRH stimulation test in post-traumatic stress disorder. *Biol Psychiatry* 1990;28:657–664.
59. Olivera AA, Fero D. Affective disorders, DST, and treatment in PTSD patients: clinical observations. *J Traum Stress* 1990;3(3):407–414.
60. Dinan TG, Barry S, Yatham Ln, Mobayed M, Brown I. A pilot study of a neuroendocrine test battery in post-traumatic stress disorder. *Biol Psychiatry* 1990;28(8): 665–672.
61. Southwick SM, Krystal JH, Morgan CA, et al. Abnormal noradrenergic function in posttraumatic stress disorders. *Arch Gen Psychiatry* 1993;50:266–274.
62. Pitman RK, van der Kolk BA, Orr SP, Greenberg MS. Naloxone-reversible analgesic response to combat-related stimuli in posttraumatic stress disorder: a pilot study. *Arch Gen Psychiatry* 1990;47(6):541–544.
63. Jensen JB, Pease JJ, ten Bensel R, Garfinkel BD. Growth hormone response patterns in sexually or physically abused boys. *J Am Acad Child Adolesc Psychiatry* 1991; 30(5):784–790.
64. Gold PW, Chrousos GP. Clinical studies with corticotropin releasing factor: implications for the diagnosis and pathophysiology of depression, Cushing's disease and adrenal insufficiency. *Psychoendocrinology* 1985;10: 401–420.
65. Holsboer F, Gerken A, Stalla GK, et al. ACTH, cortisol and corticosterone output after ovine corticotropin-releasing factor challenge during depression and after recovery. *Biol Psychiatry* 1985;20:276–286.
66. Weiss RT, Tobes M, Wertz CE, et al. The corticotropin-releasing hormone stimulation test in patients with panic disorder. *Am J Psychiatry* 1986;143:896–900.
67. Roy-Byrne P, Uhde RM, Post R, et al. The corticotropin-releasing hormone stimulation test in patients with panic disorder. *Am J Psychiatry* 1986;143:896–900.
68. Pitman RK, Orr SP. Psychophysiologic testing for post-traumatic stress disorder: forensic psychiatric application. *Bull Am Acad Psychiatry Law* 1993;21:37–52.
69. Mason JW, Giller EL, Kosten TR, Harkness L. Elevation of the urinary norepinephrine/cortisol ratio in posttraumatic stress disorder. *J Nerv Ment Dis* 1988;1746: 498–502.

Neurobiological and Clinical Consequences of Stress: From Normal Adaptation to PTSD, edited by M. J. Friedman, D. S. Charney, and A. Y. Deutch, Lippincott-Raven Publishers, Philadelphia © 1995.

25

Complications of Chemical Abuse and Dependency

Susan M. Stine and Thomas R. Kosten

Department of Psychiatry, Yale University School of Medicine, New Haven, Connecticut 06510, and National Center for Post-Traumatic Stress Disorder, Veterans Affairs Medical Center, West Haven, Connecticut 06516

Substance abuse disorders are a frequent comorbid condition associated with post-traumatic stress disorder (PTSD). Among Vietnam veterans seeking treatment for PTSD, 60%–80% exhibit concurrent diagnosis for drug or alcohol abuse or dependence (1–8). These clinical data are also supported by two major epidemiologic surveys. In the Center for Disease Control's Vietnam Experience Study (9), 39% of veterans meeting criteria for PTSD during the month before examination also meet criteria for alcohol abuse or dependence. Higher rates of comorbidity were found in the National Vietnam Veterans Readjustment Study (10), which revealed that among Vietnam veterans who met criteria for a current or lifetime PTSD diagnosis, 20% and 75%, respectively, met criteria for alcohol abuse or dependence.

On a psychological level, patients with PTSD and substance abusers both suffer because they cannot regulate their emotions, self-esteem, relationships, or behaviors. It has been observed by clinicians that substance abusers resort to using substances to relieve or to control the distress associated with their self-regulation problems. This has led to the self-medication hypothesis developed by Khantzian (11,12) to explain substance abuse behavior. Decades of clinical work with alcoholics and addicts have seemed to support such a viewpoint. As the scientific thinking in psychiatry has changed, in contrast to early psychodynamic formulations that emphasized pleasurable and aggressive drives and the unconscious meanings of drug use, the more contemporary formulations have stressed how drugs and alcohol serve substance abusers as treatment for their affect deficits, troubled self-other relationships, and developmental disabilities. An application of these ideas to PTSD is illustrated by substance use in Vietnam veterans. The Vietnam War focused attention on drug abuse as well as on PTSD in soldiers, and raised concerns that returning veterans would create an epidemic of American heroin addiction. In fact, soldiers who participated in atrocities in Vietnam reported both more severe stress symptoms and more extensive use of heroin (13). They also reported that illicit drug use acutely ameliorated these stress symptoms in Vietnam. Later, back in the United States, when these symptoms recurred as part of PTSD, some veterans again turned to illicit drugs for relief and many turned to alcohol. This phenomenon became understood as an attempt at "self-medication." The self-medication hypothesis of substance abuse provided an initial understanding of the comorbidity of PTSD with substance abuse in light of the effects of ethanol and opioids on central noradrenergic hyperactivity and endogenous opioid function. The hypothesis remains controversial, however, and has limitations especially with respect to treatment approaches that assume substance abuse is a "secondary" disorder. The self-medication model will be discussed briefly in this chapter.

Regardless of the usefulness of the self-medication model, it is clear that the interaction of PTSD and substance abuse is complex and that the direction of the causal arrow is controversial. Beyond being part of the sequelae of stress disorders, substance abuse can interact with stress in a complex way at multiple levels during both development of stress-related symptoms and subsequent maintenance of these disorders. Furthermore, there is some evidence that substance abuse may contribute to the risk of developing PTSD. Even though the direction of causality is not clearly established, it is clear that both disorders display remarkable overlap in biological systems affected and in models used to explain and investigate them.

One very important way that PTSD and substance abuse disorders converge is that they are both frequently conceptualized as disorders of both psychological and biological regulation. As biological approaches to psychiatric disorders have become increasingly influential and complex, the concept of a biological self-regulation deficit, or disorder in homeostasis, has been invoked to reconcile some of these complexities. The issue of homeostasis as it applies to PTSD is discussed in detail elsewhere in this volume. However, since this concept is also used in explaining substance abuse, it further illuminates the unique connection between these disorders. Furthermore, physiologic, behavioral, and neurochemical systems themselves are similarly affected by stress and substance abuse. Pharmacological agents used in treatment of PTSD also overlap in their biological effects with the substances abused by people with this disorder. These overlapping neurochemical and pharmacological issues will be discussed in this chapter according to specific drugs of abuse.

Since biological models for PTSD and substance abuse are addressed in detail elsewhere in this volume, this chapter will not review in detail biological and behavioral science discussed elsewhere, but will focus on the complexities of the interaction between comorbid PTSD and substance abuse conditions, and the implications that this interaction has for clinical treatment. Finally, pharmacological treatment issues that reflect the interaction of these disorders will be considered.

RISK FACTORS FOR COMORBID PTSD AND SUBSTANCE ABUSE

A variety of risk factors for comorbidity have been identified in the literature. Some clinicians have argued that the high rates of substance abuse are primarily due to the fact that alcohol and other drugs were extremely available to military personnel stationed in Vietnam during the war (14,15). This cannot explain, however, the observation that veterans with higher levels of combat exposure are more likely to abuse alcohol than those who saw considerably less combat (16), since all personnel would be equally exposed to available substances of abuse. Indeed, the latter finding suggests rather that neurobiological or other alterations associated with PTSD itself make affected individuals more susceptible to alcohol and illicit drug use. Several studies in fact suggest that severity of combat trauma exposure is correlated with higher current and lifetime rates of alcohol and drug problems (17,18). This is consistent with evidence from Helzer (19) that severity of trauma is also correlated with PTSD, and tends to also support the hypothesis that severity of substance abuse is correlated with severity of PTSD.

However, there is also evidence that preservice variables may be better predictors of postservice alcohol abuse than combat exposure (19). A variety of specific risk factors have been suggested, including family history and preservice drinking (19), relative youth at the time of combat exposure (18), and a learned helplessness attributional style (20). These issues are extensively reviewed by Kofoed et al. (21).

Additional possibilities in the causal interaction would be substance abuse behavior or vulnerability as a risk factor for PTSD, as well as the possibility that certain premorbid traits may predispose individuals to both PTSD and substance abuse disorders.

Most data on PTSD-substance abuse comorbidity is obtained from studies of Vietnam veterans with PTSD; therefore, conclusions are limited concerning possible risk of acquiring PTSD in primary substance abuse patients. There are however, some data suggesting that abusers of cocaine and opiates are more likely than other substance abusers to suffer from PTSD, and that problems with drugs and alcohol in general appear to increase the risk of PTSD.

Working on the Epidemiologic Catchment Area survey, Cottler et al. (22) reported new information on the relationship between trauma and the use of drugs and alcohol. They asked 2,663 men and women in the St. Louis area about traumatic events in their lives and symptoms of PTSD—nightmares, jumpiness, trouble concentrating, feelings of shame, and a need to avoid reminders of the traumatic event. Interviewees were also asked how much alcohol they drank and which drugs they had taken more than five times for nonmedical purposes. The PTSD was more than twice as common among the drug and alcohol users, but mainly because they had had more traumatic experiences. After a trauma, only 2% of marijuana users developed PTSD; under those circumstances they were actually slightly less vulnerable than comparison subjects who did not use drugs at all. The authors suggest that people who use illicit drugs but succeed in limiting that use to marijuana have a higher-than-average capacity to cope with stress. Cocaine/opiate users on the other hand were nearly three times more likely to have experienced a traumatic event, with physical assault the chief cause. The lifetime prevalence rate for PTSD itself was 8.3% among cocaine/opiate users versus 0.3% among comparison subjects.

Another recent study (23) supports a high prevalence of civilian PTSD in an outpatient substance abuse population. In a 6-week period, 39 of 46 patients completed the Mississippi Scale for civilian PTSD (24). Seventy-two percent of those scored 86 or greater of a 39–195 score range indicating "generalized distress." Eighteen percent scored over 120, the cutoff score for PTSD.

There is also other evidence suggesting an elevated frequency of dissociative symptoms in the substance abuse population. A study reported by Dunn et al. (25) focused on veteran patients with substance abuse problems and was designed to: 1) evaluate the base rate of self-reported dissociative experiences, and 2) assess the impact of demographic and clinical variables on scores obtained with the 28-item Dissociative Experiences Scale (26). In a cohort of 265 subjects, 41.5% had a score of 15 or more on the Dissociative Experiences Scale, which suggests that dissociative experiences are common in this population. The authors suggest that screening for dissociative disorders should be considered in chemically dependent subjects.

Although this study did not discriminate between primary dissociative symptoms with secondary substance abuse and substance abuse as a risk factor for development of dissociative symptoms, the study by Cottler et al. (22) reported that, in general, the onset of substance abuse preceded the onset of post-traumatic symptoms, suggesting that substance abuse predisposes people to exposure to traumatic events or vulnerability to PTSD after such exposure.

In the Cottler study, when other variables, including antisocial behaviors, were controlled, female gender also predicted PTSD. Women may therefore form a subpopulation especially prone to trauma and the consequences of substance abuse itself. In addition to the Cottler study, a survey of metropolitan areas carried out at the National Institute of Mental Health revealed that drug abuse/dependence was the second most common psychiatric disorder among women aged 18–24 (27). Female alcoholics may suffer a higher mortality rate than male alcoholics (28), and women opiate addicts experience poorer long-term adjustment than their male counterparts (29,30,31).

In addition to sequelae of opiate abuse, women also are more vulnerable to problems associated with cocaine abuse. A study by Mendelson et al. (32) of hospitalized cocaine abusers reported notable differences between men and women. Women began using cocaine earlier than men and entered treatment at a younger age after having used cocaine for a shorter period. Men had better overall social adjustment, including better employment and living situations. Cocaine was more likely to make men feel more guilty and to reduce guilt in women. Men had a higher incidence of antisocial personality disorder, and women were more likely to have major depression. Although men and women were almost equally depressed at admission, men improved more rapidly, according to the 24-item Hamilton Depression Rating Scale (33). The possibility of gender differences in PTSD vulnerability and in vulnerability to PTSD/substance abuse comorbidity remains an important area for future investigation.

SELF-MEDICATION MODEL AND PTSD, SUBSTANCE ABUSE, AND COMORBID CONDITIONS

The concept of "self-medication" is commonly seen in the general dual diagnosis (11,34) as well as the PTSD/substance abuse literature. Those clinicians treating patients with substance abuse and PTSD have generally assumed self-medication to be the etiology of the substance abuse (35–37). This hypothesis suggests that substance use relieves specific dysphoric symptoms of PTSD and reinforces further use. The hypothesis also touches on the controversy within the field of substance abuse concerning the relative roles of positive (e.g., drug-induced euphoria) versus negative (relief from distress) reinforcement in maintaining substance abuse behavior. This concept, espoused and developed over the years by Wikler (38,39), emphasizes physical dependence, the abstinence syndrome, correlated environmental stimuli, and primarily aversive correlates of opiate use. Currently, in most cases the role of positive reinforcement is considered to be most important in maintaining substance abuse behavior, but negative reinforcement also plays an influential role and is especially important in maintaining opiate dependence (40). Similarly, the self-medication model that can be considered a diagnosis or symptom-specific modification of more general "tension reduction" models, although widely used and popular among clinicians, has fallen out of favor in scientific literature. For example, as reviewed by Volpicelli (41), evidence for tension reduction models of substance abuse is mixed and inconsistent. A later review by Kofoed et al. (21), specifically of PTSD and substance abuse, also finds the self-medication model of little help for planning therapeutic approaches.

One argument against the self-medication model is that much of what is found in non-PTSD literature examining diagnosis/drug preference combinations seems counter-intuitive (for instance, increased rates of stimulant abuse in schizophrenic samples) (34). However, this particular example could be explained by possible relief of negative symptoms by stimulant abuse in these patients.

Another argument is that there is little specificity of drug choice in PTSD, rendering the self-medication hypothesis too nonspecific to be helpful (21). Since the main biological abnormality associated with PTSD most likely to generate susceptibility to chemical abuse/dependency is sympathetic nervous system hyperarousal and chronic lowering of endogenous opioid levels, it would be predictable that PTSD patients might successfully ameliorate intolerable symptoms with heroin, methadone, and other opiates. (Kosten and Krystal (35) have reviewed the biological basis for PTSD symptoms and substance abuse.) This is consistent with clinical experience since PTSD patients generally prefer alcohol, marijuana, central depressants or opiates, and cocaine and stimulant users with PTSD are usually also dependent on alcohol, marijuana, and opiates (8). Nevertheless, the wide range of substances used, even if understandable in terms of PTSD biology, does limit the usefulness of the self-medication hypothesis. In the section that follows, some biological grounds for use of specific substances will be mentioned.

Although the self-medication model has intuitive appeal and may be useful in understanding the etiology and persistence of comorbid PTSD/substance abuse if it is not applied too rigidly, it is certainly a misuse of this model to conclude that treating the "primary" disorder (PTSD) will lead to resolution of the "secondary" disorder. As eloquently reviewed by Kofoed et al. (21), this position has little support, and most evidence confirms the clinical necessity of simultaneous treatment of both disorders.

The self-medication hypothesis is certainly not useful if a simple one-to-one correspondence is expected between a specific comorbid psychiatric diagnosis and specific substances abused. It may be more useful if specific symptoms and affected biological systems are considered with respect to particular drugs of abuse that relieve or exacerbate those symptoms. The next section will consider specific drugs of abuse with respect to their self-medication potential for PTSD, as well as paradoxical ways in which the self-medication attempt can result in worsening of the disorder.

SPECIFIC SUBSTANCES OF ABUSE AND IMPLICATIONS FOR STRESS DISORDERS

Opiates

Opiate abuse and dependence can influence PTSD symptoms at multiple biological levels. Opiates impinge on a particular vulnerability in PTSD patients. As discussed elsewhere, PTSD is associated with abnormality in endogenous opioid systems. Most notably, plasma beta-endorphin levels have been observed to be low in PTSD patients (42). In general, acute opiate use diminishes PTSD symptoms, while chronic opiate abuse and opiate withdrawal states exacerbate such symptoms. In fact, opiate withdrawal and acute PTSD symptoms can be so similar that some authors have called PTSD a withdrawal state (43). Van der Kolk and associates (43) have also proposed that the animal model of learned helplessness in the face of inescapable shock may be directly applicable to PTSD. They hypothesize that long-term potentiation of locus coeruleus pathways to the hippocampus and amygdala may produce the hyperarousal, traumatic nightmares, and flashbacks that characterize PTSD. Such a theory also suggests that fluctuations in endogenous opioid levels will affect the response to traumatic stimuli, since the locus coeruleus is inhibited by opioids. The inescapable shock theory offers a neurobiological rationale for stress-induced analgesia, and for the "action junkie" behavior that is sometimes considered secondary to PTSD and thus has implications for opiate addiction.

Much of the interaction between PTSD and opiate abuse can be related to the adrenergic abnormalities discussed elsewhere in this volume. Kosten and Krystal (35) have suggested that adrenergic inhibition of corticotropin-releasing hormone (CRH) may account for a disturbance in the endogenous opioid system associated with PTSD. The CRH promotes release of adrenocorticotropic hormone (ACTH) from the pituitary; ACTH is coreleased with beta-endorphin, which influences the activity level of the endogenous opioid system. Therefore, Kosten and Krystal postulate that inhibition of CRH by excessive sympathetic arousal also produces an endogenous opioid deficiency in patients with PTSD. This prediction is consistent with previously mentioned clinical reports of lowered pain thresholds in PTSD patients (44) and of a possible link between chronic pain and PTSD (45,46).

Opiate withdrawal, which has been demonstrated to have an important adrenergic component (47), has even been reported to present as PTSD (48). Detoxification from opiates therefore represents a special problem in PTSD patients. Withdrawal syndromes associated with these drugs of abuse increase central noradrenergic activity and may worsen PTSD symptoms (47,49,50). Methadone-maintained patients are a special group in which chronic opiate medication may facilitate PTSD treatment in dually diagnosed patients but also ironically maintain vulnerability for adrenergic hyperactivity. Some recent evidence suggests that opiate addicts treated with methadone experience withdrawal-like symptoms when given drugs that stimulate the adrenergic system, such as cocaine (51) or yohimbine (52) (Stine et al., unpublished data).

Another recent example of overlap between PTSD and substance abuse was reported by Southwick et al. (53). In that study, a subgroup of patients with PTSD were observed to experience yohimbine-induced panic attacks (70% [14/20]) and flashbacks (40% [8/20]). In addition, in the patients with PTSD, yohimbine induced significant increases in core PTSD symptoms such as intrusive traumatic thoughts, emotional numbing, and grief.

Since yohimbine has also been reported to precipitate opiate withdrawal-like symptoms in methadone-maintained patients, it is to be expected that dually diagnosed PTSD/opiate-dependent patients have a special vulnerability to PTSD relapses. In addition to being more susceptible to adrenergic-opioid system interactions, PTSD patients are at greater risk of opiate dependence, since heroin not only dampens adrenergic hyperarousal but may also serve to replenish an endogenous opioid system that has been depleted because of the pathophysiology of PTSD.

Interestingly in this context, Pitman and associates (54) have shown that exposing Vietnam veterans with PTSD to combat scenes from the movie *Platoon* produces a naloxone-reversible 30% decrease in pain responses. This important finding of stress-induced analgesia suggests not only that PTSD is associated with dysregulation of the endogenous opioid system, but also that a possible baseline opioid deficiency might be dramatically reversed when PTSD patients are exposed to traumatic stimuli. These dramatic fluctuations of activity of the endogenous opioid system could mimic and promote the experience of opiate abuse.

Stimulants

Stimulants like cocaine and amphetamines exert their effects on the brain by increasing extracellular concentrations of dopamine (DA), noradrenaline (NA), and 5-hydroxytryptamine (5-HT). Although both cocaine and amphetamines are equally effective in blocking the reuptake of monoamines, amphetamine is more potent than cocaine in that it also promotes release (55–57).

Most studies have focused on the effects of stimulants on dopaminergic systems, due to the high correlation of this system with the rewarding efficacy and abuse liability of these drugs (58–60). Very little is known about the effects of chronic stimulant usage on other brain monoamine systems.

As mentioned earlier, PTSD has striking similarity to stimulant abuse in the way that brain neurobiology is affected. In animal research, the forebrain DA neurons show an augmented response to repeated stress, as well as to repeated administration of psychostimulants such as cocaine and amphetamines. Previous exposure to stress increases the subsequent locomotor response, subserved by DA systems of the striatum, to cocaine challenge (61–64). Thus, stress and psychostimulants can, under certain conditions, cross-sensitize. Clinical reports of severe anxiety and depression following stimulant withdrawal (65) suggest that (66,67) chronic stimulant usage may influence PTSD symptom severity through brain noradrenergic systems.

A recent study (68) in rats demonstrated that chronic exposure to amphetamine, cocaine, and the tricyclic antidepressant DMI increased activity in the noradrenergic neurons of the locus coeruleus. After 2 weeks' exposure, cells from animals in all three drug-treated groups showed a significant increase in sensitivity to the acute effects of cocaine and amphetamine. These animals also had a significant increase in behavioral stimulation (stereotypy). Autoreceptors on monoamine neurons have been reported to become subsensitive following chronic exposure to stimulants (69–71) and DMI (72,73). In the Harris and Williams study (68), there was also a significant increase in the behavioral sensitivity of the drug-treated animals to the sedative effects of clonidine.

Although a similar effect has not been demonstrated directly in humans or PTSD patients, these findings have implications for clinical observations in drug-abusing patients. Because central stimulants facilitate sympathetic hyperarousal (which would be predicted to exacerbate PTSD symptoms), PTSD patients would also be expected to exhibit less cocaine and amphetamine abuse. In fact, Friedman (8) reports that "PTSD patients do not like the heightened emotional state produced by cocaine, amphetamines, and other stimulants." It would also be expected that risk of developing PTSD and the severity of ongoing PTSD would be increased in those patients who do abuse stimulants. In fact, an epidemiological study of substance abusers discussed previously did support increased risk for development of PTSD in this population. In that study (22), cocaine/opiate users were the only group especially vulnerable to PTSD once they were exposed to trauma, thus providing interesting indirect evidence for this hypothesis. Eight percent of all cocaine/opiate users and 19% of those experiencing a trauma had the symptoms. In addition to illicit stimulant abuse, caffeine and other psychostimulants, including some found in common over-the-counter decongestant medications, also may present a risk for patients prone to anxiety disorders.

Opiate/Stimulant Interactions

The concurrent use of opiates and cocaine by addicts has interesting implications for PTSD patients because of the interaction of heroin and cocaine on opiate withdrawal. Opiate withdrawal symptoms precipitated by naloxone challenge may be attenuated in opiate-dependent individuals who have a recent history of cocaine abuse, but are not currently intoxicated with cocaine (74). This symptom relief may explain some cocaine use by patients on low doses of methadone (25–40 mg daily) and why cocaine use increases in some opiate-abusers after onset of methadone maintenance (75). However, cocaine use may also be reduced by sufficiently high doses of methadone (76). A potential mechanism for these apparently contradictory findings involves the $alpha_2$ adrenergic system. The antagonist yohimbine (which causes indirect stimulation of noradrenergic neurons of blocking feedback inhibition at the presynaptic autoreceptor) can precipitate opiate withdrawal when administered acutely (52), but it also attenuates development of withdrawal symptoms when administered chronically and repeatedly in rats dependent upon opiates (77). If cocaine-induced noradrenergic stimulation acutely precipitates or exacerbates opiate withdrawal symptoms (or adrenergic symptoms sufficiently similar to be interpreted as withdrawal), chronic cocaine use may attenuate subsequent naloxone-precipitated withdrawal in a fashion analogous to the effect of repeated yohimbine administration.

This opens the door to many complex, although speculative, interactions with the biology of PTSD, but it is clear that the attempt to self-medicate PTSD with both opiates and stimulants can both ameliorate and exacerbate PTSD symptoms.

Alcohol and Benzodiazepines

The relationship between PTSD and alcoholism may be especially important; one survey by Sierles et al. (6) showed that PTSD was highly associated with other concurrent psychiatric illnesses, particularly alcohol abuse. In a later study by Davidson et al. (78), alcohol abuse and nonbipolar depression were the most frequently noted diagnoses (41% each) in PTSD. Nonalcoholic substance abuse (16%) was also quite common. Not only is alcoholism a common diagnosis in the lifetime assessment of PTSD patients, but it is also present in more than 50% of first-degree relatives of probands with this disorder (78). Drug abuse and alcohol abuse considered together were the most common of all diagnoses in relatives of PTSD probands (60%) (79). The relative specificity of the connection is supported by the observation that alcohol and drug abuse were also significantly more common in the relatives of patients with PTSD (60%) than in relatives of patients with depression (26%) and generalized anxiety (38%).

Alcoholism also influences the neurobiologic findings of PTSD studies, and clinical experience suggests that comorbid diagnoses influence treatment response. Low platelet monoamine oxidase (MAO) activity has been suggested as a psychopathological risk factor for many conditions, including alcoholism (80–82). Davidson et al. (79) reported that the platelet MAO activity of veterans suffering from chronic PTSD was lower than that of control subjects. Patients were grouped according to whether or not they had a history of alcohol abuse and whether or not they had a concurrent diagnosis of major affective disorder. Nine patients reported a positive history and 14 patients reported a negative history for alcohol abuse. The alcohol-abusing PTSD patients also had significantly lower platelet MAO activity than the nonalcoholic PTSD patients.

All patients in the Davidson et al. (78) study had experienced another psychiatric disorder at some time in their lives. Many times these were of long duration and led to considerable morbidity, with hospitalization, loss of job, and disrupted interpersonal relationships. The authors speculate that one diagnosis sustains the other. Thus, the ability of PTSD to resolve will be hindered by a coexisting depression or other major psychiatric illness, especially substance abuse.

GENERAL OBSERVATIONS IMPLICATIONS FOR TREATMENT

General Treatment Issues

Not only do PTSD and substance abuse disorders overlap in their acute and chronic effects and influence each other at all levels—both disorders also can influence the resilience of the fundamental psychological and biological systems themselves. This results in disorders of self-regulation and compromises a patient's ability to benefit from treatment. Thus, treatment of the dually diagnosed patient must be firmly based simultaneously in both PTSD and substance abuse theory and technique.

Treatment specifically for PTSD is discussed in detail elsewhere in this volume, so this chapter will focus on the role of substance abuse treatment and its interaction with PTSD treatment. A brief mention of the history of substance abuse treatment may be useful. Dating back to the late 1960s, psychodynamic studies by investigators such as Weider and Kaplan (83), Milkman and Frosch (84), Wurmser (85), Krystal (86), Khantzian (11,12), Wilson et al. (87), Luborsky et al. (88), and Dodes (89) have indicated that substance abusers' difficulties stem from developmental handicaps and psychological deficits. Khantzian (11,12) describes four areas of self-regulation problems that predispose addicts to become dependent on and relapse to substances of abuse: 1) affect life; 2) self-esteem; 3) relationships; and 4) self care. Although the treatment of substance abuse disorders through psychodynamic methods is controversial, this formulation has been fruitful for clinicians. Other psychological approaches such as the learning model have contributed enormously in recent years as a basis for empirical research and treatment development. Learning models also deal with elements of self regulation (e.g., identifying and extinguishing automatic and impulsive responses to drug-associated cues).

Currently, most substance abuse treatment involves group or individual psychotherapy based on twelve-step programs such as Alcoholics Anonymous or relapse prevention therapies, a type of cognitive/behavioral therapy developed by Marlatt (90) and extended by others (91). Although it has been repeatedly stated that the high comorbidity of PTSD/substance abuse (91% according to Boudewyns et al. (92)) and the "inextricably intertwined" nature of these disorders require simultaneous treatment of both, there are few articles describing such a combined treatment. As reviewed by Kofoed (21), clinicians generally have combined group and individual therapies for PTSD with aspects of substance abuse treatment such as the twelve-step approach. Authors such as Abueg (93) and Jellinek (94) generally prepose a multiple-stage approach in which a stabilization and motivational period is followed by more focused PTSD and substance abuse treatment, often in a group format.

Comparisons of different therapies on outcome immediately posttreatment and at longer follow-up intervals have been limited. The Abueg (93) study suggests that relapse prevention training during a residential PTSD treatment program is superior to residential treatment alone, but that this differential effect is lost at 9 months posttreatment. This is consistent with clinicians' general assessment that treatment must be long-term, including inpatient and outpatient services.

Another study in a group of patients undergoing residential alcohol treatment (95) suggests no difference in abstinence rates for a heavy combat exposure group treated with additional readjustment group therapy and a light combat exposure group receiving no additional treatment. However, this study is limited by a nonrandomized design and no controls (21).

As discussed earlier, women present particular treatment challenges. Women remain a subgroup of noncombat substance abusers at risk for PTSD. In a recent study of women substance abusers with and without PTSD, women with PTSD were more likely to have been victims of sexual and physical abuse, had higher scores on the psychological problems and substance abuse areas of the Addiction Severity Index, and were more likely to have comorbid affective disorder. They also had lower compliance with aftercare (96). The authors recommend screening women substance abusers for victimization and PTSD in order to provide more aggressive PTSD treatment and improve treatment retention.

Clearly, more controlled studies with attention to treatment-matching issues are needed. Some progress has been made in this direction by McLellan and colleagues, who found that global severity as measured by the Addiction Severity Index (ASI) was a powerful prognostic variable (97). Psychiatric severity as measured by the ASI differentially predicted treatment response in both retrospective (98) and prospective studies (99), with low severity generally predicting greater response to treatment. Other studies have also found that psychiatric severity can predict treatment results in substance abusers, with specific diagnoses such as depression associated with differential response to specific therapies (100,101). However, no studies have examined this issue specifically in PTSD patients.

PHARMACOLOGICAL TREATMENT

Psychological techniques can be complemented by pharmacological treatment, and these pharmacological treatments can themselves constitute a psychological tool. Just as behavioral techniques have proved useful in extinguishing some of the conditioned-alarm aspects in PTSD, combined pharmacological-behavioral approaches may assist in disrupting conditioned cues that promote drug abuse. Disulfiram has been used successfully to extinguish ethanol use by negatively reinforcing its abuse. Methadone maintenance was intended to extinguish opiate abuse by blocking the rewarding properties of self-administration through cross-tolerance (102,103). Opiate maintenance may also provide some protection against PTSD symptoms. However, maintenance on opiate agonists postpones eventual detoxification, and detoxification may revive PTSD symptoms. The opiate antagonist naltrexone might extinguish previous conditioned, abuse-promoting stimuli, and although naltrexone may release peripheral catecholamines and antagonize central gamma-aminobutyric acid (GABA) systems, thereby exacerbating both PTSD and conditioned withdrawal phenomena, this occurs at much higher doses than routinely used (104).

A biological approach may complement psychological diagnostic and treatment techniques. With regard to treatment, almost every type of psychotropic agent has reported efficacy in PTSD. However, very few double-blind therapeutic trials have been published. Successful pharmacotherapy appears to diminish DSM-III-R intrusive recollections and hyperarousal, but not avoidant symptoms (8). Current information suggests that drug treatment alone is not sufficient to relieve the suffering in PTSD, but is often useful as an adjunct to psychotherapy.

Many medications have been reported helpful in treatment of PTSD, but most reports are of open treatment studies. The status of the pharmacological treatment of PTSD has been recently reviewed by Friedman (105). Controlled studies are few but have reported efficacy for tricyclic antidepressants (TCAs) (106–108), monoamine oxidase inhibitors (106,109), the beta-adrenergic antagonist propranolol (110), and alprazolam (111). Open trails have supported a possible role for the medications just mentioned, as well as carbamazepine, lithium, neuroleptics (8,112,113, 114) (see review by ver Ellen and van Kammen (115)). Other medications studied in more recent open trials are buspirone, fluoxetine, cyproheptadine, alprazolam, valproate, and TCA/clonidine combination therapy (reviewed by Friedman (105)). Most controlled studies report that successful pharmacotherapy for PTSD results in attenuation of DSM-III-R intrusive recollections (especially nightmares) and arousal (especially insomnia, startle, and irritability) symptoms, while avoidant/numbing symptoms usually do not respond to medication (8). An exciting preliminary result in this regard, however, is that fluoxetine may reduce the severity of avoidant/numbing as well as the other PTSD symptoms (see next paragraph) (116,117).

A variety of open studies has suggested the usefulness of selective serotonin reuptake inhibitors (SSRIs) in PTSD treatment (116,118–121). There is also a considerable body of evidence for the role of serotonin in the pathophysiology of PTSD. (see Chapter 26 for a more detailed review.) The only controlled trial with SSRIs, reported by van de Kolk et al. (122), confirmed the effectiveness of fluoxetine in reducing arousal and numbing but not avoidance symptoms. Fluoxetine has also been suggested as a potential pharmacotherapy for substance abuse, specifically for cocaine abuse. Fluoxetine reduces intravenous cocaine self-administration in rats (123) and has been clinically shown to reduce cocaine abuse in an open trial in opiate-dependent patients who were in methadone maintenance treatment (124).

Other pharmacological agents suggested for treatment of substance abuse also have some overlap with medications used in PTSD treatment. While there are specific medications for opiate dependence (e.g., maintenance treatment with agonists such as methadone, LAAM, and buprenorphine as well as the antagonist naltrexone), the state of the art for cocaine abusers is quite different. Similar to PTSD treatment studies, there are many encouraging open trials (e.g., medications that have appeared effective for cocaine abuse in open trials include disulfiram but not naltrexone (125), amantadine (126), bromocriptine (127), pergolide (128), methylphenidate (129), flupenthixal (130), and mazindol (131), as well as buproprion (132) and fluoxetine (124)). However, few of these pharmacological agents have been confirmed as effective in controlled trials. One medication that has shown some effectiveness for cocaine dependence in a controlled trial is desipramine (130). Although later attempts to replicate this finding have been negative or mixed (133–135), this medication may have effectiveness for a depressed subgroup (136). It may be that since desipramine has been useful for PTSD symptoms, it would be an effective medication for PTSD/cocaine abuse patients. Such specific pharmacological studies for comorbid PTSD/substance abuse patients have not yet been reported.

Carbamazepine and related medications represent a particularly interesting potential pharmacotherapy for both PTSD as well as substance abuse. The theoretical usefulness of this medication is based on the kindling model that has been independently suggested to be useful in understanding substance abuse (137) as well as PTSD (112,113). Briefly, according to this model discussed in a detailed fashion elsewhere, chronic central sympathetic arousal in PTSD, mediated by the locus coeruleus, sensitizes limbic nuclei, thereby producing a longlasting facilitation of continued abnormal arousal. As Friedman (8) comments, kindling would also explain the chronic nature of PTSD, which can continue for decades if untreated (138). Carbamazepine and valproate are two antikindling agents with reported efficacy in treatment of PTSD. The use of carbamazepine in PTSD has been reviewed by Friedman (105) and is supported by two open trials (139, 140). The usefulness of valproate in PTSD has also been supported by open trial (141) and case report (142) evidence.

Carbamazepine has also been investigated for treatment of cocaine dependence, with inconclusive results. In a short (20-day) double-blind crossover study in 32 nonmotivated crack users, carbamazepine improved early cocaine treatment retention and treatment effectiveness (143). Serum carbamazepine levels of 4 ng/ml or more were associated with greater improvement in that study. A subsequent 12-week controlled study of 183 randomized outpatients by these authors supported the usefulness of this medication (144).

Other 8-week (145) and 12-week (146) treatment studies have not replicated this effect. It is tempting to speculate that carbamazepine and related drugs may have special efficacy in a comorbid population, but this hypothesis has not been tested.

Thus, it is apparent that the options in pharmacotherapy are diverse and no treatment is a clear cure. Furthermore, increased understanding of the complex neurobiology of PTSD and substance abuse argues against the likelihood of the future discovery of any single specific pharmachotherapy for these disorders. Increased controlled clinical trials are therefore necessary to develop better approaches to matching subpopulations of patients to treatment in order to fully benefit from the recent advances in this area.

TREATMENT OF WITHDRAWAL STATES—A SPECIAL CHALLENGE

Pharmacological treatment is particularly important for certain acute problems associated with PTSD/substance abuse. As discussed earlier, drug withdrawal states, perhaps through a common noradrenergic pathway, often exacerbate post-traumatic stress syndromes. Opiate, benzodiazepine, and ethanol withdrawal are associated with central noradrenergic activation and subjective alarm states perhaps similar to that hypothesized for PTSD. These clinical states of drug withdrawal might have additive effects with PTSD on central noradrenergic function and, as a result, worsen the clinical presentation. Alcohol and opiate withdrawal alternating with abuse of these substances causes repeated adrenergic arousal, which can trigger a conditioned emotional response associated with PTSD symptoms (35). Thus, the treatment for chemical dependency, already complex, can precipitate and exacerbate PTSD symptoms, presenting a formidable combination. In other words, the normal difficulties of treating chemical dependency are multiplied by the complex risk of exacerbating PTSD symptoms. A related phenomenon during withdrawal from benzodiazepines and ethanol may be polysensory hallucinations that take the form of post-traumatic flashbacks. Clinical experience suggests that treatment of the drug withdrawal also alleviates components of the exacerbated PTSD symptoms. One could hypothesize that drug abuse whose intended purpose was to alleviate withdrawal might serve a similar protective function of reducing PTSD symptom exacerbations.

Opiate detoxification can present similar problems in PTSD patients. One approach to opiate detoxification already proving useful is to combine clonidine-assisted opiate detoxification with naltrexone in order to minimize central noradrenergic activation during the detoxification and antagonist-initiation phases of treatment (147–150). A compromise solution to minimize symptoms might be achieved through new mixed agonist-antagonist drugs such as buprenorphine (151). During maintenance phases, buprenorphine's agonist actions may suppress conditioned noradrenergic activation, while its antagonist action minimizes the severity of abstinence phenomena and may make the transition to a complete antagonist, such as naltrexone, more tolerable. This approach has been recently reviewed by Stine and Kosten (152).

Somewhat paradoxically, pure opiate antagonists may prove useful in PTSD treatment and substance abuse treatment. Intriguing preliminary results reported by Glover (153) support the potentical usefulness of this approach. In this open study of the effect of the opiate antagonist nalmefene on 18 combat veterans with PTSD, both positive and negative symptoms were observed to improve in 8 subjects. Interestingly, numbing was seen to decrease initially, while anxiety was intensified, as would be predicted in acute antagonist administration. Later responses at higher doses were improvements in both positive and negative PTSD symptoms; this is consistent with a possible protective and normalizing effect of opiate antagonists from cyclical disturbances in endogenous neurotransmission. Another report of the use of naltrexone in PTSD (to reduce flashbacks) has also been published in a brief case history of two patients (154). The use of opiate antagonist is especially interesting in light of the recent reports of naltrexone in treatment of alcoholics (155,156). Thus, naltrexone may be especially useful in patients with PTSD/alcohol abuse comorbidity.

In some chronically relapsing opiate addicts who are not good candidates for detoxification and abstinence, one might choose to medicate PTSD symptoms during opioid maintenance phases. Because of the risk for abuse in these individuals, one might choose tricyclics over benzodiazepines in this population. Reports in depressed opiate abusers suggest that proper psychopharmacological treatment may assist some individuals in maintaining abstinence from illicit drugs and relieving depressive symptoms (157,158). Control of symptoms associated with PTSD may facilitate drug abuse treatment in a parallel fashion.

An example of the synergism between PTSD and alcohol withdrawal phenomena was described by Kosten and Krystal (35). Symptoms were controlled through benzodiazepine-assisted alcohol detoxification, control of panic attacks with alprazolam, and neuroleptic administration. Control of these symptoms ultimately permitted the patient to share some of his vivid memories in therapeutic settings.

After acute treatment of intoxication and withdrawal states, longer-term relapse prevention treatment also requires that attention be paid to both PTSD and substance abuse pathology, particularly learned responses related to withdrawal phenomena. Mild protracted withdrawal as well as conditioned withdrawal symptoms in response to specific cues may exacerbate PTSD, even after detoxification from addicting substances. Individuals withdrawn from opioids and ethanol often experience withdrawal symptoms and craving for the drug of abuse when exposed to reminders of the drug abuse setting. These conditioned symptoms may reflect central noradrenergic activation and be misattributed to PTSD symptoms (47). Misattribution of the source and significance of conditioned arousal states potentially exacerbates both PTSD symptoms and drug abuse, because individuals with either PTSD or substance abuse have great difficulty labeling and expressing their feelings (159,160). Henry Krystal (161) noted that such individuals tend to experience dysphoric arousal in an undifferentiated global manner, suggesting that those with both disorders may confuse or misattribute conditioned PTSD and withdrawal phenomena. Many stimuli evoke both war experiences and prior drug use; thus, individuals may present with both drug withdrawal and PTSD symptoms. Misattributing the source of arousal, individuals might attempt to alleviate their pseudowithdrawal state rather than seek treatment for PTSD. Recognizing this overlap, the clinician can more accurately develop a treatment plan for these dually diagnosed patients. The patient who presents to drug abuse units and ignores PTSD as contributing to his symptoms, as well as the patient who presents to PTSD-oriented outreach programs and minimizes drug abuse problems, will have inadequate treatment without concurrent management of these two problems.

Benzodiazepines in PTSD/Substance Abuse Treatment

The use of alprazolam described by Kosten and Krystal in the previous section (35) leads us to consider the role of benzodiazepines in PTSD and substance comorbidity. Even the therapeutic use of benzodiazepines remains controversial. Alcoholics represent a special population within the larger PTSD/substance abuse population that may have particular problems with benzodiazepine treatment.

Benzodiazepines are frequently used transitionally to help alcoholics withdraw from alcohol. For patients withdrawing from habitual alcohol use, withdrawal itself may be anxiogenic. But for those suffering from an anxiety disorder, it may be tempting to continue benzodiazepine use indefinitely. Ciraulo et al. (162) concluded that "alcoholics as a group may be more susceptible to benzodiazepine abuse than are nonalcoholics." Also, Ciraulo et al. (163) suggest that the sons of alcoholics may be at higher risk to abuse benzodiazepines. Furthermore, even when benzodiazepines are used therapeutically and appropriately for PTSD patients, they may be problematic for patients who abuse alcohol and other drugs.

The PTSD patients may be unusually sensitive to alprazolam or its withdrawal reaction. In the case of alprazolam, these concerns are augmented by the additional risk of rebound anxiety and severe withdrawal symptoms (164,165). In fact, Risse et al. (166) reported on eight Vietnam veterans who experienced severe exacerbation of their PTSD symptoms during alprazolam withdrawal. The patients exhibited anxiety, sleep disturbance, rage reactions, hyperalertness, increased nightmares, intrusive thoughts, and homicidal ideation. Disappointing results also emerged from a trial of alprazolam in PTSD conducted by Braun and others in Israel (111). Sixteen patients entered this double-blind, random-assignment, placebo-controlled crossover trial; 10 completed 5 weeks of treatment with both alprazolam and placebo. Alprazolam demonstrated significant ($P=.02$) advantage over placebo for anxiety symptoms, but not for symptomatology specific to the PTSD syndrome itself. Although they do not provide details, the authors refer to withdrawal effects in patients who received alprazolam before placebo.

Benzodiazepines can also complicate cognitive effects of PTSD. Anthenelli et al. (167) assessed the ability of 103 healthy young white men, none of whom totally abstained from alcohol to recall information. The investigation reported a significant dose-dependent decrement in word recall during even acute diazepam challenges ($P<.0001$).

There are certainly some good arguments in favor of the use of benzodiazepines as well. As reviewed by Friedman (105), benzodiazepines in general are excellent anxiolytics, and alprazolam, in particular, has potent anxiolytic/antipanic actions. Furthermore, the kindling model of PTSD offers a theoretical reason to consider these drugs, since limbic kindling is associated with increased benzodiazepine receptor binding (168–170). There are two published reports on alprazolam in PTSD treatment. Feldman (171) conducted an open trial and found that 16 of 20 veterans with PTSD treated with alprazolam showed reduced insomnia, anxiety, irritability, and hyperarousal. Evidence for benzodiazepine-induced emotional disinhibition is indicated, however, by Feldman's report that four of these patients showed an increase in outbursts of anger. Clonazepam may be an alternative benzodiazepine that avoids some of the pitfalls of alprazolam treatment. Kofoed et al. (21) recommend clonazepam because of its efficacy, and because slow absorption and elimination reduce euphoric responses and hence abuse potential. Interestingly, a recent study by Tietz et al. (172) in rats reports that tolerance develops more slowly to clonazepam as compared with alprazolam. This effect, if also seen in humans, could support decreased abuse liability for this medication.

The issue of benzodiazepine treatment is more extensively reviewed elsewhere in this volume, but it is clear that these agents are a double-edged sword in the treatment of PTSD, especially in the presence of a substance abuse history.

CONCLUSION

Just as the concept of self-medication provides no simple one-to-one correspondence between diagnosis/treatment agent, no one-to-one correspondence can be expected between a PTSD and substance abuse diagnosis and any one method of psychological or pharmacological treatment. Some initial progress has been made in treatment matching to identify patient subgroups best treated with specific approaches, but randomized controlled drug trials and treatment-matching studies of both psychotherapy and pharmacotherapy are needed. The concept of self-medication, if used judiciously, can be useful in understanding the relationships between substance abuse and PTSD symptoms, the biological systems affected, and the drugs patients choose to abuse, and can generate useful hypotheses in the design of needed studies.

REFERENCES

1. Davidson RT, Kudler HS, Saunders WB, et al. Symptom and comorbidity patterns in World War II and Vietnam veterans with post traumatic stress disorder. *Compr Psychiatry* 1990;31:162–170.
2. Eacobar JI, Randolph ET, Puente G, et al. Post-traumatic stress disorder in Hispanic Vietnam veterans: clinical phenomenology and sociocultural characteristics. *J Nerv Ment Dis* 1983;171:585–596.
3. Friedman MJ, Kolb LC, Arnold A, et al. Third annual report of the Chief Medical Director's Special Committee on Post Traumatic Stress Disorder, 1987.
4. Keane TM, Fairbank JA. Survey analysis of combat-related stress disorders in Vietnam veterans. *Am J Psychiatry* 1983;140:348–350.
5. McFall ME, MacKay RW, Donovon DM. Combat related PTSD and psychosocial adjustment problems among substance abusing veterans. *J Nerv Ment Dis* 1991;179:33–38.
6. Sierles FS, Chen J-J, McFarland RE, Taylor MA. Post traumatic stress disorder and concurrent psychiatric illness: A preliminary report. *Am J Psychiatry* 1983; 140:1177–1179.
7. Sierles FS, Chen JJ, Messing ML, et al. Concurrent psychiatric illness in non-Hispanic outpatients diagnosed as having post traumatic stress disorder. *J Nerv Ment Dis* 1986;174:171–173.
8. Friedman MJ. Biological approaches to the diagnosis and treatment of post traumatic stress disorder. *J Traum Stress* 1991;4:67–91.
9. Health Status of Vietnam Veterans. Volume IV Psychological and Neuropsychological Evaluation. U.S. Department of Health and Human Services Public Health Service Center for Disease Control, Atlanta Georgia, pg. 62, 1989.
10. Kulka RA, Schlenger WE, Fairbank JA, et al. Trauma and the Vietnam War generation. *Bruner-Mazel* 1990: 117–126.
11. Khantzian EJ. The self-medication hypothesis of addictive disorders. *Am J Psychiatry* 1985;142:1259–1264.
12. Khantzian EJ. Self-regulation and self-medication factors in alcoholism and the addictions. In Galanter M, ed. *Recent developments in alcoholism,* vol. 8. New York: Plenum Publishing Corp; 1990.
13. Yager T, Laufer R, Gallops M. Some problems associ-

ated with war experience in men to the Vietnam generation. *Arch Gen Psychiatry* 1984;41:327–333.
14. Sapol E, Roffman RA. Marijuana in Vietnam. *J Am Pharm Assoc* 1969;NS9:615–630.
15. Wedding D. Substance abuse in the Vietnam veteran. *AAOHN J* 1987;35:74–76.
16. Keane TM, Gerardi RJ, Lyons JA, et al. The interrelationship of substance abuse and post traumatic stress disorder: epidemiological and clinical considerations. *Rec Dev Alcohol* 1988;6:27–48.
17. Jordan BK, Schlenger WE, Hough R, et al. Lifetime and current prevalence of specific psychiatric disorders among Vietnam veterans and controls. *Arch Gen Psychiatry* 1991;48:207–215.
18. Fisher VJ. Combat exposure and the etiology of post discharge substance abuse problems among Vietnam veterans. *J Traum Stress* 1991;4:251–277.
19. Helzer JE. The impact of combat on later alcohol use by Vietnam veterans. *J Psychoactive Drugs* 1984;16 (2):183–191.
20. McCormick RA, Taber JI, Knuedelback N. The relationship between attributional style and post traumatic stress disorder in addicted patients. *J Traum Stress* 1989;2:477–487.
21. Kofoed L, Friedman MJ, Peck R. Alcoholism and drug abuse in patients with PTSD. *Psychiatr Q* 1993;64 (2):151.
22. Cottler LB, Compton WM III, Mager D, et al. Post traumatic stress disorder among substance abusers from the general population. *Am J Psychiatry* May 1992;149:664–670.
23. Triffleman E, Ball S, Rounsaville B. Screening for PTSD in substance dependent outpatients. *Processings of the 56th Annual Scientific Meeting of the Committee for Problems in Drug Dependence,* National Institute on Drug Abuse Research Monograph LS Harris, ed. NIDA Monograph Pub, 1995;153(2):345.
24. Keane TM, Caddell JM, Taylor KL. Mississippi Scale for combat-related post traumatic stress disorder: three studies in reliability and validity. *J Consult Clin Psychol* 1988;56(1):85–90.
25. Dunn GE, Paolo AM, Ryan JJ, Van Fleet J. Dissociative symptoms in a substance abuse population. *Am J Psychiatry* 1993;150:1043–1047.
26. Bernstein EM, Putnam FW. Development, reliability, and validity of a dissociation scale. *J Nerv Ment Dis* 1986;174:727–735.
27. Robins LN, Helzer JE, Weissman MM, Orvaschel H, Gruenberg E, Burke JD, Regier DA. Lifetime prevalence of specific psychiatric disorders in three sites. *Arch Gen Psychiatry* 1984;41:949–958.
28. Hill SY. Physiological effects of alcohol in women. In *Women and alcohol: health-related issues.* DHHS Pub. No. (ADM) 86-1139. Washington, DC: Supt. of Docs., U.S. Govt. Print. Off., pp. 199–214, 1986.
29. Cuskey WR, Berger LH, Densen-Gerber J. Issues in the treatment of female addiction: a review and critique of the literature. *Contemp Drug Prob* 1977;6:307–371.
30. Marsh KL, Simpson DD: Sex differences in opioid addiction careers. *Am J Drug Alcohol Abuse* 1986; 12:309–329.
31. Reed T. Outcome research on treatment and on the drug abuser: An exploration. *Int J Addict* 1978;13:139–171.
32. Mendelson JH, Weiss R, Griffin M, Mirin SM, Teoh SK, Mello NK, Lex BW. Some special considerations for treatment of drug abuse and dependence in women. In *Improving Drug Abuse Treatment.* NIDA Research Monograph 106, 1991.
33. Hamilton M. A rating scale for depression. *J Neurol Neurosurg Psychiatry* 1960;23:56–62.
34. Lehman AF, Myers P, Corty E. Assessment and classification of patients with psychiatric and substance abuse syndromes. *Hosp Community Psychiatry* 1989; 40:1019–1025.
35. Kosten TR, Krystal JH. Biological mechanisms in post-traumatic stress disorder: relevance for substance abuse. In M. Galanter, ed. *Recent advances in alcoholism* vol. VI. 1988:49–68.
36. Lacoursiere RB, Godfrey KE, Ruby LM. Traumatic neurosis in the etiology of alcoholism: Vietnam combat and other trauma. *Am J Psychiatry* 1980;137:966–968.
37. Schnitt JM, Nocks JJ. Alcoholism treatment of Vietnam veterans with post traumatic stress disorder. *J Subst Abuse Treat* 1984;1:179–189.
38. Wikler A. Recent progress in research on the neurophysiological basis of morphine addiction. *Am J Psychiat* 1948;105:329–338.
39. Wikler A, Pescor FT. Classical conditioning of a morphine abstinence phenomenon: reinforcement of opioid drinking behavior and "relapse" in morphine-addicted rats. *Psychopharmacologia,* 1967;10:255–284.
40. Wise RA. The neurobiology of craving: implications for the understanding and treatment of addiction. *J Abnorm Psychol* 1988;97(2):118–132.
41. Volpicelli JR. Uncontrollable events and alcohol drinking. *Br J Addict* 1987;82:381–392.
42. Hoffman L, Watson PD, Wilson G, Montgomery J. Low plasma beta-endorphin in post traumatic stress disorder. *Aust N Z J Psychiatry* 1989;23:269–273.
43. van der Kolk BA, Greenberg M, Boyd H, Krystal J. Inescapable shock, neurotransmitters, and addiction to trauma: toward a psychobiology of post traumatic stress. *Biol Psychiatry* 1985;20:314–325.
44. Perry BD, Giller EL, Southwick SM. Altered platelet alpha 2 adrenergic binding sites in post-traumatic stress disorder. *Am J Psychiat* 1987;144:1511–1512.
45. Benedikt RA, Kolb LC. Preliminary findings on chronic pain and post traumatic stress disorder. *Am J Psychiat* 1986;143:908–910.
46. Rapaport MH. Chronic pain and post-traumatic stress disorder. *Am J Psychiat* 1987;144:120.
47. Redmond DE Jr, Krystal JH. Multiple mechanisms of withdrawal from opioid drugs. *Ann Rev Neursci* 1984; 7:443–478.
48. Salloway S, Southwick S, Sadowsky M. Opiate withdrawal presenting as PTSD. *Hosp Community Psychiatry,* vol. 41, June 1990;6:666–668.
49. Giacobini E, Izikowitz S, Wegmann A. Urinary norepinephrine and epinephrine excretion in delirium tremens. *Arch Gen Psychiatry* 1960;3:289–296.
50. Tyrer P, Owen R. Gradual withdrawal of diazepam after long-term therapy. *Lancet* 1983;2:1402–1406.
51. Stine SM, Satel S, Kosten TR. Cocaine precipitation of patient-identified opiate withdrawal. *Am J Addictions* 1993:255–258.
52. Lannon RA, Batki S, Jones BT. Yohimbine produces opioid withdrawal-like reaction in addict (abstract). In Holaday JW, Law PY, Herz A, eds. Progress in opioid research, *NIDA Research Monograph,* 1986;75:655.
53. Southwick SM, Krystal JH, Morgan CA, Johnson D,

Nagy LM, Nicolaou A, Heninger GR, Charney DS. Abnormal noradrenergic function in post traumatic stress disorder. *Arch Gen Psychiatry* 1993;50:266–274.

54. Pitman RK, van der Kolk BA, Orr SP, Greenberg MS. Naloxone-reversible analgesic response to combat-related stimuli in post traumatic stress disorder. *Arch Gen Psychiatry* 1990;47:541–544.
55. Hanson GR, Matsuda LA, Gibb JW. Effects of cocaine on methamphetamine-induced neurochemical changes: characterization of cocaine as a monoamine uptake blocker. *J Pharmacol Exp Ther* 1987;242:507–513.
56. Homan HD, Ziance RJ. The effects of *d*-amphetamine and potassium on serotonin release and metabolism in rat cerebral cortex tissue. *Res Commun Pathol Pharmacol* 1981;31:223–240.
57. Reith ME, Meisler BE, Sershem H, Lajtha A. Structural requirements for cocaine congers to interact with dopamine and serotonin uptake sites in mouse brain and to induce stereotyped behavior. *Biochem Pharmacol* 1986;35:1123–1129.
58. de Witt H, Wise RA. Blockage of cocaine reinforcement in rats with dopamine receptor blocker pimozide but not with the noradrenergic blockers phentolamine and phenoxybenzamine. *J Psychology* 1977;31:195–203.
59. Roberts DCS, Corcoran ME, Fibiger HC. On the role of ascending catecholamine systems in intravenous self-administration of cocaine. *Pharmacol Biochem Behav* 1977;6:615–620.
60. Wise RA. Neural mechanisms of the reinforcing action of cocaine. In *National Institute on Drug Abuse Research Monograph Service,* 1984;50:15–33.
61. Kalivas PW, Duffy P. Similar effects of daily cocaine and stress on mesocorticolimbic dopamine neurotransmission in the rat. *Biol Psychiatry* 1989;25:913–928.
62. Antelman SM, Eichler AJ, Black CA, Kocan D. Interchangeability of stress and amphetamine in sensitization. *Science* 1980;207:329–331.
63. Robinson TE, Angus AL, Becker JB. Sensitization to stress: the enduring effects of prior stress on amphetamine induced rotational behavior. *Life Sci* 1985;37:1039–1042.
64. Maclennan AJ, Maier SF. Coping and stress induced potentiation of stimulant stereotype in the rat. *Science* 1983;219:1091–1093.
65. Gawin FH. Cocaine addiction: psychology and neurophysiology. *Science* (Wash DC), 1991;251:1580–1586.
66. Koslow SH, Maas JW, Bowden CL, Davis JM, Hann I, Javaid J. CSF and urinary biogenic amines and metabolites in depression and mania. *Arch Gen Psychiatry* 1983;40:999–1010.
67. Redmond DE. New and old evidence for the involvement of a brain norepinephrine system in anxiety. In Fann WG, Karacan I, Pokorny D, Williams RL, eds. *Phenomenology and treatment of anxiety.* New York: Spectrum Press, 1979:153–203.
68. Harris GC, Williams JT. Sensitization of locus ceruleus neurons during withdrawal from chronic stimulants and antidepressants. *J Pharmacol Exp Ther* 1992;251 (2):476–483.
69. Henry DJ, Greene MA, White FJ. Electrophysiological effects of cocaine in the mesoaccumbens dopamine system: Repeated administration. *J Pharmacol Exp Ther* 1989;251:833–839.
70. Kalivas PW, Duffy P, DuMars LA, Skinner C. Behavioral and neurochemical effects of acute and daily cocaine administration in rats. *J Pharmacol Exp Ther* 1988;245:485–492.
71. Pitts DK, Marwah J. Chronic cocaine reduces alpha$_2$-adrenoceptor elicited mydriasis and inhibition of locus coeruleus neurons. *Eur J Pharmacol* 1989;160:201–209.
72. McMillen BA, Warnack W, Germam DC, Shore PA. Effects of chronic desipramine treatment on rat brain noradrenergic responses to alpha-adrenergic drugs. *Eur J Pharmacol* 1980;61:239–246.
73. Sugrue MF. A study of the sensitivity of rat brain alpha$_2$-adrenoceptors during chronic antidepressant treatments. *Naunyn-Schmidedberg's Arch Pharmacol* 1982;320:90–96.
74. Kosten TA. Cocaine attenuates that severity of naloxone-precipitated opioid withdrawal. *Life Sci* 1990;47:1617–1623.
75. Kosten TR, Rounsaville BJ, Kleber HD. A 2.5-year follow-up of cocaine use among treated opioid addicts: Have our treatments helped? *Arch Gen Psychiatry* 1987;44:281–284.
76. Stine SM, Burns B, Kosten TR. Methadone dose and cocaine abuse (letter). *Am J Psychiatry* 1991;148:1268.
77. Taylor JR, Lewis VO, Elsworth JD, et al. Yohimbine co-treatment during chronic morphine administration attenuates naloxone-precipated withdrawal without diminishing tail-flick analgesia in rats. *Psychopharmacology* 1991;103:407–414.
78. Davidson J, Swartz M, Storck M, Krishnan RR, Hammett E. A diagnostic and family study of post traumatic stress disorder. *Am J Psychiatry* January 1985;142:1.
79. Davidson J, Lipper S, Kilts CD, Mahorney S, Hammett E: Platelet MAO activity in post traumatic stress disorder. *Am J Psychiatry* 1985;142(11):1341–1343.
80. Haier RJ, Buchsbaum MS, Murphy DL, et al. Psychiatric vulnerability, monoamine oxidase, and the average evoked potential. *Arch Gen Psychiatry* 1980;37:340–345.
81. Wyatt RJ, Murphy DL, Balmaker RH, et al. Reduced monoamine oxidase activity in platelets: a possible genetic marker for vulnerability to schizophrenia. *Science* 1973;197:916–918.
82. Buchsbaum MS, Coursey RD, Murphy DL. The biochemical high-risk paradigm: behavioral and familial correlates of low platelet monoamine oxidase actively. *Science* 1976;194:339–341.
83. Weider H, Kaplan E. Drug use in adolescents. *Psychoanal Study Child* 1969;24:399–431.
84. Milkman H, Frousch WA. On the preferential abuse of heroin and amphetamine. *J Nerv Ment Dis* 1973;56:242–248.
85. Wurmser L. Psychoanalytic considerations of the etiology of compulsive drug use. *J Am Psychoanal Assoc* 1974;22:820–843.
86. Krystal H. Alexithymia and the effectiveness of psychoanalytic treatment. *Int J Psychoanal Psychother* 1982;9:353–378.
87. Wilson A, Passik SD, Faude J, et al. A hierarchical model of opiate addiction: failures of self-regulation as a central aspect of substance abuse. *J Nerv Ment Dis* 1989;177:390–399.
88. Luborsky L, Woody GE, Hole A, et al. A treatment manual for supportive-expressive psychoanalytically oriented psychotherapy: special adaptation for treat-

ment of drug dependence. Unpublished manual, 4th ed., 1981.
89. Dodes LM. Addiction helplessness and narcissistic rage. *Psychoanal Q* 1990;59:398–419.
90. Marlatt GA, Gordon JR. Determinants of relapse: implications for the maintenance of behavior change. In Davidson PO, Davidson DS, eds. *Behavioral medicine: changing health lifestyles*. New York: Brunner/Mazel Inc. 1980:410–415.
91. Carroll KM, Rounsaville BJ, Keller DS. Relapse prevention strategies for the treatment of cocaine abuse. *Am J Drug Alcohol Abuse* 1991;17(3):249–265.
92. Boudewyns PA, Woods GM, Hyer L, et al. Chronic combat related PTSD and concurrent substance abuse, implications for treatment of this frequent dual diagnosis. *J Traum Stress* 1991;4(4):549–560.
93. Abueg FR, Fairbank JA. Behavioral treatment of the PTSD/substance abuser: a multi-dimensional stage model. In Saigh P, ed. *Post traumatic stress disorder: a behavioral approach to assessment and treatment*. New York: Pergamon Press; 1992; Chapter 5.
94. Jellinek JM, Williams T. Post traumatic stress disorder and substance abuse in Vietnam combat veterans: treatment problems, strategies and recommendations. *J Subst Abuse Treat* 1984;1:87–97.
95. Kuhne A, Nohner W. Efficacy of chemical dependency treatment as a function of combat in Vietnam. *J Subst Abuse Treat* 1986;3:191–194.
96. Brady KT, Killeen T, Saladin ME, Dansky B, Becker S. Comorbid substance abuse and post traumatic stress disorder: characteristics of women in treatment. *Am J Addict* 1994;3(2):160–164.
97. McLellan AT, Luborsky L, Woody GE, et al. An improved diagnostic evaluation instrument for substance abuse patients: The Addiction Severity Index. *J Nerv Ment Dis* 1980;168:26–33.
98. McLellan AT, Luborsky L, Woody GE, et al. Predicting response to alcohol and drug abuse treatments. *Arch Gen Psychiatry* 1983;40:620–625.
99. McLellan AT. Psychiatric severity as a predictor of outcome from substance abuse treatments. In Alterman AI, ed. *Substance abuse and psychology*. New York: Plenum; 1985.
100. Rounsaville BJ, Dolinsky ZS, Babor TF, et al. Psychopathology as a predictor of treatment outcome in alcoholics. *Arch Gen Psychiatry* 1987;44:505–513.
101. Woody GE, McLellan AT, Luborsky L, et al. Severity of psychiatric symptoms as a predictor of benefits from psychotherapy: The Veterans Administration-Penn Study. *Am J of Psychiatry* 1984;141:1172–1177.
102. Dole VP, Nyswander ME. Heroin addiction—A metabolic disease. *Arch Intern Med* 1967;120:19–24.
103. Goldstein A. Heroin addiction and the role of methadone in its treatment. *Arch Gen Psychiatry* 1972;26: 291–298.
104. Sawynok J, Pinsky C, Labella FS. Minireview: on the specificity of naloxone as an opiate antagonist. *Life Sci* 1981;25:1221–1232.
105. Friedman MJ. Biological and pharmacological aspects of the treatment of PTSD. In Williams MB, Sommer JF, eds. *Handbook of post traumatic therapy*. Westport, CT: Greenwood Press, (in press).
106. Frank JB, Kosten TR, Giller EL, Dan E. A randomized clinical trial of phenelzine and imipramine for post traumatic stress disorder. *Am J Psychiatry* 1988;145: 1289–1291.
107. Davidson J, Kudler H, Smith R, Mahorney SL, Lipper S, Hammett E, et al. Treatment of post traumatic stress disorder with amitriptyline and placebo. *Arch Gen Psychiatry* 1990;47:259–266.
108. Reist C, Kauffmann CD, Haier RJ, Sangdahl C, et al. A controlled trial of desipramine in 18 men with post traumatic stress disorder. *Am J Psychiatry* 1989;146: 513–516.
109. Shestatzky M, Greenberg D, Lerer B. A controlled trial of phenelzine in post traumatic stress disorder. *Psychiatry Res* 1988;24:149–155.
110. Famularo R, Kinscherff R, Fenton T. Propranolol treatment for childhood post traumatic stress disorder, acute type: a pilot study. *Am J Dis Child* 1988;142:1244–1247.
111. Braun P, Greenberg D, Dasberg H, Lerer B. Core symptoms of post traumatic stress disorder unimproved by alprazolam treatment. *J Clin Psychiatry* 1990;51: 236–238.
112. van der Kolk BA. The drug treatment of post traumatic stress disorder. *J Affect Disord* 1987;13:203–213.
113. Friedman MJ. Towards rational pharmacotherapy for post traumatic stress disorder. *Am J Psychiatry* 1988; 145:281–285.
114. Silver JM, Sandberg DP, Hales RE. New approaches in the pharmacotherapy of post traumatic stress disorder. *J Clin Psychiatry* 1990;51(Supplement 10)a33-38:44–46.
115. ver Ellan P, van Kammen DP. The biological findings in post traumatic stress disorder: a review. *J App Soc Psychiatry* 1990;20(21, pt. 1):1789–1821.
116. Davidson JRT, Roth S, Newman E. Fluoxetine in post traumatic stress disorder. *Journal of Traumatic Stress* 1991;4(3):419–423.
117. McDougle C, Southwick S, St. James R, Charney D. An open trial of fluoxetine. In M Friedman, ed. *PTSD Research Quarterly*. 1990;1(2):7.
118. McDougle CJ, Southwick SM, Charney DS, St. James RL. An open trial of fluoxetine in the treatment of post traumatic stress disorder. *J Clin Psychopharmacol* 1991;11(5):325–327.
119. Nagy L, Morgan CA, Southwick S, Steven M, Charney DS. Open prospective trial of fluoxetine for post traumatic stress disorder. *J Clin Psychopharmacol* 1993;13 (2):107–113.
120. March JS. Fluoxetine and fluvoxamine in PTSD. *Am J Psychiatry* 1992;149(3):413.
121. Shay J. Fluoxetine reduces explosiveness and elevates mood of Vietnam combat veterans with PTSD. *J Trauma Stress* 1992;5:97–101.
122. van der Kolk BA, Dryfuss D, Michaels M, Berkowitz R, Saxe G, Goldenberg I. Fluoxetine in post traumatic stress disorder. *J Clin Psychiatry* 1994;55 vol II, 517–522.
123. Carroll ME, Lac ST, Asencio M, Kragh R. Fluoxetine reduces intravenous cocaine self-administration in rats. *Pharmacol Biochem Behav* 1990;35:237–244.
124. Batki SL, Manfredi LB, Jacob P 3rd, Jones RT. Fluoxetine for cocaine dependence in methadone maintenance: quantitative plasma and urine cocaine/benzoylecgonine concentrations. *J Clin Psychopharmacol* 1993;13(4):243–250.
125. Carroll K, Ziedonis D, O'Malley S, et al. Pharmaco-

logic interventions for alcohol-and cocaine-abusing individuals: a pilot study of disulfiram vs. naltrexone. *Am J Addict* 1993;2(1):77–79.

126. Morgan C, Kosten T, Gawin F, Kleber H. A pilot trial of amantadine for ambulatory withdrawal for cocaine dependence. *Processings of the 49th Annual Scientific Meeting of the Committee for Problems in Drug Dependence,* National Institute on Drug Abuse Research Monograph (1987), Washington DC: U.S. Government Printing Office.
127. Dackis CA, Said MS, Sweeney DR, et al. Single dose bromocriptine reverses cocaine craving. *Psychiatry Res* 1987;20:261–264.
128. Malcolm R, Phillips JD, Brady KT, Roberts JR. A comparison of pergolide and bromocriptine in the initial rehabilitation of cocaine dependence. *Am J Addict* 1994;3(2):144–150.
129. Khantzian EJ, Gawin FH, Riordan C, Kleber HD. Methylphenidate treatment of cocaine dependence: a preliminary report. *J Subst Abuse Treat* 1984;1:107–112.
130. Gawin FH, et al. Desipramine facilitation of initial cocaine abstinence. *Arch Gen Psych* 1989;46:117–121.
131. Berger P, Gawin F, Kosten TR. Treatment of cocaine abuse with mazindol. *Lancet* 1989;1:283.
132. Margolin A, Kosten TR, Avants SK, Petrakis I. Pre- and post-treatment cue-reactivity in cocaine addicts. *Processings of the 53rd Annual Scientific Meeting of the Committee for Problems in Drug Dependence,* National Institute on Drug Abuse Research Monograph (1991), Washington DC: U.S. Government Printing Office.
133. Kosten TR, Morgan CM, Falcione J, Schottenfeld RS. Pharmacotherapy for cocaine-abusing methadone-maintained patients using amantadine or desipramine. *Arch Gen Psychiatry* 1992;49:894–898.
134. Weddington WW, Brown BS, Haertzen CA, et al. Comparison of amantadine and desipramine combined with psychotherapy for treatment of cocaine dependence. *Am J Drug Alcohol Abuse* 1991;17:137–152.
135. Fischman MW, Foltin RW, Nestadt G, Pearlson GD. Effects of desipramine maintenance on cocaine self-administration by humans. *J Pharmacol Exp Ther* 1990; 253:760–770.
136. Carroll KM, Rounsaville BJ, Gordon LT, et al. Psychotherapy and pharmacotherapy for ambulatory cocaine abusers. *Arch Gen Psychiatry* 1994;51:177–187.
137. Post RM, Kopanda RT. Cocaine, kindling and psychosis. *Am J Psychiatry* 1976;133:627–634.
138. Archibald HC, Tuddenham RD. Persistent stress reaction after combat: a 20 year follow-up. *Arch Gen Psychiatry* 1965;12:475–481.
139. Lipper S, Davidson JRT, Grady TA, et al. Preliminary study of carbamazepine in post traumatic stress disorder. *Psychosomatics* 1986;27:849–854.
140. Wolf ME, Alavi A, Mosnaim AD. Post traumatic stress disorder in Vietnam veterans clinical and EEG findings; possible therapeutic effects of carbamazepine. *Biol Psychiatry* 1988;23:642–644.
141. Fesler FA. Valproate in combat-related post traumatic stress disorder. *J Clin Psychiatry* 1991;52(9):361–364.
142. Brodsky L, Doerman AL, Palmer LS, et al. Post traumatic stress disorder: an eclectic approach. *Int J Psychosom* 1990;37:89–95.
143. Halikas JA, Crosby RD, Carlson GA, et al. Cocaine reduction in unmotivated crack users using carbamazepine versus placebo in a short-term, double-blind crossover design. *Clin Pharmacol Ther* 1991;50:81–95.
144. Halikas JA, Crosby RD, Pearson VL, Graves NM, Nugent SM, Carlson GA. Carbamazepine improves early cocaine treatment retention and treatment effectiveness. *Processing of The 32nd Annual Meeting of the American College of Neuropsychopharmacology* NIH/NIDA/ARC (1993), Baltimore, MD.
145. Montoya ID, Levin FR, Fudala PJ, Gorelick DA. Double-blind study with carbamazepine for treatment of cocaine dependence. *Processing of The 32nd Annual Meeting of the American College of Neuropsychopharmacology,* NIH/NIDA/ARC (1993), Baltimore, MD.
146. Kranzler HR, Bauer LO. Carbamazepine for cocaine dependence. *Processing of The 32nd Annual Meeting of the American College of Neuropsychopharmacology,* NIH/NIDA/ARC (1993), Baltimore, MD.
147. Charney DS, Heninger GR, Kleber HD. The combined use of clonidine and naltrexone as a rapid, safe, and effective treatment of abrupt withdrawal from methadone. *Am J Psychiatry* 1986;143:831–837.
148. Kleber HD, Kosten TR. Naltrexone induction: Psychologic and pharmacologic strategies. *J Clin Psychiatry* 1984;9:29–38.
149. Kleber HD, Riordan CE, Rounsaville B, Kosten T, et al. Clonidine in outpatient detoxification from methadone maintenance. *Arch Gen Psychiatry* 1985;42:391–394.
150. Kleber HD, Topazian M, Gaspari J, et al. Clonidine and naltrexone in the outpatient treatment of heroin withdrawal. *Am J Drug Alcohol Abuse* 1987;13:1–18.
151. Jasinski DR, Pevnick JS, Griffith JD. Human pharmacology and abuse potential of the analgesic buprenorphine. *Arch Gen Psychiatry* 1978;35:501–516.
152. Stine SM, Kosten TR. The use of drug combinations in treatment of opioid withdrawal. *J Clin Psychopharmacol* 1992;12:203–209.
153. Glover H. A preliminary trail of nalmefene for the treatment of emotional numbing in combat veterans withess disorder. *Isr J Psychiatry Relat Sci* 1993;30 (4):255–263.
154. Bills LJ, Kreisler K. Treatments of flashback with naltrexone. *Am J Psychiatry* 1993;150:9.
155. O'Malley SS, Jaffe AJ, Chang G, Schottenfeld RS, Meyer RE, Rounsaville B. Naltrexone and coping skills therapy for alcohol dependence (a controlled study). *Arch Gen Psychiatry* 1992;49:881–887.
156. Volpicelli JR, Alterman AI, Hayashida M, O'Brien CP. Naltrexone in the treatment of alcohol dependence. *Arch Gen Psychiatry* 1992;49:876–880.
157. Kleber HD, Weissman MM, Rounsaville BJ, et al. Imipramine as treatment for depression in addicts. *Arch Gen Psychiatry* 1983;40:649–653.
158. Woody GE, O'Brien CP, McLellan AT, et al. The use of antidepressants with methadone in depressed maintenance patients. *Ann NY Acad Sci* 1982;398: 120–127.
159. Krystal H, Raskin H. *Drug dependence: aspects of ego function.* Detroit: Wayne State University Press; 1970.
160. Krystal JH, Giller EL Jr., Cicchetti DV. Assessment of alexithymia in post traumatic stress disorder and

somatic illness: introduction of a reliable measure. *Psychosom Med* 1986;48:84–94.

161. Krystal H. Trauma and the stimulus barrier. *Psychoanal Inquiry* 1985;6:131–161.
162. Ciraulo DA, et al. Critical review of liability for benzodiazepine abuse among alcoholics. *Am J Psychiat* 1988;145:1501–1506.
163. Ciraulo DA, et al. Parental alcoholism as a risk factor in benzodiazepine abuse: a pilot study. *Am J Psychiat* 1989;146:1333–1335.
164. Higgitt AC, Lader MH, Fonagy P. Clinical management of benzodiazepine dependence. *Br Med J* 1985; 291:688–690.
165. Noyes R, Clancy J, Coryell WH, Crowe RR, Chaudhry DR, Domingo DV. A withdrawal syndrome after abrupt discontinuation of alprazolam. *Am J Psychiatry* 1985;142:114–116.
166. Risse SC, Whitters A, Burke J, Chen S, Scurfield RM, Raskind . Severe withdrawal symptoms after discontinuation of alprazolam in eight patients with combat-inducted port traumatic stress disorder. *J Clin Psychiatry* 1990;51(5):206–209.
167. Anthenelli RM, Monteiro MG, Blunt B, Schuckit MA. Amnestic effects of intravenous diazepam in healthy young men. *Am J Drug Alcohol Abuse* 1991;17:129–136.
168. McNamara JO, Bonhaus DW, Shin C, Crain BJ, Gellman RL, Giacchino JL. The kindling model of epilepsy: a critical review. *CRC Critic Rev Clin Neurobiol* 1985; 1:341–391.
169. Morita K, Okamoto M, Seki K, Wada JA. Suppression of amygdala-kindled seizures in cats by enhanced GABAergic transmission in the substantial innominata. *Exp Neurol* 1985;89:225–236.
170. Tietz EI, Gomaz F, Berman RF. Amygdala kindled seizure stage is related to altered benzodiazepine binding site density. *Life Sci* 1985;36:183–190.
171. Feldmann TB. Alprazolam in the treatment of post traumatic stress disorder [letter]. *J Clin Psychiatry* 1987;48:216–217.
172. Tietz EI, Ferencak W, Aloe L, Chang H. Contingent anticonvulsant tolerance develops to the benzodiazepine (BZ) partial agonist, R016-6028, but not to clonazepam. *Processings of the 56th Annual Scientific Meeting of the Committee for Problems in Drug Dependence,* National Institute on Drug Abuse Research Monograph 1995;153 vol II, p. 238. Washington DC: U.S. Government Printing Office.

Neurobiological and Clinical Consequences of Stress: From Normal Adaptation to PTSD, edited by M. J. Friedman, D. S. Charney, and A. Y. Deutch, Lippincott-Raven Publishers, Philadelphia © 1995.

26

Towards Pharmacotherapy for Post-Traumatic Stress Disorder

Matthew J. Friedman and *Steven M. Southwick

*Departments of Psychiatry and Pharmacology, Dartmouth Medical School, Hanover, New Hampshire 03755, and National Center for Post-Traumatic Stress Disorder, Veterans Administration Medical Center, White River Junction, Vermont 05009; and *Department of Psychiatry, Yale University School of Medicine, New Haven, Connecticut 06510, and National Center for Post-Traumatic Stress Disorder, Veterans Affairs Medical Center, West Haven, Connecticut 06516*

PATHOPHYSIOLOGY OF PTSD: IMPLICATIONS FOR PHARMACOTHERAPY

Post-traumatic stress disorder (PTSD) appears to be a very complicated disorder. As many previous chapters suggest, a number of neurobiological systems are affected by exposure to catastrophic stress. There is a large body of experimental data on animal models for PTSD, including inescapable stress, conditioned fear, fear-potentiated startle, kindling, behavioral sensitization, and time-dependent sensitization. Taken together they suggest that at least seven or eight neurobiological systems that mediate cognitive, emotional, and behavioral processes involved in coping, adaptation, and survival may be dysregulated in PTSD. This impression is bolstered by human stress and PTSD research, also reviewed in previous chapters, although the data are not as comprehensive as in the animal studies. At this point in time it appears that PTSD is probably associated with abnormalities in the adrenergic, hypothalamic-pituitary-adrenocortical (HPA), opioid, dopaminergic, and thyroid systems, and possibly with alterations in the serotonergic, gamma-aminobutyric acid (GABA)-benzodiazepine and the *N*-methyl-D-aspartate (NMDA) systems. As yet, we have no evidence that neural input from these various systems is funneled through some final common pathway that may be definitively modified by pharmacological intervention.

It is important to keep this in mind as we review the current literature on the clinical psychopharmacology of PTSD. To present one of our conclusions in advance, given the complex pathophysiology of this disorder, effective pharmacotherapy for PTSD may require a multisystem approach in which several drugs, each with a unique and distinct action, are administered simultaneously.

CLINICAL PSYCHOPHARMACOLOGY—OVERVIEW

In their comprehensive review of the treatment outcome literature on PTSD, Solomon and associates (1) noted that a total of 255 English language reports on treatment of PTSD had been published, of which only 11 studies were randomized clinical trials (RCTs), 5 pharmacological, and 6 behavioral. The five drug studies included three trials of tricyclic antidepressants (TCAs) and two trials of monoamine oxidase inhibitors (MAOIs). Three other controlled trials should be mentioned: an A-B-A test of propranolol (2), an RCT of alprazolam (3), and a recently reported RCT with fluoxetine (4).

Given the paucity of RCTs and our belief that data from open trials and case reports should not be ignored, we will present a comprehensive review of the entire literature on clinical pharmacology of PTSD in this chapter. Although we give much greater weight to RCTs, we will discuss a recently published quantitative analysis that synthesizes RCT and non-RCT results regarding TCAs and MAOIs (5). We do so with the full understanding that, without an RCT, there are always serious questions about sampling, subject selection, and interpretation of results. Furthermore, there is an editorial bias such that positive rather than negative findings are much more likely to be published in an open trial or case report, whereas that is not true for randomized clinical trials.

CLINICAL PSYCHOPHARMACOLOGY—REVIEW OF THE LITERATURE

In this section we will review published data on drug trials in PTSD. Rather than present a tabulation of results with drug X or drug Y, we will offer a conceptually driven review that begins with more specific pharmacological agents and works towards more complex drugs such as TCAs or MAOIs. Therefore, we will review: 1) drugs affecting the adrenergic system (clonidine and propranolol); 2) drugs affecting the serotoninergic system (fluoxetine, buspirone, fluvoxamine, clomipramine, and cyproheptadine); 3) drugs affecting both systems (TCAs and MAOIs); 4) drugs affecting the GABA-benzodiazepine system (alprazolam and clonazepam); 5) anticonvulsants with antikindling properties (carbamazepine and valproate); 6) lithium and drugs affecting the dopamine system (neuroleptics); and 7) drugs affecting endogenous opioid activity (narcotic antagonists). A general summary of this review is shown in Table 1.

Drugs Affecting the Adrenergic System—Propranolol and Clonidine

PTSD has repeatedly been shown to be associated with excessive adrenergic activity. As reported earlier (see Chapters 18 and 19), patients with PTSD exhibit increased 24-hour urinary catecholamine levels, downregulation of alpha-2 and beta adrenergic receptors, and excessive responsitivity to the alpha-2 antagonist, yohimbine. For these reasons, one would predict that sympatholytic agents such as propranolol and clonidine would have an important place in the treatment of PTSD. Clonidine is an alpha-2 adrenergic agonist and propranolol is a postsynaptic beta adrenergic blocking agent. Both drugs reduce sympathetic arousal and adrenergic activity through different mechanisms of action. Furthermore, both drugs have demonstrated efficacy in the treatment of anxiety (6–8).

Surprisingly, there has been very little interest in either drug, despite the fact that Kolb et al. (9) reported promising results with both clonidine and propranolol as early as 1984. In open trials with nine Vietnam veteran PTSD patients, both drugs reduced traumatic nightmares, intrusive recollections, hypervigilence, insomnia, startle reactions, and angry outbursts. Perry et al. (10), in an open trial using clonidine for the treatment of 17 children with PTSD (0.05–0.1 mg twice a day), noted a profound improvement in behavioral impulsivity, anxiety, arousal, concentration, and mood. Kinzie and Leung (11) had favorable results with clonidine in combination with the TCA imipramine in Cambodian refugees who met diagnostic criteria for both PTSD and depression; in addition to improvement in depression, PTSD symptoms such as nightmares, insomnia, and startle reactions responded favorably to this treatment. There are other reports as well, suggesting that Southeast Asian refugees with PTSD can be treated successfully with the clonidine/imipramine combination (12).

There are three reports on propranolol use in PTSD patients. It successfully reduced intrusion and arousal symptoms in Vietnam veterans, as just noted (9), was ineffective in an open trial with Cambodian refugees (13), and was used successfully in American children whose PTSD

TABLE 1

Drug category	Drug	Daily dose	Hypothesized mechanism of action	# RCT	Remarks
Adrenergic Inhibitors	Clonidine	0.2–0.4 mg	Alpha 2 Agonist	0	Promising results in studies on B & D symptoms.
	Propranolol	130–160 mg	Beta Blocker	0	2 positive, 1 negative study. Successful trial with children on B & D symptoms.
Drugs Affecting Serotonergic Mechanisms	Fluoxetine	20–80 mg	SSRI	1	Only drugs consistently effective on C symptoms.
	Fluvoxamine	250–300 mg	Antidepressants	0	Effects on B & D symptoms less conclusive.
	Clomipramine	200–150 mg	TCA with SSRI Actions	0	One open trial, too little data. Reduced B symptons only.
	Buspirone	35–60 mg	5HT1A Partial Agonist	0	Case reports on 3 patients. Too little data. Only reduced B & D symptoms.
	Cyproheptidine	4–28 mg	5-HT Antagonist	0	Case reports only. Seems effective primarily on traumatic nightmares.
TCA Antidepressants	Imipramine	150–300 mg	NE/5-HT Reuptake Inhibitors	1	Quantitative analysis of RCTs and open trials suggests primary effect on B symptoms and global PTSD severity—not on C or D symptoms.
	Amitriptyline			1	
	Desipramine			1	
MAOI Antidepressants	Phenelzine	45–75 mg	Inhibits metabolism of NE and 5-HT	2	More effective than TCAs but quantitatively shows same spectrum of action.
Benzodiazepine Anxiolytic	Alprazolam	0.5–6 mg	Benzodiazepine Agonist	1	Reduce general anxiety symptoms but no conclusive actions on core PTSD symptoms. Only 3 studies.
	Clonazepam	1–6 mg		0	
Anticonvulsant	Carbamazepine	600–1000 mg	Antikindling Action	0	5 open studies. Effective on B & D symptoms.
	Valproate	750–1750 mg		0	3 open studies. Effective on C & D symptoms.
Antimanic Agent	Lithium	300–1500 mg	Multiple Actions	0	2 open trials. Appears to reduce D symptoms only.
Antipsychotic Agent	Thioridazine	200 mg	DA Antagonist	0	Not firstline drugs. 1 case report: effective on B & D symptoms.
Narcotic Antagonists	Nalmefene	200–400 mg	Opioid Antagonists	0	Open trial-mixed results: reduced C as well as B & D symptoms in some patients, but made others worse.
	Naltrexone	50 mg		0	1 case report: reduced flashbacks in two patients.

Abbreviations: PTSD B symptoms: intrusive recollections, traumatic nightmares, flashbacks, etc.
PTSD C symptoms: avoidant behavior, psychic numbing, dissociation, etc.
PTSD D symptoms: insomnia, irritability, startle, hypervigilance, hyperarousal, etc.

DA, dopamine; 5-HT, serotonin; MAOI, monoamine oxidase inhibitor; NE, norepinephrine; RCT, randomized clinical trial; SSRI, selective serotonin reuptake inhibitor; TCA, tricyclic antidepressant.

was precipitated by physical and/or sexual abuse (2). This latter study, though not a RCT, was an A-B-A design (6 weeks off—6 weeks on—6 weeks off medication) in which subjects received 2.5 mg/kg/day. Significant reductions in intrusion and arousal symptoms were observed in 8 of 11 children. In some cases, propranolol was discontinued if the pulse rate fell below 50 beats per minute or the diastolic blood pressure fell below 50 mm Hg.

Drugs Affecting the Serotonergic System—Fluoxetine, Fluvoxamine, Clomipramine, Buspirone, and Cyproheptadine

A number of the growing family of serotonin receptors (specifically the 5-HT_{1a}, 5-HT_2, and 5-HT_3) appear to mediate anxiety reactions (14). Furthermore, the pathophysiology of obsessive-compulsive disorder (OCD) clearly involves serotonergic mechanisms. This is of particular interest since the obstinate persistence of PTSD intrusive recollections sometimes resembles the pattern of obsessional thoughts in OCD, while the repetitive predictability of some avoidant behaviors is reminiscent of compulsive patterns in OCD. In addition, certain symptoms or syndromes frequently associated with PTSD (impulsivity, aggressiveness, suicidal intent, alcoholism, substance abuse, depression, and personality disorders) bring the serotonergic system to mind (15).

Despite the intriguing inferences just cited, only two preliminary studies have investigated serotonergic mechanisms in traumatized patients. Southwick et al. (16) administered the serotonergic agonist MCPP (m-chlorophenylpiperazine) to 14 Vietnam veterans with PTSD. Thirty-six percent experienced panic attacks and 25% exhibited trauma-related dissociative episodes and flashbacks, in contrast to non-PTSD combat controls who did not respond to MCPP. Corrigan et al. (17) reported significant blunting of both the prolactin and cortisol responses to intravenous clomipramine in adult female survivors of childhood sexual abuse in comparison with a nontraumatized control group. Such findings suggest that serotonergic abnormalities may be present in PTSD patients. With this in mind, we shall review published results on selective serotonin reuptake inhibitors (SSRIs), clomipramine (a TCA with SSRI-type action), buspirone (an anxiolytic 5-HT_{1A} partial agonist), and cyproheptadine (a 5-HT receptor antagonist).

SSRIs and Associated Drugs—Fluoxetine, Fluvoxamine, and Clomipramine

There are currently 5 published open trials and case reports with the SSRI fluoxetine, involving a total of 71 patients with PTSD, all but 5 of whom were American war-zone (mostly Vietnam) veterans. The remaining five were nonveteran men and women exposed to civilian trauma. Two studies were carried out at the National Center for PTSD, where marked improvement in intrusive, avoidant, and arousal symptoms was observed in 13 of 20 Vietnam veterans (18), and where marked or moderate improvement was observed in 12 of 19 military veterans (19). An interesting report by March (20) described moderate remission of PTSD symptoms in a Vietnam veteran receiving fluoxetine and later fluvoxamine who had failed to respond to other drugs. Shay (21) reported that 13 out of 18 depressed Vietnam veterans with PTSD exhibited improved mood and reduced aggressiveness after receiving fluoxetine. Finally, Davidson et al. (22) reported successful fluoxetine treatment of five nonveteran men and women with PTSD who had been exposed to sexual assault or industrial/motor vehicle trauma.

The only published RCT with SSRIs was carried out by van der Kolk et al. (4), who conducted a fluoxetine versus placebo RCT with 64 patients, 31 military veterans and 33 nonveteran adult survivors of childhood sexual abuse, with PTSD. Fluoxetine, but not placebo, significantly reduced overall PTSD symptomatology, especially with respect to numbing and arousal symptoms. Among the numbing/avoidant cluster of symptoms, fluoxetine was markedly effective in reducing numbing but not avoidance symptoms. It is also noteworthy that improvement in PTSD symptoms was not associated with reductions in depressive symptomatology, suggesting that fluoxetine had a selective anti-PTSD but not antidepressant effect in this study. Civilian childhood abuse survivors were much more responsive to medication than veterans with war-related PTSD.

Other publications on SSRIs include an open trial of fluvoxamine in 24 World War II Dutch resistance fighters with PTSD (23) and the aforementioned single case report by March (20). Results in the Dutch study were more modest and less successful than in American trials with fluoxetine. De Boer et al. (23) suggest that poorer results in their study may be due to older age and longer chronicity of PTSD among the Dutch subjects, rather than to a difference in efficacy between fluoxetine and fluvoxamine.

There is one report of an open trial of clomipramine on seven Vietnam veterans with PTSD in which OCD symptoms were also monitored with the obsession subscale of the Yale-Brown OCD Scale (24). Marked reduction in PTSD intrusive symptoms was accompanied by marked reduction in the Yale-Brown obsession score in six of the seven patients, prompting the author to speculate on the relationship between PTSD intrusive and OCD obsessional thoughts.

Several important findings worth noting differentiate clinical experience with SSRI-type drugs from other agents used in PTSD.

1. SSRIs effectively reduce avoidant/numbing symptoms as well as intrusive and arousal symptoms. For the most part, other drugs have not affected these symptoms.
2. Successful treatment usually requires 10–12 weeks of drug therapy with high doses (80 mg fluoxetine, 300 mg fluvoxamine).
3. Successful treatment is often associated with gradual and incomplete reduction of symptoms rather than complete remission. (This is often the case in OCD as well.)
4. When treatment is successful, debilitating symptoms associated with PTSD often show improvement, along with core PTSD symptoms. Such associated symptoms and syndromes include rage, impulsivity, suicidal intent, depression, OCD, and behaviors associated with alcoholism or substance abuse.
5. Reduction of PTSD symptoms was not strongly associated with significant improvement in occupational function.

Buspirone

Buspirone is an anxiolytic that acts as a 5-HT_{1A} partial agonist. Only one report appears in the literature describing the response of three combat veterans (WWII, Korea, and Vietnam) to 35–60 mg buspirone (25). All three experienced marked reduction in anxiety, insomnia, flashbacks, and depressed mood. In contrast to SSRIs, however, no patient exhibited improvement in avoidant/numbing symptoms.

Cyproheptadine

Cyproheptadine is a 5-HT antagonist that has been used specifically for PTSD-related traumatic nightmares. In case reports concerning two (26) and four (27) PTSD patients, 4–28 mg cyproheptadine at bedtime successfully suppressed recurrent traumatic nightmares. Brophy (27) has now extended his initial studies to include 80 patients whose traumatic nightmares have been suppressed by cyproheptadine (Brophy, personal communication, 1992). Furthermore, administration of methysergide, a 5-HT antagonist with action similar to cyproheptadine, has reportedly reduced nightmares and improved sleep among Vietnam veterans with PTSD (Morgan, personal communication, 1992).

Drugs Affecting Both Adrenergic and Serotonergic Systems—TCAs and MAOIs

Five of six RCTs in the PTSD literature involve TCAs and/or MAOIs. Given the high comorbidity of PTSD and major depressive disorder (MDD), and the fact that these antidepressants are also effective anxiolytic and antipanic agents (28–30), it is not surprising that they have received the lion's share of attention with regard to treatment of PTSD. Open trials and case reports on the efficacy of these drugs in PTSD are reviewed elsewhere (31,32). First, we turn our attention to the five RCTs that have been published.

Although there are three RCTs with TCAs, there are enough differences between each study to make comparisons difficult. First of all, a different drug was used in each study: imipramine (33), amitriptyline (34), and desipramine (35). Second, patients with MDD were excluded from the imipramine study but not the others. Third, the imipramine and amitriptyline studies lasted 8 weeks, whereas the desipramine study lasted only 4 weeks. These differences may explain why results from these three studies differ from one another.

Frank et al. (33) conducted an 8-week RCT comparison of imipramine versus phenelzine (a MAOI) versus placebo in 34 male Vietnam veterans with combat-related PTSD. Impressive reductions in intrusive but not avoidant symptoms were seen in the imipramine (and phenelzine) group(s) in contrast to placebo controls. This study was later extended to include a total of 60 patients (19 on phenelzine, 23 on imipramine, and 18 on placebo) by Kosten et al. (36). Results were the same as those previously reported by Frank et al. Davidson et al. (34) conducted an 8-week trial of amitriptyline versus placebo in 46 male Vietnam veterans with combat-related PTSD. In contrast to Frank et al., these investigators observed only modest improvement in PTSD symptoms. Furthermore, since patients with MDD were included in the Davidson et al. (but not the Frank et al.) study, it was found that depressed PTSD patients showed greater improvement than nondepressed patients. This raised the possibility that any amelioration of PTSD symptoms was secondary to amitriptyline's antidepressant and anxiolytic potency rather than due to a specific action on core PTSD symptoms. Davidson et al. also observed a small reduction in avoidant symptoms. This is a unique finding in the TCA literature on PTSD treatment, and may be due to the fact that amitriptyline has much greater serotonergic (SSRI-type) efficacy than most other TCAs. The final RCT was a 4-week trial of desipramine versus placebo in 18 male Vietnam veterans with combat-related PTSD (35). No differences were observed between the TCA and placebo groups in this study.

The two RCTs with MAOIs also have methodological differences and contrary results, although phenelzine was used in both studies. Frank et al. (33) and Kosten et al. (36), in the study cited previously, compared phenelzine with imipramine and placebo. Again, there were marked reductions in intrusive recollections, nightmares, and flashbacks in the phenelzine (and imipramine) group(s) in comparison to the placebo group. It should be noted that phenelzine was superior to imipramine in this study. In contrast, Shestatzky et al. (37) conducted a 4-week double-blind crossover comparison between phenelzine and placebo in 10 Israeli military veterans. There was no difference between the two groups, although it should be noted that (in contrast to most studies) the placebo group showed considerable reduction in PTSD symptoms in this study.

A final word about these RCTs of TCAs and MAOIs concerns instrumentation. In all studies, PTSD symptoms were measured by the Impact of Events Scale (IES) (38). Although this scale has been very useful in PTSD research, it is limited to measurement of intrusive and avoidant symptoms, and does not monitor arousal symptoms. We might have possibly learned more about the efficacy of TCAs and MAOIs had arousal symptoms also been measured during these RCTs.

Quantitative Analysis of Open Trials, Case Reports, and RCTs with TCAs and MAOIs

While antidepressants appear to be useful for treatment of PTSD, it is unclear from the RCTs just discussed exactly which aspects of the disorder are most responsive. For example, it could be argued that antidepressants have little or no effect on PTSD specific symptoms per se, but instead produce their overall global effects by reducing comorbid symptoms of depression and anxiety/panic. Further, even if antidepressants do reduce PTSD specific symptoms clinically, it is important to know whether they affect all three symptom clusters equally.

In order to determine the specific effects of antidepressants on PTSD symptomatology, Southwick et al. (5) performed a quantitative analysis on all available published reports using antidepressants to treat PTSD. Because most reports were not RCTs, a standard metaanalytic approach could not be used. In the quantitative analysis that was done, subjects from all 15 published reports were pooled and evaluated using a single rating scale to assess the effects on each of the three PTSD symptom clusters (reexperiencing, avoidance, hyperarousal) and on symptoms of depression and anxiety/panic. In essence, this approach allowed all subjects from the literature to be evaluated as a single sample using identical criteria for symptom improvement.

The results indicated that MAOIs were somewhat more effective than TCAs. Global response was judged to be moderate to good in 82% of subjects treated with MAOIs and in 45% of subjects taking TCAs. The response of individual symptom clusters was less pronounced, with improvement seen only in the reexperiencing cluster. Approximately 75% of subjects showed moderate or better improvement in intrusive traumatic memories, flashbacks, and nightmares, with phenelzine being superior to TCAs. Symptoms of avoidance and hyperarousal were largely unresponsive to antidepressants; the one exception was the individual symptom of insomnia that responded well to both classes of antidepressant. The quantitative analysis also suggested that a minimum of 8 weeks of treatment was needed for global improvement to occur in core PTSD symptoms.

Symptoms of depression and anxiety/panic also responded poorly to the antidepressants. For example, only 13% of phenelzine-treated and 38% of TCA-treated subjects showed a moderate or better improvement in depression. This finding is surprising given the well-known efficacy of antidepressants in treating major depression and panic disorder in non-PTSD populations. It is consistent, however, with the work of Yehuda and associates (39) suggesting that the MDD usually associated with PTSD is a very different neurobiological abnormality than endogenous depression or true melancholia (see Chapters 19 and 24).

Taken as a whole, the TCA and MAOI literature is disappointing in that so few RCTs have been carried out, no drug (except for phenelzine, which was tested twice) has been tested more than once, and, except for the recent fluoxetine study (4), no new RCTs have been published since 1991. Those early studies have helped to identify key methodological issues that must be addressed in future RCTS. Three recommendations for future research design have been proposed by Kudler et al. (40). First, any drug trial in PTSD should be carried out for a minimum of 8–10 weeks. (This recommendation is supported by quantitative analysis results indicating better outcomes in clinical trials of 8 weeks or more.) Second, observer rating instruments that are more responsive than the IES to weekly changes in symptoms should be used. And third, future research designs should include adequate controls for Axis 1 disorders that are frequently comorbid with PTSD, such as depression, alcoholism/substance abuse, and other anxiety disorders (41).

Drugs Affecting the GABA-Benzodiazepine System—Alprazolam, Clonazepam

Benzodiazepine receptors are allosterically and functionally linked to a macromolecular complex that also contains receptors for the inhibitory neurotransmitter GABA. This complex seems clearly involved in the neurobiology of stress and anxiety (42). Animal studies show that exposure to inescapable stress results in reduced benzodiazepine receptor binding (43–45). Furthermore, animals administered benzodiazepines prior to exposure to inescapable stress appear to have been protected from many adverse sequelae of this experience. The role of the GABA-benzodiazepine system may be related to alterations in other neurobiologic systems, since benzodiazepines: 1) reduce locus coeruleus activity and stress-induced increases in norepinephrine turnover 2) decrease stress-induced increases in prefrontal dopamine activity; and 3) may have their efficacy altered by endogenous steroids (46–48).

The benzodiazepine system is important to consider in PTSD because of its key involvement in human anxiety and fear states. The benzodiazepine inverse agonist, FG-7142, can produce severe anxiety and panic in healthy subjects (49), while flumazenil, a benzodiazepine receptor antagonist, can produce panic attacks in patients with panic disorders, but not in healthy subjects (50,51). There have been no published studies on the GABA-benzodiazepine system in PTSD. A final theoretical consideration for considering benzodiazepines is the kindling model of PTSD (see next paragraph), since it has been shown that limbic kindling is associated with increased benzodiazepine receptor binding (52–54).

Likewise, there are very few clinical studies with benzodiazepines in PTSD patients, although in some settings up to 71% of PTSD patients have received benzodiazepines (55). In other settings, few PTSD patients receive benzodiazepines, because clinicians are often reluctant to prescribe these drugs given the high rates of alcoholism and chemical dependency among PTSD patients. We (56) have challenged this prescribing bias against benzodiazepines, arguing that a primary objective in treatment should be symptom reduction in patients dually diagnosed with PTSD and alcoholism/substance abuse. Citing evidence that benzodiazepines may be used safely with alcoholics when they are prescribed rationally (57), we (56) have suggested that benzodiazepines may have a valuable role in the treatment of PTSD, especially in patients who also suffer from alcoholism/substance abuse.

A final clinical concern is the rebound anxiety and severe withdrawal symptoms sometimes seen in patients taking alprazolam and other triazolo-benzodiazepine derivatives. Indeed, Risse et al. (58) reported on the severe exacerbation of PTSD symptoms among eight Vietnam veterans during withdrawal from alprazolam. These patients experienced anxiety, insomnia, rage reactions, hyperalertness, increased nightmares, intrusive thoughts, and homicidal ideation.

There are three published reports on benzodiazepine treatment of PTSD, two with alprazolam and one with clonazepam. In Feldman's (59) open trial with alprazolam, 16 of 20 veterans showed reduced insomnia, anxiety, irritability, and hyperarousal. A cautionary note from this report is the observation that four patients exhibited benzodiazepine-induced disinhibition marked by increased outbursts of anger. The only published RCT on PTSD patients treated with a benzodiazepine was a 5-week crossover trial of alprazolam versus placebo in 10 Israeli patients (3). Results indicated that alprazolam treatment was no better than placebo in reducing core PTSD symptoms, although it did produce a modest reduction in general anxiety symptoms. Finally, Lowenstein et al. (60) reported on a successful trial of clonazepam in patients whose PTSD was associated with multiple personality disorder. These patients experienced marked and sustained improvement in sleep, nightmares, flashbacks, panic attacks, and other PTSD symptoms, but exhibited no improvement in avoidance or dissociative symptoms.

We believe that benzodiazepines may have a useful role in the rational treatment of carefully selected PTSD patients. We would further recommend that future trials focus on clonazepam because of its efficacy, its low abuse potential, its safety with regard to rebound anxiety/withdrawal, and possibly because of its antikindling properties.

Antikindling Agents—Carbamazepine and Valproate

PTSD can persist for decades or even a lifetime (61,62). Therefore, a theoretical model for PTSD must include a neurobiological mechanism that precipitates and sustains long-term alteration of brain function. Kindling and behavioral sensitization are such mechanisms (see Chapter 12). They are two different neurobiological processes by which limbic and other neuroanatomic structures become increasingly sensitized following exposure to electrical stimulation or stimulant (cocaine-like) drugs. Once established, kindling and behavioral sensitization remain as stable neurobiological alterations that are associated with neurophysiological abnormalities, grand mal seizures, and aberrant behavior. Kindling was first invoked as an animal model for a psychiatric syndrome by Post and Kopanda (63), who suggested that it might account for the recurrence of episodes seen in lithium-refractory bipolar affective disorder. Kindling was proposed as a possible mechanism in PTSD (64–66) because the chronic CNS sympathetic arousal associated with PTSD produced conditions that, it was hypothesized, might optimize neuronal sensitization of limbic nuclei.

A total of eight publications report beneficial effects from two anticonvulsants that were chosen specifically for their properties as antikindling agents. Five reports concern carbamazepine and three concern valproate. All were open trials or case reports.

Lipper et al. (64) administered carbamazepine to 10 Vietnam veterans with PTSD in an open trial. Their decision to test this drug was based on theoretical questions regarding kindling as a mechanism for PTSD. Seven patients reported marked reductions in both intensity and frequency of traumatic nightmares, flashbacks, intrusive recollections, and sleep disturbance. Three case reports on a total of five patients (67–69) describe a positive response to carbamazepine in traumatized patients who met diagnostic criteria for PTSD but who exhibited an EEG and clinical pattern consistent with complex partial seizures. Flashbacks, traumatic nightmares, and insomnia all diminished after anticonvulsant treatment. The authors emphasize the importance of ruling out complex partial seizures in any diagnostic workup of PTSD. It is for this reason, in particular, that Wolf et al.'s (70) open trial of carbamazepine in 10 Vietnam veterans with PTSD is so interesting. All of these patients had normal EEGs and no additional evidence for complex partial seizures before onset of carbamazepine treatment. Eight of the 10 patients showed improvement in impulsivity, irritability, and violent behavior, but there was no monitoring of PTSD symptoms during this trial.

Fesler (71) conducted an open trial of valproate in 16 Vietnam veterans with PTSD. She was motivated to conduct this trial because her patients could not tolerate carbamazepine and were therefore noncompliant. After administration of valproate, 10 of the 16 patients experienced marked improvement. It is noteworthy that improvement was greatest for hyperarousal and avoidant symptoms, and not for intrusion symptoms, which has been the case in other reports with antikindling agents. Finally, there are two case reports; Brodsky et al. (72) described one patient whose flashbacks were controlled by valproate, and Szymanski and Olympia (73) reported on two Vietnam veterans with PTSD who became less irritable and aggressive after valproate treatment.

Lithium

Lithium has a complex spectrum of actions on neuronal membranes, a number of neurotransmitter systems, phosphoinositide second messengers, and thyroid function. Its efficacy as a treatment for recurrent affective disorders is well established. Only two open trials of lithium have been reported. Van der Kolk (74) observed that 14 of 22 PTSD patients experienced marked reduction in autonomic arousal, a greater capacity to cope with stress, and reduced alcohol consumption following lithium treatment. It is noteworthy that none of these patients reportedly suffered from bipolar affective disorder. In a report of five cases of PTSD, Kitchner and Greenstein (75) described improvement following lithium treatment in irritability, inappropriate anger, anxiety, and insomnia. There have been no further publications on the efficacy of lithium in PTSD.

Drugs Affecting the Dopamine System—Neuroleptics

Mesocortical and frontal neurons exhibit increased dopaminergic activity following exposure to stress. There is additional evidence that chronic stress and repeated cocaine exposure may affect central dopaminergic functions (76); however, such responses may also be mediated by NMDA, opioid, substance P, or other neurotransmitter systems (77). Both animal research and clinical studies show that cocaine and amphetamines, which activate dopaminergic mechanisms, can produce hypervigilant and paranoid behavior. It is tempting to generalize from these findings to the hypervigilance, irritability, insomnia, and occasional psychotic behavior seen in PTSD patients. Such generalizations would be premature, however, because there has been little research on altered dopaminergic function in traumatized or PTSD patients. Two studies have shown altered dopaminergic activity in traumatized patients. Jensen et al. (78) reported alterations in the growth hormone response to levodopa (L-DOPA) in sexually abused boys, and Ende et al. (79) detected elevated urinary conjugated dopamine levels among rape victims within 24 hours of the assault. Yehuda et al. (80) have detected elevated 24-hour urinary homovanillic acid levels (HVA, the major metabolite of dopamine) in Holocaust survivors and Vietnam veterans with PTSD. Hamner et al. (81) have shown an abnormal dopamine response among PTSD patients performing a treadmill task. Clearly, the role of dopamine in the pathophysiology of PTSD needs further investigation.

Research notwithstanding, neuroleptics were widely used for treating what is now recognized as PTSD during the 1970s and into the following decade. Lacking the conceptual model provided in 1980 by DSM-III (82), and impressed by the intensity of agitation, paranoid thoughts, impulsivity, explosive behavior, potential for violence, and brief psychotic or dissociative states seen in Vietnam veterans, Department of Veterans Affairs (VA) psychiatrists prescribed large quantities of neuroleptics to achieve relief of these troubling and sometimes dangerous symptoms. As clues to the pathophysiology of PTSD began to emerge, the research focus shifted to the adrenergic, HPA, and other neurobiological systems, while the pharmacological focus shifted to TCAs MAOIs, and more recently to SSRIs. As a result, there have been no RCTs, and only a single published report regarding thioridazine, on the efficacy of neuroleptics in PTSD. Dillard et al. (76) successfully treated a Vietnam veteran who was having frequent flashbacks, nightmares, anxiety, irritability, and fear of loss of control with 200 mg thioridazine. It should be pointed out, however, that this patient probably had not received an adequate trial of a TCA, MAOI, SSRI, or any other drug before thioridazine was prescribed. Despite the paucity of empirical data, a number of writers have identified special circumstances under which neuroleptics might be used (66,84–88). These include patients with refractory PTSD who exhibit paranoid behavior, overwhelming anger, aggressivity, psychotic symptoms, fragmented ego boundaries, self-destructive behavior, and frequent flashback experiences marked by auditory or visual hallucinations of traumatic episodes. It is noteworthy, however, that Mueser and Butler (85) described a group of PTSD patients with auditory hallucinations who were refractory to neuroleptic treatment.

Drugs Affecting the Opioid System—Narcotic Antagonists

There is evidence that both laboratory stress and PTSD are associated with dysregulation of the endogenous opioid system (see Chapters 6 and 18). In 1985, van der Kolk and associates (87) proposed that chronic PTSD is associated with chronic opioid depletion. Consistent with this prediction are findings that PTSD patients exhibit: 1) lower pain thresholds (32); 2) lower plasma beta endorphin levels (88); and 3) decreased production and release of methionine-enkephalin (89). Rather than opioid depletion, however, there is evidence that, upon exposure to trauma-related stimuli, PTSD patients exhibit excessive opioid activity. Pitman et al. (90) noted a nalaxone-reversible stress-induced analgesia when they exposed Vietnam veterans with PTSD to combat scenes depicted in the movie *Platoon.*

The two pharmacological studies relevant to the opioid system have emphasized excessive rather than diminished opioid function. Glover, having hypothesized that emotional numbing in PTSD results from hypersecretion of endogenous opioids (91), predicted that emotional numbing would improve following treatment with a narcotic antagonist. He administered nalmefene, a non–FDA-approved pure opiate antagonist in an open trial to 18 Vietnam veterans with PTSD (92). Nalmefene is a potent, long-acting drug that is structurally similar to both naloxone and naltrexone but has greater binding affinity to central opiate receptors. Glover reported that nalmefene (100–200 mg b.i.d.) produced marked improvement in numbing and in other PTSD symptoms in 8 of the 18 patients. The other 10 patients showed either no improvement or a worsening of their PTSD, especially with regard to anxiety and panic symptoms. There is one additional publication, a single case report on successful treatment of PTSD flashbacks in two patients, with the narcotic antagonist, naltrexone (93). The article does not state whether other PTSD symptoms also responded to treatment.

Given laboratory results suggesting reduced opioid activity under some circumstances with PTSD patients, and excessive activity under other conditions, Glover's results showing a bidirectional action for a narcotic antagonist are intriguing. Future studies will need a better experimental design, better instrumentation, and a conceptual model that can account for the finding that some patients may benefit while others may worsen following from treatment with an opioid antagonist.

EVALUATION OF RESEARCH DATA

It is clear that our growing understanding of the pathophysiology of PTSD is much greater than our knowledge regarding effective pharmacological strategies for combating this complex disorder. Actually, there has been very little research on drug treatment. Even the few RCTs that have been published present interpretive difficulties for the following reasons.

1. Except for phenelzine (which was tested twice), no drug has been tested more than once.
2. Questions regarding dose and duration of treatment have not been adequately investigated.
3. Comorbid disorders frequently associated with PTSD (such as depression and substance abuse) need to be controlled for in future studies.
4. Most studies have relied too heavily on the Impact of Events Scale, an instrument that does not assess arousal symptoms. More comprehensive observer rating instruments have been developed in recent years that should be used in future studies (such as the Clinician Administered PTSD Scale (CAPS) and (SI-PTSD)).
5. Most studies have been conducted on American Vietnam veterans. Future studies need to explore patients whose PTSD was caused by something besides the trauma of war, such as rape, torture, natural disasters, or accidents. Future studies will also need to test women and PTSD patients from a variety of ethnocultural backgrounds.
6. American Vietnam veterans who have served as subjects in most published RCTs may be the most severely impaired, chronic, and treatment-refractory patient cohorts. The reason why they remain the most available patients for these drug trials is because they are still enrolled in VA treatment programs. It is possible, however, that patients who have remained in treatment after years of VA PTSD programs constitute a self-selected cohort of chronic patients with multiple levels of impairment who may be most refractory to drug (or any other) treatment. Future studies must include drug trials on different patient cohorts to make sure that lack of success of a specific drug is genuinely due to lack of pharmacological efficacy rather than due to a sampling strategy that inadvertently selects patients with the worst prognosis.
7. In studies in which symptom reduction is achieved, it is necessary to distinguish between statistical significance and clinical improvement. Too many studies have focused excessively on symptoms and paid scant attention to functional (marital, social, and vocational) status.

It is disturbing that valuable leads have not been pursued. For example, Kolb et al. (9) published their original report on clonidine and propranolol in 1984. Despite a promising study in 1988 (2) with propanolol, there have been no further investigations with that drug. Similarly, there has been no apparent attempt to replicate the exciting findings of Frank et al. (33) with phenelzine and imipramine. It appears, at present, that some momentum is gathering that will lead to extensive testing of SSRIs. While this is indeed exciting, such clinical trials should not lose sight of the multisystem pathology manifested in PTSD. If, for example, SSRIs prove to be efficacious, we must not forget that there are other neurobiological systems in addition to the serotonergic system that are altered in PTSD patients.

COMPLEXITY OF PTSD

As is clear from the aforementioned literature review, there is no one medication of choice for treatment of PTSD. That is, no single medication treats the entire disorder. Instead, specific symptom clusters respond best to specific medications.

The reexperiencing symptoms respond well to both MAO inhibitors and TCAs. Reports suggest that nightmares, intrusive memories, and even flashbacks are sensitive to these medications. This finding appears relatively reliable in that it has been replicated by a number of case reports, open trials, and RCTs. Other less well-studied medications also reportedly help in reducing reexperiencing symptoms. These include clonidine, propranolol, serotonin reuptake inhibitors, and carbamazepine. However, without RCTs these findings must be considered tentative.

Hyperarousal symptoms also appear to differentially respond to various medications. The most consistent and pronounced responses have been described in case reports and open trials with the antiadrenergic agents propranolol and clonidine. Other potentially useful agents include the benzodiazepines (with the cautionary note that some patients can become emotionally disinhibited) and serotonin reuptake inhibitors. Insomnia may be especially responsive to the benzodiazepines, cyproheptadine, and trazodone. Finally, although neuroleptics may be effective in treating symptoms of agitation and hyperarousal, such use is unproven and highly controversial.

Symptoms of avoidance and numbing may be the most resistant to pharmacologic intervention. The MAOIs, TCAs, clonidine, propranolol, the benzodiazepines, lithium, and carbamazepine all appear to be ineffective in treating these symptoms. It is important to note that failure to find treatment effects may be related to the relatively brief treatment trials characteristic of PTSD research to date. It may be that avoidant/numbing symptoms improve more slowly than other PTSD symptoms, making them appear unresponsive if treatment trials are too short. The most promising agents have been the serotonin reuptake inhibitors, and one RCT showed marked reduction of psychic numbing following fluoxetine treatment (4). Valproate also merits further testing in this regard, since it produces a significant decrease in avoidant/numbing symptoms.

The frequent co-occurrence of other psychiatric disorders adds to the difficulty and challenge of rationally medicating patients with PTSD. Common comorbid Axis I disorders include major depression, panic disorder, and substance abuse disorders (41). Further, high rates of Axis II disorders (borderline, antisocial, paranoid, compulsive, and schizoid personality disorders) also have been reported among treatment-seeking combat veterans with PTSD (94). It is not yet clear whether and how well these disorders respond to pharmacologic intervention. For example, in PTSD patients, preliminary evidence suggests that comorbid panic disorder and major depression may not respond well to TCAs and MAOIs.

Although no medication studied thus far has proven effective in treating the entire disorder of PTSD, even the reduction of a single symptom may be of significant help to the individual who has been severely traumatized. For example, frequent nightmares are highly disruptive for most people. They often contribute to severe sleep deprivation, irritability, and emotionally liability. Concentration decreases, interpersonal relationships deteriorate, and job performance suffers. Clearly, a reduction of this symptom alone can have a significant effect on global functioning.

PHARMACOTHERAPY AND PSYCHOTHERAPY

The addition of a psychopharmacologic agent in the context of ongoing psychotherapy with PTSD patients can be challenging. First and foremost, such patients often find it difficult to trust other people as a consequence of their severe traumatization. The world is often viewed as unsafe, unpredictable, and uncontrollable. Therefore, the stability of the therapeutic relationship is critical.

The therapist who works with traumatized patients must first provide a consistent, predictable, and safe environment. If pharmacotherapy is indicated, the specific meaning of medication to the patient must be carefully assessed and thoroughly understood. Will the patient view medication as a tool for increasing inner stability or as a threatening form of external control? Will the prescribing of medication be interpreted as a rejection, suggesting that the therapist prefers not to talk with the patient? Will the offer of medication suggest that the therapist no longer believes that the patient has the inner resources to combat PTSD symptoms and therefore must rely on external supports such as drugs? Finally, will the prescribing of medication foster regression and dependency?

It is also important for the therapist to understand what the decision to prescribe medication means to him or her, the therapist. Is medication being prescribed for specific symptoms that are known to be responsive to this particular agent, or is the medication being prescribed for other less overt or unconscious reasons? For example, is medication being prescribed in order to "do something" when therapy is progressing slowly? Is the prescribing of medication a way to divert attention away from aggressive and violent material that is upsetting to the therapist, so that treatment can focus on medication target symptoms such as appetite and sleep? Is the therapist unconsciously attempting to gain "control" over a difficult patient who constantly challenges authority figures such as him or herself? In other words, careful monitoring of countertransference and other therapy-related issues is absolutely essential when prescribing medication to traumatized individuals.

The staging of pharmacotherapy is also important. As noted earlier, some medications are effective in treating involuntary reexperiencing symptoms such as traumatic nightmares that occur spontaneously or in response to unavoidable trauma-related stimuli or situations. For some patients, the retrieval and processing of memories during therapy can cause a temporary but marked increase in distress from intense and uncontrollable intrusive or arousal symptoms. Indeed, therapy itself may come to be regarded as aversive if the focus on traumatic material becomes unbearable. If the patient is overwhelmed by such symptoms, he or she may drop out of therapy. It is under such conditions that the timely addition of a medication that suppresses intolerable PTSD symptoms may allow the patient to more freely experience, work through, and master the trauma, a process that many believe is essential to successful therapy. Many of the issues just discussed have been addressed in greater detail by Southwick and Yehuda (95).

The example cited in the previous paragraph applies to a therapeutic contract in which the goal of treatment is to process and master traumatic material, with the hope of reducing PTSD symptomatology. In this regard, pharmacotherapy is applied as an adjunct to psychotherapy, with the hope that drug-induced symptom reduction will permit greater depth and latitude for therapeutic exploration. In some cases, however, symptom reduction, in and of itself, may become an appropriate goal of treatment when the severity of PTSD symptoms has produced marked functional impairment or when the patient does not have the capacity to benefit from intensive psychotherapy. Under such conditions, the choice of medication will depend on which target symptoms are considered most debilitating at that time.

CONCLUSION

As attested by many chapters in this book, PTSD appears to have a unique pathophysiology that distinguishes it from other affective and anxiety disorders. Data suggest that this disorder is associated with abnormalities in a number of psychophysiologic, neurotransmitter, neuroendocrine, and psychoimmunologic systems. We do not know whether neural input from these various systems is funneled through some final common pathway that may be modified by pharmacological interventions. With our growing appreciation of the complexity of PTSD, however, we are beginning to understand why we may have failed to discover a pharmacological silver bullet, a single drug that will significantly reduce PTSD symptoms.

There have been too few published randomized clinical trials, and no drug has been tested more than once except for phenelzine, which has been tested only twice. Even among these few published RCTs, there are enough questions about methodology, instrumentation, and sample selection to suggest the need for expansion and extension of RCT programs to include testing and retesting of more drugs. Furthermore, since most RCTs have been conducted on chronic, and possibly treatment refractory, Vietnam veterans with PTSD, it is necessary to conduct RCTs with less severely affected patients and with patients whose PTSD has been precipitated by non–war-related traumas such as sexual assault, natural disasters, and industrial accidents.

Despite these concerns, it does appear that certain drugs may be helpful in alleviating specific PTSD symptom clusters. Reexperiencing symptoms appear to respond to TCAs and MAOIs, avoidant/numbing symptoms may respond to SSRIs (and possibly to valproate), and arousal symptoms seem to respond to antiadrenergic agents. At the present time, instead of a silver bullet, there appears to be a number of drugs that may be useful in providing specific relief for selected clusters of PTSD symptoms. Therefore, a rational approach to pharmacotherapy for PTSD may require a multisystem approach in which multiple drugs, each one with a unique and distinct action, are administered simultaneously.

At the very least, pharmacotherapy has an important role as an adjunct to psychotherapy. Reduction of disruptive PTSD symptoms makes it possible for patients to participate more productively in a variety of psychotherapeutic approaches. Symptom reduction, in and of itself, is often a legitimate and highly desirable goal of treatment. Although pharmacotherapy has already begun to prove itself as a useful clinical option, research on the potential efficacy of drug treatment for PTSD is still in its infancy.

REFERENCES

1. Solomon SD, Gerrity ET, Muff AM. Efficacy of treatments for posttraumatic stress disorder. *JAMA* 1992; 268:633–638.
2. Famularo R, Kinscherff R, Fenton T. Propranolol treatment for childhood post-traumatic stress disorder, acute type: a pilot study. *Am J Dis Child* 1988;142:1244–1247.
3. Braun P, Greenberg D, Dasberg H, Lerer B. Core symptoms of post-traumatic stress disorder unimproved by alprazolam treatment. *J Clin Psychiatry* 1990;51:236–238.
4. van der Kolk BA, Dryfuss D, Michaels M, Berkowitz R, Saxe G, Goldenberg I. Fluoxetine in post traumatic stress disorder. *J Clin Psychiatry* 1994;55:517–522.
5. Southwick SM, Yehuda R, Giller EL, Charney DS. Use of tricyclics and monoamine oxidase inhibitors in the treatment of PTSD: a quantitative review. In Murburg MM, ed. *Catecholamine function in post-traumatic stress disorder: emerging concepts.* Washington, DC: American Psychiatry Press, 1994:293–305.
6. Tanna VT, Penningroth RP, Woolson RF. Propranolol in the treatment of anxiety neurosis. *Compr Psychiatry* 1977;18:319–326.
7. Charney DS, Brier A, Jathow PI, Heninger GR. Behavioral, biochemical and blood pressure responses to alprazolam in healthy subjects: interactions with yohimbine. *Psychopharmacology* 1986;88:133–140.
8. Ravaris CL, Friedman MJ, Hauri PJ, McHugo GJ. A controlled study of alprazolam and propranolol in panic-disordered and agoraphobic outpatients. *J Clin Psychopharmacol* 1991;11:344–350.
9. Kolb LC, Burris BC, Griffiths S. Propranolol and clonidine in the treatment of the chronic post traumatic stress disorders of war. In van der Kolk BA, ed. *Post traumatic stress disorder: psychological and biological sequelae.* Washington, DC: American Psychiatric Press, 1984: 97–107.
10. Perry BD. Neurobiological sequelae of childhood trauma: PTSD in children. In Murburg MM, ed. *Catecholamine function in post-traumatic stress disorder: emerging concepts.* Washington, DC: American Psychiatric Press, 1994:233–255.
11. Kinzie JD, Leung P. Clonidine in Cambodian patients with posttraumatic stress disorder. *J Nerv Ment Dis* 1989;177:546–550.
12. Friedman MJ, Jaranson JM. The applicability of the PTSD concept of refugees. In Marsella AJ, Borneman TH, Orley J, eds. *In peril and pain: the mental health and well-being of the world's refugees.* Washington, DC: American Psychological Press, 1994:207–228.
13. Kinzie JD. Therapeutic approaches to traumatized Cambodian refugees. *J Trauma Stress* 1989;2:75–91.
14. Gonzalez-Heydrich J, Peroutka SJ. Serotonin receptor and reuptake sites: pharmacologic significance. *J Clin Psychiatry* 1990;51(4 suppl.):5–12.
15. Friedman MJ. Interrelationships between biological mechanisms and pharmacotherapy of posttraumatic stress disorder. In Wolf M, Marion E, Mosnaim A, eds. *Posttraumatic stress disorder: etiology, phenomenology, and treatment.* Washington, DC: American Psychiatric Press, 1990:204–225.

16. Southwick SM, Krystal JH, Morgan A, et al. Yohimbine and M-chlorophenyl-piperazine in PTSD. *New Research Abstracts.* American Psychiatric Association, 144th Annual Meeting, New Orleans, LA, May 11–16, 1991, Abstract NR 348.
17. Corrigan MH, Garbatt JC, Gillette GM, Mason G, Carson S, Golden RN. Serotonin and victims of childhood sexual abuse. *New Research Abstracts.* American Psychiatric Association, 143rd Annual Meeting, New York, NY, May 12–17, 1990, Abstract NR 154.
18. McDougle CJ, Southwick SM, Charney DS, Saint James RL. An open trial of fluoxetine in the treatment of posttraumatic stress disorder. *J Clin Psychopharmacol* 1991;11(5):325–327.
19. Nagy L, Morgan CA, Southwick S, Steven M, Charney DS. Open prospective trial of fluoxetine for posttraumatic stress disorder. *J Clin Psychopharmacol* 1993;13 (2):107–113.
20. March JS. Fluoxetine and fluvoxamine in PTSD. *Am J Psychiatry* 1992;149(3):413.
21. Shay J. Fluoxetine reduces explosiveness and elevates mood of Vietnam combat veterans with PTSD. *J Trauma Stress* 1992;5:97–101.
22. Davidson JRT, Roth S, Newman E. Fluoxetine in posttraumatic stress disorder. *J Trauma Stress* 1991;4(3): 419–423.
23. De Boer M, Op den Velde W, Falger PR, Hovens J, DeGroen J, Van Dujin H. Fluvoxamine treatment for chronic PTSD: a pilot study. *Psychother Psychosom* 1992;57(4):158–163.
24. Chen C-J. The obsessive quality and clomipramine treatment in PTSD. *Am J Psychiatry* 1991;148(8):1087–1088.
25. Wells GB, Chu C, Johnson R, Nasdahl C, Ayubi MA, Sewell E, Statham P. Buspirone in the treatment of posttraumatic stress disorder. *Pharmacotherapy* 1991;11(4): 340–343.
26. Harsch HH. Cyproheptadine for recurrent nightmares. *Am J Psychiatry* 1986;143:1491–1492.
27. Brophy MH. Cyproheptadine for combat nightmares in post-traumatic stress disorder and dream anxiety disorder. *Milit Med* 1991;156(2):100–101.
28. Kahn RJ, McNair DM, Lipman RS, et al. Imipramine and chlordiazepoxide in depression and anxiety disorders. *Arch Gen Psychiatry* 1986;43:79–85.
29. Sheehan DV, Ballenger J, Jacobsen G. Treatment of endogenous anxiety with phobic, hysterical and hypochondriacal symptoms. *Arch Gen Psychiatry* 1980;37: 51–59.
30. Charney DS, Menkes DB, Heninger GR. Receptor sensitivity and the mechanism of action of antidepressant treatment: implications for the etiology and therapy of depression. *Arch Gen Psychiatry* 1981;38:1160–1180.
31. Ver Ellen P, van Kammen DP. The biological findings in post-traumatic stress disorder: a review. *J Appl Soc Psychology* 1990;20(21,pt.1):1789–1821.
32. Friedman MJ. Biological approaches to the diagnosis and treatment of post-traumatic stress disorder. *J Trauma Stress* 1991;4(1):67–91.
33. Frank JB, Kosten TR, Giller EL, Dan E. A randomized clinical trial of phenelzine and imipramine for posttraumatic stress disorder. *Am J Psychiatry* 1988;145: 1289–1291.
34. Davidson J, Kudler H, Smith R, et al. Treatment of posttraumatic stress disorder with amitriptyline and placebo. *Arch Gen Psychiatry* 1990;47:259–266.
35. Reist C, Kauffmann CD, Haier RJ, Sangdahl C, Demet EM, Chicz-Demet A, Nelson JN. A controlled trial of desipramine in 18 men with post-traumatic stress disorder. *Am J Psychiatry* 1989;146:513–516.
36. Kosten TR, Frank JB, Dan E, McDougle CJ, Giller EL. Pharmacotherapy for posttraumatic stress disorder using phenelzine or imipramine. *J Nerv Ment Dis* 1991;179 (6):366–370.
37. Shestatzky M, Greenberg D, Lerer B. A controlled trial of phenelzine in post-traumatic stress disorder. *Psychiatry Res* 1988;24:149–155.
38. Horowitz M, Wilner N, Alvarez W. Impact of events scale: a measure of subjective distress. *Psychosom Med* 1979;41:209–218.
39. Yehuda R, Southwick SM, Krystal JH, Bremner D, Charney DS, Mason JW. Enhanced suppression of cortisol following dexamethasone administration in posttraumatic stress disorder. *Am J Psychiatry* 1993;150:83–86.
40. Kudler H, Davidson J, Meador K. The DST and posttraumatic stress disorder. *Am J Psychiatry* 1987;144: 1068–1071.
41. Kulka RA, Schlenger WE, Fairbank JA, Jough RL, Jordan BK, Marmar CR, Weiss DS. *Trauma and the Vietnam War generation* New York: Brunner/Mazel, 1990.
42. Guidotti A, Baraldi M, Leon A, Costa E. Benzodiazepines: a tool to explore the biochemical and neurophysiological basis of anxiety. *Fed Proc* 1980;39:1039–1042.
43. Drugan RC, Ryan SM, Minor TR, Maier SF. Librium prevents the analgesia and shuttlebox escape deficit typically observed following inescapable shock. *Pharmacol Biochem Behav* 1984;21:749–754.
44. Drugan RC, Morrow AL, Weizman R, et al. Stress induced behavioral depression in the rat is associated with a decrease in GABA receptor-mediated chloride ion flux and brain benzodiazepine receptor occupancy. *Brain Res* 1989;487:45–51.
45. Weizman R, Weizman A, Kook KA, Vocci F, Deutsch SI, Paul SM. Repeated swim stress alters brain benzodiazepine receptors measured in vivo. *J Pharmacol Ex Ther* 1989;249:701–707.
46. Ida Y, Tanaka M, Tsuda A, Tsujimaru S, Nagasaki N. Attenuating effect of diazepam on stress-induced increases in noradrenaline turnover in specific brain regions of rats: antagonism by RO15-1788. *Life Sci* 1985;37:2491–2498.
47. Redmond DE. Studies of the nucleus locus coeruleus in monkeys and hypotheses for neuropsychopharmacology. In Meltzer HY, ed. *Psychopharmacology: the third generation of progress.* New York: Raven, 967–975.
48. Majewski MD, Harrison NI, Schwartz RD, et al. Steroid metabolites are barbiturate-like modulators of the GABA receptors. *Science* 1986;232:1004–1007.
49. Dorrow R, Horowski R, Paschelka G. Severe anxiety induced by FG7142, a beta carboline ligand for benzodiazepine receptors. *Lancet* 1983;1:98–99.
50. Nutt DJ, Glue P, Lawson C, et al. Flumazenil provocation of panic attacks: evidence for altered benzodiazepine receptor in panic disorder. *Arch Gen Psychiatry* 1990;47:917–925.
51. Woods SW, Charney DS, Silver JM, et al. Behavioral, biochemical, and cardiovascular responses to the benzo-

diazepine receptor antagonist flumazenil. *Psychiatry Res* 1991;36:115–127.
52. McNamara JO, Bonhaus DS, Shin C, Crain BJ, Gellman RL, Giacchino JL. The kindling model of epilepsy: a critical review. *CRC Clin Rev Clin Neurobiol* 1985; 1:341–391.
53. Morita K, Okamoto M, Seki K, Wada JA. Suppression of amygdala-kindled seizures in cats by enhanced GABAergic transmission in the substantial innominata. *Ex Neurol* 1985;89:225–236.
54. Tietz EI, Gomaz F, Berman RF. Amygdala kindled seizure stage is related to altered benzodiazepine binding site density. *Life Sci* 1985;36:183–190.
55. Ciccone PE, Mazarek A, Weisbrot M, Greenstein RA, Olsen K, Zimmerman J. [Letter], *Am J Psychiatry* 1988; 145,1484–1485.
56. Kofoed L, Friedman M, Peck R. Alcoholism and drug abuse in patients with PTSD. *Psychiatr Q* 1993;64: 151–169.
57. Ciraulo DA, Sands BF, Shader RI. Critical review of liability for benzodiazepine abuse among alcoholics. *Am J Psychiatry* 1988;145:1501–1506.
58. Risse SC, Whitters A, Burke J, Chen S, Scurfield RM, Raskind MA. Severe withdrawal symptoms after discontinuation of alprazolam in eight patients with combat-induced post-traumatic stress disorder. *J Clin Psychiatry* 1990;51(5):206–209.
59. Feldman TB. Alprazolam in the treatment of post-traumatic stress disorder [letter]. *J Clin Psychiatry* 1987; 48:216–217.
60. Lowenstein RJ. Rational psychopharmacology in the treatment of multiple personality disorder. *Psychiatr Clin North Am* 1991;14:721–740.
61. Archibald HC, Tuddenham RD. Persistent stress reaction after combat: a 20 year follow-up. *Arch Gen Psychiatry* 1965;12:475–481.
62. Blank AS. The longitudinal course of posttraumatic stress disorder. In Davidson JRT, Foa EB, eds. *Posttraumatic stress disorder: DSM IV and beyond.* Washington, DC: American Psychiatric Press; 1993:3–22.
63. Post RM, Kopanda RT. Cocaine, kindling and psychosis. *Am J Psychiatry* 1976;133:627–634.
64. Lipper S, Davidson JRT, Grady TA, Edingar JD, Hammett EB, Mahorney SL, Cavenar JO. Preliminary study of carbamazepine in post-traumatic stress disorder. *Psychosomatics* 1986;27:849–854.
65. van der Kolk BA. The drug treatment of post-traumatic stress disorder. *J Affect Disord* 1987;13:203–213.
66. Friedman MJ. Towards rational pharmacotherapy for post-traumatic stress disorder. *Am J Psychiatry* 1988; 145:281–285.
67. Stewart JR, Bartucci RJ. Post-traumatic stress disorder and partial complex seizures. *Am J Psychiatry* 1986; 143:113–114.
68. Brodsky L, Doerman AL, Palmar LS, Slade GF, Munasifi FA. Post traumatic stress disorder: a tri-modal approach. *Psychiatr J Univ Ottawa* 1987;12(1):41–46.
69. Bennet TL, Curiel MP. Complex partial seizures presenting as a psychiatric illness. *Int J Clin Neuropsychol* 1988;10:41–44.
70. Wolf ME, Alavi A, Mosnaim AD. Post-traumatic stress disorder in Vietnam veterans clinical and EEG findings; possible therapeutic effects of carbamazepine. *Biol Psychiatry* 1988;23:642–644.
71. Fesler FA. Valproate in combat-related post-traumatic stress disorder. *J Clin Psychiatry* 1991;52(9):361–364.
72. Brodsky L, Doerman AL, Palmer LS, Slade GF, Munasifi FA. Post-traumatic stress disorder: an eclectic approach. *Int J Psychosom* 1990;37:89–95.
73. Szymanski HV, Herman V, Olympia J. Divalproex in posttraumatic stress disorder. *Am J Psychiatry* 1991; 148(8):1086–1087.
74. van der Kolk BA. Psychopharmacological issues in post traumatic stress disorder. *Hosp Community Psychiatry* 1983;34:683–691.
75. Kitchner I, Greenstein R. Low dose lithium carbonate in the treatment of post traumatic stress disorder: Brief communication. *Mil Med* 1985;150:378–381.
76. Kalivas PW, Duffy P. Similar effects of daily cocaine and stress on mesocortiolimbic dopamine neurotransmission in the rat. *Biol Psychiatry* 1989;25:913–928.
77. Southwick SM, Krystal JH, Johnson DR, Charney DS. Neurobiology of post-traumatic stress disorder. In Tasman A, ed. *Annual review of psychiatry,* vol II. Washington, DC: American Psychiatric Press, 1992:347–370.
78. Jensen JB, Pease JJ, ten Bensel R, Garfinkel BD. Growth hormone response patterns in sexually or physically abused boys. *J Am Acad Child Adolesc Psychiatry* 1991;30(5):784–790.
79. Ende N, Gertner SB, Socha B. Unexpected changes in urinary catecholamines and vanillylmandelic acid following rape assault. *Horm Behav* 1990;24(1):62–70.
80. Yehuda R, Giller EL, Southwick SM, Kahana B, Boisoneau D, Ma X, Mason JW. Relationship between catecholamine excretion and PTSD symptoms in Vietnam combat veterans and holocaust survivors. In Murburg MM, ed. *Catecholamine function in post-traumatic stress disorder: emerging concepts.* Washington, DC: American Psychiatric Press, 1994;203–219.
81. Hamner MB, Diamond BI, Hitri A. Plasma dopamine and prolactin levels in PTSD. *Biol Psychiatry* 1990; 27:72A.
82. American Psychiatric Association. Committee on Nomenclature and Statistics. *Diagnostic and statistical manual of mental disorders,* 3rd Ed. Washington, DC: American Psychiatric Association, 1980.
83. Walker JL. Chemotherapy of traumatic war stress. *Mil Med* 1982;147:1029–1033.
84. Atri PB, Gilliam JH. Comments on posttraumatic stress disorder. *Am J Psychiatry* 1989;146:128.
85. Meuser KT, Butler RW. Auditory hallucinations in combat-related chronic post-traumatic stress disorder. *Am J Psychiatry* 1987;144:299–302.
86. Dillard ML, Bendfeldt F, Jernigan P. Use of thioridazine in post-traumatic stress disorder. *South Med J* 1993; 86(11):1276–1278.
87. van der Kolk BA, Greenberg M, Boyd H, Krystal J. Inescapable shock, neurotransmitters and addiction to trauma: toward a psychobiology of post traumatic stress. *Biol Psychiatry* 1985;20:314–325.
88. Hoffman L, Watson PD, Wilson G, Montgomery J. Low plasma beta-endorphin in post-traumatic stress disorder. *Aust N Z J Psychiatry* 1989;23:268–273.
89. Wolfe E, Mosnaim AD, Puente J, Ignacio R. Plasma methionine enkephalin in PTSD. *Biol Psychiatry* 1991; 29:305–307.
90. Pitman RK, van der Kolk BA, Orr SP, Greenberg MS. Naloxone-reversible analgesic response to combat-re-

lated stimuli in post traumatic stress disorder. *Arch Gen Psychiatry* 1990;47:541–544.
91. Glover H. Emotional numbing: a possible endorphin-mediated phenomenon associated with post-traumatic stress disorder and other allied psychopathologic states. *J Trauma Stress* 1992;5:643–675.
92. Glover H. A preliminary trial of nalmefene for the treatment of emotional numbing in combat veterans with post-traumatic stress disorder. *Isr J Psychiatry Related Sci* 1993;30(4):255–263.
93. Billis LJ, Kreisler K. Treatment with flashbacks with naltrexone. *Am J Psychiatry* 1993;150(9):1430.
94. Southwick SM, Yehuda R, Giller EL. Personality disorders in treatment-seeking combat veterans with posttraumatic stress disorder. *Am J Psychiatry* 1993;150:1020–1023.
95. Southwick SM, Yehuda R. The interaction between pharmacotherapy and psychotherapy in the treatment of posttraumatic stress disorder. *Am J Psychother* 1993; 47:404–411.

Neurobiological and Clinical Consequences of Stress: From Normal Adaptation to PTSD, edited by M. J. Friedman, D. S. Charney, and A. Y. Deutch, Lippincott-Raven Publishers, Philadelphia © 1995.

27

Cognitive-Behavioral Therapy of Post-Traumatic Stress Disorder

Edna Foa, *Barbara Olasov Rothbaum, and Chris Molnar

*Department of Psychiatry, Medical College of Pennsylvania and Hahnemann University, Philadelphia, Pennsylvania 19129; and *Department of Psychiatry, Emory University School of Medicine, The Emory Clinic, Atlanta, Georgia 30322*

Epidemiological studies demonstrate that post-traumatic stress disorder (PTSD) has become a serious problem in our society. If we consider the preponderence of individuals exposed to both sexual and nonsexual assault, natural disasters such as flooding, work and motor vehicle accidents, and war, traumatic experiences are extremely common. Estimates from a national random sample of adult women indicate that almost 12 million adult women (12.9%) have been raped at least once during their lifetime (1). In a retrospective study, 35% of sexual assault victims and 39% of aggravated assault victims reported lifetime PTSD (2). In the same study, current PTSD was reported by 13% of rape victims and by 12% of aggravated assault survivors (2). A much higher rate of PTSD following assault emerges from prospective studies, with 47% of sexual assault victims and 22% of nonsexual assault victims meeting criteria for PTSD 3 months post-assault (3).

Other types of trauma also put individuals at risk for developing PTSD. Thirty-eight percent of men and 17.5% of women who experienced high combat exposure in Vietnam were diagnosed with current PTSD (4). Vicarious traumatization, such as that experienced by paramedics, law-enforcement officials, firemen, and mental health professionals, also results in post-traumatic stress reactions (5–7). Given the enormous number of individuals suffering from post-traumatic stress reactions, there is an obvious need for cost-effective, brief, and efficacious treatments.

The recognition that post-trauma reactions can cause severe disturbances led to the development of both crisis intervention programs and psychotherapy groups; they are currently the most commonly used interventions in rape crisis centers (8). Based on crisis theory (9), these interventions involve information about post-rape sequelae, active listening, and emotional support (10). Treatment with dynamic psychotherapy often has been advocated as the final stage of crisis intervention (9). Controlled studies of these interventions, however, are few. Only three reports regarding the effects of group psychotherapy for rape victims have been reported (11, 12, 83). All three studies indicated that treatment was somewhat effective in reducing post-rape distress. However, the designs of the studies did not allow for unequivocal interpretation of the results.

Several brief cognitive behavioral programs have been demonstrated to be highly effective in ameliorating PTSD symptoms and related psychopathology. However, the information about these programs has not been widely disseminated among mental health professionals who are the frontline helpers for trauma victims. This chapter offers a review of studies examining cognitive-behavioral interventions. There is a special focus on those employed with female assault survivors. Only group studies or case reports that included at least semistructured procedures for evaluating treatment outcome were included. We begin our review with a general description of cognitive-behavioral procedures. We then examine literature regarding the efficacy of these procedures with post-traumatic stress symptoms. Finally, we propose mechanisms to explain the success that cognitive behavioral procedures have had in alleviating post-traumatic stress symptoms.

COGNITIVE-BEHAVIORAL THERAPY

A cognitive-behavioral approach to pathological anxiety, such as that observed in PTSD, is founded upon two major theories: 1) learning theory and 2) cognitive theory. Learning theory encompasses both classical and operant conditioning principles. Cognitive theory emphasizes the influence of appraisal processes and their mediative role in the development of maladaptive responses to events. Two sets of treatment procedures have been developed to ameliorate anxiety disorders. The first, exposure procedures, are based primarily on a learning theory paradigm and involve having patients repeatedly confront the feared memory or situation until it ceases to evoke anxiety. Anxiety management training (AMT), which is the second set of treatment procedures and includes cognitive therapy, is founded upon both learning and cognitive paradigms. Through AMT techniques such as relaxation training, patients learn skills to directly address their anxiety responses. Through cognitive procedures, patients learn to systematically examine and change maladaptive thoughts that lead to negative responses.

Exposure Therapy

Exposure therapy is a set of techniques used when anxiety symptoms involve excessive avoidance, and treatment aims to first activate fear memories and then modify their pathological aspects. The common denominator of all exposure techniques is the confrontation of feared situations. It is helpful to view exposure therapy along three continuums that include: 1) the medium of exposure (imaginal versus in vivo); 2) the length of exposure (short versus long); and 3) the level of arousal during exposure (low versus high). When the various dimensions of exposure therapy are ordered in this manner, systematic desensitization (SD) (13), where exposure is imaginal, brief, and minimally arousing, represents one set of extremes. In vivo flooding—where exposure is to actual life events, is prolonged, and elicits high anxiety levels—represents the opposite set of extremes. It is important to note that therapists, guided by clinical needs, employ many combinations of the dimensions comprising exposure therapy.

Several studies have examined the efficacy of exposure treatments with PTSD. We will first discuss systematic desensitization and then summarize studies investigating other types of exposure.

Systematic Desensitization (SD)

In SD treatment the therapist teaches the client to relax and then describes short scenarios that focus on the feared stimulus (e.g., "you are five feet away from the *snake*"). The patient is instructed to imagine the scenarios as vividly as possible for a short period of time. Scenes are arranged hierarchically, with the least fearful one presented first. If the patient indicates feelings of anxiety, the presentation of the scene is discontinued, relaxation is reintroduced, and the scene is then presented again. Each scene is repeated until it ceases to elicit anxiety.

Systematic desensitization was employed effectively by Wolff (14) to treat the fear of a 20-year-old female who had been raped at the age of 13. When presenting for treatment, the client, fearing the assailant's return, had not spent a single night alone in 17 years. Seven sessions of SD resulted in her ability to sleep alone at night. Additional evidence for the efficacy of SD comes from a series of nine cases where SD led to improvements in measures of fear, anxiety, depression, and social adjustment (15). These findings were corroborated with a sample of 17 assault victims who received 14 sessions of a nontraditional form of SD. The SD resulted in a decrease in targeted fears, as well as an increase in social adjustment. Seventy-five percent of subjects in this study voluntarily exposed themselves in vivo to situations previously desensitized in imagination (16,17). It is important to remember that, in the absence of a control group, these positive effects cannot be unequivocally attributed to the efficacy of SD. Some of the observed improvement may have been due to the natural reduction of symptoms over the first several months following assault (18). Some support for this interpretation comes from a study indicating that SD alone was not successful in reducing assault-related anxiety (19).

Based on Celluci and Lawrence's (20) finding that SD was successful in reducing nightmares, Schindler (21) treated a 29-year-old male Vietnam veteran's recurring nightmares with SD. At a 2-week booster session, the patient reported no nightmares and decreased anxiety surrounding sleeping and dreaming. Treatment gains were maintained at a 7-month follow-up. The effect of SD on muscle tension (EMG) and heart rate (HR) was studied with 10 Veterans Administration (VA) outpatients (22). Of these, eight were Vietnam veterans, one was a World War II veteran, and one was a post-Vietnam era veteran who witnessed an airplane crash. Although all of the patients reported reexperiencing symptoms, not all met DSM-III (23) criteria for a PTSD diagnosis. The patients were assessed pre- and posttreatment on EMG and HR during verbalized descriptions of combat stress scenes and noncombat stress scenes (e.g., marital conflicts). They were subsequently treated with standard SD wherein the combat scenes were presented. Length of treatment varied across patients, with an average of 5.8 months of biweekly sessions. Results indicated that physiological responses (EMG and HR) to both types of stress scenes decreased significantly, but more decrease was evident vis-à-vis the combat scenes.

In a controlled study, 112 trauma survivors received one of three active treatments; SD was compared to hypnotherapy aiming to decrease conditioned responses, and to brief psychodynamic therapy based on Horowitz' (24) psychoanalytic theory. A wait-list group was also included. On all measures, the treatments equally decreased symptoms of PTSD, general psychopathology, and maladaptive personality traits. All treatments resulted in significantly more improvement than in the control group. The treatments were not described in detail, but all appeared to focus on the trauma and aimed to decrease conditioned discomfort surrounding the trauma, possibly accounting for the lack of a differential effect (25). Traditional SD (13 to 18 sessions), with the last 2 sessions spent in in vivo exposure, led to the resumption of normal activities in victims exposed to the trauma of automobile accidents (26). The SD was very successful in reducing PTSD symptoms, depression, fear, anxiety, and pain in this study.

Many of the studies just summarized indicate that SD can produce beneficial effects in treating PTSD in several types of trauma victims. However, methodological issues such as the lack of control groups and the failure to use standardized diagnostic measures of PTSD do not allow one to make definitive conclusions regarding the efficacy of SD alone for PTSD. In the last 20 years, anxiety disorder researchers have abandoned using SD as the only exposure treatment for PTSD. Indeed, as early as 1971 SD was found inferior to flooding in treating the complex anxiety of such disorders as agoraphobia (27). We will now examine both imaginal and in vivo exposure techniques, which initially provoke moderate to high anxiety levels in patients and involve repeated reliving of the trauma in order to facilitate one in processing it.

Imaginal and In Vivo Exposure

As mentioned earlier, when exposure begins with highly feared stimuli, is prolonged, and initially elicits high levels of anxiety, the treatment technique is called flooding. This technique can be implemented either imaginally or in vivo. Several single case studies describe the use of exposure techniques to alleviate PTSD symptoms in war victims (21,28–31). Flooding in imagination (29,32,33), as well as flooding in vivo to trauma-related events (30), appeared to be equally therapeutic. Flooding to traumatic memories has also been successful in treating adolescents (34). In addition to single case reports, studies are available that examine the effects of flooding treatments in groups of combat veterans and female assault survivors.

Six to 14 sessions of imaginal flooding were found to be slightly more effective than the "standard" PTSD treatment (weekly individual and group therapies) in treating a group of PTSD Vietnam veterans participating in a small but controlled trial (35). Flooding improved reexperiencing symptoms, sleep disturbances, state anxiety, and subjective anxiety in response to traumatic stimuli more than standard treatment alone, but had little effect on depression or trait anxiety.

In a clinical trial, 24 PTSD Vietnam veterans were randomly assigned to either 14–16 90-minute sessions of relaxation and imaginal flooding or to a waiting-list control group (33). During each session, subjects were initially instructed to relax. The subjects subsequently received 45 minutes of imaginal flooding, followed by relaxation. The treated group reported significantly less depression, state anxiety, fear, hypochondriasis, and hysteria at posttreatment and at follow-up. Therapist-rated reexperiencing of the trauma, startle reactions, memory and concentration problems, impulsivity, irritability, and legal problems were all significantly lower than in the control group. Differences were not noted on several measures, including the PTSD subscale of the Minnesota multiphasic personality inventory (MMPI), other symptoms of PTSD, and social adjustment.

Direct therapeutic exposure (12–14 sessions) was compared in a controlled investigation to traditional therapy for PTSD in 51 Vietnam combat veterans (32). All subjects also received intensive inpatient group milieu treatment for PTSD. Heart rate, EMG, and skin conductance indicated strong responding to tape-recorded individualized combat scenes for all clients, but no group differences were detected posttreatment on these measures. At follow-up, patients who displayed decreased physiological responding also demonstrated improvement on a community adjustment measure tapping anxiety/depression, alienation, vigor, and confidence in skills at follow-up. To further evaluate the differential efficacy of the two treatments, the patients who fell within the upper 25% on this community adjustment measure (the VET) were defined as successes and those in the lower 25% were designated failures. Ten of the 15 successes had received exposure treatment, whereas only 3 of the 15 failures had received exposure (36). In addition, significantly more patients who received exposure reported that they were improved as a result of therapy and remained so than patients treated with conventional therapy.

The PTSD resulting from accidents has been successfully treated with imaginal exposure, relaxation training, and self exposure in vivo (37). Although modest changes in self-reported anxiety were noted, improvement in PTSD symptoms was substantial. Exposure successfully treated the persistent post-traumatic motoric startle response (jerking the steering wheel when a car approached) of a 32-year-old woman. In this study, exposure incorporated daily exposure to highway driving and three weekly sessions of relaxation training (38). Self-monitored motoric startle responses totally remitted and self-reported anxiety decreased.

Most of the studies and case reports described in this section included additional techniques such as anger control or relaxation training, and thus the contribution of exposure itself to the overall improvement was unclear. Foa and her colleagues, aware of this confound, designed two studies to ascertain the effects of exposure in isolation from other cognitive-behavioral procedures. Imaginal exposure successfully treated the PTSD of female victims of sexual and nonsexual assault in these studies (39,40). In the first study, 45 sexual assault and nonsexual assault victims who met criteria for PTSD were randomly assigned to prolonged exposure treatment (PE), stress inoculation therapy (SIT; an anxiety management training program), supportive counseling (SC), or a no treatment wait-list control group (WL). All clients were at least 3 months postassault. Exposure treatment was delivered in nine biweekly, 90-minute sessions. Exposure treatment involved repeated imaginal exposure to the assault memory and homework. Exposure involved instructions to clients to imagine as vividly as possible the assault scene and describe it aloud in the present tense, repeating it several times for 60 minutes per session. Homework included listening to the tape-recorded narratives and practicing in vivo exposure. Supportive counseling, aimed at promoting the perception of self-control, focused on assisting clients in solving daily problems that may or may not have been rape-related. Clients were taught problem solving, and therapists engaged in active listening and support. When clients began discussing the assault itself, therapists directed them to focus on "here and now" issues. The components of stress inoculation training are described in the anxiety management section of this chapter. Victims who received either PE or SIT improved on PTSD measures immediately post-treatment and at follow-up, whereas those in the SC and WL groups did not. The PE produced superior outcome on PTSD symptoms at follow-up; clients who received PE continued to improve after treatment termination, while clients in the SIT group did not.

An ongoing study is comparing the efficacy of PE, SIT, and their combination to a wait-list control group. Again, active treatment modalities are aimed at reducing assault-related PTSD present in female sexual assault and nonsexual assault survivors. All clients are 3 months post-assault. Preliminary analyses indicate that PE results in fewer PTSD diagnoses at follow-up than SIT. The combination treatment, however, is superior to either treatment alone. This study is described in more detail later.

Eye Movement Desensitization Reprocessing (EMDR)

A new technique, Eye Movement Desensitization and Reprocessing (EMDR) (41,42) is a form of therapeutic exposure (desensitization) with a strong cognitive component accompanied by saccadic eye movements. Briefly, the technique involves the client's imagining a scene from the trauma, focusing on the accompanying cognition and arousal, and tracking the therapist's rapidly moving finger. The sequence is repeated until anxiety decreases, at which point a more adaptive cognition is introduced to the scene and eye movements. The efficacy of EMDR has received some support from case reports (41,43–48), with mixed results from controlled investigations. One controlled study of EMDR treatment for both combat veterans and sexual assault victims with PTSD resulted in decreased subjective units of discomfort ratings (SUDs) and increased positive cognitions at posttreatment and follow-up (42). Another controlled study examined the efficacy of EMDR in chronic inpatient PTSD Vietnam veterans receiving either EMDR, an exposure control treatment, or inpatient milieu treatment. The EMDR was supported by therapist and in-session anxiety ratings, but no group differences were found (49). In the final EMDR study with Vietnam veterans, only weak support was found for the contribution of eye movements to the procedure (50).

In all of the studies of EMDR just discussed, methodological concerns exist. The Shapiro (42) study, for example, measured outcome with a process variable (i.e., SUDs) rather than an outcome variable. In the remaining two studies, use of taped accounts of the trauma scenes did not allow for incorporation of cognitive changes that recounting the trauma verbally or in imagination would have. Although the efficacy of EMDR is equivocal, exposure techniques in general have received strong support that indicates they have value in the treatment of post-traumatic suffering.

Anxiety Management Training (AMT)

A cardinal symptom of PTSD is pervasive and distressing anxiety that interferes with daily functioning. Anxiety Management Training (AMT) offers a repertoire of strategies with which to combat anxiety. Relaxation training (51,52), stress inoculation training (53), cognitive restructuring (54,55), breathing retraining (56), biofeedback, social skills training (57), and distraction techniques (58) are among some of these strategies.

Stress Inoculation Training (SIT)

Stress inoculation training, an AMT package, was developed for rape victims with chronic disturbances (59). The original program included 20 therapy hours and homework assignments. It consisted of both an educational phase and a training of coping skills phase. Treatment began with a 2-hour educational phase in which the treatment program's rationale and the theoretical basis of the treatment were explained. The second phase of SIT focused on the acquisition and application of coping skills. This included deep muscle relaxation training, breathing control, role playing, covert modeling, thought stopping, and guided self-dialogue following Meichenbaum's (53) stress inoculation training. In an uncontrolled investigation, rape-related fear, anxiety, phobic anxiety, tension, and depression were reduced. Clients were 15 female rape victims who received SIT (60). Interpretation of the data is limited, however, because no control group was utilized. In a later study, Veronen and Kilpatrick (61) incorporated 10 sessions each of SIT, peer counseling, and SD. Victims were permitted to select their treatment; more than 50% of the potential subject pool rejected therapy. Of the 15 who opted for treatment, 11 selected SIT, 3 chose peer counseling, and none elected SD. Although no formal statistical analyses were conducted on the six who had completed SIT, the authors reported noticeable improvement from pre- to posttreatment on most measures. Modified SIT programs have also been successful in two case studies (62,63).

In an attempt to study several treatment programs, 37 rape victims were assigned to 1 of 3 types of group therapy: SIT, assertion training, and supportive psychotherapy plus information (64). Participants received six 2-hour sessions and were compared to those in a naturally occurring wait-list control group. The SIT was similar to that described by Kilpatrick, Veronen, and Resick (59), with two exceptions: 1) cognitive restructuring, assertiveness training, and role play were excluded since they were used in the comparative treatment, and 2) exposure in vivo was added. Results indicated that all three treatments were highly effective in reducing symptoms, with no group differences evident. Improvement was maintained at 6-month follow-up on rape-related fear measures, while improvements on depression, self-esteem, and social fears were not maintained. No improvements were found in the wait-list control group.

As was mentioned previously, Foa and colleagues (40) conducted a well-controlled study with sexual and nonsexual assault victims that compared stress inoculation training, prolonged exposure (PE), supportive counseling (SC), and a no-treatment wait-list control group (WL). The SIT included information gathering, education, treatment planning, brief breathing retraining, deep muscle relaxation, thought stopping, cognitive restructuring, guided self-dialogue, covert modeling, and role play. As discussed previously, both SIT and PE produced significantly more improvement on PTSD symptoms than WL and SC immediately following treatment. Only clients who received PE continued to improve between posttreatment and follow-up. Again, preliminary results from the continuation study in progress indicate that, in clients who received SIT, 67% no longer had PTSD when assessed immediately after treatment. The SIT, however, was not as effective as PE or SIT plus PE at follow-up (39).

Biofeedback

A few studies are available that examine the efficacy of another AMT technique: biofeedback. Seven to 14 sessions of EMG biofeedback training and progressive muscle relaxation led to improvement on almost all measures in six veterans with PTSD (66). Further support for biofeedback comes from a study that compared the effects of EMG biofeedback-assisted desensitization with a no-treatment condition in 16 Vietnam combat veterans with PTSD (67). The treatment included 48 30-minute sessions conducted 3 times per week over a 4-month interval. The EMG levels assessed at pre- and posttreatment indicated that EMG desensitization was superior to no treatment in reducing muscle tension, nightmares, and flashbacks. Additionally, none of the treated patients had been rehospitalized at a 2-year follow-up, whereas five of the eight controls were hospitalized during the 2 years preceding follow-up. Systematic biofeedback training also helped a rape victim control her tachycardia, both in the laboratory and in her natural environment (68).

Cognitive Therapy

As mentioned earlier, cognitive techniques, which are often incorporated into AMT programs, teach patients to systematically examine and change maladaptive thoughts that can lead to negative responses. Cognitive therapy (CT) appeared to be as successful as SD at leading to significant improvements in ratings of fear, anxiety, depression, and social adjustment in 25 rape victims (17). No direct comparison between SD and CT was reported, however. Furthermore, patients received treatment an average of 2 weeks post-assault. Delivering treatment at this point introduces the possibility that improvements may have been a reflection of the natural decrease in symptoms in the 3 months following an assault (3,18).

A later study investigated CT and SD in 84 subjects, some of whom had participated in previous studies (69). In each treatment modality, some subjects entered treatment soon after the rape (mean = 20 days) while others were seen several months post-assault (mean = 129 days). Analyses revealed no differences between SD and CT. Moreover, the time elapsed from rape did not appear to influence the treatment outcome. The improvement evidenced by delayed treatment seekers who were initially more symptomatic than the immediate treatment seekers suggested that both treatments had active therapeutic ingredients. The absence of a control group and the failure to exclude clients who were recently raped limit the conclusions that can be drawn from this study (3,18).

The Combination of Exposure and Anxiety Management Procedures

It is clear that exposure techniques are effective in reducing PTSD symptoms, but many PTSD sufferers are extremely avoidant. The idea of confronting traumatic memories produces much anxiety in many PTSD sufferers. The SIT techniques help reduce anxiety and are easier to tolerate; however, they may not be as effective as exposure in the long term. Therefore, a combination of the two approaches may help PTSD sufferers to confront avoided memories and situations while at the same time providing techniques to manage their anxiety. As previously discussed, Foa and her colleagues are conducting a study comparing such a combined treatment approach. They are comparing the combination of SIT plus PE to PE alone, SIT alone, and a wait-list control group.

Preliminary analyses of data from 52 female victims of sexual and nonsexual assault indicate that all active treatments significantly reduce PTSD symptoms more than WL. Immediately after treatment, 73% of the SIT/PE group, 71% of the PE group, and 67% of the SIT group no longer meet DSM-III-R criteria (70) for PTSD. At follow-up, 91% of SIT/PE clients, 71% of PE, and 58% of SIT do not meet criteria for PTSD. No significant differences, however, exist at follow-up between the three active treatment groups in their ability to alleviate PTSD suffering.

A quasi-experimental design was utilized to test cognitive processing therapy (CPT) with 19 female rape victims with PTSD (71). The CPT is based on the information processing theory of PTSD. Treatment was delivered over 12 weekly sessions in a group format and included education, exposure, and cognitive restructuring components. Treated subjects were compared to a naturally occurring wait-list control group. The CPT subjects improved significantly from pre- to posttreatment on PTSD and depression ratings, and maintained their improvement throughout the 6-month follow-up period. The wait-list subjects evidenced no change during a comparable 12-week period. Problems in this study included a failure to randomly assign clients, differences in measures for control and active treatment clients, and an absence of independent assessment. No bias, however, was demonstrated in ratings by the authors.

Summary

Overall, results from well-controlled studies and case reports converge. Results indicate that both exposure and AMT programs are effective in reducing PTSD symptoms and related psychopathology such as depression. Although these techniques appear promising, cautions must be taken regarding their use. Problems have been noted with pure exposure techniques, including negative post-trauma appraisal accompanied by shame, guilt, anger, and an increase in alcohol and drug use (72). The PTSD sufferers, especially PTSD rape victims, are a particularly avoidant group and can present a number of treatment obstacles, especially for exposure techniques (65). Pure AMT appears helpful and more easily tolerated; however, it does not appear as helpful as exposure in the long term. Combination techniques appear quite effective, possibly overcoming the pitfalls of each alone.

THEORETICAL CONSIDERATIONS

It has been proposed that PTSD arises from an inability to adequately process traumata (73). Indeed, in his classical paper about emotional processing, Rachman (74) suggested that the index of inadequate emotional processing is "the persistence or return of intrusive signs of emotional activities such as obsessions, nightmares, phobias, or inappropriate expression of emotions . . ." (p. 51). These manifestations resemble the hallmark symptoms of PTSD. If PTSD symptoms are the result of inadequate emotional processing, then therapy that aims to facilitate emotional processing can be expected to assist in the amelioration of these symptoms.

In an attempt to develop the construct of emotional processing with particular attention to the anxiety disorders, Foa and Kozak (75) integrated concepts of fear, experimental data, and clinical investigations. Adopting Lang's bioinformational theory (76,77), they viewed fear as a cognitive structure that contains three classes of elements: 1) stimuli; 2) response; and 3) the meaning associated with them. They further suggested that the fear structures in anxiety disorders include pathological elements, and that treatment should be construed as modifying these elements.

Foa and Kozak (75) suggested that two conditions are required for the reduction of fear: 1) the fear memory must be activated, and 2) new information providing elements that are incompatible with preexisting elements in the fear structure must be assimilated in order for a new memory to form. "This new information, which is at once cognitive and affective, has to be integrated into the evoked information structure for an emotional change to occur . . ." (p. 22). Indeed, exposure procedures activate the structure (i.e., elicit fear) and constitute an opportunity for corrective information to be integrated. Thus, these procedures lead to modification of the fear structure and a resultant decrease in symptomatology.

Indeed, treatment by variants of exposure to the feared situations has been applied successfully to simple phobics, obsessive-compulsives, agoraphobics, and PTSD (78). Responses associated with improvement during exposure treatment include: 1) degree of initial response (i.e., activation); 2) habituation within the exposure session; and 3) habituation across exposure sessions. Several studies lend support to these indicators (75,79). Of particular relavence to PTSD are two studies demonstrating that fear activation during treatment promotes successful outcome (32, 84). These studies demonstrated that repeated exposure to the memory of the trauma results in habituation, which allows survivors to remember the trauma without intense fear responses. When the fear elements in the structure attenuate, many stimuli that were associated with fear through generalization cease to elicit fear. The generalization gradient sharpens (i.e., rape is still perceived as dangerous and therefore would elicit fear), but other stimuli (e.g., being home alone) no longer elicit fear.

Foa and Kozak (75) suggested that in addition to correcting distorted stimulus-response associations via habituation, (i.e., which severs the associations between anxiety responses and safe stimuli), treatment also offers corrective affective and cognitive information that assists in replacing mistaken beliefs with more accurate ones. Trauma memories of victims with PTSD, it is suggested, are distinguished by representations of the world as indiscriminantly dangerous and of the self as an inadequate coper (80). The observation that victims with PTSD engage in this "black and white" type of thinking was also discussed by Alford, Mahone, and Fielstein (81). They further suggested that traumatic memories are more disorganized than other memories, and that these disorganized memories are more difficult to modify. If this is true, then treatments should be directed toward both organizing and correcting the negative beliefs that can result in maladaptive responses. How can the efficacy of SIT and PE be understood within this framework?

Foa and Riggs (80) suggested that repeated reliving of the rape memories during PE treatment decreased, via habituation, the anxiety associated with trauma memories (via habituation) and, thus, allowed for a reevaluation of meaning representations in the memory. It is also suggested that repeated reliving generates a more organized memory record, one more readily integrated into existing schemas. Support for this comes from a study analyzing the narratives of the trauma memory during exposure (85). A decrease, from the first to the last descriptions of the trauma, in expressions of disorganization (i.e., unfinished thoughts and repetitions) was correlated with improvement on indices of PTSD and related pathology. While PE does not directly address the victim's world and self schemas, one result of emotionally processing the trauma memory is the victim's ability to view the rape as a distinct event rather than representing the world as a whole. This more positive world view assists the victim in discriminating between danger and safety. Moreover, the successful processing of the traumatic memories leads to a reduction in PTSD symptoms, particularly those of intrusion and avoidance. Such processing results in the victim's perception of herself as an adequate coper. Indirectly, exposure treatments may also foster positive social interactions; if the victim no longer perceives the world as indiscriminantly dangerous and herself as an inadequate coper, then she may more readily seek social support that can facilitate her confrontation of stressors.

The SIT techniques seem to impact more directly on the self-schema of the victims. While SIT does not organize the trauma memory structure, it can assist the client in managing the anxiety that exposure evokes. In addition, teaching the victim techniques for coping with stress and anxiety may lead to the development of a more positive self-image (e.g., the image of self as a successful coper). The SIT may impact indirectly in two ways on the victim's perception of the world: 1) the increased perception of control may allow the victim to tolerate exposure to the trauma memories for longer periods of time, which may serve as self-directed prolonged exposure, and 2) the perception of oneself as able to cope with stress reduces one's sense of vulnerability; if the victim perceives herself as an "adequate coper," she will see herself as more capable of averting potential dangers. Just as exposure treatments may lead to increased social support, so too does SIT. Positive social interactions, in turn, serve to further strengthen functional schemas.

It is not surprising that a combination of SIT and PE is proving more effective than either procedure alone in reducing PTSD symptoms. The PE both organizes the trauma memory and leads to an accompanying change in world schema, while SIT directly impacts self-schema. It follows that any treatment that organizes the trauma memory, and either directly or indirectly alters the victim's schemas of self and world will help to ameliorate symptoms of PTSD. Indeed, the effectiveness of treatments such as supportive counseling focusing on problem solving (40,64) may be due to changes that these techniques produce in the victim's self-schema.

Identification of specific impairments that underlie panic disorder has led to very powerful treatments (82). Elucidating mechanisms that underlie PTSD and incorporating into treatment programs techniques that address such mechanisms will render treatment more effective in alleviating the symptoms of PTSD.

REFERENCES

1. National Victim Center and Crime Victim Research and Treatment Center, 1992.
2. Kilpatrick DG, Resnick HS. Posttraumatic stress disorder associated with exposure to criminal victimization in clinical and community populations. In Davidson JRT, Foa EB, eds. *Posttraumatic stress disorder in review: recent research and future development.* Washington, DC: American Psychiatric Press; 1993.
3. Rothbaum BO, Foa EB, Riggs DS, Murdock T, Walsh W. A prospective examination of post-traumatic stress disorder in rape victims. *J Traum Stress* 1992;5:455–475.
4. Kulka RA, Schlenger WE, Fairbank JA, et al. *Trauma and the Vietnam War generation.* New York: Brunner/Mazel; 1990.
5. McCann L, Pearlman A. Vicarious traumatization: a framework for understanding the psychological effects of working with victims. *J Traum Stress* 1990;3:131–149.
6. McFarlane AC. The aetiology of post traumatic morbidity: predisposing, precipitating, and perpetuating factors. *Br J Psychol* 1989;154:221–228.
7. Mitchell JT, Bray GP. *Emergency services stress: guidelines for preserving the health and careers of emergency services personnel.* Englewood Cliffs, NJ: Prentice Hall, 1990.
8. Koss MP, Harvey MR. *The rape victim: clinical and community approaches to treatment.* Lexington, MA: The Stephen Greene Press; 1987.
9. Burgess AW, Holstrom LL. Coping behavior of the rape victim. *Am J Psychiatry* 1976;133:413–418.
10. Forman BD. Cognitive modification of obsessive thinking in a rape victim: a preliminary study. *Psychol Rep* 1980;47:819–822.
11. Cryer L, Beutler L. Group therapy: An alternative treatment approach for rape victims. *J Sex Marital Ther* 1980;6:40–46.
12. Perl M, Westin AB, Peterson LG. The female rape survivor: time-limited group therapy with female/male co-therapists. *J Psychosom Obstet Gyneacol* 1985;4:197–205.
13. Wolpe, J. *Psychotherapy by reciprocal inhibition.* Stanford, CA: Stanford University Press; 1958.
14. Wolff R. Systematic desensitization and negative practice to alter the aftereffects of a rape attempt. *J Behav Exp* 1977;8:423–425.
15. Turner SM. *Systematic desensitization of fears and anxiety in rape victims.* Paper presented at the Association for the Advancement of Behavior Therapy, San Francisco, CA; 1979.
16. Frank E, Stewart BD. Physical aggression: treating the victims. In Bleckman EA, ed. *Behavior modification with women.* New York: Guilford Press, 1983:245–272.
17. Frank E, Stewart BD. Depressive symptoms in rape victims. *J Affect Disord* 1984;1:269–277.
18. Kilpatrick DG, Calhoun KS. Early behavioral treatment for rape trauma: efficacy or artifact? *Behav Ther* 1988;19:421–427.
19. Becker JV, Abel GG. Behavioral treatment of victims of sexual assault. In Turner SM, Calhoun KS, Adams HE, eds. *Handbook of clinical behavior therapy.* New York: Wiley; 1981:347–379.
20. Celluci AJ, Lawrence PS. The efficacy of systematic desensitization in reducing nightmares. *J Behav Exp* 1978;9:109–114.
21. Schindler FE. Treatment of systematic desensitization of a recurring nightmare of a real life trauma. *J Behav Exp* 1980;11:53–54.
22. Bowen GR, Lambert JA. Systematic desensitization therapy with post-traumatic stress disorder cases. In Figley CR, ed. *Trauma and its wake: Vol. II.* New York: Brunner/Mazel, 1986:280–291.
23. American Psychiatric Association. *Diagnostic and statistical manual of mental disorders,* 3rd ed. Washington, DC: APA, 1980.
24. Horowitz, MD. *Stress response syndromes.* New York: Aronson, 1976.
25. Brom D, Kleber RJ, Defare PB. Brief psychotherapy for post-traumatic stress disorder. *J Cons Clin* 1989;57:607–612.
26. Muse M. Stress-related, posttraumatic chronic pain syndrome: behavioral treatment approach. *Pain* 1986;25:389–394.
27. Boulougouris J, Marks I, Marset P. Superiority of flooding (implosion) to desensitization for reducing pathological fear. *Behav Res Ther* 1971;9:7–16.
28. Fairbank JA, Gross RT, Keane TM. Treatment of post-traumatic stress disorder: evaluation of outcome with a behavioral code. *Behav Modif* 1983;7:557–568.
29. Fairbank JA, Keane TM. Flooding for combat-related stress disorders: assessment of anxiety reduction across traumatic memories. *Behav Ther* 1982;13:499–510.
30. Johnson CH, Gilmore JD, Shenoy RZ. Use of a feeding procedure in the treatment of a stress-related anxiety disorder. *J Behav Exp* 1982;13:235–237.
31. Keane TM, Kaloupek DG. Imaginal flooding in the treatment of post-traumatic stress disorder. *J Cons Clin* 1982;50:138–140.
32. Boudewyns PA, Hyer L. Physiological response to combat memories and preliminary treatment outcome in Vietnam veteran PTSD patients treated with direct therapeutic exposure. *Behav Ther* 1990;21:63–87.
33. Keane TM, Fairbank JA, Caddell JM, Zimering RT. Implosive (flooding) therapy reduces symptoms of PTSD in Vietnam combat veterans. *Behav Ther* 1989;20:245–260.
34. Saigh P. The use of an in vitro flooding package in the treatment of traumatized adolescents. *Dev Behav Pediatrics* 1989;10:17–21.
35. Cooper NA, Clum GA. Imaginal flooding as a supplementary treatment for PTSD in combat veterans: a controlled study. *Behav Ther* 1989;20:381–391.
36. Boudewyns PA, Hyer L, Woods MG, Harrison WR, McCranie E. PTSD among Vietnam veterans: an early look at treatment outcome using direct therapeutic exposure. *J Traum Stress* 1990;3:359–368.
37. McCaffrey RJ, Fairbank JA. Post-traumatic stress disorder associated with transportation accidents: two case studies. *Behav Ther* 1985;16:406–416.

38. Fairbank JA, DeGood DE, Jenkins CW. Behavioral treatment of a persistent post-traumatic startle response. *J Behav Exp* 1981;12:321–324.
39. Foa EB. *Cognitive behavioral treatment for rape: victims with chronic PTSD.* Paper presented at the International Society for Traumatic Stress Studies, San Antonio, TX; October 1993.
40. Foa EB, Rothbaum BO, Riggs DS, Murdock T. Treatment of post-traumatic stress disorder in rape victims: a comparison between cognitive-behavioral procedures and counseling. *J Cons Clin* 1991;59:715–723.
41. Shapiro F. Efficacy of eye movement desensitization procedure in the treatment of traumatic memories. *J Traum Stress* 1989;2:199–223.
42. Shapiro F. Eye movement desensitization procedure: a new treatment for post-traumatic stress disorder. *J Behav Exp* 1989;20:211–217.
43. Lipke HJ, Botkin AL. Case studies of eye movement desensitization and reprocessing (EMDR) with chronic post-traumatic stress disorder. *Psychotherapy* 1992;4: 591–595.
44. Marquis JN. A report on seventy-eight cases treated by eye movement desensitization. *J Behav Exp* 1991;22: 187–192.
45. Puk G. Treating traumatic memories: a case report on the eye movement desensitization procedure. *J Behav Exp* 1991:149–151.
46. Rothbaum BO. *Eye movement desensitization and reprocessing in the treatment of PTSD and painful memories.* Paper presented at the Association for the Advancement of Behavior Therapy Annual Convention, Atlanta, GA; November 1993.
47. Shapiro F. Eye movement desensitization and reprocessing procedure: from EMD to EMD/R—a new treatment model for anxiety and related traumata. *Behav Ther* 1991;12:133–135.
48. Wolpe J, Abrams J. Post-traumatic stress disorder overcome by eye movement desensitization: a case report. *J Behav Exp* 1991;22:39–43.
49. Boudewyns PA, Swertka SA, Hyer LA, Albrecht JW, Sperr EV. Eye movement desensitization for PTSD of combat: a treatment outcome pilot study. *Behav Ther* 1993;16:29–33.
50. Pitman RK, Orr S, Altman B, Longpre RE, Poire RE, Lasko NB. A controlled study of eye movement desensitization/reprocessing (EMDR) treatment for post-traumatic stress disorder. Paper presented at the Annual Meeting of the International Society for Traumatic Stress Studies, San Antonio, TX; October 1993.
51. Bernstein, DA, Borkovec, TD. *Progressive relaxation training.* Champaign, Il: Research Press; 1973.
52. Jacobson, E. *Progressive relaxation.* Chicago: University of Chicago Press; 1938.
53. Meichenbaum, D. *Cognitive behavior modification.* Morristown, NJ: General Learning Press; 1974.
54. Beck, AT. *Depression: causes and treatment.* Philadelphia: University of Pennsylvania Press; 1972.
55. Ellis A. The basic clinical theory and rational-emotive therapy. In Ellis A, Grieger R, eds. *Handbook of rational-emotive therapy,* New York: Springer, 1977:3–34.
56. Clark DM, Salkovskis PM, Chaukley AJ. Respiratory control as a treatment for panic attacks. *J Behav Exp* 1985;16:23–30.
57. Becker RE, Heimberg RG, Bellack AS. *Social skills training treatment for depression.* New York: Pergamon Press; 1987.
58. Wolpe, J. *The practice of behavior therapy.* New York: Pergamon Press; 1973.
59. Kilpatrick DG, Veronen LJ, Resick PA. Psychological sequelae to rape: assessment and treatment strategies. In Dolays DM, Meredith RL, eds. *Behavioral medicine: assessment and treatment strategies.* New York: Plenum Press, 1982:473–497.
60. Veronen LJ, Kilpatrick DG. *Stress inoculation training for victims of rape: efficacy and differential findings.* Paper presented at the Sixteenth Annual Convention of the Association for the Advancement of Behavior Therapy, Los Angeles, CA; November 1982.
61. Veronen LJ, Kilpatrick DG. Stress management for rape victims. In Meichenbaum D, Jaremko ME, eds. *Stress reduction and prevention.* New York: Plenum Press; 1983:341–374.
62. Kilpatrick DG, Amick AE. Rape trauma. In Hersen M, Last CG, eds. *Behavior therapy casebook.* New York: Springer; 1985:86–103.
63. Pearson MA, Poquette BM, Wasden RE. Stress inoculation and the treatment of post-rape trauma: a case report. *Behav Ther* 1983;6:58–59.
64. Resick PA, Jordan CG, Girelli SA, Hutter CK, Marhoefer-Dvorak S. A comparative outcome study of behavioral group therapy for sexual assault victims. *Behav Ther* 1988;19:385–401.
65. Rothbaum BO, Foa EB. Exposure therapy for rape victims with post-traumatic stress disorder. *Behav Ther* 1992;15:219–222.
66. Hickling EJ, Sison GFP, Vanderploeg RD. Treatment of posttraumatic stress disorder with relaxation and biofeedback training. *Behav Ther* 1986;16:406–416.
67. Peniston EG. EMG biofeedback-assisted desensitization treatment for Vietnam combat veterans post-traumatic stress disorder. *Clin Biofeedback Health* 1986;9: 35–41.
68. Blanchard EB, Abel GG. An experimental case study of the biofeedback treatment of a rape induced psychophysiological cardiovascular disorder. *Behav Ther* 1976; 7:113–119.
69. Frank E, Anderson B, Stewart BD, Dancu C, Hughes C, West D. Efficacy of cognitive behavior therapy and systematic desensitization in the treatment of rape trauma. *Behav Ther* 1988;19:403–420.
70. American Psychiatric Association. *Diagnostic and statistical manual of mental disorders,* 3rd ed. revised. Washington, DC: APA, 1987.
71. Resick PA, Schnicke MK. Cognitive processing therapy for sexual assault victims. *J Consult Clin Psychol* 1992; 60:748–756.
72. Pitman RK, Altman B, Greenwald E, Longpre RE, Macklin ML, Poire RE, Steketee G. Psychiatric complications during flooding therapy for posttraumatic stress disorder. *J Clin Psychiatry* 1991;52:17–20.
73. Foa EB, Steketee G, Rothbaum BO. Behavioral/cognitive conceptualizations of post-traumatic stress disorder. *Behav Ther* 1989;20:155–176.
74. Rachman S. Emotional processing. *Behav Res Ther* 1980; 18:51–60.
75. Foa EB, Kozak MJ. Emotional processing of fear: exposure to corrective information. *Psychol Bull* 1986;99: 20–35.

76. Lang PJ. Imagery in therapy: an information processing analysis of fear. *Behav Ther* 1977;8:862–886.
77. Lang PJ. A bio-informational theory of emotional imagery. *Psychophysiology* 1979;6:495–511.
78. Marks, I. *Fears, phobias, and rituals: panic, anxiety, and their disorders.* Oxford: Oxford University Press; 1987.
79. Kozak MJ, Foa EB, Steketee G. Process and outcome of exposure treatment with obsessive-compulsives: psychophysiological indicators of emotional processing. *Behav Ther* 1988;19:157–169.
80. Foa EB, Riggs DS. Post-traumatic stress disorder in rape victims. In Oldham J, Riba MB, Tasman A, eds. *American psychiatric press review of psychiatry, Vol. 12.* Washington, DC: American Psychiatric Press, 1993: 273–303.
81. Alford JD, Mahone C, Fielstein EM. Cognitive and behavioral sequelae of combat: conceptualization and implications for treatment. *J Traum Stress* 1988;1:489–501.
82. Margraf J, Barlow DH, Clark D, Telch M. Psychological treatment of panic: work in progress on outcome, active ingredients, and follow-up. *Behav Res Ther* 1993;31: 1–8.
83. Roth S, Dye E, Lebowitz L. Group therapy for sexual-assault victims. *Psychotherapy* 1988;25:82–93.
84. Foa EB, Riggs DS, Massie ED, Yarczower M. The impact of fear activation and anger on the efficacy of exposure treatment for PTSD. *Behavior Therapy,* in press.
85. Foa EB, Molnar C, Cashman L. Change in rape narratives during exposure therapy for PTSD. *Journal of Traumatic Stress,* in press.

Neurobiological and Clinical Consequences of Stress: From Normal Adaptation to PTSD, edited by M. J. Friedman, D. S. Charney, and A. Y. Deutch, Lippincott-Raven Publishers, Philadelphia © 1995.

28

Dynamic Psychotherapy of Post-Traumatic Stress Disorder

Charles R. Marmar, Daniel S. Weiss, and *Robert S. Pynoos

*Department of Psychiatry, University of California, San Francisco, San Francisco, California 94143; and *Department of Psychiatry and Biobehavioral Sciences, University of California, Los Angeles, Neuropsychiatric Institute, Los Angeles, California 90024*

PSYCHODYNAMIC MODELS OF TRAUMATIC STRESS RESPONSE

The past decade has seen increasing clinical and research interest in dynamic approaches to the treatment of civilian and combat-related post-traumatic stress disorder (PTSD). These approaches build upon earlier developments in the field of psychodynamic psychotherapy for anxiety and mood disorders, with specific nuances for the management of responses to traumatic events.

Psychodynamic approaches to the understanding of traumatic stress emphasize the impact of a traumatic event on the person's self concepts and view of others, affects resulting when: 1) conscious and unconscious representations of the self and others triggered by the trauma are discrepant with usual views, and 2) defenses are mobilized to cope with the discrepant meanings and painful emotions. Psychodynamic approaches emphasize meaningful conscious and unconscious associations linked to specific moments within the traumatic event and its aftermath. These associations or themes may recruit earlier conflicts from preexisting developmental periods. Traumatic events activate earlier mental schemas concerning danger, injury, and protection, activating concerns from childhood and adolescence regarding safety, trust, risk, injury, loss, parental protection, dependency, and autonomy. Adult trauma may activate specific pre-oedipal or oedipal constellations, particularly those concerning maternal protection and nurturance, control of emotions and bodily functions, and conflicts about potency, rivalry, aggression, and fears of retaliation. The trauma-activated themes are seen as bridges from current concerns to self representations, representations of other, affect states, and defenses arising during early developmental periods.

When these post-traumatic shifts in self concept are unaddressed over time, a deterioration in character functioning may result. Shifting views of the self as victim or victimizer interfere with mastery of the trauma. The adult who is functioning reasonably well before the trauma may develop post-traumatic narcissistic or borderline personality features with: 1) fracturing of a coherent positive sense of self; 2) traumatic disillusionment in self, others, and institutions; 3) use of more primitive defenses, including splitting (1) and dissociation (2); 4) poorly controlled affect states; 5) heightened impulsivity; and 6) increased vulnerability to future narcissistic injuries (3–5). As part of this regression in psychological functioning, trauma victims have a tendency to repeat situationally inappropriate, immature, or conflicted relationship patterns that have been internalized during formative relationships (6). These relationship patterns prominently feature shifting views of the self and others as victims, victimizers, and rescuers (7,8). The result is more turbulent current interpersonal relationships and intensified transference reactions during treatment that can threaten the patient's capacity and willingness to engage in the therapeutic process (1,9,10).

FACTORS INFLUENCING THE RESPONSE TO TRAUMA

The course of response following traumatic stress exposure is impacted by multiple factors. Four broad factors with relevance to psychodynamic treatments are discussed in the section that follows: 1) the circumstances of the trauma; 2) reactivation of latent trauma from earlier developmental periods; 3) the phase of development at which trauma occurs; and 4) the social context during and after the trauma.

Circumstances of the Trauma

The psychological sequelae following traumatic life events are in part determined by the circumstances of the trauma. In combat trauma, repeated endangerment to the self, witnessing the death and dismemberment of buddies, witnessing or participating in atrocities, and physical injury with permanent disability increase the risk for more chronic and severe traumatic stress reactions (11). For rape trauma victims, use of physical force, display of a weapon, and injury to the victim are associated with increased risk of PTSD (12). For battered women, the risk for PTSD has also been shown to be associated with the characteristics of the battery experience (13). Not only are different levels of exposure within a similar traumatic event associated with different outcomes, but different classes of traumatic events are associated with different psychological sequelae. For example, rape trauma victims suffer more acute guilt, shame, and suicidal ideation compared to individuals experiencing civilian accidents or industrial disasters (14).

Traumatic grief responses are complicated when there have been mutilative injuries to the deceased that have been witnessed by the survivor, and because of frequent protracted court proceedings that lead to compulsive inquiry with frustrated efforts to provide a frame of meaning for a tragic death (7,15). These preoccupations with the circumstances of the death reinforce the occurrence of painful, unbidden visual imagery, cause preoccupation with the brutality and the meaningless of the death, and interfere with the survivor's capacity to modulate the loss by reminiscing about positive memories of the deceased.

The course of traumatic stress response may be further complicated by a domino-like progression of secondary adversities that frequently follow violent experience. For example, rape, assault, and accident victims may require repeated hospitalization and reconstructive surgery, and suffer variable lengths of disability. The repeated medical procedures may be traumatic in and of themselves, serve as inevitable reminders of the original trauma, and, in combination with scarring or other physical stigmata, constitute unavoidable reminders of the trauma (16). Financial compensation for the trauma in the form of disability payments is frequently a mixed blessing, providing needed financial help during a transition back to health but risking the development of chronic dependency and ultimately a profound self-concept disturbance in a reorganized identity as a victim. Once established, such identity shifts constitute powerful resistances to recovery.

Reactivation of Latent Trauma

Adult trauma can transiently, and in some cases permanently, alter the gains made in building a healthy self concept over the course of development. For those adult trauma victims without a history of prior traumatization, the trauma in adulthood frequently recruits two classes of problematic self concepts. Consciously felt and expressed concerns relate to problematic weak and defective feelings about the self, including fear, helplessness, vulnerability to repetition of the trauma, grief over losses, and shameful feelings about being unable to control one's emotional reactions or fate. There is a second set of problematic self concepts, usually warded off and operating outside of the person's conscious awareness, as in which the self is viewed as dangerously powerful and responsible for victimizing the self or others during the trauma (10). The latter problematic and powerful self concepts are associated with exaggerated feelings of personal responsibility, intense survivor guilt, and self-defeating behaviors linked to the magical belief of personal responsibility for the trauma.

The picture is further complicated when feelings of helplessness and victimization at the time of adult trauma reinvoke memories of childhood experiences of victimization. Similarly, views of the self as dangerously powerful and responsible for inflicting harm during the adult trauma may recruit real or fantasized views of the self as having victimized others during childhood or adolescence (6). These changes recapitulate cognitive, affective, and physiological reactions experienced at the time of the original trauma. As a result, the trauma victim is flooded with intrusive memories of the current adult trauma, as well as thoughts, images, memories, and dreams of trauma occurring years or decades earlier, and compounded levels of hyperarousal.

Impact of Developmental Phase on Traumatic Stress Response

Pynoos (16) has summarized the high risks of childhood physical and sexual abuse, and exposure to family violence. The representational world of the normal developing child is in flux, with constant revisions of inner models of the self, others, and the world. Compared to adults, identity is more fluid, with positive and negative representations of the self and others less well integrated. The child is dependent upon the social, emotional, and physical environment of the family, which may itself be the source of the abuse or become severely strained in an effort to cope with the consequences of trauma affecting the family unit.

Abuse of infants and toddlers may have far-reaching influence on attachment behavior. Preschool children are limited in their capacity to verbally represent trauma, and even with increasing maturity these early experiences are not readily assimilated into verbal representation. For these reasons, the trauma may be represented for the very young child primarily at the level of nonverbal communication, bodily representation, and play with compulsive behavioral reinactments. Chronic or repetitive trauma in childhood is associated with profound restrictions in emotional responsiveness, social isolation and detachment, difficulties with impulse control, self-injurious behaviors, traumatic reenactments, and increased frequency of dissociative responses (7,17). In extreme cases, fragmentations of self representation occur, with longlasting or permanent dissociation of part self concepts expressed later in development as multiple personality (2, 18) or borderline personality (19).

Trauma exposure during early adolescence interferes with stage-appropriate tasks of 1) integrating past, present, and future orientations; 2) separation from parents; 3) disturbances in moral development; and 4) interference with development of symbolic thinking (7). Repetitive severe traumatization during later adolescence, as seen in young combat veterans, can result in severe personality disorders. These effects are not limited to troubled adolescents who are subsequently traumatized, but may occur in late-adolescent military personnel whose prior academic, social, and moral development were in the normal range (10,20).

Models of traumatization have not always adequately addressed the impact of adult developmental phases on trauma response. The traumatic event may impact at varying time periods with respect to the adult's achievement of interpersonal and occupational goals, and health status (16). As an example, the young adult who is sexually assaulted by a stranger at the time of forming an important love relationship may suffer consequences not limited to the nightmares, flashbacks, and arousal reactions of PTSD. The current love relationship is at risk of disruption, subsequent love relationships may be delayed, and marital and family goals may be derailed. Avoidance of intimacy, including sexuality, may lead over time to impaired social skills.

For traumatized adults, the profound experience of helplessness associated with the traumatic event is at variance with adult self conceptualizations as competent, in control, and invulnerable, leading to subsequent changes in ambition, assertiveness, and initiative. Representations of the world as a violent, unpredictable, and uncontrollable place are not readily assimilated into existing schema, requiring modified views of the self and others that take account of the reality of trauma without swamping trusting, safer, and more predictable views of relationships and life experience. These strains, particularly in vulnerable individuals, can intensify borderline (1), narcissistic (5), and other severe personality disturbances (21).

In addition, there is often the concomitant internal experience of social isolation, with the sense that no one can understand how the victim feels. This sense may be linked to the internal working models of attachment between self and others described by Bowlby (22). It is quite likely that the experience of trauma as an adult reactivates earlier internal models of disruptions of safe attachment that were experienced in childhood. The feelings of isolation may be the product of these reliving experiences, or the result of defensive flight from attachment in order to protect against the dangers of reexperiencing the traumatic childhood losses. Even when the trauma does not include betrayal of trust by others, such experiences are commonly evoked. Moreover, the social withdrawal is often most strongly felt towards loved ones who did not forsake the traumatized individual but are nonetheless indicted. Reactivation of childhood feelings of abandonment and betrayal in adult trauma victims implicates the structural framework of the attachment paradigm in this aspect of the reactions to traumatic events. Inadequate attention has been given to the nuances of traumatic stress response in older adults. Physical, material, medical, and emotional abuse of elders has been identified (23). When natural and manmade disasters strike older adults, they occur at a sensitive developmental period in which the individual is attempting to come to terms with declining social status, loss of relatives and friends, diminishing income, and increasing dependency on others (16). The older adults' perception of self efficacy in addressing and coping with traumatic events may be low. A large percentage of older adults are widows living alone, without spouses, and with compromised financial and social support.

Older adults physically traumatized from assaults or natural disasters may experience relatively greater physical injury and slower recovery. Trauma may necessitate changes in living situations and social routines that are more difficult for older adults. The combination of physical vulnerability, difficulty adjusting to change, and limited financial resources increase the risk of depression secondary to traumatization in the elderly (24).

Mediating Role of the Interpersonal Environment

Victims of natural and manmade disasters have a heightened need for reliance on appropriate social support, within the context of the immediate family and community, and the more narrow group of those sharing a common traumatic event. The availability of social support from family, friends, and community may be compromised when those called upon for support have had their own lives adversely changed by the traumatic event, a phenomenon referred to as indirect victimization. Family members and friends of trauma survivors may directly witness the traumatic event or receive graphic information about the circumstances of the trauma, resulting in both intrusive and avoidant reactions in members of the support system. The support system may be greatly strained, for example, when waiting to hear the outcome of hostage or kidnapping events, or rescue and recovery operations, or when awaiting the outcome of post-trauma emergency care. Family and friends may be further depleted when called upon to assume supportive roles for a longer period of time.

Even when supports are available, the trauma victim may decline help, in part because of the invitation for disclosure of disturbing memories, and in part because of realistic or exaggerated worries about burdening others with emotional reactions to the trauma. The trauma victim may erect a "trauma membrane," allowing intimacy and disclosure only with those who have shared the immediate trauma or survived a similar past trauma (25). The barrier of the trauma membrane is erected to prevent access to the trauma victim by any individual who is not conferred the status of a safe and protective figure by virtue of having shared a similar life experience. The assumption is that a similar life experience will automatically confer empathy. In erecting such a barrier, trauma victims may deprive themselves of much-needed help.

Further adding to the traumatized individual's ambivalent approach towards support is a conflict between the need to address increased dependency following victimization and a heightened need for autonomy. Well-meaning family and friends may offer help that undermines rather than builds the traumatized individual's need to reassert autonomy, mastery being more difficult in those with a history of difficult childhood and adolescent attachment, separation, and autonomy experiences.

The family environment provides a fundamental context for reconferring a sense of safety and security following trauma. A well-functioning family serves as a natural buffer, helping a traumatized member to legitimize and validate his or her experience, express painful reactions, clarify issues of realistic responsibility, and return to full functioning in work and love relations (26). These normal self-maintaining functions of the intact family may be compromised when all family members are exposed to a common traumatic event or when dysfunctional patterns of the family unit preexist the traumatic event. In the former case, family members may be at different phases of traumatic stress response. One member may experience a compulsive need to discuss the trauma while another rigidly wishes to deny the event and avoid reminders, creating strains in the support system. In the latter case, dysfunctional patterns increase the risk of regression in family functioning following family trauma, with greater divisiveness, blaming, unrealistic expectations for rapid recovery, distorted communication patterns, isolation, and risk of rupture of an already fragile family cohesion, resulting in greater rigidity or chaos in family functioning.

Group support structures such as mutual help groups, community support groups, and formal group therapy may play a decisive role for the trauma victim. A well-functioning support group will provide practical support for the individual temporarily overwhelmed by trauma, reinforce the normative nature of stress reactions and recovery patterns, share mutual concerns, address common fears and traumatic memories, increase the capacity to tolerate disturbing emotions, and share strategies for coping (27).

DYNAMIC APPROACHES FOR TREATING POST-TRAUMATIC STRESS

Normal Stress Response

The majority of adults who are relatively well-functioning prior to a traumatic life event, have adequate social supports, do not have a history of childhood or adolescent trauma or psychiatric illness, and experience a single discrete traumatic event in adulthood will have a normal stress response. The reaction usually lasts from several weeks to months, and leads to full recovery. As described by Horowitz (9), the typical phases are as follows: 1) an immediate painful registration and outpouring of terror, grief, rage, and other emotions, lasting minutes to hours; 2) several weeks of denial, numbing, and unreality that yield to reexperiencing the trauma in the form of daytime unbidden thoughts and images, and nightmares; and 3) working through with recovery by 3–5 months.

Interventions that have been demonstrated to facilitate normal recovery from traumatic stressors and reduce the occurrence of post-traumatic stress disorder and other post-traumatic psychological problems include education and support, avoidance of maladaptive coping through isolation, alcohol and drugs, compulsive work, and individual or group debriefing. Armstrong, O'Callahan, and Marmar (28) have described a

TABLE 1. *Dynamic treatment approaches for PTSD*

Traumatic stress category	Dynamic approaches
Normal Stress Response	debriefing disclosure of events expression of reactions sharing coping strategies transition after event
Acute Catastrophic Stress Reaction	debriefing abreaction support self-cohesion adjunctive pharmacotherapy
PTSD without Comorbidity	time-limited dynamic psychotherapy establish therapeutic alliance focus on self concepts working through conflicts linkage to prior trauma attention to transference and countertransference
PTSD with Axis I Comorbidity	time-limited dynamic therapy treat alcohol/substance abuse first treat other comorbidities concurrently
Chronic Secondary PTSD with Axis II Comorbidity	multimodal longterm dynamic group and pharmacologic inpatient at times for uncovering or crises individual, marital, and family treatment vocational rehabilitation, and social skills training

PTSD, post-trauma stress disorder.

dynamically informed multiple-stressor debriefing model aimed at reducing complications following traumatic stress exposure. The model is derived in part from earlier work by Mitchell (29) and Raphael (30), and was modified for use in debriefing emergency services personnel following multiple repeated stress events in the aftermath of the 1989 Loma Prieta earthquake and the 1991 Oakland Hills firestorm. The multiple-stressor debriefing model provides a single 2-hour debriefing session in four stages: disclosure of events, exploration of troubling reactions, identification of practical coping strategies; and exploration of feelings about leaving the disaster, plan of action regarding transition, and requests for referrals.

Acute Catastrophic Stress Reactions

Acute catastrophic stress reactions occur within hours or a few days following traumatic stress exposure and are characterized by panic reactions, cognitive disorganization, disorientation, dissociation, severe insomnia, tics and other movement disorders, paranoid reactions, and incapacity to manage even basic self care, work, and interpersonal functions (10,31).

Treatment strategies for acute catastrophic stress reactions include immediate support, removal from the scene of the trauma, time-limited use of antianxiety medications for daytime anxiety and insomnia, brief support-expressive dynamic psychotherapy aimed at abreaction and integration of the trauma, and careful follow-up (32). Brief supportive-expressive dynamic psychotherapy is conducted within a crisis intervention framework, emphasizing: 1) anxiety reduction; 2) shoring up of defenses; 3) effort to restabilize the feelings of safety, predictability, and coherence of self that have been fragmented by catastrophic stress exposure; and 4) provision of caring and empathic support in the framework of a holding environment aimed at countering post-traumatic narcissistic regression (5). Therapists offer assistance in processing the event, modulating overwhelming affect, making decisions, and related supportive functions that allow the acute catastrophic stress victim to identify with the therapist's calm step-by-step problem-solving approach.

Uncomplicated Post-Traumatic Stress Disorders

Post-traumatic stress disorder, in a pure form uncomplicated by comorbid conditions and extensive secondary adversities, is most frequently seen 6–8 months following exposure to catastrophic events. The individual's mental state is dominated by reexperiencing, avoidant, and arousal symptoms that persist, do not remit with the passage of time and support of family and friends, and interfere with work and interpersonal functioning. For individuals whose PTSD problems persist beyond 3–5 months following the trauma, treatment is strongly indicated. Left untreated, the majority of this group will not remit with further passage of time, and are at risk for depression, alcohol/substance abuse, panic disorder and multiple phobic avoidances, marital and family disturbances, and work impairment (33).

Brief dynamic psychotherapy (6,9,10,34) is effective for uncomplicated PTSD (35) and for pathological grief reactions (27). At the beginning of treatment, the patient is encouraged to retell the story of the traumatic event. The opportunity to retell the event in the presence of an expert healer who remains calm, empathic, compassionate, and nonjudgmental begins to depressurize the trauma response for the patient and facilitate a working collaboration with the therapist.

The rapid establishment of a therapeutic alliance, central to brief therapy approaches, may take on special nuances with recently traumatized individuals. The threat that emotions may escalate to unmanageable proportions, leaving the patient disorganized and overwhelmed, or that the therapist may not tolerate the expression of rage, sorrow, panic, and other intense emotions, may inhibit self disclosure. A further threat to the therapeutic alliance derives from the tendency of the trauma victim to repeat situationally inappropriate, immature, or conflicted relationship models that have been internalized during sensitive developmental periods and heightened following traumatic stress exposure. Common transferential views of the therapist include victim, victimizer, and rescuer representations (8, 36,37).

For individuals with PTSD, the impact of a traumatic event on self concepts serves as an organizer for the unfolding treatment process. The initial focus of treatment centers on the therapist's empathic response to the patient's feelings of helplessness, shame, and vulnerability. Later, when the patient reexperiences a greater sense of self cohesion, utilizes better-functioning defense and coping strategies, and more successfully modulates the powerful affects related to uncovering of the trauma, the therapist addresses the warded off conflict-laden meanings of the traumatic event (e.g., angry reactions and feelings of exaggerated responsibility (5)). The pace and depth of the uncovering processes are also adjusted to take account of the patient's level of functioning prior to the trauma (38).

For the trauma victim who has experienced a failure of control resulting in intrusive breakthrough of imagery and undermodulated feeling states, the therapeutic effort is directed at resurrecting adaptive defenses. Interventions include provision of support, asking questions or repeating comments, helping the patient to compartmentalize elements of the trauma experience to examine them in tolerable doses, and helping to organize information by reconstructing sequences. Further support is provided by recommending reduced interpersonal and work demands, removal of environmental cues that trigger memories of the trauma, and encouraging the adaptive use of suppression and avoidance techniques so that the person can function while struggling to master the trauma.

For individuals who present with overcontrol, avoidant coping, and emotional numbing, therapists' efforts are directed at creating a safe environment for the exploration of warded-off images and feelings related to the traumatic event. The therapist directs the patient's attention to characteristic avoidant coping strategies, including use of alcohol and drugs, workaholism, compulsive spending, eating or sexual behavior, thrill-seeking behavior, and avoidance of events associated with the trauma. Transference-based fantasies of injuring the therapist or self if feelings emerge in the treatment are addressed. The therapist encourages the patient to use imagery rather than words in reconstructing the trauma, associate to the images, express the feelings experienced at the time of the trauma, and elaborate conflicted meanings.

Patients with PTSD most commonly present with rigid, overcontrolled states alternating with poorly controlled undermodulated states. The result is an approach-avoidance struggle with difficulty staying with feelings without taking defensive flight or being overwhelmed. The working-through process is marked by the patient's increasing capacity to review memories of the trauma, express feelings, revive distorted meanings, and plan for the future in a way that takes account of the consequences of the trauma. The capacity to dose the intensity and duration of work on the trauma depends heavily on the therapist's intervention early in treatment. Readiness for termination is signaled when the patient can talk about the trauma without being overwhelmed and can also put the trauma out of mind to cope with interpersonal and work responsibilities.

Just as weak and strong problematic self concepts characterize trauma responses in patients, parallel weak and strong views of the self can occur as countertransference reactions. As trauma victims relate the details of their experiences, therapists frequently experience helplessness, fear, disgust, and grief. In a process of secondary traumatization, therapists may experience intrusive images and dreams in response to material disclosed by patients. Traumatized patients may, for conscious or unconscious reasons, attempt to shock their therapist with recapitulation of the trauma, in part to test the safety of the relationship and in part to evacuate painful memories of the trauma. Pynoos and Nader (7) describe modifications of brief dynamic psychotherapy for acutely traumatized children and adolescents. Debriefing, designated psychological "first aid," is adjusted for the developmental level of the child. Attention is given to both proximal and distal impact on developmental stages. For example, a sexually abused 3-year-old who responds to treatment at age 4 is at risk for reactivation of PTSD during early adolescence. For this reason, a postintervention model is recommended, with reassessment and booster treatments as necessary to prevent derailment at later developmental crossroads.

Post-Traumatic Stress Disorder Comorbid with Other Disorders

Individuals with PTSD are at risk for developing secondary affective, alcohol/substance abuse, panic, and phobic disorders. The occurrence of these secondary conditions is greater in those with pre-traumatic histories of these conditions. Treatment of the comorbid conditions is essential to management of PTSD. Alcohol and substance abuse treatment generally precedes psychotherapy for PTSD. Relapse prevention for alcohol and substance abuse is provided concurrent with PTSD treatment. Use of tricyclic antidepressants or serotonin reuptake inhibitors, for treatment of major depression or panic disorder that develops in the course of PTSD, has the advantage of reducing reexperiencing and arousal symptoms in addition to targeted depressive and panic disorders.

Post-Traumatic Personality Disorder

A variant of PTSD is seen many years or decades after unresolved trauma seen in adults who present for treatment with histories of childhood sexual or physical abuse, Holocaust survivors, and combat veterans. These individuals may show all the prototypical features of PTSD, multiple secondary adversities with poor physical health, unstable work identities, and turbulent interpersonal relations, resulting over time in a clinical picture complicated by avoidant, narcissistic, or borderline personality disturbances. The invasion of the sphere of personality functioning by the long-term maladaptive consequences of chronic PTSD produces a condition recently designated as post-traumatic personality disorder (20). While genetic and early developmental factors contribute, development of personality disorders in these cases is greatly influenced by the impact of decades of unresolved PTSD and associated cumulative secondary adversities on character functioning. High rates of comorbidity with alcohol and substance abuse, depression, and psychosomatic disorders are seen in individuals with post-traumatic personality disorder.

Treatment is long term and multimodal for this difficult population. Long term group therapy (39) and pharmacotherapy with antidepressant medication provide a cost-effective treatment approach. Treatment must be tailored to the needs of the individual, with consideration given to long term supportive-expressive individual psychotherapy, family therapy, and mood-stabilizing agents. Inpatient treatment may be required during periods of regression to permit a safe uncovering of traumatic memories. As the individual with severe post-traumatic personality disorder recovers vocational rehabilitation, social skills training, and alcohol and drug abuse relapse prevention improve functioning and stabilize gains.

Group treatment provides an opportunity to reconstruct and master trauma, and addresses multiple secondary adversities. These approaches have formed the cornerstone for treatment of chronic PTSD in war veterans (39,42). Koller, Marmar, and Kanas (43) developed a model of treating post-traumatic personality disorder in combat veterans in group therapy incorporating psychodynamic principals. Eight to 12 combat veterans with PTSD are accepted into each group, with the goal of retaining 6 to 10 working members. The group is conducted on a weekly basis for 1.5 hour sessions. A 1-year total period of treatment allows adequate time to address trauma in the war zone, and pre- and postmilitary service problems affecting the course of response. A 1-year period has the further advantage of recapitulating the typical veteran's tour of duty, heightening anniversary reactions, and enriching the symbolic meaning of the group experience as an effort to master the war zone experience. Over the course of treatment, the first 6-month block is directed primarily at processing traumatic experiences related to military service. The second 6-month block focuses on developmental issues that proceeded combat and post-military experiences, including maladaptive coping strategies and secondary adversities arising out of failed efforts to deal with combat trauma, and that influenced the development of post-traumatic personality disorder.

MIND-BRAIN RELATIONSHIPS IN PTSD: IMPLICATIONS FOR DYNAMIC PSYCHOTHERAPY

Individuals suffering from post-traumatic stress disorder have been shown to have stable neurobiologic changes in central and peripheral catecholamines, as well as hypothalamic-pituitary-adrenocortical axis, dopaminergic, opioid, and thyroid systems (44–46). These neurobiologic alterations are expressed clinically in a state of chronic psychophysiological hyperarousal, both baseline and in response to reminders of the traumatic events. This autonomic arousal is expressed in altered heart rate, skin conductance, and facial muscle activity (47). Several theories have been proposed to account for the impact of traumatic exposure on biologic functioning, resulting in changes in neurotransmitter relationships and autonomic activity. These models include a fear-enhanced startle paradigm (48), inescapable shock (8), and kindling (44).

Successful psychodynamic treatment, especially when provided early in the course of the development of post-traumatic stress disorder, may ameliorate the development of chronic stable biologic changes resulting from traumatic stress exposure. For example, fear-enhanced startle appears to be mediated by a relatively small number of brain stem pathways involving the thalamus and amygdala, maintained by an associative learning process that results in what may be considered multiple false alarms for danger. Early preventative psychotherapeutic intervention, including such techniques as the dynamically informed multiple stressor debriefing model just described, may alter the meanings of the traumatic event and curtail the fear networks that may activate or reinforce fear-enhanced startle.

Exposing animals to inescapable shock produces a syndrome of "learned helplessness" characterized by behavioral depression, impaired learning, decreased appetite, decreased immunity, and decreased reproductive behavior. Chronic PTSD patients exhibit features of this learned helplessness paradigm, reflected in passive rather than active problem-coping strategies. For example, emergency services personnel with chronic stress response symptoms cope through escape avoidance, including by: 1) having fantasies or wishes about how things might turn out better; 2) trying to make themselves feel better by eating, drinking, smoking, or using drugs or medication; 3) sleeping more than usual; 4) refusing to believe what had happened; 5) avoiding people in general; and 6) using related avoidant coping strategies. In contrast, emergency services personnel who make successful adaptations following traumatic exposure preferentially use planned solving strategies, including making a plan of action and following it, changing some things so the situation will turn our alright, coming up with different solutions to the problem, and knowing what has to be done and intensifying their efforts to make things work (49). Psychodynamic psychotherapy specifically addresses avoidant coping strategies as pathological defenses, encourages active problem solving, and allows tolerable confrontation with painful feelings and memories that are warded off through avoidant coping. Psychodynamic approaches, as well as cognitive-behavioral interventions, can reduce the intensity of deficits induced by learned hopelessness once established. It is also possible that dynamic and cognitive interventions directed at maladaptive coping that reflect learned helplessness in PTSD might result in changes in the noradrenergic system thought to underlie learned helplessness.

Repeated traumatization through actual events or reexperiencing phenomena may be the triggers for kindling in PTSD. Psychodynamic intervention directed at mastery and working through of traumatic memories counters repeated cycles of intrusive reexperiencing of traumatic life events, and in so doing may well reduce the toxic kindling effects associated with such reexposure. Biological evidence suggests that chronic intrusive symptomatology is associated with the release of stress hormones, including cortisol, which may underlie kindling and other neurotoxic effects. Early preventive interventions with the dynamically informed multiple stressor debriefing model, or active treatment of acute or chronic post-traumatic stress disorder with psychodynamic individual or group psychotherapy, may reduce the release of stress hormones and protect against or even reverse kindling and other neurotoxic phenomena.

COMBINED TREATMENT APPROACHES

Integrated approaches to treating PTSD take account of the forms of traumatic stress response and interactions of dynamic, cognitive-behavioral, and pharmacotherapeutic approaches (50). Management of normal stress response includes psychoeducation, support, caution against maladaptive coping, debriefing, and support groups. The management of acute catastrophic stress reactions emphasizes brief supportive-expressive dynamic psychotherapy conducted within a crisis intervention framework, time-limited cognitive-behavioral techniques aimed at anxiety reduction, and shortterm pharmacotherapy with adrenergic blockers or benzodiazepines for overwhelming daytime anxiety. Insomnia in acute catastrophic stress reactions can be safely managed with sedative antidepressants, including low-dose trazadone, or short-term use of benzodiazepines for those without a history of alcohol or sedative dependency.

For uncomplicated post-traumatic stress disorders, brief dynamic psychotherapy, or time-limited cognitive-behavioral treatments emphasizing exposure strategies and cognitive restructuring, are effective treatments. Both the dynamic and the cognitive-behavioral methods emphasize revision of pathogenic beliefs, as well as dosing of exposure to traumatic images and affects. When PTSD is comorbid with panic disorder, partial agoraphobia, major depressive episode, and related disorders, the addition of specific psychotherapeutic and pharmacotherapeutic treatments for these conditions is indicated.

For post-traumatic personality disorder in trauma victims with decades of unresolved traumatic stress symptomatology, extensive secondary adversities, and complex comorbidities, long-term multimodal integrated treatment offers the greatest hope. Long-term psychodynamic group therapy, in combination with long-term pharmacotherapy emphasizing tricyclic antidepressants, serotonin reuptake inhibitors, adrenergic blockers, and mood-stabilizing agents in an individualized approach, provides a cost-effective treatment. The initial phase of treatment is directed at exploration of the trauma and modulation of intrusive, avoidant, and arousal symptoms. Inpatient treatment may be required in severe cases to permit a safe uncovering of traumatic memories. As the individual with severe post-traumatic personality disorders is able to tolerate reconstruction of the trauma and reduce conditioned emotional arousal to traumatic cues, long-term recovery depends upon: 1) the integration of pharmacotherapy to reduce irritability, arousal, and insomnia; 2) communication and social skills training; 3) vocational rehabilitation; and 4) alcohol and drug abuse relapse prevention.

REFERENCES

1. Marmar CR, Horowitz MJ. Phenomenological analysis of splitting. *Psychotherapy* 1986;23:21–29.
2. Spiegel D, Cardena E. Disintegrated experience: the dissociative disorders revisited. *J Abnorm Psychol* 1991; 100:366–378.
3. Fox PP. Narcissistic rage and the problem of combat aggression. *Arch Gen Psychiatry* 1974;31:807–811.
4. Horowitz M, Zilberg N. Regressive alterations in the self-concept. *Am J Psychiatry* 1983;140:284–289.
5. Marmar CR, Freeman M. Brief dynamic psychotherapy of post-traumatic stress disorders: management of narcissistic regression. *J Traumatic Stress* 1988;1:323–337.
6. Weiss DS. Psychodynamic treatment of PTSD: an accident victim with childhood trauma. *J Traumatic Stress,* in press.
7. Pynoos RS, Nader K. Issues in the treatment of post-traumatic stress in children and adolescents. In Wilson JP, Raphael B, eds. *The international handbook of traumatic stress syndromes.* New York: Plenum Press, 1993; 535–549.
8. van der Kolk B. *Advances in the psychotherapy of post-traumatic stress disorder.* Paper presented at the World Conference of the International Society for Traumatic Stress Studies, Amsterdam, The Netherlands; June 1992.
9. Horowitz MJ. *Stress response syndromes.* Northvale, NJ: Jason Aronson, 1986.
10. Marmar CR. Brief dynamic psychotherapy of post-traumatic stress disorder. *Psychiatric Annals* 1991;21:405–414.
11. Kulka RA, Schlenger WE, Fairbank JA, Hough RL, Jordan BK, Marmar CR, Weiss DS. *Trauma and the Vietnam War generation.* New York: Brunner/Mazel; 1990.
12. Bownes IT, O'Gorman EC, Sayers A. Assault characteristics in post-traumatic stress disorder in rape victims. *Acta Psychiatrica Scandivica* 1991;83:27–30.
13. Kemp A, Rawlings EI, Green BL. Post-traumatic stress disorder (PTSD) in battered women: a shelter sample. *J Traumatic Stress* 1991;4:137–148.
14. Dahl S. Acute responses to rape—a PTSD variant. *Acta Physiol Scand* 1989;80(suppl 355):56–62.
15. Rynearson EK. Psychological adjustment to unnatural dying. In Zisnook S, ed. *Biopsychosocial aspects of bereavement.* 1990:77–93.
16. Pynoos RS, Sorenson SB, Steinberg AM. Interpersonal violence and traumatic stress reactions. In Goldberger L, Breznitz S, eds. *Handbook of stress: theoretical and clinical aspects,* 2nd ed.(in press).
17. Terr LC. Family anxiety after traumatic events. *J Clin Psychiatry* 1989;50(11,suppl):15–19.
18. Putnam FW. Dissociation as a response to extreme trauma. In Kulft RP, ed. *The childhood antecedents of multiple personality.* Washington, DC: American Psychiatric Press; 1985:65–97.
19. Herman JL, Perry JC, van der Kolk BA. Childhood trauma in borderline personality disorder. *Am J Psychiatry* 1989;146:490–495.
20. Horowitz MJ, Weiss DS, Marmar CR. Diagnosis of post-traumatic stress disorder. *J Nerv Ment Dis* 1987; 175:267–268.
21. Horowitz MJ, Marmar C, Weiss D, DeWitt K, Rosenbaum R. Brief psychotherapy of bereavement reactions. *Arch Gen Psychiatry* 1984;41:438–448.
22. Bowlby J. *A secure base: parent-child attachment and healthy human development.* New York: Basic Books; 1988.
23. Finklehor D, Hotaling GT. Sexual abuse in the National Incidence Study of Child Abuse and Neglect: an appraisal. *Childhood Abuse Negl* 1984;8:23–32.
24. Maida C, Gordon N, Steinberg A, Gordon G. Psychosocial impact of disasters: victims of the Baldwin Hills fire. *J Traum Stress* 1989;2:37–48.
25. Lindy JD. The trauma membrane and other clinical concepts derived from psychotherapy work with survivors of natural disasters. *Psychiatric Annals* 1985;15:153–160.
26. McCubbin MA, McCubbin HI. Theoretical orientations to family stress and coping. In Figley CR, ed. *Treating stress in families.* New York: Brunner/Mazel, 1989:3–43.
27. Marmar CR, Horowitz MJ, Weiss D, Wilner N, Kaltreider N. A controlled trial of brief psychotherapy and mutual help treatment of conjugal bereavement. *Am J Psychiatry* 1988;145:203–209.
28. Armstrong K, O'Callahan W, Marmar CR. Debriefing red cross disaster personnel: the multiple stressor debriefing model. *J Traum Stress* 1991;4:481–491.

29. Mitchell J. When disaster strikes . . . The critical incident stress debriefing process. *J Emerg Med Serv* 1983;8:36–39.
30. Raphael B. *When disaster strikes.* New York: Basic Books Inc.; 1986.
31. Rahe RH. Acute versus chronic psychological reactions to combat. *Mil Med* 1988;153:365–371.
32. Forster P, Marmar CR. Benzodiazepines in acute stress reactions: benefits, risks, and controversies. In Roy-Byrne PP, Cowley BS, eds. *Benzodiazopines in clinical practice: risks and benefits.* Washington, DC: American Psychiatric Press; 1991.
33. Windholz MJ, Marmar CR, Horowitz MJ. A review of the research on conjugal bereavement: impact on health and efficacy of interventions. *Compr Psychiatry* 1985;26:433–447.
34. Weiss DS, Marmar CR. Teaching time-limited dynamic psychotherapy for post-traumatic stress disorder and pathological grief. *Psychotherapy* 1993;30:587–591.
35. Brom D, Kleber RJ, Defares PB. Brief psychotherapy for posttraumatic stress disorders. *J Consul Clin Psychol* 1989;57(5):607–612.
36. Lindy JD. An outline for the psychoanalytic psychotherapy of post-traumatic stress disorder. In Figley CR, ed. *Trauma and its wake,* vol 2. New York: Brunner/Mazel, 1986:195–202.
37. Rose DS. A model for psychodynamic psychotherapy with the rape victim. *Psychotherapy* 1991;28:85–95.
38. Horowitz MJ, Marmar CR, Weiss DW, et al. Brief psychotherapy of bereavement reactions: the relationship of process to outcome. *Arch Gen Psychiatry* 1984;41:438–448.
39. Egendorf A. Vietnam veterans rap groups and themes of postwar life. *J Soc Issues* 1975;4:111–124.
40. Lifton R. *Home from the war.* New York: Simon & Schuster; 1973.
41. Scurfield R, Corker T, Gongla P, Hough R. Three post-Vietnam "rap/therapy" groups: an analysis. *Group* 1984;4:3–24.
42. Rozynko V, Dondershine HE. Trauma focus group therapy for Vietnam veterans with PTSD. *Psychotherapy* 1991;28:157–161.
43. Koller P, Marmar CR, Kanas N. Psychodynamic group treatment of post-traumatic stress disorder in Vietnam veterans. *Int J Group Psychoth* 1992;42:225–246.
44. Friedman MJ. Biological approaches to the diagnosis and treatment of post-traumatic stress disorder. *J Traum Stress* 1991;4:67–92.
45. Mason JW, Kosten TR, Southwick SM, Giller EL. The use of psychoendocrine strategies in post-traumatic stress disorder. *J Appl Psychol* 1990;20:1822–1846.
46. Mason JW, Bremner JD, Yehuda R, Southwick SM, Ma X, Johnson D, Morgan, Charney D. *Alterations in thyroid function in PTSD.* Presented at the Seventh Annual Meeting of the Society for Psychotherapy Research, Washington, DC; 1991.
47. Pitman RK, Orr SP, Forgue DF, de Jong JB, Clairborn JM. Psychophysiologic assessment of posttraumatic stress disorder imagery in Vietnam combat veterans. *Arch Gen Psychiatry* 1987;44:970–975.
48. Krystal JH, Kosten TR, Southwick S, et al. Neurobiological aspects of PTSD: review of clinical and preclinical studies. *Behav Ther* 1989;20:177–198.
49. Marmar CR, Weiss DW, Metzler T, Ronfeldt H, Foreman C. *Responses of emergency services personnel to the Interstate 880 Nimitz freeway collapse during the 1989 Loma Prieta Earthquake.* Presented at the Annual Meeting of the Anxiety Disorders of America Association, Santa Monica, 1994.
50. Marmar CR, Foy D, Kagan B, Pynoos RS. An integrated approach for treating post-traumatic stress. In Oldham JM, Riba MB, Tasman A, eds. *American psychiatric press review of psychiatry* 1993;12:239–272.

Neurobiological and Clinical Consequences of Stress: From Normal Adaptation to PTSD, edited by M. J. Friedman, D. S. Charney, and A. Y. Deutch, Lippincott-Raven Publishers, Philadelphia © 1995.

29

The Relationship Between Trauma, Post-Traumatic Stress Disorder, and Physical Health

*Matthew J. Friedman and †Paula P. Schnurr

*Departments of *†Psychiatry and *Pharmacology, Dartmouth Medical School, Hanover, New Hampshire 03755 and †Lebanon, New Hampshire 03756, and National Center for Post-Traumatic Stress Disorder, Veterans Administration Medical Center, White River Junction, Vermont 05009*

Since publication of Selye's seminal book, *The Stress of Life* (1), a great deal of research has been aimed at furthering our understanding of the relationship between stress and physical health. Several important reviews have addressed etiological mechanisms through which exposure to stress might affect biological systems that are critical for maintaining health. Chrousos and Gold (2) define a "stress system" that has as its main components the hypothalamic-pituitary-adrenocortical (HPA) system, and the central and peripheral adrenergic system. According to these authors, the stress system can produce a variety of pathophysiologic states that may precipitate vulnerability to or actual expression of a wide range of psychiatric, endocrine, and inflammatory disorders. Specific reviews have focused on literature pertaining to stress-induced impairment of immunologic function (3,4) and increased susceptibility to infectious disease following exposure to stress (5). There also is a relevant literature on the relationship between stress-induced cardiovascular abnormalities and somatic anxiety (6). Kaplan and associates (7) have reviewed studies, many of them their own work, that show how social stress can precipitate atherosclerotic pathology in primates, and Saab and Schneiderman (8) have reviewed laboratory and clinical studies showing a linear correlation between intensity of stress-induced cardiovascular reactivity and later development of hypertension. Kiecolt-Glaser and associates (9) concluded that there appears to be "a convergence among cardiovascular, neuroendocrine and psychoneuroimmunological research and the evaluation of differences among people who vary in autonomic activation. A better understanding of these individual differences in response to stress could help identify those individuals who may be more prone to long-term health changes . . ." (p. 684).

From the perspective of the field of trauma research, it is surprising that so few of the thousands of studies that have been devoted to the topic of stress and physical health have examined the effects associated with extreme stress. As Cohen and Williamson (5) suggest in their review of the literature on stress and infectious disease in humans, "Impact of severe events would provide the fairest test of a stress-disease relationship . . ." (p. 18). Perhaps equally surprising are the examples of severe events offered by Cohen and Williamson: divorce, bereavement, and job loss. While not denying the emotional impact of such events, we believe that an even fairer test of the stress-disease relationship would be provided by the study of extremely traumatic events, including war-zone exposure, sexual and other criminal victimization, natural and human-made disasters, and serious accidents.

First we review the literature on the physical health outcomes associated with traumatic events. Despite the extensive literature suggesting that exposure to stressful events may be associated with adverse health outcomes, much less has been written on the medical and somatic consequences of exposure to extreme stress. Nonetheless, reviewers have suggested that physical health may be severely and chronically impaired following traumatic experiences such as concentration camp incarceration (10,11), confinement as a prisoner of war (POW) (12–14), war zone exposure during military hostilities (15,16), sexual assault as a child or adult (17–19), or exposure to natural disasters (20). Second, we review the literature on the physical health outcomes associated with post-traumatic stress disorder (PTSD). Like Wolfe and colleagues (21), we argue that PTSD is an important mediator through which trauma may be related to adverse outcomes. Third, we review biological and psychological correlates of PTSD that might predispose affected individuals toward increased risk for medical problems.

METHODOLOGICAL CONSIDERATIONS

Our review includes four categories of health outcome: 1) self-reports; 2) utilization of medical services; 3) morbidity as indicated by physician diagnosis or laboratory tests; and 4) mortality. Self-reports, by far the most commonly measured outcomes, include ratings of health status, symptoms, or illness conditions. There has been much controversy about the validity of self-reports as indicators of an individual's "true" health status, because self-reports may not be confirmed by laboratory tests or diagnostic assessments. For example, male Vietnam theater veterans were found to be more likely than Vietnam era veterans to report poorer health and increased health problems in a variety of domains, yet medical tests failed to corroborate virtually all of the self-reported differences (22,23). We think that it is important to remember that self-reports may not always provide valid information about health or disease, and, specifically, that they may be strongly influenced by psychological states and processes (see further discussion (24,25)). We also think that self-reports should not be dismissed as necessarily or categorically invalid. Self-reports can, at least to some extent, be valid indicators of an individual's physical health status. Self-reported health as rated on a five-point single-item scale is an excellent predictor of mortality (26), and self-reported health indicators can distinguish groups that vary in terms of disease type and severity (27). Accordingly, we have chosen to include self-reports in our review.

Another methodological issue that must be considered when interpreting the literature on stress and physical health is that all of the human studies are correlational, which makes it difficult to infer a causal role for stress or PTSD. As in any correlational problem, it is necessary to specify (i.e., measure and control for) other potential causes. At the same time, it is necessary to interpret the results of most multivariate procedures with extreme caution; we say "most" because we believe that structural modeling offers a more appropriate way to analyze the data in this area. Imagine that an investigator wishes to examine the relationship between amount of exposure to a flood and utilization of medical services in the year following the flood, while controlling for other factors known to influence utilization such as age. The investigator finds that amount of exposure is positively correlated with utilization, but that the relationship is no longer statistically significant when age is taken into account. Should the investigator conclude that amount of exposure has no relationship to utilization? The answer is "no" if age is correlated with exposure because, for example, older individuals tend to live in less sturdy homes that are closer to flood zones. The investigator would make a Type II error by failing to statistically model this relationship in a path or structural model, or failing to pay careful attention to the correlations among independent variables such as exposure and control variables such as age. We provide an actual example of such a complex relationship in our introduction to the section on PTSD and health that follows.

A further methodological issue is that longitudinal studies are rare, and usually complicated by an absence of pre-exposure or pre-PTSD baseline health data. Baseline information often must be inferred from retrospective self-reports, which may be biased or incomplete. Given this and the aforementioned problems, we believe that it is especially important to examine the literature on trauma, PTSD, and health as a whole, and also to look to the experimental animal literature as reviewed in Part I of this volume.

THE ROLE OF PHYSICAL INJURY

People are sometimes killed, injured, or medically disabled as a direct result of their exposure to traumatic events. Military veterans who survive battlefield wounds may develop permanent medical problems from spinal cord injuries, loss of a limb, and other deficits. Female survivors of rape, criminal victimization, or domestic violence may suffer from acute bodily injury as well as long-term consequences of such damage (19). Ex-POW survivors of prolonged captivity often experience malnutrition, infectious diseases (especially tuberculosis), beatings, torture, prolonged physical exertion (on work details, forced marches, etc.), frostbite, and other stressors. Therefore, it is sometimes difficult, if not impossible, to identify a specific psychological or physical stressor as the major cause of a subsequent medical problem. Furthermore, it is neither useful nor realistic to separate psychological from physical stress in this regard, because each may interact with the other and both undoubtedly contribute to the suffering and despair associated with traumatic exposure.

We readily acknowledge that long-term health consequences may result from injuries or from medical conditions associated with a traumatic event. We do not believe, however, that this alone can account for the many findings (see review that follows) showing that traumatic exposure in general and PTSD in particular are associated with increased risk of adverse health outcomes. Although there is a relationship between severity of injury (28) and risk of PTSD, it is unlikely that injury is the sole cause of the relationship between PTSD and adverse health outcomes. Most people who develop PTSD are not injured as a result of the trauma that caused their PTSD, as illustrated in a study of the discharge summaries of 543 veterans with PTSD. Sixty percent had an identified medical problem and 42% had multiple problems, yet only 8% had physical sequelae of combat-related trauma (29). Also, injury to a specific site or system is unlikely to explain the multiple nature of the physical health complaints and physical morbidity seen in PTSD.

To summarize, it appears that trauma-related injuries and medical problems contribute to subsequent health problems, but their contribution is minor and cannot explain much of the data suggesting that traumatic exposure is associated with adverse health outcomes. Although we will not reiterate this point, it should be kept in mind as a potentially important variable underlying research findings summarized in the following review of the literature.

TRAUMA AND HEALTH

Self-Reports

Most publications in this category concern Vietnam veterans. In a national epidemiological survey, the National Vietnam Veterans Readjustment Study (NVVRS), male and female Vietnam veterans with high war-zone exposure reported poorer health and more health problems relative to era veterans or civilians (15). Other studies have independently shown that combat exposure predicted increased reporting of a variety of health problems among 6,800 male Midwestern Vietnam combat veterans who belonged to the American Legion (16) and among Vietnam combat veterans who believed that they had been exposed to the herbicide Agent Orange, whether or not such exposure had actually occurred (30). Combat exposure also predicted adverse health reports in an elegant study of 2,260 male-male monozygotic twin pairs who participated in mail or telephone interviews. In comparison with their twins who were not exposed to war-zone stress, the siblings who served in Southeast Asia reported significantly more hearing and skin problems (31). In their Vietnam Experience Study, the Centers for Disease Control (CDC) (22) conducted a telephone survey of 7,924 Vietnam theater veterans and 7,364 Vietnam era veterans. The combat veterans were almost twice as likely to assess their health as "poor" or "fair," and to report a wide variety of medical problems, as the era veterans. However, a further component of the study illustrates how self-reports may not be corroborated by diagnostic tests. A random subsample of 2,490 Vietnam veterans and 1,972 non-Vietnam veterans underwent a comprehensive medical examination that detected few current objective differences in physical health between the two groups (23). Wolfe et al. (21) found a relationship between the magnitude of war-zone exposure and current adverse health reports among female Vietnam veteran nurses, although subsequent analyses suggested that PTSD was an important predictor of the association (see section on PTSD and health that follows).

Many of the publications on adverse health effects among survivors of childhood or adult sexual abuse suggest that women who have a history of sexual victimization are more likely than controls who do not have a history of such exposure to report a variety of medical problems. Female rape victims were more likely to report lower perceived health status, more somatic symptoms, and more negative health behaviors than nonvictim controls (32–34). Somatic symptoms, especially regarding gastrointestinal distress and recurrent headaches (35), as well as dysuria, vaginal discharge, and chronic abdominal pain (36), were reported more frequently by rape or incest survivors than by nonabused controls.

An interesting facet of the sexual trauma and health literature is the suggestion that a history of sexual abuse, especially during childhood, is associated with the presence of chronic pelvic pain (37–39). Walker et al. (40) performed laparoscopic exams on 50 women with chronic pelvic pain and 50 women seeking tubal ligation or infertility evaluation. There was a significant correlation between a history of severe sexual abuse and medically unexplained symptoms; in other words, patients with chronic pelvic pain were more likely than patients in the comparison group to have a positive history of sexual abuse and a normal laparoscopic examination. A different conclusion was reached by Rapkin et al. (41), who found no difference in childhood abuse histories between women with chronic pelvic pain and those with chronic pain in other locations. In fact, childhood physical abuse rather than sexual abuse predicted adult pain syndromes, and was greater for both chronic pain groups than for controls. An important difference between Rapkin et al.'s study and other investigations of chronic pelvic pain is that it was the only study to include a chronic pain (nonpelvic) control group. This may account for the different results reported by these investigators, and underscores the necessity for designing studies with appropriate control groups in order to interpret data pertaining to this complex area.

A review by Hovanitz (20) provides an excellent synthesis of literature on physical health outcomes associated with the aftermath of disasters. Most of the studies reviewed concern the aftermath of floods, and Hovanitz concludes that, in all of these studies, severity of exposure to flood trauma was associated with increased reports of health problems. Two studies of individuals who had been assessed before exposure to the severe Puerto Rican flood of 1985 and were reinterviewed in 1987 indicate a small but significant increase in somatic symptoms (42, 43). Similarly, among older adults exposed to the 1984 flood in southeastern Kentucky who were interviewed 18 months afterwards, severity of flood exposure was related to reported magnitude of somatic symptoms (44).

Two self-report studies concern survivors of a volcanic eruption. Murphy (45) failed to find a relationship between severity of trauma and physical health reports among survivors of the 1980 Mount Saint Helens eruption, although she looked only at change from 11 months after the eruption to 35 months after. Among survivors of the 1987 volcano that destroyed Armero, Colombia, those who reported emotional distress 7 months after the disaster were more likely than those who were not distressed to report nonspecific physical problems and to have multiple physical complaints (46).

Uba and Chung (47) found that, among Cambodian refugees living in the United States, the amount of traumatic exposure during the war in Cambodia was positively related to poor health outcomes more than 10 years later. Mollica et al. (11) found that therapeutic intervention actually worsened somatic symptoms but improved psychological symptoms among Cambodian, Hmong, Laotian, and Vietnamese refugees previously exposed to the war in Southeast Asia. Van der Ploeg and Kleijn (48) compared Dutch men and women previously held hostage during terrorist hijackings with family members not abducted. Six to 9 years after captivity, the ex-hostages reported significantly higher rates of migraine or severe headache, intestinal problems, rheumatic problems and pains, skin diseases (eczema), and stomach problems. Bartone et al. (49) interviewed military family assistance officers 6 months and 1 year after a U.S. Army airplane crash in Gander, Newfoundland, in 1985 that killed 248 soldiers. Results showed a robust dose-response relationship between severity of exposure to survivor stress (the intensity and duration of contact with surviving family members) and magnitude of self-reported physical symptoms or illnesses. Two studies concern the relationship between exposure to motor vehicle accidents and the subsequent development of chronic pain and PTSD (50,51).

Utilization

The National Academy of Sciences follow-up of 2,500 U.S. Army World War II and Korean conflict ex-POWs has generated an important dataset regarding utilization of medical resources following the extreme stress of POW captivity (12,13). The most notable increases in hospital admission rates for POWs versus controls were for psychiatric disorders, nutritional problems, tuberculosis, and other infectious and parasitic diseases. In addition, POWs had higher hospitalization rates for cardiovascular and gastrointestinal diseases in contrast to non-POW veteran controls. Similar results were found by Eitinger (10) in his morbidity studies of Norwegian survivors of Nazi concentration camps. Survivors had more sick periods, longer sick leaves, and more frequent and longer-lasting hospitalization periods than controls.

A number of studies have shown increased utilization of medical resources among women with histories of recent or lifetime sexual assault, physical assault, or criminal victimization. Kimerling and Calhoun (32) prospectively compared utilization rates among women who had been recently raped with a matched control group for 1 year after the victims' assault. By 4 months post assault, the victim group reported increased medical (but not psychological) service utilization that persisted at the 1-year follow-up. Waigandt et al. (34) also found that sexual assault victims reported significantly more doctor visits for a prolonged period (2 years) after the rape than a matched control group.

Koss et al. (33) found twice as many physician visits (and 2.5 times greater medical expenses) among women who had been exposed to criminal victimization in comparison with controls, during the first year following the assault. Severity of victimization was the most powerful predictor of physician visits and outpatient costs. Similarly, Norris et al. (52) interviewed men and women exposed to violent crime 6 and 12 months after the assault and compared them with adults who had been subjected to property crimes but who had not been physically attacked. Medical (as well as mental health, legal, and pastoral) utilization rates were significantly elevated among the physically victimized cohort in contrast to victims of property crimes. In a study of Swedish battered women, Bergman and Brismar (53) examined medical records for 10 years preceding and 5 years after the battering, and compared them with records of a matched control group. In addition to hospital admissions for injuries directly related to the battering, battered women had significantly increased admission rates for nontraumatic surgical disorders, gynecological disorders, induced abortion, medical disorders, suicide attempts, and inpatient observation.

With regard to female adult survivors of childhood or adult sexual abuse seeking treatment at an outpatient clinic for gastrointestinal disorders, Drossman et al. (37) found that victimized women reported more lifetime surgeries than nonabused controls, although there was no difference in number of physician visits per 6 months or number of hospitalizations per 2 years. Finally, Felitti (35), reviewing utilization patterns among lifetime sexually abused patients receiving treatment at an HMO, found that they were more likely than a nonabused comparison group to exhibit a high utilization pattern (10 or more doctor office visits per year).

Increased medical utilization also has been observed following natural disasters. Hospital referrals more than doubled following the Bristol, England, flood in 1968. There was no relationship between flood-related injuries and referral rates, but a clear dose-response relationship between depth of flooding and medical utilization was found (54). A similar increase in visits to health-care practitioners, especially among males, was reported following the 1974 flood in Brisbane, Australia (55). During the 7 months following the 1980 eruption of Mount Saint Helens and subsequent ashfall, there was a significant increase, relative to the 7 months prior to the eruption, in emergency room visits but not in hospital admission or occupancy rates (56).

Morbidity

The largest body of information in which adverse health outcomes are documented by a medical examination concerns the aftermath of war trauma. Although most of the studies focus on military veterans, including POWs, there are important publications on the medical status of civilians exposed to concentration camp and/or war-zone trauma.

Longitudinal studies of World War II and Korean POWs have been conducted in Canada, Australia, and the U.S. A number of reports have appeared on a National Academy of Sciences follow-up of over 2,500 U.S. Army repatriated POWs following World War II and the Korean War who were compared with approximately 2,500 U.S. Army non-POW veteran controls. In general, POWs had higher rates of medical illness in most organ systems, but interpretation is difficult, as noted previously. The POWs also had greater psychiatric morbidity than non-POW controls (12,13,57). Goulston et al. (58) assessed gastrointestinal symptoms among a random sample of surviving Australian World War II POWs and non-POW veteran controls 40 years after the stress of internment. Duodenal ulcers and strongyloidiasis were more prevalent in the POW group. Venn and Guest (14) reviewed studies of former Australian POWs and found five conditions consistently associated with imprisonment: 1) strongyloidiasis; 2) peptic ulcer; 3) anxiety states; 4) depression; and 5) hepatitis B. Among Canadian POWs, higher prevalence rates of gastrointestinal, neurological, and cardiovascular disease were found in comparison with non-POW controls (59). A second Canadian study found no difference in rates of heart disease between POWs and non-POW siblings, although POWs had higher cardiovascular morbidity than expected for the Canadian male population (60).

Eitinger's (10) classic mortality and morbidity study of 498 Norwegian World War II concentration camp survivors and matched controls has the same interpretive problems as the POW studies because of the terrible physical and hygienic conditions in the camps. Ex-prisoners had higher rates of tuberculosis, cardiovascular and respiratory disease, and both gastric and duodenal ulcers.

The CDC (23) Vietnam Experience Study mentioned earlier, in which subjects received comprehensive medical exams, detected few differences in physical health between Vietnam war-zone victims and Vietnam era veterans. The exceptions were hearing loss and decreased sperm count among war-zone veterans. These findings, which differ markedly from the investigators' self-report data (22), suggest that there may not be a relationship between traumatic exposure and the development of medical problems. However, veterans enrolled in the CDC study were approximately 37 years old at the time that they were examined. This may be too early for a stress-induced vulnerability to produce a detectable medical abnormality. We hope that the CDC will carry out a follow-up evaluation when these veterans are in their 50s.

Perhaps the most interesting reports in this group of studies concern coronary heart disease among civilians exposed to war-zone stress. Sibai et al. (61) studied patients who underwent coronary angiography at the American University of Beirut Medical Center during the Lebanon Civil War. Exposure to acute and chronic war stress was higher in patients with abnormal coronary angiographies than in patients with normal angiographic findings. A similar report documented a higher incidence of acute myocardial infarction during the civil war in Croatia than during the same time period during the previous year before the outbreak of hostilities (62).

Two reviews of literature on morbidity following sexual trauma (17,63) emphasize that, since most of the few studies in this area lack methodological rigor, it is difficult to link medical illnesses to childhood sexual abuse. Each reviewer, however, comes to a different conclusion. On the one hand, Laws (63) states that a history of sexual abuse is common in women with a history of chronic pain (especially pelvic pain), functional bowel disorders, eating disorders, obesity, and alcohol abuse. On the other hand, Fry (17) concludes that the findings from more recent studies have called into question results from some of the earlier reports suggesting links between childhood sexual abuse and specific physical symptoms.

Among consecutive admissions to an interdisciplinary inpatient pain rehabilitation program, Wurtele et al. (64) observed that 39% of women and 7% of men reported childhood sexual abuse. Felitti (35) found that sexually abused patients seen in general practice were more likely than controls to have morbid obesity. Drossman et al. (37) observed that 89% of consecutive patients admitted to a GI clinic reported childhood or adult sexual abuse. Patients with a functional disease (irritable bowel syndrome, nonulcer dyspepsia, chronic abdominal pain, constipation, etc.) were more likely than those with organic disease (Crohn's disease, ulcerative colitis, acid peptic disease, liver diseases, etc.) to report a history of forced intercourse or frequent physical abuse.

Two weeks after the 1980 earthquake in southern Italy, subjects exhibited elevations in heart rate, serum cholesterol, and triglyceride levels. These increases were no longer present 7 years later (65). Eight years after a North Sea oil rig disaster that occurred in 1980, Norwegian survivors exhibited significant increases in "psychosomatic diagnoses" such as cardiovascular disease (mostly hypertension), musculoskeletal problems, lower back pain, and dermatological problems in comparison to control subjects (66). Increased morbidity was found among survivors of the floods in both Bristol, England, in 1968 and Brisbane, Australia, in 1974 (54,55). Higher rates of leukemia, lymphoma, and spontaneous abortion were observed after the flooding in western New York that followed Hurricane Agnes in 1972 (67). Adams and Adams (56), who used archival data to study a community before and after the eruption of Mount Saint Helens, reported a greater number of stress-aggravated and psychosomatic illnesses, as well as increased employee sick leave. Murphy (45) failed to find effects due to amount of trauma among survivors who were studied 11 and 35 months after the eruption. As noted by Hovanitz (20), there are enough methodological differences between the last two studies to make direct comparison difficult.

Mortality

Several reports concerning American World War II and Korean War ex-POWs have reported excess mortality in comparison to non-POW controls due mostly to accidental trauma, tuberculosis, suicide, and cirrhosis of the liver. Highest mortality rates occurred during the years immediately following repatriation and converged to control rates by the 9th to 13th year. Other causes of death cited in some, but not all, follow-up studies include atherosclerotic cardiovascular disease, hypertension, and GI problems (12,13, 57,68,69). Similar results were found regarding Australian World War II former POWs who had significantly elevated death rates from liver cirrhosis, tuberculosis, arteriosclerosis and degenerative heart disease, motor vehicle accidents, and suicide (70). Accidents and suicide tended to occur within the first 5 years after imprisonment, whereas other causes of death remained elevated up to 18 years after repatriation. Perhaps the best-documented evidence of an association between war-related trauma and mortality from acute myocardial infarction is the report on Croatian civilians exposed to the recent civil war in that country (62).

In a study by the CDC, Vietnam theater veterans also exhibited increased mortality due to motor vehicle accidents, suicide, homicide, and accidental poisoning during the first 5 years postdischarge (71). It is noteworthy that death due to circulatory disease was significantly lower among Vietnam theater veterans than among era veteran controls. In a comparison of drafted Vietnam theater veterans versus not-drafted draft-eligible men, Hearst et al. (72) also found higher mortality rates due to suicide, motor vehicle accidents, and all causes of death. The final study on Vietnam veterans concerns women, mostly nurses, who served in the war zone. Thomas et al. (73) found an increased mortality due to uterine and pancreatic cancer among the war zone veterans in contrast to era veteran controls; as in the CDC's study of male veterans (71), there was lower mortality due to circulatory disease in the theater veterans.

Following the Newcastle earthquake in New South Wales, Australia, there was a significant increase in fatal myocardial infarctions and coronary deaths among people aged less than 70 years. This increase in fatal cardiac deaths lasted for the 4 days immediately after the earthquake and returned to normal levels during the next 4 months (74). For the first 12 months following the Bristol, England, flood in 1968 there was a 50% increase in deaths compared to the previous year (54). In contrast, there was no increased mortality following the Brisbane, Australia, flood in 1974 (55). Increased mortality rates, especially due to malignancies and spontaneous abortions, occurred in western New York following the flooding of the river valley by Hurricane Agnes in 1972 (67). Based on obituaries in a local paper, Adams and Adams (56) reported a substantial increase in the death rate (during the 7 months following, in contrast to the 7 months preceding the disaster) in a community that was traumatized by the Mount Saint Helens eruption and ashfall.

Summary

The trauma and health literature is impressive for the consistency of results showing that exposure to catastrophic stress is associated with adverse health reports, medical utilization, morbidity, and mortality among survivors (see Table 1).

The trauma and health literature also is a large, uneven, and often methodologically flawed body of work. Readers who are skeptical about whether trauma has any "real" effects on physical health can find plenty of support for their position. As indicated in Table 1, the self-report and utilization data are more extensive and conclusive than studies on morbidity or mortality. Also, the self-report findings of one large study did not correlate with physical exam and laboratory findings (22,23). Moreover, as previously stated, morbidity data concerning survivors of long-term imprisonment need to be interpreted cautiously, because the many privations endured during such captivity increased the risk of adverse health outcomes above and beyond the possible independent contributions of PTSD.

To address these kinds of concerns, we again must emphasize the general consistency of findings across diverse trauma populations and outcomes and, in particular, the fact that the self-report and utilization data are supported by the

TABLE 1. *Summary of results concerning the relationship between traumatic exposure and adverse health outcomes*

	Outcome			
Trauma	Self-report	Utilization	Morbidity	Mortality
Military	++	+	+/−	+
Sexual	++	++	+/−	NA
Disaster	+	+	+	+/−
Other[a]	++	NA	+	+

[a]Includes war refugees, hostages, and motor vehicle accident survivors.

++, clear association; +, probable association; +/−, inconsistent information; NA, information not available.

morbidity and mortality data. For example, studies showing a greater occurrence of atherosclerotic heart disease among Lebanese (61) and Croatian (62) civilians during civil war are quite persuasive, as are documented increases in mortality due to cardiovascular events among civilians during wartime (62) and following natural disasters (54,74). Interpreting the self-report and utilization literature in light of such findings brings up the question of parsimony; we believe that a true effect of trauma on physical health is the best single explanation that could underlie all of the data.

PTSD AND HEALTH

PTSD as a Mediator of the Relationship between Trauma and Physical Health

The foregoing section documents evidence of poor health outcomes in individuals who have experienced an extremely traumatic event. A question that has been largely unaddressed in this literature is the role of PTSD in the relationship between trauma and health. Cohen and Williamson (5) discuss the distinction between stress (i.e., a stressor) and distress (the reaction to a stressor) in their review of the literature on stress and infectious disease. We believe that a similar event-reaction distinction is in order when considering traumatic events and, further, that PTSD is a major mediator of the relationship between trauma and health. The reasoning behind our position is illustrated in the following example. Wolfe et al. (21) used multiple regression to examine the effects of war-zone exposure and PTSD on self-reported health outcomes in 109 female Vietnam veterans, first by examining each variable separately and then by examining the effects of each variable while controlling for the other. They found robust effects of both variables on all of the health outcomes that they studied when each variable was considered separately, but when both exposure and PTSD were simultaneously included in the models (all of which included age, education, and health before Vietnam as covariates), only PTSD robustly predicted poor health outcomes. In other words, the association between war-zone exposure and poor health was dramatically reduced when PTSD was taken into account.

This marked decrease in the effects attributable to war-zone exposure after adjusting for PTSD suggests an important mediational role for PTSD in the association between exposure and perceived health. We tested this hypothesis by performing path analyses to predict two of Wolfe's primary outcomes, self-reported current health and number of current health problems. As in Wolfe et al.'s multiple regression analyses, war-zone exposure and PTSD served as predictors, and age, education, and health before Vietnam were included as covariates. In each model, we tested the direct effects of all variables on each outcome and the indirect effects of exposure, age, education, and health before Vietnam as mediated by PTSD; only the results for PTSD and exposure are reported here. (Full details of the analyses are available from the second author.) An advantage of this approach is that it allowed us to determine how much of the relationship between trauma and health is direct, as opposed to predicted by the association between trauma and PTSD.

We found that war-zone exposure had statistically significant total effects on both current health and number of problems (standardized path coefficients of .511 and .432, respectively, $ps < .001$). For current health, 56% of the total effect was indirectly mediated through PTSD (standardized path coefficient $= -.295$, $P < .001$), although there also was a direct effect of exposure (standardized path coefficient $= -.216$, $P < .01$). For number of current health problems, only the indirect effect of exposure, 76% of the total, was statistically significant (standardized path coefficient $= .326$, $P < .001$). These analyses illustrate the importance of PTSD—the reaction to a trauma—in mediating the relationship between trauma and physical health.

Self-Reports

Several studies have found that PTSD is associated with complaints of poor health outcomes in military veterans (15,75,76) and active duty personnel (76). This association does not seem to be either gender- or nationality-specific. Similar findings have been reported for Israeli (77,78), New Zealander (79), and Canadian male veterans (80), as well as for female veterans (15,21, 76). The health problems associated with PTSD include multiple physical symptoms and systems. For example, Wolfe et al. (21) found that PTSD (controlling for amount of war-zone exposure) was associated with increased likelihood of cardiovascular, gastrointestinal, gynecological, dermatological, opthamological/otolaryngological, and pain problems. Shalev et al. (77) found that combat veterans with PTSD were more likely than combat controls to report cardiovascular, neurological, gastrointestinal, audiological, headache, and back pain problems. Litz et al. (75) examined self-reports of symptoms and physician-diagnosed disorders, and found that PTSD was associated with elevations for 17 of the 22 symptoms studied, although not with more physician-diagnosed disorders. In contrast, Long et al. (79) found that PTSD was associated with increased numbers of both symptoms and chronic disorder.

To our knowledge, there is only one study of the relationship between PTSD and health complaints among non-veterans. Breslau and Davis (81) found that young community-residing adults whose PTSD lasted at least 1 year reported more medical conditions than those whose PTSD lasted less than a year.

Utilization

It would be reasonable to expect that PTSD is associated with increased use of medical services, given that complaints of poor health are increased in PTSD. The NVVRS found that Vietnam veterans with PTSD reported increased use of medical health services (15); similar findings were reported by Long et al. (79) for New Zealand Vietnam veterans. Long et al. (79), as well as Card (82), found that PTSD was associated with increased disability days due to illness, which suggests that the problems experienced by individuals with PTSD may affect occupational and social function. One study suggests that the relationship between PTSD and utilization of medical services may be additionally influenced by other factors. Garfein et al. (83) examined the likelihood of receiving VA inpatient treatment for cardiovascular, cerebrovascular, and gastrointestinal problems among patients who had either PTSD, major depressive disorder, or schizophrenia. The hyperarousal and hyperreactivity in PTSD could be argued to make these disorders more likely among the PTSD group (75). The PTSD group was more likely than the schizophrenic group to have used inpatient services for cardiovascular or gastrointestinal problems, but did not differ from the depressed group. This finding highlights the need for psychiatric controls, especially those that share overlapping symptoms with PTSD patients (such as depression and alcohol abuse), to be included in studies of the relationship between PTSD and physical health outcomes.

Morbidity and Mortality

Only a few investigators have examined the relationship between PTSD and morbidity, and no one has examined mortality. Yet, as just noted (29), 60% of a sample of PTSD inpatients in a VA hospital were found to have an identified medical problem, and 42% had multiple problems. Lipton and Schaffer (84) reported multiple, severe medical problems among patients receiving psychotherapy in a PTSD treatment program.

We have found three studies that report cardiovascular morbidity information for individuals with and without PTSD, and all show increased morbidity associated with PTSD. Falger et al. (85) compared veterans of the Dutch Resistance in World War II who currently had PTSD with men of comparable age who had recently had either surgery or a myocardial infarction (MI). The PTSD group was comparable to the MI group in prevalence of angina (31% versus 26%, respectively), and higher than the surgery group (7%). Litz et al. (86) studied male Vietnam combat veterans and found that PTSD predicted poorer effort tolerance in a laboratory stress test.

Similar results were reported by Shalev et al. (77) for Israeli combat veterans with PTSD who did not differ from non-PTSD controls on physical exam and laboratory test findings. Although the PTSD veterans were more likely to smoke, Shalev et al. observed a difference in effort tolerance even when PTSD and no PTSD groups were stratified based on smoking status (yes/no); in other words, effort tolerance was poorer in veterans who had PTSD regardless of whether they smoked.

Whereas Litz et al. (86) found no association between PTSD symptoms and total cholesterol, Zimering et al. (87) have recently reported that male Vietnam veterans with PTSD have unfavorable blood lipid profiles, relative to combat veterans without PTSD. The groups in Zimering et al.'s study were equivalent in terms of hypertension and family history of heart disease, and smoking was controlled, so the effects cannot be attributed to these risk factors for cardiovascular disease. Still, as we are about to discuss, PTSD is associated with poor health habits (77,86). Therefore, other risk factors may have contributed to the difference between groups.

Summary

PTSD is associated with increased complaints of poor health, utilization of medical services, and cardiovascular morbidity. Caution is warranted when interpreting the data because there are relatively few studies in the area, and only one of the cardiovascular morbidity studies used controls who had a major psychiatric disorder other than depression (83). Despite the similarity observed in that study between PTSD and major depressive disorder (in utilization of inpatient medical services), we believe that PTSD is distinctive among psychiatric disorders in terms of its potential to promote poor health. As a reaction to extreme trauma, PTSD is an important mediator of the relationship between trauma and health. Next we discuss this mediational role in light of known biological, psychophysiological, and psychological components of PTSD.

FACTORS THAT MAY EXPLAIN THE RELATIONSHIP BETWEEN PTSD AND HEALTH OUTCOMES

Neurobiological Factors

Rosen and Fields (88) were the first to propose that neurochemical changes in the brain associated with PTSD might be etiologically related to the long-term morbidity observed in trauma victims. They focused specifically on the cardiovascular and gastrointestinal disorders observed in World War II POWs (12,13), which they hypothesized had resulted from PTSD-related alterations in brain catecholaminergic activity. Friedman (89) expanded this conceptual approach, and cited literature dating back to 1864 that presumes a relationship between combat trauma and cardiovascular illness. He speculated that the pathophysiology of PTSD might be etiologically responsible for cardiovascular abnormalities observed among war veterans who have received a number of diagnostic labels, including soldier's heart, Da Costa's Syndrome, and neurocirculatory asthenia.

As emphasized throughout this book, many of the biological systems affected by exposure of animals to stressful laboratory paradigms exhibit stable alterations in humans with PTSD. The best documented abnormalities associated with PTSD include enhanced cardiovascular reactivity, autonomic hyperarousal, disturbed sleep physiology, adrenergic dysregulation, enhanced thyroid function, and altered HPA activity. Taken together, these findings suggest that the pathophysiology of PTSD is such as to increase the risk for a number of medical illnesses. Specifically, it appears that excessive sympathetic reactivity and adrenergic dysregulation will increase vulnerability to cardiovascular disorders such as hypertension (8) and atherosclerotic heart disease (7). Endocrinological abnormalities such as HPA dysregulation, altered thyroid function, and possible alteration in testosterone and growth hormone (see Chapters 13, 18, 20, and 24) might increase vulnerability to specific endocrinopathies. In addition, such abnormalities, especially in the HPA system, may increase susceptibility to infections and immunologic disorders. Opioid dysregulation would certainly be expected to affect pain perception, pain tolerance, and chronic pain syndromes (90), as well as to impair endogenous cardiovascular regulation (91).

If certain abnormalities shown on animal stress paradigms are also found in humans with PTSD, we predict that there will be increased risk for medical illness among these patients. In particular, we predict that immunologic abnormalities will be shown to be reliably associated with PTSD. We base this prediction on psychoimmunologic research findings (see Chapter 21), on the stress and health literature (3,5), and on known interactions between immunologic function, the HPA axis, and the central and peripheral adrenergic systems (2,92). Finally, in addition to affecting susceptibility to infectious disease, a dysregulated immune system might also increase the risk for autoimmune disorders among PTSD patients. We predict that additional disease vulnerabilities will be identified among PTSD patients after other laboratory stress-induced abnormalities are found in human patients. These include alterations in corticotropin-releasing hormone (CRH), peptidergic, serotonergic, cholinergic, gamma-aminobutyric acid (GABA)-benzodiazepine, and dopaminergic systems.

Elsewhere in this volume, Post and colleagues discuss the applicability of sensitization models such as kindling and behavioral sensitization to PTSD. It is possible that a sensitization mechanism might cause individuals with PTSD to be at increased risk for medical disorders, for which a limbic sensitization model has been proposed (e.g., multiple chemical sensitivity syndrome (MCS) (93)). This latter speculation (for which the authors accept full responsibility) may be pertinent to preliminary data suggesting that American veterans of the Persian Gulf War who have PTSD may be at greater risk than colleagues without PTSD for symptoms suggestive of MCS.

To summarize, there appears to be good agreement between adverse health outcomes that one would predict based on our emerging understanding of the pathophysiology of PTSD and adverse health outcomes that have been reported in the literature concerning patients with PTSD.

Psychological and Behavioral Factors

A number of psychological and behavioral correlates of PTSD are themselves known risk factors for negative health outcomes. These correlates, and their relationship to PTSD, are so interrelated as to make separate discussion of each one virtually impossible. They may operate directly by creating psychological states that are potentially unhealthy (e.g., hyperreactivity), and indirectly by increasing the likelihood of unhealthy habits.

The literature reviewed by Williams (see Chapter 22) shows that hostility predicts poor cardiovascular health. Hostility and anger are increased in PTSD (94,95); thus, individuals who have PTSD could be at risk for cardiovascular disease. This risk could be related to the same mechanism by which trauma-related hyperreactivity is postulated to operate (7,8). Additionally, the risk could be related to poor health behaviors. Siegler et al. (96) found that hostility measured in college students predicted increased adult risk factors for coronary artery disease (including caffeine intake, body mass index, unfavorable lipid ratios, and smoking). In the same study, adult hostility was associated with increased risk on all of these factors plus alcohol consumption and hypertension. Individuals with PTSD have increased health risk assessment scores (86) and are more likely to smoke (77,87) and have alcohol problems (15). In fact, one longitudinal study of Israeli combat veterans found that increases in PTSD were associated with increases in alcohol consumption and smoking (97). Smith (98) discusses increased interpersonal stress—low social support and high conflict—as another explanation for the relationship between hostility and health. Poor social support has been found in several studies to be related to PTSD (99).

Health outcomes associated with depression also may be responsible for any association between PTSD and physical health. Depression is a common comorbid diagnosis among individuals with PTSD (15). Depressed individuals report more physical symptoms and use more medical treatment than nondepressed individuals (see review (100)). One possibility is that physical symptom inventories often include symptoms that are components of anxious and depressed states, such as headaches, nausea, and breathlessness. This is a plausible, although only partial, explanation for Litz et al.'s (75) observation of increased symptom complaints in combat-related PTSD. Another possibility is that depressed individuals are simply presenting their psychological distress in a somatic context (100). The relationship between depression and health outcomes is not only observed for self-reports and utilization, however; like hostility, depression has been linked to increased mortality in cardiac patients (101) and to poorer immune function (102). Moreover, depression interferes with good health habits (103). For example, depression is associated with reduced success at quitting smoking (104,105) and with increased alcohol consumption (106).

Like depression, alcohol and substance abuse are common comorbid diagnoses in PTSD (15,107). Both are significant public health problems. Although smoking is related to excessive drinking (106,108), heavy drinking is associated with increased cardiovascular morbidity and mortality when smoking is taken into account (108,109).

Our discussion would be incomplete if we did not include coping as a potential mediator of the relationship between PTSD and physical health. As noted by Aldwin (110), a major reason for studying coping is that coping is thought to moderate the negative effects of stress on health. Although much remains to be learned about coping in PTSD, it appears that PTSD is related to avoidant and emotion-focused coping (111–113), strategies that often are associated with psychological distress and physical symptoms (but see evidence that the efficacy of a particular strategy may depend on the perceived controllability of a stressor (114)). Aldwin mentions direct, mediational, and moderating (buffering) pathways by which coping could influence physical health. To our knowledge, only one study has included PTSD in tests of these pathways (115). The authors used a longitudinal design and hierarchical multiple regression to predict time 2 somatic symptoms as a function of variables that were significantly correlated with these symptoms: time 1 somatic symptoms, time 2 PTSD, negative life events, coping, and social resources. After controlling for initial symptom levels, only PTSD and social resources predicted symptoms at time 2; interactions involving coping, which would have provided tests of moderating pathways, were not statistically significant. Still, we suspect that the mediating and moderating roles of coping are likely to be important in understanding the relationship between PTSD and physical health. For example, if one adopts avoidant strategies such as alcoholism and substance abuse in order to cope with intrusive symptoms, the effects of avoidant coping are actually being mediated by these behaviors.

CONCLUDING COMMENTS

A number of authors have reviewed the relationship between health and sexual trauma (17–19,63), natural disasters (20), and POW captivity (13). There also have been large-scale studies on the adverse health consequences of military trauma, including war-zone exposure (15,16), POW captivity (12–14), and civilian incarceration in a concentration camp (10). This review, to our knowledge, is the most comprehensive effort at identifying and synthesizing the entire English language literature on trauma and health, and at extending the focus to question the relationship between PTSD and health.

We have taken as our starting point the massive stress and health literature, which, beginning with Selye's classic work (1), has shown that exposure to laboratory stress, everyday hassles, and severe events such as divorce, bereavement, and job loss can produce physiological abnormalities, impaired immunologic function, and increased susceptibility to infection (2–5, 7–9). We agree with Cohen and Williamson (5), who suggested that a review of the impact of traumatic events on health would be the "fairest test of the stress-disease relationship" (p. 18), and believe that the literature strongly suggests a clear association between exposure to extreme stress and poor health.

Following the suggestions of Rosen and Fields (88), and Friedman (89) that the adverse health consequences of POW and military trauma are due substantially to PTSD, we have reviewed the PTSD and health literature in its entirety. This is a much smaller body of work than either the stress and health or even the trauma and health literature, but it too supports this conclusion. Most of the existing studies are based on self-reports of physical health, and to a lesser extent on utilization, but the few studies of morbidity also found PTSD to be related to poor health outcomes.

When viewing the literature on trauma, PTSD, and physical health, it is important to recognize the general consistency of findings across self-reports and more objective outcome domains. The consistency bolsters our interpretation of the self-report data as a valid indicator of health status. We raise this issue because of a concern in the everyday stress literature that self-reported physical symptoms substantially reflect neuroticism, or "negative affectivity" (NA) (116). NA is associated with increased self-reports of physical symptoms, presumably because individuals high in NA experience exaggerated perception of and reaction to mild physical symptoms. One significant characteristic of NA is that it is related to self-reported symptoms, but not to utilization, morbidity, mortality, or disease risk factors (116). This pattern of relationships stands in marked contrast to patterns observed for trauma and PTSD. Thus, although NA may partially be the mechanism whereby trauma and PTSD are related to self-reported symptoms, it is unlikely to be the primary mechanism. Furthermore, it is even less likely that NA is the primary mechanism relating trauma and PTSD to other health outcomes.

We believe that PTSD is distinctive among psychiatric disorders in terms of its potential to promote poor health because of both the physiological and psychological abnormalities associated with this disorder. Specifically, it appears that medical problems may result from demonstrated biological alterations such as sympathetic hyperreactivity, adrenergic dysregulation, endocrinological abnormalities, opioid dysregulations, and probably (we speculate from the stress and health literature) altered immunologic mechanisms. Of equal importance, psychological and behavioral abnormalities associated with PTSD are likely to promote poor health. These include hostility, depression, poor health habits (smoking, drinking, drug use, and abnormal eating behavior), and poor coping skills. We believe that there are crucial cumulative and interaction effects between each of these biological and psychological factors through which PTSD promotes poor health.

There are several implications of this conclusion. From a scientific perspective, future research on stress and physical health should pay much more attention to the physical health outcomes associated with extreme stress and PTSD. Furthermore, PTSD investigators should pay more attention to the medical consequences of PTSD. In both cases, researchers should attempt to include multiple physical health outcome domains whenever possible. From a clinical perspective, medical practitioners should take a careful trauma history in their overall assessment of patients. Mental health practitioners should monitor medical complaints and consider the possibility that their patients' physical health status may be related to PTSD. Finally, the strictest test of this proposed relationship between PTSD and poor health would be to provide PTSD treatment to PTSD patients with medical problems. Controlling for equal medical interventions, we speculate that patients whose PTSD responds favorably to treatment might benefit medically. Such benefits might take several forms. In the case of early intervention, successful PTSD treatment might promote improvement in physical health. In the case of chronic PTSD and long-term medical problems, PTSD treatment might slow or halt the progression of well-established disorders.

Powerful institutional and conceptual factors interfere with the medical/mental health collaboration that we deem essential for diagnosis and treatment of PTSD-related medical problems. The major institutional factor is the segregation of clinicians into medical, surgical, and mental health divisions. The major conceptual factor, which we consider more powerful and insidious, is a pervasive eighteenth-century mind-body dualism that remains influential in many medical circles.

Our assertion that PTSD promotes poor health through a complex interaction between biological and psychological mechanisms flies in the face of such dualism. We hope that the present literature review and discussion will help to foster the crucial medical/mental health collaborative initiatives needed to address the problem of trauma-related medical illness, and lead to better diagnosis and treatment of such disorders in the future.

ACKNOWLEDGMENTS

We would like to thank Drs. Carolyn Aldwin and Ilene Siegler for their helpful comments on an earlier version of the manuscript.

REFERENCES

1. Selye H. *The stress of life*. New York: McGraw-Hill; 1956.
2. Chrousos GP, Gold PW. The concepts of stress and stress system disorders: overview of physical and behavioral homeostasis. *JAMA* 1992;267:1244–1252.
3. Kiecolt-Glaser JK, Glaser R. Psychoneuroimmunology: can psychological interventions modulate immunity? *J Consult Clin Psychol* 1992;60:1–6.
4. Schneiderman L, Baum A. Acute and chronic stress and the immune system. In Schneiderman N, McCabe P, Baum A, eds. *Stress and disease processes*. Hillsdale, NJ: Erlbaum Associates, 1992;1–25.
5. Cohen S, Williamson GM. Stress and infectious disease in humans. *Psychol Bull* 1991;109:5–24.
6. Hoehn-Saric R, MacLeod DR. The peripheral sympathetic nervous system. *Psychiatr Clin North Am* 1988; 11:375–386.
7. Kaplan JR, Manuck SB, Williams JK, Strawn W. Psychosocial influences on atherosclerosis: evidence for effects and mechanisms in nonhuman primates. In Blascovich J, Katkin E, eds. *Cardiovascular reactivity to psychological stress and disease*. Washington, DC: American Psychological Association, 1993:3–26.
8. Saab PG, Schneiderman N. Biobehavioral stressors, laboratory investigation, and the risk of hypertension. In Blascovich J, Katkin E, eds. *Cardiovascular reactivity to psychological stress and disease*. Washington, DC: American Psychological Association, 1993;49–82.
9. Kiecolt-Glaser JK, Cacioppo JT, Glaser R. Acute psychological stressors and short-term immune changes: what, why, for whom, and to what extent? *Psychosom Med* 1992;54:680–685.
10. Eitinger L. A follow-up study of the Norwegian concentration camp survivors' mortality and morbidity. *Isr Annals Psychiatry Related Disciplines* 1973;11: 199–209.
11. Mollica RF, Wyshak G, Lavelle J, Truong T, Tor S, Yang T. Assessing symptom change in Southeast Asian refugee survivors of mass violence and torture. *Am J Psychiatry* 1990;147:83–88.
12. Beebe GW. Follow-up studies of World War II and Korean war prisoners: II. Morbidity, disability, and maladjustments. *Am J Epidemiol* 1975;101:400–422.
13. Page WF. *The health of former prisoners of war*. Washington, DC: National Academy Press; 1992.
14. Venn AJ, Guest CS. Chronic morbidity of former prisoners of war and other Australian veterans. *Med J Aust* 1991;155:705–712.
15. Kulka RA, Schlenger WE, Fairbank JA, et al. *Trauma and the Vietnam War generation*. New York: Brunner/ Mazel; 1990.
16. Stellman SD, Stellman JM, Sommer JF. Health and reproductive outcomes among American Legionnaires in relation to combat and herbicide exposure in Vietnam. *Environ Res* 1988;47:150–174.
17. Fry R. Adult physical illness and childhood sexual abuse. *J Psychosom Res* 1993;37:89–103.
18. Goodman LA, Koss MP, Russo NF. Violence against women: physical and mental health effects. Part I: Research findings. *Appl Preven Psychol* 1993;2:79–89.
19. Koss MP, Heslet L. Somatic consequences of violence against women. *Arch Fam Med* 1992;1:53–59.
20. Hovanitz CA. Physical health risks associated with aftermath of disaster: basic paths of influence and their implications for prevention intervention. *J Soc Behav Personality* 1993;8:213–254.
21. Wolfe J, Schnurr PP, Brown PJ, Furey J. PTSD and war-zone exposure as correlates of perceived health in female Vietnam veterans. *J Consult Clin Psychol* 1994;62:1235–1240.
22. Centers for Disease Control. Health status of Vietnam veterans: II. Physical health. *JAMA* 1988a;259:2708–2714.
23. Centers for Disease Control. Health status of Vietnam veterans: III. Reproductive outcomes and child health. *JAMA* 1988b;259:2715–2719.
24. Cioffi D. Beyond attentional strategies: a cognitive-perceptual model of somatic interpretation. *Psychol Bull* 1991;109:25–41.
25. Pennebaker JW. *The psychology of physical symptoms*. New York: Springer-Verlag; 1982.
26. Piljs LTK, Feskens EJM, Kromhout D. Self-rated health, mortality, and chronic diseases in elderly men: the Zutphen Study, 1985–1990. *Am J Epidemiol* 1993; 138:840–848.
27. McHorney CA, Ware JE, Rogers W, Raczek AE, Lu JFR. The validity and relative precision of MOS short- and long-form health status scales and Dartmouth COOP charts. *Med Care* 1992;30:MS253–MS265.
28. Norman EM, Getek DM, Griffin C. Post-traumatic stress disorder in an urban trauma population. *Appl Nurs Res* 1991;4:171–176.
29. White PA, Faustman WO. Coexisting physical conditions among inpatients with post-traumatic stress disorder. *Mil Med* 1989;154:66–71.
30. Decouflé P, Holmgreen P, Boyle CA, Stroup NE. Self-reported health status of Vietnam veterans in relation to perceived exposure to herbicides and combat. *Am J Epidemiol* 1992;135:312–323.

31. Eisen SA, Goldberg J, True WR, Henderson WG. A co-twin study of the effects of the Vietnam War on the self-reported physical health of veterans. *Am J Epidemiol* 1991;134:49–58.
32. Kimerling R, Calhoun KS. Somatic symptoms, social support and treatment seeking among sexual assault victims. *J Consult Clin Psychol,* 1994;62:333–340.
33. Koss MP, Woodruff WJ, Koss PG. Relation of criminal victimization to health perceptions among women medical patients. *J Consult Clin Psychol* 1990;58: 147–152.
34. Waigandt A, Wallace DL, Phelps L, Miller D. The impact of sexual assault on physical health status. *J Traum Stress* 1990;3:93–102.
35. Felitti VJ. Long-term medical consequences of incest, rape, and molestation. *South Med J* 1991;84:328–331.
36. Rimsza ME, Berg RA. Sexual abuse: somatic and emotional reactions. *Child Abuse Negl* 1988;12:201–208.
37. Drossman DA, Leserman J, Nachman G, et al. Sexual and physical abuse in women with functional or organic gastrointestinal disorders. *Ann Int Med* 1990;113:828–833.
38. Schei B. Sexual factors in pelvic pain: a study of women living in physically abusive relationships and of randomly selected controls. *J Psychosom Obstet Gynaecol* 1991;12:99–108.
39. Toomey TC, Hernandez JT, Gittelman DF, Hulka JF. Relationship of sexual and physical abuse to pain and psychological assessment variables in chronic pelvic pain patients. *Pain* 1993;53:105–109.
40. Walker EA, Katon WJ, Hansom J, et al. Medical and psychiatric symptoms in women with childhood sexual abuse. *Psychosom Med* 1992;54:638–664.
41. Rapkin AJ, Kames LD, Darke LL, Stampler FM, Nabiloff BD. History of physical and sexual abuse in women with chronic pelvic pain. *Obstet Gynecol* 1990;76: 92–96.
42. Bravo M, Rubio-Stepec M, Canino GJ, Woodbury MA, Ribera JC. The psychological sequelae of disaster stress prospectively and retrospectively evaluated. *Am J Community Psychol* 1990;18:661–680.
43. Escobar JI, Canino G, Rubio-Stipec MS, Bravo M. Somatic symptoms after a natural disaster: a prospective study. *Am J Psychiatry* 1992;149:965–967.
44. Phifer JF. Psychological distress and somatic symptoms after natural disaster: differential vulnerability among older veterans. *Psychol Aging* 1990;5:412–420.
45. Murphy SA. Health and recovery status of victims one and three years following a natural disaster. In Figley CR, ed. *Trauma and its wake,* vol 2. New York: Brunner/Mazel; 1985:133–155.
46. Lima BR, Pai S, Santacruz H, Lozano J, Luna J. Screening for the psychological consequences of a major disaster in a developing country: Armero, Colombia. *Acta Psychiatr Scand* 1987;76:561–567.
47. Uba L, Chung RC. The relationship between trauma and financial and physical well-being among Cambodians in the United States. *J Gen Psychol* 1991;118: 215–225.
48. Van der Ploeg HM, Kleijn WC. Being held hostage in the Netherlands: a study of long-term aftereffects. *J Traum Stress* 1989;2:153–169.
49. Bartone PT, Ursano RJ, Wright KM, Ingraham LH. The impact of a military air disaster on the health of assistance workers. *J Nerv Ment Dis* 1989;177:317–328.
50. Kuch K, Evans RJ, Watson PC, Bubela C. Road vehicle accidents and phobias in 60 patients with fibromyalgia. *J Anx Disord* 1991;5:273–280.
51. Muse M. Stress-related, posttraumatic chronic pain syndrome: behavioral treatment approach. *Pain* 1986; 25:389–394.
52. Norris FH, Kaniasty KZ, Scheer DA. Use of mental health services among victims of crime: frequency, correlates, and subsequent recovery. *J Consult Clin Psychol* 1990;58:538–547.
53. Bergman B, Brismar B. A 5-year follow-up study of 117 battered women. *Am J Public Health* 1991;81: 1486–1489.
54. Bennet G. Bristol floods 1968: controlled survey of effects on health of local community disaster. *Br Med J* 1968;3:454–458.
55. Abrahams MJ, Price J, Whitlock FA, Williams G. The Brisbane floods, January 1974: their impact on health. *Med J Aust* 1976;2:936–939.
56. Adams PR, Adams GR. Mount St. Helens ashfall: evidence for a disaster stress reaction. *Am Psychol* 1984;39:252–260.
57. Cohen BM, Cooper MZ. A follow-up study of World War II Prisoners of War. (Veterans Administration Medical Monograph). Washington, DC: Government Printing Office, 1955.
58. Goulston KJ, Dent OF, Chapuis PH, et al. Gastrointestinal morbidity among World War II prisoners of war. *Med J Aust* 1985;143:6–10.
59. Adamson JD, Brerton D. Ultimate disabilities in the Hong Kong repatriates. *Treatment Serv Bull* 1948;3: 5–10 (cited in 88).
60. Richardson HJ. Report of a study of disabilities and problems of Hong Kong veterans, 1964–1965. Ottawa, Canada: Ottawa Pensions Commission, 1965 (cited in 13).
61. Sibai AM, Armenian HK, Alam S. Wartime determinants of arteriographically confirmed coronary artery disease in Beirut. *Am J Epidemiol* 1989;130:623–631.
62. Bergovec M, Mihatov S, Prpic H, et al. Acute myocardial infarction among civilians in Zagreb city area. *Lancet* 1992;339:303.
63. Laws AMI. Does a history of sexual abuse in childhood play a role in women's medical problems? A review. *J Women's Health* 1993;2:165–172.
64. Wurtele SK, Kaplan GM, Keairnes M. Childhood sexual abuse among chronic pain patients. *Clin J Pain* 1990;6:110–113.
65. Trevisan M, Jossa F, Farinaro E, et al. Earthquake and coronary heart disease risk factors: a longitudinal study. *Am J Epidemiol* 1986;135:632–637.
66. Holen A. A longitudinal study of the occurrence and persistence of post-traumatic health problems in disaster survivors. *Stress Med* 1989;7:11–17.
67. Janerich DT, Stark AD, Greenwald P, Burnett WS, Jacobson HI, McCusker J. Increased leukemia, lymphoma, and spontaneous abortion in western New York following a flood disaster. *Public Health Rep* 1981; 96:350–356.
68. Keehn RJ. Follow-up studies of World War II and Korean war prisoners: III. Mortality to January 1, 1976. *Am J Epidemiol* 1980;111:194–211.

69. Nefzger MD. Follow-up studies of World War II and Korean war prisoners: I. Study plan and mortality findings. *Am J Epidemiol* 1970;91:123–138.
70. Guest CS, Venn AJ. Mortality of former prisoners of war and other Australian veterans. *Med J Aust* 1992; 157:132–135.
71. Centers for Disease Control. Postservice mortality among Vietnam veterans. *JAMA* 1987;257:790–795.
72. Hearst N, Newman TB, Hulley SB. Delayed effects of the military draft on mortality: a randomized natural experiment. *N Engl J Med* 1986;314:620–624.
73. Thomas TL, Kang HK, Dalager NA. Mortality among women Vietnam veterans, 1973–1987. *Am J Epidemiol* 1991;134:973–980.
74. Dobson AJ, Alexander HM, Malcolm JA, Steele PL, Miles TA. Heart attacks and the Newcastle earthquake. *Med J Aust* 1991;155:757–760.
75. Litz BT, Keane TM, Fisher L, Marx B, Monaco V. Physical health complaints in combat-related post-traumatic stress disorder: a preliminary report. *J Traum Stress* 1992;5:131–141.
76. Stretch R. Post-traumatic stress disorder among Vietnam and Vietnam era veterans. In Figley CR, ed. *Trauma and its wake,* vol 2. New York: Brunner/Mazel, 1986;156–192.
77. Shalev A, Bleich A, Ursano R. Posttraumatic stress disorder: somatic comorbidity and effort tolerance. *Psychosomatics* 1990;31:197–203.
78. Solomon Z. Somatic complaints, stress reaction, and post-traumatic stress disorder: a three-year follow-up study. *Behav Med* 1988;14:179–185.
79. Long N, Chamberlain K, Vincent C. The health and mental health of New Zealand Vietnam war veterans with posttraumatic stress disorder. *N Z Med J* 1992; 105:417–419.
80. Stretch R. Psychosocial readjustment of Canadian Vietnam veterans. *J Consult Clin Psychol* 1991;59: 188–189.
81. Breslau N, Davis GC. Posttraumatic stress disorder in an urban population of young adults: risk factors for chronicity. *Am J Psychiatry* 1992;149:671–675.
82. Card JJ. *Lives after Vietnam.* Lexington, MA: Heath; 1983.
83. Garfein AJ, Friedman MJ, Bates EW, Ronis DL, Falcon SP, Furey J. *On the relationship of PTSD to cardiovascular, cerebrovascular, and gastrointestinal problems: Use of VA inpatient services.* Paper presented at the annual meeting of the American Public Health Association, San Francisco, CA; October 1993.
84. Lipton M, Schaffer WR. Physical symptoms related to post-traumatic stress disorder (PTSD) in an aging population. *Mil Med* 1988;153:316–318.
85. Falger PRJ, Op den Velde W, Hovens JE, Schouten EGW, DeGroen JHM, Van Duijn H. Current posttraumatic stress disorder and cardiovascular disease risk factors in Dutch Resistance veterans from World War II. *Psychother Psychosom* 1992;57:164–171.
86. Litz BT, Fisher L, Keane TK. *Physical health problems in post-traumatic stress disorder: a risk factor analysis.* In Keane T, Wolfe J (Co-chairs), Components of Health Risk in Veterans with Post-traumatic Stress Disorder. Symposium presented at the annual American Psychological Association Meeting, San Francisco, CA; August 1991.
87. Zimering R, Norris V, Lurie A, Hearst D. Elevated lipid profiles in Vietnam combat veterans with PTSD. Submitted, 1994.
88. Rosen J, Fields R. The long-term effects of extraordinary trauma: a look beyond PTSD. *J Anx Disord* 1988; 2:179–181.
89. Friedman MJ. Biological approaches to the diagnosis and treatment of post-traumatic stress disorder. *J Traum Stress* 1991;4:67–91.
90. Pitman RK, Van der Kolk BA, Orr SP, Greenberg MS. Naloxone-reversible analgesic response to combat-related stimuli in posttraumatic stress disorder: a pilot study. *Arch Gen Psychiatry* 1990;47:541–544.
91. McCubbin JA. Stress and endogenous opioids: behavioral and circulatory interactions. *Biol Psychiatry* 1993; 35:91–122.
92. Cacioppo JT. Social neuroscience: autonomic, neuroendocrine, and immune responses to stress. *Psychophysiology* 1993;31:113–128.
93. Bell IR, Miller CS, Schwartz GE. An olfactory-limbic model of multiple chemical sensitivity syndrome: possible relationships to kindling and affective spectrum disorders. *Biol Psychiatry* 1992;32:218–242.
94. Hovens JE, Op den Velde W, Falger PRJ, Schouten EGW, DeGroen JHM, Van Duijn H. Anxiety, depression, and anger in Dutch Resistance veterans from World War II. *Psychother Psychosom* 1992;57:172–179.
95. Zimering R, Norris G, Davenport P. *Anger in Vietnam veterans with PTSD: implications for long-term cardiovascular health.* In Zimering R (Chair), Empirical advances in the study of PTSD in Vietnam combat veterans. Symposium presented at the annual meeting of the Society for Traumatic Stress Studies, Dallas, TX; October 1988.
96. Siegler IC, Peterson BL, Barefoot JC, Williams RB. Hostility during late adolescence predicts coronary risk factors at mid-life. *Am J Epidemiol* 1992;136:146–154.
97. Solomon Z, Mikulincer M, Kotler M. A two-year follow-up of somatic complaints among Israeli combat stress reaction casualties. *J Psychosom Res* 1987;31: 463–469.
98. Smith T. Hostility and health: current status of a psychosomatic hypothesis. *Health Psychology* 1992;11: 139–150.
99. Keane TM, Scott WO, Chavoya GA, Lamparski DM, Fairbank JA. Social support in Vietnam veterans with Posttraumatic Stress Disorder: a comparative analysis. *J Consult Clin Psychol* 1985;53:95–102.
100. Schulberg HC, McClelland M, Burns BJ. Depression and physical illness: the prevalence, causation, and diagnosis of comorbidity. *Clin Psychol Rev* 1987;7: 145–167.
101. Frasure-Smith N, Lespérance F, Talajic M. Depression following myocardial infarction: impact on 6-month survival. *JAMA* 1993;270:1819–1825.
102. Evans DL, Folds JD, Pettito JM, et al. Circulating natural killer cell phenotypes in men and women with major depression: relation to cytotoxic activity and severity of depression. *Arch Gen Psychiatry* 1992;49: 388–395.
103. Frederick T, Frerichs RR, Clark VA. Personal health habits and symptoms of depression at the community level. *Prev Med* 1988;17:173–182.

104. Anda RF, Williamson DF, Escobedo LG, Mast EE, Giovino GA, Remington PL. Depression and the dynamics of smoking: a national perspective. *JAMA* 1990;264:1541–1545.
105. Glassman AH, Helzer JE, Covey LS, et al. Smoking, smoking cessation, and major depression. *JAMA* 1990;264:1546–1549.
106. Hartka E, Johnstone B, Leino EV, Motoyoshi M, Temple MT, Fillmore KM. A meta-analysis of depressive symptomatology and alcohol consumption over time. *Br J Addict* 1991;86:1283–1298.
107. Kofoed L, Friedman MJ, Peck R. Alcoholism and drug abuse in patients with PTSD. *Psychiatr Q* 1993;64: 151–171.
108. Rehm J, Fichter MM, Elton M. Effects on mortality of alcohol consumption, smoking, physical activity, and close personal relationships. *Addiction* 1993;88: 101–112.
109. Klatsky AL, Armstrong MA, Friedman GD. Risk of cardiovascular mortality in alcohol drinkers, ex-drinkers and nondrinkers. *Am J Cardiol* 1990;66:1237–1242.
110. Aldwin CM. *Stress, coping and development: an integrated perspective.* New York: Guilford, 1994.
111. Blake DD, Cook J, Keane TM. Post-traumatic stress disorder and coping in veterans who are seeking medical treatment. *J Clin Psychol* 1992;48:695–704.
112. Fairbank J, Hansen DJ, Fitterling JM. Patterns of appraisal and coping across different stressor conditions among former prisoners of war with and without posttraumatic stress disorder. *J Consult Clin Psychol* 1991; 59:274–281.
113. Wolfe J, Keane TM, Kaloupek DG, Mora CA, Wine P. Patterns of positive readjustment in Vietnam combat veterans. *J Traum Stress* 1993;6:179–193.
114. Solomon SD, Regier DA, Burke JD. Role of perceived control in coping with disaster. *J Soc Clin Psychol* 1989;8:376–392.
115. Solomon Z, Mikulincer M, Habershaim N. Life events, coping strategies, social resources, and somatic complaints among combat stress reaction casualties. *Br J Med Psychol* 1990;63:137–148.
116. Watson D, Pennebaker JW. Health complaints, stress, and distress: exploring the central role of negative affectivity. *Psychol Rev* 1989;96:234–254.

PART V

Summary

Neurobiological and Clinical Consequences of Stress: From Normal Adaptation to PTSD, edited by M. J. Friedman, D. S. Charney, and A. Y. Deutch, Lippincott-Raven Publishers, Philadelphia © 1995.

30

Key Questions and a Research Agenda for the Future

Matthew J. Friedman, *Dennis S. Charney, and *†Ariel Y. Deutch

*Departments of Psychiatry and Pharmacology, Dartmouth Medical School, Hanover, New Hampshire 03755, and National Center for Post-Traumatic Stress Disorder, Veterans Administration Medical Center, White River Junction, Vermont 05009; and Departments of *Psychiatry and *†Pharmacology, Yale University School of Medicine, New Haven, Connecticut 06510, and National Center for Post-Traumatic Stress Disorder, Veterans Affairs Medical Center, West Haven, Connecticut 06516*

To our knowledge, this book represents the first comprehensive effort to conceptualize post-traumatic stress disorder (PTSD) in terms of the basic neurobiological mechanisms that promote normal adaptation to stress. We believe that PTSD may be distinct among psychiatric disorders because of the wealth of laboratory research and animal models that can help us understand the pathophysiology of this disorder. In our own effort to synthesize some of the major themes presented throughout this book, we offer the following questions.

1. What defines ''normal'' as opposed to ''traumatic'' stress? What differentiates normal from traumatic stress?
2. What are the major similarities and differences between the neurobiology of the stress response generally observed experimentally, and that observed among PTSD patients?
3. What neurobiological aspects of the stress response have yet to be explored in PTSD?
4. To what degree do peripheral events influence central neuronal function in PTSD?
5. Do the somatic consequences of stress help us understand the adverse health findings associated with exposure to traumatic events and PTSD?
6. Which neural mechanisms seem most applicable to PTSD?
7. Which neural mechanisms of memory seem most applicable to PTSD?
8. Is allostasis a useful concept for PTSD?
9. What animal models that are applicable to PTSD?
10. Are there any useful clinical applications of laboratory paradigms currently used in PTSD research with humans?
11. Where should we direct future research on the clinical pharmacology of PTSD?
12. How can we understand, from a neurobiological perspective, the fact that PTSD is usually associated with at least one other comorbid psychiatric disorder?
13. What should be the future agenda for research on PTSD?

1. WHAT DEFINES ''NORMAL'' AS OPPOSED TO ''TRAUMATIC'' STRESS? WHAT DIFFERENTIATES NORMAL FROM TRAUMATIC STRESS?

The definition of stress has always been somewhat circular. The event that precipitates stress is a stressor, but the response to a given stressor, designated stress, varies greatly across individuals. Stressors that are part of the normal vicissitudes of modern life, and that give rise to no demonstrable psycho- or physio-pathology in most people, may be severe stressors for other (predisposed) individuals and thus lead to overt symptomatology. Thus, the response (stress) differs widely despite the same evoking stimulus (stressor). In addition to the difference in response to a given stressor that depends upon individual vulnerability, there are also differences in responsiveness to the same stressor that depend upon context (controllable versus uncontrollable stress). Finally, the type of stressor (e.g., footshock versus restraint) may also produce large response differences from one individual to another.

Part of the difficulty in working on the neurobiology of stress and PTSD is the fact that stress needs to be defined operationally for each investigation. This holds for both preclinical and clinical studies. It will be increasingly important to provide full documentation of the defining characteristics of stressors employed in preclinical studies if we hope to compare and contrast findings in a rational manner. Similar issues plague clinical studies. For example, epidemiological studies of the effects of past stress exposure have been hampered by difficulties in defining stress and its various presentations; not only must major life stressors be categorized, but so must those events comprising the "social stress" that varies as a function of family, community, and social support.

In summary, investigations of stress have been hampered by a lack of clear definition. While many might substitute "stress" for "pornography" in the now famous statement of Justice Potter Stewart—"I shall not attempt today to further define the kinds of [pornographic] material . . . but I know pornography when I see it"—it is apparent that this attitude will not be useful in research endeavors on the effects of stress.

2. WHAT ARE THE MAJOR SIMILARITIES AND DIFFERENCES BETWEEN THE NEUROBIOLOGY OF THE STRESS RESPONSE GENERALLY OBSERVED EXPERIMENTALLY, AND THAT OBSERVED AMONG PTSD PATIENTS?

Among the most consistent observations in PTSD patients and in laboratory animals exposed to stress are changes in catecholaminergic systems. For example, both exposure of PTSD patients to traumatic reminders and exposure of rats to a neutral stimulus previously paired with stress increase sympathetic nervous system activity. Many PTSD patients have increased urinary catecholamine levels and decreased platelet α_2 and lymphocyte β adrenergic receptors; PTSD patients also exhibit potentiated behavioral, biochemical, and cardiovascular responses to yohimbine. These findings suggest that the pathophysiology of PTSD involves altered peripheral catecholaminergic function.

It remains unclear to what degree these changes actually reflect central events, or are derived from changes in peripheral catecholamines. A considerable body of data from nonhuman animal studies suggests that changes in peripheral levels of norepinephrine and epinephrine may contribute to changes in acquisition and consolidation of memory, as well as contribute to anxiety states. Future studies should be directed to unraveling the relative contributions of peripheral and central catecholaminergic alterations in the pathophysiology of PTSD.

A key component of the acute response to stress is activation of the hypothalamic-pituitary-adrenal (HPA) axis. Several recent animal studies have demonstrated that exposure to stressful events early in life can have profound and longlasting consequences; these sequelae may be qualitatively different and differ quantitatively from the changes elicited by acute stress exposure. For example, acute stress exposure increases HPA function. In contrast, PTSD patients paradoxically exhibit decreased HPA function, including decreased 24-hour urinary cortisol levels, increased glucocorticoid receptor density, supersuppression in response to dexamethasone, and blunting of the corticotropin-releasing–hormone (CRH)–adrenocorticotropic-hormone (ACTH) response. However, these same parameters of altered HPA function can also be observed in adult rats subjected to mild stressors during critical developmental periods (such as transient maternal separation); in contrast, more "severe" stressors typically induce increased HPA tone. There are several other areas of similarity between PTSD and effects of stress in rodents and other animals. The neural mechanisms of fear conditioning, extinction, and sensitization have been extensively investigated in laboratory animals, and appear to hold clues as to the mechanisms underlying the morbidity and chronicity of PTSD. In addition, there are some similarities in certain aspects of the adrenergic and opioid response to both normal and traumatic stress.

The similarities and differences between animal studies and changes observed in PTSD indicate that there is a compelling need for more work in laboratory animals to determine the role of different stress intensities, duration, and contingencies (i.e., escapable versus inescapable) in various central and peripheral systems; only recently have contemporary approaches been brought to bear on these issues. In addition, studies of the effect of different stressors during critical developmental periods are clearly indicated.

3. WHAT NEUROBIOLOGICAL ASPECTS OF THE STRESS RESPONSE HAVE YET TO BE EXPLORED IN PTSD?

Studies in laboratory animals have shown myriad responses to stressors. Stress has been shown to alter the function of virtually every neurotransmitter system; similarly, a wide array of nuclei in the CNS are involved. Clinical studies of PTSD have only started to assess the function of a few of the relevant brain regions and transmitter systems that basic science studies implicate.

The amygdala appears to mediate or modulate many of the emotional aspects of stress, such as fear conditioning, extinction, sensitization, and recall of traumatic memories. There have been, to our knowledge, no studies of amygdala function in PTSD. Similarly, the amygdala can be considered as part of an extended neural network of structures that subserve emotion, including prefrontal cortices, hippocampus, and reticular formation. Studies examining the relationship of amygdala function to other areas connected with the amygdala are lacking, particularly in primates. In part, the paucity of data on these regions is due to a lack of appropriate methodologies for studying these regions in man. Certain in vivo imaging techniques, particularly studies of blood flow by echo planer and other fast-scan MRI, may be well suited to this task.

Other imaging techniques that can detect significant changes in regional volumes may also be useful in defining the sites of pathology in PTSD. For example, several recent MRI studies have indicated a reduction in hippocampal volume in PTSD patients; this fits well with the effects of chronic stress on hippocampal neurons. However, the animal data indicate that there is significant potential for recovery after chronic stress- or glucocorticoid-induced morphological changes in hippocampal neurons. This discrepancy emphasizes the need for long-term studies of the effects of chronic stress in laboratory animals.

In vivo imaging methods, however, are probably of limited utility for examining such changes as the dystrophic alterations in hippocampal pyramidal cells observed after chronic stress. These changes (as opposed to overt cell loss or atrophy) are decreased dendritic spine density and changes in branching patterns of processes. Detection of such changes currently requires sophisticated computer-based assessments of histological material; it does not seem likely that current in vivo imaging methods will be able to determine such changes in man at any time in the near future. Since neuronal plasticity is much more common than once thought, it seems appropriate to actively investigate morphological changes in PTSD patients. We believe that it is time to consider the development and implementation of a brain bank focusing on PTSD.

In addition to needing a greater focus on structural abnormalities associated with PTSD, we also believe it necessary to apply sophisticated clinical neuropharmacological techniques to the study of neurotransmitters that have received insufficient attention in PTSD. This is particularly true for CRH, neuropeptide Y, oxytocin, excitatory amino acids, and the gamma-aminobutyric acid (GABA)-benzodiazepine system.

4. TO WHAT DEGREE DO PERIPHERAL EVENTS INFLUENCE CENTRAL NEURONAL FUNCTION IN PTSD?

The relative involvement of peripheral versus central mechanisms in emotion and stress has been an ongoing area of study and controversy for over a century. As noted in the section addressing question 2, there are several changes in catecholaminergic parameters in PTSD; many of these changes are probably attributable to changes in peripheral catecholamine systems. The difficulty in monitoring central neurochemical events has plagued biological psychiatry. Since there are robust changes in peripheral markers of catecholamine function, and since peripheral events can theoretically influence central processes both directly and indirectly, it is appropriate to examine the degree to which central systems are impacted by peripheral events. Such studies may conceivably lead to pharmacological interventions that may be of some utility in PTSD.

5. DO THE SOMATIC CONSEQUENCES OF STRESS HELP US UNDERSTAND THE ADVERSE HEALTH FINDINGS ASSOCIATED WITH EXPOSURE TO TRAUMATIC EVENTS AND PTSD?

Most studies of the somatic consequences of stress are correlational in nature. There has been very little hypothesis-driven research concerning the somatic consequences of chronic stress in PTSD. Moreover, the factors that may increase the risk of adverse health outcomes for PTSD patients are not specific to this disorder. However, given the high incidence of comorbidity in PTSD, it seems appropriate to investigate the somatic consequences of stress in any attempt to reveal potential interventions that are specifically useful in PTSD.

Among individuals exposed to ''nontraumatic'' stress, there are a number of characteristics that predispose individuals to adverse health outcomes. These include altered sympathetic tone and concomitant cardiovascular hyperreactivity, smoking, drug abuse (including alcohol abuse), and a host of other factors. Many of these attributes appear to apply to PTSD patients. For example, many PTSD patients exhibit increased sympathetic tone, while alcohol and substance abuse are very frequently present. In general, available epidemiological studies indicate that exposure to trauma in general and the development of PTSD in particular are important risk factors for adverse health outcomes. In view of the paucity of rigorous, hypothesis-driven research published to date, further research is clearly needed to address the relationship between trauma, PTSD, and medical illness.

6. WHICH NEURAL MECHANISMS SEEM MOST APPLICABLE TO PTSD?

Fear conditioning, extinction, and sensitization are the three neural mechanisms that seem most applicable to PTSD. All these mechanisms provide major links through which a large body of laboratory findings can be reconceptualized and applied to PTSD.

Fear conditioning may account for the common observation in PTSD that sensory and cognitive stimuli associated with or resembling the original frightening experience elicit panic attacks, flashbacks, and a variety of autonomic symptoms. Pervasive anxiety may be due to contextual fear conditioning.

A failure in the neural mechanisms underlying extinction may relate to treatment-resistant avoidance behavior and the persistence of traumatic memories.

Sensitization may explain the adverse effects of early life trauma on subsequent responses to stressful life events. Sensitization may also play a role in the chronic course of PTSD and, in some cases, the worsening of the illness over time.

7. WHICH NEURAL MECHANISMS OF MEMORY SEEM MOST APPLICABLE TO PTSD?

In many respects, PTSD is a disorder of memory. Research into the neural mechanisms subserving the acquisition and retrieval of traumatic events will enhance our understanding of many of the clinical features of PTSD, and may facilitate the development of novel treatment approaches. It will be important to determine if there are differences in the brain mechanisms involved in the incorporation of traumatic as compared with nontraumatic events into longterm memory, and if the retrieval mechanisms differ. In particular, pharmacological studies are indicated. This need is amplified by the recent presentation of data demonstrating that pretreatment with the β adrenoceptor antagonist propranolol decreased components of the memory of an emotionally-charged traumatic story. This study suggests that similar pharmacological interventions may be useful adjuncts for persons in whom exposure to traumatic events, such as military personnel or emergency medical professionals, is expected.

Recent studies of long-term potentiation (LTP) and long-term depression (LTD) have emphasized the relation of these effects to memory, and established the role of certain neuronal messengers—including glutamate and other excitatory amino acids, and nitric oxide—in memory. The phenomenon of LTP is much more widespread than originally suspected, and now has been demonstrated in a wide variety of brain areas. Future studies may shed light on trauma-induced changes in memory, and the elucidation of the transmitter systems involved may uncover useful therapeutic approaches for prevention or treatment of PTSD.

8. IS ALLOSTASIS A USEFUL MODEL FOR UNDERSTANDING PTSD?

We believe that allostasis is an excellent model for understanding PTSD. The chronicity, stability, and refractory nature of this disorder suggest that a pathologic, allostatic balance has been achieved. The best evidence to suggest this conclusion comes from research on both the adrenergic and HPA systems. Excessive adrenergic function is balanced by downregulation of alpha-2 and beta adrenergic receptors whereas allostatic HPA equilibrium is marked by lower cortisol levels, upregulation of glucocorticoid receptors, and blunting of the CRH-ACTH response. The high price of allostasis, however, is dysregulation of these key systems. As a result, PTSD patients exhibit major biobehavioral deficits in their ability to function and cope with the normal vicissitudes of life.

It is not clear when a pathological equilibrium is achieved in response to chronic stress (i.e., when allostasis appears as opposed to transient deviations from homeostasis). Similarly, is there a decay of allostasis with reversion to a true homeostatic equilibrium, or is allostasis permanent? Studies of the adaptive changes to chronic stress in laboratory animals are clearly needed, with an emphasis on longterm (months to years) studies of chronic stress.

9. WHAT ANIMAL MODELS ARE APPLICABLE TO PTSD?

An effort needs to be made to develop animal models that are more analogous to the human situation than most current laboratory paradigms. Criteria that need to be satisfied in any animal model of PTSD must include factors that are essential to the initiation of PTSD-like symptoms, as well as factors that influence their expression.

Important preclinical studies of nonhuman primates and other animals in this regard include investigations on the consequences of maternal neglect, protracted social and hierarchical stress, and the impact of uncontrollable and unpredictable stress. Determination of the effects of these stressors on physiological, somatic, and behavioral functions is critical. Further attempts to discover "treatments" that reverse stress effects may provide new clinical insights.

10. ARE THERE ANY USEFUL CLINICAL APPLICATIONS OF LABORATORY PARADIGMS RECENTLY USED IN PTSD RESEARCH WITH HUMANS?

Of the many psychobiological strategies employed to investigate the pathophysiology of PTSD, we believe that several approaches have potential applicability as clinical tools for distinguishing PTSD from other disorders. Several techniques appear quite promising at this time, especially measurement of psychophysiological reactivity, the dexamethasone suppression test, the acoustic startle response, and the 24-hour urinary neurohormone profile. This is a short list; hopefully it will be lengthened and fortified by future research.

11. WHERE SHOULD WE DIRECT FUTURE RESEARCH ON THE CLINICAL PHARMACOLOGY OF PTSD?

We have mentioned several suggestions for future research on PTSD, including clinical pharmacological issues. We wish to emphasize several additional points.

Clinical reports concerning military and disaster trauma indicate that clinical interventions within the first 24–72 hours of traumatic exposure, such as critical incident stress debriefing, may not only result in rapid symptom relief but may also prevent the later development of PTSD. We believe that acute pharmacological intervention may achieve similar results. There is clearly a need to identify drugs that, when administered shortly after acute exposure to traumatic events, will prevent autonomic instability and the encoding of horrific traumatic memories.

Treatment of established cases of PTSD with adrenergic antagonists (such as clonidine and propanolol) and agents that reduce panic attacks (tricyclic antidepressants and monomine oxidase inhibitors) has been disappointing, since even the most successful trials have achieved only modest symptomatic relief. Preliminary trials with selective serotonin reuptake inhibitors (SSRIs) have been promising, as have an array of case reports and open trials with other drugs. Our major concern in this regard is the low number of randomized clinical trials with PTSD patients. This is a crucial area of research that has not received appropriate attention.

From a theoretical perspective, there are several classes of drugs that would be particularly interesting to test in the treatment of PTSD. These include novel peptidergic (CRH, neuropeptide Y (NPY)) antagonists, allosteric modulators of the *N*-methyl-D-aspartate (NMDA) receptor, and certain anticonvulsant medications. These choices are suggested by preclinical research findings.

It is important to emphasize that we are entering a new period in the treatment of PTSD, one in which rationally-based therapeutic interventions may be envisioned. It is imperative that we focus on carefully executed, randomized clinical trials that are based on clearly stated hypotheses.

12. HOW CAN WE UNDERSTAND, FROM A NEUROBIOLOGICAL PERSPECTIVE, THE FACT THAT PTSD IS USUALLY ASSOCIATED WITH AT LEAST ONE OTHER COMORBID PSYCHIATRIC DISORDER?

As with most other psychiatric disorders, PTSD rarely occurs in "pure form," but is most often associated with other DSM-III-R diagnoses. In view of the considerable symptom overlap between PTSD and other diagnoses, comorbidity may reflect an artifact of a nosological system that cannot make fine distinctions between different disorders based on phenomenology alone. On the other hand, it may indicate that there is a common pathophysiology underlying a number of distinct psychiatric disorders that may exhibit different clinical presentations. The only thing that is not in doubt is that phenomenology cannot take us much further. The key to distinguishing true comorbidity from apparent comorbidity lies in biological research. Only when we understand the differences and similarities between various psychiatric diagnoses in terms of underlying pathophysiology will we be able to put the clinical question of comorbidity in its proper context. At that time, we may be able to move to hypothesis-driven research on intervention strategies and treatment approaches that are informed by a pathophysiological understanding of the psychiatric disorder(s) in question.

13. WHAT SHOULD BE THE FUTURE AGENDA FOR RESEARCH ON PTSD?

We have addressed this question throughout this chapter. The field is wide open, since we are really only now entering the realm of contemporary PTSD research, despite the long history of the disorder. The following list summarizes areas that we believe are of importance to our understanding of PTSD. Since stress is so pervasive, and affects all of us in various ways, it is likely that the benefits derived from PTSD research will not be limited to PTSD. We anticipate that the findings derived from studies of the following areas will have wide relevance to a large number of conditions:

a. further exploration of basic neurobiological and psychological mechanisms known to play a central role in the stress response;
b. greater emphasis on theory-driven research derived from important animal models in this field;
c. intensification of efforts to understand the pathophysiology of PTSD;
d. application of laboratory paradigms to psychobiological diagnostic protocols;
e. greater emphasis on randomized clinical tests of drugs in the current pharmacopoeia, as well as initiatives to test new classes of drugs that are suggested by extrapolation from basic stress research;
f. attention to the impact of traumatic stress and PTSD on physical health.

It is our hope that this book will serve as a starting point from which to implement this research agenda during the next few years. We hope that clinical researchers will investigate questions about PTSD from the broader perspective of coping and adaptation, and that their research initiatives will be driven by theoretical considerations. We also hope that basic biobehavioral scientists will test laboratory paradigms that are informed by clinical observations and questions.

In closing, we wish to emphasize that PTSD is a widespread and prevalent disorder. The diagnosis of PTSD is frequently missed. Current treatment strategies for PTSD have generally been attempts at symptomatic relief, with rather poor outcomes. However, with the rapid advances in treatment strategies, aimed at treating symptoms, acutely intervening to ameliorate or retard the emergence of subsequent symptoms, or actively engaging in preventive approaches, it is imperative that PTSD be appropriately diagnosed and that persons at risk be identified. More clinical training is needed. More recognition is needed regarding PTSD's profound long-term effects on brain function. Future success with regard to improved recognition and treatment will require ongoing collaboration between neuroscientists, clinicians, public policy makers, and the community at large. We hope that this book is a start in that direction.

Subject Index

Numerals in *italic* indicate a figure; "t" following a page number indicates a table.

B

C

D

E

O

P

T

U

V

W

Y